P9-CEB-842

Modern Control Engineering

Fifth Edition

Katsuhiko Ogata

Prentice Hall

Boston   Columbus   Indianapolis   New York   San Francisco   Upper Saddle River
Amsterdam   Cape Town   Dubai   London   Madrid   Milan   Munich   Paris   Montreal   Toronto
Delhi   Mexico City   Sao Paulo   Sydney   Hong Kong   Seoul   Singapore   Taipei   Tokyo

VP/Editorial Director, Engineering/Computer Science: Marcia J. Horton
Assistant/Supervisor: Dolores Mars
Senior Editor: Andrew Gilfillan
Associate Editor: Alice Dworkin
Editorial Assistant: William Opaluch
Director of Marketing: Margaret Waples
Senior Marketing Manager: Tim Galligan
Marketing Assistant: Mack Patterson
Senior Managing Editor: Scott Disanno
Art Editor: Greg Dulles
Senior Operations Supervisor: Alan Fischer
Operations Specialist: Lisa McDowell
Art Director: Kenny Beck
Cover Designer: Carole Anson
Media Editor: Daniel Sandin

Credits and acknowledgments borrowed from other sources and reproduced, with permission, in this textbook appear on appropriate page within text.

MATLAB is a registered trademark of The MathWorks, Inc., 3 Apple Hill Drive, Natick MA 01760-2098.

Copyright © 2010, 2002, 1997, 1990, 1970 Pearson Education, Inc., publishing as Prentice Hall, One Lake Street, Upper Saddle River, New Jersey 07458. All rights reserved. Manufactured in the United States of America. This publication is protected by Copyright, and permission should be obtained from the publisher prior to any prohibited reproduction, storage in a retrieval system, or transmission in any form or by any means, electronic, mechanical, photocopying, recording, or likewise. To obtain permission(s) to use material from this work, please submit a written request to Pearson Education, Inc., Permissions Department, One Lake Street, Upper Saddle River, New Jersey 07458.

Many of the designations by manufacturers and seller to distinguish their products are claimed as trademarks. Where those designations appear in this book, and the publisher was aware of a trademark claim, the designations have been printed in initial caps or all caps.

Library of Congress Cataloging-in-Publication Data on File

Prentice Hall
is an imprint of

www.pearsonhighered.com

ISBN 10: 0-13-615673-8
ISBN 13: 978-0-13-615673-4

Contents

Preface   ix

Chapter 1   Introduction to Control Systems   1

1–1   Introduction   1
1–2   Examples of Control Systems   4
1–3   Closed-Loop Control Versus Open-Loop Control   7
1–4   Design and Compensation of Control Systems   9
1–5   Outline of the Book   10

Chapter 2   Mathematical Modeling of Control Systems   13

2–1   Introduction   13
2–2   Transfer Function and Impulse-Response Function   15
2–3   Automatic Control Systems   17
2–4   Modeling in State Space   29
2–5   State-Space Representation of Scalar Differential
       Equation Systems   35
2–6   Transformation of Mathematical Models with MATLAB   39

2–7   Linearization of Nonlinear Mathematical Models   43

Example Problems and Solutions   46

Problems   60

**Chapter 3   Mathematical Modeling of Mechanical Systems
and Electrical Systems**   **63**

3–1   Introduction   63

3–2   Mathematical Modeling of Mechanical Systems   63

3–3   Mathematical Modeling of Electrical Systems   72

Example Problems and Solutions   86

Problems   97

**Chapter 4   Mathematical Modeling of Fluid Systems
and Thermal Systems**   **100**

4–1   Introduction   100

4–2   Liquid-Level Systems   101

4–3   Pneumatic Systems   106

4–4   Hydraulic Systems   123

4–5   Thermal Systems   136

Example Problems and Solutions   140

Problems   152

Chapter 5   Transient and Steady-State Response Analyses   **159**

5–1   Introduction   159

5–2   First-Order Systems   161

5–3   Second-Order Systems   164

5–4   Higher-Order Systems   179

5–5   Transient-Response Analysis with MATLAB   183

5–6   Routh's Stability Criterion   212

5–7   Effects of Integral and Derivative Control Actions
on System Performance   218

5–8   Steady-State Errors in Unity-Feedback Control Systems   225

Example Problems and Solutions   231

Problems   263

Chapter 6   Control Systems Analysis and Design by the Root-Locus Method   269

6–1   Introduction   269
6–2   Root-Locus Plots   270
6–3   Plotting Root Loci with MATLAB   290
6–4   Root-Locus Plots of Positive Feedback Systems   303
6–5   Root-Locus Approach to Control-Systems Design   308
6–6   Lead Compensation   311
6–7   Lag Compensation   321
6–8   Lag–Lead Compensation   330
6–9   Parallel Compensation   342
       Example Problems and Solutions   347
       Problems   394

Chapter 7   Control Systems Analysis and Design by the Frequency-Response Method   398

7–1   Introduction   398
7–2   Bode Diagrams   403
7–3   Polar Plots   427
7–4   Log-Magnitude-versus-Phase Plots   443
7–5   Nyquist Stability Criterion   445
7–6   Stability Analysis   454
7–7   Relative Stability Analysis   462
7–8   Closed-Loop Frequency Response of Unity-Feedback Systems   477
7–9   Experimental Determination of Transfer Functions   486
7–10   Control Systems Design by Frequency-Response Approach   491
7–11   Lead Compensation   493
7–12   Lag Compensation   502
7–13   Lag–Lead Compensation   511
       Example Problems and Solutions   521
       Problems   561

Chapter 8   PID Controllers and Modified PID Controllers   567

8–1   Introduction   567
8–2   Ziegler–Nichols Rules for Tuning PID Controllers   568

8–3 Design of PID Controllers with Frequency-Response Approach   577

8–4 Design of PID Controllers with Computational Optimization Approach   583

8–5 Modifications of PID Control Schemes   590

8–6 Two-Degrees-of-Freedom Control   592

8–7 Zero-Placement Approach to Improve Response Characteristics   595

Example Problems and Solutions   614

Problems   641

Chapter 9   Control Systems Analysis in State Space   648

9–1 Introduction   648

9–2 State-Space Representations of Transfer-Function Systems   649

9–3 Transformation of System Models with MATLAB   656

9–4 Solving the Time-Invariant State Equation   660

9–5 Some Useful Results in Vector-Matrix Analysis   668

9–6 Controllability   675

9–7 Observability   682

Example Problems and Solutions   688

Problems   720

Chapter 10   Control Systems Design in State Space   722

10–1 Introduction   722

10–2 Pole Placement   723

10–3 Solving Pole-Placement Problems with MATLAB   735

10–4 Design of Servo Systems   739

10–5 State Observers   751

10–6 Design of Regulator Systems with Observers   778

10–7 Design of Control Systems with Observers   786

10–8 Quadratic Optimal Regulator Systems   793

10–9 Robust Control Systems   806

Example Problems and Solutions   817

Problems   855

Appendix A   Laplace Transform Tables   859

Appendix B   Partial-Fraction Expansion   867

Appendix C   Vector-Matrix Algebra   874

References   882

Index   886

Preface

This book introduces important concepts in the analysis and design of control systems. Readers will find it to be a clear and understandable textbook for control system courses at colleges and universities. It is written for senior electrical, mechanical, aerospace, or chemical engineering students. The reader is expected to have fulfilled the following prerequisites: introductory courses on differential equations, Laplace transforms, vector-matrix analysis, circuit analysis, mechanics, and introductory thermodynamics.

The main revisions made in this edition are as follows:

- The use of MATLAB for obtaining responses of control systems to various inputs has been increased.
- The usefulness of the computational optimization approach with MATLAB has been demonstrated.
- New example problems have been added throughout the book.
- Materials in the previous edition that are of secondary importance have been deleted in order to provide space for more important subjects. Signal flow graphs were dropped from the book. A chapter on Laplace transform was deleted. Instead, Laplace transform tables, and partial-fraction expansion with MATLAB are presented in Appendix A and Appendix B, respectively.
- A short summary of vector-matrix analysis is presented in Appendix C; this will help the reader to find the inverses of n x n matrices that may be involved in the analysis and design of control systems.

This edition of *Modern Control Engineering* is organized into ten chapters. The outline of this book is as follows: Chapter 1 presents an introduction to control systems. Chapter 2 deals with mathematical modeling of control systems. A linearization technique for non-linear mathematical models is presented in this chapter. Chapter 3 derives mathematical models of mechanical systems and electrical systems. Chapter 4 discusses mathematical modeling of fluid systems (such as liquid-level systems, pneumatic systems, and hydraulic systems) and thermal systems.

Chapter 5 treats transient response and steady-state analyses of control systems. MATLAB is used extensively for obtaining transient response curves. Routh's stability criterion is presented for stability analysis of control systems. Hurwitz stability criterion is also presented.

Chapter 6 discusses the root-locus analysis and design of control systems, including positive feedback systems and conditionally stable systems. Plotting root loci with MATLAB is discussed in detail. Design of lead, lag, and lag-lead compensators with the root-locus method is included.

Chapter 7 treats the frequency-response analysis and design of control systems. The Nyquist stability criterion is presented in an easily understandable manner. The Bode diagram approach to the design of lead, lag, and lag-lead compensators is discussed.

Chapter 8 deals with basic and modified PID controllers. Computational approaches for obtaining optimal parameter values for PID controllers are discussed in detail, particularly with respect to satisfying requirements for step-response characteristics.

Chapter 9 treats basic analyses of control systems in state space. Concepts of controllability and observability are discussed in detail.

Chapter 10 deals with control systems design in state space. The discussions include pole placement, state observers, and quadratic optimal control. An introductory discussion of robust control systems is presented at the end of Chapter 10.

The book has been arranged toward facilitating the student's gradual understanding of control theory. Highly mathematical arguments are carefully avoided in the presentation of the materials. Statement proofs are provided whenever they contribute to the understanding of the subject matter presented.

Special effort has been made to provide example problems at strategic points so that the reader will have a clear understanding of the subject matter discussed. In addition, a number of solved problems (A-problems) are provided at the end of each chapter, except Chapter 1. The reader is encouraged to study all such solved problems carefully; this will allow the reader to obtain a deeper understanding of the topics discussed. In addition, many problems (without solutions) are provided at the end of each chapter, except Chapter 1. The unsolved problems (B-problems) may be used as homework or quiz problems.

If this book is used as a text for a semester course (with 56 or so lecture hours), a good portion of the material may be covered by skipping certain subjects. Because of the abundance of example problems and solved problems (A-problems) that might answer many possible questions that the reader might have, this book can also serve as a self-study book for practicing engineers who wish to study basic control theories.

I would like to thank the following reviewers for this edition of the book: Mark Campbell, Cornell University; Henry Sodano, Arizona State University; and Atul G. Kelkar, Iowa State University. Finally, I wish to offer my deep appreciation to Ms. Alice Dworkin, Associate Editor, Mr. Scott Disanno, Senior Managing Editor, and all the people involved in this publishing project, for the speedy yet superb production of this book.

Katsuhiko Ogata

Introduction
to Control Systems

1–1 INTRODUCTION

Control theories commonly used today are classical control theory (also called conventional control theory), modern control theory, and robust control theory. This book presents comprehensive treatments of the analysis and design of control systems based on the classical control theory and modern control theory. A brief introduction of robust control theory is included in Chapter 10.

Automatic control is essential in any field of engineering and science. Automatic control is an important and integral part of space-vehicle systems, robotic systems, modern manufacturing systems, and any industrial operations involving control of temperature, pressure, humidity, flow, etc. It is desirable that most engineers and scientists are familiar with theory and practice of automatic control.

This book is intended to be a text book on control systems at the senior level at a college or university. All necessary background materials are included in the book. Mathematical background materials related to Laplace transforms and vector-matrix analysis are presented separately in appendixes.

Brief Review of Historical Developments of Control Theories and Practices. The first significant work in automatic control was James Watt's centrifugal governor for the speed control of a steam engine in the eighteenth century. Other significant works in the early stages of development of control theory were due to

Minorsky, Hazen, and Nyquist, among many others. In 1922, Minorsky worked on automatic controllers for steering ships and showed how stability could be determined from the differential equations describing the system. In 1932, Nyquist developed a relatively simple procedure for determining the stability of closed-loop systems on the basis of open-loop response to steady-state sinusoidal inputs. In 1934, Hazen, who introduced the term *servomechanisms* for position control systems, discussed the design of relay servomechanisms capable of closely following a changing input.

During the decade of the 1940s, frequency-response methods (especially the Bode diagram methods due to Bode) made it possible for engineers to design linear closed-loop control systems that satisfied performance requirements. Many industrial control systems in 1940s and 1950s used PID controllers to control pressure, temperature, etc. In the early 1940s Ziegler and Nichols suggested rules for tuning PID controllers, called Ziegler–Nichols tuning rules. From the end of the 1940s to the 1950s, the root-locus method due to Evans was fully developed.

The frequency-response and root-locus methods, which are the core of classical control theory, lead to systems that are stable and satisfy a set of more or less arbitrary performance requirements. Such systems are, in general, acceptable but not optimal in any meaningful sense. Since the late 1950s, the emphasis in control design problems has been shifted from the design of one of many systems that work to the design of one optimal system in some meaningful sense.

As modern plants with many inputs and outputs become more and more complex, the description of a modern control system requires a large number of equations. Classical control theory, which deals only with single-input, single-output systems, becomes powerless for multiple-input, multiple-output systems. Since about 1960, because the availability of digital computers made possible time-domain analysis of complex systems, modern control theory, based on time-domain analysis and synthesis using state variables, has been developed to cope with the increased complexity of modern plants and the stringent requirements on accuracy, weight, and cost in military, space, and industrial applications.

During the years from 1960 to 1980, optimal control of both deterministic and stochastic systems, as well as adaptive and learning control of complex systems, were fully investigated. From 1980s to 1990s, developments in modern control theory were centered around robust control and associated topics.

Modern control theory is based on time-domain analysis of differential equation systems. Modern control theory made the design of control systems simpler because the theory is based on a model of an actual control system. However, the system's stability is sensitive to the error between the actual system and its model. This means that when the designed controller based on a model is applied to the actual system, the system may not be stable. To avoid this situation, we design the control system by first setting up the range of possible errors and then designing the controller in such a way that, if the error of the system stays within the assumed range, the designed control system will stay stable. The design method based on this principle is called robust control theory. This theory incorporates both the frequency-response approach and the time-domain approach. The theory is mathematically very complex.

Because this theory requires mathematical background at the graduate level, inclusion of robust control theory in this book is limited to introductory aspects only. The reader interested in details of robust control theory should take a graduate-level control course at an established college or university.

Definitions. Before we can discuss control systems, some basic terminologies must be defined.

Controlled Variable and Control Signal or Manipulated Variable. The *controlled* variable is the quantity or condition that is measured and controlled. The *control signal* or *manipulated* variable is the quantity or condition that is varied by the controller so as to affect the value of the controlled variable. Normally, the controlled variable is the output of the system. *Control* means measuring the value of the controlled variable of the system and applying the control signal to the system to correct or limit deviation of the measured value from a desired value.

In studying control engineering, we need to define additional terms that are necessary to describe control systems.

Plants. A plant may be a piece of equipment, perhaps just a set of machine parts functioning together, the purpose of which is to perform a particular operation. In this book, we shall call any physical object to be controlled (such as a mechanical device, a heating furnace, a chemical reactor, or a spacecraft) a plant.

Processes. The *Merriam–Webster Dictionary* defines a process to be a natural, progressively continuing operation or development marked by a series of gradual changes that succeed one another in a relatively fixed way and lead toward a particular result or end; or an artificial or voluntary, progressively continuing operation that consists of a series of controlled actions or movements systematically directed toward a particular result or end. In this book we shall call any operation to be controlled a *process.* Examples are chemical, economic, and biological processes.

Systems. A system is a combination of components that act together and perform a certain objective. A system need not be physical. The concept of the system can be applied to abstract, dynamic phenomena such as those encountered in economics. The word system should, therefore, be interpreted to imply physical, biological, economic, and the like, systems.

Disturbances. A disturbance is a signal that tends to adversely affect the value of the output of a system. If a disturbance is generated within the system, it is called *internal,* while an *external* disturbance is generated outside the system and is an input.

Feedback Control. Feedback control refers to an operation that, in the presence of disturbances, tends to reduce the difference between the output of a system and some reference input and does so on the basis of this difference. Here only unpredictable disturbances are so specified, since predictable or known disturbances can always be compensated for within the system.

In this section we shall present a few examples of control systems.

Speed Control System. The basic principle of a Watt's speed governor for an engine is illustrated in the schematic diagram of Figure 1–1. The amount of fuel admitted to the engine is adjusted according to the difference between the desired and the actual engine speeds.

The sequence of actions may be stated as follows: The speed governor is adjusted such that, at the desired speed, no pressured oil will flow into either side of the power cylinder. If the actual speed drops below the desired value due to disturbance, then the decrease in the centrifugal force of the speed governor causes the control valve to move downward, supplying more fuel, and the speed of the engine increases until the desired value is reached. On the other hand, if the speed of the engine increases above the desired value, then the increase in the centrifugal force of the governor causes the control valve to move upward. This decreases the supply of fuel, and the speed of the engine decreases until the desired value is reached.

In this speed control system, the plant (controlled system) is the engine and the controlled variable is the speed of the engine. The difference between the desired speed and the actual speed is the error signal. The control signal (the amount of fuel) to be applied to the plant (engine) is the actuating signal. The external input to disturb the controlled variable is the disturbance. An unexpected change in the load is a disturbance.

Temperature Control System. Figure 1–2 shows a schematic diagram of temperature control of an electric furnace. The temperature in the electric furnace is measured by a thermometer, which is an analog device. The analog temperature is converted

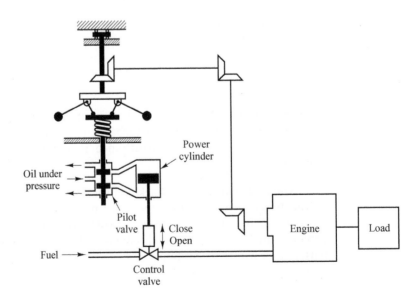

Figure 1–1
Speed control
system.

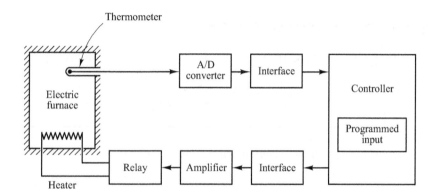

Figure 1–2
Temperature control system.

to a digital temperature by an A/D converter. The digital temperature is fed to a controller through an interface. This digital temperature is compared with the programmed input temperature, and if there is any discrepancy (error), the controller sends out a signal to the heater, through an interface, amplifier, and relay, to bring the furnace temperature to a desired value.

Business Systems.   A business system may consist of many groups. Each task assigned to a group will represent a dynamic element of the system. Feedback methods of reporting the accomplishments of each group must be established in such a system for proper operation. The cross-coupling between functional groups must be made a minimum in order to reduce undesirable delay times in the system. The smaller this cross-coupling, the smoother the flow of work signals and materials will be.

A business system is a closed-loop system. A good design will reduce the managerial control required. Note that disturbances in this system are the lack of personnel or materials, interruption of communication, human errors, and the like.

The establishment of a well-founded estimating system based on statistics is mandatory to proper management. It is a well-known fact that the performance of such a system can be improved by the use of lead time, or *anticipation*.

To apply control theory to improve the performance of such a system, we must represent the dynamic characteristic of the component groups of the system by a relatively simple set of equations.

Although it is certainly a difficult problem to derive mathematical representations of the component groups, the application of optimization techniques to business systems significantly improves the performance of the business system.

Consider, as an example, an engineering organizational system that is composed of major groups such as management, research and development, preliminary design, experiments, product design and drafting, fabrication and assembling, and tesing. These groups are interconnected to make up the whole operation.

Such a system may be analyzed by reducing it to the most elementary set of components necessary that can provide the analytical detail required and by representing the dynamic characteristics of each component by a set of simple equations. (The dynamic performance of such a system may be determined from the relation between progressive accomplishment and time.)

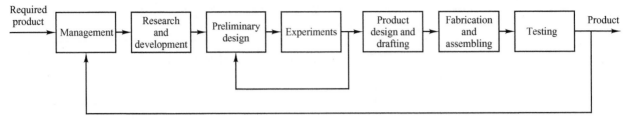

Figure 1–3
Block diagram of an engineering organizational system.

A functional block diagram may be drawn by using blocks to represent the functional activities and interconnecting signal lines to represent the information or product output of the system operation. Figure 1–3 is a possible block diagram for this system.

Robust Control System.   The first step in the design of a control system is to obtain a mathematical model of the plant or control object. In reality, any model of a plant we want to control will include an error in the modeling process. That is, the actual plant differs from the model to be used in the design of the control system.

To ensure the controller designed based on a model will work satisfactorily when this controller is used with the actual plant, one reasonable approach is to assume from the start that there is an uncertainty or error between the actual plant and its mathematical model and include such uncertainty or error in the design process of the control system. The control system designed based on this approach is called a robust control system.

Suppose that the actual plant we want to control is $\widetilde{G}(s)$ and the mathematical model of the actual plant is $G(s)$, that is,

$$\widetilde{G}(s) = \text{actual plant model that has uncertainty } \Delta(s)$$

$$G(s) = \text{nominal plant model to be used for designing the control system}$$

$\widetilde{G}(s)$ and $G(s)$ may be related by a multiplicative factor such as

$$\widetilde{G}(s) = G(s)[1 + \Delta(s)]$$

or an additive factor

$$\widetilde{G}(s) = G(s) + \Delta(s)$$

or in other forms.

Since the exact description of the uncertainty or error $\Delta(s)$ is unknown, we use an estimate of $\Delta(s)$ and use this estimate, $W(s)$, in the design of the controller. $W(s)$ is a scalar transfer function such that

$$\|\Delta(s)\|_\infty < \|W(s)\|_\infty = \max_{0 \le \omega \le \infty} |W(j\omega)|$$

where $\|W(s)\|_\infty$ is the maximum value of $|W(j\omega)|$ for $0 \le \omega \le \infty$ and is called the H infinity norm of $W(s)$.

Chapter 1   /   **Introduction to Control Systems**

Using the small gain theorem, the design procedure here boils down to the determination of the controller $K(s)$ such that the inequality

$$\left\| \frac{W(s)}{1 + K(s)G(s)} \right\|_\infty < 1$$

is satisfied, where $G(s)$ is the transfer function of the model used in the design process, $K(s)$ is the transfer function of the controller, and $W(s)$ is the chosen transfer function to approximate $\Delta(s)$. In most practical cases, we must satisfy more than one such inequality that involves $G(s)$, $K(s)$, and $W(s)$'s. For example, to guarantee robust stability and robust performance we may require two inequalities, such as

$$\left\| \frac{W_m(s)K(s)G(s)}{1 + K(s)G(s)} \right\|_\infty < 1 \quad \text{for robust stability}$$

$$\left\| \frac{W_s(s)}{1 + K(s)G(s)} \right\|_\infty < 1 \quad \text{for robust performance}$$

be satisfied. (These inequalities are derived in Section 10–9.) There are many different such inequalities that need to be satisfied in many different robust control systems. (Robust stability means that the controller $K(s)$ guarantees internal stability of all systems that belong to a group of systems that include the system with the actual plant. Robust performance means the specified performance is satisfied in all systems that belong to the group.) In this book all the plants of control systems we discuss are assumed to be known precisely, except the plants we discuss in Section 10–9 where an introductory aspect of robust control theory is presented.

1–3 CLOSED-LOOP CONTROL VERSUS OPEN-LOOP CONTROL

Feedback Control Systems.　A system that maintains a prescribed relationship between the output and the reference input by comparing them and using the difference as a means of control is called a *feedback control system*. An example would be a room-temperature control system. By measuring the actual room temperature and comparing it with the reference temperature (desired temperature), the thermostat turns the heating or cooling equipment on or off in such a way as to ensure that the room temperature remains at a comfortable level regardless of outside conditions.

Feedback control systems are not limited to engineering but can be found in various nonengineering fields as well. The human body, for instance, is a highly advanced feedback control system. Both body temperature and blood pressure are kept constant by means of physiological feedback. In fact, feedback performs a vital function: It makes the human body relatively insensitive to external disturbances, thus enabling it to function properly in a changing environment.

Closed-Loop Control Systems. Feedback control systems are often referred to as *closed-loop control* systems. In practice, the terms feedback control and closed-loop control are used interchangeably. In a closed-loop control system the actuating error signal, which is the difference between the input signal and the feedback signal (which may be the output signal itself or a function of the output signal and its derivatives and/or integrals), is fed to the controller so as to reduce the error and bring the output of the system to a desired value. The term closed-loop control always implies the use of feedback control action in order to reduce system error.

Open-Loop Control Systems. Those systems in which the output has no effect on the control action are called *open-loop control systems.* In other words, in an open-loop control system the output is neither measured nor fed back for comparison with the input. One practical example is a washing machine. Soaking, washing, and rinsing in the washer operate on a time basis. The machine does not measure the output signal, that is, the cleanliness of the clothes.

In any open-loop control system the output is not compared with the reference input. Thus, to each reference input there corresponds a fixed operating condition; as a result, the accuracy of the system depends on calibration. In the presence of disturbances, an open-loop control system will not perform the desired task. Open-loop control can be used, in practice, only if the relationship between the input and output is known and if there are neither internal nor external disturbances. Clearly, such systems are not feedback control systems. Note that any control system that operates on a time basis is open loop. For instance, traffic control by means of signals operated on a time basis is another example of open-loop control.

Closed-Loop versus Open-Loop Control Systems. An advantage of the closed-loop control system is the fact that the use of feedback makes the system response relatively insensitive to external disturbances and internal variations in system parameters. It is thus possible to use relatively inaccurate and inexpensive components to obtain the accurate control of a given plant, whereas doing so is impossible in the open-loop case.

From the point of view of stability, the open-loop control system is easier to build because system stability is not a major problem. On the other hand, stability is a major problem in the closed-loop control system, which may tend to overcorrect errors and thereby can cause oscillations of constant or changing amplitude.

It should be emphasized that for systems in which the inputs are known ahead of time and in which there are no disturbances it is advisable to use open-loop control. Closed-loop control systems have advantages only when unpredictable disturbances and/or unpredictable variations in system components are present. Note that the output power rating partially determines the cost, weight, and size of a control system. The number of components used in a closed-loop control system is more than that for a corresponding open-loop control system. Thus, the closed-loop control system is generally higher in cost and power. To decrease the required power of a system, open-loop control may be used where applicable. A proper combination of open-loop and closed-loop controls is usually less expensive and will give satisfactory overall system performance.

Most analyses and designs of control systems presented in this book are concerned with closed-loop control systems. Under certain circumstances (such as where no disturbances exist or the output is hard to measure) open-loop control systems may be desired. Therefore, it is worthwhile to summarize the advantages and disadvantages of using open-loop control systems.

The major advantages of open-loop control systems are as follows:

1. Simple construction and ease of maintenance.
2. Less expensive than a corresponding closed-loop system.
3. There is no stability problem.
4. Convenient when output is hard to measure or measuring the output precisely is economically not feasible. (For example, in the washer system, it would be quite expensive to provide a device to measure the quality of the washer's output, cleanliness of the clothes.)

The major disadvantages of open-loop control systems are as follows:

1. Disturbances and changes in calibration cause errors, and the output may be different from what is desired.
2. To maintain the required quality in the output, recalibration is necessary from time to time.

1–4 DESIGN AND COMPENSATION OF CONTROL SYSTEMS

This book discusses basic aspects of the design and compensation of control systems. Compensation is the modification of the system dynamics to satisfy the given specifications. The approaches to control system design and compensation used in this book are the root-locus approach, frequency-response approach, and the state-space approach. Such control systems design and compensation will be presented in Chapters 6, 7, 9 and 10. The PID-based compensational approach to control systems design is given in Chapter 8.

In the actual design of a control system, whether to use an electronic, pneumatic, or hydraulic compensator is a matter that must be decided partially based on the nature of the controlled plant. For example, if the controlled plant involves flammable fluid, then we have to choose pneumatic components (both a compensator and an actuator) to avoid the possibility of sparks. If, however, no fire hazard exists, then electronic compensators are most commonly used. (In fact, we often transform nonelectrical signals into electrical signals because of the simplicity of transmission, increased accuracy, increased reliability, ease of compensation, and the like.)

Performance Specifications. Control systems are designed to perform specific tasks. The requirements imposed on the control system are usually spelled out as performance specifications. The specifications may be given in terms of transient response requirements (such as the maximum overshoot and settling time in step response) and of steady-state requirements (such as steady-state error in following ramp input) or may be given in frequency-response terms. The specifications of a control system must be given before the design process begins.

For routine design problems, the performance specifications (which relate to accuracy, relative stability, and speed of response) may be given in terms of precise numerical values. In other cases they may be given partially in terms of precise numerical values and partially in terms of qualitative statements. In the latter case the specifications may have to be modified during the course of design, since the given specifications may never be satisfied (because of conflicting requirements) or may lead to a very expensive system.

Generally, the performance specifications should not be more stringent than necessary to perform the given task. If the accuracy at steady-state operation is of prime importance in a given control system, then we should not require unnecessarily rigid performance specifications on the transient response, since such specifications will require expensive components. Remember that the most important part of control system design is to state the performance specifications precisely so that they will yield an optimal control system for the given purpose.

System Compensation. Setting the gain is the first step in adjusting the system for satisfactory performance. In many practical cases, however, the adjustment of the gain alone may not provide sufficient alteration of the system behavior to meet the given specifications. As is frequently the case, increasing the gain value will improve the steady-state behavior but will result in poor stability or even instability. It is then necessary to redesign the system (by modifying the structure or by incorporating additional devices or components) to alter the overall behavior so that the system will behave as desired. Such a redesign or addition of a suitable device is called *compensation*. A device inserted into the system for the purpose of satisfying the specifications is called a *compensator*. The compensator compensates for deficient performance of the original system.

Design Procedures. In the process of designing a control system, we set up a mathematical model of the control system and adjust the parameters of a compensator. The most time-consuming part of the work is the checking of the system performance by analysis with each adjustment of the parameters. The designer should use MATLAB or other available computer package to avoid much of the numerical drudgery necessary for this checking.

Once a satisfactory mathematical model has been obtained, the designer must construct a prototype and test the open-loop system. If absolute stability of the closed loop is assured, the designer closes the loop and tests the performance of the resulting closed-loop system. Because of the neglected loading effects among the components, nonlinearities, distributed parameters, and so on, which were not taken into consideration in the original design work, the actual performance of the prototype system will probably differ from the theoretical predictions. Thus the first design may not satisfy all the requirements on performance. The designer must adjust system parameters and make changes in the prototype until the system meets the specificications. In doing this, he or she must analyze each trial, and the results of the analysis must be incorporated into the next trial. The designer must see that the final system meets the performance apecifications and, at the same time, is reliable and economical.

1–5 OUTLINE OF THE BOOK

This text is organized into 10 chapters. The outline of each chapter may be summarized as follows:

Chapter 1 presents an introduction to this book.

Chapter 2 deals with mathematical modeling of control systems that are described by linear differential equations. Specifically, transfer function expressions of differential equation systems are derived. Also, state-space expressions of differential equation systems are derived. MATLAB is used to transform mathematical models from transfer functions to state-space equations and vice versa. This book treats linear systems in detail. If the mathematical model of any system is nonlinear, it needs to be linearized before applying theories presented in this book. A technique to linearize nonlinear mathematical models is presented in this chapter.

Chapter 3 derives mathematical models of various mechanical and electrical systems that appear frequently in control systems.

Chapter 4 discusses various fluid systems and thermal systems, that appear in control systems. Fluid systems here include liquid-level systems, pneumatic systems, and hydraulic systems. Thermal systems such as temperature control systems are also discussed here. Control engineers must be familiar with all of these systems discussed in this chapter.

Chapter 5 presents transient and steady-state response analyses of control systems defined in terms of transfer functions. MATLAB approach to obtain transient and steady-state response analyses is presented in detail. MATLAB approach to obtain three-dimensional plots is also presented. Stability analysis based on Routh's stability criterion is included in this chapter and the Hurwitz stability criterion is briefly discussed.

Chapter 6 treats the root-locus method of analysis and design of control systems. It is a graphical method for determining the locations of all closed-loop poles from the knowledge of the locations of the open-loop poles and zeros of a closed-loop system as a parameter (usually the gain) is varied from zero to infinity. This method was developed by W. R. Evans around 1950. These days MATLAB can produce root-locus plots easily and quickly. This chapter presents both a manual approach and a MATLAB approach to generate root-locus plots. Details of the design of control systems using lead compensators, lag compensators, are lag–lead compensators are presented in this chapter.

Chapter 7 presents the frequency-response method of analysis and design of control systems. This is the oldest method of control systems analysis and design and was developed during 1940–1950 by Nyquist, Bode, Nichols, Hazen, among others. This chapter presents details of the frequency-response approach to control systems design using lead compensation technique, lag compensation technique, and lag–lead compensation technique. The frequency-response method was the most frequently used analysis and design method until the state-space method became popular. However, since H-infinity control for designing robust control systems has become popular, frequency response is gaining popularity again.

Chapter 8 discusses PID controllers and modified ones such as multidegrees-of-freedom PID controllers. The PID controller has three parameters; proportional gain, integral gain, and derivative gain. In industrial control systems more than half of the controllers used have been PID controllers. The performance of PID controllers depends on the relative magnitudes of those three parameters. Determination of the relative magnitudes of the three parameters is called tuning of PID controllers.

Ziegler and Nichols proposed so-called "Ziegler–Nichols tuning rules" as early as 1942. Since then numerous tuning rules have been proposed. These days manufacturers of PID controllers have their own tuning rules. In this chapter we present a computer optimization approach using MATLAB to determine the three parameters to satisfy given transient response characteristics. The approach can be expanded to determine the three parameters to satisfy any specific given characteristics.

Chapter 9 presents basic analysis of state-space equations. Concepts of controllability and observability, most important concepts in modern control theory, due to Kalman are discussed in full. In this chapter, solutions of state-space equations are derived in detail.

Chapter 10 discusses state-space designs of control systems. This chapter first deals with pole placement problems and state observers. In control engineering, it is frequently desirable to set up a meaningful performance index and try to minimize it (or maximize it, as the case may be). If the performance index selected has a clear physical meaning, then this approach is quite useful to determine the optimal control variable. This chapter discusses the quadratic optimal regulator problem where we use a performance index which is an integral of a quadratic function of the state variables and the control variable. The integration is performed from $t = 0$ to $t = \infty$. This chapter concludes with a brief discussion of robust control systems.

Mathematical Modeling of Control Systems

2-1 INTRODUCTION

In studying control systems the reader must be able to model dynamic systems in mathematical terms and analyze their dynamic characteristics. A mathematical model of a dynamic system is defined as a set of equations that represents the dynamics of the system accurately, or at least fairly well. Note that a mathematical model is not unique to a given system. A system may be represented in many different ways and, therefore, may have many mathematical models, depending on one's perspective.

The dynamics of many systems, whether they are mechanical, electrical, thermal, economic, biological, and so on, may be described in terms of differential equations. Such differential equations may be obtained by using physical laws governing a particular system—for example, Newton's laws for mechanical systems and Kirchhoff's laws for electrical systems. We must always keep in mind that deriving reasonable mathematical models is the most important part of the entire analysis of control systems.

Throughout this book we assume that the principle of causality applies to the systems considered. This means that the current output of the system (the output at time $t = 0$) depends on the past input (the input for $t < 0$) but does not depend on the future input (the input for $t > 0$).

Mathematical Models. Mathematical models may assume many different forms. Depending on the particular system and the particular circumstances, one mathematical model may be better suited than other models. For example, in optimal control problems, it is advantageous to use state-space representations. On the other hand, for the transient-response or frequency-response analysis of single-input, single-output, linear, time-invariant systems, the transfer-function representation may be more convenient than any other. Once a mathematical model of a system is obtained, various analytical and computer tools can be used for analysis and synthesis purposes.

Simplicity Versus Accuracy.   In obtaining a mathematical model, we must make a compromise between the simplicity of the model and the accuracy of the results of the analysis. In deriving a reasonably simplified mathematical model, we frequently find it necessary to ignore certain inherent physical properties of the system. In particular, if a linear lumped-parameter mathematical model (that is, one employing ordinary differential equations) is desired, it is always necessary to ignore certain nonlinearities and distributed parameters that may be present in the physical system. If the effects that these ignored properties have on the response are small, good agreement will be obtained between the results of the analysis of a mathematical model and the results of the experimental study of the physical system.

In general, in solving a new problem, it is desirable to build a simplified model so that we can get a general feeling for the solution. A more complete mathematical model may then be built and used for a more accurate analysis.

We must be well aware that a linear lumped-parameter model, which may be valid in low-frequency operations, may not be valid at sufficiently high frequencies, since the neglected property of distributed parameters may become an important factor in the dynamic behavior of the system. For example, the mass of a spring may be neglected in low-frequency operations, but it becomes an important property of the system at high frequencies. (For the case where a mathematical model involves considerable errors, robust control theory may be applied. Robust control theory is presented in Chapter 10.)

Linear Systems.   A system is called linear if the principle of superposition applies. The principle of superposition states that the response produced by the simultaneous application of two different forcing functions is the sum of the two individual responses. Hence, for the linear system, the response to several inputs can be calculated by treating one input at a time and adding the results. It is this principle that allows one to build up complicated solutions to the linear differential equation from simple solutions.

In an experimental investigation of a dynamic system, if cause and effect are proportional, thus implying that the principle of superposition holds, then the system can be considered linear.

Linear Time-Invariant Systems and Linear Time-Varying Systems.   A differential equation is linear if the coefficients are constants or functions only of the independent variable. Dynamic systems that are composed of linear time-invariant lumped-parameter components may be described by linear time-invariant differential equations—that is, constant-coefficient differential equations. Such systems are called *linear time-invariant* (or *linear constant-coefficient*) systems. Systems that are represented by differential equations whose coefficients are functions of time are called *linear time-varying* systems. An example of a time-varying control system is a spacecraft control system. (The mass of a spacecraft changes due to fuel consumption.)

Outline of the Chapter. Section 2–1 has presented an introduction to the mathematical modeling of dynamic systems. Section 2–2 presents the transfer function and impulse-response function. Section 2–3 introduces automatic control systems and Section 2–4 discusses concepts of modeling in state space. Section 2–5 presents state-space representation of dynamic systems. Section 2–6 discusses transformation of mathematical models with MATLAB. Finally, Section 2–7 discusses linearization of nonlinear mathematical models.

2–2 TRANSFER FUNCTION AND IMPULSE-RESPONSE FUNCTION

In control theory, functions called transfer functions are commonly used to characterize the input-output relationships of components or systems that can be described by linear, time-invariant, differential equations. We begin by defining the transfer function and follow with a derivation of the transfer function of a differential equation system. Then we discuss the impulse-response function.

Transfer Function. The *transfer function* of a linear, time-invariant, differential equation system is defined as the ratio of the Laplace transform of the output (response function) to the Laplace transform of the input (driving function) under the assumption that all initial conditions are zero.

Consider the linear time-invariant system defined by the following differential equation:

$$a_0 \overset{(n)}{y} + a_1 \overset{(n-1)}{y} + \cdots + a_{n-1} \dot{y} + a_n y$$

$$= b_0 \overset{(m)}{x} + b_1 \overset{(m-1)}{x} + \cdots + b_{m-1} \dot{x} + b_m x \qquad (n \geq m)$$

where y is the output of the system and x is the input. The transfer function of this system is the ratio of the Laplace transformed output to the Laplace transformed input when all initial conditions are zero, or

$$\text{Transfer function} = G(s) = \frac{\mathscr{L}[\text{output}]}{\mathscr{L}[\text{input}]}\Bigg|_{\text{zero initial conditions}}$$

$$= \frac{Y(s)}{X(s)} = \frac{b_0 s^m + b_1 s^{m-1} + \cdots + b_{m-1} s + b_m}{a_0 s^n + a_1 s^{n-1} + \cdots + a_{n-1} s + a_n}$$

By using the concept of transfer function, it is possible to represent system dynamics by algebraic equations in s. If the highest power of s in the denominator of the transfer function is equal to n, the system is called an *nth-order system*.

Comments on Transfer Function. The applicability of the concept of the transfer function is limited to linear, time-invariant, differential equation systems. The transfer function approach, however, is extensively used in the analysis and design of such systems. In what follows, we shall list important comments concerning the transfer function. (Note that a system referred to in the list is one described by a linear, time-invariant, differential equation.)

1. The transfer function of a system is a mathematical model in that it is an operational method of expressing the differential equation that relates the output variable to the input variable.

2. The transfer function is a property of a system itself, independent of the magnitude and nature of the input or driving function.

3. The transfer function includes the units necessary to relate the input to the output; however, it does not provide any information concerning the physical structure of the system. (The transfer functions of many physically different systems can be identical.)

4. If the transfer function of a system is known, the output or response can be studied for various forms of inputs with a view toward understanding the nature of the system.

5. If the transfer function of a system is unknown, it may be established experimentally by introducing known inputs and studying the output of the system. Once established, a transfer function gives a full description of the dynamic characteristics of the system, as distinct from its physical description.

Convolution Integral. For a linear, time-invariant system the transfer function $G(s)$ is

$$G(s) = \frac{Y(s)}{X(s)}$$

where $X(s)$ is the Laplace transform of the input to the system and $Y(s)$ is the Laplace transform of the output of the system, where we assume that all initial conditions involved are zero. It follows that the output $Y(s)$ can be written as the product of $G(s)$ and $X(s)$, or

$$Y(s) = G(s)X(s) \qquad (2\text{--}1)$$

Note that multiplication in the complex domain is equivalent to convolution in the time domain (see Appendix A), so the inverse Laplace transform of Equation (2–1) is given by the following convolution integral:

$$y(t) = \int_0^t x(\tau)g(t - \tau)\,d\tau$$

$$= \int_0^t g(\tau)x(t - \tau)\,d\tau$$

where both $g(t)$ and $x(t)$ are 0 for $t < 0$.

Impulse-Response Function. Consider the output (response) of a linear time-invariant system to a unit-impulse input when the initial conditions are zero. Since the Laplace transform of the unit-impulse function is unity, the Laplace transform of the output of the system is

$$Y(s) = G(s) \qquad (2\text{--}2)$$

The inverse Laplace transform of the output given by Equation (2–2) gives the impulse response of the system. The inverse Laplace transform of $G(s)$, or

$$\mathcal{L}^{-1}[G(s)] = g(t)$$

is called the impulse-response function. This function $g(t)$ is also called the weighting function of the system.

The impulse-response function $g(t)$ is thus the response of a linear time-invariant system to a unit-impulse input when the initial conditions are zero. The Laplace transform of this function gives the transfer function. Therefore, the transfer function and impulse-response function of a linear, time-invariant system contain the same information about the system dynamics. It is hence possible to obtain complete information about the dynamic characteristics of the system by exciting it with an impulse input and measuring the response. (In practice, a pulse input with a very short duration compared with the significant time constants of the system can be considered an impulse.)

2–3 AUTOMATIC CONTROL SYSTEMS

A control system may consist of a number of components. To show the functions performed by each component, in control engineering, we commonly use a diagram called the *block diagram*. This section first explains what a block diagram is. Next, it discusses introductory aspects of automatic control systems, including various control actions. Then, it presents a method for obtaining block diagrams for physical systems, and, finally, discusses techniques to simplify such diagrams.

Block Diagrams. A *block diagram* of a system is a pictorial representation of the functions performed by each component and of the flow of signals. Such a diagram depicts the interrelationships that exist among the various components. Differing from a purely abstract mathematical representation, a block diagram has the advantage of indicating more realistically the signal flows of the actual system.

In a block diagram all system variables are linked to each other through functional blocks. The *functional* block or simply *block* is a symbol for the mathematical operation on the input signal to the block that produces the output. The transfer functions of the components are usually entered in the corresponding blocks, which are connected by arrows to indicate the direction of the flow of signals. Note that the signal can pass only in the direction of the arrows. Thus a block diagram of a control system explicitly shows a unilateral property.

Figure 2–1 shows an element of the block diagram. The arrowhead pointing toward the block indicates the input, and the arrowhead leading away from the block represents the output. Such arrows are referred to as *signals*.

Figure 2–1
Element of a block diagram.

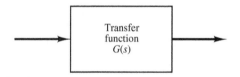

Note that the dimension of the output signal from the block is the dimension of the input signal multiplied by the dimension of the transfer function in the block.

The advantages of the block diagram representation of a system are that it is easy to form the overall block diagram for the entire system by merely connecting the blocks of the components according to the signal flow and that it is possible to evaluate the contribution of each component to the overall performance of the system.

In general, the functional operation of the system can be visualized more readily by examining the block diagram than by examining the physical system itself. A block diagram contains information concerning dynamic behavior, but it does not include any information on the physical construction of the system. Consequently, many dissimilar and unrelated systems can be represented by the same block diagram.

It should be noted that in a block diagram the main source of energy is not explicitly shown and that the block diagram of a given system is not unique. A number of different block diagrams can be drawn for a system, depending on the point of view of the analysis.

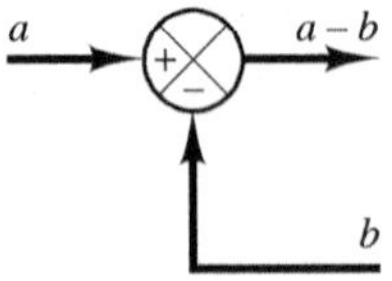

Summing Point. Referring to Figure 2–2, a circle with a cross is the symbol that indicates a summing operation. The plus or minus sign at each arrowhead indicates whether that signal is to be added or subtracted. It is important that the quantities being added or subtracted have the same dimensions and the same units.

Branch Point. A *branch point* is a point from which the signal from a block goes concurrently to other blocks or summing points.

Figure 2–2
Summing point.

Block Diagram of a Closed-Loop System. Figure 2–3 shows an example of a block diagram of a closed-loop system. The output $C(s)$ is fed back to the summing point, where it is compared with the reference input $R(s)$. The closed-loop nature of the system is clearly indicated by the figure. The output of the block, $C(s)$ in this case, is obtained by multiplying the transfer function $G(s)$ by the input to the block, $E(s)$. Any linear control system may be represented by a block diagram consisting of blocks, summing points, and branch points.

When the output is fed back to the summing point for comparison with the input, it is necessary to convert the form of the output signal to that of the input signal. For example, in a temperature control system, the output signal is usually the controlled temperature. The output signal, which has the dimension of temperature, must be converted to a force or position or voltage before it can be compared with the input signal. This conversion is accomplished by the feedback element whose transfer function is $H(s)$, as shown in Figure 2–4. The role of the feedback element is to modify the output before it is compared with the input. (In most cases the feedback element is a sensor that measures

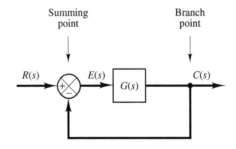

Figure 2–3
Block diagram of a closed-loop system.

Chapter 2 / **Mathematical Modeling of Control Systems**

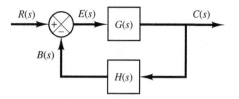

Figure 2–4
Closed-loop system.

the output of the plant. The output of the sensor is compared with the system input, and the actuating error signal is generated.) In the present example, the feedback signal that is fed back to the summing point for comparison with the input is $B(s) = H(s)C(s)$.

Open-Loop Transfer Function and Feedforward Transfer Function. Referring to Figure 2–4, the ratio of the feedback signal $B(s)$ to the actuating error signal $E(s)$ is called the *open-loop transfer function*. That is,

$$\text{Open-loop transfer function} = \frac{B(s)}{E(s)} = G(s)H(s)$$

The ratio of the output $C(s)$ to the actuating error signal $E(s)$ is called the *feedforward transfer function*, so that

$$\text{Feedforward transfer function} = \frac{C(s)}{E(s)} = G(s)$$

If the feedback transfer function $H(s)$ is unity, then the open-loop transfer function and the feedforward transfer function are the same.

Closed-Loop Transfer Function. For the system shown in Figure 2–4, the output $C(s)$ and input $R(s)$ are related as follows: since

$$C(s) = G(s)E(s)$$
$$E(s) = R(s) - B(s)$$
$$= R(s) - H(s)C(s)$$

eliminating $E(s)$ from these equations gives

$$C(s) = G(s)[R(s) - H(s)C(s)]$$

or

$$\frac{C(s)}{R(s)} = \frac{G(s)}{1 + G(s)H(s)} \tag{2–3}$$

The transfer function relating $C(s)$ to $R(s)$ is called the *closed-loop transfer function*. It relates the closed-loop system dynamics to the dynamics of the feedforward elements and feedback elements.

From Equation (2–3), $C(s)$ is given by

$$C(s) = \frac{G(s)}{1 + G(s)H(s)} R(s)$$

Thus the output of the closed-loop system clearly depends on both the closed-loop transfer function and the nature of the input.

Obtaining Cascaded, Parallel, and Feedback (Closed-Loop) Transfer Functions with MATLAB. In control-systems analysis, we frequently need to calculate the cascaded transfer functions, parallel-connected transfer functions, and feedback-connected (closed-loop) transfer functions. MATLAB has convenient commands to obtain the cascaded, parallel, and feedback (closed-loop) transfer functions.

Suppose that there are two components $G_1(s)$ and $G_2(s)$ connected differently as shown in Figure 2–5 (a), (b), and (c), where

$$G_1(s) = \frac{\text{num1}}{\text{den1}}, \qquad G_2(s) = \frac{\text{num2}}{\text{den2}}$$

To obtain the transfer functions of the cascaded system, parallel system, or feedback (closed-loop) system, the following commands may be used:

[num, den] = series(num1,den1,num2,den2)
[num, den] = parallel(num1,den1,num2,den2)
[num, den] = feedback(num1,den1,num2,den2)

As an example, consider the case where

$$G_1(s) = \frac{10}{s^2 + 2s + 10} = \frac{\text{num1}}{\text{den1}}, \qquad G_2(s) = \frac{5}{s + 5} = \frac{\text{num2}}{\text{den2}}$$

MATLAB Program 2–1 gives $C(s)/R(s) = $ num/den for each arrangement of $G_1(s)$ and $G_2(s)$. Note that the command

printsys(num,den)

displays the num/den [that is, the transfer function $C(s)/R(s)$] of the system considered.

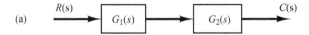

(a)

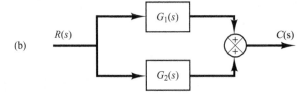

(b)

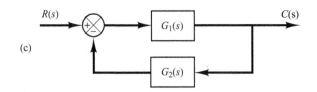

(c)

Figure 2–5
(a) Cascaded system;
(b) parallel system;
(c) feedback (closed-loop) system.

Chapter 2 / Mathematical Modeling of Control Systems

```
MATLAB Program 2-1

num1 = [10];
den1 = [1  2  10];
num2 = [5];
den2 = [1  5];
[num, den] = series(num1,den1,num2,den2);
printsys(num,den)

num/den =

                  50
      ─────────────────────────
      s^3 + 7s^2 + 20s + 50

[num, den] = parallel(num1,den1,num2,den2);
printsys(num,den)

num/den =

         5s^2 + 20s + 100
      ─────────────────────────
      s^3 + 7s^2 + 20s + 50

[num, den] = feedback(num1,den1,num2,den2);
printsys(num,den)

num/den =

              10s + 50
      ─────────────────────────
      s^3 + 7s^2 + 20s + 100
```

Automatic Controllers. An automatic controller compares the actual value of the plant output with the reference input (desired value), determines the deviation, and produces a control signal that will reduce the deviation to zero or to a small value. The manner in which the automatic controller produces the control signal is called the *control action.* Figure 2–6 is a block diagram of an industrial control system, which

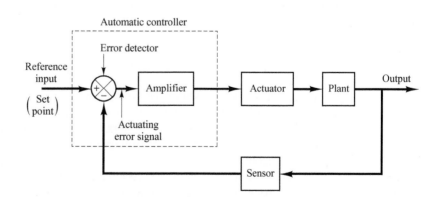

Figure 2–6
Block diagram of an industrial control system, which consists of an automatic controller, an actuator, a plant, and a sensor (measuring element).

consists of an automatic controller, an actuator, a plant, and a sensor (measuring element). The controller detects the actuating error signal, which is usually at a very low power level, and amplifies it to a sufficiently high level. The output of an automatic controller is fed to an actuator, such as an electric motor, a hydraulic motor, or a pneumatic motor or valve. (The actuator is a power device that produces the input to the plant according to the control signal so that the output signal will approach the reference input signal.)

The sensor or measuring element is a device that converts the output variable into another suitable variable, such as a displacement, pressure, voltage, etc., that can be used to compare the output to the reference input signal. This element is in the feedback path of the closed-loop system. The set point of the controller must be converted to a reference input with the same units as the feedback signal from the sensor or measuring element.

Classifications of Industrial Controllers. Most industrial controllers may be classified according to their control actions as:

1. Two-position or on–off controllers
2. Proportional controllers
3. Integral controllers
4. Proportional-plus-integral controllers
5. Proportional-plus-derivative controllers
6. Proportional-plus-integral-plus-derivative controllers

Most industrial controllers use electricity or pressurized fluid such as oil or air as power sources. Consequently, controllers may also be classified according to the kind of power employed in the operation, such as pneumatic controllers, hydraulic controllers, or electronic controllers. What kind of controller to use must be decided based on the nature of the plant and the operating conditions, including such considerations as safety, cost, availability, reliability, accuracy, weight, and size.

Two-Position or On–Off Control Action. In a two-position control system, the actuating element has only two fixed positions, which are, in many cases, simply on and off. Two-position or on–off control is relatively simple and inexpensive and, for this reason, is very widely used in both industrial and domestic control systems.

Let the output signal from the controller be $u(t)$ and the actuating error signal be $e(t)$. In two-position control, the signal $u(t)$ remains at either a maximum or minimum value, depending on whether the actuating error signal is positive or negative, so that

$$u(t) = U_1, \quad \text{for } e(t) > 0$$
$$= U_2, \quad \text{for } e(t) < 0$$

where U_1 and U_2 are constants. The minimum value U_2 is usually either zero or $-U_1$. Two-position controllers are generally electrical devices, and an electric solenoid-operated valve is widely used in such controllers. Pneumatic proportional controllers with very high gains act as two-position controllers and are sometimes called pneumatic two-position controllers.

Figures 2–7(a) and (b) show the block diagrams for two-position or on–off controllers. The range through which the actuating error signal must move before the switching occurs

Figure 2–7
(a) Block diagram of
an on–off controller;
(b) block diagram of
an on–off controller
with differential gap.

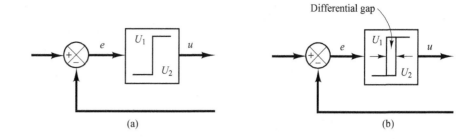

(a)   (b)

is called the *differential gap*. A differential gap is indicated in Figure 2–7(b). Such a differential gap causes the controller output $u(t)$ to maintain its present value until the actuating error signal has moved slightly beyond the zero value. In some cases, the differential gap is a result of unintentional friction and lost motion; however, quite often it is intentionally provided in order to prevent too-frequent operation of the on–off mechanism.

Consider the liquid-level control system shown in Figure 2–8(a), where the electromagnetic valve shown in Figure 2–8(b) is used for controlling the inflow rate. This valve is either open or closed. With this two-position control, the water inflow rate is either a positive constant or zero. As shown in Figure 2–9, the output signal continuously moves between the two limits required to cause the actuating element to move from one fixed position to the other. Notice that the output curve follows one of two exponential curves, one corresponding to the filling curve and the other to the emptying curve. Such output oscillation between two limits is a typical response characteristic of a system under two-position control.

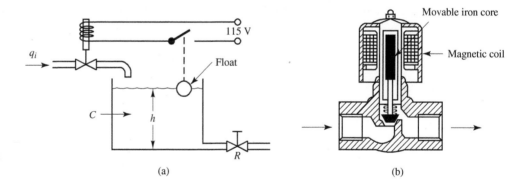

Figure 2–8
(a) Liquid-level
control system;
(b) electromagnetic
valve.

(a)   (b)

Figure 2–9
Level $h(t)$-versus-t
curve for the system
shown in Figure 2–8(a).

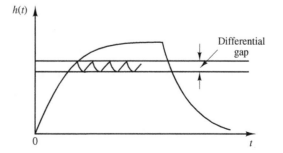

From Figure 2–9, we notice that the amplitude of the output oscillation can be reduced by decreasing the differential gap. The decrease in the differential gap, however, increases the number of on–off switchings per minute and reduces the useful life of the component. The magnitude of the differential gap must be determined from such considerations as the accuracy required and the life of the component.

Proportional Control Action. For a controller with proportional control action, the relationship between the output of the controller $u(t)$ and the actuating error signal $e(t)$ is

$$u(t) = K_p e(t)$$

or, in Laplace-transformed quantities,

$$\frac{U(s)}{E(s)} = K_p$$

where K_p is termed the proportional gain.

Whatever the actual mechanism may be and whatever the form of the operating power, the proportional controller is essentially an amplifier with an adjustable gain.

Integral Control Action. In a controller with integral control action, the value of the controller output $u(t)$ is changed at a rate proportional to the actuating error signal $e(t)$. That is,

$$\frac{du(t)}{dt} = K_i e(t)$$

or

$$u(t) = K_i \int_0^t e(t)\, dt$$

where K_i is an adjustable constant. The transfer function of the integral controller is

$$\frac{U(s)}{E(s)} = \frac{K_i}{s}$$

Proportional-Plus-Integral Control Action. The control action of a proportional-plus-integral controller is defined by

$$u(t) = K_p e(t) + \frac{K_p}{T_i} \int_0^t e(t)\, dt$$

or the transfer function of the controller is

$$\frac{U(s)}{E(s)} = K_p\left(1 + \frac{1}{T_i s}\right)$$

where T_i is called the *integral time*.

Proportional-Plus-Derivative Control Action. The control action of a proportional-plus-derivative controller is defined by

$$u(t) = K_p e(t) + K_p T_d \frac{de(t)}{dt}$$

and the transfer function is

$$\frac{U(s)}{E(s)} = K_p(1 + T_d s)$$

where T_d is called the *derivative time*.

Proportional-Plus-Integral-Plus-Derivative Control Action. The combination of proportional control action, integral control action, and derivative control action is termed proportional-plus-integral-plus-derivative control action. It has the advantages of each of the three individual control actions. The equation of a controller with this combined action is given by

$$u(t) = K_p e(t) + \frac{K_p}{T_i} \int_0^t e(t)\,dt + K_p T_d \frac{de(t)}{dt}$$

or the transfer function is

$$\frac{U(s)}{E(s)} = K_p\left(1 + \frac{1}{T_i s} + T_d s\right)$$

where K_p is the proportional gain, T_i is the integral time, and T_d is the derivative time. The block diagram of a proportional-plus-integral-plus-derivative controller is shown in Figure 2–10.

Figure 2–10
Block diagram of a proportional-plus-integral-plus-derivative controller.

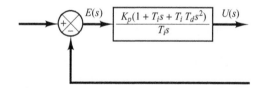

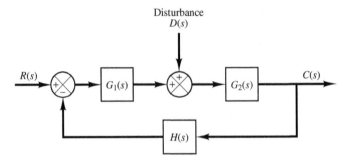

Figure 2–11
Closed-loop system
subjected to a
disturbance.

Closed-Loop System Subjected to a Disturbance.　Figure 2–11 shows a closed-loop system subjected to a disturbance. When two inputs (the reference input and disturbance) are present in a linear time-invariant system, each input can be treated independently of the other; and the outputs corresponding to each input alone can be added to give the complete output. The way each input is introduced into the system is shown at the summing point by either a plus or minus sign.

Consider the system shown in Figure 2–11. In examining the effect of the disturbance $D(s)$, we may assume that the reference input is zero; we may then calculate the response $C_D(s)$ to the disturbance only. This response can be found from

$$\frac{C_D(s)}{D(s)} = \frac{G_2(s)}{1 + G_1(s)G_2(s)H(s)}$$

On the other hand, in considering the response to the reference input $R(s)$, we may assume that the disturbance is zero. Then the response $C_R(s)$ to the reference input $R(s)$ can be obtained from

$$\frac{C_R(s)}{R(s)} = \frac{G_1(s)G_2(s)}{1 + G_1(s)G_2(s)H(s)}$$

The response to the simultaneous application of the reference input and disturbance can be obtained by adding the two individual responses. In other words, the response $C(s)$ due to the simultaneous application of the reference input $R(s)$ and disturbance $D(s)$ is given by

$$C(s) = C_R(s) + C_D(s)$$

$$= \frac{G_2(s)}{1 + G_1(s)G_2(s)H(s)} \left[G_1(s)R(s) + D(s) \right]$$

Consider now the case where $|G_1(s)H(s)| \gg 1$ and $|G_1(s)G_2(s)H(s)| \gg 1$. In this case, the closed-loop transfer function $C_D(s)/D(s)$ becomes almost zero, and the effect of the disturbance is suppressed. This is an advantage of the closed-loop system.

On the other hand, the closed-loop transfer function $C_R(s)/R(s)$ approaches $1/H(s)$ as the gain of $G_1(s)G_2(s)H(s)$ increases. This means that if $|G_1(s)G_2(s)H(s)| \gg 1$, then the closed-loop transfer function $C_R(s)/R(s)$ becomes independent of $G_1(s)$ and $G_2(s)$ and inversely proportional to $H(s)$, so that the variations of $G_1(s)$ and $G_2(s)$ do not affect the closed-loop transfer function $C_R(s)/R(s)$. This is another advantage of the closed-loop system. It can easily be seen that any closed-loop system with unity feedback, $H(s) = 1$, tends to equalize the input and output.

Procedures for Drawing a Block Diagram. To draw a block diagram for a system, first write the equations that describe the dynamic behavior of each component. Then take the Laplace transforms of these equations, assuming zero initial conditions, and represent each Laplace-transformed equation individually in block form. Finally, assemble the elements into a complete block diagram.

As an example, consider the RC circuit shown in Figure 2–12(a). The equations for this circuit are

$$i = \frac{e_i - e_o}{R} \tag{2-4}$$

$$e_o = \frac{\int i\, dt}{C} \tag{2-5}$$

The Laplace transforms of Equations (2–4) and (2–5), with zero initial condition, become

$$I(s) = \frac{E_i(s) - E_o(s)}{R} \tag{2-6}$$

$$E_o(s) = \frac{I(s)}{Cs} \tag{2-7}$$

Equation (2–6) represents a summing operation, and the corresponding diagram is shown in Figure 2–12(b). Equation (2–7) represents the block as shown in Figure 2–12(c). Assembling these two elements, we obtain the overall block diagram for the system as shown in Figure 2–12(d).

Block Diagram Reduction. It is important to note that blocks can be connected in series only if the output of one block is not affected by the next following block. If there are any loading effects between the components, it is necessary to combine these components into a single block.

Any number of cascaded blocks representing nonloading components can be replaced by a single block, the transfer function of which is simply the product of the individual transfer functions.

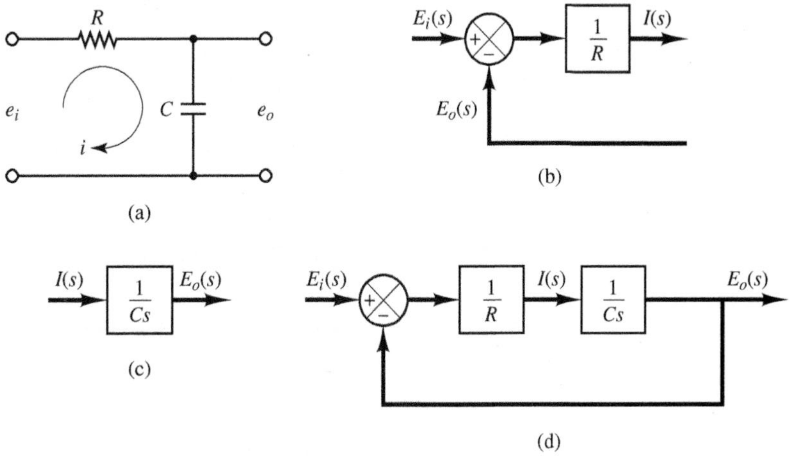

Figure 2–12
(a) RC circuit;
(b) block diagram representing Equation (2–6);
(c) block diagram representing Equation (2–7);
(d) block diagram of the RC circuit.

A complicated block diagram involving many feedback loops can be simplified by a step-by-step rearrangement. Simplification of the block diagram by rearrangements considerably reduces the labor needed for subsequent mathematical analysis. It should be noted, however, that as the block diagram is simplified, the transfer functions in new blocks become more complex because new poles and new zeros are generated.

EXAMPLE 2–1 Consider the system shown in Figure 2–13(a). Simplify this diagram.

By moving the summing point of the negative feedback loop containing H_2 outside the positive feedback loop containing H_1, we obtain Figure 2–13(b). Eliminating the positive feedback loop, we have Figure 2–13(c). The elimination of the loop containing H_2/G_1 gives Figure 2–13(d). Finally, eliminating the feedback loop results in Figure 2–13(e).

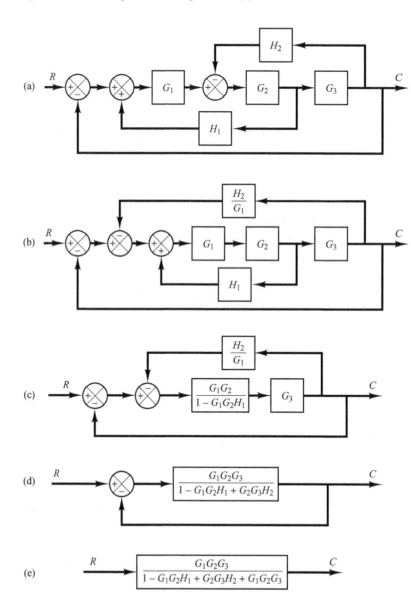

Figure 2–13
(a) Multiple-loop system;
(b)–(e) successive reductions of the block diagram shown in (a).

Chapter 2 / Mathematical Modeling of Control Systems

Notice that the numerator of the closed-loop transfer function $C(s)/R(s)$ is the product of the transfer functions of the feedforward path. The denominator of $C(s)/R(s)$ is equal to

$$1 + \sum (\text{product of the transfer functions around each loop})$$

$$= 1 + (-G_1 G_2 H_1 + G_2 G_3 H_2 + G_1 G_2 G_3)$$

$$= 1 - G_1 G_2 H_1 + G_2 G_3 H_2 + G_1 G_2 G_3$$

(The positive feedback loop yields a negative term in the denominator.)

2–4 MODELING IN STATE SPACE

In this section we shall present introductory material on state-space analysis of control systems.

Modern Control Theory. The modern trend in engineering systems is toward greater complexity, due mainly to the requirements of complex tasks and good accuracy. Complex systems may have multiple inputs and multiple outputs and may be time varying. Because of the necessity of meeting increasingly stringent requirements on the performance of control systems, the increase in system complexity, and easy access to large scale computers, modern control theory, which is a new approach to the analysis and design of complex control systems, has been developed since around 1960. This new approach is based on the concept of state. The concept of state by itself is not new, since it has been in existence for a long time in the field of classical dynamics and other fields.

Modern Control Theory Versus Conventional Control Theory. Modern control theory is contrasted with conventional control theory in that the former is applicable to multiple-input, multiple-output systems, which may be linear or nonlinear, time invariant or time varying, while the latter is applicable only to linear time-invariant single-input, single-output systems. Also, modern control theory is essentially time-domain approach and frequency domain approach (in certain cases such as H-infinity control), while conventional control theory is a complex frequency-domain approach. Before we proceed further, we must define state, state variables, state vector, and state space.

State. The state of a dynamic system is the smallest set of variables (called *state variables*) such that knowledge of these variables at $t = t_0$, together with knowledge of the input for $t \geq t_0$, completely determines the behavior of the system for any time $t \geq t_0$.

Note that the concept of state is by no means limited to physical systems. It is applicable to biological systems, economic systems, social systems, and others.

State Variables. The state variables of a dynamic system are the variables making up the smallest set of variables that determine the state of the dynamic system. If at least n variables $x_1, x_2, \ldots, x_n$ are needed to completely describe the behavior of a dynamic system (so that once the input is given for $t \geq t_0$ and the initial state at $t = t_0$ is specified, the future state of the system is completely determined), then such n variables are a set of state variables.

Note that state variables need not be physically measurable or observable quantities. Variables that do not represent physical quantities and those that are neither measurable nor observable can be chosen as state variables. Such freedom in choosing state variables is an advantage of the state-space methods. Practically, however, it is convenient to choose easily measurable quantities for the state variables, if this is possible at all, because optimal control laws will require the feedback of all state variables with suitable weighting.

State Vector. If n state variables are needed to completely describe the behavior of a given system, then these n state variables can be considered the n components of a vector $\mathbf{x}$. Such a vector is called a *state vector*. A state vector is thus a vector that determines uniquely the system state $\mathbf{x}(t)$ for any time $t \geq t_0$, once the state at $t = t_0$ is given and the input $u(t)$ for $t \geq t_0$ is specified.

State Space. The n-dimensional space whose coordinate axes consist of the x_1 axis, x_2 axis, $\ldots, x_n$ axis, where $x_1, x_2, \ldots, x_n$ are state variables, is called a *state space*. Any state can be represented by a point in the state space.

State-Space Equations. In state-space analysis we are concerned with three types of variables that are involved in the modeling of dynamic systems: input variables, output variables, and state variables. As we shall see in Section 2–5, the state-space representation for a given system is not unique, except that the number of state variables is the same for any of the different state-space representations of the same system.

The dynamic system must involve elements that memorize the values of the input for $t \geq t_1$. Since integrators in a continuous-time control system serve as memory devices, the outputs of such integrators can be considered as the variables that define the internal state of the dynamic system. Thus the outputs of integrators serve as state variables. The number of state variables to completely define the dynamics of the system is equal to the number of integrators involved in the system.

Assume that a multiple-input, multiple-output system involves n integrators. Assume also that there are r inputs $u_1(t), u_2(t), \ldots, u_r(t)$ and m outputs $y_1(t), y_2(t), \ldots, y_m(t)$. Define n outputs of the integrators as state variables: $x_1(t), x_2(t), \ldots, x_n(t)$ Then the system may be described by

$$\dot{x}_1(t) = f_1(x_1, x_2, \ldots, x_n; u_1, u_2, \ldots, u_r; t)$$
$$\dot{x}_2(t) = f_2(x_1, x_2, \ldots, x_n; u_1, u_2, \ldots, u_r; t)$$

$$\cdot$$
$$\cdot \qquad\qquad\qquad\qquad\qquad\qquad\qquad (2\text{--}8)$$
$$\cdot$$

$$\dot{x}_n(t) = f_n(x_1, x_2, \ldots, x_n; u_1, u_2, \ldots, u_r; t)$$

The outputs $y_1(t), y_2(t), \ldots, y_m(t)$ of the system may be given by

$$y_1(t) = g_1(x_1, x_2, \ldots, x_n; u_1, u_2, \ldots, u_r; t)$$
$$y_2(t) = g_2(x_1, x_2, \ldots, x_n; u_1, u_2, \ldots, u_r; t)$$

$$\cdot$$
$$\cdot \qquad\qquad (2\text{--}9)$$
$$\cdot$$

$$y_m(t) = g_m(x_1, x_2, \ldots, x_n; u_1, u_2, \ldots, u_r; t)$$

If we define

$$
\mathbf{x}(t) = \begin{bmatrix} x_1(t) \\ x_2(t) \\ \cdot \\ \cdot \\ \cdot \\ x_n(t) \end{bmatrix}, \quad
\mathbf{f}(\mathbf{x}, \mathbf{u}, t) = \begin{bmatrix} f_1(x_1, x_2, \ldots, x_n; u_1, u_2, \ldots, u_r; t) \\ f_2(x_1, x_2, \ldots, x_n; u_1, u_2, \ldots, u_r; t) \\ \cdot \\ \cdot \\ \cdot \\ f_n(x_1, x_2, \ldots, x_n; u_1, u_2, \ldots, u_r; t) \end{bmatrix},
$$

$$
\mathbf{y}(t) = \begin{bmatrix} y_1(t) \\ y_2(t) \\ \cdot \\ \cdot \\ \cdot \\ y_m(t) \end{bmatrix}, \quad
\mathbf{g}(\mathbf{x}, \mathbf{u}, t) = \begin{bmatrix} g_1(x_1, x_2, \ldots, x_n; u_1, u_2, \ldots, u_r; t) \\ g_2(x_1, x_2, \ldots, x_n; u_1, u_2, \ldots, u_r; t) \\ \cdot \\ \cdot \\ \cdot \\ g_m(x_1, x_2, \ldots, x_n; u_1, u_2, \ldots, u_r; t) \end{bmatrix}, \quad
\mathbf{u}(t) = \begin{bmatrix} u_1(t) \\ u_2(t) \\ \cdot \\ \cdot \\ \cdot \\ u_r(t) \end{bmatrix}
$$

then Equations (2–8) and (2–9) become

$$\dot{\mathbf{x}}(t) = \mathbf{f}(\mathbf{x}, \mathbf{u}, t) \qquad\qquad (2\text{--}10)$$

$$\mathbf{y}(t) = \mathbf{g}(\mathbf{x}, \mathbf{u}, t) \qquad\qquad (2\text{--}11)$$

where Equation (2–10) is the state equation and Equation (2–11) is the output equation. If vector functions $\mathbf{f}$ and/or $\mathbf{g}$ involve time t explicitly, then the system is called a time-varying system.

If Equations (2–10) and (2–11) are linearized about the operating state, then we have the following linearized state equation and output equation:

$$\dot{\mathbf{x}}(t) = \mathbf{A}(t)\mathbf{x}(t) + \mathbf{B}(t)\mathbf{u}(t) \qquad\qquad (2\text{--}12)$$

$$\mathbf{y}(t) = \mathbf{C}(t)\mathbf{x}(t) + \mathbf{D}(t)\mathbf{u}(t) \qquad\qquad (2\text{--}13)$$

where $\mathbf{A}(t)$ is called the state matrix, $\mathbf{B}(t)$ the input matrix, $\mathbf{C}(t)$ the output matrix, and $\mathbf{D}(t)$ the direct transmission matrix. (Details of linearization of nonlinear systems about

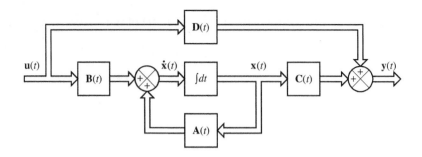

Figure 2–14
Block diagram of the linear, continuous-time control system represented in state space.

the operating state are discussed in Section 2–7.) A block diagram representation of Equations (2–12) and (2–13) is shown in Figure 2–14.

If vector functions **f** and **g** do not involve time t explicitly then the system is called a time-invariant system. In this case, Equations (2–12) and (2–13) can be simplified to

$$\dot{\mathbf{x}}(t) = \mathbf{A}\mathbf{x}(t) + \mathbf{B}\mathbf{u}(t) \qquad (2\text{–}14)$$

$$\mathbf{y}(t) = \mathbf{C}\mathbf{x}(t) + \mathbf{D}\mathbf{u}(t) \qquad (2\text{–}15)$$

Equation (2–14) is the state equation of the linear, time-invariant system and Equation (2–15) is the output equation for the same system. In this book we shall be concerned mostly with systems described by Equations (2–14) and (2–15).

In what follows we shall present an example for deriving a state equation and output equation.

EXAMPLE 2–2 Consider the mechanical system shown in Figure 2–15. We assume that the system is linear. The external force $u(t)$ is the input to the system, and the displacement $y(t)$ of the mass is the output. The displacement $y(t)$ is measured from the equilibrium position in the absence of the external force. This system is a single-input, single-output system.

From the diagram, the system equation is

$$m\ddot{y} + b\dot{y} + ky = u \qquad (2\text{–}16)$$

This system is of second order. This means that the system involves two integrators. Let us define state variables $x_1(t)$ and $x_2(t)$ as

$$x_1(t) = y(t)$$
$$x_2(t) = \dot{y}(t)$$

Then we obtain

$$\dot{x}_1 = x_2$$

$$\dot{x}_2 = \frac{1}{m}(-ky - b\dot{y}) + \frac{1}{m}u$$

or

$$\dot{x}_1 = x_2 \qquad (2\text{–}17)$$

$$\dot{x}_2 = -\frac{k}{m}x_1 - \frac{b}{m}x_2 + \frac{1}{m}u \qquad (2\text{–}18)$$

The output equation is

$$y = x_1 \qquad (2\text{–}19)$$

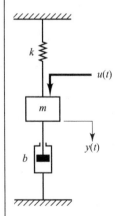

Figure 2–15
Mechanical system.

Chapter 2 / Mathematical Modeling of Control Systems

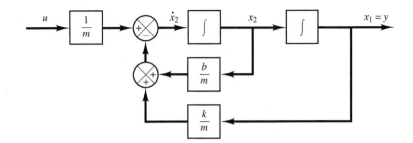

Figure 2–16
Block diagram of the
mechanical system
shown in Figure 2–15.

In a vector-matrix form, Equations (2–17) and (2–18) can be written as

$$\begin{bmatrix} \dot{x}_1 \\ \dot{x}_2 \end{bmatrix} = \begin{bmatrix} 0 & 1 \\ -\dfrac{k}{m} & -\dfrac{b}{m} \end{bmatrix} \begin{bmatrix} x_1 \\ x_2 \end{bmatrix} + \begin{bmatrix} 0 \\ \dfrac{1}{m} \end{bmatrix} u \qquad (2\text{--}20)$$

The output equation, Equation (2–19), can be written as

$$y = \begin{bmatrix} 1 & 0 \end{bmatrix} \begin{bmatrix} x_1 \\ x_2 \end{bmatrix} \qquad (2\text{--}21)$$

Equation (2–20) is a state equation and Equation (2–21) is an output equation for the system.
They are in the standard form:

$$\dot{\mathbf{x}} = \mathbf{A}\mathbf{x} + \mathbf{B}u$$
$$y = \mathbf{C}\mathbf{x} + Du$$

where

$$\mathbf{A} = \begin{bmatrix} 0 & 1 \\ -\dfrac{k}{m} & -\dfrac{b}{m} \end{bmatrix}, \qquad \mathbf{B} = \begin{bmatrix} 0 \\ \dfrac{1}{m} \end{bmatrix}, \qquad \mathbf{C} = \begin{bmatrix} 1 & 0 \end{bmatrix}, \qquad D = 0$$

Figure 2–16 is a block diagram for the system. Notice that the outputs of the integrators are state
variables.

Correlation Between Transfer Functions and State-Space Equations. In what
follows we shall show how to derive the transfer function of a single-input, single-output
system from the state-space equations.

Let us consider the system whose transfer function is given by

$$\frac{Y(s)}{U(s)} = G(s) \qquad (2\text{--}22)$$

This system may be represented in state space by the following equations:

$$\dot{\mathbf{x}} = \mathbf{A}\mathbf{x} + \mathbf{B}u \qquad (2\text{--}23)$$

$$y = \mathbf{C}\mathbf{x} + Du \qquad (2\text{--}24)$$

where $\mathbf{x}$ is the state vector, u is the input, and y is the output. The Laplace transforms of Equations (2–23) and (2–24) are given by

$$s\mathbf{X}(s) - \mathbf{x}(0) = \mathbf{A}\mathbf{X}(s) + \mathbf{B}U(s) \tag{2–25}$$

$$Y(s) = \mathbf{C}\mathbf{X}(s) + DU(s) \tag{2–26}$$

Since the transfer function was previously defined as the ratio of the Laplace transform of the output to the Laplace transform of the input when the initial conditions were zero, we set $\mathbf{x}(0)$ in Equation (2–25) to be zero. Then we have

$$s\mathbf{X}(s) - \mathbf{A}\mathbf{X}(s) = \mathbf{B}U(s)$$

or

$$(s\mathbf{I} - \mathbf{A})\mathbf{X}(s) = \mathbf{B}U(s)$$

By premultiplying $(s\mathbf{I} - \mathbf{A})^{-1}$ to both sides of this last equation, we obtain

$$\mathbf{X}(s) = (s\mathbf{I} - \mathbf{A})^{-1}\mathbf{B}U(s) \tag{2–27}$$

By substituting Equation (2–27) into Equation (2–26), we get

$$Y(s) = \left[\mathbf{C}(s\mathbf{I} - \mathbf{A})^{-1}\mathbf{B} + D\right]U(s) \tag{2–28}$$

Upon comparing Equation (2–28) with Equation (2–22), we see that

$$G(s) = \mathbf{C}(s\mathbf{I} - \mathbf{A})^{-1}\mathbf{B} + D \tag{2–29}$$

This is the transfer-function expression of the system in terms of $\mathbf{A}, \mathbf{B}, \mathbf{C}$, and D.

Note that the right-hand side of Equation (2–29) involves $(s\mathbf{I} - \mathbf{A})^{-1}$. Hence $G(s)$ can be written as

$$G(s) = \frac{Q(s)}{|s\mathbf{I} - \mathbf{A}|}$$

where $Q(s)$ is a polynomial in s. Notice that $|s\mathbf{I} - \mathbf{A}|$ is equal to the characteristic polynomial of $G(s)$. In other words, the eigenvalues of $\mathbf{A}$ are identical to the poles of $G(s)$.

EXAMPLE 2–3 Consider again the mechanical system shown in Figure 2–15. State-space equations for the system are given by Equations (2–20) and (2–21). We shall obtain the transfer function for the system from the state-space equations.

By substituting $\mathbf{A}, \mathbf{B}, \mathbf{C}$, and D into Equation (2–29), we obtain

$$G(s) = \mathbf{C}(s\mathbf{I} - \mathbf{A})^{-1}\mathbf{B} + D$$

$$= \begin{bmatrix} 1 & 0 \end{bmatrix} \left\{ \begin{bmatrix} s & 0 \\ 0 & s \end{bmatrix} - \begin{bmatrix} 0 & 1 \\ -\dfrac{k}{m} & -\dfrac{b}{m} \end{bmatrix} \right\}^{-1} \begin{bmatrix} 0 \\ \dfrac{1}{m} \end{bmatrix} + 0$$

$$= \begin{bmatrix} 1 & 0 \end{bmatrix} \begin{bmatrix} s & -1 \\ \dfrac{k}{m} & s + \dfrac{b}{m} \end{bmatrix}^{-1} \begin{bmatrix} 0 \\ \dfrac{1}{m} \end{bmatrix}$$

Note that

$$
\begin{bmatrix} s & -1 \\ \dfrac{k}{m} & s + \dfrac{b}{m} \end{bmatrix}^{-1} = \dfrac{1}{s^2 + \dfrac{b}{m}s + \dfrac{k}{m}} \begin{bmatrix} s + \dfrac{b}{m} & 1 \\ -\dfrac{k}{m} & s \end{bmatrix}
$$

(Refer to Appendix C for the inverse of the 2 × 2 matrix.)
Thus, we have

$$
G(s) = \begin{bmatrix} 1 & 0 \end{bmatrix} \dfrac{1}{s^2 + \dfrac{b}{m}s + \dfrac{k}{m}} \begin{bmatrix} s + \dfrac{b}{m} & 1 \\ -\dfrac{k}{m} & s \end{bmatrix} \begin{bmatrix} 0 \\ \dfrac{1}{m} \end{bmatrix}
$$

$$
= \dfrac{1}{ms^2 + bs + k}
$$

which is the transfer function of the system. The same transfer function can be obtained from Equation (2–16).

Transfer Matrix. Next, consider a multiple-input, multiple-output system. Assume that there are r inputs $u_1, u_2, \ldots, u_r$, and m outputs $y_1, y_2, \ldots, y_m$. Define

$$
\mathbf{y} = \begin{bmatrix} y_1 \\ y_2 \\ \cdot \\ \cdot \\ \cdot \\ y_m \end{bmatrix}, \qquad \mathbf{u} = \begin{bmatrix} u_1 \\ u_2 \\ \cdot \\ \cdot \\ \cdot \\ u_r \end{bmatrix}
$$

The transfer matrix $\mathbf{G}(s)$ relates the output $\mathbf{Y}(s)$ to the input $\mathbf{U}(s)$, or

$$
\mathbf{Y}(s) = \mathbf{G}(s)\mathbf{U}(s)
$$

where $\mathbf{G}(s)$ is given by

$$
\mathbf{G}(s) = \mathbf{C}(s\mathbf{I} - \mathbf{A})^{-1}\mathbf{B} + \mathbf{D}
$$

[The derivation for this equation is the same as that for Equation (2–29).] Since the input vector $\mathbf{u}$ is r dimensional and the output vector $\mathbf{y}$ is m dimensional, the transfer matrix $\mathbf{G}(s)$ is an $m \times r$ matrix.

2–5 STATE-SPACE REPRESENTATION OF SCALAR DIFFERENTIAL EQUATION SYSTEMS

A dynamic system consisting of a finite number of lumped elements may be described by ordinary differential equations in which time is the independent variable. By use of vector-matrix notation, an nth-order differential equation may be expressed by a first-order vector-matrix differential equation. If n elements of the vector are a set of state variables, then the vector-matrix differential equation is a *state* equation. In this section we shall present methods for obtaining state-space representations of continuous-time systems.

State-Space Representation of nth-Order Systems of Linear Differential Equations in which the Forcing Function Does Not Involve Derivative Terms. Consider the following nth-order system:

$$\overset{(n)}{y} + a_1 \overset{(n-1)}{y} + \cdots + a_{n-1}\dot{y} + a_n y = u \qquad (2\text{--}30)$$

Noting that the knowledge of $y(0), \dot{y}(0), \ldots, \overset{(n-1)}{y}(0)$, together with the input $u(t)$ for $t \geq 0$, determines completely the future behavior of the system, we may take $y(t), \dot{y}(t), \ldots, \overset{(n-1)}{y}(t)$ as a set of n state variables. (Mathematically, such a choice of state variables is quite convenient. Practically, however, because higher-order derivative terms are inaccurate, due to the noise effects inherent in any practical situations, such a choice of the state variables may not be desirable.)

Let us define

$$x_1 = y$$
$$x_2 = \dot{y}$$
$$\cdot$$
$$\cdot$$
$$\cdot$$
$$x_n = \overset{(n-1)}{y}$$

Then Equation (2–30) can be written as

$$\dot{x}_1 = x_2$$
$$\dot{x}_2 = x_3$$
$$\cdot$$
$$\cdot$$
$$\cdot$$
$$\dot{x}_{n-1} = x_n$$
$$\dot{x}_n = -a_n x_1 - \cdots - a_1 x_n + u$$

or

$$\dot{\mathbf{x}} = \mathbf{A}\mathbf{x} + \mathbf{B}u \qquad (2\text{--}31)$$

where

$$\mathbf{x} = \begin{bmatrix} x_1 \\ x_2 \\ \cdot \\ \cdot \\ \cdot \\ x_n \end{bmatrix}, \quad \mathbf{A} = \begin{bmatrix} 0 & 1 & 0 & \cdots & 0 \\ 0 & 0 & 1 & \cdots & 0 \\ \cdot & \cdot & \cdot & & \cdot \\ \cdot & \cdot & \cdot & & \cdot \\ 0 & 0 & 0 & \cdots & 1 \\ -a_n & -a_{n-1} & -a_{n-2} & \cdots & -a_1 \end{bmatrix}, \quad \mathbf{B} = \begin{bmatrix} 0 \\ 0 \\ \cdot \\ \cdot \\ \cdot \\ 0 \\ 1 \end{bmatrix}$$

The output can be given by

$$
y = \begin{bmatrix} 1 & 0 & \cdots & 0 \end{bmatrix}
\begin{bmatrix} x_1 \\ x_2 \\ \cdot \\ \cdot \\ \cdot \\ x_n \end{bmatrix}
$$

or

$$
y = \mathbf{Cx} \qquad (2\text{--}32)
$$

where

$$
\mathbf{C} = \begin{bmatrix} 1 & 0 & \cdots & 0 \end{bmatrix}
$$

[Note that D in Equation (2–24) is zero.] The first-order differential equation, Equation (2–31), is the state equation, and the algebraic equation, Equation (2–32), is the output equation.

Note that the state-space representation for the transfer function system

$$
\frac{Y(s)}{U(s)} = \frac{1}{s^n + a_1 s^{n-1} + \cdots + a_{n-1} s + a_n}
$$

is given also by Equations (2–31) and (2–32).

State-Space Representation of nth-Order Systems of Linear Differential Equations in which the Forcing Function Involves Derivative Terms.   Consider the differential equation system that involves derivatives of the forcing function, such as

$$
\overset{(n)}{y} + a_1 \overset{(n-1)}{y} + \cdots + a_{n-1} \dot{y} + a_n y = b_0 \overset{(n)}{u} + b_1 \overset{(n-1)}{u} + \cdots + b_{n-1} \dot{u} + b_n u \qquad (2\text{--}33)
$$

The main problem in defining the state variables for this case lies in the derivative terms of the input u. The state variables must be such that they will eliminate the derivatives of u in the state equation.

One way to obtain a state equation and output equation for this case is to define the following n variables as a set of n state variables:

$$
\begin{aligned}
x_1 &= y - \beta_0 u \\
x_2 &= \dot{y} - \beta_0 \dot{u} - \beta_1 u = \dot{x}_1 - \beta_1 u \\
x_3 &= \ddot{y} - \beta_0 \ddot{u} - \beta_1 \dot{u} - \beta_2 u = \dot{x}_2 - \beta_2 u
\end{aligned}
$$

$$ \qquad (2\text{--}34) $$

$$
\begin{aligned}
&\cdot \\
&\cdot \\
&\cdot
\end{aligned}
$$

$$
x_n = \overset{(n-1)}{y} - \beta_0 \overset{(n-1)}{u} - \beta_1 \overset{(n-2)}{u} - \cdots - \beta_{n-2} \dot{u} - \beta_{n-1} u = \dot{x}_{n-1} - \beta_{n-1} u
$$

where $\beta_0, \beta_1, \beta_2, \ldots, \beta_{n-1}$ are determined from

$$\beta_0 = b_0$$
$$\beta_1 = b_1 - a_1\beta_0$$
$$\beta_2 = b_2 - a_1\beta_1 - a_2\beta_0$$
$$\beta_3 = b_3 - a_1\beta_2 - a_2\beta_1 - a_3\beta_0$$

$$\cdot$$
$$\cdot$$
$$\cdot$$

(2–35)

$$\beta_{n-1} = b_{n-1} - a_1\beta_{n-2} - \cdots - a_{n-2}\beta_1 - a_{n-1}\beta_0$$

With this choice of state variables the existence and uniqueness of the solution of the state equation is guaranteed. (Note that this is not the only choice of a set of state variables.) With the present choice of state variables, we obtain

$$\dot{x}_1 = x_2 + \beta_1 u$$
$$\dot{x}_2 = x_3 + \beta_2 u$$

$$\cdot$$
$$\cdot$$
$$\cdot$$

(2–36)

$$\dot{x}_{n-1} = x_n + \beta_{n-1} u$$
$$\dot{x}_n = -a_n x_1 - a_{n-1} x_2 - \cdots - a_1 x_n + \beta_n u$$

where β_n is given by

$$\beta_n = b_n - a_1\beta_{n-1} - \cdots - a_{n-1}\beta_1 - a_{n-1}\beta_0$$

[To derive Equation (2–36), see Problem **A–2–6**.] In terms of vector-matrix equations, Equation (2–36) and the output equation can be written as

$$
\begin{bmatrix} \dot{x}_1 \\ \dot{x}_2 \\ \cdot \\ \cdot \\ \cdot \\ \dot{x}_{n-1} \\ \dot{x}_n \end{bmatrix}
=
\begin{bmatrix}
0 & 1 & 0 & \cdots & 0 \\
0 & 0 & 1 & \cdots & 0 \\
\cdot & \cdot & \cdot & & \cdot \\
\cdot & \cdot & \cdot & & \cdot \\
\cdot & \cdot & \cdot & & \cdot \\
0 & 0 & 0 & \cdots & 1 \\
-a_n & -a_{n-1} & -a_{n-2} & \cdots & -a_1
\end{bmatrix}
\begin{bmatrix} x_1 \\ x_2 \\ \cdot \\ \cdot \\ \cdot \\ x_{n-1} \\ x_n \end{bmatrix}
+
\begin{bmatrix} \beta_1 \\ \beta_2 \\ \cdot \\ \cdot \\ \cdot \\ \beta_{n-1} \\ \beta_n \end{bmatrix} u
$$

$$
y = \begin{bmatrix} 1 & 0 & \cdots & 0 \end{bmatrix}
\begin{bmatrix} x_1 \\ x_2 \\ \cdot \\ \cdot \\ \cdot \\ x_n \end{bmatrix}
+ \beta_0 u
$$

or

$$\dot{\mathbf{x}} = \mathbf{A}\mathbf{x} + \mathbf{B}u \qquad (2\text{--}37)$$

$$y = \mathbf{C}\mathbf{x} + Du \qquad (2\text{--}38)$$

where

$$\mathbf{x} = \begin{bmatrix} x_1 \\ x_2 \\ \cdot \\ \cdot \\ \cdot \\ x_{n-1} \\ x_n \end{bmatrix}, \qquad \mathbf{A} = \begin{bmatrix} 0 & 1 & 0 & \cdots & 0 \\ 0 & 0 & 1 & \cdots & 0 \\ \cdot & \cdot & \cdot & & \cdot \\ \cdot & \cdot & \cdot & & \cdot \\ \cdot & \cdot & \cdot & & \cdot \\ 0 & 0 & 0 & \cdots & 1 \\ -a_n & -a_{n-1} & -a_{n-2} & \cdots & -a_1 \end{bmatrix}$$

$$\mathbf{B} = \begin{bmatrix} \beta_1 \\ \beta_2 \\ \cdot \\ \cdot \\ \cdot \\ \beta_{n-1} \\ \beta_n \end{bmatrix}, \qquad \mathbf{C} = \begin{bmatrix} 1 & 0 & \cdots & 0 \end{bmatrix}, \qquad D = \beta_0 = b_0$$

In this state-space representation, matrices $\mathbf{A}$ and $\mathbf{C}$ are exactly the same as those for the system of Equation (2–30). The derivatives on the right-hand side of Equation (2–33) affect only the elements of the $\mathbf{B}$ matrix.

Note that the state-space representation for the transfer function

$$\frac{Y(s)}{U(s)} = \frac{b_0 s^n + b_1 s^{n-1} + \cdots + b_{n-1} s + b_n}{s^n + a_1 s^{n-1} + \cdots + a_{n-1} s + a_n}$$

is given also by Equations (2–37) and (2–38).

There are many ways to obtain state-space representations of systems. Methods for obtaining canonical representations of systems in state space (such as controllable canonical form, observable canonical form, diagonal canonical form, and Jordan canonical form) are presented in Chapter 9.

MATLAB can also be used to obtain state-space representations of systems from transfer-function representations, and vice versa. This subject is presented in Section 2–6.

2–6 TRANSFORMATION OF MATHEMATICAL MODELS WITH MATLAB

MATLAB is quite useful to transform the system model from transfer function to state space, and vice versa. We shall begin our discussion with transformation from transfer function to state space.

Let us write the closed-loop transfer function as

$$\frac{Y(s)}{U(s)} = \frac{\text{numerator polynomial in } s}{\text{denominator polynomial in } s} = \frac{\text{num}}{\text{den}}$$

Once we have this transfer-function expression, the MATLAB command

$$[A,B,C,D] = \text{tf2ss(num,den)}$$

will give a state-space representation. It is important to note that the state-space representation for any system is not unique. There are many (infinitely many) state-space representations for the same system. The MATLAB command gives one possible such state-space representation.

Transformation from Transfer Function to State Space Representation. Consider the transfer-function system

$$\frac{Y(s)}{U(s)} = \frac{s}{(s + 10)(s^2 + 4s + 16)}$$

$$= \frac{s}{s^3 + 14s^2 + 56s + 160} \qquad (2\text{--}39)$$

There are many (infinitely many) possible state-space representations for this system. One possible state-space representation is

$$\begin{bmatrix} \dot{x}_1 \\ \dot{x}_2 \\ \dot{x}_3 \end{bmatrix} = \begin{bmatrix} 0 & 1 & 0 \\ 0 & 0 & 1 \\ -160 & -56 & -14 \end{bmatrix} \begin{bmatrix} x_1 \\ x_2 \\ x_3 \end{bmatrix} + \begin{bmatrix} 0 \\ 1 \\ -14 \end{bmatrix} u$$

$$y = \begin{bmatrix} 1 & 0 & 0 \end{bmatrix} \begin{bmatrix} x_1 \\ x_2 \\ x_3 \end{bmatrix} + [0]u$$

Another possible state-space representation (among infinitely many alternatives) is

$$\begin{bmatrix} \dot{x}_1 \\ \dot{x}_2 \\ \dot{x}_3 \end{bmatrix} = \begin{bmatrix} -14 & -56 & -160 \\ 1 & 0 & 0 \\ 0 & 1 & 0 \end{bmatrix} \begin{bmatrix} x_1 \\ x_2 \\ x_3 \end{bmatrix} + \begin{bmatrix} 1 \\ 0 \\ 0 \end{bmatrix} u \qquad (2\text{--}40)$$

$$y = \begin{bmatrix} 0 & 1 & 0 \end{bmatrix} \begin{bmatrix} x_1 \\ x_2 \\ x_3 \end{bmatrix} + [0]u \qquad (2\text{--}41)$$

MATLAB transforms the transfer function given by Equation (2–39) into the state-space representation given by Equations (2–40) and (2–41). For the example system considered here, MATLAB Program 2–2 will produce matrices **A**, **B**, **C**, and *D*.

MATLAB Program 2–2

```
num = [1    0];
den = [1   14   56   160];
[A,B,C,D] = tf2ss(num,den)

A =

   -14   -56  -160
     1     0     0
     0     1     0

B =

     1
     0
     0

C =

     0     1     0

D =

     0
```

Transformation from State Space Representation to Transfer Function.   To obtain the transfer function from state-space equations, use the following command:

$$[num,den] = ss2tf(A,B,C,D,iu)$$

iu must be specified for systems with more than one input. For example, if the system has three inputs ($u1$, $u2$, $u3$), then iu must be either 1, 2, or 3, where 1 implies $u1$, 2 implies $u2$, and 3 implies $u3$.

If the system has only one input, then either

$$[num,den] = ss2tf(A,B,C,D)$$

or

$$[\text{num,den}] = \text{ss2tf}(A,B,C,D,1)$$

may be used. For the case where the system has multiple inputs and multiple outputs, see Problem A–2–12.

EXAMPLE 2–4   Obtain the transfer function of the system defined by the following state-space equations:

$$\begin{bmatrix} \dot{x}_1 \\ \dot{x}_2 \\ \dot{x}_3 \end{bmatrix} = \begin{bmatrix} 0 & 1 & 0 \\ 0 & 0 & 1 \\ -5 & -25 & -5 \end{bmatrix} \begin{bmatrix} x_1 \\ x_2 \\ x_3 \end{bmatrix} + \begin{bmatrix} 0 \\ 25 \\ -120 \end{bmatrix} u$$

$$y = \begin{bmatrix} 1 & 0 & 0 \end{bmatrix} \begin{bmatrix} x_1 \\ x_2 \\ x_3 \end{bmatrix}$$

MATLAB Program 2-3 will produce the transfer function for the given system. The transfer function obtained is given by

$$\frac{Y(s)}{U(s)} = \frac{25s + 5}{s^3 + 5s^2 + 25s + 5}$$

MATLAB Program 2–3

```
A = [0  1  0; 0  0  1; -5  -25  -5];
B = [0; 25; -120];
C = [1  0   0];
D = [0];
[num,den] = ss2tf(A,B,C,D)

num =

  0  0.0000  25.0000  5.0000

den

  1.0000  5.0000  25.0000  5.0000
```

% ***** The same result can be obtained by entering the following command: *****

```
[num,den] = ss2tf(A,B,C,D,1)

num =

  0  0.0000  25.0000  5.0000

den =

  1.0000  5.0000  25.0000  5.0000
```

Nonlinear Systems. A system is nonlinear if the principle of superposition does not apply. Thus, for a nonlinear system the response to two inputs cannot be calculated by treating one input at a time and adding the results.

Although many physical relationships are often represented by linear equations, in most cases actual relationships are not quite linear. In fact, a careful study of physical systems reveals that even so-called "linear systems" are really linear only in limited operating ranges. In practice, many electromechanical systems, hydraulic systems, pneumatic systems, and so on, involve nonlinear relationships among the variables. For example, the output of a component may saturate for large input signals. There may be a dead space that affects small signals. (The dead space of a component is a small range of input variations to which the component is insensitive.) Square-law nonlinearity may occur in some components. For instance, dampers used in physical systems may be linear for low-velocity operations but may become nonlinear at high velocities, and the damping force may become proportional to the square of the operating velocity.

Linearization of Nonlinear Systems. In control engineering a normal operation of the system may be around an equilibrium point, and the signals may be considered small signals around the equilibrium. (It should be pointed out that there are many exceptions to such a case.) However, if the system operates around an equilibrium point and if the signals involved are small signals, then it is possible to approximate the nonlinear system by a linear system. Such a linear system is equivalent to the nonlinear system considered within a limited operating range. Such a linearized model (linear, time-invariant model) is very important in control engineering.

The linearization procedure to be presented in the following is based on the expansion of nonlinear function into a Taylor series about the operating point and the retention of only the linear term. Because we neglect higher-order terms of the Taylor series expansion, these neglected terms must be small enough; that is, the variables deviate only slightly from the operating condition. (Otherwise, the result will be inaccurate.)

Linear Approximation of Nonlinear Mathematical Models. To obtain a linear mathematical model for a nonlinear system, we assume that the variables deviate only slightly from some operating condition. Consider a system whose input is $x(t)$ and output is $y(t)$. The relationship between $y(t)$ and $x(t)$ is given by

$$y = f(x) \tag{2–42}$$

If the normal operating condition corresponds to $\bar{x}$, $\bar{y}$, then Equation (2–42) may be expanded into a Taylor series about this point as follows:

$$y = f(x)$$

$$= f(\bar{x}) + \frac{df}{dx}(x - \bar{x}) + \frac{1}{2!}\frac{d^2f}{dx^2}(x - \bar{x})^2 + \cdots \tag{2–43}$$

where the derivatives df/dx, d^2f/dx^2, ... are evaluated at $x = \bar{x}$. If the variation $x - \bar{x}$ is small, we may neglect the higher-order terms in $x - \bar{x}$. Then Equation (2–43) may be written as

$$y = \bar{y} + K(x - \bar{x}) \tag{2–44}$$

where

$$\bar{y} = f(\bar{x})$$

$$K = \left. \frac{df}{dx} \right|_{x=\bar{x}}$$

Equation (2–44) may be rewritten as

$$y - \bar{y} = K(x - \bar{x}) \tag{2–45}$$

which indicates that $y - \bar{y}$ is proportional to $x - \bar{x}$. Equation (2–45) gives a linear mathematical model for the nonlinear system given by Equation (2–42) near the operating point $x = \bar{x}$, $y = \bar{y}$.

Next, consider a nonlinear system whose output y is a function of two inputs x_1 and x_2, so that

$$y = f(x_1, x_2) \tag{2–46}$$

To obtain a linear approximation to this nonlinear system, we may expand Equation (2–46) into a Taylor series about the normal operating point $\bar{x}_1, \bar{x}_2$. Then Equation (2–46) becomes

$$y = f(\bar{x}_1, \bar{x}_2) + \left[\frac{\partial f}{\partial x_1}(x_1 - \bar{x}_1) + \frac{\partial f}{\partial x_2}(x_2 - \bar{x}_2) \right]$$

$$+ \frac{1}{2!} \left[\frac{\partial^2 f}{\partial x_1^2}(x_1 - \bar{x}_1)^2 + 2\frac{\partial^2 f}{\partial x_1 \partial x_2}(x_1 - \bar{x}_1)(x_2 - \bar{x}_2) \right.$$

$$\left. + \frac{\partial^2 f}{\partial x_2^2}(x_2 - \bar{x}_2)^2 \right] + \cdots$$

where the partial derivatives are evaluated at $x_1 = \bar{x}_1$, $x_2 = \bar{x}_2$. Near the normal operating point, the higher-order terms may be neglected. The linear mathematical model of this nonlinear system in the neighborhood of the normal operating condition is then given by

$$y - \bar{y} = K_1(x_1 - \bar{x}_1) + K_2(x_2 - \bar{x}_2)$$

where

$$\bar{y} = f(\bar{x}_1, \bar{x}_2)$$

$$K_1 = \frac{\partial f}{\partial x_1}\bigg|_{x_1=\bar{x}_1,\, x_2=\bar{x}_2}$$

$$K_2 = \frac{\partial f}{\partial x_2}\bigg|_{x_1=\bar{x}_1,\, x_2=\bar{x}_2}$$

The linearization technique presented here is valid in the vicinity of the operating condition. If the operating conditions vary widely, however, such linearized equations are not adequate, and nonlinear equations must be dealt with. It is important to remember that a particular mathematical model used in analysis and design may accurately represent the dynamics of an actual system for certain operating conditions, but may not be accurate for other operating conditions.

EXAMPLE 2–5 Linearize the nonlinear equation

$$z = xy$$

in the region $5 \le x \le 7$, $10 \le y \le 12$. Find the error if the linearized equation is used to calculate the value of z when $x = 5$, $y = 10$.

Since the region considered is given by $5 \le x \le 7$, $10 \le y \le 12$, choose $\bar{x} = 6$, $\bar{y} = 11$. Then $\bar{z} = \bar{x}\bar{y} = 66$. Let us obtain a linearized equation for the nonlinear equation near a point $\bar{x} = 6$, $\bar{y} = 11$.

Expanding the nonlinear equation into a Taylor series about point $x = \bar{x}$, $y = \bar{y}$ and neglecting the higher-order terms, we have

$$z - \bar{z} = a(x - \bar{x}) + b(y - \bar{y})$$

where

$$a = \frac{\partial(xy)}{\partial x}\bigg|_{x=\bar{x},\, y=\bar{y}} = \bar{y} = 11$$

$$b = \frac{\partial(xy)}{\partial y}\bigg|_{x=\bar{x},\, y=\bar{y}} = \bar{x} = 6$$

Hence the linearized equation is

$$z - 66 = 11(x - 6) + 6(y - 11)$$

or

$$z = 11x + 6y - 66$$

When $x = 5$, $y = 10$, the value of z given by the linearized equation is

$$z = 11x + 6y - 66 = 55 + 60 - 66 = 49$$

The exact value of z is $z = xy = 50$. The error is thus $50 - 49 = 1$. In terms of percentage, the error is 2%.

EXAMPLE PROBLEMS AND SOLUTIONS

A–2–1. Simplify the block diagram shown in Figure 2–17.

Solution. First, move the branch point of the path involving H_1 outside the loop involving H_2, as shown in Figure 2–18(a). Then eliminating two loops results in Figure 2–18(b). Combining two blocks into one gives Figure 2–18(c).

A–2–2. Simplify the block diagram shown in Figure 2–19. Obtain the transfer function relating $C(s)$ and $R(s)$.

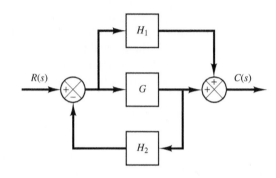

Figure 2–17
Block diagram of a system.

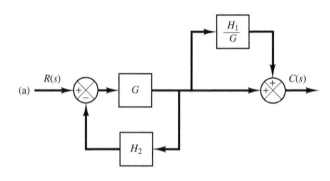

(a)

(b)

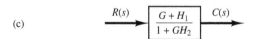

Figure 2–18
Simplified block diagrams for the system shown in Figure 2–17.

(c)

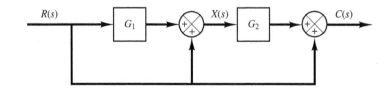

Figure 2–19
Block diagram of a system.

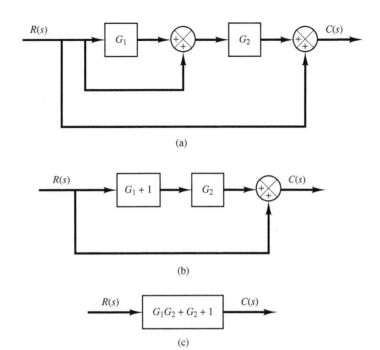

Figure 2–20
Reduction of the
block diagram shown
in Figure 2–19.

Solution. The block diagram of Figure 2–19 can be modified to that shown in Figure 2–20(a). Eliminating the minor feedforward path, we obtain Figure 2–20(b), which can be simplified to Figure 2–20(c). The transfer function $C(s)/R(s)$ is thus given by

$$\frac{C(s)}{R(s)} = G_1 G_2 + G_2 + 1$$

The same result can also be obtained by proceeding as follows: Since signal $X(s)$ is the sum of two signals $G_1 R(s)$ and $R(s)$, we have

$$X(s) = G_1 R(s) + R(s)$$

The output signal $C(s)$ is the sum of $G_2 X(s)$ and $R(s)$. Hence

$$C(s) = G_2 X(s) + R(s) = G_2 \big[G_1 R(s) + R(s)\big] + R(s)$$

And so we have the same result as before:

$$\frac{C(s)}{R(s)} = G_1 G_2 + G_2 + 1$$

A–2–3. Simplify the block diagram shown in Figure 2–21. Then obtain the closed-loop transfer function $C(s)/R(s)$.

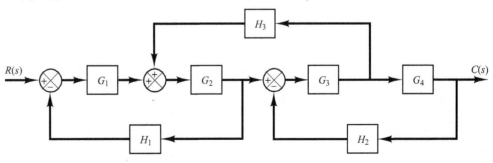

Figure 2–21
Block diagram of a
system.

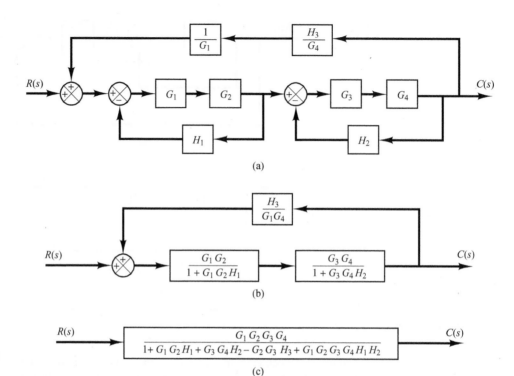

Figure 2–22
Successive
reductions of the
block diagram shown
in Figure 2–21.

Solution. First move the branch point between G_3 and G_4 to the right-hand side of the loop containing G_3, G_4, and H_2. Then move the summing point between G_1 and G_2 to the left-hand side of the first summing point. See Figure 2–22(a). By simplifying each loop, the block diagram can be modified as shown in Figure 2–22(b). Further simplification results in Figure 2–22(c), from which the closed-loop transfer function $C(s)/R(s)$ is obtained as

$$\frac{C(s)}{R(s)} = \frac{G_1 G_2 G_3 G_4}{1 + G_1 G_2 H_1 + G_3 G_4 H_2 - G_2 G_3 H_3 + G_1 G_2 G_3 G_4 H_1 H_2}$$

A–2–4. Obtain transfer functions $C(s)/R(s)$ and $C(s)/D(s)$ of the system shown in Figure 2–23.

Solution. From Figure 2–23 we have

$$U(s) = G_f R(s) + G_c E(s) \tag{2–47}$$
$$C(s) = G_p[D(s) + G_1 U(s)] \tag{2–48}$$
$$E(s) = R(s) - HC(s) \tag{2–49}$$

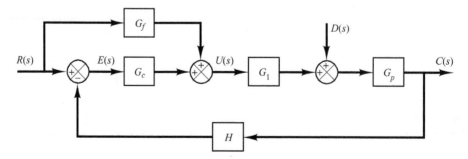

Figure 2–23
Control system with
reference input and
disturbance input.

Chapter 2 / Mathematical Modeling of Control Systems

By substituting Equation (2–47) into Equation (2–48), we get

$$C(s) = G_p D(s) + G_1 G_p \left[G_f R(s) + G_c E(s) \right] \tag{2–50}$$

By substituting Equation (2–49) into Equation (2–50), we obtain

$$C(s) = G_p D(s) + G_1 G_p \left\{ G_f R(s) + G_c \left[R(s) - H C(s) \right] \right\}$$

Solving this last equation for $C(s)$, we get

$$C(s) + G_1 G_p G_c H C(s) = G_p D(s) + G_1 G_p (G_f + G_c) R(s)$$

Hence

$$C(s) = \frac{G_p D(s) + G_1 G_p (G_f + G_c) R(s)}{1 + G_1 G_p G_c H} \tag{2–51}$$

Note that Equation (2–51) gives the response $C(s)$ when both reference input $R(s)$ and disturbance input $D(s)$ are present.

To find transfer function $C(s)/R(s)$, we let $D(s) = 0$ in Equation (2–51). Then we obtain

$$\frac{C(s)}{R(s)} = \frac{G_1 G_p (G_f + G_c)}{1 + G_1 G_p G_c H}$$

Similarly, to obtain transfer function $C(s)/D(s)$, we let $R(s) = 0$ in Equation (2–51). Then $C(s)/D(s)$ can be given by

$$\frac{C(s)}{D(s)} = \frac{G_p}{1 + G_1 G_p G_c H}$$

A–2–5. Figure 2–24 shows a system with two inputs and two outputs. Derive $C_1(s)/R_1(s)$, $C_1(s)/R_2(s)$, $C_2(s)/R_1(s)$, and $C_2(s)/R_2(s)$. (In deriving outputs for $R_1(s)$, assume that $R_2(s)$ is zero, and vice versa.)

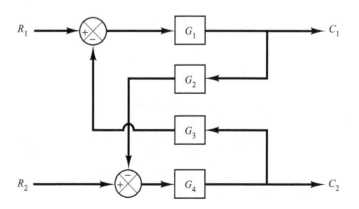

Figure 2–24
System with two inputs and two outputs.

Solution. From the figure, we obtain

$$C_1 = G_1(R_1 - G_3 C_2) \tag{2-52}$$

$$C_2 = G_4(R_2 - G_2 C_1) \tag{2-53}$$

By substituting Equation (2–53) into Equation (2–52), we obtain

$$C_1 = G_1[R_1 - G_3 G_4(R_2 - G_2 C_1)] \tag{2-54}$$

By substituting Equation (2–52) into Equation (2–53), we get

$$C_2 = G_4[R_2 - G_2 G_1(R_1 - G_3 C_2)] \tag{2-55}$$

Solving Equation (2–54) for C_1, we obtain

$$C_1 = \frac{G_1 R_1 - G_1 G_3 G_4 R_2}{1 - G_1 G_2 G_3 G_4} \tag{2-56}$$

Solving Equation (2–55) for C_2 gives

$$C_2 = \frac{-G_1 G_2 G_4 R_1 + G_4 R_2}{1 - G_1 G_2 G_3 G_4} \tag{2-57}$$

Equations (2–56) and (2–57) can be combined in the form of the transfer matrix as follows:

$$\begin{bmatrix} C_1 \\ C_2 \end{bmatrix} = \begin{bmatrix} \dfrac{G_1}{1 - G_1 G_2 G_3 G_4} & -\dfrac{G_1 G_3 G_4}{1 - G_1 G_2 G_3 G_4} \\ -\dfrac{G_1 G_2 G_4}{1 - G_1 G_2 G_3 G_4} & \dfrac{G_4}{1 - G_1 G_2 G_3 G_4} \end{bmatrix} \begin{bmatrix} R_1 \\ R_2 \end{bmatrix}$$

Then the transfer functions $C_1(s)/R_1(s)$, $C_1(s)/R_2(s)$, $C_2(s)/R_1(s)$ and $C_2(s)/R_2(s)$ can be obtained as follows:

$$\frac{C_1(s)}{R_1(s)} = \frac{G_1}{1 - G_1 G_2 G_3 G_4}, \qquad \frac{C_1(s)}{R_2(s)} = -\frac{G_1 G_3 G_4}{1 - G_1 G_2 G_3 G_4}$$

$$\frac{C_2(s)}{R_1(s)} = -\frac{G_1 G_2 G_4}{1 - G_1 G_2 G_3 G_4}, \qquad \frac{C_2(s)}{R_2(s)} = \frac{G_4}{1 - G_1 G_2 G_3 G_4}$$

Note that Equations (2–56) and (2–57) give responses C_1 and C_2, respectively, when both inputs R_1 and R_2 are present.

Notice that when $R_2(s) = 0$, the original block diagram can be simplified to those shown in Figures 2–25(a) and (b). Similarly, when $R_1(s) = 0$, the original block diagram can be simplified to those shown in Figures 2–25(c) and (d). From these simplified block diagrams we can also obtain $C_1(s)/R_1(s)$, $C_2(s)/R_1(s)$, $C_1(s)/R_2(s)$, and $C_2(s)/R_2(s)$, as shown to the right of each corresponding block diagram.

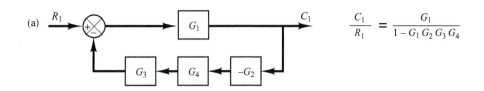

(a) R_1 G_1 C_1 G_3 G_4 $-G_2$

$$\frac{C_1}{R_1} = \frac{G_1}{1 - G_1 G_2 G_3 G_4}$$

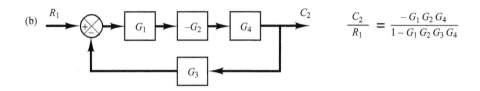

(b) R_1 G_1 $-G_2$ G_4 C_2 G_3

$$\frac{C_2}{R_1} = \frac{-G_1 G_2 G_4}{1 - G_1 G_2 G_3 G_4}$$

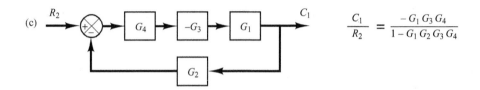

(c) R_2 G_4 $-G_3$ G_1 C_1 G_2

$$\frac{C_1}{R_2} = \frac{-G_1 G_3 G_4}{1 - G_1 G_2 G_3 G_4}$$

Figure 2–25
Simplified block
diagrams and
corresponding
closed-loop transfer
functions.

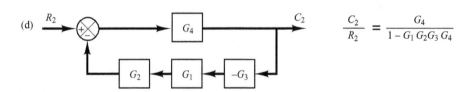

(d) R_2 G_4 C_2 G_2 G_1 $-G_3$

$$\frac{C_2}{R_2} = \frac{G_4}{1 - G_1 G_2 G_3 G_4}$$

A–2–6. Show that for the differential equation system

$$\dddot{y} + a_1 \ddot{y} + a_2 \dot{y} + a_3 y = b_0 \dddot{u} + b_1 \ddot{u} + b_2 \dot{u} + b_3 u \tag{2–58}$$

state and output equations can be given, respectively, by

$$\begin{bmatrix} \dot{x}_1 \\ \dot{x}_2 \\ \dot{x}_3 \end{bmatrix} = \begin{bmatrix} 0 & 1 & 0 \\ 0 & 0 & 1 \\ -a_3 & -a_2 & -a_1 \end{bmatrix} \begin{bmatrix} x_1 \\ x_2 \\ x_3 \end{bmatrix} + \begin{bmatrix} \beta_1 \\ \beta_2 \\ \beta_3 \end{bmatrix} u \tag{2–59}$$

and

$$y = \begin{bmatrix} 1 & 0 & 0 \end{bmatrix} \begin{bmatrix} x_1 \\ x_2 \\ x_3 \end{bmatrix} + \beta_0 u \tag{2–60}$$

where state variables are defined by

$$x_1 = y - \beta_0 u$$

$$x_2 = \dot{y} - \beta_0 \dot{u} - \beta_1 u = \dot{x}_1 - \beta_1 u$$

$$x_3 = \ddot{y} - \beta_0 \ddot{u} - \beta_1 \dot{u} - \beta_2 u = \dot{x}_2 - \beta_2 u$$

Example Problems and Solutions

and

$$\beta_0 = b_0$$
$$\beta_1 = b_1 - a_1\beta_0$$
$$\beta_2 = b_2 - a_1\beta_1 - a_2\beta_0$$
$$\beta_3 = b_3 - a_1\beta_2 - a_2\beta_1 - a_3\beta_0$$

Solution. From the definition of state variables x_2 and x_3, we have

$$\dot{x}_1 = x_2 + \beta_1 u \tag{2–61}$$
$$\dot{x}_2 = x_3 + \beta_2 u \tag{2–62}$$

To derive the equation for $\dot{x}_3$, we first note from Equation (2–58) that

$$\dddot{y} = -a_1\ddot{y} - a_2\dot{y} - a_3 y + b_0\dddot{u} + b_1\ddot{u} + b_2\dot{u} + b_3 u$$

Since

$$x_3 = \ddot{y} - \beta_0\ddot{u} - \beta_1\dot{u} - \beta_2 u$$

we have

$$
\begin{aligned}
\dot{x}_3 &= \dddot{y} - \beta_0\dddot{u} - \beta_1\ddot{u} - \beta_2\dot{u} \\
&= \left(-a_1\ddot{y} - a_2\dot{y} - a_3 y\right) + b_0\dddot{u} + b_1\ddot{u} + b_2\dot{u} + b_3 u - \beta_0\dddot{u} - \beta_1\ddot{u} - \beta_2\dot{u} \\
&= -a_1\left(\ddot{y} - \beta_0\ddot{u} - \beta_1\dot{u} - \beta_2 u\right) - a_1\beta_0\ddot{u} - a_1\beta_1\dot{u} - a_1\beta_2 u \\
&\quad -a_2\left(\dot{y} - \beta_0\dot{u} - \beta_1 u\right) - a_2\beta_0\dot{u} - a_2\beta_1 u - a_3\left(y - \beta_0 u\right) - a_3\beta_0 u \\
&\quad + b_0\dddot{u} + b_1\ddot{u} + b_2\dot{u} + b_3 u - \beta_0\dddot{u} - \beta_1\ddot{u} - \beta_2\dot{u} \\
&= -a_1 x_3 - a_2 x_2 - a_3 x_1 + \left(b_0 - \beta_0\right)\dddot{u} + \left(b_1 - \beta_1 - a_1\beta_0\right)\ddot{u} \\
&\quad + \left(b_2 - \beta_2 - a_1\beta_1 - a_2\beta_0\right)\dot{u} + \left(b_3 - a_1\beta_2 - a_2\beta_1 - a_3\beta_0\right)u \\
&= -a_1 x_3 - a_2 x_2 - a_3 x_1 + \left(b_3 - a_1\beta_2 - a_2\beta_1 - a_3\beta_0\right)u \\
&= -a_1 x_3 - a_2 x_2 - a_3 x_1 + \beta_3 u
\end{aligned}
$$

Hence, we get

$$\dot{x}_3 = -a_3 x_1 - a_2 x_2 - a_1 x_3 + \beta_3 u \tag{2–63}$$

Combining Equations (2–61), (2–62), and (2–63) into a vector-matrix equation, we obtain Equation (2–59). Also, from the definition of state variable x_1, we get the output equation given by Equation (2–60).

A–2–7. Obtain a state-space equation and output equation for the system defined by

$$\frac{Y(s)}{U(s)} = \frac{2s^3 + s^2 + s + 2}{s^3 + 4s^2 + 5s + 2}$$

Solution. From the given transfer function, the differential equation for the system is

$$\dddot{y} + 4\ddot{y} + 5\dot{y} + 2y = 2\dddot{u} + \ddot{u} + \dot{u} + 2u$$

Comparing this equation with the standard equation given by Equation (2–33), rewritten

$$\dddot{y} + a_1\ddot{y} + a_2\dot{y} + a_3 y = b_0\dddot{u} + b_1\ddot{u} + b_2\dot{u} + b_3 u$$

we find

$$a_1 = 4, \qquad a_2 = 5, \qquad a_3 = 2$$
$$b_0 = 2, \qquad b_1 = 1, \qquad b_2 = 1, \qquad b_3 = 2$$

Referring to Equation (2–35), we have

$$\beta_0 = b_0 = 2$$
$$\beta_1 = b_1 - a_1\beta_0 = 1 - 4 \times 2 = -7$$
$$\beta_2 = b_2 - a_1\beta_1 - a_2\beta_0 = 1 - 4 \times (-7) - 5 \times 2 = 19$$
$$\beta_3 = b_3 - a_1\beta_2 - a_2\beta_1 - a_3\beta_0$$
$$= 2 - 4 \times 19 - 5 \times (-7) - 2 \times 2 = -43$$

Referring to Equation (2–34), we define

$$x_1 = y - \beta_0 u = y - 2u$$
$$x_2 = \dot{x}_1 - \beta_1 u = \dot{x}_1 + 7u$$
$$x_3 = \dot{x}_2 - \beta_2 u = \dot{x}_2 - 19u$$

Then referring to Equation (2–36),

$$\dot{x}_1 = x_2 - 7u$$
$$\dot{x}_2 = x_3 + 19u$$
$$\dot{x}_3 = -a_3 x_1 - a_2 x_2 - a_1 x_3 + \beta_3 u$$
$$= -2x_1 - 5x_2 - 4x_3 - 43u$$

Hence, the state-space representation of the system is

$$\begin{bmatrix} \dot{x}_1 \\ \dot{x}_2 \\ \dot{x}_3 \end{bmatrix} = \begin{bmatrix} 0 & 1 & 0 \\ 0 & 0 & 1 \\ -2 & -5 & -4 \end{bmatrix} \begin{bmatrix} x_1 \\ x_2 \\ x_3 \end{bmatrix} + \begin{bmatrix} -7 \\ 19 \\ -43 \end{bmatrix} u$$

$$y = \begin{bmatrix} 1 & 0 & 0 \end{bmatrix} \begin{bmatrix} x_1 \\ x_2 \\ x_3 \end{bmatrix} + 2u$$

This is one possible state-space representation of the system. There are many (infinitely many) others. If we use MATLAB, it produces the following state-space representation:

$$\begin{bmatrix} \dot{x}_1 \\ \dot{x}_2 \\ \dot{x}_3 \end{bmatrix} = \begin{bmatrix} -4 & -5 & -2 \\ 1 & 0 & 0 \\ 0 & 1 & 0 \end{bmatrix} \begin{bmatrix} x_1 \\ x_2 \\ x_3 \end{bmatrix} + \begin{bmatrix} 1 \\ 0 \\ 0 \end{bmatrix} u$$

$$y = \begin{bmatrix} -7 & -9 & -2 \end{bmatrix} \begin{bmatrix} x_1 \\ x_2 \\ x_3 \end{bmatrix} + 2u$$

See MATLAB Program 2-4. (Note that all state-space representations for the same system are equivalent.)

Example Problems and Solutions

MATLAB Program 2–4

```
num = [2 1 1 2];
den = [1 4 5 2];
[A,B,C,D] = tf2ss(num,den)
```

A =

```
    -4  -5  -2
     1   0   0
     0   1   0
```

B =

```
    1
    0
    0
```

C =

```
    -7  -9  -2
```

D =

```
    2
```

A–2–8. Obtain a state-space model of the system shown in Figure 2–26.

Solution. The system involves one integrator and two delayed integrators. The output of each integrator or delayed integrator can be a state variable. Let us define the output of the plant as x_1, the output of the controller as x_2, and the output of the sensor as x_3. Then we obtain

$$\frac{X_1(s)}{X_2(s)} = \frac{10}{s + 5}$$

$$\frac{X_2(s)}{U(s) - X_3(s)} = \frac{1}{s}$$

$$\frac{X_3(s)}{X_1(s)} = \frac{1}{s + 1}$$

$$Y(s) = X_1(s)$$

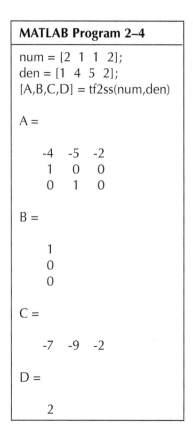

Figure 2–26
Control system.

which can be rewritten as

$$sX_1(s) = -5X_1(s) + 10X_2(s)$$
$$sX_2(s) = -X_3(s) + U(s)$$
$$sX_3(s) = X_1(s) - X_3(s)$$
$$Y(s) = X_1(s)$$

By taking the inverse Laplace transforms of the preceding four equations, we obtain

$$\dot{x}_1 = -5x_1 + 10x_2$$
$$\dot{x}_2 = -x_3 + u$$
$$\dot{x}_3 = x_1 - x_3$$
$$y = x_1$$

Thus, a state-space model of the system in the standard form is given by

$$\begin{bmatrix} \dot{x}_1 \\ \dot{x}_2 \\ \dot{x}_3 \end{bmatrix} = \begin{bmatrix} -5 & 10 & 0 \\ 0 & 0 & -1 \\ 1 & 0 & -1 \end{bmatrix} \begin{bmatrix} x_1 \\ x_2 \\ x_3 \end{bmatrix} + \begin{bmatrix} 0 \\ 1 \\ 0 \end{bmatrix} u$$

$$y = \begin{bmatrix} 1 & 0 & 0 \end{bmatrix} \begin{bmatrix} x_1 \\ x_2 \\ x_3 \end{bmatrix}$$

It is important to note that this is not the only state-space representation of the system. Infinitely many other state-space representations are possible. However, the number of state variables is the same in any state-space representation of the same system. In the present system, the number of state variables is three, regardless of what variables are chosen as state variables.

A–2–9. Obtain a state-space model for the system shown in Figure 2–27(a).

Solution. First, notice that $(as + b)/s^2$ involves a derivative term. Such a derivative term may be avoided if we modify $(as + b)/s^2$ as

$$\frac{as + b}{s^2} = \left(a + \frac{b}{s}\right)\frac{1}{s}$$

Using this modification, the block diagram of Figure 2–27(a) can be modified to that shown in Figure 2–27(b).

Define the outputs of the integrators as state variables, as shown in Figure 2–27(b). Then from Figure 2–27(b) we obtain

$$\frac{X_1(s)}{X_2(s) + a[U(s) - X_1(s)]} = \frac{1}{s}$$

$$\frac{X_2(s)}{U(s) - X_1(s)} = \frac{b}{s}$$

$$Y(s) = X_1(s)$$

which may be modified to

$$sX_1(s) = X_2(s) + a[U(s) - X_1(s)]$$
$$sX_2(s) = -bX_1(s) + bU(s)$$
$$Y(s) = X_1(s)$$

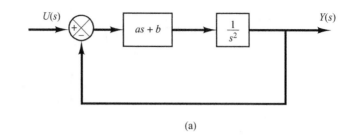

(a)

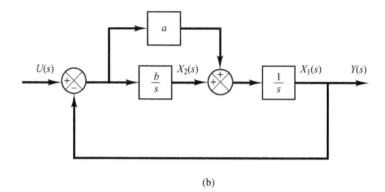

Figure 2–27
(a) Control system;
(b) modified block
diagram.

(b)

Taking the inverse Laplace transforms of the preceding three equations, we obtain

$$\dot{x}_1 = -ax_1 + x_2 + au$$

$$\dot{x}_2 = -bx_1 + bu$$

$$y = x_1$$

Rewriting the state and output equations in the standard vector-matrix form, we obtain

$$\begin{bmatrix} \dot{x}_1 \\ \dot{x}_2 \end{bmatrix} = \begin{bmatrix} -a & 1 \\ -b & 0 \end{bmatrix} \begin{bmatrix} x_1 \\ x_2 \end{bmatrix} + \begin{bmatrix} a \\ b \end{bmatrix} u$$

$$y = \begin{bmatrix} 1 & 0 \end{bmatrix} \begin{bmatrix} x_1 \\ x_2 \end{bmatrix}$$

A–2–10. Obtain a state-space representation of the system shown in Figure 2–28(a).

Solution. In this problem, first expand $(s + z)/(s + p)$ into partial fractions.

$$\frac{s + z}{s + p} = 1 + \frac{z - p}{s + p}$$

Next, convert $K/[s(s + a)]$ into the product of K/s and $1/(s + a)$. Then redraw the block diagram, as shown in Figure 2–28(b). Defining a set of state variables, as shown in Figure 2–28(b), we obtain the following equations:

$$\dot{x}_1 = -ax_1 + x_2$$

$$\dot{x}_2 = -Kx_1 + Kx_3 + Ku$$

$$\dot{x}_3 = -(z - p)x_1 - px_3 + (z - p)u$$

$$y = x_1$$

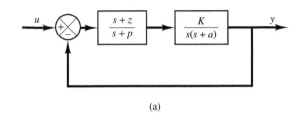

(a)

Figure 2–28
(a) Control system;
(b) block diagram
defining state
variables for the
system.

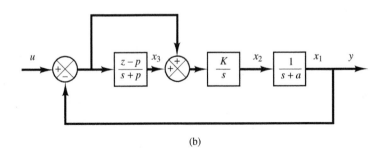

(b)

Rewriting gives

$$\begin{bmatrix} \dot{x}_1 \\ \dot{x}_2 \\ \dot{x}_3 \end{bmatrix} = \begin{bmatrix} -a & 1 & 0 \\ -K & 0 & K \\ -(z-p) & 0 & -p \end{bmatrix} \begin{bmatrix} x_1 \\ x_2 \\ x_3 \end{bmatrix} + \begin{bmatrix} 0 \\ K \\ z-p \end{bmatrix} u$$

$$y = \begin{bmatrix} 1 & 0 & 0 \end{bmatrix} \begin{bmatrix} x_1 \\ x_2 \\ x_3 \end{bmatrix}$$

Notice that the output of the integrator and the outputs of the first-order delayed integrators $[1/(s+a)$ and $(z-p)/(s+p)]$ are chosen as state variables. It is important to remember that the output of the block $(s+z)/(s+p)$ in Figure 2–28(a) cannot be a state variable, because this block involves a derivative term, $s+z$.

A–2–11. Obtain the transfer function of the system defined by

$$\begin{bmatrix} \dot{x}_1 \\ \dot{x}_2 \\ \dot{x}_3 \end{bmatrix} = \begin{bmatrix} -1 & 1 & 0 \\ 0 & -1 & 1 \\ 0 & 0 & -2 \end{bmatrix} \begin{bmatrix} x_1 \\ x_2 \\ x_3 \end{bmatrix} + \begin{bmatrix} 0 \\ 0 \\ 1 \end{bmatrix} u$$

$$y = \begin{bmatrix} 1 & 0 & 0 \end{bmatrix} \begin{bmatrix} x_1 \\ x_2 \\ x_3 \end{bmatrix}$$

Solution. Referring to Equation (2–29), the transfer function $G(s)$ is given by

$$G(s) = \mathbf{C}(s\mathbf{I} - \mathbf{A})^{-1}\mathbf{B} + D$$

In this problem, matrices $\mathbf{A}, \mathbf{B}, \mathbf{C},$ and D are

$$\mathbf{A} = \begin{bmatrix} -1 & 1 & 0 \\ 0 & -1 & 1 \\ 0 & 0 & -2 \end{bmatrix}, \qquad \mathbf{B} = \begin{bmatrix} 0 \\ 0 \\ 1 \end{bmatrix}, \qquad \mathbf{C} = \begin{bmatrix} 1 & 0 & 0 \end{bmatrix}, \qquad D = 0$$

Hence

$$G(s) = \begin{bmatrix} 1 & 0 & 0 \end{bmatrix} \begin{bmatrix} s + 1 & -1 & 0 \\ 0 & s + 1 & -1 \\ 0 & 0 & s + 2 \end{bmatrix}^{-1} \begin{bmatrix} 0 \\ 0 \\ 1 \end{bmatrix}$$

$$= \begin{bmatrix} 1 & 0 & 0 \end{bmatrix} \begin{bmatrix} \dfrac{1}{s + 1} & \dfrac{1}{(s + 1)^2} & \dfrac{1}{(s + 1)^2(s + 2)} \\ 0 & \dfrac{1}{s + 1} & \dfrac{1}{(s + 1)(s + 2)} \\ 0 & 0 & \dfrac{1}{s + 2} \end{bmatrix} \begin{bmatrix} 0 \\ 0 \\ 1 \end{bmatrix}$$

$$= \frac{1}{(s + 1)^2(s + 2)} = \frac{1}{s^3 + 4s^2 + 5s + 2}$$

A–2–12. Consider a system with multiple inputs and multiple outputs. When the system has more than one output, the MATLAB command

$$[\text{NUM,den}] = \text{ss2tf(A,B,C,D,iu)}$$

produces transfer functions for all outputs to each input. (The numerator coefficients are returned to matrix NUM with as many rows as there are outputs.)

Consider the system defined by

$$\begin{bmatrix} \dot{x}_1 \\ \dot{x}_2 \end{bmatrix} = \begin{bmatrix} 0 & 1 \\ -25 & -4 \end{bmatrix} \begin{bmatrix} x_1 \\ x_2 \end{bmatrix} + \begin{bmatrix} 1 & 1 \\ 0 & 1 \end{bmatrix} \begin{bmatrix} u_1 \\ u_2 \end{bmatrix}$$

$$\begin{bmatrix} y_1 \\ y_2 \end{bmatrix} = \begin{bmatrix} 1 & 0 \\ 0 & 1 \end{bmatrix} \begin{bmatrix} x_1 \\ x_2 \end{bmatrix} + \begin{bmatrix} 0 & 0 \\ 0 & 0 \end{bmatrix} \begin{bmatrix} u_1 \\ u_2 \end{bmatrix}$$

This system involves two inputs and two outputs. Four transfer functions are involved: $Y_1(s)/U_1(s)$, $Y_2(s)/U_1(s)$, $Y_1(s)/U_2(s)$, and $Y_2(s)/U_2(s)$. (When considering input u_1, we assume that input u_2 is zero and vice versa.)

Solution. MATLAB Program 2-5 produces four transfer functions.
This is the MATLAB representation of the following four transfer functions:

$$\frac{Y_1(s)}{U_1(s)} = \frac{s + 4}{s^2 + 4s + 25}, \qquad \frac{Y_2(s)}{U_1(s)} = \frac{-25}{s^2 + 4s + 25}$$

$$\frac{Y_1(s)}{U_2(s)} = \frac{s + 5}{s^2 + 4s + 25}, \qquad \frac{Y_2(s)}{U_2(s)} = \frac{s - 25}{s^2 + 4s + 25}$$

```
MATLAB Program 2–5
A = [0    1;-25   -4];
B = [1    1;0    1];
C = [1    0;0    1];
D = [0    0;0    0];
[NUM,den] = ss2tf(A,B,C,D,1)

NUM =
          0    1    4
          0    0   -25

den =
          1    4    25

[NUM,den] = ss2tf(A,B,C,D,2)

NUM =

     0  1.0000      5.0000
     0  1.0000    -25.0000

den =

          1    4    25
```

A–2–13. Linearize the nonlinear equation

$$z = x^2 + 4xy + 6y^2$$

in the region defined by $8 \le x \le 10, 2 \le y \le 4$.

Solution. Define

$$f(x, y) = z = x^2 + 4xy + 6y^2$$

Then

$$z = f(x, y) = f(\bar{x}, \bar{y}) + \left[\frac{\partial f}{\partial x}(x - \bar{x}) + \frac{\partial f}{\partial y}(y - \bar{y}) \right]_{x=\bar{x}, y=\bar{y}} + \cdots$$

where we choose $\bar{x} = 9, \bar{y} = 3$.

Since the higher-order terms in the expanded equation are small, neglecting these higher-order terms, we obtain

$$z - \bar{z} = K_1(x - \bar{x}) + K_2(y - \bar{y})$$

where

$$K_1 = \frac{\partial f}{\partial x}\bigg|_{x=\bar{x}, y=\bar{y}} = 2\bar{x} + 4\bar{y} = 2 \times 9 + 4 \times 3 = 30$$

$$K_2 = \frac{\partial f}{\partial y}\bigg|_{x=\bar{x}, y=\bar{y}} = 4\bar{x} + 12\bar{y} = 4 \times 9 + 12 \times 3 = 72$$

$$\bar{z} = \bar{x}^2 + 4\bar{x}\bar{y} + 6\bar{y}^2 = 9^2 + 4 \times 9 \times 3 + 6 \times 9 = 243$$

Thus

$$z - 243 = 30(x - 9) + 72(y - 3)$$

Hence a linear approximation of the given nonlinear equation near the operating point is

$$z - 30x - 72y + 243 = 0$$

PROBLEMS

B-2-1. Simplify the block diagram shown in Figure 2-29 and obtain the closed-loop transfer function $C(s)/R(s)$.

B-2-2. Simplify the block diagram shown in Figure 2-30 and obtain the closed-loop transfer function $C(s)/R(s)$.

B-2-3. Simplify the block diagram shown in Figure 2-31 and obtain the closed-loop transfer function $C(s)/R(s)$.

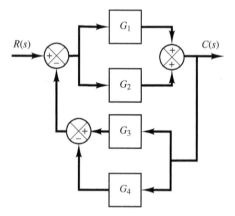

Figure 2-29
Block diagram of a system.

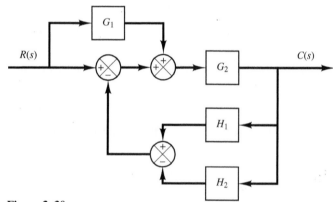

Figure 2-30
Block diagram of a system.

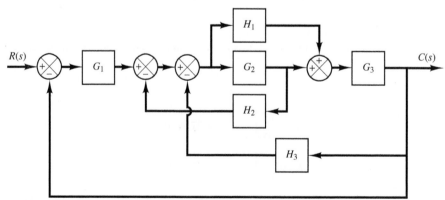

Figure 2-31
Block diagram of a system.

B–2–4. Consider industrial automatic controllers whose control actions are proportional, integral, proportional-plus-integral, proportional-plus-derivative, and proportional-plus-integral-plus-derivative. The transfer functions of these controllers can be given, respectively, by

$$\frac{U(s)}{E(s)} = K_p$$

$$\frac{U(s)}{E(s)} = \frac{K_i}{s}$$

$$\frac{U(s)}{E(s)} = K_p\left(1 + \frac{1}{T_i s}\right)$$

$$\frac{U(s)}{E(s)} = K_p(1 + T_d s)$$

$$\frac{U(s)}{E(s)} = K_p\left(1 + \frac{1}{T_i s} + T_d s\right)$$

where $U(s)$ is the Laplace transform of $u(t)$, the controller output, and $E(s)$ the Laplace transform of $e(t)$, the actuating error signal. Sketch $u(t)$-versus-t curves for each of the five types of controllers when the actuating error signal is

(a) $e(t) =$ unit-step function
(b) $e(t) =$ unit-ramp function

In sketching curves, assume that the numerical values of K_p, K_i, T_i, and T_d are given as

$$K_p = \text{proportional gain} = 4$$
$$K_i = \text{integral gain} = 2$$
$$T_i = \text{integral time} = 2 \text{ sec}$$
$$T_d = \text{derivative time} = 0.8 \text{ sec}$$

B–2–5. Figure 2–32 shows a closed-loop system with a reference input and disturbance input. Obtain the expression for the output $C(s)$ when both the reference input and disturbance input are present.

B–2–6. Consider the system shown in Figure 2–33. Derive the expression for the steady-state error when both the reference input $R(s)$ and disturbance input $D(s)$ are present.

B–2–7. Obtain the transfer functions $C(s)/R(s)$ and $C(s)/D(s)$ of the system shown in Figure 2–34.

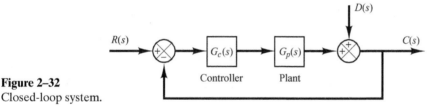

Figure 2–32
Closed-loop system.

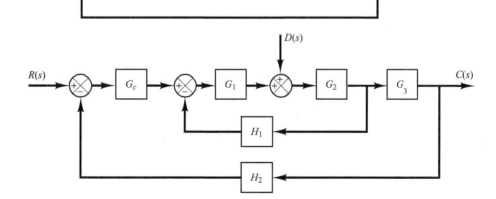

Figure 2–33
Control system.

Figure 2–34
Control system.

B–2–8. Obtain a state-space representation of the system shown in Figure 2–35.

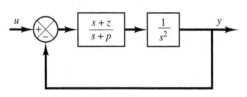

Figure 2–35
Control system.

B–2–9. Consider the system described by

$$\ddot{y} + 3\ddot{y} + 2\dot{y} = u$$

Derive a state-space representation of the system.

B–2–10. Consider the system described by

$$\begin{bmatrix} \dot{x}_1 \\ \dot{x}_2 \end{bmatrix} = \begin{bmatrix} -4 & -1 \\ 3 & -1 \end{bmatrix} \begin{bmatrix} x_1 \\ x_2 \end{bmatrix} + \begin{bmatrix} 1 \\ 1 \end{bmatrix} u$$

$$y = \begin{bmatrix} 1 & 0 \end{bmatrix} \begin{bmatrix} x_1 \\ x_2 \end{bmatrix}$$

Obtain the transfer function of the system.

B–2–11. Consider a system defined by the following state-space equations:

$$\begin{bmatrix} \dot{x}_1 \\ \dot{x}_2 \end{bmatrix} = \begin{bmatrix} -5 & -1 \\ 3 & -1 \end{bmatrix} \begin{bmatrix} x_1 \\ x_2 \end{bmatrix} + \begin{bmatrix} 2 \\ 5 \end{bmatrix} u$$

$$y = \begin{bmatrix} 1 & 2 \end{bmatrix} \begin{bmatrix} x_1 \\ x_2 \end{bmatrix}$$

Obtain the transfer function $G(s)$ of the system.

B–2–12. Obtain the transfer matrix of the system defined by

$$\begin{bmatrix} \dot{x}_1 \\ \dot{x}_2 \\ \dot{x}_3 \end{bmatrix} = \begin{bmatrix} 0 & 1 & 0 \\ 0 & 0 & 1 \\ -2 & -4 & -6 \end{bmatrix} \begin{bmatrix} x_1 \\ x_2 \\ x_3 \end{bmatrix} + \begin{bmatrix} 0 & 0 \\ 0 & 1 \\ 1 & 0 \end{bmatrix} \begin{bmatrix} u_1 \\ u_2 \end{bmatrix}$$

$$\begin{bmatrix} y_1 \\ y_2 \end{bmatrix} = \begin{bmatrix} 1 & 0 & 0 \\ 0 & 1 & 0 \end{bmatrix} \begin{bmatrix} x_1 \\ x_2 \\ x_3 \end{bmatrix}$$

B–2–13. Linearize the nonlinear equation

$$z = x^2 + 8xy + 3y^2$$

in the region defined by $2 \le x \le 4, 10 \le y \le 12$.

B–2–14. Find a linearized equation for

$$y = 0.2x^3$$

about a point $x = 2$.

Mathematical Modeling of Mechanical Systems and Electrical Systems

3–1 INTRODUCTION

This chapter presents mathematical modeling of mechanical systems and electrical systems. In Chapter 2 we obtained mathematical models of a simple electrical circuit and a simple mechanical system. In this chapter we consider mathematical modeling of a variety of mechanical systems and electrical systems that may appear in control systems.

The fundamental law govering mechanical systems is Newton's second law. In Section 3–2 we apply this law to various mechanical systems and derive transfer-function models and state-space models.

The basic laws governing electrical circuits are Kirchhoff's laws. In Section 3–3 we obtain transfer-function models and state-space models of various electrical circuits and operational amplifier systems that may appear in many control systems.

3–2 MATHEMATICAL MODELING OF MECHANICAL SYSTEMS

This section first discusses simple spring systems and simple damper systems. Then we derive transfer-function models and state-space models of various mechanical systems.

Figure 3–1
(a) System consisting of two springs in parallel;
(b) system consisting of two springs in series.

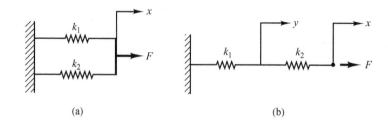

(a)   (b)

EXAMPLE 3–1   Let us obtain the equivalent spring constants for the systems shown in Figures 3–1(a) and (b), respectively.

For the springs in parallel [Figure 3–1(a)] the equivalent spring constant k_{eq} is obtained from

$$k_1 x + k_2 x = F = k_{eq} x$$

or

$$k_{eq} = k_1 + k_2$$

For the springs in series [Figure–3–1(b)], the force in each spring is the same. Thus

$$k_1 y = F, \qquad k_2(x - y) = F$$

Elimination of y from these two equations results in

$$k_2\left(x - \frac{F}{k_1}\right) = F$$

or

$$k_2 x = F + \frac{k_2}{k_1} F = \frac{k_1 + k_2}{k_1} F$$

The equivalent spring constant k_{eq} for this case is then found as

$$k_{eq} = \frac{F}{x} = \frac{k_1 k_2}{k_1 + k_2} = \frac{1}{\dfrac{1}{k_1} + \dfrac{1}{k_2}}$$

EXAMPLE 3–2   Let us obtain the equivalent viscous-friction coefficient b_{eq} for each of the damper systems shown in Figures 3–2(a) and (b). An oil-filled damper is often called a dashpot. A dashpot is a device that provides viscous friction, or damping. It consists of a piston and oil-filled cylinder. Any relative motion between the piston rod and the cylinder is resisted by the oil because the oil must flow around the piston (or through orifices provided in the piston) from one side of the piston to the other. The dashpot essentially absorbs energy. This absorbed energy is dissipated as heat, and the dashpot does not store any kinetic or potential energy.

Figure 3–2
(a) Two dampers
connected in parallel;
(b) two dampers
connected in series.

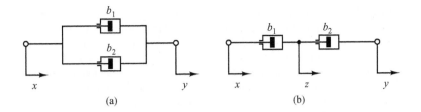

(a)   (b)

(a) The force f due to the dampers is

$$f = b_1(\dot{y} - \dot{x}) + b_2(\dot{y} - \dot{x}) = (b_1 + b_2)(\dot{y} - \dot{x})$$

In terms of the equivalent viscous-friction coefficient b_{eq}, force f is given by

$$f = b_{eq}(\dot{y} - \dot{x})$$

Hence

$$b_{eq} = b_1 + b_2$$

(b) The force f due to the dampers is

$$f = b_1(\dot{z} - \dot{x}) = b_2(\dot{y} - \dot{z}) \tag{3-1}$$

where z is the displacement of a point between damper b_1 and damper b_2. (Note that the same force is transmitted through the shaft.) From Equation (3–1), we have

$$(b_1 + b_2)\dot{z} = b_2\dot{y} + b_1\dot{x}$$

or

$$\dot{z} = \frac{1}{b_1 + b_2}(b_2\dot{y} + b_1\dot{x}) \tag{3-2}$$

In terms of the equivalent viscous-friction coefficient b_{eq}, force f is given by

$$f = b_{eq}(\dot{y} - \dot{x})$$

By substituting Equation (3–2) into Equation (3–1), we have

$$f = b_2(\dot{y} - \dot{z}) = b_2\left[\dot{y} - \frac{1}{b_1 + b_2}(b_2\dot{y} + b_1\dot{x})\right]$$

$$= \frac{b_1 b_2}{b_1 + b_2}(\dot{y} - \dot{x})$$

Thus,

$$f = b_{eq}(\dot{y} - \dot{x}) = \frac{b_1 b_2}{b_1 + b_2}(\dot{y} - \dot{x})$$

Hence,

$$b_{eq} = \frac{b_1 b_2}{b_1 + b_2} = \frac{1}{\dfrac{1}{b_1} + \dfrac{1}{b_2}}$$

EXAMPLE 3–3   Consider the spring-mass-dashpot system mounted on a massless cart as shown in Figure 3–3. Let us obtain mathematical models of this system by assuming that the cart is standing still for $t < 0$ and the spring-mass-dashpot system on the cart is also standing still for $t < 0$. In this system, $u(t)$ is the displacement of the cart and is the input to the system. At $t = 0$, the cart is moved at a constant speed, or $\dot{u} = $ constant. The displacement $y(t)$ of the mass is the output. (The displacement is relative to the ground.) In this system, m denotes the mass, b denotes the viscous-friction coefficient, and k denotes the spring constant. We assume that the friction force of the dashpot is proportional to $\dot{y} - \dot{u}$ and that the spring is a linear spring; that is, the spring force is proportional to $y - u$.

For translational systems, Newton's second law states that

$$ma = \sum F$$

where m is a mass, a is the acceleration of the mass, and $\sum F$ is the sum of the forces acting on the mass in the direction of the acceleration a. Applying Newton's second law to the present system and noting that the cart is massless, we obtain

$$m\frac{d^2y}{dt^2} = -b\left(\frac{dy}{dt} - \frac{du}{dt}\right) - k(y - u)$$

or

$$m\frac{d^2y}{dt^2} + b\frac{dy}{dt} + ky = b\frac{du}{dt} + ku$$

This equation represents a mathematical model of the system considered. Taking the Laplace transform of this last equation, assuming zero initial condition, gives

$$\left(ms^2 + bs + k\right)Y(s) = (bs + k)U(s)$$

Taking the ratio of $Y(s)$ to $U(s)$, we find the transfer function of the system to be

$$\text{Transfer function} = G(s) = \frac{Y(s)}{U(s)} = \frac{bs + k}{ms^2 + bs + k}$$

Such a transfer-function representation of a mathematical model is used very frequently in control engineering.

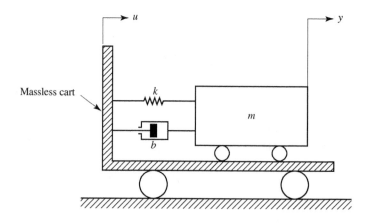

Figure 3–3
Spring-mass-dashpot system mounted on a cart.

Next we shall obtain a state-space model of this system. We shall first compare the differential equation for this system

$$\ddot{y} + \frac{b}{m}\dot{y} + \frac{k}{m}y = \frac{b}{m}\dot{u} + \frac{k}{m}u$$

with the standard form

$$\ddot{y} + a_1\dot{y} + a_2 y = b_0\ddot{u} + b_1\dot{u} + b_2 u$$

and identify a_1, a_2, b_0, b_1, and b_2 as follows:

$$a_1 = \frac{b}{m}, \qquad a_2 = \frac{k}{m}, \qquad b_0 = 0, \qquad b_1 = \frac{b}{m}, \qquad b_2 = \frac{k}{m}$$

Referring to Equation (2–35), we have

$$\beta_0 = b_0 = 0$$

$$\beta_1 = b_1 - a_1\beta_0 = \frac{b}{m}$$

$$\beta_2 = b_2 - a_1\beta_1 - a_2\beta_0 = \frac{k}{m} - \left(\frac{b}{m}\right)^2$$

Then, referring to Equation (2–34), define

$$x_1 = y - \beta_0 u = y$$

$$x_2 = \dot{x}_1 - \beta_1 u = \dot{x}_1 - \frac{b}{m}u$$

From Equation (2–36) we have

$$\dot{x}_1 = x_2 + \beta_1 u = x_2 + \frac{b}{m}u$$

$$\dot{x}_2 = -a_2 x_1 - a_1 x_2 + \beta_2 u = -\frac{k}{m}x_1 - \frac{b}{m}x_2 + \left[\frac{k}{m} - \left(\frac{b}{m}\right)^2\right]u$$

and the output equation becomes

$$y = x_1$$

or

$$\begin{bmatrix} \dot{x}_1 \\ \dot{x}_2 \end{bmatrix} = \begin{bmatrix} 0 & 1 \\ -\dfrac{k}{m} & -\dfrac{b}{m} \end{bmatrix} \begin{bmatrix} x_1 \\ x_2 \end{bmatrix} + \begin{bmatrix} \dfrac{b}{m} \\ \dfrac{k}{m} - \left(\dfrac{b}{m}\right)^2 \end{bmatrix} u \qquad (3\text{--}3)$$

and

$$y = \begin{bmatrix} 1 & 0 \end{bmatrix} \begin{bmatrix} x_1 \\ x_2 \end{bmatrix} \qquad (3\text{--}4)$$

Equations (3–3) and (3–4) give a state-space representation of the system. (Note that this is not the only state-space representation. There are infinitely many state-space representations for the system.)

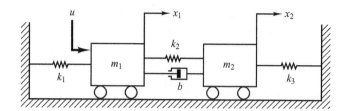

Figure 3–4
Mechanical system.

EXAMPLE 3–4 Obtain the transfer functions $X_1(s)/U(s)$ and $X_2(s)/U(s)$ of the mechanical system shown in Figure 3–4.

The equations of motion for the system shown in Figure 3–4 are

$$m_1 \ddot{x}_1 = -k_1 x_1 - k_2(x_1 - x_2) - b(\dot{x}_1 - \dot{x}_2) + u$$

$$m_2 \ddot{x}_2 = -k_3 x_2 - k_2(x_2 - x_1) - b(\dot{x}_2 - \dot{x}_1)$$

Simplifying, we obtain

$$m_1 \ddot{x}_1 + b\dot{x}_1 + (k_1 + k_2)x_1 = b\dot{x}_2 + k_2 x_2 + u$$

$$m_2 \ddot{x}_2 + b\dot{x}_2 + (k_2 + k_3)x_2 = b\dot{x}_1 + k_2 x_1$$

Taking the Laplace transforms of these two equations, assuming zero initial conditions, we obtain

$$[m_1 s^2 + bs + (k_1 + k_2)]X_1(s) = (bs + k_2)X_2(s) + U(s) \tag{3–5}$$

$$[m_2 s^2 + bs + (k_2 + k_3)]X_2(s) = (bs + k_2)X_1(s) \tag{3–6}$$

Solving Equation (3–6) for $X_2(s)$ and substituting it into Equation (3–5) and simplifying, we get

$$[(m_1 s^2 + bs + k_1 + k_2)(m_2 s^2 + bs + k_2 + k_3) - (bs + k_2)^2]X_1(s)$$
$$= (m_2 s^2 + bs + k_2 + k_3)U(s)$$

from which we obtain

$$\frac{X_1(s)}{U(s)} = \frac{m_2 s^2 + bs + k_2 + k_3}{(m_1 s^2 + bs + k_1 + k_2)(m_2 s^2 + bs + k_2 + k_3) - (bs + k_2)^2} \tag{3–7}$$

From Equations (3–6) and (3–7) we have

$$\frac{X_2(s)}{U(s)} = \frac{bs + k_2}{(m_1 s^2 + bs + k_1 + k_2)(m_2 s^2 + bs + k_2 + k_3) - (bs + k_2)^2} \tag{3–8}$$

Equations (3–7) and (3–8) are the transfer functions $X_1(s)/U(s)$ and $X_2(s)/U(s)$, respectively.

EXAMPLE 3–5 An inverted pendulum mounted on a motor-driven cart is shown in Figure 3–5(a). This is a model of the attitude control of a space booster on takeoff. (The objective of the attitude control problem is to keep the space booster in a vertical position.) The inverted pendulum is unstable in that it may fall over any time in any direction unless a suitable control force is applied. Here we consider

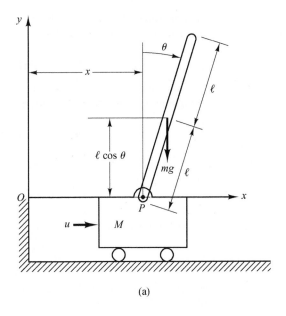

(a)

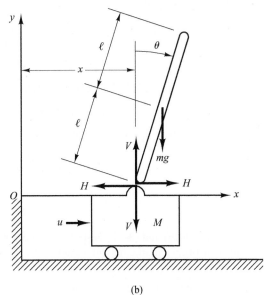

(b)

Figure 3–5
(a) Inverted
pendulum system;
(b) free-body
diagram.

only a two-dimensional problem in which the pendulum moves only in the plane of the page. The control force u is applied to the cart. Assume that the center of gravity of the pendulum rod is at its geometric center. Obtain a mathematical model for the system.

Define the angle of the rod from the vertical line as θ. Define also the (x, y) coordinates of the center of gravity of the pendulum rod as (x_G, y_G). Then

$$x_G = x + l \sin \theta$$

$$y_G = l \cos \theta$$

To derive the equations of motion for the system, consider the free-body diagram shown in Figure 3–5(b). The rotational motion of the pendulum rod about its center of gravity can be described by

$$I\ddot{\theta} = Vl\sin\theta - Hl\cos\theta \tag{3–9}$$

where I is the moment of inertia of the rod about its center of gravity.

The horizontal motion of center of gravity of pendulum rod is given by

$$m\frac{d^2}{dt^2}(x + l\sin\theta) = H \tag{3–10}$$

The vertical motion of center of gravity of pendulum rod is

$$m\frac{d^2}{dt^2}(l\cos\theta) = V - mg \tag{3–11}$$

The horizontal motion of cart is described by

$$M\frac{d^2x}{dt^2} = u - H \tag{3–12}$$

Since we must keep the inverted pendulum vertical, we can assume that $\theta(t)$ and $\dot{\theta}(t)$ are small quantities such that $\sin\theta \doteq \theta, \cos\theta = 1$, and $\theta\dot{\theta}^2 = 0$. Then, Equations (3–9) through (3–11) can be linearized. The linearized equations are

$$I\ddot{\theta} = Vl\theta - Hl \tag{3–13}$$

$$m(\ddot{x} + l\ddot{\theta}) = H \tag{3–14}$$

$$0 = V - mg \tag{3–15}$$

From Equations (3–12) and (3–14), we obtain

$$(M + m)\ddot{x} + ml\ddot{\theta} = u \tag{3–16}$$

From Equations (3–13), (3–14), and (3–15), we have

$$I\ddot{\theta} = mgl\theta - Hl$$

$$= mgl\theta - l(m\ddot{x} + ml\ddot{\theta})$$

or

$$(I + ml^2)\ddot{\theta} + ml\ddot{x} = mgl\theta \tag{3–17}$$

Equations (3–16) and (3–17) describe the motion of the inverted-pendulum-on-the-cart system. They constitute a mathematical model of the system.

EXAMPLE 3–6   Consider the inverted-pendulum system shown in Figure 3–6. Since in this system the mass is concentrated at the top of the rod, the center of gravity is the center of the pendulum ball. For this case, the moment of inertia of the pendulum about its center of gravity is small, and we assume $I = 0$ in Equation (3–17). Then the mathematical model for this system becomes as follows:

$$(M + m)\ddot{x} + ml\ddot{\theta} = u \tag{3–18}$$

$$ml^2\ddot{\theta} + ml\ddot{x} = mgl\theta \tag{3–19}$$

Equations (3–18) and (3–19) can be modified to

$$Ml\ddot{\theta} = (M + m)g\theta - u \tag{3–20}$$

$$M\ddot{x} = u - mg\theta \tag{3–21}$$

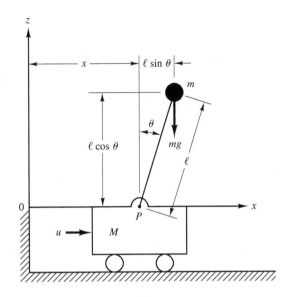

Figure 3–6
Inverted-pendulum
system.

Equation (3–20) was obtained by eliminating $\ddot{x}$ from Equations (3–18) and (3–19). Equation (3–21) was obtained by eliminating $\ddot{\theta}$ from Equations (3–18) and (3–19). From Equation (3–20) we obtain the plant transfer function to be

$$\frac{\Theta(s)}{-U(s)} = \frac{1}{Mls^2 - (M + m)g}$$

$$= \frac{1}{Ml\left(s + \sqrt{\dfrac{M + m}{Ml}g}\right)\left(s - \sqrt{\dfrac{M + m}{Ml}g}\right)}$$

The inverted-pendulum plant has one pole on the negative real axis $\left[s = -(\sqrt{M + m}/\sqrt{Ml})\sqrt{g}\right]$ and another on the positive real axis $\left[s = (\sqrt{M + m}/\sqrt{Ml})\sqrt{g}\right]$. Hence, the plant is open-loop unstable.

Define state variables x_1, x_2, x_3, and x_4 by

$$x_1 = \theta$$
$$x_2 = \dot{\theta}$$
$$x_3 = x$$
$$x_4 = \dot{x}$$

Note that angle θ indicates the rotation of the pendulum rod about point P, and x is the location of the cart. If we consider θ and x as the outputs of the system, then

$$\mathbf{y} = \begin{bmatrix} y_1 \\ y_2 \end{bmatrix} = \begin{bmatrix} \theta \\ x \end{bmatrix} = \begin{bmatrix} x_1 \\ x_3 \end{bmatrix}$$

(Notice that both θ and x are easily measurable quantities.) Then, from the definition of the state variables and Equations (3–20) and (3–21), we obtain

$$\dot{x}_1 = x_2$$
$$\dot{x}_2 = \frac{M + m}{Ml}gx_1 - \frac{1}{Ml}u$$
$$\dot{x}_3 = x_4$$
$$\dot{x}_4 = -\frac{m}{M}gx_1 + \frac{1}{M}u$$

In terms of vector-matrix equations, we have

$$
\begin{bmatrix} \dot{x}_1 \\ \dot{x}_2 \\ \dot{x}_3 \\ \dot{x}_4 \end{bmatrix} = \begin{bmatrix} 0 & 1 & 0 & 0 \\ \dfrac{M+m}{Ml}g & 0 & 0 & 0 \\ 0 & 0 & 0 & 1 \\ -\dfrac{m}{M}g & 0 & 0 & 0 \end{bmatrix} \begin{bmatrix} x_1 \\ x_2 \\ x_3 \\ x_4 \end{bmatrix} + \begin{bmatrix} 0 \\ -\dfrac{1}{Ml} \\ 0 \\ \dfrac{1}{M} \end{bmatrix} u \qquad (3\text{--}22)
$$

$$
\begin{bmatrix} y_1 \\ y_2 \end{bmatrix} = \begin{bmatrix} 1 & 0 & 0 & 0 \\ 0 & 0 & 1 & 0 \end{bmatrix} \begin{bmatrix} x_1 \\ x_2 \\ x_3 \\ x_4 \end{bmatrix} \qquad (3\text{--}23)
$$

Equations (3–22) and (3–23) give a state-space representation of the inverted-pendulum system. (Note that state-space representation of the system is not unique. There are infinitely many such representations for this system.)

3–3 MATHEMATICAL MODELING OF ELECTRICAL SYSTEMS

Basic laws governing electrical circuits are Kirchhoff's current law and voltage law. Kirchhoff's current law (node law) states that the algebraic sum of all currents entering and leaving a node is zero. (This law can also be stated as follows: The sum of currents entering a node is equal to the sum of currents leaving the same node.) Kirchhoff's voltage law (loop law) states that at any given instant the algebraic sum of the voltages around any loop in an electrical circuit is zero. (This law can also be stated as follows: The sum of the voltage drops is equal to the sum of the voltage rises around a loop.) A mathematical model of an electrical circuit can be obtained by applying one or both of Kirchhoff's laws to it.

This section first deals with simple electrical circuits and then treats mathematical modeling of operational amplifier systems.

***LRC* Circuit.** Consider the electrical circuit shown in Figure 3–7. The circuit consists of an inductance L (henry), a resistance R (ohm), and a capacitance C (farad). Applying Kirchhoff's voltage law to the system, we obtain the following equations:

$$
L\frac{di}{dt} + Ri + \frac{1}{C}\int i\, dt = e_i \qquad (3\text{--}24)
$$

$$
\frac{1}{C}\int i\, dt = e_o \qquad (3\text{--}25)
$$

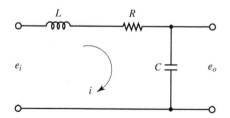

Figure 3–7
Electrical circuit.

Equations (3–24) and (3–25) give a mathematical model of the circuit.

A transfer-function model of the circuit can also be obtained as follows: Taking the Laplace transforms of Equations (3–24) and (3–25), assuming zero initial conditions, we obtain

$$LsI(s) + RI(s) + \frac{1}{C}\frac{1}{s}I(s) = E_i(s)$$

$$\frac{1}{C}\frac{1}{s}I(s) = E_o(s)$$

If e_i is assumed to be the input and e_o the output, then the transfer function of this system is found to be

$$\frac{E_o(s)}{E_i(s)} = \frac{1}{LCs^2 + RCs + 1} \tag{3–26}$$

A state-space model of the system shown in Figure 3–7 may be obtained as follows: First, note that the differential equation for the system can be obtained from Equation (3–26) as

$$\ddot{e}_o + \frac{R}{L}\dot{e}_o + \frac{1}{LC}e_o = \frac{1}{LC}e_i$$

Then by defining state variables by

$$x_1 = e_o$$
$$x_2 = \dot{e}_o$$

and the input and output variables by

$$u = e_i$$
$$y = e_o = x_1$$

we obtain

$$\begin{bmatrix} \dot{x}_1 \\ \dot{x}_2 \end{bmatrix} = \begin{bmatrix} 0 & 1 \\ -\dfrac{1}{LC} & -\dfrac{R}{L} \end{bmatrix}\begin{bmatrix} x_1 \\ x_2 \end{bmatrix} + \begin{bmatrix} 0 \\ \dfrac{1}{LC} \end{bmatrix} u$$

and

$$y = \begin{bmatrix} 1 & 0 \end{bmatrix}\begin{bmatrix} x_1 \\ x_2 \end{bmatrix}$$

These two equations give a mathematical model of the system in state space.

Transfer Functions of Cascaded Elements. Many feedback systems have components that load each other. Consider the system shown in Figure 3–8. Assume that e_i is the input and e_o is the output. The capacitances C_1 and C_2 are not charged initially.

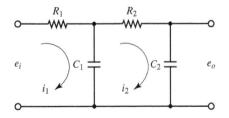

Figure 3–8
Electrical system.

It will be shown that the second stage of the circuit ($R_2 C_2$ portion) produces a loading effect on the first stage ($R_1 C_1$ portion). The equations for this system are

$$\frac{1}{C_1} \int (i_1 - i_2)\, dt + R_1 i_1 = e_i \tag{3-27}$$

and

$$\frac{1}{C_1} \int (i_2 - i_1)\, dt + R_2 i_2 + \frac{1}{C_2} \int i_2\, dt = 0 \tag{3-28}$$

$$\frac{1}{C_2} \int i_2\, dt = e_o \tag{3-29}$$

Taking the Laplace transforms of Equations (3-27) through (3-29), respectively, using zero initial conditions, we obtain

$$\frac{1}{C_1 s}[I_1(s) - I_2(s)] + R_1 I_1(s) = E_i(s) \tag{3-30}$$

$$\frac{1}{C_1 s}[I_2(s) - I_1(s)] + R_2 I_2(s) + \frac{1}{C_2 s} I_2(s) = 0 \tag{3-31}$$

$$\frac{1}{C_2 s} I_2(s) = E_o(s) \tag{3-32}$$

Eliminating $I_1(s)$ from Equations (3-30) and (3-31) and writing $E_i(s)$ in terms of $I_2(s)$, we find the transfer function between $E_o(s)$ and $E_i(s)$ to be

$$\frac{E_o(s)}{E_i(s)} = \frac{1}{(R_1 C_1 s + 1)(R_2 C_2 s + 1) + R_1 C_2 s}$$

$$= \frac{1}{R_1 C_1 R_2 C_2 s^2 + (R_1 C_1 + R_2 C_2 + R_1 C_2)s + 1} \tag{3-33}$$

The term $R_1 C_2 s$ in the denominator of the transfer function represents the interaction of two simple RC circuits. Since $(R_1 C_1 + R_2 C_2 + R_1 C_2)^2 > 4 R_1 C_1 R_2 C_2$, the two roots of the denominator of Equation (3-33) are real.

The present analysis shows that, if two RC circuits are connected in cascade so that the output from the first circuit is the input to the second, the overall transfer function is not the product of $1/(R_1 C_1 s + 1)$ and $1/(R_2 C_2 s + 1)$. The reason for this is that, when we derive the transfer function for an isolated circuit, we implicitly assume that the output is unloaded. In other words, the load impedance is assumed to be infinite, which means that no power is being withdrawn at the output. When the second circuit is connected to the output of the first, however, a certain amount of power is withdrawn, and thus the assumption of no loading is violated. Therefore, if the transfer function of this system is obtained under the assumption of no loading, then it is not valid. The degree of the loading effect determines the amount of modification of the transfer function.

Complex Impedances. In deriving transfer functions for electrical circuits, we frequently find it convenient to write the Laplace-transformed equations directly, without writing the differential equations. Consider the system shown in Figure 3–9(a). In this system, Z_1 and Z_2 represent complex impedances. The complex impedance $Z(s)$ of a two-terminal circuit is the ratio of $E(s)$, the Laplace transform of the voltage across the terminals, to $I(s)$, the Laplace transform of the current through the element, under the assumption that the initial conditions are zero, so that $Z(s) = E(s)/I(s)$. If the two-terminal element is a resistance R, capacitance C, or inductance L, then the complex impedance is given by $R, 1/Cs$, or Ls, respectively. If complex impedances are connected in series, the total impedance is the sum of the individual complex impedances.

Remember that the impedance approach is valid only if the initial conditions involved are all zeros. Since the transfer function requires zero initial conditions, the impedance approach can be applied to obtain the transfer function of the electrical circuit. This approach greatly simplifies the derivation of transfer functions of electrical circuits.

Consider the circuit shown in Figure 3–9(b). Assume that the voltages e_i and e_o are the input and output of the circuit, respectively. Then the transfer function of this circuit is

$$\frac{E_o(s)}{E_i(s)} = \frac{Z_2(s)}{Z_1(s) + Z_2(s)}$$

For the system shown in Figure 3–7,

$$Z_1 = Ls + R, \qquad Z_2 = \frac{1}{Cs}$$

Hence the transfer function $E_o(s)/E_i(s)$ can be found as follows:

$$\frac{E_o(s)}{E_i(s)} = \frac{\dfrac{1}{Cs}}{Ls + R + \dfrac{1}{Cs}} = \frac{1}{LCs^2 + RCs + 1}$$

which is, of course, identical to Equation (3–26).

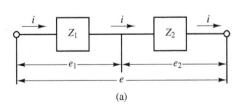

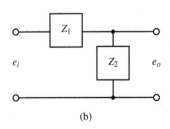

Figure 3–9
Electrical circuits.

(a)

(b)

EXAMPLE 3–7 Consider again the system shown in Figure 3–8. Obtain the transfer function $E_o(s)/E_i(s)$ by use of the complex impedance approach. (Capacitors C_1 and C_2 are not charged initially.)

The circuit shown in Figure 3–8 can be redrawn as that shown in Figure 3–10(a), which can be further modified to Figure 3–10(b).

In the system shown in Figure 3–10(b) the current I is divided into two currents I_1 and I_2. Noting that

$$Z_2 I_1 = (Z_3 + Z_4)I_2, \qquad I_1 + I_2 = I$$

we obtain

$$I_1 = \frac{Z_3 + Z_4}{Z_2 + Z_3 + Z_4} I, \qquad I_2 = \frac{Z_2}{Z_2 + Z_3 + Z_4} I$$

Noting that

$$E_i(s) = Z_1 I + Z_2 I_1 = \left[Z_1 + \frac{Z_2(Z_3 + Z_4)}{Z_2 + Z_3 + Z_4} \right] I$$

$$E_o(s) = Z_4 I_2 = \frac{Z_2 Z_4}{Z_2 + Z_3 + Z_4} I$$

we obtain

$$\frac{E_o(s)}{E_i(s)} = \frac{Z_2 Z_4}{Z_1(Z_2 + Z_3 + Z_4) + Z_2(Z_3 + Z_4)}$$

Substituting $Z_1 = R_1$, $Z_2 = 1/(C_1 s)$, $Z_3 = R_2$, and $Z_4 = 1/(C_2 s)$ into this last equation, we get

$$\frac{E_o(s)}{E_i(s)} = \frac{\dfrac{1}{C_1 s}\dfrac{1}{C_2 s}}{R_1 \left(\dfrac{1}{C_1 s} + R_2 + \dfrac{1}{C_2 s} \right) + \dfrac{1}{C_1 s}\left(R_2 + \dfrac{1}{C_2 s} \right)}$$

$$= \frac{1}{R_1 C_1 R_2 C_2 s^2 + (R_1 C_1 + R_2 C_2 + R_1 C_2)s + 1}$$

which is the same as that given by Equation (3–33).

Figure 3–10
(a) The circuit of Figure 3–8 shown in terms of impedances; (b) equivalent circuit diagram.

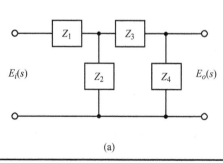

(a)

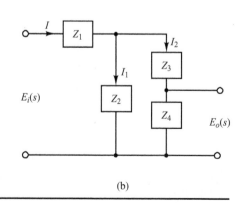

(b)

(a)   (b)

Figure 3–11
(a) System consisting of two nonloading cascaded elements; (b) an equivalent system.

Transfer Functions of Nonloading Cascaded Elements. The transfer function of a system consisting of two nonloading cascaded elements can be obtained by eliminating the intermediate input and output. For example, consider the system shown in Figure 3–11(a). The transfer functions of the elements are

$$G_1(s) = \frac{X_2(s)}{X_1(s)} \quad \text{and} \quad G_2(s) = \frac{X_3(s)}{X_2(s)}$$

If the input impedance of the second element is infinite, the output of the first element is not affected by connecting it to the second element. Then the transfer function of the whole system becomes

$$G(s) = \frac{X_3(s)}{X_1(s)} = \frac{X_2(s)X_3(s)}{X_1(s)X_2(s)} = G_1(s)G_2(s)$$

The transfer function of the whole system is thus the product of the transfer functions of the individual elements. This is shown in Figure 3–11(b).

As an example, consider the system shown in Figure 3–12. The insertion of an isolating amplifier between the circuits to obtain nonloading characteristics is frequently used in combining circuits. Since amplifiers have very high input impedances, an isolation amplifier inserted between the two circuits justifies the nonloading assumption.

The two simple *RC* circuits, isolated by an amplifier as shown in Figure 3–12, have negligible loading effects, and the transfer function of the entire circuit equals the product of the individual transfer functions. Thus, in this case,

$$\frac{E_o(s)}{E_i(s)} = \left(\frac{1}{R_1C_1s + 1}\right)(K)\left(\frac{1}{R_2C_2s + 1}\right)$$

$$= \frac{K}{(R_1C_1s + 1)(R_2C_2s + 1)}$$

Electronic Controllers. In what follows we shall discuss electronic controllers using operational amplifiers. We begin by deriving the transfer functions of simple operational-amplifier circuits. Then we derive the transfer functions of some of the operational-amplifier controllers. Finally, we give operational-amplifier controllers and their transfer functions in the form of a table.

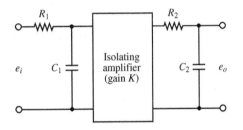

Figure 3–12
Electrical system.

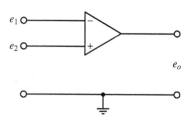

Figure 3–13
Operational
amplifier.

Operational Amplifiers.   Operational amplifiers, often called op amps, are frequently used to amplify signals in sensor circuits. Op amps are also frequently used in filters used for compensation purposes. Figure 3–13 shows an op amp. It is a common practice to choose the ground as 0 volt and measure the input voltages e_1 and e_2 relative to the ground. The input e_1 to the minus terminal of the amplifier is inverted, and the input e_2 to the plus terminal is not inverted. The total input to the amplifier thus becomes $e_2 - e_1$. Hence, for the circuit shown in Figure 3–13, we have

$$e_o = K(e_2 - e_1) = -K(e_1 - e_2)$$

where the inputs e_1 and e_2 may be dc or ac signals and K is the differential gain (voltage gain). The magnitude of K is approximately $10^5 \sim 10^6$ for dc signals and ac signals with frequencies less than approximately 10 Hz. (The differential gain K decreases with the signal frequency and becomes about unity for frequencies of 1 MHz $\sim$ 50 MHz.) Note that the op amp amplifies the difference in voltages e_1 and e_2. Such an amplifier is commonly called a differential amplifier. Since the gain of the op amp is very high, it is necessary to have a negative feedback from the output to the input to make the amplifier stable. (The feedback is made from the output to the inverted input so that the feedback is a negative feedback.)

In the ideal op amp, no current flows into the input terminals, and the output voltage is not affected by the load connected to the output terminal. In other words, the input impedance is infinity and the output impedance is zero. In an actual op amp, a very small (almost negligible) current flows into an input terminal and the output cannot be loaded too much. In our analysis here, we make the assumption that the op amps are ideal.

Inverting Amplifier.   Consider the operational-amplifier circuit shown in Figure 3–14. Let us obtain the output voltage e_o.

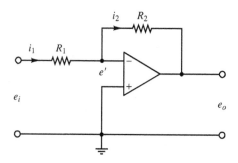

Figure 3–14
Inverting amplifier.

The equation for this circuit can be obtained as follows: Define

$$i_1 = \frac{e_i - e'}{R_1}, \qquad i_2 = \frac{e' - e_o}{R_2}$$

Since only a negligible current flows into the amplifier, the current i_1 must be equal to current i_2. Thus

$$\frac{e_i - e'}{R_1} = \frac{e' - e_o}{R_2}$$

Since $K(0 - e') = e_o$ and $K \gg 1$, e' must be almost zero, or $e' \doteq 0$. Hence we have

$$\frac{e_i}{R_1} = \frac{-e_o}{R_2}$$

or

$$e_o = -\frac{R_2}{R_1} e_i$$

Thus the circuit shown is an inverting amplifier. If $R_1 = R_2$, then the op-amp circuit shown acts as a sign inverter.

Noninverting Amplifier. Figure 3–15(a) shows a noninverting amplifier. A circuit equivalent to this one is shown in Figure 3–15(b). For the circuit of Figure 3–15(b), we have

$$e_o = K\left(e_i - \frac{R_1}{R_1 + R_2} e_o \right)$$

where K is the differential gain of the amplifier. From this last equation, we get

$$e_i = \left(\frac{R_1}{R_1 + R_2} + \frac{1}{K} \right) e_o$$

Since $K \gg 1$, if $R_1/(R_1 + R_2) \gg 1/K$, then

$$e_o = \left(1 + \frac{R_2}{R_1} \right) e_i$$

This equation gives the output voltage e_o. Since e_o and e_i have the same signs, the op-amp circuit shown in Figure 3–15(a) is noninverting.

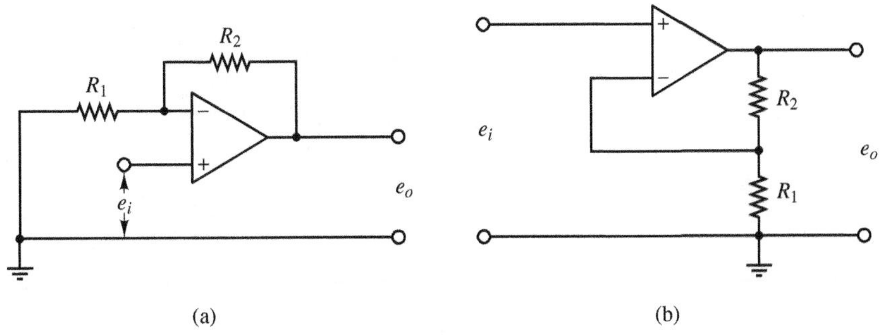

Figure 3–15
(a) Noninverting
operational
amplifier;
(b) equivalent
circuit.

(a)

(b)

EXAMPLE 3–8   Figure 3–16 shows an electrical circuit involving an operational amplifier. Obtain the output e_o. Let us define

$$i_1 = \frac{e_i - e'}{R_1}, \qquad i_2 = C\frac{d(e' - e_o)}{dt}, \qquad i_3 = \frac{e' - e_o}{R_2}$$

Noting that the current flowing into the amplifier is negligible, we have

$$i_1 = i_2 + i_3$$

Hence

$$\frac{e_i - e'}{R_1} = C\frac{d(e' - e_o)}{dt} + \frac{e' - e_o}{R_2}$$

Since $e' \doteq 0$, we have

$$\frac{e_i}{R_1} = -C\frac{de_o}{dt} - \frac{e_o}{R_2}$$

Taking the Laplace transform of this last equation, assuming the zero initial condition, we have

$$\frac{E_i(s)}{R_1} = -\frac{R_2 Cs + 1}{R_2} E_o(s)$$

which can be written as

$$\frac{E_o(s)}{E_i(s)} = -\frac{R_2}{R_1}\frac{1}{R_2 Cs + 1}$$

The op-amp circuit shown in Figure 3–16 is a first-order lag circuit. (Several other circuits involving op amps are shown in Table 3–1 together with their transfer functions. Table 3–1 is given on page 85.)

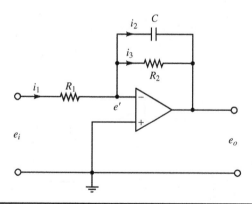

Figure 3–16
First-order lag circuit using operational amplifier.

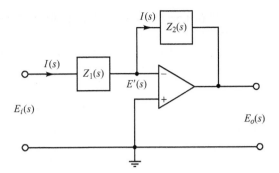

Figure 3–17
Operational-
amplifier circuit.

Impedance Approach to Obtaining Transfer Functions. Consider the op-amp circuit shown in Figure 3–17. Similar to the case of electrical circuits we discussed earlier, the impedance approach can be applied to op-amp circuits to obtain their transfer functions. For the circuit shown in Figure 3–17, we have

$$\frac{E_i(s) - E'(s)}{Z_1} = \frac{E'(s) - E_o(s)}{Z_2}$$

Since $E'(s) \doteq 0$, we have

$$\frac{E_o(s)}{E_i(s)} = -\frac{Z_2(s)}{Z_1(s)} \tag{3-34}$$

EXAMPLE 3–9 Referring to the op-amp circuit shown in Figure 3–16, obtain the transfer function $E_o(s)/E_i(s)$ by use of the impedance approach.

The complex impedances $Z_1(s)$ and $Z_2(s)$ for this circuit are

$$Z_1(s) = R_1 \quad \text{and} \quad Z_2(s) = \frac{1}{Cs + \dfrac{1}{R_2}} = \frac{R_2}{R_2 Cs + 1}$$

The transfer function $E_o(s)/E_i(s)$ is, therefore, obtained as

$$\frac{E_o(s)}{E_i(s)} = -\frac{Z_2(s)}{Z_1(s)} = -\frac{R_2}{R_1}\frac{1}{R_2 Cs + 1}$$

which is, of course, the same as that obtained in Example 3-8.

Lead or Lag Networks Using Operational Amplifiers. Figure 3–18(a) shows an electronic circuit using an operational amplifier. The transfer function for this circuit can be obtained as follows: Define the input impedance and feedback impedance as Z_1 and Z_2, respectively. Then

$$Z_1 = \frac{R_1}{R_1 C_1 s + 1}, \qquad Z_2 = \frac{R_2}{R_2 C_2 s + 1}$$

Hence, referring to Equation (3–34), we have

$$\frac{E(s)}{E_i(s)} = -\frac{Z_2}{Z_1} = -\frac{R_2}{R_1} \frac{R_1 C_1 s + 1}{R_2 C_2 s + 1} = -\frac{C_1}{C_2} \frac{s + \dfrac{1}{R_1 C_1}}{s + \dfrac{1}{R_2 C_2}} \tag{3–35}$$

Notice that the transfer function in Equation (3–35) contains a minus sign. Thus, this circuit is sign inverting. If such a sign inversion is not convenient in the actual application, a sign inverter may be connected to either the input or the output of the circuit of Figure 3–18(a). An example is shown in Figure 3–18(b). The sign inverter has the transfer function of

$$\frac{E_o(s)}{E(s)} = -\frac{R_4}{R_3}$$

The sign inverter has the gain of $-R_4/R_3$. Hence the network shown in Figure 3–18(b) has the following transfer function:

$$\frac{E_o(s)}{E_i(s)} = \frac{R_2 R_4}{R_1 R_3} \frac{R_1 C_1 s + 1}{R_2 C_2 s + 1} = \frac{R_4 C_1}{R_3 C_2} \frac{s + \dfrac{1}{R_1 C_1}}{s + \dfrac{1}{R_2 C_2}}$$

$$= K_c \alpha \frac{Ts + 1}{\alpha Ts + 1} = K_c \frac{s + \dfrac{1}{T}}{s + \dfrac{1}{\alpha T}} \tag{3–36}$$

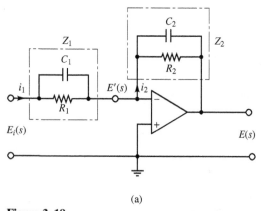

(a)

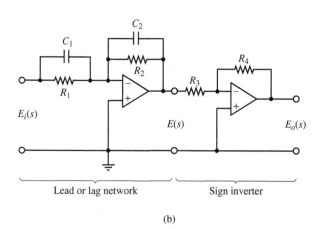

(b)

Figure 3–18
(a) Operational-amplifier circuit; (b) operational-amplifier circuit used as a lead or lag compensator.

where

$$T = R_1 C_1, \qquad \alpha T = R_2 C_2, \qquad K_c = \frac{R_4 C_1}{R_3 C_2}$$

Notice that

$$K_c \alpha = \frac{R_4 C_1}{R_3 C_2} \frac{R_2 C_2}{R_1 C_1} = \frac{R_2 R_4}{R_1 R_3}, \qquad \alpha = \frac{R_2 C_2}{R_1 C_1}$$

This network has a dc gain of $K_c \alpha = R_2 R_4 / (R_1 R_3)$.

Note that this network, whose transfer function is given by Equation (3–36), is a lead network if $R_1 C_1 > R_2 C_2$, or $\alpha < 1$. It is a lag network if $R_1 C_1 < R_2 C_2$.

PID Controller Using Operational Amplifiers.   Figure 3–19 shows an electronic proportional-plus-integral-plus-derivative controller (a PID controller) using operational amplifiers. The transfer function $E(s)/E_i(s)$ is given by

$$\frac{E(s)}{E_i(s)} = -\frac{Z_2}{Z_1}$$

where

$$Z_1 = \frac{R_1}{R_1 C_1 s + 1}, \qquad Z_2 = \frac{R_2 C_2 s + 1}{C_2 s}$$

Thus

$$\frac{E(s)}{E_i(s)} = -\left(\frac{R_2 C_2 s + 1}{C_2 s} \right)\left(\frac{R_1 C_1 s + 1}{R_1} \right)$$

Noting that

$$\frac{E_o(s)}{E(s)} = -\frac{R_4}{R_3}$$

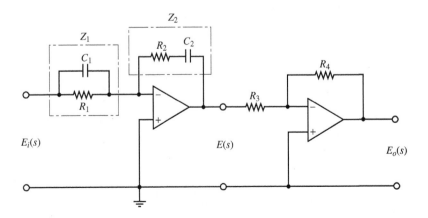

Figure 3–19
Electronic PID
controller.

we have

$$\frac{E_o(s)}{E_i(s)} = \frac{E_o(s)}{E(s)} \frac{E(s)}{E_i(s)} = \frac{R_4 R_2}{R_3 R_1} \frac{(R_1 C_1 s + 1)(R_2 C_2 s + 1)}{R_2 C_2 s}$$

$$= \frac{R_4 R_2}{R_3 R_1} \left(\frac{R_1 C_1 + R_2 C_2}{R_2 C_2} + \frac{1}{R_2 C_2 s} + R_1 C_1 s \right)$$

$$= \frac{R_4(R_1 C_1 + R_2 C_2)}{R_3 R_1 C_2} \left[1 + \frac{1}{(R_1 C_1 + R_2 C_2)s} + \frac{R_1 C_1 R_2 C_2}{R_1 C_1 + R_2 C_2} s \right] \qquad (3\text{–}37)$$

Notice that the second operational-amplifier circuit acts as a sign inverter as well as a gain adjuster.

When a PID controller is expressed as

$$\frac{E_o(s)}{E_i(s)} = K_p \left(1 + \frac{T_i}{s} + T_d s \right)$$

K_p is called the proportional gain, T_i is called the integral time, and T_d is called the derivative time. From Equation (3–37) we obtain the proportional gain K_p, integral time T_i, and derivative time T_d to be

$$K_p = \frac{R_4(R_1 C_1 + R_2 C_2)}{R_3 R_1 C_2}$$

$$T_i = \frac{1}{R_1 C_1 + R_2 C_2}$$

$$T_d = \frac{R_1 C_1 R_2 C_2}{R_1 C_1 + R_2 C_2}$$

When a PID controller is expressed as

$$\frac{E_o(s)}{E_i(s)} = K_p + \frac{K_i}{s} + K_d s$$

K_p is called the proportional gain, K_i is called the integral gain, and K_d is called the derivative gain. For this controller

$$K_p = \frac{R_4(R_1 C_1 + R_2 C_2)}{R_3 R_1 C_2}$$

$$K_i = \frac{R_4}{R_3 R_1 C_2}$$

$$K_d = \frac{R_4 R_2 C_1}{R_3}$$

Table 3–1 shows a list of operational-amplifier circuits that may be used as controllers or compensators.

Table 3–1 Operational-Amplifier Circuits That May Be Used as Compensators

	Control Action	$G(s) = \dfrac{E_o(s)}{E_i(s)}$	Operational-Amplifier Circuits
1	P	$\dfrac{R_4}{R_3}\dfrac{R_2}{R_1}$	
2	I	$\dfrac{R_4}{R_3}\dfrac{1}{R_1 C_2 s}$	
3	PD	$\dfrac{R_4}{R_3}\dfrac{R_2}{R_1}(R_1 C_1 s + 1)$	
4	PI	$\dfrac{R_4}{R_3}\dfrac{R_2}{R_1}\dfrac{R_2 C_2 s + 1}{R_2 C_2 s}$	
5	PID	$\dfrac{R_4}{R_3}\dfrac{R_2}{R_1}\dfrac{(R_1 C_1 s + 1)(R_2 C_2 s + 1)}{R_2 C_2 s}$	
6	Lead or lag	$\dfrac{R_4}{R_3}\dfrac{R_2}{R_1}\dfrac{R_1 C_1 s + 1}{R_2 C_2 s + 1}$	
7	Lag–lead	$\dfrac{R_6}{R_5}\dfrac{R_4}{R_3}\dfrac{[(R_1 + R_3)\,C_1 s + 1]\,(R_2 C_2 s + 1)}{(R_1 C_1 s + 1)\,[(R_2 + R_4)\,C_2 s + 1]}$	

EXAMPLE PROBLEMS AND SOLUTIONS

A–3–1. Figure 3–20(a) shows a schematic diagram of an automobile suspension system. As the car moves along the road, the vertical displacements at the tires act as the motion excitation to the automobile suspension system. The motion of this system consists of a translational motion of the center of mass and a rotational motion about the center of mass. Mathematical modeling of the complete system is quite complicated.

A very simplified version of the suspension system is shown in Figure 3–20(b). Assuming that the motion x_i at point P is the input to the system and the vertical motion x_o of the body is the output, obtain the transfer function $X_o(s)/X_i(s)$. (Consider the motion of the body only in the vertical direction.) Displacement x_o is measured from the equilibrium position in the absence of input x_i.

Solution. The equation of motion for the system shown in Figure 3–20(b) is

$$m\ddot{x}_o + b(\dot{x}_o - \dot{x}_i) + k(x_o - x_i) = 0$$

or

$$m\ddot{x}_o + b\dot{x}_o + kx_o = b\dot{x}_i + kx_i$$

Taking the Laplace transform of this last equation, assuming zero initial conditions, we obtain

$$(ms^2 + bs + k)X_o(s) = (bs + k)X_i(s)$$

Hence the transfer function $X_o(s)/X_i(s)$ is given by

$$\frac{X_o(s)}{X_i(s)} = \frac{bs + k}{ms^2 + bs + k}$$

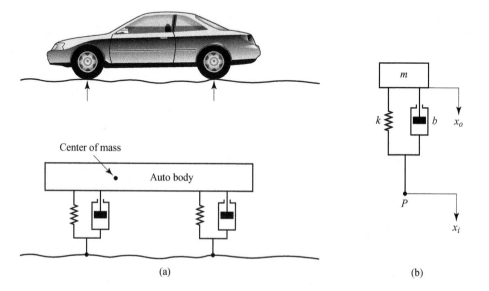

Figure 3–20
(a) Automobile suspension system;
(b) simplified suspension system.

(a)

(b)

A–3–2. Obtain the transfer function $Y(s)/U(s)$ of the system shown in Figure 3–21. The input u is a displacement input. (Like the system of Problem **A–3–1**, this is also a simplified version of an automobile or motorcycle suspension system.)

Solution. Assume that displacements x and y are measured from respective steady-state positions in the absence of the input u. Applying the Newton's second law to this system, we obtain

$$m_1 \ddot{x} = k_2(y - x) + b(\dot{y} - \dot{x}) + k_1(u - x)$$

$$m_2 \ddot{y} = -k_2(y - x) - b(\dot{y} - \dot{x})$$

Hence, we have

$$m_1 \ddot{x} + b\dot{x} + (k_1 + k_2)x = b\dot{y} + k_2 y + k_1 u$$

$$m_2 \ddot{y} + b\dot{y} + k_2 y = b\dot{x} + k_2 x$$

Taking Laplace transforms of these two equations, assuming zero initial conditions, we obtain

$$\left[m_1 s^2 + bs + (k_1 + k_2)\right]X(s) = (bs + k_2)Y(s) + k_1 U(s)$$

$$\left[m_2 s^2 + bs + k_2\right]Y(s) = (bs + k_2)X(s)$$

Eliminating $X(s)$ from the last two equations, we have

$$\left(m_1 s^2 + bs + k_1 + k_2\right)\frac{m_2 s^2 + bs + k_2}{bs + k_2}Y(s) = (bs + k_2)Y(s) + k_1 U(s)$$

which yields

$$\frac{Y(s)}{U(s)} = \frac{k_1(bs + k_2)}{m_1 m_2 s^4 + (m_1 + m_2)bs^3 + \left[k_1 m_2 + (m_1 + m_2)k_2\right]s^2 + k_1 bs + k_1 k_2}$$

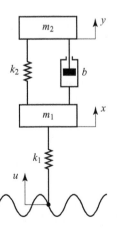

Figure 3–21
Suspension system.

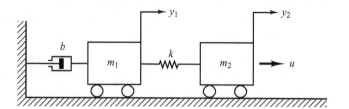

Figure 3–22
Mechanical system.

A–3–3. Obtain a state-space representation of the system shown in Figure 3–22.

Solution. The system equations are

$$m_1\ddot{y}_1 + b\dot{y}_1 + k(y_1 - y_2) = 0$$
$$m_2\ddot{y}_2 + k(y_2 - y_1) = u$$

The output variables for this system are y_1 and y_2. Define state variables as

$$x_1 = y_1$$
$$x_2 = \dot{y}_1$$
$$x_3 = y_2$$
$$x_4 = \dot{y}_2$$

Then we obtain the following equations:

$$\dot{x}_1 = x_2$$
$$\dot{x}_2 = \frac{1}{m_1}[-b\dot{y}_1 - k(y_1 - y_2)] = -\frac{k}{m_1}x_1 - \frac{b}{m_1}x_2 + \frac{k}{m_1}x_3$$
$$\dot{x}_3 = x_4$$
$$\dot{x}_4 = \frac{1}{m_2}[-k(y_2 - y_1) + u] = \frac{k}{m_2}x_1 - \frac{k}{m_2}x_3 + \frac{1}{m_2}u$$

Hence, the state equation is

$$
\begin{bmatrix} \dot{x}_1 \\ \dot{x}_2 \\ \dot{x}_3 \\ \dot{x}_4 \end{bmatrix}
=
\begin{bmatrix}
0 & 1 & 0 & 0 \\
-\dfrac{k}{m_1} & -\dfrac{b}{m_1} & \dfrac{k}{m_1} & 0 \\
0 & 0 & 0 & 1 \\
\dfrac{k}{m_2} & 0 & -\dfrac{k}{m_2} & 0
\end{bmatrix}
\begin{bmatrix} x_1 \\ x_2 \\ x_3 \\ x_4 \end{bmatrix}
+
\begin{bmatrix} 0 \\ 0 \\ 0 \\ \dfrac{1}{m_2} \end{bmatrix} u
$$

and the output equation is

$$
\begin{bmatrix} y_1 \\ y_2 \end{bmatrix}
=
\begin{bmatrix}
1 & 0 & 0 & 0 \\
0 & 0 & 1 & 0
\end{bmatrix}
\begin{bmatrix} x_1 \\ x_2 \\ x_3 \\ x_4 \end{bmatrix}
$$

A–3–4. Obtain the transfer function $X_o(s)/X_i(s)$ of the mechanical system shown in Figure 3–23(a). Also obtain the transfer function $E_o(s)/E_i(s)$ of the electrical system shown in Figure 3–23(b). Show that these transfer functions of the two systems are of identical form and thus they are analogous systems.

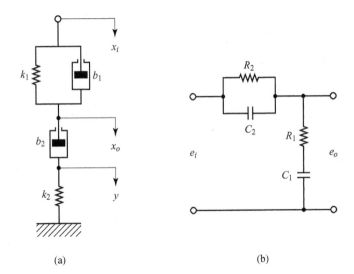

Figure 3–23
(a) Mechanical
system;
(b) analogous
electrical system.

(a)   (b)

Solution. In Figure 3–23(a) we assume that displacements x_i, x_o, and y are measured from their respective steady-state positions. Then the equations of motion for the mechanical system shown in Figure 3–23(a) are

$$b_1(\dot{x}_i - \dot{x}_o) + k_1(x_i - x_o) = b_2(\dot{x}_o - \dot{y})$$
$$b_2(\dot{x}_o - \dot{y}) = k_2 y$$

By taking the Laplace transforms of these two equations, assuming zero initial conditions, we have

$$b_1[sX_i(s) - sX_o(s)] + k_1[X_i(s) - X_o(s)] = b_2[sX_o(s) - sY(s)]$$
$$b_2[sX_o(s) - sY(s)] = k_2 Y(s)$$

If we eliminate $Y(s)$ from the last two equations, then we obtain

$$b_1[sX_i(s) - sX_o(s)] + k_1[X_i(s) - X_o(s)] = b_2 s X_o(s) - b_2 s \frac{b_2 s X_o(s)}{b_2 s + k_2}$$

or

$$(b_1 s + k_1)X_i(s) = \left(b_1 s + k_1 + b_2 s - b_2 s \frac{b_2 s}{b_2 s + k_2}\right)X_o(s)$$

Hence the transfer function $X_o(s)/X_i(s)$ can be obtained as

$$\frac{X_o(s)}{X_i(s)} = \frac{\left(\dfrac{b_1}{k_1}s + 1\right)\left(\dfrac{b_2}{k_2}s + 1\right)}{\left(\dfrac{b_1}{k_1}s + 1\right)\left(\dfrac{b_2}{k_2}s + 1\right) + \dfrac{b_2}{k_1}s}$$

For the electrical system shown in Figure 3–23(b), the transfer function $E_o(s)/E_i(s)$ is found to be

$$\frac{E_o(s)}{E_i(s)} = \frac{R_1 + \dfrac{1}{C_1 s}}{\dfrac{1}{(1/R_2) + C_2 s} + R_1 + \dfrac{1}{C_1 s}}$$

$$= \frac{(R_1 C_1 s + 1)(R_2 C_2 s + 1)}{(R_1 C_1 s + 1)(R_2 C_2 s + 1) + R_2 C_1 s}$$

Example Problems and Solutions

89

A comparison of the transfer functions shows that the systems shown in Figures 3–23(a) and (b) are analogous.

A–3–5. Obtain the transfer functions $E_o(s)/E_i(s)$ of the bridged T networks shown in Figures 3–24(a) and (b).

Solution. The bridged T networks shown can both be represented by the network of Figure 3–25(a), where we used complex impedances. This network may be modified to that shown in Figure 3–25(b).

In Figure 3–25(b), note that

$$I_1 = I_2 + I_3, \qquad I_2Z_1 = (Z_3 + Z_4)I_3$$

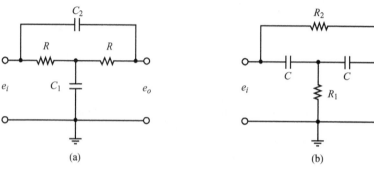

Figure 3–24
Bridged T networks.

(a)   (b)

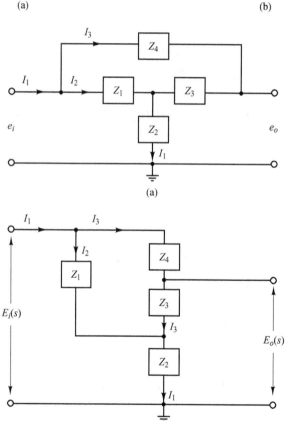

(a)

Figure 3–25
(a) Bridged T
network in terms of
complex impedances;
(b) equivalent
network.

(b)

Hence

$$I_2 = \frac{Z_3 + Z_4}{Z_1 + Z_3 + Z_4} I_1, \qquad I_3 = \frac{Z_1}{Z_1 + Z_3 + Z_4} I_1$$

Then the voltages $E_i(s)$ and $E_o(s)$ can be obtained as

$$E_i(s) = Z_1 I_2 + Z_2 I_1$$

$$= \left[Z_2 + \frac{Z_1(Z_3 + Z_4)}{Z_1 + Z_3 + Z_4} \right] I_1$$

$$= \frac{Z_2(Z_1 + Z_3 + Z_4) + Z_1(Z_3 + Z_4)}{Z_1 + Z_3 + Z_4} I_1$$

$$E_o(s) = Z_3 I_3 + Z_2 I_1$$

$$= \frac{Z_3 Z_1}{Z_1 + Z_3 + Z_4} I_1 + Z_2 I_1$$

$$= \frac{Z_3 Z_1 + Z_2(Z_1 + Z_3 + Z_4)}{Z_1 + Z_3 + Z_4} I_1$$

Hence, the transfer function $E_o(s)/E_i(s)$ of the network shown in Figure 3–25(a) is obtained as

$$\frac{E_o(s)}{E_i(s)} = \frac{Z_3 Z_1 + Z_2(Z_1 + Z_3 + Z_4)}{Z_2(Z_1 + Z_3 + Z_4) + Z_1 Z_3 + Z_1 Z_4} \qquad (3\text{–}38)$$

For the bridged T network shown in Figure 3–24(a), substitute

$$Z_1 = R, \qquad Z_2 = \frac{1}{C_1 s}, \qquad Z_3 = R, \qquad Z_4 = \frac{1}{C_2 s}$$

into Equation (3–38). Then we obtain the transfer function $E_o(s)/E_i(s)$ to be

$$\frac{E_o(s)}{E_i(s)} = \frac{R^2 + \dfrac{1}{C_1 s}\left(R + R + \dfrac{1}{C_2 s} \right)}{\dfrac{1}{C_1 s}\left(R + R + \dfrac{1}{C_2 s} \right) + R^2 + R\dfrac{1}{C_2 s}}$$

$$= \frac{R C_1 R C_2 s^2 + 2R C_2 s + 1}{R C_1 R C_2 s^2 + (2R C_2 + R C_1)s + 1}$$

Similarly, for the bridged T network shown in Figure 3–24(b), we substitute

$$Z_1 = \frac{1}{Cs}, \qquad Z_2 = R_1, \qquad Z_3 = \frac{1}{Cs}, \qquad Z_4 = R_2$$

into Equation (3–38). Then the transfer function $E_o(s)/E_i(s)$ can be obtained as follows:

$$\frac{E_o(s)}{E_i(s)} = \frac{\dfrac{1}{Cs}\dfrac{1}{Cs} + R_1\left(\dfrac{1}{Cs} + \dfrac{1}{Cs} + R_2 \right)}{R_1\left(\dfrac{1}{Cs} + \dfrac{1}{Cs} + R_2 \right) + \dfrac{1}{Cs}\dfrac{1}{Cs} + R_2\dfrac{1}{Cs}}$$

$$= \frac{R_1 C R_2 C s^2 + 2R_1 C s + 1}{R_1 C R_2 C s^2 + (2R_1 C + R_2 C)s + 1}$$

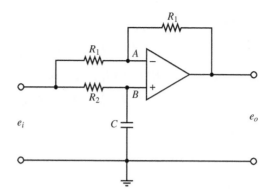

Figure 3–26
Operational-
amplifier circuit.

A–3–6. Obtain the transfer function $E_o(s)/E_i(s)$ of the op-amp circuit shown in Figure 3–26.

Solution. The voltage at point A is

$$e_A = \frac{1}{2}(e_i - e_o) + e_o$$

The Laplace-transformed version of this last equation is

$$E_A(s) = \frac{1}{2}[E_i(s) + E_o(s)]$$

The voltage at point B is

$$E_B(s) = \frac{\dfrac{1}{Cs}}{R_2 + \dfrac{1}{Cs}} E_i(s) = \frac{1}{R_2Cs + 1} E_i(s)$$

Since $[E_B(s) - E_A(s)]K = E_o(s)$ and $K \gg 1$, we must have $E_A(s) = E_B(s)$. Thus

$$\frac{1}{2}[E_i(s) + E_o(s)] = \frac{1}{R_2Cs + 1} E_i(s)$$

Hence

$$\frac{E_o(s)}{E_i(s)} = -\frac{R_2Cs - 1}{R_2Cs + 1} = -\frac{s - \dfrac{1}{R_2C}}{s + \dfrac{1}{R_2C}}$$

A–3–7. Obtain the transfer function $E_o(s)/E_i(s)$ of the op-amp system shown in Figure 3–27 in terms of complex impedances $Z_1, Z_2, Z_3,$ and Z_4. Using the equation derived, obtain the transfer function $E_o(s)/E_i(s)$ of the op-amp system shown in Figure 3–26.

Solution. From Figure 3–27, we find

$$\frac{E_i(s) - E_A(s)}{Z_3} = \frac{E_A(s) - E_o(s)}{Z_4}$$

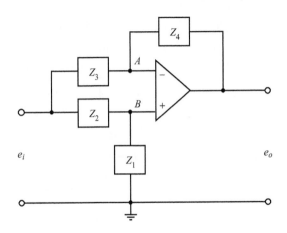

Figure 3–27
Operational-
amplifier circuit.

or

$$E_i(s) - \left(1 + \frac{Z_3}{Z_4}\right)E_A(s) = -\frac{Z_3}{Z_4}E_o(s) \tag{3-39}$$

Since

$$E_A(s) = E_B(s) = \frac{Z_1}{Z_1 + Z_2}E_i(s) \tag{3-40}$$

by substituting Equation (3–40) into Equation (3–39), we obtain

$$\left[\frac{Z_4Z_1 + Z_4Z_2 - Z_4Z_1 - Z_3Z_1}{Z_4(Z_1 + Z_2)}\right]E_i(s) = -\frac{Z_3}{Z_4}E_o(s)$$

from which we get the transfer function $E_o(s)/E_i(s)$ to be

$$\frac{E_o(s)}{E_i(s)} = -\frac{Z_4Z_2 - Z_3Z_1}{Z_3(Z_1 + Z_2)} \tag{3-41}$$

To find the transfer function $E_o(s)/E_i(s)$ of the circuit shown in Figure 3–26, we substitute

$$Z_1 = \frac{1}{Cs}, \quad Z_2 = R_2, \quad Z_3 = R_1, \quad Z_4 = R_1$$

into Equation (3–41). The result is

$$\frac{E_o(s)}{E_i(s)} = -\frac{R_1R_2 - R_1\dfrac{1}{Cs}}{R_1\left(\dfrac{1}{Cs} + R_2\right)} = -\frac{R_2Cs - 1}{R_2Cs + 1}.$$

which is, as a matter of course, the same as that obtained in Problem **A–3–6**.

Example Problems and Solutions

A–3–8. Obtain the transfer function $E_o(s)/E_i(s)$ of the operational-amplifier circuit shown in Figure 3–28.

Solution. We will first obtain currents i_1, i_2, i_3, i_4, and i_5. Then we will use node equations at nodes A and B.

$$i_1 = \frac{e_i - e_A}{R_1}; \qquad i_2 = \frac{e_A - e_o}{R_3}, \qquad i_3 = C_1 \frac{de_A}{dt}$$

$$i_4 = \frac{e_A}{R_2}, \qquad i_5 = C_2 \frac{-de_o}{dt}$$

At node A, we have $i_1 = i_2 + i_3 + i_4$, or

$$\frac{e_i - e_A}{R_1} = \frac{e_A - e_o}{R_3} + C_1 \frac{de_A}{dt} + \frac{e_A}{R_2} \tag{3–42}$$

At node B, we get $i_4 = i_5$, or

$$\frac{e_A}{R_2} = C_2 \frac{-de_o}{dt} \tag{3–43}$$

By rewriting Equation (3–42), we have

$$C_1 \frac{de_A}{dt} + \left(\frac{1}{R_1} + \frac{1}{R_2} + \frac{1}{R_3} \right) e_A = \frac{e_i}{R_1} + \frac{e_o}{R_3} \tag{3–44}$$

From Equation (3–43), we get

$$e_A = -R_2 C_2 \frac{de_o}{dt} \tag{3–45}$$

By substituting Equation (3–45) into Equation (3–44), we obtain

$$C_1 \left(-R_2 C_2 \frac{d^2 e_o}{dt^2} \right) + \left(\frac{1}{R_1} + \frac{1}{R_2} + \frac{1}{R_3} \right) (-R_2 C_2) \frac{de_o}{dt} = \frac{e_i}{R_1} + \frac{e_o}{R_3}$$

Taking the Laplace transform of this last equation, assuming zero initial conditions, we obtain

$$-C_1 C_2 R_2 s^2 E_o(s) + \left(\frac{1}{R_1} + \frac{1}{R_2} + \frac{1}{R_3} \right) (-R_2 C_2) s E_o(s) - \frac{1}{R_3} E_o(s) = \frac{E_i(s)}{R_1}$$

from which we get the transfer function $E_o(s)/E_i(s)$ as follows:

$$\frac{E_o(s)}{E_i(s)} = -\frac{1}{R_1 C_1 R_2 C_2 s^2 + \left[R_2 C_2 + R_1 C_2 + (R_1/R_3) R_2 C_2 \right] s + (R_1/R_3)}$$

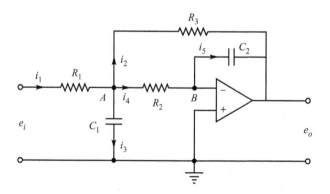

Figure 3–28
Operational-
amplifier circuit.

A–3–9. Consider the servo system shown in Figure 3–29(a). The motor shown is a servomotor, a dc motor designed specifically to be used in a control system. The operation of this system is as follows: A pair of potentiometers acts as an error-measuring device. They convert the input and output positions into proportional electric signals. The command input signal determines the angular position r of the wiper arm of the input potentiometer. The angular position r is the reference input to the system, and the electric potential of the arm is proportional to the angular position of the arm. The output shaft position determines the angular position c of the wiper arm of the output potentiometer. The difference between the input angular position r and the output angular position c is the error signal e, or

$$e = r - c$$

The potential difference $e_r - e_c = e_v$ is the error voltage, where e_r is proportional to r and e_c is proportional to c; that is, $e_r = K_0 r$ and $e_c = K_0 c$, where K_0 is a proportionality constant. The error voltage that appears at the potentiometer terminals is amplified by the amplifier whose gain constant is K_1. The output voltage of this amplifier is applied to the armature circuit of the dc motor. A fixed voltage is applied to the field winding. If an error exists, the motor develops a torque to rotate the output load in such a way as to reduce the error to zero. For constant field current, the torque developed by the motor is

$$T = K_2 i_a$$

where K_2 is the motor torque constant and i_a is the armature current.

When the armature is rotating, a voltage proportional to the product of the flux and angular velocity is induced in the armature. For a constant flux, the induced voltage e_b is directly proportional to the angular velocity $d\theta/dt$, or

$$e_b = K_3 \frac{d\theta}{dt}$$

where e_b is the back emf, K_3 is the back emf constant of the motor, and θ is the angular displacement of the motor shaft.

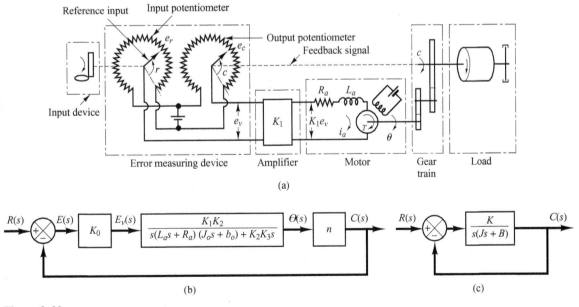

(a)

(b)   (c)

Figure 3–29
(a) Schematic diagram of servo system; (b) block diagram for the system; (c) simplified block diagram.

Example Problems and Solutions

Obtain the transfer function between the motor shaft angular displacement θ and the error voltage e_v. Obtain also a block diagram for this system and a simplified block diagram when L_a is negligible.

Solution. The speed of an armature-controlled dc servomotor is controlled by the armature voltage e_a. (The armature voltage $e_a = K_1 e_v$ is the output of the amplifier.) The differential equation for the armature circuit is

$$L_a \frac{di_a}{dt} + R_a i_a + e_b = e_a$$

or

$$L_a \frac{di_a}{dt} + R_a i_a + K_3 \frac{d\theta}{dt} = K_1 e_v \tag{3-46}$$

The equation for torque equilibrium is

$$J_0 \frac{d^2\theta}{dt^2} + b_0 \frac{d\theta}{dt} = T = K_2 i_a \tag{3-47}$$

where J_0 is the inertia of the combination of the motor, load, and gear train referred to the motor shaft and b_0 is the viscous-friction coefficient of the combination of the motor, load, and gear train referred to the motor shaft.

By eliminating i_a from Equations (3–46) and (3–47), we obtain

$$\frac{\Theta(s)}{E_v(s)} = \frac{K_1 K_2}{s(L_a s + R_a)(J_0 s + b_0) + K_2 K_3 s} \tag{3-48}$$

We assume that the gear ratio of the gear train is such that the output shaft rotates n times for each revolution of the motor shaft. Thus,

$$C(s) = n\Theta(s) \tag{3-49}$$

The relationship among $E_v(s)$, $R(s)$, and $C(s)$ is

$$E_v(s) = K_0[R(s) - C(s)] = K_0 E(s) \tag{3-50}$$

The block diagram of this system can be constructed from Equations (3–48), (3–49), and (3–50), as shown in Figure 3–29(b). The transfer function in the feedforward path of this system is

$$G(s) = \frac{C(s)}{\Theta(s)} \frac{\Theta(s)}{E_v(s)} \frac{E_v(s)}{E(s)} = \frac{K_0 K_1 K_2 n}{s[(L_a s + R_a)(J_0 s + b_0) + K_2 K_3]}$$

When L_a is small, it can be neglected, and the transfer function $G(s)$ in the feedforward path becomes

$$G(s) = \frac{K_0 K_1 K_2 n}{s[R_a(J_0 s + b_0) + K_2 K_3]}$$

$$= \frac{K_0 K_1 K_2 n / R_a}{J_0 s^2 + \left(b_0 + \frac{K_2 K_3}{R_a}\right)s} \tag{3-51}$$

The term $[b_0 + (K_2 K_3 / R_a)]s$ indicates that the back emf of the motor effectively increases the viscous friction of the system. The inertia J_0 and viscous friction coefficient $b_0 + (K_2 K_3 / R_a)$ are referred to the motor shaft. When J_0 and $b_0 + (K_2 K_3/R_a)$ are multiplied by $1/n^2$, the inertia and viscous-friction coefficient are expressed in terms of the output shaft. Introducing new parameters defined by

$$J = J_0/n^2 = \text{moment of inertia referred to the output shaft}$$

$$B = [b_0 + (K_2 K_3/R_a)]/n^2 = \text{viscous-friction coefficient referred to the output shaft}$$

$$K = K_0 K_1 K_2/n R_a$$

the transfer function $G(s)$ given by Equation (3–51) can be simplified, yielding

$$G(s) = \frac{K}{Js^2 + Bs}$$

or

$$G(s) = \frac{K_m}{s(T_m s + 1)}$$

where

$$K_m = \frac{K}{B}, \qquad T_m = \frac{J}{B} = \frac{R_a J_0}{R_a b_0 + K_2 K_3}$$

The block diagram of the system shown in Figure 3–29(b) can thus be simplified as shown in Figure 3–29(c).

PROBLEMS

B–3–1. Obtain the equivalent viscous-friction coefficient b_{eq} of the system shown in Figure 3–30.

B–3–2. Obtain mathematical models of the mechanical systems shown in Figures 3–31(a) and (b).

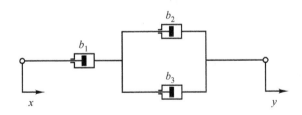

Figure 3–30
Damper system.

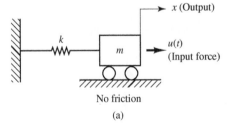

(a)

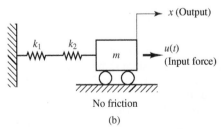

(b)

Figure 3–31
Mechanical systems.

B–3–3. Obtain a state-space representation of the mechanical system shown in Figure 3–32, where u_1 and u_2 are the inputs and y_1 and y_2 are the outputs.

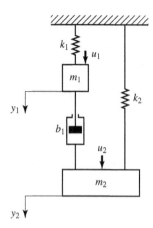

Figure 3–32 Mechanical system.

B–3–4. Consider the spring-loaded pendulum system shown in Figure 3–33. Assume that the spring force acting on the pendulum is zero when the pendulum is vertical, or $\theta = 0$. Assume also that the friction involved is negligible and the angle of oscillation θ is small. Obtain a mathematical model of the system.

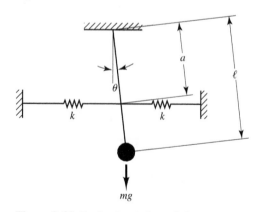

Figure 3–33 Spring-loaded pendulum system.

B–3–5. Referring to Examples 3–5 and 3–6, consider the inverted-pendulum system shown in Figure 3–34. Assume that the mass of the inverted pendulum is m and is evenly distributed along the length of the rod. (The center of gravity of the pendulum is located at the center of the rod.) Assuming that θ is small, derive mathematical models for the system in the forms of differential equations, transfer functions, and state-space equations.

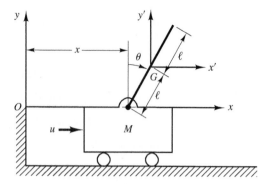

Figure 3–34 Inverted-pendulum system.

B–3–6. Obtain the transfer functions $X_1(s)/U(s)$ and $X_2(s)/U(s)$ of the mechanical system shown in Figure 3–35.

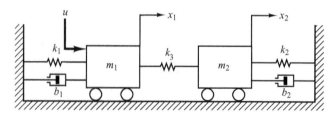

Figure 3–35 Mechanical system.

B–3–7. Obtain the transfer function $E_o(s)/E_i(s)$ of the electrical circuit shown in Figure 3–36.

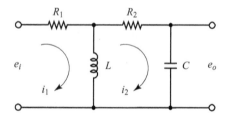

Figure 3–36 Electrical circuit.

B–3–8. Consider the electrical circuit shown in Figure 3–37. Obtain the transfer function $E_o(s)/E_i(s)$ by use of the block diagram approach.

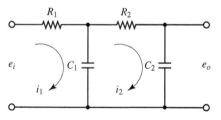

Figure 3–37 Electrical circuit.

B–3–9. Derive the transfer function of the electrical circuit shown in Figure 3–38. Draw a schematic diagram of an analogous mechanical system.

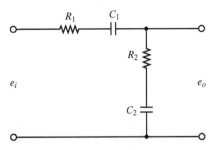

Figure 3–38 Electrical circuit.

B–3–10. Obtain the transfer function $E_o(s)/E_i(s)$ of the op-amp circuit shown in Figure 3–39.

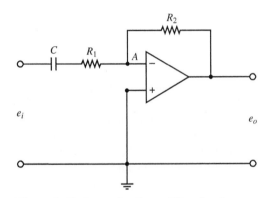

Figure 3–39 Operational-amplifier circuit.

B–3–11. Obtain the transfer function $E_o(s)/E_i(s)$ of the op-amp circuit shown in Figure 3–40.

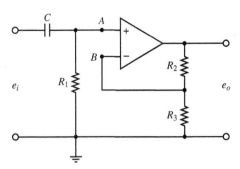

Figure 3–40 Operational-amplifier circuit.

B–3–12. Using the impedance approach, obtain the transfer function $E_o(s)/E_i(s)$ of the op-amp circuit shown in Figure 3–41.

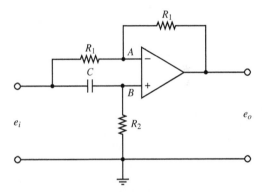

Figure 3–41 Operational-amplifier circuit.

B–3–13. Consider the system shown in Figure 3–42. An armature-controlled dc servomotor drives a load consisting of the moment of inertia J_L. The torque developed by the motor is T. The moment of inertia of the motor rotor is J_m. The angular displacements of the motor rotor and the load element are θ_m and θ, respectively. The gear ratio is $n = \theta/\theta_m$. Obtain the transfer function $\Theta(s)/E_i(s)$.

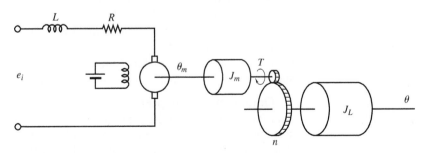

Figure 3–42 Armature-controlled dc servomotor system.

Mathematical Modeling of Fluid Systems and Thermal Systems

4-1 INTRODUCTION

This chapter treats mathematical modeling of fluid systems and thermal systems. As the most versatile medium for transmitting signals and power, fluids—liquids and gases— have wide usage in industry. Liquids and gases can be distinguished basically by their relative incompressibilities and the fact that a liquid may have a free surface, whereas a gas expands to fill its vessel. In the engineering field the term *pneumatic* describes fluid systems that use air or gases and *hydraulic* applies to those using oil.

We first discuss liquid-level systems that are frequently used in process control. Here we introduce the concepts of resistance and capacitance to describe the dynamics of such systems. Then we treat pneumatic systems. Such systems are extensively used in the automation of production machinery and in the field of automatic controllers. For instance, pneumatic circuits that convert the energy of compressed air into mechanical energy enjoy wide usage. Also, various types of pneumatic controllers are widely used in industry. Next, we present hydraulic servo systems. These are widely used in machine tool systems, aircraft control systems, etc. We discuss basic aspects of hydraulic servo systems and hydraulic controllers. Both pneumatic systems and hydraulic systems can be modeled easily by using the concepts of resistance and capacitance. Finally, we treat simple thermal systems. Such systems involve heat transfer from one substance to another. Mathematical models of such systems can be obtained by using thermal resistance and thermal capacitance.

Outline of the Chapter.   Section 4-1 has presented introductory material for the chapter. Section 4-2 discusses liquid-level systems. Section 4-3 treats pneumatic systems—in particular, the basic principles of pneumatic controllers. Section 4-4 first discusses hydraulic servo systems and then presents hydraulic controllers. Finally, Section 4-5 analyzes thermal systems and obtains mathematical models of such systems.

In analyzing systems involving fluid flow, we find it necessary to divide flow regimes into laminar flow and turbulent flow, according to the magnitude of the Reynolds number. If the Reynolds number is greater than about 3000 to 4000, then the flow is turbulent. The flow is laminar if the Reynolds number is less than about 2000. In the laminar case, fluid flow occurs in streamlines with no turbulence. Systems involving laminar flow may be represented by linear differential equations.

Industrial processes often involve flow of liquids through connecting pipes and tanks. The flow in such processes is often turbulent and not laminar. Systems involving turbulent flow often have to be represented by nonlinear differential equations. If the region of operation is limited, however, such nonlinear differential equations can be linearized. We shall discuss such linearized mathematical models of liquid-level systems in this section. Note that the introduction of concepts of resistance and capacitance for such liquid-level systems enables us to describe their dynamic characteristics in simple forms.

Resistance and Capacitance of Liquid-Level Systems. Consider the flow through a short pipe connecting two tanks. The resistance R for liquid flow in such a pipe or restriction is defined as the change in the level difference (the difference of the liquid levels of the two tanks) necessary to cause a unit change in flow rate; that is,

$$R = \frac{\text{change in level difference, m}}{\text{change in flow rate, m}^3/\text{sec}}$$

Since the relationship between the flow rate and level difference differs for the laminar flow and turbulent flow, we shall consider both cases in the following.

Consider the liquid-level system shown in Figure 4–1(a). In this system the liquid spouts through the load valve in the side of the tank. If the flow through this restriction is laminar, the relationship between the steady-state flow rate and steady-state head at the level of the restriction is given by

$$Q = KH$$

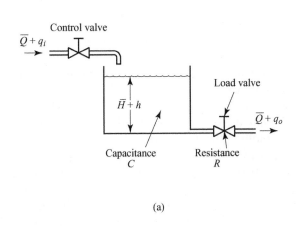

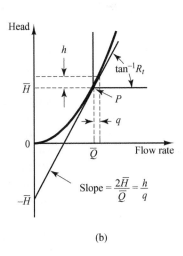

Figure 4–1
(a) Liquid-level system; (b) head-versus-flow-rate curve.

(a)

(b)

where Q = steady-state liquid flow rate, m^3/sec
K = coefficient, m^2/sec
H = steady-state head, m

For laminar flow, the resistance R_l is obtained as

$$R_l = \frac{dH}{dQ} = \frac{H}{Q}$$

The laminar-flow resistance is constant and is analogous to the electrical resistance.

If the flow through the restriction is turbulent, the steady-state flow rate is given by

$$Q = K\sqrt{H} \tag{4-1}$$

where Q = steady-state liquid flow rate, m^3/sec
K = coefficient, m$^{2.5}$/sec
H = steady-state head, m

The resistance R_t for turbulent flow is obtained from

$$R_t = \frac{dH}{dQ}$$

Since from Equation (4-1) we obtain

$$dQ = \frac{K}{2\sqrt{H}}\,dH$$

we have

$$\frac{dH}{dQ} = \frac{2\sqrt{H}}{K} = \frac{2\sqrt{H}\,\sqrt{H}}{Q} = \frac{2H}{Q}$$

Thus,

$$R_t = \frac{2H}{Q}$$

The value of the turbulent-flow resistance R_t depends on the flow rate and the head. The value of R_t, however, may be considered constant if the changes in head and flow rate are small.

By use of the turbulent-flow resistance, the relationship between Q and H can be given by

$$Q = \frac{2H}{R_t}$$

Such linearization is valid, provided that changes in the head and flow rate from their respective steady-state values are small.

In many practical cases, the value of the coefficient K in Equation (4-1), which depends on the flow coefficient and the area of restriction, is not known. Then the resistance may be determined by plotting the head-versus-flow-rate curve based on experimental data and measuring the slope of the curve at the operating condition. An example of such a plot is shown in Figure 4-1(b). In the figure, point P is the steady-state operating point. The tangent line to the curve at point P intersects the ordinate at point $(0, -\bar{H})$. Thus, the slope of this tangent line is $2\bar{H}/\bar{Q}$. Since the resistance R_t at the operating point P is given by $2\bar{H}/\bar{Q}$, the resistance R_t is the slope of the curve at the operating point.

Consider the operating condition in the neighborhood of point P. Define a small deviation of the head from the steady-state value as h and the corresponding small change of the flow rate as q. Then the slope of the curve at point P can be given by

$$\text{Slope of curve at point } P = \frac{h}{q} = \frac{2\bar{H}}{\bar{Q}} = R_t$$

The linear approximation is based on the fact that the actual curve does not differ much from its tangent line if the operating condition does not vary too much.

The capacitance C of a tank is defined to be the change in quantity of stored liquid necessary to cause a unit change in the potential (head). (The potential is the quantity that indicates the energy level of the system.)

$$C = \frac{\text{change in liquid stored, m}^3}{\text{change in head, m}}$$

It should be noted that the capacity (m³) and the capacitance (m²) are different. The capacitance of the tank is equal to its cross-sectional area. If this is constant, the capacitance is constant for any head.

Liquid-Level Systems. Consider the system shown in Figure 4–1(a). The variables are defined as follows:

$\bar{Q}$ = steady-state flow rate (before any change has occurred), m³/sec

q_i = small deviation of inflow rate from its steady-state value, m³/sec

q_o = small deviation of outflow rate from its steady-state value, m³/sec

$\bar{H}$ = steady-state head (before any change has occurred), m

h = small deviation of head from its steady-state value, m

As stated previously, a system can be considered linear if the flow is laminar. Even if the flow is turbulent, the system can be linearized if changes in the variables are kept small. Based on the assumption that the system is either linear or linearized, the differential equation of this system can be obtained as follows: Since the inflow minus outflow during the small time interval dt is equal to the additional amount stored in the tank, we see that

$$C\,dh = (q_i - q_o)\,dt$$

From the definition of resistance, the relationship between q_o and h is given by

$$q_o = \frac{h}{R}$$

The differential equation for this system for a constant value of R becomes

$$RC\frac{dh}{dt} + h = Rq_i \tag{4–2}$$

Note that RC is the time constant of the system. Taking the Laplace transforms of both sides of Equation (4–2), assuming the zero initial condition, we obtain

$$(RCs + 1)H(s) = RQ_i(s)$$

where

$$H(s) = \mathscr{L}[h] \quad \text{and} \quad Q_i(s) = \mathscr{L}[q_i]$$

If q_i is considered the input and h the output, the transfer function of the system is

$$\frac{H(s)}{Q_i(s)} = \frac{R}{RCs + 1}$$

If, however, q_o is taken as the output, the input being the same, then the transfer function is

$$\frac{Q_o(s)}{Q_i(s)} = \frac{1}{RCs + 1}$$

where we have used the relationship

$$Q_o(s) = \frac{1}{R} H(s)$$

Liquid-Level Systems with Interaction. Consider the system shown in Figure 4–2. In this system, the two tanks interact. Thus the transfer function of the system is not the product of two first-order transfer functions.

In the following, we shall assume only small variations of the variables from the steady-state values. Using the symbols as defined in Figure 4–2, we can obtain the following equations for this system:

$$\frac{h_1 - h_2}{R_1} = q_1 \tag{4–3}$$

$$C_1 \frac{dh_1}{dt} = q - q_1 \tag{4–4}$$

$$\frac{h_2}{R_2} = q_2 \tag{4–5}$$

$$C_2 \frac{dh_2}{dt} = q_1 - q_2 \tag{4–6}$$

If q is considered the input and q_2 the output, the transfer function of the system is

$$\frac{Q_2(s)}{Q(s)} = \frac{1}{R_1 C_1 R_2 C_2 s^2 + (R_1 C_1 + R_2 C_2 + R_2 C_1)s + 1} \tag{4–7}$$

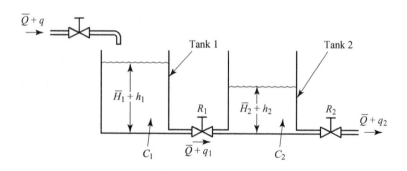

Figure 4–2
Liquid-level system
with interaction.

$\bar{Q}$: Steady-state flow rate
$\bar{H}_1$: Steady-state liquid level of tank 1
$\bar{H}_2$: Steady-state liquid level of tank 2

Chapter 4 / **Mathematical Modeling of Fluid Systems and Thermal Systems**

It is instructive to obtain Equation (4–7), the transfer function of the interacted system, by block diagram reduction. From Equations (4–3) through (4–6), we obtain the elements of the block diagram, as shown in Figure 4–3(a). By connecting signals properly, we can construct a block diagram, as shown in Figure 4–3(b). This block diagram can be simplified, as shown in Figure 4–3(c). Further simplifications result in Figures 4–3(d) and (e). Figure 4–3(e) is equivalent to Equation (4–7).

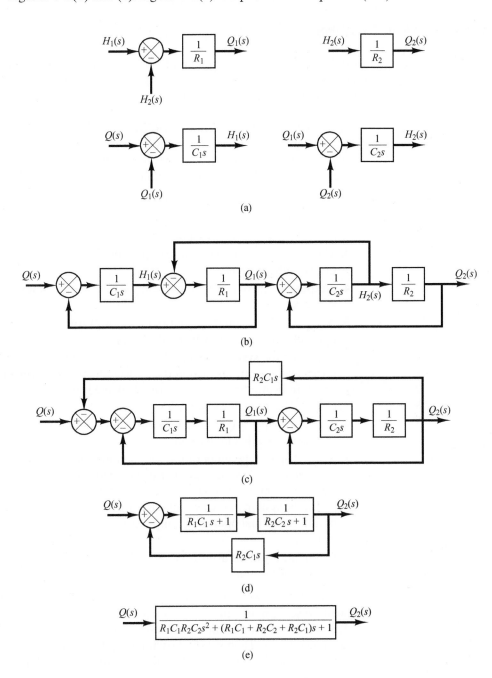

Figure 4–3
(a) Elements of the block diagram of the system shown in Figure 4–2; (b) block diagram of the system; (c)–(e) successive reductions of the block diagram.

Notice the similarity and difference between the transfer function given by Equation (4–7) and that given by Equation (3–33). The term $R_2 C_1 s$ that appears in the denominator of Equation (4–7) exemplifies the interaction between the two tanks. Similarly, the term $R_1 C_2 s$ in the denominator of Equation (3–33) represents the interaction between the two RC circuits shown in Figure 3–8.

4–3 PNEUMATIC SYSTEMS

In industrial applications pneumatic systems and hydraulic systems are frequently compared. Therefore, before we discuss pneumatic systems in detail, we shall give a brief comparison of these two kinds of systems.

Comparison Between Pneumatic Systems and Hydraulic Systems. The fluid generally found in pneumatic systems is air; in hydraulic systems it is oil. And it is primarily the different properties of the fluids involved that characterize the differences between the two systems. These differences can be listed as follows:

1. Air and gases are compressible, whereas oil is incompressible (except at high pressure).
2. Air lacks lubricating property and always contains water vapor. Oil functions as a hydraulic fluid as well as a lubricator.
3. The normal operating pressure of pneumatic systems is very much lower than that of hydraulic systems.
4. Output powers of pneumatic systems are considerably less than those of hydraulic systems.
5. Accuracy of pneumatic actuators is poor at low velocities, whereas accuracy of hydraulic actuators may be made satisfactory at all velocities.
6. In pneumatic systems, external leakage is permissible to a certain extent, but internal leakage must be avoided because the effective pressure difference is rather small. In hydraulic systems internal leakage is permissible to a certain extent, but external leakage must be avoided.
7. No return pipes are required in pneumatic systems when air is used, whereas they are always needed in hydraulic systems.
8. Normal operating temperature for pneumatic systems is 5° to 60°C (41° to 140°F). The pneumatic system, however, can be operated in the 0° to 200°C (32° to 392°F) range. Pneumatic systems are insensitive to temperature changes, in contrast to hydraulic systems, in which fluid friction due to viscosity depends greatly on temperature. Normal operating temperature for hydraulic systems is 20° to 70°C (68° to 158°F).
9. Pneumatic systems are fire- and explosion-proof, whereas hydraulic systems are not, unless nonflammable liquid is used.

In what follows we begin with a mathematical modeling of pneumatic systems. Then we shall present pneumatic proportional controllers.

We shall first give detailed discussions of the principle by which proportional controllers operate. Then we shall treat methods for obtaining derivative and integral control actions. Throughout the discussions, we shall place emphasis on the fundamental principles, rather than on the details of the operation of the actual mechanisms.

Pneumatic Systems. The past decades have seen a great development in low-pressure pneumatic controllers for industrial control systems, and today they are used extensively in industrial processes. Reasons for their broad appeal include an explosion-proof character, simplicity, and ease of maintenance.

Resistance and Capacitance of Pressure Systems. Many industrial processes and pneumatic controllers involve the flow of a gas or air through connected pipelines and pressure vessels.

Consider the pressure system shown in Figure 4–4(a). The gas flow through the restriction is a function of the gas pressure difference $p_i - p_o$. Such a pressure system may be characterized in terms of a resistance and a capacitance.

The gas flow resistance R may be defined as follows:

$$R = \frac{\text{change in gas pressure difference, lb}_f/\text{ft}^2}{\text{change in gas flow rate, lb/sec}}$$

or

$$R = \frac{d(\Delta P)}{dq} \tag{4–8}$$

where $d(\Delta P)$ is a small change in the gas pressure difference and dq is a small change in the gas flow rate. Computation of the value of the gas flow resistance R may be quite time consuming. Experimentally, however, it can be easily determined from a plot of the pressure difference versus flow rate by calculating the slope of the curve at a given operating condition, as shown in Figure 4–4(b).

The capacitance of the pressure vessel may be defined by

$$C = \frac{\text{change in gas stored, lb}}{\text{change in gas pressure, lb}_f/\text{ft}^2}$$

or

$$C = \frac{dm}{dp} = V\frac{d\rho}{dp} \tag{4–9}$$

Figure 4–4
(a) Schematic diagram of a pressure system; (b) pressure-difference-versus-flow-rate curve.

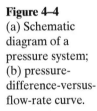
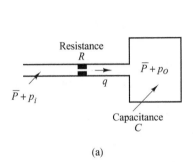
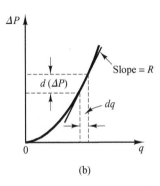

(a)

(b)

where C = capacitance, lb-ft^2/lb$_f$
$\quad\quad m$ = mass of gas in vessel, lb
$\quad\quad p$ = gas pressure, lb$_f$/ft^2
$\quad\quad V$ = volume of vessel, ft^3
$\quad\quad \rho$ = density, lb/ft^3

The capacitance of the pressure system depends on the type of expansion process involved. The capacitance can be calculated by use of the ideal gas law. If the gas expansion process is polytropic and the change of state of the gas is between isothermal and adiabatic, then

$$p\left(\frac{V}{m}\right)^n = \frac{p}{\rho^n} = \text{constant} = K \tag{4-10}$$

where n = polytropic exponent.

For ideal gases,

$$p\bar{v} = \bar{R}T \quad\quad \text{or} \quad\quad pv = \frac{\bar{R}}{M}T$$

where $\quad p$ = absolute pressure, lb$_f$/ft^2
$\quad\quad \bar{v}$ = volume occupied by 1 mole of a gas, ft^3/lb-mole
$\quad\quad \bar{R}$ = universal gas constant, ft-lb$_f$/lb-mole °R
$\quad\quad T$ = absolute temperature, °R
$\quad\quad v$ = specific volume of gas, ft^3/lb
$\quad\quad M$ = molecular weight of gas per mole, lb/lb-mole

Thus

$$pv = \frac{p}{\rho} = \frac{\bar{R}}{M}T = R_{gas}T \tag{4-11}$$

where R_{gas} = gas constant, ft-lb$_f$/lb °R.

The polytropic exponent n is unity for isothermal expansion. For adiabatic expansion, n is equal to the ratio of specific heats c_p/c_v, where c_p is the specific heat at constant pressure and c_v is the specific heat at constant volume. In many practical cases, the value of n is approximately constant, and thus the capacitance may be considered constant.

The value of $d\rho/dp$ is obtained from Equations (4-10) and (4-11). From Equation (4-10) we have

$$dp = Kn\rho^{n-1}\,d\rho$$

or

$$\frac{d\rho}{dp} = \frac{1}{Kn\rho^{n-1}} = \frac{\rho^n}{pn\rho^{n-1}} = \frac{\rho}{pn}$$

Substituting Equation (4-11) into this last equation, we get

$$\frac{d\rho}{dp} = \frac{1}{nR_{gas}T}$$

The capacitance C is then obtained as

$$C = \frac{V}{nR_{gas}T} \tag{4–12}$$

The capacitance of a given vessel is constant if the temperature stays constant. (In many practical cases, the polytropic exponent n is approximately $1.0 \sim 1.2$ for gases in uninsulated metal vessels.)

Pressure Systems. Consider the system shown in Figure 4–4(a). If we assume only small deviations in the variables from their respective steady-state values, then this system may be considered linear.

Let us define

$\bar{P}$ = gas pressure in the vessel at steady state (before changes in pressure have occurred), lb_f/ft^2

p_i = small change in inflow gas pressure, lb_f/ft^2

p_o = small change in gas pressure in the vessel, lb_f/ft^2

V = volume of the vessel, ft^3

m = mass of gas in the vessel, lb

q = gas flow rate, lb/sec

ρ = density of gas, lb/ft^3

For small values of p_i and p_o, the resistance R given by Equation (4–8) becomes constant and may be written as

$$R = \frac{p_i - p_o}{q}$$

The capacitance C is given by Equation (4–9), or

$$C = \frac{dm}{dp}$$

Since the pressure change dp_o times the capacitance C is equal to the gas added to the vessel during dt seconds, we obtain

$$C\, dp_o = q\, dt$$

or

$$C \frac{dp_o}{dt} = \frac{p_i - p_o}{R}$$

which can be written as

$$RC \frac{dp_o}{dt} + p_o = p_i$$

If p_i and p_o are considered the input and output, respectively, then the transfer function of the system is

$$\frac{P_o(s)}{P_i(s)} = \frac{1}{RCs + 1}$$

where RC has the dimension of time and is the time constant of the system.

Pneumatic Nozzle–Flapper Amplifiers. A schematic diagram of a pneumatic nozzle–flapper amplifier is shown in Figure 4–5(a). The power source for this amplifier is a supply of air at constant pressure. The nozzle–flapper amplifier converts small changes in the position of the flapper into large changes in the back pressure in the nozzle. Thus a large power output can be controlled by the very little power that is needed to position the flapper.

In Figure 4–5(a), pressurized air is fed through the orifice, and the air is ejected from the nozzle toward the flapper. Generally, the supply pressure P_s for such a controller is 20 psig (1.4 kg_f/cm^2 gage). The diameter of the orifice is on the order of 0.01 in. (0.25 mm) and that of the nozzle is on the order of 0.016 in. (0.4 mm). To ensure proper functioning of the amplifier, the nozzle diameter must be larger than the orifice diameter.

In operating this system, the flapper is positioned against the nozzle opening. The nozzle back pressure P_b is controlled by the nozzle–flapper distance X. As the flapper approaches the nozzle, the opposition to the flow of air through the nozzle increases, with the result that the nozzle back pressure P_b increases. If the nozzle is completely closed by the flapper, the nozzle back pressure P_b becomes equal to the supply pressure P_s. If the flapper is moved away from the nozzle, so that the nozzle–flapper distance is wide (on the order of 0.01 in.), then there is practically no restriction to flow, and the nozzle back pressure P_b takes on a minimum value that depends on the nozzle–flapper device. (The lowest possible pressure will be the ambient pressure P_a.)

Note that, because the air jet puts a force against the flapper, it is necessary to make the nozzle diameter as small as possible.

A typical curve relating the nozzle back pressure P_b to the nozzle–flapper distance X is shown in Figure 4–5(b). The steep and almost linear part of the curve is utilized in the actual operation of the nozzle–flapper amplifier. Because the range of flapper displacements is restricted to a small value, the change in output pressure is also small, unless the curve is very steep.

The nozzle–flapper amplifier converts displacement into a pressure signal. Since industrial process control systems require large output power to operate large pneumatic actuating valves, the power amplification of the nozzle–flapper amplifier is usually insufficient. Consequently, a pneumatic relay is often needed as a power amplifier in connection with the nozzle–flapper amplifier.

Figure 4–5
(a) Schematic diagram of a pneumatic nozzle–flapper amplifier; (b) characteristic curve relating nozzle back pressure and nozzle–flapper distance.

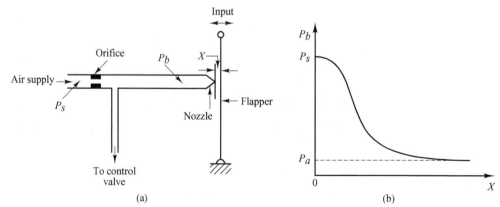

Pneumatic Relays. In practice, in a pneumatic controller, a nozzle–flapper amplifier acts as the first-stage amplifier and a pneumatic relay as the second-stage amplifier. The pneumatic relay is capable of handling a large quantity of airflow.

A schematic diagram of a pneumatic relay is shown in Figure 4–6(a). As the nozzle back pressure P_b increases, the diaphragm valve moves downward. The opening to the atmosphere decreases and the opening to the pneumatic valve increases, thereby increasing the control pressure P_c. When the diaphragm valve closes the opening to the atmosphere, the control pressure P_c becomes equal to the supply pressure P_s. When the nozzle back pressure P_b decreases and the diaphragm valve moves upward and shuts off the air supply, the control pressure P_c drops to the ambient pressure P_a. The control pressure P_c can thus be made to vary from 0 psig to full supply pressure, usually 20 psig.

The total movement of the diaphragm valve is very small. In all positions of the valve, except at the position to shut off the air supply, air continues to bleed into the atmosphere, even after the equilibrium condition is attained between the nozzle back pressure and the control pressure. Thus the relay shown in Figure 4–6(a) is called a bleed-type relay.

There is another type of relay, the nonbleed type. In this one the air bleed stops when the equilibrium condition is obtained and, therefore, there is no loss of pressurized air at steady-state operation. Note, however, that the nonbleed-type relay must have an atmospheric relief to release the control pressure P_c from the pneumatic actuating valve. A schematic diagram of a nonbleed-type relay is shown in Figure 4–6(b).

In either type of relay, the air supply is controlled by a valve, which is in turn controlled by the nozzle back pressure. Thus, the nozzle back pressure is converted into the control pressure with power amplification.

Since the control pressure P_c changes almost instantaneously with changes in the nozzle back pressure P_b, the time constant of the pneumatic relay is negligible compared with the other larger time constants of the pneumatic controller and the plant.

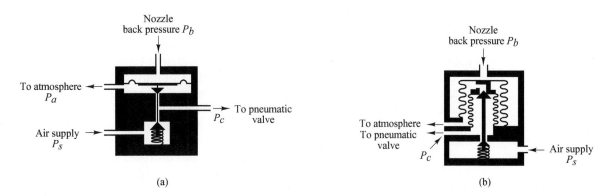

Figure 4–6
(a) Schematic diagram of a bleed-type relay; (b) schematic diagram of a nonbleed-type relay.

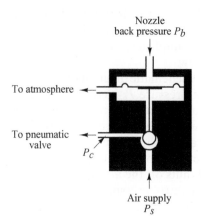

Figure 4–7
Reverse-acting relay.

It is noted that some pneumatic relays are reverse acting. For example, the relay shown in Figure 4–7 is a reverse-acting relay. Here, as the nozzle back pressure P_b increases, the ball valve is forced toward the lower seat, thereby decreasing the control pressure P_c. Thus, this relay is a reverse-acting relay.

Pneumatic Proportional Controllers (Force-Distance Type). Two types of pneumatic controllers, one called the force-distance type and the other the force-balance type, are used extensively in industry. Regardless of how differently industrial pneumatic controllers may appear, careful study will show the close similarity in the functions of the pneumatic circuit. Here we shall consider the force-distance type of pneumatic controllers.

Figure 4–8(a) shows a schematic diagram of such a proportional controller. The nozzle–flapper amplifier constitutes the first-stage amplifier, and the nozzle back pressure is controlled by the nozzle–flapper distance. The relay-type amplifier constitutes the second-stage amplifier. The nozzle back pressure determines the position of the diaphragm valve for the second-stage amplifier, which is capable of handling a large quantity of airflow.

In most pneumatic controllers, some type of pneumatic feedback is employed. Feedback of the pneumatic output reduces the amount of actual movement of the flapper. Instead of mounting the flapper on a fixed point, as shown in Figure 4–8(b), it is often pivoted on the feedback bellows, as shown in Figure 4–8(c). The amount of feedback can be regulated by introducing a variable linkage between the feedback bellows and the flapper connecting point. The flapper then becomes a floating link. It can be moved by both the error signal and the feedback signal.

The operation of the controller shown in Figure 4–8(a) is as follows. The input signal to the two-stage pneumatic amplifier is the actuating error signal. Increasing the actuating error signal moves the flapper to the left. This move will, in turn, increase the nozzle back pressure, and the diaphragm valve moves downward. This results in an increase of the control pressure. This increase will cause bellows F to expand and move the flapper to the right, thus opening the nozzle. Because of this feedback, the nozzle–flapper displacement is very small, but the change in the control pressure can be large.

It should be noted that proper operation of the controller requires that the feedback bellows move the flapper less than that movement caused by the error signal alone. (If these two movements were equal, no control action would result.)

Equations for this controller can be derived as follows. When the actuating error is zero, or $e = 0$, an equilibrium state exists with the nozzle–flapper distance equal to $\bar{X}$, the

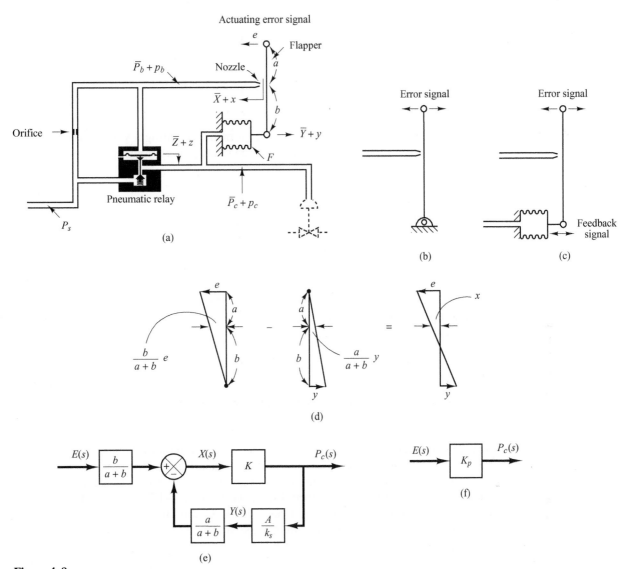

Figure 4–8
(a) Schematic diagram of a force-distance type of pneumatic proportional controller;
(b) flapper mounted on a fixed point; (c) flapper mounted on a feedback bellows;
(d) displacement x as a result of addition of two small displacements;
(e) block diagram for the controller; (f) simplified block diagram for the controller.

displacement of bellows equal to $\bar{Y}$, the displacement of the diaphragm equal to $\bar{Z}$, the nozzle back pressure equal to $\bar{P}_b$, and the control pressure equal to $\bar{P}_c$. When an actuating error exists, the nozzle–flapper distance, the displacement of the bellows, the displacement of the diaphragm, the nozzle back pressure, and the control pressure deviate from their respective equilibrium values. Let these deviations be x, y, z, p_b, and p_c, respectively. (The positive direction for each displacement variable is indicated by an arrowhead in the diagram.)

Assuming that the relationship between the variation in the nozzle back pressure and the variation in the nozzle–flapper distance is linear, we have

$$p_b = K_1 x \tag{4–13}$$

where K_1 is a positive constant. For the diaphragm valve,

$$p_b = K_2 z \tag{4–14}$$

where K_2 is a positive constant. The position of the diaphragm valve determines the control pressure. If the diaphragm valve is such that the relationship between p_c and z is linear, then

$$p_c = K_3 z \tag{4–15}$$

where K_3 is a positive constant. From Equations (4–13), (4–14), and (4–15), we obtain

$$p_c = \frac{K_3}{K_2} p_b = \frac{K_1 K_3}{K_2} x = K x \tag{4–16}$$

where $K = K_1 K_3 / K_2$ is a positive constant. For the flapper, since there are two small movements (e and y) in opposite directions, we can consider such movements separately and add up the results of two movements into one displacement x. See Figure 4–8(d). Thus, for the flapper movement, we have

$$x = \frac{b}{a + b} e - \frac{a}{a + b} y \tag{4–17}$$

The bellows acts like a spring, and the following equation holds true:

$$A p_c = k_s y \tag{4–18}$$

where A is the effective area of the bellows and k_s is the equivalent spring constant—that is, the stiffness due to the action of the corrugated side of the bellows.

Assuming that all variations in the variables are within a linear range, we can obtain a block diagram for this system from Equations (4–16), (4–17), and (4–18) as shown in Figure 4–8(e). From Figure 4–8(e), it can be clearly seen that the pneumatic controller shown in Figure 4–8(a) itself is a feedback system. The transfer function between p_c and e is given by

$$\frac{P_c(s)}{E(s)} = \frac{\dfrac{b}{a + b} K}{1 + K \dfrac{a}{a + b} \dfrac{A}{k_s}} = K_p \tag{4–19}$$

A simplified block diagram is shown in Figure 4–8(f). Since p_c and e are proportional, the pneumatic controller shown in Figure 4–8(a) is a *pneumatic proportional controller.* As seen from Equation (4–19), the gain of the pneumatic proportional controller can be widely varied by adjusting the flapper connecting linkage. [The flapper connecting linkage is not shown in Figure 4–8(a).] In most commercial proportional controllers an adjusting knob or other mechanism is provided for varying the gain by adjusting this linkage.

As noted earlier, the actuating error signal moved the flapper in one direction, and the feedback bellows moved the flapper in the opposite direction, but to a smaller degree.

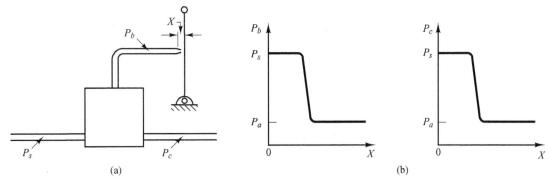

Figure 4–9
(a) Pneumatic controller without a feedback mechanism; (b) curves P_b versus X and P_c versus X.

The effect of the feedback bellows is thus to reduce the sensitivity of the controller. The principle of feedback is commonly used to obtain wide proportional-band controllers.

Pneumatic controllers that do not have feedback mechanisms [which means that one end of the flapper is fixed, as shown in Figure 4–9(a)] have high sensitivity and are called *pneumatic two-position controllers* or *pneumatic on–off controllers*. In such a controller, only a small motion between the nozzle and the flapper is required to give a complete change from the maximum to the minimum control pressure. The curves relating P_b to X and P_c to X are shown in Figure 4–9(b). Notice that a small change in X can cause a large change in P_b, which causes the diaphragm valve to be completely open or completely closed.

Pneumatic Proportional Controllers (Force-Balance Type). Figure 4–10 shows a schematic diagram of a force-balance type pneumatic proportional controller. Force-balance type controllers are in extensive use in industry. Such controllers are called stack controllers. The basic principle of operation does not differ from that of the force-distance type controller. The main advantage of the force-balance type controller is that it eliminates many mechanical linkages and pivot joints, thereby reducing the effects of friction.

In what follows, we shall consider the principle of the force-balance type controller. In the controller shown in Figure 4–10, the reference input pressure P_r and the output pressure P_o are fed to large diaphragm chambers. Note that a force-balance type pneumatic controller operates only on pressure signals. Therefore, it is necessary to convert the reference input and system output to corresponding pressure signals.

Figure 4–10
Schematic diagram of a force-balance type pneumatic proportional controller.

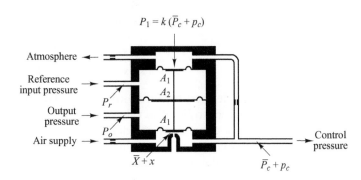

As in the case of the force-distance type controller, this controller employs a flapper, nozzle, and orifices. In Figure 4–10, the drilled opening in the bottom chamber is the nozzle. The diaphragm just above the nozzle acts as a flapper.

The operation of the force-balance type controller shown in Figure 4–10 may be summarized as follows: 20-psig air from an air supply flows through an orifice, causing a reduced pressure in the bottom chamber. Air in this chamber escapes to the atmosphere through the nozzle. The flow through the nozzle depends on the gap and the pressure drop across it. An increase in the reference input pressure P_r, while the output pressure P_o remains the same, causes the valve stem to move down, decreasing the gap between the nozzle and the flapper diaphragm. This causes the control pressure P_c to increase. Let

$$p_e = P_r - P_o \tag{4-20}$$

If $p_e = 0$, there is an equilibrium state with the nozzle–flapper distance equal to $\bar{X}$ and the control pressure equal to $\bar{P}_c$. At this equilibrium state, $P_1 = \bar{P}_c k$ (where $k < 1$) and

$$\bar{X} = \alpha(\bar{P}_c A_1 - \bar{P}_c k A_1) \tag{4-21}$$

where α is a constant.

Let us assume that $p_e \neq 0$ and define small variations in the nozzle–flapper distance and control pressure as x and p_c, respectively. Then we obtain the following equation:

$$\bar{X} + x = \alpha[(\bar{P}_c + p_c)A_1 - (\bar{P}_c + p_c)kA_1 - p_e(A_2 - A_1)] \tag{4-22}$$

From Equations (4–21) and (4–22), we obtain

$$x = \alpha[p_c(1 - k)A_1 - p_e(A_2 - A_1)] \tag{4-23}$$

At this point, we must examine the quantity x. In the design of pneumatic controllers, the nozzle–flapper distance is made quite small. In view of the fact that x/α is very much smaller than $p_c(1 - k)A_1$ or $p_e(A_2 - A_1)$—that is, for $p_e \neq 0$

$$\frac{x}{\alpha} \ll p_c(1 - k)A_1$$

$$\frac{x}{\alpha} \ll p_e(A_2 - A_1)$$

we may neglect the term x in our analysis. Equation (4–23) can then be rewritten to reflect this assumption as follows:

$$p_c(1 - k)A_1 = p_e(A_2 - A_1)$$

and the transfer function between p_c and p_e becomes

$$\frac{P_c(s)}{P_e(s)} = \frac{A_2 - A_1}{A_1} \frac{1}{1 - k} = K_p$$

where p_e is defined by Equation (4–20). The controller shown in Figure 4–10 is a proportional controller. The value of gain K_p increases as k approaches unity. Note that the value of k depends on the diameters of the orifices in the inlet and outlet pipes of the feedback chamber. (The value of k approaches unity as the resistance to flow in the orifice of the inlet pipe is made smaller.)

Pneumatic Actuating Valves. One characteristic of pneumatic controls is that they almost exclusively employ pneumatic actuating valves. A pneumatic actuating valve can provide a large power output. (Since a pneumatic actuator requires a large power input to produce a large power output, it is necessary that a sufficient quantity of pressurized air be available.) In practical pneumatic actuating valves, the valve characteristics may not be linear; that is, the flow may not be directly proportional to the valve stem position, and also there may be other nonlinear effects, such as hysteresis.

Consider the schematic diagram of a pneumatic actuating valve shown in Figure 4–11. Assume that the area of the diaphragm is A. Assume also that when the actuating error is zero, the control pressure is equal to $\bar{P}_c$ and the valve displacement is equal to $\bar{X}$.

In the following analysis, we shall consider small variations in the variables and linearize the pneumatic actuating valve. Let us define the small variation in the control pressure and the corresponding valve displacement to be p_c and x, respectively. Since a small change in the pneumatic pressure force applied to the diaphragm repositions the load, consisting of the spring, viscous friction, and mass, the force-balance equation becomes

$$A p_c = m\ddot{x} + b\dot{x} + kx$$

where m = mass of the valve and valve stem
  b = viscous-friction coefficient
  k = spring constant

If the force due to the mass and viscous friction are negligibly small, then this last equation can be simplified to

$$A p_c = kx$$

The transfer function between x and p_c thus becomes

$$\frac{X(s)}{P_c(s)} = \frac{A}{k} = K_c$$

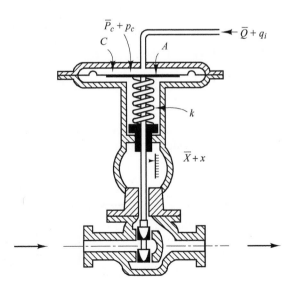

Figure 4–11
Schematic diagram of a pneumatic actuating valve.

where $X(s) = \mathcal{L}[x]$ and $P_c(s) = \mathcal{L}[p_c]$. If q_i, the change in flow through the pneumatic actuating valve, is proportional to x, the change in the valve-stem displacement, then

$$\frac{Q_i(s)}{X(s)} = K_q$$

where $Q_i(s) = \mathcal{L}[q_i]$ and K_q is a constant. The transfer function between q_i and p_c becomes

$$\frac{Q_i(s)}{P_c(s)} = K_c K_q = K_v$$

where K_v is a constant.

The standard control pressure for this kind of a pneumatic actuating valve is between 3 and 15 psig. The valve-stem displacement is limited by the allowable stroke of the diaphragm and is only a few inches. If a longer stroke is needed, a piston–spring combination may be employed.

In pneumatic actuating valves, the static-friction force must be limited to a low value so that excessive hysteresis does not result. Because of the compressibility of air, the control action may not be positive; that is, an error may exist in the valve-stem position. The use of a valve positioner results in improvements in the performance of a pneumatic actuating valve.

Basic Principle for Obtaining Derivative Control Action.   We shall now present methods for obtaining derivative control action. We shall again place the emphasis on the principle and not on the details of the actual mechanisms.

The basic principle for generating a desired control action is to insert the inverse of the desired transfer function in the feedback path. For the system shown in Figure 4–12, the closed-loop transfer function is

$$\frac{C(s)}{R(s)} = \frac{G(s)}{1 + G(s)H(s)}$$

If $|G(s)H(s)| \gg 1$, then $C(s)/R(s)$ can be modified to

$$\frac{C(s)}{R(s)} = \frac{1}{H(s)}$$

Thus, if proportional-plus-derivative control action is desired, we insert an element having the transfer function $1/(Ts + 1)$ in the feedback path.

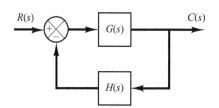

Figure 4–12
Control system.

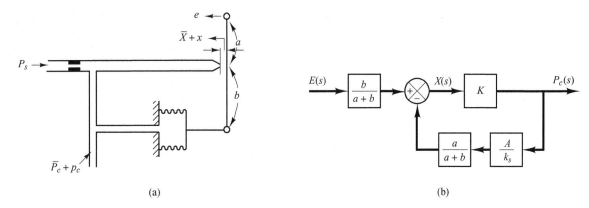

(a) (b)

Figure 4–13
(a) Pneumatic proportional controller; (b) block diagram of the controller.

Consider the pneumatic controller shown in Figure 4–13(a). Considering small changes in the variables, we can draw a block diagram of this controller as shown in Figure 4–13(b). From the block diagram we see that the controller is of proportional type.

We shall now show that the addition of a restriction in the negative feedback path will modify the proportional controller to a proportional-plus-derivative controller, or a PD controller.

Consider the pneumatic controller shown in Figure 4–14(a). Assuming again small changes in the actuating error, nozzle–flapper distance, and control pressure, we can summarize the operation of this controller as follows: Let us first assume a small step change in e.

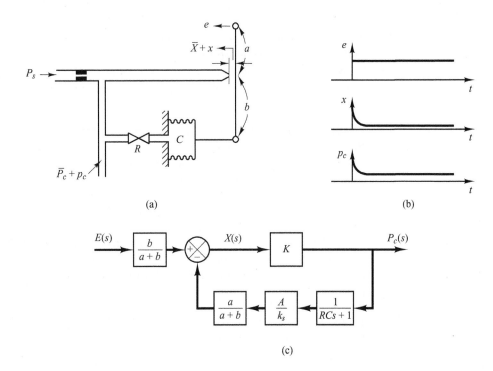

Figure 4–14
(a) Pneumatic proportional-plus-derivative controller; (b) step change in e and the corresponding changes in x and p_c plotted versus t; (c) block diagram of the controller.

(a)   (b)

(c)

Then the change in the control pressure p_c will be instantaneous. The restriction R will momentarily prevent the feedback bellows from sensing the pressure change p_c. Thus the feedback bellows will not respond momentarily, and the pneumatic actuating valve will feel the full effect of the movement of the flapper. As time goes on, the feedback bellows will expand. The change in the nozzle–flapper distance x and the change in the control pressure p_c can be plotted against time t, as shown in Figure 4–14(b). At steady state, the feedback bellows acts like an ordinary feedback mechanism. The curve p_c versus t clearly shows that this controller is of the proportional-plus-derivative type.

A block diagram corresponding to this pneumatic controller is shown in Figure 4–14(c). In the block diagram, K is a constant, A is the area of the bellows, and k_s is the equivalent spring constant of the bellows. The transfer function between p_c and e can be obtained from the block diagram as follows:

$$\frac{P_c(s)}{E(s)} = \frac{\dfrac{b}{a+b} K}{1 + \dfrac{Ka}{a+b}\dfrac{A}{k_s}\dfrac{1}{RCs+1}}$$

In such a controller the loop gain $\left| KaA/[(a+b)k_s(RCs+1)] \right|$ is made much greater than unity. Thus the transfer function $P_c(s)/E(s)$ can be simplified to give

$$\frac{P_c(s)}{E(s)} = K_p(1 + T_d s)$$

where

$$K_p = \frac{bk_s}{aA}, \qquad T_d = RC$$

Thus, delayed negative feedback, or the transfer function $1/(RCs+1)$ in the feedback path, modifies the proportional controller to a proportional-plus-derivative controller.

Note that if the feedback valve is fully opened, the control action becomes proportional. If the feedback valve is fully closed, the control action becomes narrow-band proportional (on–off).

Obtaining Pneumatic Proportional-Plus-Integral Control Action. Consider the proportional controller shown in Figure 4–13(a). Considering small changes in the variables, we can show that the addition of delayed positive feedback will modify this proportional controller to a proportional-plus-integral controller, or a PI controller.

Consider the pneumatic controller shown in Figure 4–15(a). The operation of this controller is as follows: The bellows denoted by I is connected to the control pressure source without any restriction. The bellows denoted by II is connected to the control pressure source through a restriction. Let us assume a small step change in the actuating error. This will cause the back pressure in the nozzle to change instantaneously. Thus a change in the control pressure p_c also occurs instantaneously. Due to the restriction of the valve in the path to bellows II, there will be a pressure drop across the valve. As time goes on, air will flow across the valve in such a way that the change in pressure in bellows II attains the value p_c. Thus bellows II will expand or contract as time elapses in such a way as to move the flapper an additional amount in the direction of the original displacement e. This will cause the back pressure p_c in the nozzle to change continuously, as shown in Figure 4–15(b).

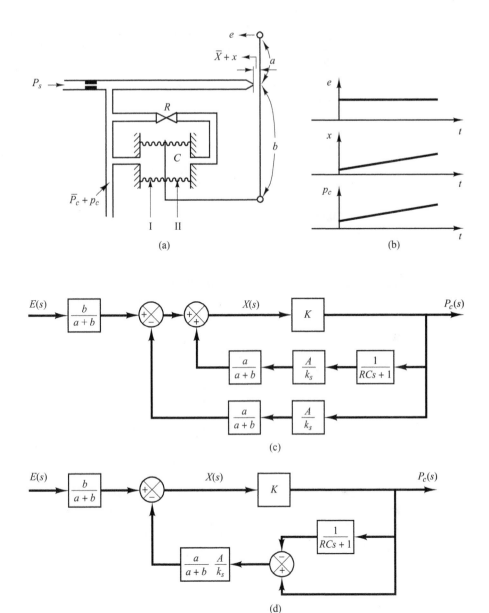

Figure 4–15
(a) Pneumatic proportional-plus-integral controller; (b) step change in e and the corresponding changes in x and p_c plotted versus t; (c) block diagram of the controller; (d) simplified block diagram.

Note that the integral control action in the controller takes the form of slowly canceling the feedback that the proportional control originally provided.

A block diagram of this controller under the assumption of small variations in the variables is shown in Figure 4–15(c). A simplification of this block diagram yields Figure 4–15(d). The transfer function of this controller is

$$\frac{P_c(s)}{E(s)} = \frac{\dfrac{b}{a+b} K}{1 + \dfrac{Ka}{a+b} \dfrac{A}{k_s} \left(1 - \dfrac{1}{RCs+1}\right)}$$

where K is a constant, A is the area of the bellows, and k_s is the equivalent spring constant of the combined bellows. If $|KaARCs/[(a + b)k_s(RCs + 1)]| \gg 1$, which is usually the case, the transfer function can be simplified to

$$\frac{P_c(s)}{E(s)} = K_p\left(1 + \frac{1}{T_i s}\right)$$

where

$$K_p = \frac{bk_s}{aA}, \qquad T_i = RC$$

Obtaining Pneumatic Proportional-Plus-Integral-Plus-Derivative Control Action.   A combination of the pneumatic controllers shown in Figures 4–14(a) and 4–15(a) yields a proportional-plus-integral-plus-derivative controller, or a PID controller. Figure 4–16(a) shows a schematic diagram of such a controller. Figure 4–16(b) shows a block diagram of this controller under the assumption of small variations in the variables.

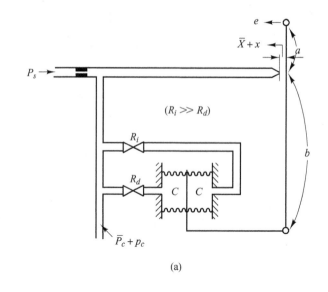

(a)

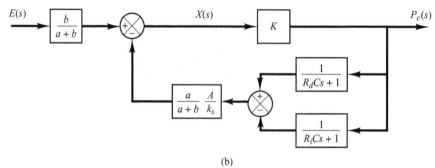

(b)

Figure 4–16
(a) Pneumatic proportional-plus-integral-plus-derivative controller; (b) block diagram of the controller.

The transfer function of this controller is

$$\frac{P_c(s)}{E(s)} = \frac{\dfrac{bK}{a+b}}{1 + \dfrac{Ka}{a+b}\dfrac{A}{k_s}\dfrac{(R_iC - R_dC)s}{(R_dCs + 1)(R_iCs + 1)}}$$

By defining

$$T_i = R_iC, \qquad T_d = R_dC$$

and noting that under normal operation $\left| KaA(T_i - T_d)s / \left[(a+b)k_s(T_ds + 1)(T_is + 1) \right] \right| \gg 1$
and $T_i \gg T_d$, we obtain

$$\frac{P_c(s)}{E(s)} \doteq \frac{bk_s}{aA}\frac{(T_ds + 1)(T_is + 1)}{(T_i - T_d)s}$$

$$\doteq \frac{bk_s}{aA}\frac{T_dT_is^2 + T_is + 1}{T_is}$$

$$= K_p\left(1 + \frac{1}{T_is} + T_ds \right) \qquad (4\text{-}24)$$

where

$$K_p = \frac{bk_s}{aA}$$

Equation (4–24) indicates that the controller shown in Figure 4–16(a) is a proportional-plus-integral-plus-derivative controller or a PID controller.

4–4 HYDRAULIC SYSTEMS

Except for low-pressure pneumatic controllers, compressed air has seldom been used for the continuous control of the motion of devices having significant mass under external load forces. For such a case, hydraulic controllers are generally preferred.

Hydraulic Systems. The widespread use of hydraulic circuitry in machine tool applications, aircraft control systems, and similar operations occurs because of such factors as positiveness, accuracy, flexibility, high horsepower-to-weight ratio, fast starting, stopping, and reversal with smoothness and precision, and simplicity of operations.

The operating pressure in hydraulic systems is somewhere between 145 and 5000 lb$_f$/in.^{2} (between 1 and 35 MPa). In some special applications, the operating pressure may go up to 10,000 lb$_f$/in.^{2} (70 MPa). For the same power requirement, the weight and size of the hydraulic unit can be made smaller by increasing the supply pressure. With high-pressure hydraulic systems, very large force can be obtained. Rapid-acting, accurate positioning of heavy loads is possible with hydraulic systems. A combination of electronic and hydraulic systems is widely used because it combines the advantages of both electronic control and hydraulic power.

Advantages and Disadvantages of Hydraulic Systems. There are certain advantages and disadvantages in using hydraulic systems rather than other systems. Some of the advantages are the following:

1. Hydraulic fluid acts as a lubricant, in addition to carrying away heat generated in the system to a convenient heat exchanger.
2. Comparatively small-sized hydraulic actuators can develop large forces or torques.
3. Hydraulic actuators have a higher speed of response with fast starts, stops, and speed reversals.
4. Hydraulic actuators can be operated under continuous, intermittent, reversing, and stalled conditions without damage.
5. Availability of both linear and rotary actuators gives flexibility in design.
6. Because of low leakages in hydraulic actuators, speed drop when loads are applied is small.

On the other hand, several disadvantages tend to limit their use.

1. Hydraulic power is not readily available compared to electric power.
2. Cost of a hydraulic system may be higher than that of a comparable electrical system performing a similar function.
3. Fire and explosion hazards exist unless fire-resistant fluids are used.
4. Because it is difficult to maintain a hydraulic system that is free from leaks, the system tends to be messy.
5. Contaminated oil may cause failure in the proper functioning of a hydraulic system.
6. As a result of the nonlinear and other complex characteristics involved, the design of sophisticated hydraulic systems is quite involved.
7. Hydraulic circuits have generally poor damping characteristics. If a hydraulic circuit is not designed properly, some unstable phenomena may occur or disappear, depending on the operating condition.

Comments. Particular attention is necessary to ensure that the hydraulic system is stable and satisfactory under all operating conditions. Since the viscosity of hydraulic fluid can greatly affect damping and friction effects of the hydraulic circuits, stability tests must be carried out at the highest possible operating temperature.

Note that most hydraulic systems are nonlinear. Sometimes, however, it is possible to linearize nonlinear systems so as to reduce their complexity and permit solutions that are sufficiently accurate for most purposes. A useful linearization technique for dealing with nonlinear systems was presented in Section 2–7.

Hydraulic Servo System. Figure 4–17(a) shows a hydraulic servomotor. It is essentially a pilot-valve-controlled hydraulic power amplifier and actuator. The pilot valve is a balanced valve, in the sense that the pressure forces acting on it are all balanced. A very large power output can be controlled by a pilot valve, which can be positioned with very little power.

In practice, the ports shown in Figure 4–17(a) are often made wider than the corresponding valves. In such a case, there is always leakage through the valves. Such leak-

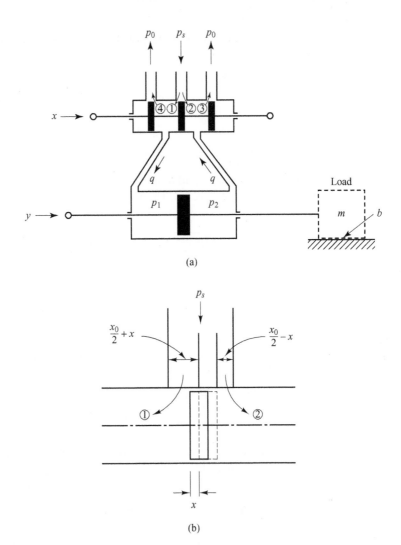

Figure 4–17
(a) Hydraulic servo
system; (b) enlarged
diagram of the valve
orifice area.

age improves both the sensitivity and the linearity of the hydraulic servomotor. In the
following analysis we shall make the assumption that the ports are made wider than
the valves—that is, the valves are underlapped. [Note that sometimes a dither signal, a
high-frequency signal of very small amplitude (with respect to the maximum
displacement of the valve), is superimposed on the motion of the pilot valve. This also
improves the sensitivity and linearity. In this case also there is leakage through the valve.]

We shall apply the linearization technique presented in Section 2–7 to obtain a linearized mathematical model of the hydraulic servomotor. We assume that the valve is
underlapped and symmetrical and admits hydraulic fluid under high pressure into a
power cylinder that contains a large piston, so that a large hydraulic force is established
to move a load.

In Figure 4–17(b) we have an enlarged diagram of the valve orifice area. Let us
define the valve orifice areas of ports $1, 2, 3, 4$ as A_1, A_2, A_3, A_4, respectively. Also, define
the flow rates through ports $1, 2, 3, 4$ as q_1, q_2, q_3, q_4, respectively. Note that, since the

valve is symmetrical, $A_1 = A_3$ and $A_2 = A_4$. Assuming the displacement x to be small, we obtain

$$A_1 = A_3 = k\left(\frac{x_0}{2} + x\right)$$

$$A_2 = A_4 = k\left(\frac{x_0}{2} - x\right)$$

where k is a constant.

Furthermore, we shall assume that the return pressure p_o in the return line is small and thus can be neglected. Then, referring to Figure 4–17(a), flow rates through valve orifices are

$$q_1 = c_1 A_1 \sqrt{\frac{2g}{\gamma}(p_s - p_1)} = C_1\sqrt{p_s - p_1}\left(\frac{x_0}{2} + x\right)$$

$$q_2 = c_2 A_2 \sqrt{\frac{2g}{\gamma}(p_s - p_2)} = C_2\sqrt{p_s - p_2}\left(\frac{x_0}{2} - x\right)$$

$$q_3 = c_1 A_3 \sqrt{\frac{2g}{\gamma}(p_2 - p_0)} = C_1\sqrt{p_2 - p_0}\left(\frac{x_0}{2} + x\right) = C_1\sqrt{p_2}\left(\frac{x_0}{2} + x\right)$$

$$q_4 = c_2 A_4 \sqrt{\frac{2g}{\gamma}(p_1 - p_0)} = C_2\sqrt{p_1 - p_0}\left(\frac{x_0}{2} - x\right) = C_2\sqrt{p_1}\left(\frac{x_0}{2} - x\right)$$

where $C_1 = c_1 k\sqrt{2g/\gamma}$ and $C_2 = c_2 k\sqrt{2g/\gamma}$, and γ is the specific weight and is given by $\gamma = \rho g$, where ρ is mass density and g is the acceleration of gravity. The flow rate q to the left-hand side of the power piston is

$$q = q_1 - q_4 = C_1\sqrt{p_s - p_1}\left(\frac{x_0}{2} + x\right) - C_2\sqrt{p_1}\left(\frac{x_0}{2} - x\right) \qquad (4\text{--}25)$$

The flow rate from the right-hand side of the power piston to the drain is the same as this q and is given by

$$q = q_3 - q_2 = C_1\sqrt{p_2}\left(\frac{x_0}{2} + x\right) - C_2\sqrt{p_s - p_2}\left(\frac{x_0}{2} - x\right)$$

In the present analysis we assume that the fluid is incompressible. Since the valve is symmetrical, we have $q_1 = q_3$ and $q_2 = q_4$. By equating q_1 and q_3, we obtain

$$p_s - p_1 = p_2$$

or

$$p_s = p_1 + p_2$$

If we define the pressure difference across the power piston as Δp or

$$\Delta p = p_1 - p_2$$

then

$$p_1 = \frac{p_s + \Delta p}{2}, \qquad p_2 = \frac{p_s - \Delta p}{2}$$

For the symmetrical valve shown in Figure 4–17(a), the pressure in each side of the power piston is $(1/2)p_s$ when no load is applied, or $\Delta p = 0$. As the spool valve is displaced, the pressure in one line increases as the pressure in the other line decreases by the same amount.

In terms of p_s and Δp, we can rewrite the flow rate q given by Equation (4–25) as

$$q = q_1 - q_4 = C_1 \sqrt{\frac{p_s - \Delta p}{2}} \left(\frac{x_0}{2} + x \right) - C_2 \sqrt{\frac{p_s + \Delta p}{2}} \left(\frac{x_0}{2} - x \right)$$

Noting that the supply pressure p_s is constant. the flow rate q can be written as a function of the valve displacement x and pressure difference Δp, or

$$q = C_1 \sqrt{\frac{p_s - \Delta p}{2}} \left(\frac{x_0}{2} + x \right) - C_2 \sqrt{\frac{p_s + \Delta p}{2}} \left(\frac{x_0}{2} - x \right) = f(x, \Delta p)$$

By applying the linearization technique presented in Section 2–7 to this case, the linearized equation about point $x = \bar{x}, \Delta p = \Delta \bar{p}, q = \bar{q}$ is

$$q - \bar{q} = a(x - \bar{x}) + b(\Delta p - \Delta \bar{p}) \tag{4–26}$$

where

$$\bar{q} = f(\bar{x}, \Delta \bar{p})$$

$$a = \left. \frac{\partial f}{\partial x} \right|_{x = \bar{x}, \, \Delta p = \Delta \bar{p}} = C_1 \sqrt{\frac{p_s - \Delta \bar{p}}{2}} + C_2 \sqrt{\frac{p_s + \Delta \bar{p}}{2}}$$

$$b = \left. \frac{\partial f}{\partial \Delta p} \right|_{x = \bar{x}, \, \Delta p = \Delta \bar{p}} = -\left[\frac{C_1}{2\sqrt{2}\sqrt{p_s - \Delta \bar{p}}} \left(\frac{x_0}{2} + \bar{x} \right) \right.$$

$$\left. + \frac{C_2}{2\sqrt{2}\sqrt{p_s + \Delta \bar{p}}} \left(\frac{x_0}{2} - \bar{x} \right) \right] < 0$$

Coefficients a and b here are called *valve coefficients*. Equation (4–26) is a linearized mathematical model of the spool valve near an operating point $x = \bar{x}, \Delta p = \Delta \bar{p}, q = \bar{q}$. The values of valve coefficients a and b vary with the operating point. Note that $\partial f / \partial \Delta p$ is negative and so b is negative.

Since the normal operating point is the point where $\bar{x} = 0, \Delta \bar{p} = 0, \bar{q} = 0$, near the normal operating point Equation (4–26) becomes

$$q = K_1 x - K_2 \Delta p \tag{4–27}$$

where

$$K_1 = (C_1 + C_2) \sqrt{\frac{p_s}{2}} > 0$$

$$K_2 = (C_1 + C_2) \frac{x_0}{4\sqrt{2}\sqrt{p_s}} > 0$$

Equation (4–27) is a linearized mathematical model of the spool valve near the origin ($\bar{x} = 0$, $\Delta \bar{p} = 0$, $\bar{q} = 0$.) Note that the region near the origin is most important in this kind of system, because the system operation usually occurs near this point.

Figure 4–18 shows this linearized relationship among q, x, and ΔP. The straight lines shown are the characteristic curves of the linearized hydraulic servomotor. This family of curves consists of equidistant parallel straight lines, parametrized by x.

In the present analysis we assume that the load reactive forces are small, so that the leakage flow rate and oil compressibility can be ignored.

Referring to Figure 4–17(a), we see that the rate of flow of oil q times dt is equal to the power-piston displacement dy times the piston area A times the density of oil ρ. Thus, we obtain

$$A\rho \, dy = q \, dt$$

Notice that for a given flow rate q the larger the piston area A is, the lower will be the velocity dy/dt. Hence, if the piston area A is made smaller, the other variables remaining constant, the velocity dy/dt will become higher. Also, an increased flow rate q will cause an increased velocity of the power piston and will make the response time shorter.

Equation (4–27) can now be written as

$$\Delta P = \frac{1}{K_2} \left(K_1 x - A\rho \frac{dy}{dt} \right)$$

The force developed by the power piston is equal to the pressure difference ΔP times the piston area A or

$$\text{Force developed by the power piston} = A \, \Delta P$$

$$= \frac{A}{K_2} \left(K_1 x - A\rho \frac{dy}{dt} \right)$$

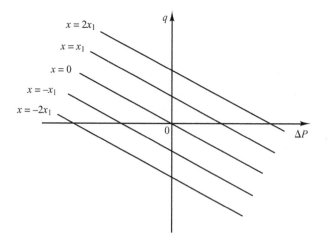

Figure 4–18
Characteristic curves of the linearized hydraulic servomotor.

Chapter 4 / Mathematical Modeling of Fluid Systems and Thermal Systems

For a given maximum force, if the pressure difference is sufficiently high, the piston area, or the volume of oil in the cylinder, can be made small. Consequently, to minimize the weight of the controller, we must make the supply pressure sufficiently high.

Assume that the power piston moves a load consisting of a mass and viscous friction. Then the force developed by the power piston is applied to the load mass and friction, and we obtain

$$m\ddot{y} + b\dot{y} = \frac{A}{K_2}\left(K_1 x - A\rho\dot{y}\right)$$

or

$$m\ddot{y} + \left(b + \frac{A^2\rho}{K_2}\right)\dot{y} = \frac{AK_1}{K_2}x \qquad (4\text{--}28)$$

where m is the mass of the load and b is the viscous-friction coefficient.

Assuming that the pilot-valve displacement x is the input and the power-piston displacement y is the output, we find that the transfer function for the hydraulic servo-motor is, from Equation (4–28),

$$\frac{Y(s)}{X(s)} = \frac{1}{s\left[\left(\dfrac{mK_2}{AK_1}\right)s + \dfrac{bK_2}{AK_1} + \dfrac{A\rho}{K_1}\right]}$$

$$= \frac{K}{s(Ts + 1)} \qquad (4\text{--}29)$$

where

$$K = \frac{1}{\dfrac{bK_2}{AK_1} + \dfrac{A\rho}{K_1}} \qquad \text{and} \qquad T = \frac{mK_2}{bK_2 + A^2\rho}$$

From Equation (4–29) we see that this transfer function is of the second order. If the ratio $mK_2/(bK_2 + A^2\rho)$ is negligibly small or the time constant T is negligible, the transfer function $Y(s)/X(s)$ can be simplified to give

$$\frac{Y(s)}{X(s)} = \frac{K}{s}$$

It is noted that a more detailed analysis shows that if oil leakage, compressibility (including the effects of dissolved air), expansion of pipelines, and the like are taken into consideration, the transfer function becomes

$$\frac{Y(s)}{X(s)} = \frac{K}{s(T_1 s + 1)(T_2 s + 1)}$$

where T_1 and T_2 are time constants. As a matter of fact, these time constants depend on the volume of oil in the operating circuit. The smaller the volume, the smaller the time constants.

Hydraulic Integral Controller. The hydraulic servomotor shown in Figure 4–19 is a pilot-valve-controlled hydraulic power amplifier and actuator. Similar to the hydraulic servo system shown in Figure 4–17, for negligibly small load mass the servomotor shown in Figure 4–19 acts as an integrator or an integral controller. Such a servomotor constitutes the basis of the hydraulic control circuit.

In the hydraulic servomotor shown in Figure 4–19, the pilot valve (a four-way valve) has two lands on the spool. If the width of the land is smaller than the port in the valve sleeve, the valve is said to be *underlapped*. *Overlapped* valves have a land width greater than the port width. A *zero-lapped* valve has a land width that is identical to the port width. (If the pilot valve is a zero-lapped valve, analyses of hydraulic servomotors become simpler.)

In the present analysis, we assume that hydraulic fluid is incompressible and that the inertia force of the power piston and load is negligible compared to the hydraulic force at the power piston. We also assume that the pilot valve is a zero-lapped valve, and the oil flow rate is proportional to the pilot valve displacement.

Operation of this hydraulic servomotor is as follows. If input x moves the pilot valve to the right, port II is uncovered, and so high-pressure oil enters the right-hand side of the power piston. Since port I is connected to the drain port, the oil in the left-hand side of the power piston is returned to the drain. The oil flowing into the power cylinder is at high pressure; the oil flowing out from the power cylinder into the drain is at low pressure. The resulting difference in pressure on both sides of the power piston will cause it to move to the left.

Note that the rate of flow of oil q (kg/sec) times dt (sec) is equal to the power-piston displacement dy (m) times the piston area A (m²) times the density of oil ρ (kg/m³). Therefore,

$$A\rho \, dy = q \, dt \qquad (4\text{--}30)$$

Because of the assumption that the oil flow rate q is proportional to the pilot-valve displacement x, we have

$$q = K_1 x \qquad (4\text{--}31)$$

where K_1 is a positive constant. From Equations (4–30) and (4–31) we obtain

$$A\rho \frac{dy}{dt} = K_1 x$$

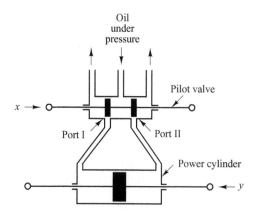

Figure 4–19
Hydraulic
servomotor.

The Laplace transform of this last equation, assuming a zero initial condition, gives

$$A\rho s Y(s) = K_1 X(s)$$

or

$$\frac{Y(s)}{X(s)} = \frac{K_1}{A\rho s} = \frac{K}{s}$$

where $K = K_1/(A\rho)$. Thus the hydraulic servomotor shown in Figure 4–19 acts as an integral controller.

Hydraulic Proportional Controller. It has been shown that the servomotor in Figure 4–19 acts as an integral controller. This servomotor can be modified to a proportional controller by means of a feedback link. Consider the hydraulic controller shown in Figure 4–20(a). The left-hand side of the pilot valve is joined to the left-hand side of the power piston by a link ABC. This link is a floating link rather than one moving about a fixed pivot.

The controller here operates in the following way. If input e moves the pilot valve to the right, port II will be uncovered and high-pressure oil will flow through port II into the right-hand side of the power piston and force this piston to the left. The power piston, in moving to the left, will carry the feedback link ABC with it, thereby moving the pilot valve to the left. This action continues until the pilot piston again covers ports I and II. A block diagram of the system can be drawn as in Figure 4–20(b). The transfer function between $Y(s)$ and $E(s)$ is given by

$$\frac{Y(s)}{E(s)} = \frac{\dfrac{b}{a+b}\dfrac{K}{s}}{1 + \dfrac{K}{s}\dfrac{a}{a+b}}$$

Noting that under the normal operating conditions we have $\left|Ka/[s(a + b)]\right| \gg 1$, this last equation can be simplified to

$$\frac{Y(s)}{E(s)} = \frac{b}{a} = K_p$$

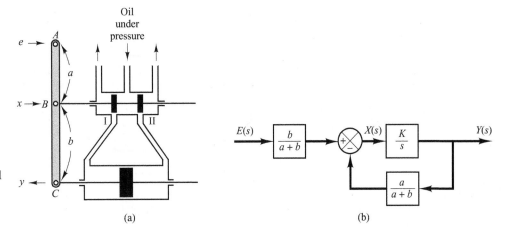

Figure 4–20
(a) Servomotor that acts as a proportional controller; (b) block diagram of the servomotor.

(a)

(b)

The transfer function between y and e becomes a constant. Thus, the hydraulic controller shown in Figure 4–20(a) acts as a proportional controller, the gain of which is K_p. This gain can be adjusted by effectively changing the lever ratio b/a. (The adjusting mechanism is not shown in the diagram.)

We have thus seen that the addition of a feedback link will cause the hydraulic servomotor to act as a proportional controller.

Dashpots. The dashpot (also called a damper) shown in Figure 4–21(a) acts as a differentiating element. Suppose that we introduce a step displacement to the piston position y. Then the displacement z becomes equal to y momentarily. Because of the spring force, however, the oil will flow through the resistance R and the cylinder will come back to the original position. The curves y versus t and z versus t are shown in Figure 4–21(b).

Let us derive the transfer function between the displacement z and displacement y. Define the pressures existing on the right and left sides of the piston as $P_1(\text{lb}_f/\text{in.}^2)$ and $P_2(\text{lb}_f/\text{in.}^2)$, respectively. Suppose that the inertia force involved is negligible. Then the force acting on the piston must balance the spring force. Thus

$$A(P_1 - P_2) = kz$$

where A = piston area, in.^{2}
k = spring constant, $\text{lb}_f/\text{in.}$

The flow rate q is given by

$$q = \frac{P_1 - P_2}{R}$$

where q = flow rate through the restriction, lb/sec
R = resistance to flow at the restriction, $\text{lb}_f\text{-sec}/\text{in.}^2\text{-lb}$

Since the flow through the restriction during dt seconds must equal the change in the mass of oil to the left of the piston during the same dt seconds, we obtain

$$q \, dt = A\rho(dy - dz)$$

where ρ = density, lb/in.^{3}. (We assume that the fluid is incompressible or ρ = constant.) This last equation can be rewritten as

$$\frac{dy}{dt} - \frac{dz}{dt} = \frac{q}{A\rho} = \frac{P_1 - P_2}{RA\rho} = \frac{kz}{RA^2\rho}$$

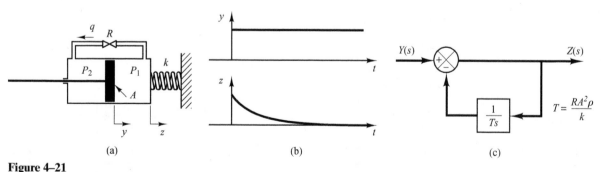

| (a) | (b) | (c) |

Figure 4–21
(a) Dashpot; (b) step change in y and the corresponding change in z plotted versus t; (c) block diagram of the dashpot.

Chapter 4 / Mathematical Modeling of Fluid Systems and Thermal Systems

or

$$\frac{dy}{dt} = \frac{dz}{dt} + \frac{kz}{RA^2\rho}$$

Taking the Laplace transforms of both sides of this last equation, assuming zero initial conditions, we obtain

$$sY(s) = sZ(s) + \frac{k}{RA^2\rho} Z(s)$$

The transfer function of this system thus becomes

$$\frac{Z(s)}{Y(s)} = \frac{s}{s + \dfrac{k}{RA^2\rho}}$$

Let us define $RA^2\rho/k = T$. (Note that $RA^2\rho/k$ has the dimension of time.) Then

$$\frac{Z(s)}{Y(s)} = \frac{Ts}{Ts + 1} = \frac{1}{1 + \dfrac{1}{Ts}}$$

Clearly, the dashpot is a differentiating element. Figure 4–21(c) shows a block diagram representation for this system.

Obtaining Hydraulic Proportional-Plus-Integral Control Action. Figure 4–22(a) shows a schematic diagram of a hydraulic proportional-plus-integral controller. A block diagram of this controller is shown in Figure 4–22(b). The transfer function $Y(s)/E(s)$ is given by

$$\frac{Y(s)}{E(s)} = \frac{\dfrac{b}{a+b}\dfrac{K}{s}}{1 + \dfrac{Ka}{a+b}\dfrac{T}{Ts+1}}$$

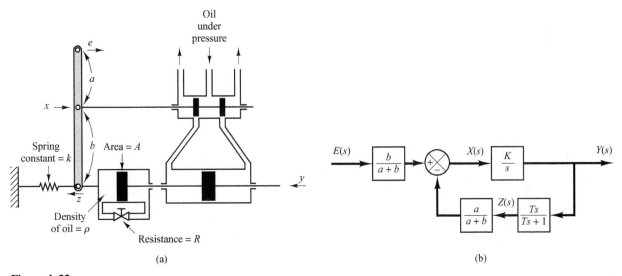

(a)   (b)

Figure 4–22
(a) Schematic diagram of a hydraulic proportional-plus-integral controller; (b) block diagram of the controller.

In such a controller, under normal operation $|KaT/[(a + b)(Ts + 1)]| \gg 1$, with the result that

$$\frac{Y(s)}{E(s)} = K_p\left(1 + \frac{1}{T_i s}\right)$$

where

$$K_p = \frac{b}{a}, \qquad T_i = T = \frac{RA^2\rho}{k}$$

Thus the controller shown in Figure 4–22(a) is a proportional-plus-integral controller (PI controller).

Obtaining Hydraulic Proportional-Plus-Derivative Control Action. Figure 4–23(a) shows a schematic diagram of a hydraulic proportional-plus-derivative controller. The cylinders are fixed in space and the pistons can move. For this system, notice that

$$k(y - z) = A(P_2 - P_1)$$

$$q = \frac{P_2 - P_1}{R}$$

$$q\,dt = \rho A\,dz$$

Hence

$$y = z + \frac{A}{k}qR = z + \frac{RA^2\rho}{k}\frac{dz}{dt}$$

or

$$\frac{Z(s)}{Y(s)} = \frac{1}{Ts + 1}$$

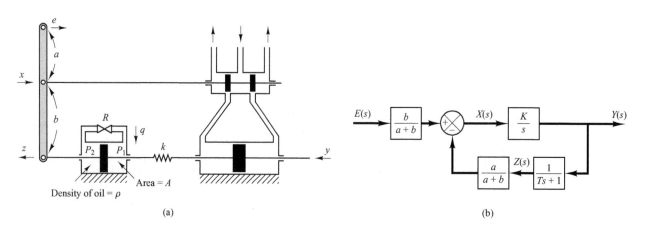

(a)   (b)

Figure 4–23
(a) Schematic diagram of a hydraulic proportional-plus-derivative controller; (b) block diagram of the controller.

where

$$T = \frac{RA^2\rho}{k}$$

A block diagram for this system is shown in Figure 4–23(b). From the block diagram the transfer function $Y(s)/E(s)$ can be obtained as

$$\frac{Y(s)}{E(s)} = \frac{\dfrac{b}{a+b}\dfrac{K}{s}}{1 + \dfrac{a}{a+b}\dfrac{K}{s}\dfrac{1}{Ts+1}}$$

Under normal operation we have $|aK/[(a+b)s(Ts+1)]| \gg 1$. Hence

$$\frac{Y(s)}{E(s)} = K_p(1 + Ts)$$

where

$$K_p = \frac{b}{a}, \qquad T = \frac{RA^2\rho}{k}$$

Thus the controller shown in Figure 4–23(a) is a proportional-plus-derivative controller (PD controller).

Obtaining Hydraulic Proportional-Plus-Integral-Plus-Derivative Control Action.
Figure 4–24 shows a schematic diagram of a hydraulic proportional-plus-integral-plus-derivative controller. It is a combination of the proportional-plus-integral controller and proportional-plus derivative controller.

If the two dashpots are identical except the piston shafts, the transfer function $Z(s)/Y(s)$ can be obtained as follows:

$$\frac{Z(s)}{Y(s)} = \frac{T_1 s}{T_1 T_2 s^2 + (T_1 + 2T_2)s + 1}$$

(For the derivation of this transfer function, refer to Problem **A–4–9**.)

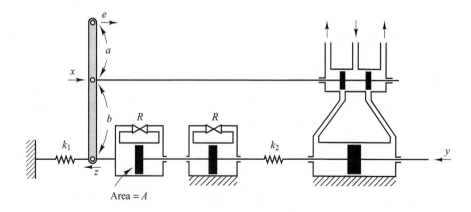

Figure 4–24
Schematic diagram of a hydraulic proportional-plus-integral-plus-derivative controller.

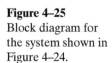

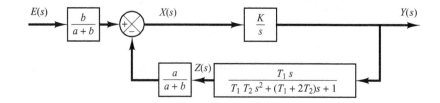

Figure 4–25
Block diagram for
the system shown in
Figure 4–24.

A block diagram for this system is shown in Figure 4–25. The transfer function $Y(s)/E(s)$ can be obtained as

$$\frac{Y(s)}{E(s)} = \frac{b}{a+b} \frac{\dfrac{K}{s}}{1 + \dfrac{a}{a+b} \dfrac{K}{s} \dfrac{T_1 s}{T_1 T_2 s^2 + (T_1 + 2T_2)s + 1}}$$

Under normal circumstances we design the system such that

$$\left| \frac{a}{a+b} \frac{K}{s} \frac{T_1 s}{T_1 T_2 s^2 + (T_1 + 2T_2)s + 1} \right| \gg 1$$

then

$$\frac{Y(s)}{E(s)} = \frac{b}{a} \frac{T_1 T_2 s^2 + (T_1 + 2T_2)s + 1}{T_1 s}$$

$$= K_p + \frac{K_i}{s} + K_d s$$

where

$$K_p = \frac{b}{a} \frac{T_1 + 2T_2}{T_1}, \qquad K_i = \frac{b}{a} \frac{1}{T_1}, \qquad K_d = \frac{b}{a} T_2$$

Thus, the controller shown in Figure 4–24 is a proportional-plus-integral-plus-derivative controller (PID controller).

4–5 THERMAL SYSTEMS

Thermal systems are those that involve the transfer of heat from one substance to another. Thermal systems may be analyzed in terms of resistance and capacitance, although the thermal capacitance and thermal resistance may not be represented accurately as lumped parameters, since they are usually distributed throughout the substance. For precise analysis, distributed-parameter models must be used. Here, however, to simplify the analysis we shall assume that a thermal system can be represented by a lumped-parameter model, that substances that are characterized by resistance to heat flow have negligible heat capacitance, and that substances that are characterized by heat capacitance have negligible resistance to heat flow.

There are three different ways heat can flow from one substance to another: conduction, convection, and radiation. Here we consider only conduction and convection. (Radiation heat transfer is appreciable only if the temperature of the emitter is very high compared to that of the receiver. Most thermal processes in process control systems do not involve radiation heat transfer.)

For conduction or convection heat transfer,

$$q = K \, \Delta\theta$$

where q = heat flow rate, kcal/sec
$\Delta\theta$ = temperature difference, °C
K = coefficient, kcal/sec °C

The coefficient K is given by

$$K = \frac{kA}{\Delta X}, \qquad \text{for conduction}$$

$$= HA, \qquad \text{for convection}$$

where k = thermal conductivity, kcal/m sec °C
A = area normal to heat flow, m^2
ΔX = thickness of conductor, m
H = convection coefficient, kcal/m^2 sec °C

Thermal Resistance and Thermal Capacitance. The thermal resistance R for heat transfer between two substances may be defined as follows:

$$R = \frac{\text{change in temperature difference, °C}}{\text{change in heat flow rate, kcal/sec}}$$

The thermal resistance for conduction or convection heat transfer is given by

$$R = \frac{d(\Delta\theta)}{dq} = \frac{1}{K}$$

Since the thermal conductivity and convection coefficients are almost constant, the thermal resistance for either conduction or convection is constant.

The thermal capacitance C is defined by

$$C = \frac{\text{change in heat stored, kcal}}{\text{change in temperature, °C}}$$

or

$$C = mc$$

where m = mass of substance considered, kg
c = specific heat of substance, kcal/kg °C

Thermal System. Consider the system shown in Figure 4–26(a). It is assumed that the tank is insulated to eliminate heat loss to the surrounding air. It is also assumed that there is no heat storage in the insulation and that the liquid in the tank is perfectly mixed so that it is at a uniform temperature. Thus, a single temperature is used to describe the temperature of the liquid in the tank and of the outflowing liquid.

Let us define

$$\bar{\Theta}_i = \text{steady-state temperature of inflowing liquid, °C}$$

$$\bar{\Theta}_o = \text{steady-state temperature of outflowing liquid, °C}$$

$$G = \text{steady-state liquid flow rate, kg/sec}$$

$$M = \text{mass of liquid in tank, kg}$$

$$c = \text{specific heat of liquid, kcal/kg °C}$$

$$R = \text{thermal resistance, °C sec/kcal}$$

$$C = \text{thermal capacitance, kcal/°C}$$

$$\bar{H} = \text{steady-state heat input rate, kcal/sec}$$

Assume that the temperature of the inflowing liquid is kept constant and that the heat input rate to the system (heat supplied by the heater) is suddenly changed from $\bar{H}$ to $\bar{H} + h_i$, where h_i represents a small change in the heat input rate. The heat outflow rate will then change gradually from $\bar{H}$ to $\bar{H} + h_o$. The temperature of the outflowing liquid will also be changed from $\bar{\Theta}_o$ to $\bar{\Theta}_o + \theta$. For this case, h_o, C, and R are obtained, respectively, as

$$h_o = Gc\theta$$

$$C = Mc$$

$$R = \frac{\theta}{h_o} = \frac{1}{Gc}$$

The heat-balance equation for this system is

$$C\,d\theta = (h_i - h_o)\,dt$$

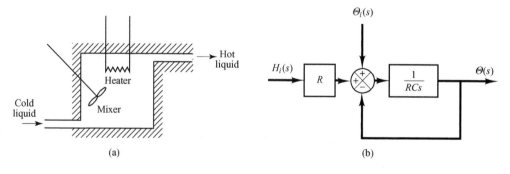

Figure 4–26
(a) Thermal system:
(b) block diagram of
the system.

138   Chapter 4 / Mathematical Modeling of Fluid Systems and Thermal Systems

or

$$C \frac{d\theta}{dt} = h_i - h_o$$

which may be rewritten as

$$RC \frac{d\theta}{dt} + \theta = Rh_i$$

Note that the time constant of the system is equal to RC or M/G seconds. The transfer function relating θ and h_i is given by

$$\frac{\Theta(s)}{H_i(s)} = \frac{R}{RCs + 1}$$

where $\Theta(s) = \mathcal{L}[\theta(t)]$ and $H_i(s) = \mathcal{L}[h_i(t)]$.

In practice, the temperature of the inflowing liquid may fluctuate and may act as a load disturbance. (If a constant outflow temperature is desired, an automatic controller may be installed to adjust the heat inflow rate to compensate for the fluctuations in the temperature of the inflowing liquid.) If the temperature of the inflowing liquid is suddenly changed from $\bar{\Theta}_i$ to $\bar{\Theta}_i + \theta_i$ while the heat input rate H and the liquid flow rate G are kept constant, then the heat outflow rate will be changed from $\bar{H}$ to $\bar{H} + h_o$, and the temperature of the outflowing liquid will be changed from $\bar{\Theta}_o$ to $\bar{\Theta}_o + \theta$. The heat-balance equation for this case is

$$C \, d\theta = (Gc\theta_i - h_o) \, dt$$

or

$$C \frac{d\theta}{dt} = Gc\theta_i - h_o$$

which may be rewritten

$$RC \frac{d\theta}{dt} + \theta = \theta_i$$

The transfer function relating θ and θ_i is given by

$$\frac{\Theta(s)}{\Theta_i(s)} = \frac{1}{RCs + 1}$$

where $\Theta(s) = \mathcal{L}[\theta(t)]$ and $\Theta_i(s) = \mathcal{L}[\theta_i(t)]$.

If the present thermal system is subjected to changes in both the temperature of the inflowing liquid and the heat input rate, while the liquid flow rate is kept constant, the change θ in the temperature of the outflowing liquid can be given by the following equation:

$$RC \frac{d\theta}{dt} + \theta = \theta_i + Rh_i$$

A block diagram corresponding to this case is shown in Figure 4–26(b). Notice that the system involves two inputs.

A–4–1. In the liquid-level system of Figure 4–27 assume that the outflow rate Q m^3/sec through the out-flow valve is related to the head H m by

$$Q = K\sqrt{H} = 0.01\sqrt{H}$$

Assume also that when the inflow rate Q_i is 0.015 m^3/sec the head stays constant. For $t < 0$ the system is at steady state $(Q_i = 0.015$ m^3/sec$)$. At $t = 0$ the inflow valve is closed and so there is no inflow for $t \geq 0$. Find the time necessary to empty the tank to half the original head. The capacitance C of the tank is 2 m^2.

Solution. When the head is stationary, the inflow rate equals the outflow rate. Thus head H_o at $t = 0$ is obtained from

$$0.015 = 0.01\sqrt{H_o}$$

or

$$H_o = 2.25 \text{ m}$$

The equation for the system for $t > 0$ is

$$-C\,dH = Q\,dt$$

or

$$\frac{dH}{dt} = -\frac{Q}{C} = \frac{-0.01\sqrt{H}}{2}$$

Hence

$$\frac{dH}{\sqrt{H}} = -0.005\,dt$$

Assume that, at $t = t_1$, $H = 1.125$ m. Integrating both sides of this last equation, we obtain

$$\int_{2.25}^{1.125} \frac{dH}{\sqrt{H}} = \int_0^{t_1}(-0.005)\,dt = -0.005t_1$$

It follows that

$$2\sqrt{H}\,\Big|_{2.25}^{1.125} = 2\sqrt{1.125} - 2\sqrt{2.25} = -0.005t_1$$

or

$$t_1 = 175.7$$

Thus, the head becomes half the original value (2.25 m) in 175.7 sec.

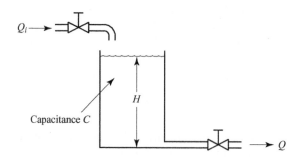

Figure 4–27
Liquid-level system.

A–4–2. Consider the liquid-level system shown in Figure 4–28. In the system, $\bar{Q}_1$ and $\bar{Q}_2$ are steady-state inflow rates and $\bar{H}_1$ and $\bar{H}_2$ are steady-state heads. The quantities $q_{i1}, q_{i2}, h_1, h_2, q_1$, and q_o are considered small. Obtain a state-space representation for the system when h_1 and h_2 are the outputs and q_{i1} and q_{i2} are the inputs.

Solution. The equations for the system are

$$C_1\, dh_1 = \left(q_{i1} - q_1\right) dt \tag{4–32}$$

$$\frac{h_1 - h_2}{R_1} = q_1 \tag{4–33}$$

$$C_2\, dh_2 = \left(q_1 + q_{i2} - q_o\right) dt \tag{4–34}$$

$$\frac{h_2}{R_2} = q_o \tag{4–35}$$

Elimination of q_1 from Equation (4–32) using Equation (4–33) results in

$$\frac{dh_1}{dt} = \frac{1}{C_1}\left(q_{i1} - \frac{h_1 - h_2}{R_1}\right) \tag{4–36}$$

Eliminating q_1 and q_o from Equation (4–34) by using Equations (4–33) and (4–35) gives

$$\frac{dh_2}{dt} = \frac{1}{C_2}\left(\frac{h_1 - h_2}{R_1} + q_{i2} - \frac{h_2}{R_2}\right) \tag{4–37}$$

Define state variables x_1 and x_2 by

$$x_1 = h_1$$

$$x_2 = h_2$$

the input variables u_1 and u_2 by

$$u_1 = q_{i1}$$

$$u_2 = q_{i2}$$

and the output variables y_1 and y_2 by

$$y_1 = h_1 = x_1$$

$$y_2 = h_2 = x_2$$

Then Equations (4–36) and (4–37) can be written as

$$\dot{x}_1 = -\frac{1}{R_1 C_1} x_1 + \frac{1}{R_1 C_1} x_2 + \frac{1}{C_1} u_1$$

$$\dot{x}_2 = \frac{1}{R_1 C_2} x_1 - \left(\frac{1}{R_1 C_2} + \frac{1}{R_2 C_2}\right) x_2 + \frac{1}{C_2} u_2$$

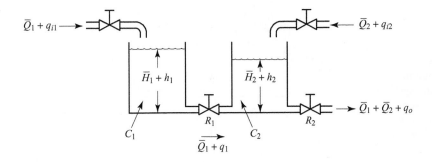

Figure 4–28
Liquid-level system.

In the form of the standard vector-matrix representation, we have

$$\begin{bmatrix} \dot{x}_1 \\ \dot{x}_2 \end{bmatrix} = \begin{bmatrix} -\dfrac{1}{R_1 C_1} & \dfrac{1}{R_1 C_1} \\ \dfrac{1}{R_1 C_2} & -\left(\dfrac{1}{R_1 C_2} + \dfrac{1}{R_2 C_2}\right) \end{bmatrix} \begin{bmatrix} x_1 \\ x_2 \end{bmatrix} + \begin{bmatrix} \dfrac{1}{C_1} & 0 \\ 0 & \dfrac{1}{C_2} \end{bmatrix} \begin{bmatrix} u_1 \\ u_2 \end{bmatrix}$$

which is the state equation, and

$$\begin{bmatrix} y_1 \\ y_2 \end{bmatrix} = \begin{bmatrix} 1 & 0 \\ 0 & 1 \end{bmatrix} \begin{bmatrix} x_1 \\ x_2 \end{bmatrix}$$

which is the output equation.

A–4–3. The value of the gas constant for any gas may be determined from accurate experimental observations of simultaneous values of p, v, and T.

Obtain the gas constant R_{air} for air. Note that at 32°F and 14.7 psia the specific volume of air is 12.39 ft³/lb. Then obtain the capacitance of a 20-ft³ pressure vessel that contains air at 160°F. Assume that the expansion process is isothermal.

Solution.

$$R_{air} = \frac{pv}{T} = \frac{14.7 \times 144 \times 12.39}{460 + 32} = 53.3 \text{ ft-lb}_f/\text{lb}°\text{R}$$

Referring to Equation (4–12), the capacitance of a 20-ft³ pressure vessel is

$$C = \frac{V}{n R_{air} T} = \frac{20}{1 \times 53.3 \times 620} = 6.05 \times 10^{-4} \frac{\text{lb}}{\text{lb}_f/\text{ft}^2}$$

Note that in terms of SI units, R_{air} is given by

$$R_{air} = 287 \text{ N-m/kg K}$$

A–4–4. In the pneumatic pressure system of Figure 4–29(a), assume that, for $t < 0$, the system is at steady state and that the pressure of the entire system is $\bar{P}$. Also, assume that the two bellows are identical. At $t = 0$, the input pressure is changed from $\bar{P}$ to $\bar{P} + p_i$. Then the pressures in bellows 1 and 2 will change from $\bar{P}$ to $\bar{P} + p_1$ and from $\bar{P}$ to $\bar{P} + p_2$, respectively. The capacity (volume) of each bellows is 5×10^{-4} m³, and the operating-pressure difference Δp (difference between p_i and p_1 or difference between p_i and p_2) is between -0.5×10^5 N/m² and 0.5×10^5 N/m². The corresponding mass flow rates (kg/sec) through the valves are shown in Figure 4–29(b). Assume that the bellows expand or contract linearly with the air pressures applied to them, that the equivalent spring constant of the bellows system is $k = 1 \times 10^5$ N/m, and that each bellows has area $A = 15 \times 10^{-4}$ m².

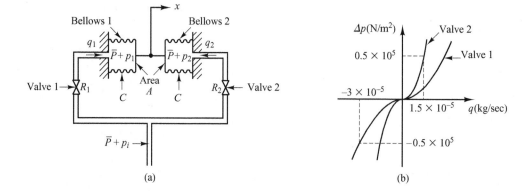

Figure 4–29
(a) Pneumatic pressure system;
(b) pressure-difference-versus-mass-flow-rate curves.

Defining the displacement of the midpoint of the rod that connects two bellows as x, find the transfer function $X(s)/P_i(s)$. Assume that the expansion process is isothermal and that the temperature of the entire system stays at 30°C. Assume also that the polytropic exponent n is 1.

Solution. Referring to Section 4–3, transfer function $P_1(s)/P_i(s)$ can be obtained as

$$\frac{P_1(s)}{P_i(s)} = \frac{1}{R_1Cs + 1} \tag{4-38}$$

Similarly, transfer function $P_2(s)/P_i(s)$ is

$$\frac{P_2(s)}{P_i(s)} = \frac{1}{R_2Cs + 1} \tag{4-39}$$

The force acting on bellows 1 in the x direction is $A(\bar{P} + p_1)$, and the force acting on bellows 2 in the negative x direction is $A(\bar{P} + p_2)$. The resultant force balances with kx, the equivalent spring force of the corrugated sides of the bellows.

$$A(p_1 - p_2) = kx$$

or

$$A[P_1(s) - P_2(s)] = kX(s) \tag{4-40}$$

Referring to Equations (4–38) and (4–39), we see that

$$P_1(s) - P_2(s) = \left(\frac{1}{R_1Cs + 1} - \frac{1}{R_2Cs + 1} \right) P_i(s)$$

$$= \frac{R_2Cs - R_1Cs}{(R_1Cs + 1)(R_2Cs + 1)} P_i(s)$$

By substituting this last equation into Equation (4–40) and rewriting, the transfer function $X(s)/P_i(s)$ is obtained as

$$\frac{X(s)}{P_i(s)} = \frac{A}{k} \frac{(R_2C - R_1C)s}{(R_1Cs + 1)(R_2Cs + 1)} \tag{4-41}$$

The numerical values of average resistances R_1 and R_2 are

$$R_1 = \frac{d\,\Delta p}{dq_1} = \frac{0.5 \times 10^5}{3 \times 10^{-5}} = 0.167 \times 10^{10} \frac{\text{N/m}^2}{\text{kg/sec}}$$

$$R_2 = \frac{d\,\Delta p}{dq_2} = \frac{0.5 \times 10^5}{1.5 \times 10^{-5}} = 0.333 \times 10^{10} \frac{\text{N/m}^2}{\text{kg/sec}}$$

The numerical value of capacitance C of each bellows is

$$C = \frac{V}{nR_{\text{air}}T} = \frac{5 \times 10^{-4}}{1 \times 287 \times (273 + 30)} = 5.75 \times 10^{-9} \frac{\text{kg}}{\text{N/m}^2}$$

where $R_{\text{air}} = 287$ N-m/kg K. (See Problem **A–4–3.**) Consequently,

$$R_1C = 0.167 \times 10^{10} \times 5.75 \times 10^{-9} = 9.60 \text{ sec}$$

$$R_2C = 0.333 \times 10^{10} \times 5.75 \times 10^{-9} = 19.2 \text{ sec}$$

By substituting the numerical values for A, k, R_1C, and R_2C into Equation (4–41), we obtain

$$\frac{X(s)}{P_i(s)} = \frac{1.44 \times 10^{-7}s}{(9.6s + 1)(19.2s + 1)}$$

Example Problems and Solutions

143

A–4–5. Draw a block diagram of the pneumatic controller shown in Figure 4–30. Then derive the transfer function of this controller. Assume that $R_d \ll R_i$. Assume also that the two bellows are identical.

If the resistance R_d is removed (replaced by the line-sized tubing), what control action do we get? If the resistance R_i is removed (replaced by the line-sized tubing), what control action do we get?

Solution. Let us assume that when $e = 0$ the nozzle–flapper distance is equal to $\bar{X}$ and the control pressure is equal to $\bar{P}_c$. In the present analysis, we shall assume small deviations from the respective reference values as follows:

$\quad e = $ small error signal

$\quad x = $ small change in the nozzle–flapper distance

$\quad p_c = $ small change in the control pressure

$\quad p_I = $ small pressure change in bellows I due to small change in the control pressure

$\quad p_{II} = $ small pressure change in bellows II due to small change in the control pressure

$\quad y = $ small displacement at the lower end of the flapper

In this controller, p_c is transmitted to bellows I through the resistance R_d. Similarly, p_c is transmitted to bellows II through the series of resistances R_d and R_i. The relationship between p_I and p_c is

$$\frac{P_I(s)}{P_c(s)} = \frac{1}{R_d C s + 1} = \frac{1}{T_d s + 1}$$

where $T_d = R_d C = $ derivative time. Similarly, p_{II} and p_I are related by the transfer function

$$\frac{P_{II}(s)}{P_I(s)} = \frac{1}{R_i C s + 1} = \frac{1}{T_i s + 1}$$

where $T_i = R_i C = $ integral time. The force-balance equation for the two bellows is

$$(p_I - p_{II})A = k_s y$$

where k_s is the stiffness of the two connected bellows and A is the cross-sectional area of the bellows. The relationship among the variables e, x, and y is

$$x = \frac{b}{a + b}e - \frac{a}{a + b}y$$

The relationship between p_c and x is

$$p_c = Kx \qquad (K > 0)$$

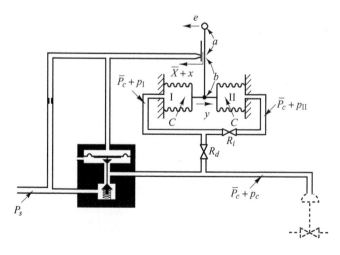

Figure 4–30
Schematic diagram
of a pneumatic
controller.

From the equations just derived, a block diagram of the controller can be drawn, as shown in Figure 4–31(a). Simplification of this block diagram results in Figure 4–31(b).

The transfer function between $P_c(s)$ and $E(s)$ is

$$\frac{P_c(s)}{E(s)} = \frac{\dfrac{b}{a+b}K}{1 + K\dfrac{a}{a+b}\dfrac{A}{k_s}\left(\dfrac{T_i s}{T_i s + 1}\right)\left(\dfrac{1}{T_d s + 1}\right)}$$

For a practical controller, under normal operation $|KaAT_i s/[(a+b)k_s(T_i s + 1)(T_d s + 1)]|$ is very much greater than unity and $T_i \gg T_d$. Therefore, the transfer function can be simplified as follows:

$$\frac{P_c(s)}{E(s)} \doteq \frac{bk_s(T_i s + 1)(T_d s + 1)}{aAT_i s}$$

$$= \frac{bk_s}{aA}\left(\frac{T_i + T_d}{T_i} + \frac{1}{T_i s} + T_d s\right)$$

$$\doteq K_p\left(1 + \frac{1}{T_i s} + T_d s\right)$$

where

$$K_p = \frac{bk_s}{aA}$$

Thus the controller shown in Figure 4–30 is a proportional-plus-integral-plus-derivative one.

If the resistance R_d is removed, or $R_d = 0$, the action becomes that of a proportional-plus-integral controller.

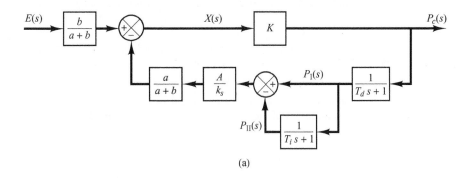

(a)

Figure 4–31
(a) Block diagram of the pneumatic controller shown in Figure 4–30;
(b) simplified block diagram.

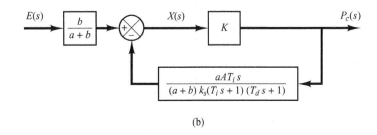

(b)

Example Problems and Solutions

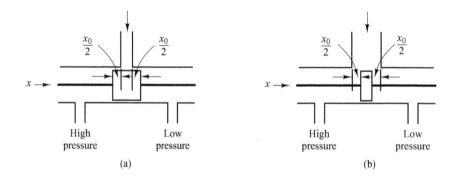

Figure 4–32
(a) Overlapped spool valve;
(b) underlapped spool valve.

High pressure   Low pressure   High pressure   Low pressure

(a)   (b)

If the resistance R_i is removed, or $R_i = 0$, the action becomes that of a narrow-band proportional, or two-position, controller. (Note that the actions of two feedback bellows cancel each other, and there is no feedback.)

A–4–6. Actual spool valves are either overlapped or underlapped because of manufacturing tolerances. Consider the overlapped and underlapped spool valves shown in Figures 4–32(a) and (b). Sketch curves relating the uncovered port area A versus displacement x.

Solution. For the overlapped valve, a dead zone exists between $-\frac{1}{2}x_0$ and $\frac{1}{2}x_0$, or $-\frac{1}{2}x_0 < x < \frac{1}{2}x_0$. The curve for uncovered port area A versus displacement x is shown in Figure 4–33(a). Such an overlapped valve is unfit as a control valve.

For the underlapped valve, the curve for port area A versus displacement x is shown in Figure 4–33(b). The effective curve for the underlapped region has a higher slope, meaning a higher sensitivity. Valves used for controls are usually underlapped.

A–4–7. Figure 4–34 shows a hydraulic jet-pipe controller. Hydraulic fluid is ejected from the jet pipe. If the jet pipe is shifted to the right from the neutral position, the power piston moves to the left, and vice versa. The jet-pipe valve is not used as much as the flapper valve because of large null flow, slower response, and rather unpredictable characteristics. Its main advantage lies in its insensitivity to dirty fluids.

Suppose that the power piston is connected to a light load so that the inertia force of the load element is negligible compared to the hydraulic force developed by the power piston. What type of control action does this controller produce?

Solution. Define the displacement of the jet nozzle from the neutral position as x and the displacement of the power piston as y. If the jet nozzle is moved to the right by a small displace-

Figure 4–33
(a) Uncovered-port-area-A-versus displacement-x curve for the overlapped valve; (b) uncovered-port-area-A-versus-displacement-x curve for the underlapped valve.

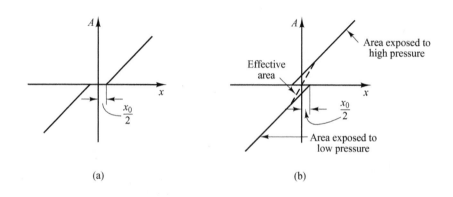

Effective area

Area exposed to high pressure

Area exposed to low pressure

(a)   (b)

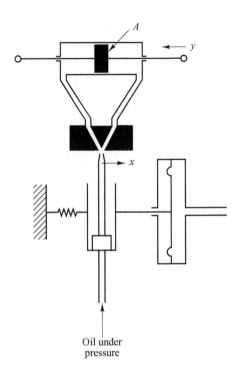

Figure 4–34
Hydraulic jet-pipe
controller.

Oil under
pressure

ment x, the oil flows to the right side of the power piston, and the oil in the left side of the power piston is returned to the drain. The oil flowing into the power cylinder is at high pressure; the oil flowing out from the power cylinder into the drain is at low pressure. The resulting pressure difference causes the power piston to move to the left.

For a small jet-nozzle displacement x, the flow rate q to the power cylinder is proportional to x; that is,

$$q = K_1 x$$

For the power cylinder,

$$A\rho\, dy = q\, dt$$

where A is the power-piston area and ρ is the density of oil. Hence

$$\frac{dy}{dt} = \frac{q}{A\rho} = \frac{K_1}{A\rho} x = Kx$$

where $K = K_1/(A\rho)$ = constant. The transfer function $Y(s)/X(s)$ is thus

$$\frac{Y(s)}{X(s)} = \frac{K}{s}$$

The controller produces the integral control action.

Example Problems and Solutions

147

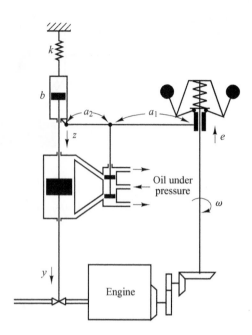

Figure 4–35
Speed control
system.

A–4–8. Explain the operation of the speed control system shown in Figure 4–35.

Solution. If the engine speed increases, the sleeve of the fly-ball governor moves upward. This movement acts as the input to the hydraulic controller. A positive error signal (upward motion of the sleeve) causes the power piston to move downward, reduces the fuel-valve opening, and decreases the engine speed. A block diagram for the system is shown in Figure 4–36.

From the block diagram the transfer function $Y(s)/E(s)$ can be obtained as

$$\frac{Y(s)}{E(s)} = \frac{a_2}{a_1 + a_2} \frac{\dfrac{K}{s}}{1 + \dfrac{a_1}{a_1 + a_2} \dfrac{bs}{bs + k} \dfrac{K}{s}}$$

If the following condition applies,

$$\left| \frac{a_1}{a_1 + a_2} \frac{bs}{bs + k} \frac{K}{s} \right| \gg 1$$

the transfer function $Y(s)/E(s)$ becomes

$$\frac{Y(s)}{E(s)} \doteq \frac{a_2}{a_1 + a_2} \frac{a_1 + a_2}{a_1} \frac{bs + k}{bs} = \frac{a_2}{a_1} \left(1 + \frac{k}{bs} \right)$$

The speed controller has proportional-plus-integral control action.

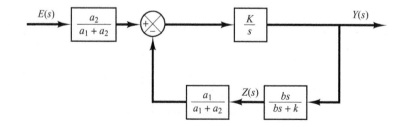

Figure 4–36
Block diagram for
the speed control
system shown in
Figure 4–35.

A-4-9. Derive the transfer function $Z(s)/Y(s)$ of the hydraulic system shown in Figure 4–37. Assume that the two dashpots in the system are identical ones except the piston shafts.

Solution. In deriving the equations for the system, we assume that force F is applied at the right end of the shaft causing displacement y. (All displacements y, w, and z are measured from respective equilibrium positions when no force is applied at the right end of the shaft.) When force F is applied, pressure P_1 becomes higher than pressure P_1', or $P_1 > P_1'$. Similarly, $P_2 > P_2'$.

For the force balance, we have the following equation:

$$k_2(y - w) = A(P_1 - P_1') + A(P_2 - P_2') \tag{4-42}$$

Since

$$k_1 z = A(P_1 - P_1') \tag{4-43}$$

and

$$q_1 = \frac{P_1 - P_1'}{R}$$

we have

$$k_1 z = ARq_1$$

Also, since

$$q_1 \, dt = A(dw - dz)\rho$$

we have

$$q_1 = A(\dot{w} - \dot{z})\rho$$

or

$$\dot{w} - \dot{z} = \frac{k_1 z}{A^2 R\rho}$$

Define $A^2 R\rho = B$. (B is the viscous-friction coefficient.) Then

$$\dot{w} - \dot{z} = \frac{k_1}{B} z \tag{4-44}$$

Also, for the right-hand-side dashpot we have

$$q_2 \, dt = A\rho \, dw$$

Since $q_2 = (P_2 - P_2')/R$, we obtain

$$\dot{w} = \frac{q_2}{A\rho} = \frac{A(P_2 - P_2')}{A^2 R\rho}$$

or

$$A(P_2 - P_2') = B\dot{w} \tag{4-45}$$

Substituting Equations (4–43) and (4–45) into Equation (4–42), we have

$$k_2 y - k_2 w = k_1 z + B\dot{w}$$

Taking the Laplace transform of this last equation, assuming zero initial condition, we obtain

$$k_2 Y(s) = (k_2 + Bs)W(s) + k_1 Z(s) \tag{4-46}$$

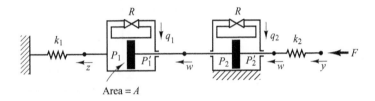

Figure 4–37
Hydraulic system.

Area = A

Taking the Laplace transform of Equation (4–44), assuming zero initial condition, we obtain

$$W(s) = \frac{k_1 + Bs}{Bs} Z(s) \tag{4–47}$$

By using Equation (4–47) to eliminate $W(s)$ from Equation (4–46), we obtain

$$k_2 Y(s) = (k_2 + Bs)\frac{k_1 + Bs}{Bs} Z(s) + k_1 Z(s)$$

from which we obtain the transfer function $Z(s)/Y(s)$ to be

$$\frac{Z(s)}{Y(s)} = \frac{k_2 s}{Bs^2 + (2k_1 + k_2)s + \dfrac{k_1 k_2}{B}}$$

Multiplying $B/(k_1 k_2)$ to both the numerator and denominator of this last equation, we get

$$\frac{Z(s)}{Y(s)} = \frac{\dfrac{B}{k_1} s}{\dfrac{B^2}{k_1 k_2} s^2 + \left(\dfrac{2B}{k_2} + \dfrac{B}{k_1}\right)s + 1}$$

Define $B/k_1 = T_1$, $B/k_2 = T_2$. Then the transfer function $Z(s)/Y(s)$ becomes as follows:

$$\frac{Z(s)}{Y(s)} = \frac{T_1 s}{T_1 T_2 s^2 + (T_1 + 2T_2)s + 1}$$

A–4–10. Considering small deviations from steady-state operation, draw a block diagram of the air heating system shown in Figure 4–38. Assume that the heat loss to the surroundings and the heat capacitance of the metal parts of the heater are negligible.

Solution. Let us define

$\bar{\Theta}_i$ = steady-state temperature of inlet air, °C
$\bar{\Theta}_o$ = steady-state temperature of outlet air, °C
G = mass flow rate of air through the heating chamber, kg/sec
M = mass of air contained in the heating chamber, kg
c = specific heat of air, kcal/kg °C
R = thermal resistance, °C sec/kcal
C = thermal capacitance of air contained in the heating chamber = Mc, kcal/°C
$\bar{H}$ = steady-state heat input, kcal/sec

Let us assume that the heat input is suddenly changed from $\bar{H}$ to $\bar{H} + h$ and the inlet air temperature is suddenly changed from $\bar{\Theta}_i$ to $\bar{\Theta}_i + \theta_i$. Then the outlet air temperature will be changed from $\bar{\Theta}_o$ to $\bar{\Theta}_o + \theta_o$.

The equation describing the system behavior is

$$C \, d\theta_o = \left[h + Gc(\theta_i - \theta_o)\right] dt$$

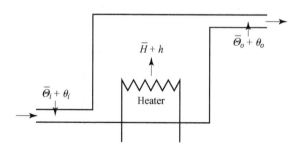

Figure 4–38
Air heating system.

Chapter 4 / Mathematical Modeling of Fluid Systems and Thermal Systems

Figure 4–39
Block diagram of the
air heating system
shown in
Figure 4–38.

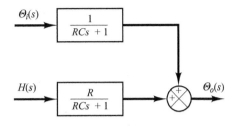

or

$$C\frac{d\theta_o}{dt} = h + Gc(\theta_i - \theta_o)$$

Noting that

$$Gc = \frac{1}{R}$$

we obtain

$$C\frac{d\theta_o}{dt} = h + \frac{1}{R}(\theta_i - \theta_o)$$

or

$$RC\frac{d\theta_o}{dt} + \theta_o = Rh + \theta_i$$

Taking the Laplace transforms of both sides of this last equation and substituting the initial condition that $\theta_o(0) = 0$, we obtain

$$\Theta_o(s) = \frac{R}{RCs + 1}H(s) + \frac{1}{RCs + 1}\Theta_i(s)$$

The block diagram of the system corresponding to this equation is shown in Figure 4–39.

A–4–11. Consider the thin, glass-wall, mercury thermometer system shown in Figure 4–40. Assume that the thermometer is at a uniform temperature $\bar{\Theta}$ (ambient temperature) and that at $t = 0$ it is immersed in a bath of temperature $\bar{\Theta} + \theta_b$, where θ_b is the bath temperature (which may be constant or changing) measured from the ambient temperature $\bar{\Theta}$. Define the instantaneous thermometer temperature by $\bar{\Theta} + \theta$, so that θ is the change in the thermometer temperature satisfying the condition that $\theta(0) = 0$. Obtain a mathematical model for the system. Also obtain an electrical analog of the thermometer system.

Solution. A mathematical model for the system can be derived by considering heat balance as follows: The heat entering the thermometer during dt sec is $q\,dt$, where q is the heat flow rate to the thermometer. This heat is stored in the thermal capacitance C of the thermometer, thereby raising its temperature by $d\theta$. Thus the heat-balance equation is

$$C\,d\theta = q\,dt \tag{4-48}$$

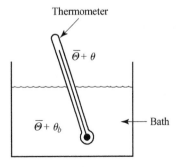

Figure 4–40
Thin, glass-wall,
mercury thermo-
meter system.

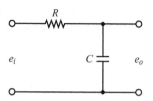

Figure 4–41
Electrical analog of
the thermometer
system shown in
Figure 4–40.

Since thermal resistance R may be written as

$$R = \frac{d(\Delta\theta)}{dq} = \frac{\Delta\theta}{q}$$

heat flow rate q may be given, in terms of thermal resistance R, as

$$q = \frac{(\bar{\Theta} + \theta_b) - (\bar{\Theta} + \theta)}{R} = \frac{\theta_b - \theta}{R}$$

where $\bar{\Theta} + \theta_b$ is the bath temperature and $\bar{\Theta} + \theta$ is the thermometer temperature. Hence, we can rewrite Equation (4–48) as

$$C\frac{d\theta}{dt} = \frac{\theta_b - \theta}{R}$$

or

$$RC\frac{d\theta}{dt} + \theta = \theta_b \qquad (4\text{–}49)$$

Equation (4–49) is a mathematical model of the thermometer system.

   Referring to Equation (4–49), an electrical analog for the thermometer system can be written as

$$RC\frac{de_o}{dt} + e_o = e_i$$

An electrical circuit represented by this last equation is shown in Figure 4–41.

PROBLEMS

B–4–1. Consider the conical water-tank system shown in Figure 4–42. The flow through the valve is turbulent and is related to the head H by

$$Q = 0.005\sqrt{H}$$

where Q is the flow rate measured in m³/sec and H is in meters.

   Suppose that the head is 2 m at $t = 0$. What will be the head at $t = 60$ sec?

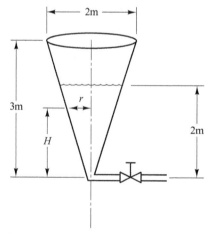

Figure 4–42 Conical water-tank system.

B–4–2. Consider the liquid-level control system shown in Figure 4–43. The controller is of the proportional type. The set point of the controller is fixed.

Draw a block diagram of the system, assuming that changes in the variables are small. Obtain the transfer function between the level of the second tank and the disturbance input q_d. Obtain the steady-state error when the disturbance q_d is a unit-step function.

B–4–3. For the pneumatic system shown in Figure 4–44, assume that steady-state values of the air pressure and the displacement of the bellows are $\bar{P}$ and $\bar{X}$, respectively. Assume also that the input pressure is changed from $\bar{P}$ to $\bar{P} + p_i$, where p_i is a small change in the input pressure. This change will cause the displacement of the bellows to change a small amount x. Assuming that the capacitance of the bellows is C and the resistance of the valve is R, obtain the transfer function relating x and p_i.

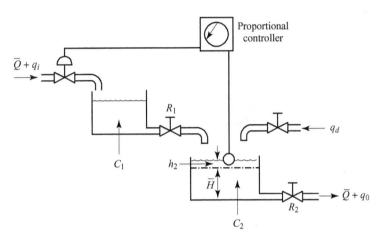

Figure 4–43
Liquid-level control system.

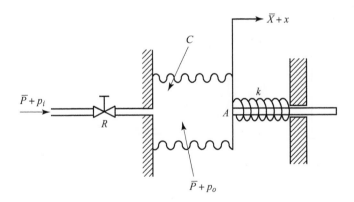

Figure 4–44
Pneumatic system.

B–4–4. Figure 4–45 shows a pneumatic controller. The pneumatic relay has the characteristic that $p_c = Kp_b$, where $K > 0$. What kind of control action does this controller produce? Derive the transfer function $P_c(s)/E(s)$.

B–4–5. Consider the pneumatic controller shown in Figure 4–46. Assuming that the pneumatic relay has the characteristics that $p_c = Kp_b$ (where $K > 0$), determine the control action of this controller. The input to the controller is e and the output is p_c.

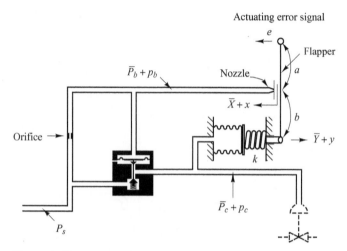

Figure 4–45
Pneumatic controller.

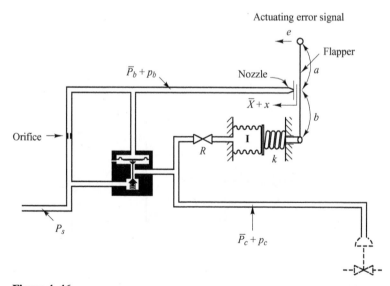

Figure 4–46
Pneumatic controller.

B–4–6. Figure 4–47 shows a pneumatic controller. The signal e is the input and the change in the control pressure p_c is the output. Obtain the transfer function $P_c(s)/E(s)$. Assume that the pneumatic relay has the characteristics that $p_c = Kp_b$, where $K > 0$.

B–4–7. Consider the pneumatic controller shown in Figure 4–48. What control action does this controller produce? Assume that the pneumatic relay has the characteristics that $p_c = Kp_b$, where $K > 0$.

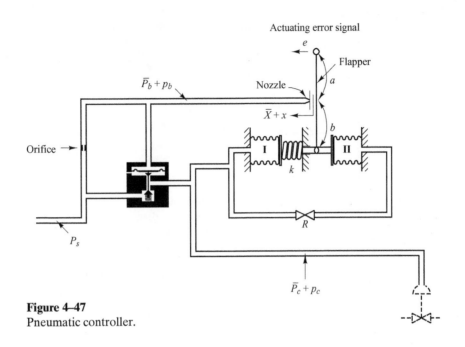

Figure 4–47
Pneumatic controller.

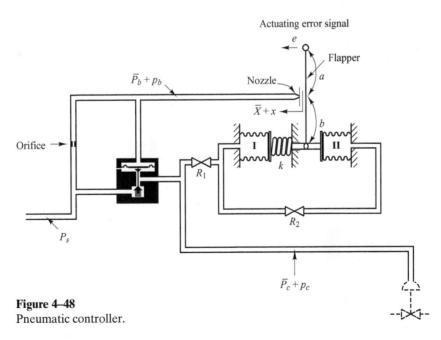

Figure 4–48
Pneumatic controller.

B–4–8. Figure 4–49 shows a flapper valve. It is placed between two opposing nozzles. If the flapper is moved slightly to the right, the pressure unbalance occurs in the nozzles and the power piston moves to the left, and vice versa. Such a device is frequently used in hydraulic servos as the first-stage valve in two-stage servovalves. This usage occurs because considerable force may be needed to stroke larger spool valves that result from the steady-state flow force. To reduce or compensate this force, two-stage valve configuration is often employed; a flapper valve or jet pipe is used as the first-stage valve to provide a necessary force to stroke the second-stage spool valve.

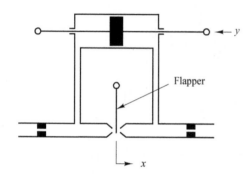

Figure 4–49 Flapper valve.

Figure 4–50 shows a schematic diagram of a hydraulic servomotor in which the error signal is amplified in two stages using a jet pipe and a pilot valve. Draw a block diagram of the system of Figure 4–50 and then find the transfer function between y and x, where x is the air pressure and y is the displacement of the power piston.

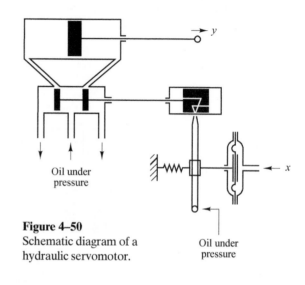

Figure 4–50
Schematic diagram of a
hydraulic servomotor.

B–4–9. Figure 4–51 is a schematic diagram of an aircraft elevator control system. The input to the system is the deflection angle θ of the control lever, and the output is the elevator angle ϕ. Assume that angles θ and ϕ are relatively small. Show that for each angle θ of the control lever there is a corresponding (steady-state) elevator angle ϕ.

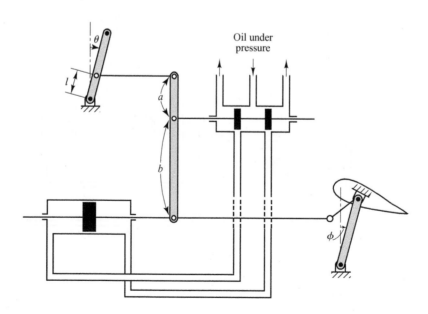

Figure 4–51
Aircraft elevator
control system.

B–4–10. Consider the liquid-level control system shown in Figure 4–52. The inlet valve is controlled by a hydraulic integral controller. Assume that the steady-state inflow rate is $\bar{Q}$ and steady-state outflow rate is also $\bar{Q}$, the steady-state head is $\bar{H}$, steady-state pilot valve displacement is $\bar{X} = 0$, and steady-state valve position is $\bar{Y}$. We assume that the set point $\bar{R}$ corresponds to the steady-state head $\bar{H}$. The set point is fixed. Assume also that the disturbance inflow rate q_d, which is a small quantity, is applied to the water tank at $t = 0$. This disturbance causes the head to change from $\bar{H}$ to $\bar{H} + h$. This change results in a change in the outflow rate by q_o. Through the hydraulic controller, the change in head causes a change in the inflow rate from $\bar{Q}$ to $\bar{Q} + q_i$. (The integral controller tends to keep the head constant as much as possible in the presence of disturbances.) We assume that all changes are of small quantities.

We assume that the velocity of the power piston (valve) is proportional to pilot-valve displacement x, or

$$\frac{dy}{dt} = K_1 x$$

where K_1 is a positive constant. We also assume that the change in the inflow rate q_i is negatively proportional to the change in the valve opening y, or

$$q_i = -K_v y$$

where K_v is a positive constant.

Assuming the following numerical values for the system,

$$C = 2 \text{ m}^2, \qquad R = 0.5 \text{ sec/m}^2, \qquad K_v = 1 \text{ m}^2/\text{sec}$$

$$a = 0.25 \text{ m}, \qquad b = 0.75 \text{ m}, \qquad K_1 = 4 \text{ sec}^{-1}$$

obtain the transfer function $H(s)/Q_d(s)$.

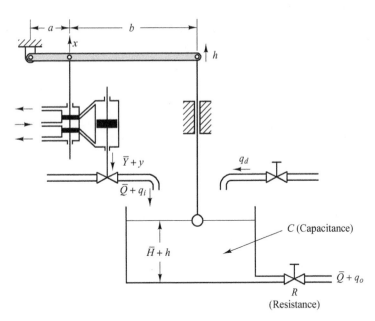

Figure 4–52
Liquid-level control system.

B–4–11. Consider the controller shown in Figure 4–53. The input is the air pressure p_i measured from some steady-state reference pressure $\bar{P}$ and the output is the displacement y of the power piston. Obtain the transfer function $Y(s)/P_i(s)$.

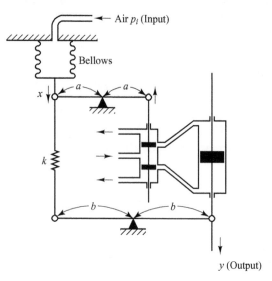

Figure 4–53
Controller.

B–4–12. A thermocouple has a time constant of 2 sec. A thermal well has a time constant of 30 sec. When the thermocouple is inserted into the well, this temperature-measuring device can be considered a two-capacitance system.

Determine the time constants of the combined thermocouple–thermal-well system. Assume that the weight of the thermocouple is 8 g and the weight of the thermal well is 40 g. Assume also that the specific heats of the thermocouple and thermal well are the same.

Transient and Steady-State Response Analyses

5-1 INTRODUCTION

In early chapters it was stated that the first step in analyzing a control system was to derive a mathematical model of the system. Once such a model is obtained, various methods are available for the analysis of system performance.

In practice, the input signal to a control system is not known ahead of time but is random in nature, and the instantaneous input cannot be expressed analytically. Only in some special cases is the input signal known in advance and expressible analytically or by curves, such as in the case of the automatic control of cutting tools.

In analyzing and designing control systems, we must have a basis of comparison of performance of various control systems. This basis may be set up by specifying particular test input signals and by comparing the responses of various systems to these input signals.

Many design criteria are based on the response to such test signals or on the response of systems to changes in initial conditions (without any test signals). The use of test signals can be justified because of a correlation existing between the response characteristics of a system to a typical test input signal and the capability of the system to cope with actual input signals.

Typical Test Signals. The commonly used test input signals are step functions, ramp functions, acceleration functions, impulse functions, sinusoidal functions, and white noise. In this chapter we use test signals such as step, ramp, acceleration and impulse signals. With these test signals, mathematical and experimental analyses of control systems can be carried out easily, since the signals are very simple functions of time.

Which of these typical input signals to use for analyzing system characteristics may be determined by the form of the input that the system will be subjected to most frequently under normal operation. If the inputs to a control system are gradually changing functions of time, then a ramp function of time may be a good test signal. Similarly, if a system is subjected to sudden disturbances, a step function of time may be a good test signal; and for a system subjected to shock inputs, an impulse function may be best. Once a control system is designed on the basis of test signals, the performance of the system in response to actual inputs is generally satisfactory. The use of such test signals enables one to compare the performance of many systems on the same basis.

Transient Response and Steady-State Response.   The time response of a control system consists of two parts: the transient response and the steady-state response. By transient response, we mean that which goes from the initial state to the final state. By steady-state response, we mean the manner in which the system output behaves as t approaches infinity. Thus the system response $c(t)$ may be written as

$$c(t) = c_{tr}(t) + c_{ss}(t)$$

where the first term on the right-hand side of the equation is the transient response and the second term is the steady-state response.

Absolute Stability, Relative Stability, and Steady-State Error.   In designing a control system, we must be able to predict the dynamic behavior of the system from a knowledge of the components. The most important characteristic of the dynamic behavior of a control system is absolute stability—that is, whether the system is stable or unstable. A control system is in equilibrium if, in the absence of any disturbance or input, the output stays in the same state. A linear time-invariant control system is stable if the output eventually comes back to its equilibrium state when the system is subjected to an initial condition. A linear time-invariant control system is critically stable if oscillations of the output continue forever. It is unstable if the output diverges without bound from its equilibrium state when the system is subjected to an initial condition. Actually, the output of a physical system may increase to a certain extent but may be limited by mechanical "stops," or the system may break down or become nonlinear after the output exceeds a certain magnitude so that the linear differential equations no longer apply.

Important system behavior (other than absolute stability) to which we must give careful consideration includes relative stability and steady-state error. Since a physical control system involves energy storage, the output of the system, when subjected to an input, cannot follow the input immediately but exhibits a transient response before a steady state can be reached. The transient response of a practical control system often exhibits damped oscillations before reaching a steady state. If the output of a system at steady state does not exactly agree with the input, the system is said to have steady-state error. This error is indicative of the accuracy of the system. In analyzing a control system, we must examine transient-response behavior and steady-state behavior.

Outline of the Chapter.   This chapter is concerned with system responses to aperiodic signals (such as step, ramp, acceleration, and impulse functions of time). The outline of the chapter is as follows: Section 5–1 has presented introductory material for the chapter. Section 5–2 treats the response of first-order systems to aperiodic inputs. Section 5–3 deals with the transient response of the second-order systems. Detailed analyses of the step response, ramp response, and impulse response of the second-order systems are presented. Section 5–4 discusses the transient-response analysis of higher-order systems. Section 5–5 gives an introduction to the MATLAB approach to the solution of transient-response problems. Section 5–6 gives an example of a transient-response problem solved with MATLAB. Section 5–7 presents Routh's stability criterion. Section 5–8 discusses effects of integral and derivative control actions on system performance. Finally, Section 5–9 treats steady-state errors in unity-feedback control systems.

5–2 FIRST-ORDER SYSTEMS

Consider the first-order system shown in Figure 5–1(a). Physically, this system may represent an RC circuit, thermal system, or the like. A simplified block diagram is shown in Figure 5–1(b). The input-output relationship is given by

$$\frac{C(s)}{R(s)} = \frac{1}{Ts + 1} \tag{5-1}$$

In the following, we shall analyze the system responses to such inputs as the unit-step, unit-ramp, and unit-impulse functions. The initial conditions are assumed to be zero.

Note that all systems having the same transfer function will exhibit the same output in response to the same input. For any given physical system, the mathematical response can be given a physical interpretation.

Unit-Step Response of First-Order Systems.   Since the Laplace transform of the unit-step function is $1/s$, substituting $R(s) = 1/s$ into Equation (5–1), we obtain

$$C(s) = \frac{1}{Ts + 1}\frac{1}{s}$$

Expanding $C(s)$ into partial fractions gives

$$C(s) = \frac{1}{s} - \frac{T}{Ts + 1} = \frac{1}{s} - \frac{1}{s + (1/T)} \tag{5-2}$$

Taking the inverse Laplace transform of Equation (5–2), we obtain

$$c(t) = 1 - e^{-t/T}, \qquad \text{for } t \geq 0 \tag{5-3}$$

Equation (5–3) states that initially the output $c(t)$ is zero and finally it becomes unity. One important characteristic of such an exponential response curve $c(t)$ is that at $t = T$ the value of $c(t)$ is 0.632, or the response $c(t)$ has reached 63.2% of its total change. This may be easily seen by substituting $t = T$ in $c(t)$. That is,

$$c(T) = 1 - e^{-1} = 0.632$$

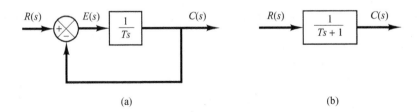

Figure 5–1
(a) Block diagram of a first-order system; (b) simplified block diagram.

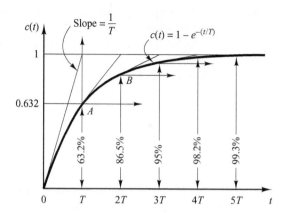

Figure 5–2
Exponential
response curve.

Note that the smaller the time constant T, the faster the system response. Another important characteristic of the exponential response curve is that the slope of the tangent line at $t = 0$ is $1/T$, since

$$\left.\frac{dc}{dt}\right|_{t=0} = \left.\frac{1}{T} e^{-t/T}\right|_{t=0} = \frac{1}{T} \tag{5–4}$$

The output would reach the final value at $t = T$ if it maintained its initial speed of response. From Equation (5–4) we see that the slope of the response curve $c(t)$ decreases monotonically from $1/T$ at $t = 0$ to zero at $t = \infty$.

The exponential response curve $c(t)$ given by Equation (5–3) is shown in Figure 5–2. In one time constant, the exponential response curve has gone from 0 to 63.2% of the final value. In two time constants, the response reaches 86.5% of the final value. At $t = 3T, 4T$, and $5T$, the response reaches 95%, 98.2%, and 99.3%, respectively, of the final value. Thus, for $t \geq 4T$, the response remains within 2% of the final value. As seen from Equation (5–3), the steady state is reached mathematically only after an infinite time. In practice, however, a reasonable estimate of the response time is the length of time the response curve needs to reach and stay within the 2% line of the final value, or four time constants.

Unit-Ramp Response of First-Order Systems. Since the Laplace transform of the unit-ramp function is $1/s^2$, we obtain the output of the system of Figure 5–1(a) as

$$C(s) = \frac{1}{Ts + 1} \frac{1}{s^2}$$

Expanding $C(s)$ into partial fractions gives

$$C(s) = \frac{1}{s^2} - \frac{T}{s} + \frac{T^2}{Ts + 1} \tag{5–5}$$

Taking the inverse Laplace transform of Equation (5–5), we obtain

$$c(t) = t - T + Te^{-t/T}, \qquad \text{for } t \geq 0 \tag{5–6}$$

The error signal $e(t)$ is then

$$e(t) = r(t) - c(t)$$
$$= T\left(1 - e^{-t/T}\right)$$

Chapter 5 / Transient and Steady-State Response Analyses

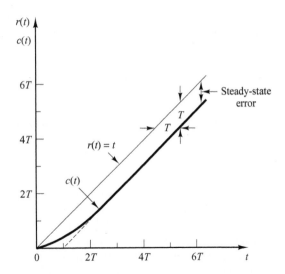

Figure 5–3
Unit-ramp response
of the system shown
in Figure 5–1(a).

As t approaches infinity, $e^{-t/T}$ approaches zero, and thus the error signal $e(t)$ approaches T or

$$e(\infty) = T$$

The unit-ramp input and the system output are shown in Figure 5–3. The error in following the unit-ramp input is equal to T for sufficiently large t. The smaller the time constant T, the smaller the steady-state error in following the ramp input.

Unit-Impulse Response of First-Order Systems. For the unit-impulse input, $R(s) = 1$ and the output of the system of Figure 5–1(a) can be obtained as

$$C(s) = \frac{1}{Ts + 1} \tag{5–7}$$

The inverse Laplace transform of Equation (5–7) gives

$$c(t) = \frac{1}{T} e^{-t/T}, \qquad \text{for } t \geq 0 \tag{5–8}$$

The response curve given by Equation (5–8) is shown in Figure 5–4.

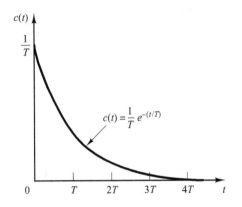

Figure 5–4
Unit-impulse
response of the
system shown in
Figure 5–1(a).

An Important Property of Linear Time-Invariant Systems. In the analysis above, it has been shown that for the unit-ramp input the output $c(t)$ is

$$c(t) = t - T + Te^{-t/T}, \quad \text{for } t \geq 0 \quad \text{[See Equation (5-6).]}$$

For the unit-step input, which is the derivative of unit-ramp input, the output $c(t)$ is

$$c(t) = 1 - e^{-t/T}, \quad \text{for } t \geq 0 \quad \text{[See Equation (5-3).]}$$

Finally, for the unit-impulse input, which is the derivative of unit-step input, the output $c(t)$ is

$$c(t) = \frac{1}{T} e^{-t/T}, \quad \text{for } t \geq 0 \quad \text{[See Equation (5-8).]}$$

Comparing the system responses to these three inputs clearly indicates that the response to the derivative of an input signal can be obtained by differentiating the response of the system to the original signal. It can also be seen that the response to the integral of the original signal can be obtained by integrating the response of the system to the original signal and by determining the integration constant from the zero-output initial condition. This is a property of linear time-invariant systems. Linear time-varying systems and nonlinear systems do not possess this property.

5-3 SECOND-ORDER SYSTEMS

In this section, we shall obtain the response of a typical second-order control system to a step input, ramp input, and impulse input. Here we consider a servo system as an example of a second-order system.

Servo System. The servo system shown in Figure 5-5(a) consists of a proportional controller and load elements (inertia and viscous-friction elements). Suppose that we wish to control the output position c in accordance with the input position r.

The equation for the load elements is

$$J\ddot{c} + B\dot{c} = T$$

where T is the torque produced by the proportional controller whose gain is K. By taking Laplace transforms of both sides of this last equation, assuming the zero initial conditions, we obtain

$$Js^2 C(s) + BsC(s) = T(s)$$

So the transfer function between $C(s)$ and $T(s)$ is

$$\frac{C(s)}{T(s)} = \frac{1}{s(Js + B)}$$

By using this transfer function, Figure 5-5(a) can be redrawn as in Figure 5-5(b), which can be modified to that shown in Figure 5-5(c). The closed-loop transfer function is then obtained as

$$\frac{C(s)}{R(s)} = \frac{K}{Js^2 + Bs + K} = \frac{K/J}{s^2 + (B/J)s + (K/J)}$$

Such a system where the closed-loop transfer function possesses two poles is called a second-order system. (Some second-order systems may involve one or two zeros.)

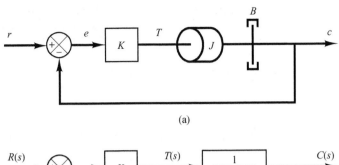

(a)

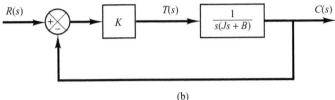

(b)

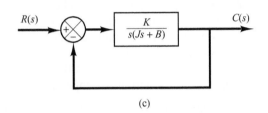

(c)

Figure 5–5
(a) Servo system;
(b) block diagram;
(c) simplified block
diagram.

Step Response of Second-Order System. The closed-loop transfer function of the system shown in Figure 5–5(c) is

$$\frac{C(s)}{R(s)} = \frac{K}{Js^2 + Bs + K} \tag{5–9}$$

which can be rewritten as

$$\frac{C(s)}{R(s)} = \frac{\dfrac{K}{J}}{\left[s + \dfrac{B}{2J} + \sqrt{\left(\dfrac{B}{2J}\right)^2 - \dfrac{K}{J}} \right]\left[s + \dfrac{B}{2J} - \sqrt{\left(\dfrac{B}{2J}\right)^2 - \dfrac{K}{J}} \right]}$$

The closed-loop poles are complex conjugates if $B^2 - 4JK < 0$ and they are real if $B^2 - 4JK \geq 0$. In the transient-response analysis, it is convenient to write

$$\frac{K}{J} = \omega_n^2, \qquad \frac{B}{J} = 2\zeta\omega_n = 2\sigma$$

where σ is called the *attenuation*; ω_n, the *undamped natural frequency*; and ζ, the *damping ratio* of the system. The damping ratio ζ is the ratio of the actual damping B to the critical damping $B_c = 2\sqrt{JK}$ or

$$\zeta = \frac{B}{B_c} = \frac{B}{2\sqrt{JK}}$$

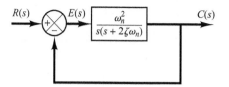

Figure 5–6
Second-order system.

In terms of ζ and ω_n, the system shown in Figure 5–5(c) can be modified to that shown in Figure 5–6, and the closed-loop transfer function $C(s)/R(s)$ given by Equation (5–9) can be written

$$\frac{C(s)}{R(s)} = \frac{\omega_n^2}{s^2 + 2\zeta\omega_n s + \omega_n^2} \tag{5–10}$$

This form is called the *standard form* of the second-order system.

The dynamic behavior of the second-order system can then be described in terms of two parameters ζ and ω_n. If $0 < \zeta < 1$, the closed-loop poles are complex conjugates and lie in the left-half s plane. The system is then called underdamped, and the transient response is oscillatory. If $\zeta = 0$, the transient response does not die out. If $\zeta = 1$, the system is called critically damped. Overdamped systems correspond to $\zeta > 1$.

We shall now solve for the response of the system shown in Figure 5–6 to a unit-step input. We shall consider three different cases: the underdamped ($0 < \zeta < 1$), critically damped ($\zeta = 1$), and overdamped ($\zeta > 1$) cases.

(1) *Underdamped case* $(0 < \zeta < 1)$: In this case, $C(s)/R(s)$ can be written

$$\frac{C(s)}{R(s)} = \frac{\omega_n^2}{(s + \zeta\omega_n + j\omega_d)(s + \zeta\omega_n - j\omega_d)}$$

where $\omega_d = \omega_n\sqrt{1 - \zeta^2}$. The frequency ω_d is called the *damped natural frequency*. For a unit-step input, $C(s)$ can be written

$$C(s) = \frac{\omega_n^2}{(s^2 + 2\zeta\omega_n s + \omega_n^2)s} \tag{5–11}$$

The inverse Laplace transform of Equation (5–11) can be obtained easily if $C(s)$ is written in the following form:

$$C(s) = \frac{1}{s} - \frac{s + 2\zeta\omega_n}{s^2 + 2\zeta\omega_n s + \omega_n^2}$$

$$= \frac{1}{s} - \frac{s + \zeta\omega_n}{(s + \zeta\omega_n)^2 + \omega_d^2} - \frac{\zeta\omega_n}{(s + \zeta\omega_n)^2 + \omega_d^2}$$

Referring to the Laplace transform table in Appendix A, it can be shown that

$$\mathcal{L}^{-1}\left[\frac{s + \zeta\omega_n}{(s + \zeta\omega_n)^2 + \omega_d^2}\right] = e^{-\zeta\omega_n t}\cos\omega_d t$$

$$\mathcal{L}^{-1}\left[\frac{\omega_d}{(s + \zeta\omega_n)^2 + \omega_d^2}\right] = e^{-\zeta\omega_n t}\sin\omega_d t$$

Hence the inverse Laplace transform of Equation (5–11) is obtained as

$$\mathcal{L}^{-1}[C(s)] = c(t)$$

$$= 1 - e^{-\zeta\omega_n t}\left(\cos\omega_d t + \frac{\zeta}{\sqrt{1-\zeta^2}}\sin\omega_d t\right)$$

$$= 1 - \frac{e^{-\zeta\omega_n t}}{\sqrt{1-\zeta^2}}\sin\left(\omega_d t + \tan^{-1}\frac{\sqrt{1-\zeta^2}}{\zeta}\right), \qquad \text{for } t \geq 0 \quad (5\text{--}12)$$

From Equation (5–12), it can be seen that the frequency of transient oscillation is the damped natural frequency ω_d and thus varies with the damping ratio ζ. The error signal for this system is the difference between the input and output and is

$$e(t) = r(t) - c(t)$$

$$= e^{-\zeta\omega_n t}\left(\cos\omega_d t + \frac{\zeta}{\sqrt{1-\zeta^2}}\sin\omega_d t\right), \qquad \text{for } t \geq 0$$

This error signal exhibits a damped sinusoidal oscillation. At steady state, or at $t = \infty$, no error exists between the input and output.

If the damping ratio ζ is equal to zero, the response becomes undamped and oscillations continue indefinitely. The response $c(t)$ for the zero damping case may be obtained by substituting $\zeta = 0$ in Equation (5–12), yielding

$$c(t) = 1 - \cos\omega_n t, \qquad \text{for } t \geq 0 \qquad (5\text{--}13)$$

Thus, from Equation (5–13), we see that ω_n represents the undamped natural frequency of the system. That is, ω_n is that frequency at which the system output would oscillate if the damping were decreased to zero. If the linear system has any amount of damping, the undamped natural frequency cannot be observed experimentally. The frequency that may be observed is the damped natural frequency ω_d, which is equal to $\omega_n\sqrt{1-\zeta^2}$. This frequency is always lower than the undamped natural frequency. An increase in ζ would reduce the damped natural frequency ω_d. If ζ is increased beyond unity, the response becomes overdamped and will not oscillate.

(2) *Critically damped case* ($\zeta = 1$): If the two poles of $C(s)/R(s)$ are equal, the system is said to be a critically damped one.

For a unit-step input, $R(s) = 1/s$ and $C(s)$ can be written

$$C(s) = \frac{\omega_n^2}{(s + \omega_n)^2 s} \qquad (5\text{--}14)$$

The inverse Laplace transform of Equation (5–14) may be found as

$$c(t) = 1 - e^{-\omega_n t}(1 + \omega_n t), \qquad \text{for } t \geq 0 \qquad (5\text{--}15)$$

This result can also be obtained by letting ζ approach unity in Equation (5–12) and by using the following limit:

$$\lim_{\zeta \to 1}\frac{\sin\omega_d t}{\sqrt{1-\zeta^2}} = \lim_{\zeta \to 1}\frac{\sin\omega_n\sqrt{1-\zeta^2}\,t}{\sqrt{1-\zeta^2}} = \omega_n t$$

(3) Overdamped case ($\zeta > 1$): In this case, the two poles of $C(s)/R(s)$ are negative real and unequal. For a unit-step input, $R(s) = 1/s$ and $C(s)$ can be written

$$C(s) = \frac{\omega_n^2}{\left(s + \zeta\omega_n + \omega_n\sqrt{\zeta^2 - 1}\right)\left(s + \zeta\omega_n - \omega_n\sqrt{\zeta^2 - 1}\right)s} \qquad (5\text{–}16)$$

The inverse Laplace transform of Equation (5–16) is

$$c(t) = 1 + \frac{1}{2\sqrt{\zeta^2 - 1}\left(\zeta + \sqrt{\zeta^2 - 1}\right)} e^{-\left(\zeta + \sqrt{\zeta^2 - 1}\right)\omega_n t}$$

$$- \frac{1}{2\sqrt{\zeta^2 - 1}\left(\zeta - \sqrt{\zeta^2 - 1}\right)} e^{-\left(\zeta - \sqrt{\zeta^2 - 1}\right)\omega_n t}$$

$$= 1 + \frac{\omega_n}{2\sqrt{\zeta^2 - 1}}\left(\frac{e^{-s_1 t}}{s_1} - \frac{e^{-s_2 t}}{s_2}\right), \qquad \text{for } t \geq 0 \qquad (5\text{–}17)$$

where $s_1 = \left(\zeta + \sqrt{\zeta^2 - 1}\right)\omega_n$ and $s_2 = \left(\zeta - \sqrt{\zeta^2 - 1}\right)\omega_n$. Thus, the response $c(t)$ includes two decaying exponential terms.

When ζ is appreciably greater than unity, one of the two decaying exponentials decreases much faster than the other, so the faster-decaying exponential term (which corresponds to a smaller time constant) may be neglected. That is, if $-s_2$ is located very much closer to the $j\omega$ axis than $-s_1$ (which means $|s_2| \ll |s_1|$), then for an approximate solution we may neglect $-s_1$. This is permissible because the effect of $-s_1$ on the response is much smaller than that of $-s_2$, since the term involving s_1 in Equation (5–17) decays much faster than the term involving s_2. Once the faster-decaying exponential term has disappeared, the response is similar to that of a first-order system, and $C(s)/R(s)$ may be approximated by

$$\frac{C(s)}{R(s)} = \frac{\zeta\omega_n - \omega_n\sqrt{\zeta^2 - 1}}{s + \zeta\omega_n - \omega_n\sqrt{\zeta^2 - 1}} = \frac{s_2}{s + s_2}$$

This approximate form is a direct consequence of the fact that the initial values and final values of both the original $C(s)/R(s)$ and the approximate one agree with each other.

With the approximate transfer function $C(s)/R(s)$, the unit-step response can be obtained as

$$C(s) = \frac{\zeta\omega_n - \omega_n\sqrt{\zeta^2 - 1}}{\left(s + \zeta\omega_n - \omega_n\sqrt{\zeta^2 - 1}\right)s}$$

The time response $c(t)$ is then

$$c(t) = 1 - e^{-\left(\zeta - \sqrt{\zeta^2 - 1}\right)\omega_n t}, \qquad \text{for } t \geq 0$$

This gives an approximate unit-step response when one of the poles of $C(s)/R(s)$ can be neglected.

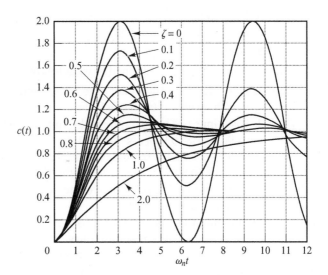

Figure 5–7
Unit-step response curves of the system shown in Figure 5–6.

A family of unit-step response curves $c(t)$ with various values of ζ is shown in Figure 5–7, where the abscissa is the dimensionless variable $\omega_n t$. The curves are functions only of ζ. These curves are obtained from Equations (5–12), (5–15), and (5–17). The system described by these equations was initially at rest.

Note that two second-order systems having the same ζ but different ω_n will exhibit the same overshoot and the same oscillatory pattern. Such systems are said to have the same relative stability.

From Figure 5–7, we see that an underdamped system with ζ between 0.5 and 0.8 gets close to the final value more rapidly than a critically damped or overdamped system. Among the systems responding without oscillation, a critically damped system exhibits the fastest response. An overdamped system is always sluggish in responding to any inputs.

It is important to note that, for second-order systems whose closed-loop transfer functions are different from that given by Equation (5–10), the step-response curves may look quite different from those shown in Figure 5–7.

Definitions of Transient-Response Specifications.   Frequently, the performance characteristics of a control system are specified in terms of the transient response to a unit-step input, since it is easy to generate and is sufficiently drastic. (If the response to a step input is known, it is mathematically possible to compute the response to any input.)

The transient response of a system to a unit-step input depends on the initial conditions. For convenience in comparing transient responses of various systems, it is a common practice to use the standard initial condition that the system is at rest initially with the output and all time derivatives thereof zero. Then the response characteristics of many systems can be easily compared.

The transient response of a practical control system often exhibits damped oscillations before reaching steady state. In specifying the transient-response characteristics of a control system to a unit-step input, it is common to specify the following:

1. Delay time, t_d
2. Rise time, t_r

3. Peak time, t_p

4. Maximum overshoot, M_p

5. Settling time, t_s

These specifications are defined in what follows and are shown graphically in Figure 5–8.

1. Delay time, t_d: The delay time is the time required for the response to reach half the final value the very first time.

2. Rise time, t_r: The rise time is the time required for the response to rise from 10% to 90%, 5% to 95%, or 0% to 100% of its final value. For underdamped second-order systems, the 0% to 100% rise time is normally used. For overdamped systems, the 10% to 90% rise time is commonly used.

3. Peak time, t_p: The peak time is the time required for the response to reach the first peak of the overshoot.

4. Maximum (percent) overshoot, M_p: The maximum overshoot is the maximum peak value of the response curve measured from unity. If the final steady-state value of the response differs from unity, then it is common to use the maximum percent overshoot. It is defined by

$$\text{Maximum percent overshoot} = \frac{c(t_p) - c(\infty)}{c(\infty)} \times 100\%$$

The amount of the maximum (percent) overshoot directly indicates the relative stability of the system.

5. Settling time, t_s: The settling time is the time required for the response curve to reach and stay within a range about the final value of size specified by absolute percentage of the final value (usually 2% or 5%). The settling time is related to the largest time constant of the control system. Which percentage error criterion to use may be determined from the objectives of the system design in question.

The time-domain specifications just given are quite important, since most control systems are time-domain systems; that is, they must exhibit acceptable time responses. (This means that, the control system must be modified until the transient response is satisfactory.)

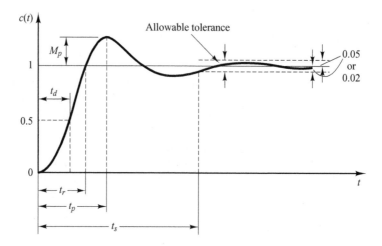

Figure 5–8
Unit-step response
curve showing t_d, t_r,
t_p, M_p, and t_s.

Note that not all these specifications necessarily apply to any given case. For example, for an overdamped system, the terms peak time and maximum overshoot do not apply. (For systems that yield steady-state errors for step inputs, this error must be kept within a specified percentage level. Detailed discussions of steady-state errors are postponed until Section 5–8.)

A Few Comments on Transient-Response Specifications. Except for certain applications where oscillations cannot be tolerated, it is desirable that the transient response be sufficiently fast and be sufficiently damped. Thus, for a desirable transient response of a second-order system, the damping ratio must be between 0.4 and 0.8. Small values of ζ (that is, $\zeta < 0.4$) yield excessive overshoot in the transient response, and a system with a large value of ζ (that is, $\zeta > 0.8$) responds sluggishly.

We shall see later that the maximum overshoot and the rise time conflict with each other. In other words, both the maximum overshoot and the rise time cannot be made smaller simultaneously. If one of them is made smaller, the other necessarily becomes larger.

Second-Order Systems and Transient-Response Specifications. In the following, we shall obtain the rise time, peak time, maximum overshoot, and settling time of the second-order system given by Equation (5–10). These values will be obtained in terms of ζ and ω_n. The system is assumed to be underdamped.

Rise time t_r: Referring to Equation (5–12), we obtain the rise time t_r by letting $c(t_r) = 1$.

$$c(t_r) = 1 = 1 - e^{-\zeta \omega_n t_r}\left(\cos \omega_d t_r + \frac{\zeta}{\sqrt{1 - \zeta^2}} \sin \omega_d t_r \right) \tag{5–18}$$

Since $e^{-\zeta \omega_n t_r} \neq 0$, we obtain from Equation (5–18) the following equation:

$$\cos \omega_d t_r + \frac{\zeta}{\sqrt{1 - \zeta^2}} \sin \omega_d t_r = 0$$

Since $\omega_n \sqrt{1 - \zeta^2} = \omega_d$ and $\zeta \omega_n = \sigma$, we have

$$\tan \omega_d t_r = -\frac{\sqrt{1 - \zeta^2}}{\zeta} = -\frac{\omega_d}{\sigma}$$

Thus, the rise time t_r is

$$t_r = \frac{1}{\omega_d} \tan^{-1}\left(\frac{\omega_d}{-\sigma}\right) = \frac{\pi - \beta}{\omega_d} \tag{5–19}$$

where angle β is defined in Figure 5–9. Clearly, for a small value of t_r, ω_d must be large.

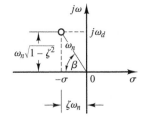

Figure 5–9
Definition of the angle β.

Peak time t_p: Referring to Equation (5–12), we may obtain the peak time by differentiating $c(t)$ with respect to time and letting this derivative equal zero. Since

$$\frac{dc}{dt} = \zeta \omega_n e^{-\zeta \omega_n t} \left(\cos \omega_d t + \frac{\zeta}{\sqrt{1 - \zeta^2}} \sin \omega_d t \right)$$

$$+ e^{-\zeta \omega_n t} \left(\omega_d \sin \omega_d t - \frac{\zeta \omega_d}{\sqrt{1 - \zeta^2}} \cos \omega_d t \right)$$

and the cosine terms in this last equation cancel each other, dc/dt, evaluated at $t = t_p$, can be simplified to

$$\frac{dc}{dt}\bigg|_{t=t_p} = (\sin \omega_d t_p) \frac{\omega_n}{\sqrt{1 - \zeta^2}} e^{-\zeta \omega_n t_p} = 0$$

This last equation yields the following equation:

$$\sin \omega_d t_p = 0$$

or

$$\omega_d t_p = 0, \pi, 2\pi, 3\pi, \ldots$$

Since the peak time corresponds to the first peak overshoot, $\omega_d t_p = \pi$. Hence

$$t_p = \frac{\pi}{\omega_d} \tag{5–20}$$

The peak time t_p corresponds to one-half cycle of the frequency of damped oscillation.

Maximum overshoot M_p: The maximum overshoot occurs at the peak time or at $t = t_p = \pi/\omega_d$. Assuming that the final value of the output is unity, M_p is obtained from Equation (5–12) as

$$M_p = c(t_p) - 1$$

$$= -e^{-\zeta \omega_n (\pi/\omega_d)} \left(\cos \pi + \frac{\zeta}{\sqrt{1 - \zeta^2}} \sin \pi \right)$$

$$= e^{-(\sigma/\omega_d)\pi} = e^{-(\zeta/\sqrt{1-\zeta^2})\pi} \tag{5–21}$$

The maximum percent overshoot is $e^{-(\sigma/\omega_d)\pi} \times 100\%$.

If the final value $c(\infty)$ of the output is not unity, then we need to use the following equation:

$$M_p = \frac{c(t_p) - c(\infty)}{c(\infty)}$$

Settling time t_s: For an underdamped second-order system, the transient response is obtained from Equation (5–12) as

$$c(t) = 1 - \frac{e^{-\zeta \omega_n t}}{\sqrt{1 - \zeta^2}} \sin \left(\omega_d t + \tan^{-1} \frac{\sqrt{1 - \zeta^2}}{\zeta} \right), \qquad \text{for } t \geq 0$$

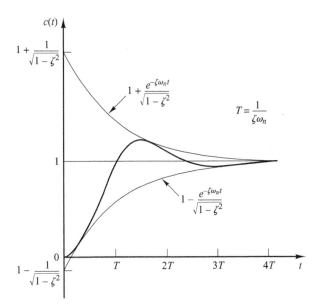

Figure 5–10
Pair of envelope curves for the unit-step response curve of the system shown in Figure 5–6.

The curves $1 \pm \left(e^{-\zeta\omega_n t}/\sqrt{1-\zeta^2}\right)$ are the envelope curves of the transient response to a unit-step input. The response curve $c(t)$ always remains within a pair of the envelope curves, as shown in Figure 5–10. The time constant of these envelope curves is $1/\zeta\omega_n$.

The speed of decay of the transient response depends on the value of the time constant $1/\zeta\omega_n$. For a given ω_n, the settling time t_s is a function of the damping ratio ζ. From Figure 5–7, we see that for the same ω_n and for a range of ζ between 0 and 1 the settling time t_s for a very lightly damped system is larger than that for a properly damped system. For an overdamped system, the settling time t_s becomes large because of the sluggish response.

The settling time corresponding to a $\pm2\%$ or $\pm5\%$ tolerance band may be measured in terms of the time constant $T = 1/\zeta\omega_n$ from the curves of Figure 5–7 for different values of ζ. The results are shown in Figure 5–11. For $0 < \zeta < 0.9$, if the 2% criterion is used, t_s is approximately four times the time constant of the system. If the 5% criterion is used, then t_s is approximately three times the time constant. Note that the settling time reaches a minimum value around $\zeta = 0.76$ (for the 2% criterion) or $\zeta = 0.68$ (for the 5% criterion) and then increases almost linearly for large values of ζ. The discontinuities in the curves of Figure 5–11 arise because an infinitesimal change in the value of ζ can cause a finite change in the settling time.

For convenience in comparing the responses of systems, we commonly define the settling time t_s to be

$$t_s = 4T = \frac{4}{\sigma} = \frac{4}{\zeta\omega_n} \qquad \text{(2\% criterion)} \qquad (5\text{--}22)$$

or

$$t_s = 3T = \frac{3}{\sigma} = \frac{3}{\zeta\omega_n} \qquad \text{(5\% criterion)} \qquad (5\text{--}23)$$

Note that the settling time is inversely proportional to the product of the damping ratio and the undamped natural frequency of the system. Since the value of ζ is usually determined from the requirement of permissible maximum overshoot, the settling time

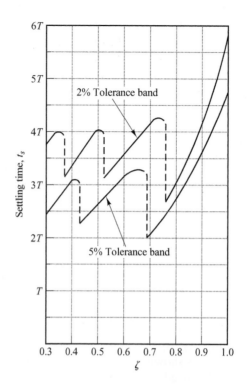

Figure 5–11
Settling time t_s
versus ζ curves.

is determined primarily by the undamped natural frequency ω_n. This means that the duration of the transient period may be varied, without changing the maximum over-shoot, by adjusting the undamped natural frequency ω_n.

From the preceding analysis, it is evident that for rapid response ω_n must be large. To limit the maximum overshoot M_p and to make the settling time small, the damping ratio ζ should not be too small. The relationship between the maximum percent overshoot M_p and the damping ratio ζ is presented in Figure 5–12. Note that if the damping ratio is between 0.4 and 0.7, then the maximum percent overshoot for step response is between 25% and 4%.

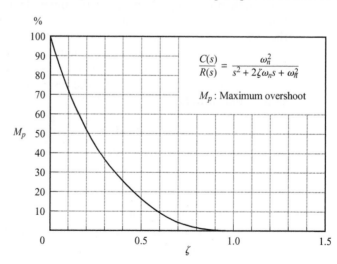

Figure 5–12
M_p versus ζ curve.

It is important to note that the equations for obtaining the rise time, peak time, maximum overshoot, and settling time are valid only for the standard second-order system defined by Equation (5–10). If the second-order system involves a zero or two zeros, the shape of the unit-step response curve will be quite different from those shown in Figure 5–7.

EXAMPLE 5–1 Consider the system shown in Figure 5–6, where $\zeta = 0.6$ and $\omega_n = 5$ rad/sec. Let us obtain the rise time t_r, peak time t_p, maximum overshoot M_p, and settling time t_s when the system is subjected to a unit-step input.

From the given values of ζ and ω_n, we obtain $\omega_d = \omega_n \sqrt{1 - \zeta^2} = 4$ and $\sigma = \zeta \omega_n = 3$.

Rise time t_r: The rise time is

$$t_r = \frac{\pi - \beta}{\omega_d} = \frac{3.14 - \beta}{4}$$

where β is given by

$$\beta = \tan^{-1} \frac{\omega_d}{\sigma} = \tan^{-1} \frac{4}{3} = 0.93 \text{ rad}$$

The rise time t_r is thus

$$t_r = \frac{3.14 - 0.93}{4} = 0.55 \text{ sec}$$

Peak time t_p: The peak time is

$$t_p = \frac{\pi}{\omega_d} = \frac{3.14}{4} = 0.785 \text{ sec}$$

Maximum overshoot M_p: The maximum overshoot is

$$M_p = e^{-(\sigma/\omega_d)\pi} = e^{-(3/4)\times 3.14} = 0.095$$

The maximum percent overshoot is thus 9.5%.

Settling time t_s: For the 2% criterion, the settling time is

$$t_s = \frac{4}{\sigma} = \frac{4}{3} = 1.33 \text{ sec}$$

For the 5% criterion,

$$t_s = \frac{3}{\sigma} = \frac{3}{3} = 1 \text{ sec}$$

Servo System with Velocity Feedback. The derivative of the output signal can be used to improve system performance. In obtaining the derivative of the output position signal, it is desirable to use a tachometer instead of physically differentiating the output signal. (Note that the differentiation amplifies noise effects. In fact, if discontinuous noises are present, differentiation amplifies the discontinuous noises more than the useful signal. For example, the output of a potentiometer is a discontinuous voltage signal because, as the potentiometer brush is moving on the windings, voltages are induced in the switchover turns and thus generate transients. The output of the potentiometer therefore should not be followed by a differentiating element.)

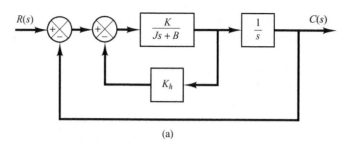

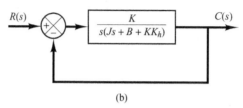

Figure 5–13
(a) Block diagram of
a servo system;
(b) simplified block
diagram.

The tachometer, a special dc generator, is frequently used to measure velocity without differentiation process. The output of a tachometer is proportional to the angular velocity of the motor.

Consider the servo system shown in Figure 5–13(a). In this device, the velocity signal, together with the positional signal, is fed back to the input to produce the actuating error signal. In any servo system, such a velocity signal can be easily generated by a tachometer. The block diagram shown in Figure 5–13(a) can be simplified, as shown in Figure 5–13(b), giving

$$\frac{C(s)}{R(s)} = \frac{K}{Js^2 + (B + KK_h)s + K} \tag{5–24}$$

Comparing Equation (5–24) with Equation (5–9), notice that the velocity feedback has the effect of increasing damping. The damping ratio ζ becomes

$$\zeta = \frac{B + KK_h}{2\sqrt{KJ}} \tag{5–25}$$

The undamped natural frequency $\omega_n = \sqrt{K/J}$ is not affected by velocity feedback. Noting that the maximum overshoot for a unit-step input can be controlled by controlling the value of the damping ratio ζ, we can reduce the maximum overshoot by adjusting the velocity-feedback constant K_h so that ζ is between 0.4 and 0.7.

It is important to remember that velocity feedback has the effect of increasing the damping ratio without affecting the undamped natural frequency of the system.

EXAMPLE 5–2 For the system shown in Figure 5–13(a), determine the values of gain K and velocity-feedback constant K_h so that the maximum overshoot in the unit-step response is 0.2 and the peak time is 1 sec. With these values of K and K_h, obtain the rise time and settling time. Assume that $J = 1$ kg-m² and $B = 1$ N-m/rad/sec.

Determination of the values of K and K_h: The maximum overshoot M_p is given by Equation (5–21) as

$$M_p = e^{-(\zeta/\sqrt{1-\zeta^2})\pi}$$

This value must be 0.2. Thus,

$$e^{-(\zeta/\sqrt{1-\zeta^2})\pi} = 0.2$$

or

$$\frac{\zeta\pi}{\sqrt{1-\zeta^2}} = 1.61$$

which yields

$$\zeta = 0.456$$

The peak time t_p is specified as 1 sec; therefore, from Equation (5–20),

$$t_p = \frac{\pi}{\omega_d} = 1$$

or

$$\omega_d = 3.14$$

Since ζ is 0.456, ω_n is

$$\omega_n = \frac{\omega_d}{\sqrt{1-\zeta^2}} = 3.53$$

Since the natural frequency ω_n is equal to $\sqrt{K/J}$,

$$K = J\omega_n^2 = \omega_n^2 = 12.5 \text{ N-m}$$

Then K_h is, from Equation (5–25),

$$K_h = \frac{2\sqrt{KJ}\zeta - B}{K} = \frac{2\sqrt{K}\zeta - 1}{K} = 0.178 \text{ sec}$$

Rise time t_r: From Equation (5–19), the rise time t_r is

$$t_r = \frac{\pi - \beta}{\omega_d}$$

where

$$\beta = \tan^{-1}\frac{\omega_d}{\sigma} = \tan^{-1}1.95 = 1.10$$

Thus, t_r is

$$t_r = 0.65 \text{ sec}$$

Settling time t_s: For the 2% criterion,

$$t_s = \frac{4}{\sigma} = 2.48 \text{ sec}$$

For the 5% criterion,

$$t_s = \frac{3}{\sigma} = 1.86 \text{ sec}$$

Impulse Response of Second-Order Systems. For a unit-impulse input $r(t)$, the corresponding Laplace transform is unity, or $R(s) = 1$. The unit-impulse response $C(s)$ of the second-order system shown in Figure 5-6 is

$$C(s) = \frac{\omega_n^2}{s^2 + 2\zeta\omega_n s + \omega_n^2}$$

The inverse Laplace transform of this equation yields the time solution for the response $c(t)$ as follows:

For $0 \le \zeta < 1$,

$$c(t) = \frac{\omega_n}{\sqrt{1 - \zeta^2}} e^{-\zeta\omega_n t} \sin \omega_n \sqrt{1 - \zeta^2}\, t, \qquad \text{for } t \ge 0 \qquad (5\text{–}26)$$

For $\zeta = 1$,

$$c(t) = \omega_n^2 t e^{-\omega_n t}, \qquad \text{for } t \ge 0 \qquad (5\text{–}27)$$

For $\zeta > 1$,

$$c(t) = \frac{\omega_n}{2\sqrt{\zeta^2 - 1}} e^{-(\zeta - \sqrt{\zeta^2-1})\omega_n t} - \frac{\omega_n}{2\sqrt{\zeta^2 - 1}} e^{-(\zeta + \sqrt{\zeta^2-1})\omega_n t}, \qquad \text{for } t \ge 0 \qquad (5\text{–}28)$$

Note that without taking the inverse Laplace transform of $C(s)$ we can also obtain the time response $c(t)$ by differentiating the corresponding unit-step response, since the unit-impulse function is the time derivative of the unit-step function. A family of unit-impulse response curves given by Equations (5–26) and (5–27) with various values of ζ is shown in Figure 5–14. The curves $c(t)/\omega_n$ are plotted against the dimensionless variable $\omega_n t$, and thus they are functions only of ζ. For the critically damped and overdamped cases, the unit-impulse response is always positive or zero; that is, $c(t) \ge 0$. This can be seen from Equations (5–27) and (5–28). For the underdamped case, the unit-impulse response $c(t)$ oscillates about zero and takes both positive and negative values.

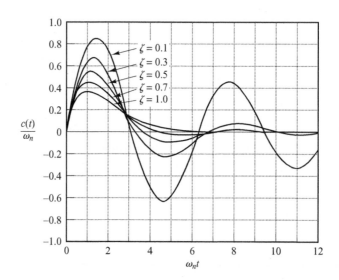

Figure 5–14
Unit-impulse response curves of the system shown in Figure 5–6.

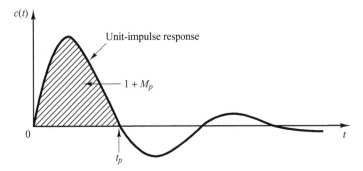

Figure 5–15
Unit-impulse
response curve of the
system shown in
Figure 5–6.

From the foregoing analysis, we may conclude that if the impulse response $c(t)$ does not change sign, the system is either critically damped or overdamped, in which case the corresponding step response does not overshoot but increases or decreases monotonically and approaches a constant value.

The maximum overshoot for the unit-impulse response of the underdamped system occurs at

$$t = \frac{\tan^{-1}\dfrac{\sqrt{1-\zeta^2}}{\zeta}}{\omega_n\sqrt{1-\zeta^2}}, \qquad \text{where } 0 < \zeta < 1 \tag{5–29}$$

[Equation (5–29) can be obtained by equating dc/dt to zero and solving for t.] The maximum overshoot is

$$c(t)_{max} = \omega_n\exp\left(-\frac{\zeta}{\sqrt{1-\zeta^2}}\tan^{-1}\frac{\sqrt{1-\zeta^2}}{\zeta}\right), \qquad \text{where } 0 < \zeta < 1 \tag{5–30}$$

[Equation (5–30) can be obtained by substituting Equation (5–29) into Equation (5–26).]

Since the unit-impulse response function is the time derivative of the unit-step response function, the maximum overshoot M_p for the unit-step response can be found from the corresponding unit-impulse response. That is, the area under the unit-impulse response curve from $t = 0$ to the time of the first zero, as shown in Figure 5–15, is $1 + M_p$, where M_p is the maximum overshoot (for the unit-step response) given by Equation (5–21). The peak time t_p (for the unit-step response) given by Equation (5–20) corresponds to the time that the unit-impulse response first crosses the time axis.

5–4 HIGHER-ORDER SYSTEMS

In this section we shall present a transient-response analysis of higher-order systems in general terms. It will be seen that the response of a higher-order system is the sum of the responses of first-order and second-order systems.

Transient Response of Higher-Order Systems. Consider the system shown in Figure 5–16. The closed-loop transfer function is

$$\frac{C(s)}{R(s)} = \frac{G(s)}{1 + G(s)H(s)} \tag{5-31}$$

In general, $G(s)$ and $H(s)$ are given as ratios of polynomials in s, or

$$G(s) = \frac{p(s)}{q(s)} \quad \text{and} \quad H(s) = \frac{n(s)}{d(s)}$$

where $p(s)$, $q(s)$, $n(s)$, and $d(s)$ are polynomials in s. The closed-loop transfer function given by Equation (5–31) may then be written

$$\frac{C(s)}{R(s)} = \frac{p(s)d(s)}{q(s)d(s) + p(s)n(s)}$$

$$= \frac{b_0 s^m + b_1 s^{m-1} + \cdots + b_{m-1}s + b_m}{a_0 s^n + a_1 s^{n-1} + \cdots + a_{n-1}s + a_n} \quad (m \leq n)$$

The transient response of this system to any given input can be obtained by a computer simulation. (See Section 5–5.) If an analytical expression for the transient response is desired, then it is necessary to factor the denominator polynomial. [MATLAB may be used for finding the roots of the denominator polynomial. Use the command roots(den).] Once the numerator and the denominator have been factored, $C(s)/R(s)$ can be written in the form

$$\frac{C(s)}{R(s)} = \frac{K(s + z_1)(s + z_2)\cdots(s + z_m)}{(s + p_1)(s + p_2)\cdots(s + p_n)} \tag{5-32}$$

Let us examine the response behavior of this system to a unit-step input. Consider first the case where the closed-loop poles are all real and distinct. For a unit-step input, Equation (5–32) can be written

$$C(s) = \frac{a}{s} + \sum_{i=1}^{n} \frac{a_i}{s + p_i} \tag{5-33}$$

where a_i is the residue of the pole at $s = -p_i$. (If the system involves multiple poles, then $C(s)$ will have multiple-pole terms.) [The partial-fraction expansion of $C(s)$, as given by Equation (5–33), can be obtained easily with MATLAB. Use the residue command. (See Appendix B.)]

If all closed-loop poles lie in the left-half s plane, the relative magnitudes of the residues determine the relative importance of the components in the expanded form of

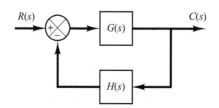

Figure 5–16
Control system.

$C(s)$. If there is a closed-loop zero close to a closed-loop pole, then the residue at this pole is small and the coefficient of the transient-response term corresponding to this pole becomes small. A pair of closely located poles and zeros will effectively cancel each other. If a pole is located very far from the origin, the residue at this pole may be small. The transients corresponding to such a remote pole are small and last a short time. Terms in the expanded form of $C(s)$ having very small residues contribute little to the transient response, and these terms may be neglected. If this is done, the higher-order system may be approximated by a lower-order one. (Such an approximation often enables us to estimate the response characteristics of a higher-order system from those of a simplified one.)

Next, consider the case where the poles of $C(s)$ consist of real poles and pairs of complex-conjugate poles. A pair of complex-conjugate poles yields a second-order term in s. Since the factored form of the higher-order characteristic equation consists of first- and second-order terms, Equation (5–33) can be rewritten

$$C(s) = \frac{a}{s} + \sum_{j=1}^{q} \frac{a_j}{s + p_j} + \sum_{k=1}^{r} \frac{b_k(s + \zeta_k\omega_k) + c_k\omega_k\sqrt{1 - \zeta_k^2}}{s^2 + 2\zeta_k\omega_k s + \omega_k^2} \qquad (q + 2r = n)$$

where we assumed all closed-loop poles are distinct. [If the closed-loop poles involve multiple poles, $C(s)$ must have multiple-pole terms.] From this last equation, we see that the response of a higher-order system is composed of a number of terms involving the simple functions found in the responses of first- and second-order systems. The unit-step response $c(t)$, the inverse Laplace transform of $C(s)$, is then

$$c(t) = a + \sum_{j=1}^{q} a_j e^{-p_j t} + \sum_{k=1}^{r} b_k e^{-\zeta_k\omega_k t} \cos \omega_k \sqrt{1 - \zeta_k^2}\, t$$

$$+ \sum_{k=1}^{r} c_k e^{-\zeta_k\omega_k t} \sin \omega_k \sqrt{1 - \zeta_k^2}\, t, \qquad \text{for } t \geq 0 \qquad (5–34)$$

Thus the response curve of a stable higher-order system is the sum of a number of exponential curves and damped sinusoidal curves.

If all closed-loop poles lie in the left-half s plane, then the exponential terms and the damped exponential terms in Equation (5–34) will approach zero as time t increases. The steady-state output is then $c(\infty) = a$.

Let us assume that the system considered is a stable one. Then the closed-loop poles that are located far from the $j\omega$ axis have large negative real parts. The exponential terms that correspond to these poles decay very rapidly to zero. (Note that the horizontal distance from a closed-loop pole to the $j\omega$ axis determines the settling time of transients due to that pole. The smaller the distance is, the longer the settling time.)

Remember that the type of transient response is determined by the closed-loop poles, while the shape of the transient response is primarily determined by the closed-loop zeros. As we have seen earlier, the poles of the input $R(s)$ yield the steady-state response terms in the solution, while the poles of $C(s)/R(s)$ enter into the exponential transient-response terms and/or damped sinusoidal transient-response terms. The zeros of $C(s)/R(s)$ do not affect the exponents in the exponential terms, but they do affect the magnitudes and signs of the residues.

Dominant Closed-Loop Poles.   The relative dominance of closed-loop poles is determined by the ratio of the real parts of the closed-loop poles, as well as by the relative magnitudes of the residues evaluated at the closed-loop poles. The magnitudes of the residues depend on both the closed-loop poles and zeros.

If the ratios of the real parts of the closed-loop poles exceed 5 and there are no zeros nearby, then the closed-loop poles nearest the $j\omega$ axis will dominate in the transient-response behavior because these poles correspond to transient-response terms that decay slowly. Those closed-loop poles that have dominant effects on the transient-response behavior are called *dominant closed-loop* poles. Quite often the dominant closed-loop poles occur in the form of a complex-conjugate pair. The dominant closed-loop poles are most important among all closed-loop poles.

Note that the gain of a higher-order system is often adjusted so that there will exist a pair of dominant complex-conjugate closed-loop poles. The presence of such poles in a stable system reduces the effects of such nonlinearities as dead zone, backlash, and coulomb-friction.

Stability Analysis in the Complex Plane.   The stability of a linear closed-loop system can be determined from the location of the closed-loop poles in the s plane. If any of these poles lie in the right-half s plane, then with increasing time they give rise to the dominant mode, and the transient response increases monotonically or oscillates with increasing amplitude. This represents an unstable system. For such a system, as soon as the power is turned on, the output may increase with time. If no saturation takes place in the system and no mechanical stop is provided, then the system may eventually be subjected to damage and fail, since the response of a real physical system cannot increase indefinitely. Therefore, closed-loop poles in the right-half s plane are not permissible in the usual linear control system. If all closed-loop poles lie to the left of the $j\omega$ axis, any transient response eventually reaches equilibrium. This represents a stable system.

Whether a linear system is stable or unstable is a property of the system itself and does not depend on the input or driving function of the system. The poles of the input, or driving function, do not affect the property of stability of the system, but they contribute only to steady-state response terms in the solution. Thus, the problem of absolute stability can be solved readily by choosing no closed-loop poles in the right-half s plane, including the $j\omega$ axis. (Mathematically, closed-loop poles on the $j\omega$ axis will yield oscillations, the amplitude of which is neither decaying nor growing with time. In practical cases, where noise is present, however, the amplitude of oscillations may increase at a rate determined by the noise power level. Therefore, a control system should not have closed-loop poles on the $j\omega$ axis.)

Note that the mere fact that all closed-loop poles lie in the left-half s plane does not guarantee satisfactory transient-response characteristics. If dominant complex-conjugate closed-loop poles lie close to the $j\omega$ axis, the transient response may exhibit excessive oscillations or may be very slow. Therefore, to guarantee fast, yet well-damped, transient-response characteristics, it is necessary that the closed-loop poles of the system lie in a particular region in the complex plane, such as the region bounded by the shaded area in Figure 5–17.

Since the relative stability and transient-response performance of a closed-loop control system are directly related to the closed-loop pole-zero configuration in the s plane,

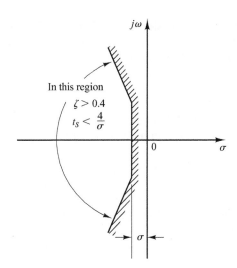

Figure 5–17
Region in the complex plane satisfying the conditions $\zeta > 0.4$ and $t_s < 4/\sigma$.

it is frequently necessary to adjust one or more system parameters in order to obtain suitable configurations. The effects of varying system parameters on the closed-loop poles will be discussed in detail in Chapter 6.

5–5 TRANSIENT-RESPONSE ANALYSIS WITH MATLAB

Introduction.   The practical procedure for plotting time response curves of systems higher than second order is through computer simulation. In this section we present the computational approach to the transient-response analysis with MATLAB. In particular, we discuss step response, impulse response, ramp response, and responses to other simple inputs.

MATLAB Representation of Linear Systems.   The transfer function of a system is represented by two arrays of numbers. Consider the system

$$\frac{C(s)}{R(s)} = \frac{2s + 25}{s^2 + 4s + 25} \tag{5–35}$$

This system can be represented as two arrays, each containing the coefficients of the polynomials in decreasing powers of s as follows:

$$\text{num} = [2 \quad 25]$$
$$\text{den} = [1 \quad 4 \quad 25]$$

An alternative representation is

$$\text{num} = [0 \quad 2 \quad 25]$$
$$\text{den} = [1 \quad 4 \quad 25]$$

In this expression a zero is padded. Note that if zeros are padded, the dimensions of "num" vector and "den" vector become the same. An advantage of padding zeros is that the "num" vector and "den" vector can be directly added. For example,

$$\text{num} + \text{den} = [0 \ 2 \ 25] + [1 \ 4 \ 25]$$
$$= [1 \ 6 \ 50]$$

If num and den (the numerator and denominator of the closed-loop transfer function) are known, commands such as

$$\text{step(num,den)}, \quad \text{step(num,den,t)}$$

will generate plots of unit-step responses (t in the step command is the user-specified time.) For a control system defined in a state-space form, where state matrix $\mathbf{A}$, control matrix $\mathbf{B}$, output matrix $\mathbf{C}$, and direct transmission matrix $\mathbf{D}$ of state-space equations are known, the command

$$\text{step(A,B,C,D)}, \quad \text{step(A,B,C,D,t)}$$

will generate plots of unit-step responses. When t is not explicitly included in the step commands, the time vector is automatically determined.

Note that the command step(sys) may be used to obtain the unit-step response of a system. First, define the system by

$$\text{sys} = \text{tf(num,den)}$$

or

$$\text{sys} = \text{ss(A,B,C,D)}$$

Then, to obtain, for example, the unit-step response, enter

$$\text{step(sys)}$$

into the computer.

When step commands have left-hand arguments such as

$$[y,x,t] = \text{step(num,den,t)}$$
$$[y,x,t] = \text{step(A,B,C,D,iu)}$$
$$[y,x,t] = \text{step(A,B,C,D,iu,t)} \tag{5-36}$$

no plot is shown on the screen. Hence it is necessary to use a plot command to see the response curves. The matrices y and x contain the output and state response of the system, respectively, evaluated at the computation time points t. (y has as many columns as outputs and one row for each element in t. x has as many columns as states and one row for each element in t.)

Note in Equation (5–36) that the scalar iu is an index into the inputs of the system and specifies which input is to be used for the response, and t is the user-specified time. If the system involves multiple inputs and multiple outputs, the step command, such as given by Equation (5–36), produces a series of step-response plots, one for each input and output combination of

$$\dot{\mathbf{x}} = \mathbf{Ax} + \mathbf{Bu}$$
$$\mathbf{y} = \mathbf{Cx} + \mathbf{Du}$$

(For details, see Example 5–3.)

EXAMPLE 5–3 Consider the following system:

$$\begin{bmatrix} \dot{x}_1 \\ \dot{x}_2 \end{bmatrix} = \begin{bmatrix} -1 & -1 \\ 6.5 & 0 \end{bmatrix} \begin{bmatrix} x_1 \\ x_2 \end{bmatrix} + \begin{bmatrix} 1 & 1 \\ 1 & 0 \end{bmatrix} \begin{bmatrix} u_1 \\ u_2 \end{bmatrix}$$

$$\begin{bmatrix} y_1 \\ y_2 \end{bmatrix} = \begin{bmatrix} 1 & 0 \\ 0 & 1 \end{bmatrix} \begin{bmatrix} x_1 \\ x_2 \end{bmatrix} + \begin{bmatrix} 0 & 0 \\ 0 & 0 \end{bmatrix} \begin{bmatrix} u_1 \\ u_2 \end{bmatrix}$$

Obtain the unit-step response curves.

Although it is not necessary to obtain the transfer-matrix expression for the system to obtain the unit-step response curves with MATLAB, we shall derive such an expression for reference. For the system defined by

$$\dot{x} = Ax + Bu$$

$$y = Cx + Du$$

the transfer matrix $G(s)$ is a matrix that relates $Y(s)$ and $U(s)$ as follows:

$$Y(s) = G(s)U(s)$$

Taking Laplace transforms of the state-space equations, we obtain

$$sX(s) - x(0) = AX(s) + BU(s) \tag{5–37}$$

$$Y(s) = CX(s) + DU(s) \tag{5–38}$$

In deriving the transfer matrix, we assume that $x(0) = 0$. Then, from Equation (5–37), we get

$$X(s) = (sI - A)^{-1}BU(s) \tag{5–39}$$

Substituting Equation (5–39) into Equation (5–38), we obtain

$$Y(s) = [C(sI - A)^{-1}B + D]U(s)$$

Thus the transfer matrix $G(s)$ is given by

$$G(s) = C(sI - A)^{-1}B + D$$

The transfer matrix $G(s)$ for the given system becomes

$$G(s) = C(sI - A)^{-1}B$$

$$= \begin{bmatrix} 1 & 0 \\ 0 & 1 \end{bmatrix} \begin{bmatrix} s+1 & 1 \\ -6.5 & s \end{bmatrix}^{-1} \begin{bmatrix} 1 & 1 \\ 1 & 0 \end{bmatrix}$$

$$= \frac{1}{s^2 + s + 6.5} \begin{bmatrix} s & -1 \\ 6.5 & s+1 \end{bmatrix} \begin{bmatrix} 1 & 1 \\ 1 & 0 \end{bmatrix}$$

$$= \frac{1}{s^2 + s + 6.5} \begin{bmatrix} s-1 & s \\ s+7.5 & 6.5 \end{bmatrix}$$

Hence

$$\begin{bmatrix} Y_1(s) \\ Y_2(s) \end{bmatrix} = \begin{bmatrix} \dfrac{s-1}{s^2 + s + 6.5} & \dfrac{s}{s^2 + s + 6.5} \\ \dfrac{s+7.5}{s^2 + s + 6.5} & \dfrac{6.5}{s^2 + s + 6.5} \end{bmatrix} \begin{bmatrix} U_1(s) \\ U_2(s) \end{bmatrix}$$

Since this system involves two inputs and two outputs, four transfer functions may be defined, depending on which signals are considered as input and output. Note that, when considering the signal u_1 as the input, we assume that signal u_2 is zero, and vice versa. The four transfer functions are

$$\frac{Y_1(s)}{U_1(s)} = \frac{s-1}{s^2 + s + 6.5}, \qquad \frac{Y_1(s)}{U_2(s)} = \frac{s}{s^2 + s + 6.5}$$

$$\frac{Y_2(s)}{U_1(s)} = \frac{s+7.5}{s^2 + s + 6.5}, \qquad \frac{Y_2(s)}{U_2(s)} = \frac{6.5}{s^2 + s + 6.5}$$

Assume that u_1 and u_2 are unit-step functions. The four individual step-response curves can then be plotted by use of the command

$$step(A,B,C,D)$$

MATLAB Program 5–1 produces four such step-response curves. The curves are shown in Figure 5–18. (Note that the time vector t is automatically determined, since the command does not include t.)

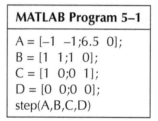

MATLAB Program 5–1

```
A = [-1  -1;6.5  0];
B = [1  1;1  0];
C = [1  0;0  1];
D = [0  0;0  0];
step(A,B,C,D)
```

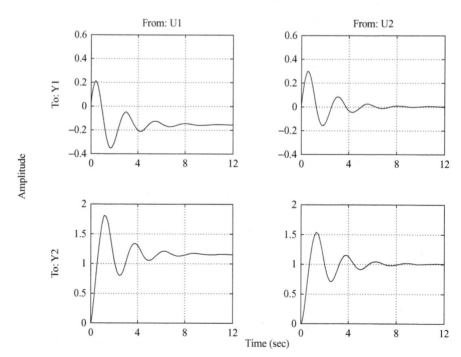

Figure 5–18
Unit-step response curves.

To plot two step-response curves for the input u_1 in one diagram and two step-response curves for the input u_2 in another diagram, we may use the commands

$$\text{step(A,B,C,D,1)}$$

and

$$\text{step(A,B,C,D,2)}$$

respectively. MATLAB Program 5–2 is a program to plot two step-response curves for the input u_1 in one diagram and two step-response curves for the input u_2 in another diagram. Figure 5–19 shows the two diagrams, each consisting of two step-response curves. (This MATLAB program uses text commands. For such commands, refer to the paragraph following this example.)

MATLAB Program 5–2

```
% ***** In this program we plot step-response curves of a system
% having two inputs (u1 and u2) and two outputs (y1 and y2) *****

% ***** We shall first plot step-response curves when the input is
% u1. Then we shall plot step-response curves when the input is
% u2 *****

% ***** Enter matrices A, B, C, and D *****

A = [-1  -1;6.5  0];
B = [1  1;1  0];
C = [1  0;0  1];
D = [0  0;0  0];

% ***** To plot step-response curves when the input is u1, enter
% the command 'step(A,B,C,D,1)' *****

step(A,B,C,D,1)
grid
title ('Step-Response Plots: Input = u1 (u2 = 0)')
text(3.4, -0.06,'Y1')
text(3.4, 1.4,'Y2')

% ***** Next, we shall plot step-response curves when the input
% is u2. Enter the command 'step(A,B,C,D,2)' *****

step(A,B,C,D,2)
grid
title ('Step-Response Plots: Input = u2 (u1 = 0)')
text(3,0.14,'Y1')
text(2.8,1.1,'Y2')
```

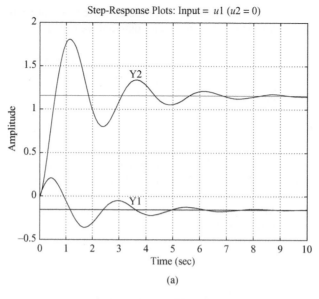

Step-Response Plots: Input = $u1$ ($u2 = 0$)

(a)

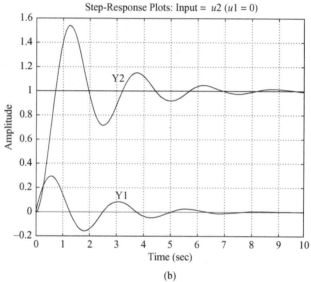

Step-Response Plots: Input = $u2$ ($u1 = 0$)

Figure 5–19
Unit-step response
curves. (a) u_1 is the
input $(u_2 = 0)$; (b) u_2
is the input $(u_1 = 0)$.

(b)

Writing Text on the Graphics Screen. To write text on the graphics screen, enter, for example, the following statements:

text(3.4, -0.06,'Y1')

and

text(3.4,1.4,'Y2')

The first statement tells the computer to write 'Y1' beginning at the coordinates $x = 3.4$, $y = -0.06$. Similarly, the second statement tells the computer to write 'Y2' beginning at the coordinates $x = 3.4$, $y = 1.4$. [See MATLAB Program 5–2 and Figure 5–19(a).]

Another way to write a text or texts in the plot is to use the gtext command. The syntax is

gtext('text')

When gtext is executed, the computer waits until the cursor is positioned (using a mouse) at the desired position in the screen. When the left mouse button is pressed, the text enclosed in simple quotes is written on the plot at the cursor's position. Any number of gtext commands can be used in a plot. (See, for example, MATLAB Program 5–15.)

MATLAB Description of Standard Second-Order System. As noted earlier, the second-order system

$$G(s) = \frac{\omega_n^2}{s^2 + 2\zeta\omega_n s + \omega_n^2} \tag{5-40}$$

is called the standard second-order system. Given ω_n and ζ, the command

printsys(num,den)   or   printsys(num,den,s)

prints num/den as a ratio of polynomials in s.

Consider, for example, the case where $\omega_n = 5$ rad/sec and $\zeta = 0.4$. MATLAB Program 5–3 generates the standard second-order system, where $\omega_n = 5$ rad/sec and $\zeta = 0.4$. Note that in MATLAB Program 5–3, "num 0" is 1.

MATLAB Program 5–3

```
wn = 5;
damping_ratio = 0.4;
[num0,den] = ord2(wn,damping_ratio);
num = 5^2*num0;
printsys(num,den,'s')
num/den =
```

$$\frac{25}{S\char94 2 + 4s + 25}$$

Obtaining the Unit-Step Response of the Transfer-Function System. Let us consider the unit-step response of the system given by

$$G(s) = \frac{25}{s^2 + 4s + 25}$$

MATLAB Program 5–4 will yield a plot of the unit-step response of this system. A plot of the unit-step response curve is shown in Figure 5–20.

MATLAB Program 5–4

% ------------- Unit-step response -------------

% ***** Enter the numerator and denominator of the transfer
% function *****

num = [25];
 den = [1 4 25];

% ***** Enter the following step-response command *****

step(num,den)

% ***** Enter grid and title of the plot *****

grid
title (' Unit-Step Response of G(s) = 25/(s^2+4s+25)')

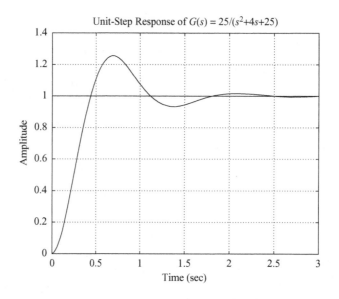

Unit-Step Response of $G(s) = 25/(s^2+4s+25)$

Figure 5–20
Unit-step response
curve.

Notice in Figure 5–20 (and many others) that the x-axis and y-axis labels are automatically determined. If it is desired to label the x axis and y axis differently, we need to modify the step command. For example, if it is desired to label the x axis as 't Sec' and the y axis as 'Output,' then use step-response commands with left-hand arguments, such as

$$c = step(num,den,t)$$

or, more generally,

$$[y,x,t] = step(num,den,t)$$

and use plot(t,y) command. See, for example, MATLAB Program 5–5 and Figure 5–21.

MATLAB Program 5–5

```
% ------------ Unit-step response ------------
num = [25];
den = [1  4  25];
t = 0:0.01:3;
[y,x,t] = step(num,den,t);
plot(t,y)
grid
title('Unit-Step Response of G(s)=25/(s^2+4s+25)')
xlabel('t Sec')
ylabel('Output')
```

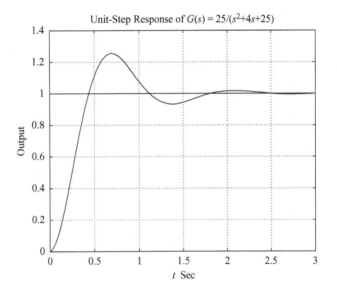

Figure 5–21
Unit-step response curve.

Obtaining Three-Dimensional Plot of Unit-Step Response Curves with MATLAB.

MATLAB enables us to plot three-dimensional plots easily. The commands to obtain three-dimensional plots are "mesh" and "surf." The difference between the "mesh" plot and "surf" plot is that in the former only the lines are drawn and in the latter the spaces between the lines are filled in by colors. In this book we use only the "mesh" command.

EXAMPLE 5–4   Consider the closed-loop system defined by

$$\frac{C(s)}{R(s)} = \frac{1}{s^2 + 2\zeta s + 1}$$

(The undamped natural frequency ω_n is normalized to 1.) Plot unit-step response curves $c(t)$ when ζ assumes the following values:

$$\zeta = 0,\ 0.2,\ 0.4,\ 0.6.\ 0.8,\ 1.0$$

Also plot a three-dimensional plot.

Section 5–5 / Transient-Response Analysis with MATLAB

An illustrative MATLAB Program for plotting a two-dimensional diagram and a three-dimensional diagram of unit-step response curves of this second-order system is given in MATLAB Program 5–6. The resulting plots are shown in Figures 5–22(a) and (b), respectively. Notice that we used the command mesh(t,zeta,y') to plot the three-dimensional plot. We may use a command mesh(y') to get the same result. [Note that command mesh(t,zeta,y) or mesh(y) will produce a three-dimensional plot the same as Figure 5–22(b), except that x axis and y axis are interchanged. See Problem A–5–15.]

When we want to solve a problem using MATLAB and if the solution process involves many repetitive computations, various approaches may be conceived to simplify the MATLAB program. A frequently used approach to simplify the computation is to use "for loops." MATLAB Program 5–6 uses such a "for loop." In this book many MATLAB programs using "for loops" are presented for solving a variety of problems. Readers are advised to study all those problems carefully to familiarize themselves with the approach.

MATLAB Program 5–6

```
% ------- Two-dimensional plot and three-dimensional plot of unit-step
% response curves for the standard second-order system with wn = 1
% and zeta = 0, 0.2, 0.4, 0.6, 0.8, and 1. -------

t = 0:0.2:10;
zeta = [0   0.2   0.4   0.6   0.8   1];
    for n = 1:6;
    num = [1];
    den = [1   2*zeta(n)   1];
    [y(1:51,n),x,t] = step(num,den,t);
    end

% To plot a two-dimensional diagram, enter the command plot(t,y).

plot(t,y)
grid
title('Plot of Unit-Step Response Curves with \omega_n = 1 and \zeta = 0, 0.2, 0.4, 0.6, 0.8, 1')
xlabel('t (sec)')
ylabel('Response')
text(4.1,1.86,'\zeta = 0')
text(3.5,1.5,'0.2')
text(3 .5,1.24,'0.4')
text(3.5,1.08,'0.6')
text(3.5,0.95,'0.8')
text(3.5,0.86,'1.0')

% To plot a three-dimensional diagram, enter the command mesh(t,zeta,y').

mesh(t,zeta,y')
title('Three-Dimensional Plot of Unit-Step Response Curves')
xlabel('t Sec')
ylabel('\zeta')
zlabel('Response')
```

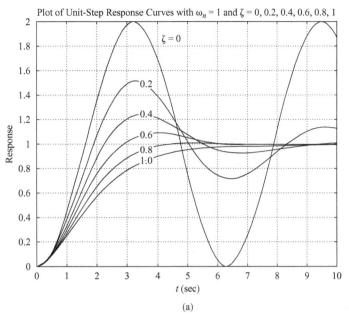

Plot of Unit-Step Response Curves with $\omega_n = 1$ and $\zeta = 0, 0.2, 0.4, 0.6, 0.8, 1$

(a)

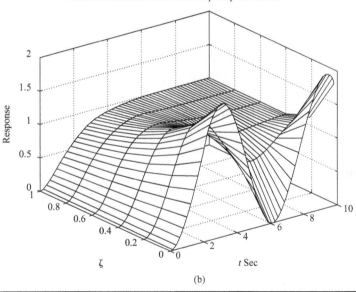

Three-Dimensional Plot of Unit-Step Response Curves

(b)

Figure 5–22
(a) Two-dimensional plot of unit-step response curves for $\zeta = 0, 0.2, 0.4, 0.6, 0.8,$ and 1.0; (b) three-dimensional plot of unit-step response curves.

Obtaining Rise Time, Peak Time, Maximum Overshoot, and Settling Time with MATLAB. MATLAB can conveniently be used to obtain the rise time, peak time, maximum overshoot, and settling time. Consider the system defined by

$$\frac{C(s)}{R(s)} = \frac{25}{s^2 + 6s + 25}$$

MATLAB Program 5–7 yields the rise time, peak time, maximum overshoot, and settling time. A unit-step response curve for this system is given in Figure 5–23 to verify the results obtained with MATLAB Program 5–7. (Note that this program can also be applied to higher-order systems. See Problem **A–5–10**.)

MATLAB Program 5–7

```
% ------- This is a MATLAB program to find the rise time, peak time,
% maximum overshoot, and settling time of the second-order system
% and higher-order system -------
% ------- In this example, we assume zeta = 0.6 and wn = 5 -------
num = [25];
den = [1  6  25];
t = 0:0.005:5;
[y,x,t] = step(num,den,t);
r = 1; while y(r) < 1.0001; r = r + 1; end;
rise_time = (r - 1)*0.005

rise_time =

   0.5550

[ymax,tp] = max(y);
peak_time = (tp - 1)*0.005

peak_time =

   0.7850

max_overshoot = ymax-1

max_overshoot =

   0.0948

s = 1001; while y(s) > 0.98 & y(s) < 1.02; s = s - 1; end;
settling_time = (s - 1)*0.005

settling_time =

   1.1850
```

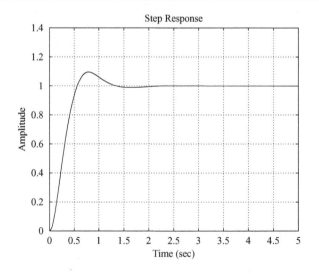

Figure 5–23
Unit-step response curve.

Impulse Response. The unit-impulse response of a control system may be obtained by using any of the impulse commands such as

$$\text{impulse(num,den)}$$

$$\text{impulse(A,B,C,D)}$$

$$[y,x,t] = \text{impulse(num,den)}$$

$$[y,x,t] = \text{impulse(num,den,t)} \tag{5-41}$$

$$[y,x,t] = \text{impulse(A,B,C,D)}$$

$$[y,x,t] = \text{impulse(A,B,C,D,iu)} \tag{5-42}$$

$$[y,x,t] = \text{impulse(A,B,C,D,iu,t)} \tag{5-43}$$

The command impulse(num,den) plots the unit-impulse response on the screen. The command impulse(A,B,C,D) produces a series of unit-impulse-response plots, one for each input and output combination of the system

$$\dot{\mathbf{x}} = \mathbf{Ax} + \mathbf{Bu}$$

$$\mathbf{y} = \mathbf{Cx} + \mathbf{Du}$$

Note that in Equations (5–42) and (5–43) the scalar iu is an index into the inputs of the system and specifies which input to be used for the impulse response.

Note also that if the command used does not include "t" explicitly, the time vector is automatically determined. If the command includes the user-supplied time vector "t", as do the commands given by Equations (5–41) and (5–43)], this vector specifies the times at which the impulse response is to be computed.

If MATLAB is invoked with the left-hand argument [y,x,t], such as in the case of [y,x,t] = impulse(A,B,C,D), the command returns the output and state responses of the system and the time vector t. No plot is drawn on the screen. The matrices y and x contain the output and state responses of the system evaluated at the time points t. (y has as many columns as outputs and one row for each element in t. x has as many columns as state variables and one row for each element in t.) To plot the response curve, we must include a plot command, such as plot(t,y).

EXAMPLE 5–5 Obtain the unit-impulse response of the following system:

$$\frac{C(s)}{R(s)} = G(s) = \frac{1}{s^2 + 0.2s + 1}$$

MATLAB Program 5–8 will produce the unit-impulse response. The resulting plot is shown in Figure 5–24.

MATLAB Program 5–8

```
num = [1];
den = [1  0.2  1];
impulse(num,den);
grid
title('Unit-Impulse Response of G(s) = 1/(s^2 + 0.2s + 1)')
```

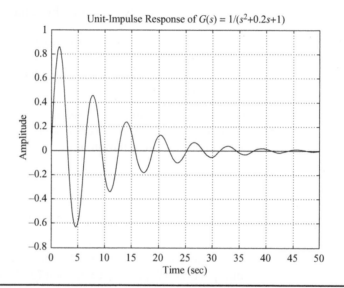

Unit-Impulse Response of $G(s) = 1/(s^2+0.2s+1)$

Figure 5–24
Unit-impulse-
response curve.

Alternative Approach to Obtain Impulse Response. Note that when the initial conditions are zero, the unit-impulse response of $G(s)$ is the same as the unit-step response of $sG(s)$.

Consider the unit-impulse response of the system considered in Example 5–5. Since $R(s) = 1$ for the unit-impulse input, we have

$$\frac{C(s)}{R(s)} = C(s) = G(s) = \frac{1}{s^2 + 0.2s + 1}$$

$$= \frac{s}{s^2 + 0.2s + 1} \frac{1}{s}$$

We can thus convert the unit-impulse response of $G(s)$ to the unit-step response of $sG(s)$.

If we enter the following num and den into MATLAB,

$$num = [0 \ 1 \ 0]$$

$$den = [1 \ 0.2 \ 1]$$

and use the step-response command; as given in MATLAB Program 5–9, we obtain a plot of the unit-impulse response of the system as shown in Figure 5–25.

MATLAB Program 5–9

```
num = [1  0];
den = [1  0.2  1];
step(num,den);
grid
title('Unit-Step Response of sG(s) = s/(s^2 + 0.2s + 1)')
```

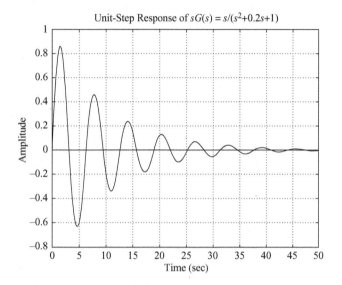

Figure 5–25
Unit-impulse-response curve obtained as the unit-step response of $sG(s) = s/(s^2 + 0.2s + 1)$.

Ramp Response.   There is no ramp command in MATLAB. Therefore, we need to use the step command or the lsim command (presented later) to obtain the ramp response. Specifically, to obtain the ramp response of the transfer-function system $G(s)$, divide $G(s)$ by s and use the step-response command. For example, consider the closed-loop system

$$\frac{C(s)}{R(s)} = \frac{2s + 1}{s^2 + s + 1}$$

For a unit-ramp input, $R(s) = 1/s^2$. Hence

$$C(s) = \frac{2s + 1}{s^2 + s + 1}\frac{1}{s^2} = \frac{2s + 1}{(s^2 + s + 1)s}\frac{1}{s}$$

To obtain the unit-ramp response of this system, enter the following numerator and denominator into the MATLAB program:

$$num = [2\ \ 1];$$

$$den = [1\ \ 1\ \ 1\ \ 0];$$

and use the step-response command. See MATLAB Program 5–10. The plot obtained by using this program is shown in Figure 5–26.

MATLAB Program 5–10

```
% -------------- Unit-ramp response --------------

% ***** The unit-ramp response is obtained as the unit-step
% response of G(s)/s *****

% ***** Enter the numerator and denominator of G(s)/s *****

num = [2  1];
den = [1  1  1  0];

% ***** Specify the computing time points (such as t = 0:0.1:10)
% and then enter step-response command: c = step(num,den,t) *****

t = 0:0.1:10;
c = step(num,den,t);

% ***** In plotting the ramp-response curve, add the reference
% input to the plot. The reference input is t. Add to the
% argument of the plot command with the following: t,t,'-'. Thus
% the plot command becomes as follows: plot(t,c,'o',t,t,'-') *****

plot(t,c,'o',t,t,'-')

% ***** Add grid, title, xlabel, and ylabel *****

grid
title('Unit-Ramp Response Curve for System G(s) = (2s + 1)/(s^2 + s + 1)')
xlabel('t Sec')
ylabel('Input and Output')
```

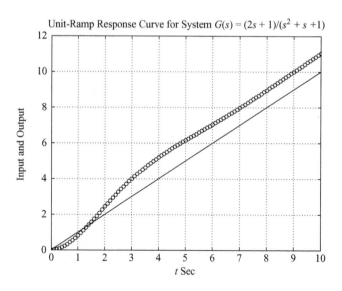

Figure 5–26
Unit-ramp response curve.

Chapter 5 / Transient and Steady-State Response Analyses

Unit-Ramp Response of a System Defined in State Space.   Next, we shall treat the unit-ramp response of the system in state-space form. Consider the system described by

$$\dot{x} = Ax + Bu$$
$$y = Cx + Du$$

where u is the unit-ramp function. In what follows, we shall consider a simple example to explain the method. Consider the case where

$$A = \begin{bmatrix} 0 & 1 \\ -1 & -1 \end{bmatrix}, \qquad B = \begin{bmatrix} 0 \\ 1 \end{bmatrix}, \qquad x(0) = 0$$
$$C = [1 \quad 0], \qquad D = [0]$$

When the initial conditions are zeros, the unit-ramp response is the integral of the unit-step response. Hence the unit-ramp response can be given by

$$z = \int_0^t y \, dt \tag{5-44}$$

From Equation (5–44), we obtain

$$\dot{z} = y = x_1 \tag{5-45}$$

Let us define

$$z = x_3$$

Then Equation (5–45) becomes

$$\dot{x}_3 = x_1 \tag{5-46}$$

Combining Equation (5–46) with the original state-space equation, we obtain

$$\begin{bmatrix} \dot{x}_1 \\ \dot{x}_2 \\ \dot{x}_3 \end{bmatrix} = \begin{bmatrix} 0 & 1 & 0 \\ -1 & -1 & 0 \\ 1 & 0 & 0 \end{bmatrix} \begin{bmatrix} x_1 \\ x_2 \\ x_3 \end{bmatrix} + \begin{bmatrix} 0 \\ 1 \\ 0 \end{bmatrix} u \tag{5-47}$$

$$z = [0 \quad 0 \quad 1] \begin{bmatrix} x_1 \\ x_2 \\ x_3 \end{bmatrix} \tag{5-48}$$

where u appearing in Equation (5–47) is the unit-step function. These equations can be written as

$$\dot{x} = AAx + BBu$$
$$z = CCx + DDu$$

where

$$AA = \begin{bmatrix} 0 & 1 & 0 \\ -1 & -1 & 0 \\ 1 & 0 & 0 \end{bmatrix} = \begin{bmatrix} A & 0 \\ \hline C & 0 \end{bmatrix}$$

$$BB = \begin{bmatrix} 0 \\ 1 \\ 0 \end{bmatrix} = \begin{bmatrix} B \\ 0 \end{bmatrix}, \qquad CC = [0 \quad 0 \quad 1], \qquad DD = [0]$$

Note that x_3 is the third element of x. A plot of the unit-ramp response curve $z(t)$ can be obtained by entering MATLAB Program 5–11 into the computer. A plot of the unit-ramp response curve obtained from this MATLAB program is shown in Figure 5–27.

MATLAB Program 5–11

```
% --------------- Unit-ramp response ---------------

% ***** The unit-ramp response is obtained by adding a new
% state variable x3. The dimension of the state equation
% is enlarged by one *****

% ***** Enter matrices A, B, C, and D of the original state
% equation and output equation *****

A = [0  1;-1  -1];
B = [0;  1];
C = [1  0];
D = [0];

% ***** Enter matrices AA, BB, CC, and DD of the new,
% enlarged state equation and output equation *****

AA = [A  zeros(2,1);C  0];
BB = [B;0];
CC = [0  0  1];
DD = [0];

% ***** Enter step-response command: [z,x,t] = step(AA,BB,CC,DD) *****

[z,x,t] = step(AA,BB,CC,DD);

% ***** In plotting x3 add the unit-ramp input t in the plot
% by entering the following command: plot(t,x3,'o',t,t,'-') *****

x3 = [0  0  1]*x'; plot(t,x3,'o',t,t,'-')
grid
title('Unit-Ramp Response')
xlabel('t Sec')
ylabel('Input and Output')
```

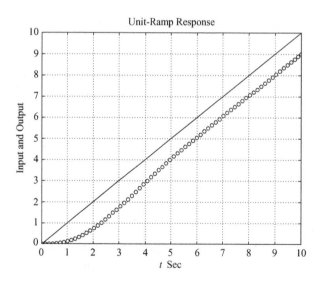

Figure 5–27
Unit-ramp response
curve.

Obtaining Response to Arbitrary Input. To obtain the response to an arbitrary input, the command lsim may be used. The commands like

$$lsim(num,den,r,t)$$
$$lsim(A,B,C,D,u,t)$$
$$y = lsim(num,den,r,t)$$
$$y = lsim(A,B,C,D,u,t)$$

will generate the response to input time function r or u. See the following two examples. (Also, see Problems **A–5–14** through **A–5–16**.)

EXAMPLE 5–6   Using the lsim command, obtain the unit-ramp response of the following system:

$$\frac{C(s)}{R(s)} = \frac{2s + 1}{s^2 + s + 1}$$

We may enter MATLAB Program 5–12 into the computer to obtain the unit-ramp response. The resulting plot is shown in Figure 5–28.

MATLAB Program 5–12

```
% ------- Ramp Response -------
num = [2  1];
 den = [1  1  1];
t = 0:0.1:10;
r = t;
y = lsim(num,den,r,t);
plot(t,r,'-',t,y,'o')
grid
title('Unit-Ramp Response Obtained by Use of Command "lsim"')
xlabel('t Sec')
ylabel('Unit-Ramp Input and System Output')
text(6.3,4.6,'Unit-Ramp Input')
text(4.75,9.0,'Output')
```

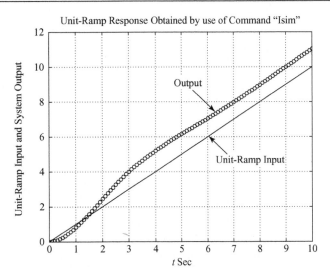

Figure 5–28
Unit-ramp response.

EXAMPLE 5–7 Consider the system

$$\begin{bmatrix} \dot{x}_1 \\ \dot{x}_2 \end{bmatrix} = \begin{bmatrix} -1 & 0.5 \\ -1 & 0 \end{bmatrix} \begin{bmatrix} x_1 \\ x_2 \end{bmatrix} + \begin{bmatrix} 0 \\ 1 \end{bmatrix} u$$

$$y = \begin{bmatrix} 1 & 0 \end{bmatrix} \begin{bmatrix} x_1 \\ x_2 \end{bmatrix}$$

Using MATLAB, obtain the response curves $y(t)$ when the input u is given by

1. u = unit-step input
2. $u = e^{-t}$

Assume that the initial state is $\mathbf{x}(0) = \mathbf{0}$.

A possible MATLAB program to produce the responses of this system to the unit-step input $[u = 1(t)]$ and the exponential input $[u = e^{-t}]$ is shown in MATLAB Program 5–13. The resulting response curves are shown in Figures 5–29(a) and (b), respectively.

MATLAB Program 5–13

```
t = 0:0.1:12;
A = [-1  0.5;-1  0];
B = [0;1];
C = [1  0];
D = [0];

% For the unit-step input u = 1(t), use the command "y = step(A,B,C,D,1,t)".

y = step(A,B,C,D,1,t);
plot(t,y)
grid
title('Unit-Step Response')
xlabel('t Sec')
ylabel('Output')

% For the response to exponential input u = exp(-t), use the command
% "z = lsim(A,B,C,D,u,t)".

u = exp(-t);
z = lsim(A,B,C,D,u,t);
plot(t,u,'-',t,z,'o')
grid
title('Response to Exponential Input u = exp(-t)')
xlabel('t Sec')
ylabel('Exponential Input and System Output')
text(2.3,0.49,'Exponential input')
text(6.4,0.28,'Output')
```

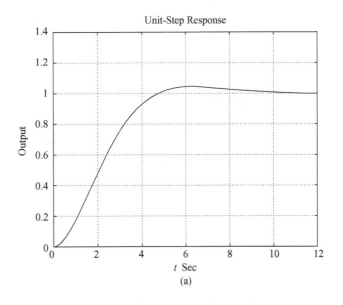

Unit-Step Response

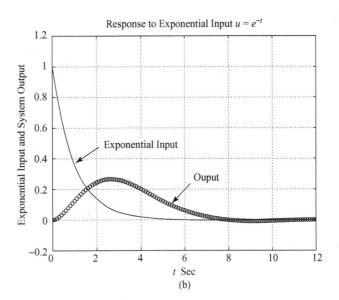

Figure 5–29
(a) Unit-step
response;
(b) response to input
$u = e^{-t}$.

Response to Exponential Input $u = e^{-t}$

Response to Initial Condition. In what follows we shall present a few methods for obtaining the response to an initial condition. Commands that we may use are "step" or "initial". We shall first present a method to obtain the response to the initial condition using a simple example. Then we shall discuss the response to the initial condition when the system is given in state-space form. Finally, we shall present a command initial to obtain the response of a system given in a state-space form.

EXAMPLE 5–8 Consider the mechanical system shown in Figure 5–30, where $m = 1$ kg, $b = 3$ N-sec/m, and $k = 2$ N/m. Assume that at $t = 0$ the mass m is pulled downward such that $x(0) = 0.1$ m and $\dot{x}(0) = 0.05$ m/sec. The displacement $x(t)$ is measured from the equilibrium position before the mass is pulled down. Obtain the motion of the mass subjected to the initial condition. (Assume no external forcing function.)

The system equation is

$$m\ddot{x} + b\dot{x} + kx = 0$$

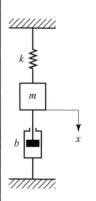

Figure 5–30
Mechanical system.

with the initial conditions $x(0) = 0.1$ m and $\dot{x}(0) = 0.05$ m/sec. (x is measured from the equilibrium position.) The Laplace transform of the system equation gives

$$m\left[s^2X(s) - sx(0) - \dot{x}(0)\right] + b\left[sX(s) - x(0)\right] + kX(s) = 0$$

or

$$(ms^2 + bs + k)X(s) = mx(0)s + m\dot{x}(0) + bx(0)$$

Solving this last equation for $X(s)$ and substituting the given numerical values, we obtain

$$X(s) = \frac{mx(0)s + m\dot{x}(0) + bx(0)}{ms^2 + bs + k}$$

$$= \frac{0.1s + 0.35}{s^2 + 3s + 2}$$

This equation can be written as

$$X(s) = \frac{0.1s^2 + 0.35s}{s^2 + 3s + 2}\frac{1}{s}$$

Hence the motion of the mass m may be obtained as the unit-step response of the following system:

$$G(s) = \frac{0.1s^2 + 0.35s}{s^2 + 3s + 2}$$

MATLAB Program 5–14 will give a plot of the motion of the mass. The plot is shown in Figure 5–31.

MATLAB Program 5–14

```
% -------------- Response to initial condition ---------------

% ***** System response to initial condition is converted to
% a unit-step response by modifying the numerator polynomial *****

% ***** Enter the numerator and denominator of the transfer
% function G(s) *****

num = [0.1  0.35  0];
den = [1  3  2];

% ***** Enter the following step-response command *****

step(num,den)

% ***** Enter grid and title of the plot *****

grid
title('Response of Spring-Mass-Damper System to Initial Condition')
```

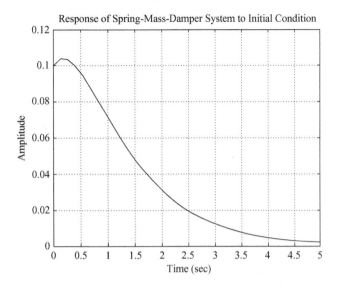

Figure 5–31
Response of the
mechanical system
considered in
Example 5–8.

Response of Spring-Mass-Damper System to Initial Condition

Response to Initial Condition (State-Space Approach, Case 1).   Consider the system defined by

$$\dot{\mathbf{x}} = \mathbf{Ax}, \qquad \mathbf{x}(0) = \mathbf{x}_0 \tag{5–49}$$

Let us obtain the response $\mathbf{x}(t)$ when the initial condition $\mathbf{x}(0)$ is specified. Assume that there is no external input function acting on this system. Assume also that $\mathbf{x}$ is an n-vector.

First, take Laplace transforms of both sides of Equation (5–49).

$$s\mathbf{X}(s) - \mathbf{x}(0) = \mathbf{AX}(s)$$

This equation can be rewritten as

$$s\mathbf{X}(s) = \mathbf{AX}(s) + \mathbf{x}(0) \tag{5–50}$$

Taking the inverse Laplace transform of Equation (5–50), we obtain

$$\dot{\mathbf{x}} = \mathbf{Ax} + \mathbf{x}(0)\,\delta(t) \tag{5–51}$$

(Notice that by taking the Laplace transform of a differential equation and then by taking the inverse Laplace transform of the Laplace-transformed equation we generate a differential equation that involves the initial condition.)

Now define

$$\dot{\mathbf{z}} = \mathbf{x} \tag{5–52}$$

Then Equation (5–51) can be written as

$$\ddot{\mathbf{z}} = \mathbf{A\dot{z}} + \mathbf{x}(0)\,\delta(t) \tag{5–53}$$

By integrating Equation (5–53) with respect to t, we obtain

$$\dot{\mathbf{z}} = \mathbf{Az} + \mathbf{x}(0)1(t) = \mathbf{Az} + \mathbf{B}u \tag{5–54}$$

where

$$\mathbf{B} = \mathbf{x}(0), \qquad u = 1(t)$$

Section 5–5 / Transient-Response Analysis with MATLAB

205

Referring to Equation (5–52), the state $\mathbf{x}(t)$ is given by $\dot{\mathbf{z}}(t)$. Thus,

$$\mathbf{x} = \dot{\mathbf{z}} = \mathbf{Az} + \mathbf{B}u \tag{5–55}$$

The solution of Equations (5–54) and (5–55) gives the response to the initial condition.

Summarizing, the response of Equation (5–49) to the initial condition $\mathbf{x}(0)$ is obtained by solving the following state-space equations:

$$\dot{\mathbf{z}} = \mathbf{Az} + \mathbf{B}u$$

$$\mathbf{x} = \mathbf{Az} + \mathbf{B}u$$

where

$$\mathbf{B} = \mathbf{x}(0), \qquad u = 1(t)$$

MATLAB commands to obtain the response curves, where we do not specify the time vector t (that is, we let the time vector be determined automatically by MATLAB), are given next.

```
% Specify matrices A and B
[x,z,t] = step(A,B,A,B);
x1 = [1  0  0 ... 0]*x';
x2 = [0  1  0 ... 0]*x';

        .

        .

        .

xn = [0  0  0 ... 1]*x';
plot(t,x1,t,x2, ... ,t,xn)
```

If we choose the time vector t (for example, let the computation time duration be from t = 0 to t = tp with the computing time increment of Δt), then we use the following MATLAB commands:

```
t = 0: Δt: tp;
% Specify matrices A and B
[x,z,t] = step(A,B,A,B,1,t);
x1 = [1  0  0 ... 0]*x';
x2 = [0  1  0 ... 0]*x';

        .

        .

        .

xn = [0  0  0 ... 1]*x';
plot(t,x1,t,x2, ... ,t,xn)
```

(See, for example, Example 5–9.)

Response to Initial Condition (State-Space Approach, Case 2). Consider the system defined by

$$\dot{\mathbf{x}} = \mathbf{Ax}, \qquad \mathbf{x}(0) = \mathbf{x}_0 \tag{5–56}$$

$$\mathbf{y} = \mathbf{Cx} \tag{5–57}$$

(Assume that $\mathbf{x}$ is an n-vector and $\mathbf{y}$ is an m-vector.)

Similar to case 1, by defining

$$\dot{\mathbf{z}} = \mathbf{x}$$

we can obtain the following equation:

$$\dot{\mathbf{z}} = \mathbf{Az} + \mathbf{x}(0)1(t) = \mathbf{Az} + \mathbf{B}u \tag{5–58}$$

where

$$\mathbf{B} = \mathbf{x}(0), \qquad u = 1(t)$$

Noting that $\mathbf{x} = \dot{\mathbf{z}}$, Equation (5–57) can be written as

$$\mathbf{y} = \mathbf{C}\dot{\mathbf{z}} \tag{5–59}$$

By substituting Equation (5–58) into Equation (5–59), we obtain

$$\mathbf{y} = \mathbf{C}(\mathbf{Az} + \mathbf{B}u) = \mathbf{CAz} + \mathbf{CB}u \tag{5–60}$$

The solution of Equations (5–58) and (5–60), rewritten here

$$\dot{\mathbf{z}} = \mathbf{Az} + \mathbf{B}u$$

$$\mathbf{y} = \mathbf{CAz} + \mathbf{CB}u$$

where $\mathbf{B} = \mathbf{x}(0)$ and $u = 1(t)$, gives the response of the system to a given initial condition. MATLAB commands to obtain the response curves (output curves y1 versus t, y2 versus t, ... , ym versus t) are shown next for two cases:

Case A.   When the time vector t is not specified (that is, the time vector t is to be determined automatically by MATLAB):

```
% Specify matrices A, B, and C
[y,z,t] = step(A,B,C*A,C*B);
y1 = [1  0  0 ... 0]*y';
y2 = [0  1  0 ... 0]*y';
       .

       .

       .

ym = [0  0  0 ... 1]*y';
plot(t,y1,t,y2, ... ,t,ym)
```

When the time vector t is specified:

$$t = 0: \Delta t: tp;$$

% Specify matrices A, B, and C

[y,z,t] = step(A,B,C*A,C*B,1,t)

y1 = [1 0 0 ... 0]*y';

y2 = [0 1 0 ... 0]*y';

.

.

.

ym = [0 0 0 ... 1]*y';

plot(t,y1,t,y2, ... ,t,ym)

EXAMPLE 5–9 Obtain the response of the system subjected to the given initial condition.

$$\begin{bmatrix} \dot{x}_1 \\ \dot{x}_2 \end{bmatrix} = \begin{bmatrix} 0 & 1 \\ -10 & -5 \end{bmatrix} \begin{bmatrix} x_1 \\ x_2 \end{bmatrix}, \quad \begin{bmatrix} x_1(0) \\ x_2(0) \end{bmatrix} = \begin{bmatrix} 2 \\ 1 \end{bmatrix}$$

or

$$\dot{\mathbf{x}} = \mathbf{A}\mathbf{x}, \quad \mathbf{x}(0) = \mathbf{x}_0$$

Obtaining the response of the system to the given initial condition resolves to solving the unit-step response of the following system:

$$\dot{\mathbf{z}} = \mathbf{A}\mathbf{z} + \mathbf{B}u$$

$$\mathbf{x} = \mathbf{A}\mathbf{z} + \mathbf{B}u$$

where

$$\mathbf{B} = \mathbf{x}(0), \quad u = 1(t)$$

Hence a possible MATLAB program for obtaining the response may be given as shown in MATLAB Program 5–15. The resulting response curves are shown in Figure 5–32.

MATLAB Program 5–15

```
t = 0:0.01:3;
A = [0  1;-10  -5];
B = [2;1];
[x,z,t] = step(A,B,A,B,1,t);
x1 = [1  0]*x';
x2 = [0  1]*x';
plot(t,x1,'x',t,x2,'-')
grid
title('Response to Initial Condition')
xlabel('t Sec')
ylabel('State Variables x1 and x2')
gtext('x1')
gtext('x2')
```

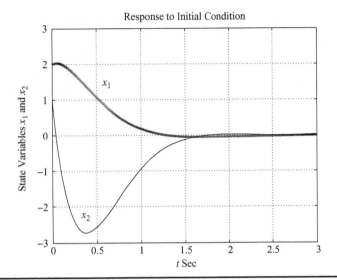

Response to Initial Condition

Figure 5–32
Response of system
in Example 5–9 to
initial condition.

For an illustrative example of how to use Equations (5–58) and (5–60) to find the response to the initial condition, see Problem **A–5–16**.

Obtaining Response to Initial Condition by Use of Command Initial. If the system is given in the state-space form, then the following command

$$initial(A,B,C,D,[initial\ condition],t)$$

will produce the response to the initial condition.
Suppose that we have the system defined by

$$\dot{\mathbf{x}} = \mathbf{A}\mathbf{x} + \mathbf{B}u, \qquad \mathbf{x}(0) = \mathbf{x}_0$$

$$y = \mathbf{C}\mathbf{x} + Du$$

where

$$\mathbf{A} = \begin{bmatrix} 0 & 1 \\ -10 & -5 \end{bmatrix}, \qquad \mathbf{B} = \begin{bmatrix} 0 \\ 0 \end{bmatrix}, \qquad \mathbf{C} = \begin{bmatrix} 0 & 0 \end{bmatrix}, \qquad D = 0$$

$$\mathbf{x}_0 = \begin{bmatrix} 2 \\ 1 \end{bmatrix}$$

Then the command "initial" can be used as shown in MATLAB Program 5–16 to obtain the response to the initial condition. The response curves $x_1(t)$ and $x_2(t)$ are shown in Figure 5–33. They are the same as those shown in Figure 5–32.

MATLAB Program 5–16

```
t = 0:0.05:3;
A = [0  1;-10  -5];
B = [0;0];
C = [0  0];
D = [0];
[y,x] = initial(A,B,C,D,[2;1],t);
x1 = [1  0]*x';
x2 = [0  1]*x';
plot(t,x1,'o',t,x1,t,x2,'x',t,x2)
grid
title('Response to Initial Condition')
xlabel('t Sec')
ylabel('State Variables x1 and x2')
gtext('x1')
gtext('x2')
```

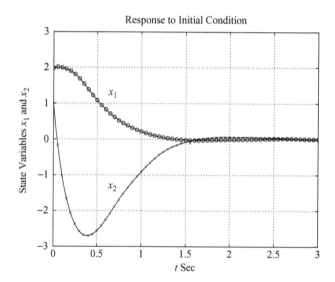

Figure 5–33
Response curves to
initial condition.

EXAMPLE 5–10 Consider the following system that is subjected to the initial condition. (No external forcing function is present.)

$$\dddot{y} + 8\ddot{y} + 17\dot{y} + 10y = 0$$

$$y(0) = 2, \qquad \dot{y}(0) = 1, \qquad \ddot{y}(0) = 0.5$$

Obtain the response $y(t)$ to the given initial condition.

Chapter 5 / Transient and Steady-State Response Analyses

By defining the state variables as

$$x_1 = y$$
$$x_2 = \dot{y}$$
$$x_3 = \ddot{y}$$

we obtain the following state-space representation for the system:

$$\begin{bmatrix} \dot{x}_1 \\ \dot{x}_2 \\ \dot{x}_3 \end{bmatrix} = \begin{bmatrix} 0 & 1 & 0 \\ 0 & 0 & 1 \\ -10 & -17 & -8 \end{bmatrix} \begin{bmatrix} x_1 \\ x_2 \\ x_3 \end{bmatrix}, \qquad \begin{bmatrix} x_1(0) \\ x_2(0) \\ x_3(0) \end{bmatrix} = \begin{bmatrix} 2 \\ 1 \\ 0.5 \end{bmatrix}$$

$$y = \begin{bmatrix} 1 & 0 & 0 \end{bmatrix} \begin{bmatrix} x_1 \\ x_2 \\ x_3 \end{bmatrix}$$

A possible MATLAB program to obtain the response $y(t)$ is given in MATLAB Program 5–17. The resulting response curve is shown in Figure 5–34.

MATLAB Program 5–17

```
t = 0:0.05:10;
A = [0 1 0;0 0 1;-10 -17 -8];
B = [0;0;0];
C = [1 0 0];
D = [0];
y = initial(A,B,C,D,[2;1;0.5],t);
plot(t,y)
grid
title('Response to Initial Condition')
xlabel('t (sec)')
ylabel('Output y')
```

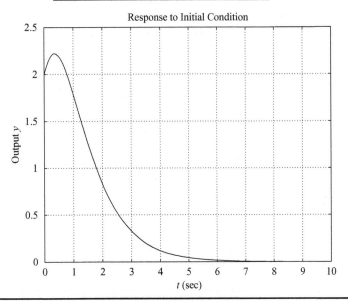

Figure 5–34
Response $y(t)$ to
initial condition.

The most important problem in linear control systems concerns stability. That is, under what conditions will a system become unstable? If it is unstable, how should we stabilize the system? In Section 5–4 it was stated that a control system is stable if and only if all closed-loop poles lie in the left-half s plane. Most linear closed-loop systems have closed-loop transfer functions of the form

$$\frac{C(s)}{R(s)} = \frac{b_0 s^m + b_1 s^{m-1} + \cdots + b_{m-1} s + b_m}{a_0 s^n + a_1 s^{n-1} + \cdots + a_{n-1} s + a_n} = \frac{B(s)}{A(s)}$$

where the a's and b's are constants and $m \leq n$. A simple criterion, known as Routh's stability criterion, enables us to determine the number of closed-loop poles that lie in the right-half s plane without having to factor the denominator polynomial. (The polynomial may include parameters that MATLAB cannot handle.)

Routh's Stability Criterion. Routh's stability criterion tells us whether or not there are unstable roots in a polynomial equation without actually solving for them. This stability criterion applies to polynomials with only a finite number of terms. When the criterion is applied to a control system, information about absolute stability can be obtained directly from the coefficients of the characteristic equation.

The procedure in Routh's stability criterion is as follows:

1. Write the polynomial in s in the following form:

$$a_0 s^n + a_1 s^{n-1} + \cdots + a_{n-1} s + a_n = 0 \tag{5–61}$$

where the coefficients are real quantities. We assume that $a_n \neq 0$; that is, any zero root has been removed.

2. If any of the coefficients are zero or negative in the presence of at least one positive coefficient, a root or roots exist that are imaginary or that have positive real parts. Therefore, in such a case, the system is not stable. If we are interested in only the absolute stability, there is no need to follow the procedure further. Note that all the coefficients must be positive. This is a necessary condition, as may be seen from the following argument: A polynomial in s having real coefficients can always be factored into linear and quadratic factors, such as $(s + a)$ and $(s^2 + bs + c)$, where a, b, and c are real. The linear factors yield real roots and the quadratic factors yield complex-conjugate roots of the polynomial. The factor $(s^2 + bs + c)$ yields roots having negative real parts only if b and c are both positive. For all roots to have negative real parts, the constants a, b, c, and so on, in all factors must be positive. The product of any number of linear and quadratic factors containing only positive coefficients always yields a polynomial with positive coefficients. It is important to note that the condition that all the coefficients be positive is not sufficient to assure stability. The necessary but not sufficient condition for stability is that the coefficients of Equation (5–61) all be present and all have a positive sign. (If all a's are negative, they can be made positive by multiplying both sides of the equation by -1.)

3. If all coefficients are positive, arrange the coefficients of the polynomial in rows and columns according to the following pattern:

$$
\begin{array}{ll}
s^n   & a_0 \quad a_2 \quad a_4 \quad a_6 \quad \cdots \\
s^{n-1} & a_1 \quad a_3 \quad a_5 \quad a_7 \quad \cdots \\
s^{n-2} & b_1 \quad b_2 \quad b_3 \quad b_4 \quad \cdots \\
s^{n-3} & c_1 \quad c_2 \quad c_3 \quad c_4 \quad \cdots \\
s^{n-4} & d_1 \quad d_2 \quad d_3 \quad d_4 \quad \cdots \\
\cdot & \quad\; \cdot \quad\; \cdot \\
\cdot & \quad\; \cdot \quad\; \cdot \\
\cdot & \quad\; \cdot \quad\; \cdot \\
s^2   & e_1 \quad e_2 \\
s^1   & f_1 \\
s^0   & g_1
\end{array}
$$

The process of forming rows continues until we run out of elements. (The total number of rows is $n + 1$.) The coefficients b_1, b_2, b_3, and so on, are evaluated as follows:

$$
b_1 = \frac{a_1 a_2 - a_0 a_3}{a_1}
$$

$$
b_2 = \frac{a_1 a_4 - a_0 a_5}{a_1}
$$

$$
b_3 = \frac{a_1 a_6 - a_0 a_7}{a_1}
$$

$$
\cdot
$$
$$
\cdot
$$
$$
\cdot
$$

The evaluation of the b's is continued until the remaining ones are all zero. The same pattern of cross-multiplying the coefficients of the two previous rows is followed in evaluating the c's, d's, e's, and so on. That is,

$$
c_1 = \frac{b_1 a_3 - a_1 b_2}{b_1}
$$

$$
c_2 = \frac{b_1 a_5 - a_1 b_3}{b_1}
$$

$$
c_3 = \frac{b_1 a_7 - a_1 b_4}{b_1}
$$

$$
\cdot
$$
$$
\cdot
$$
$$
\cdot
$$

and

$$d_1 = \frac{c_1 b_2 - b_1 c_2}{c_1}$$

$$d_2 = \frac{c_1 b_3 - b_1 c_3}{c_1}$$

.

.

.

This process is continued until the nth row has been completed. The complete array of coefficients is triangular. Note that in developing the array an entire row may be divided or multiplied by a positive number in order to simplify the subsequent numerical calculation without altering the stability conclusion.

Routh's stability criterion states that the number of roots of Equation (5–61) with positive real parts is equal to the number of changes in sign of the coefficients of the first column of the array. It should be noted that the exact values of the terms in the first column need not be known; instead, only the signs are needed. The necessary and sufficient condition that all roots of Equation (5–61) lie in the left-half s plane is that all the coefficients of Equation (5–61) be positive and all terms in the first column of the array have positive signs.

EXAMPLE 5–11   Let us apply Routh's stability criterion to the following third-order polynomial:

$$a_0 s^3 + a_1 s^2 + a_2 s + a_3 = 0$$

where all the coefficients are positive numbers. The array of coefficients becomes

$$
\begin{array}{ccc}
s^3 & a_0 & a_2 \\
s^2 & a_1 & a_3 \\
s^1 & \dfrac{a_1 a_2 - a_0 a_3}{a_1} & \\
s^0 & a_3 &
\end{array}
$$

The condition that all roots have negative real parts is given by

$$a_1 a_2 > a_0 a_3$$

EXAMPLE 5–12   Consider the following polynomial:

$$s^4 + 2s^3 + 3s^2 + 4s + 5 = 0$$

Let us follow the procedure just presented and construct the array of coefficients. (The first two rows can be obtained directly from the given polynomial. The remaining terms are obtained from these. If any coefficients are missing, they may be replaced by zeros in the array.)

$$
\begin{array}{cccc}
s^4 & 1 & 3 & 5 \\
s^3 & 2 & 4 & 0 \\
\\
s^2 & 1 & 5 \\
s^1 & -6 \\
s^0 & 5
\end{array}
\qquad
\begin{array}{cccc}
s^4 & 1 & 3 & 5 \\
s^3 & \cancel{2} & \cancel{4} & \cancel{0} & \text{The second row is divided} \\
 & 1 & 2 & 0 & \text{by 2.} \\
s^2 & 1 & 5 \\
s^1 & -3 \\
s^0 & 5
\end{array}
$$

In this example, the number of changes in sign of the coefficients in the first column is 2. This means that there are two roots with positive real parts. Note that the result is unchanged when the coefficients of any row are multiplied or divided by a positive number in order to simplify the computation.

Special Cases. If a first-column term in any row is zero, but the remaining terms are not zero or there is no remaining term, then the zero term is replaced by a very small positive number ϵ and the rest of the array is evaluated. For example, consider the following equation:

$$
s^3 + 2s^2 + s + 2 = 0 \tag{5–62}
$$

The array of coefficients is

$$
\begin{array}{ccc}
s^3 & 1 & 1 \\
s^2 & 2 & 2 \\
s^1 & 0 \approx \epsilon \\
s^0 & 2
\end{array}
$$

If the sign of the coefficient above the zero (ϵ) is the same as that below it, it indicates that there are a pair of imaginary roots. Actually, Equation (5–62) has two roots at $s = \pm j$.

If, however, the sign of the coefficient above the zero (ϵ) is opposite that below it, it indicates that there is one sign change. For example, for the equation

$$
s^3 - 3s + 2 = (s - 1)^2(s + 2) = 0
$$

the array of coefficients is

One sign change:

$$
\begin{array}{ccc}
s^3 & 1 & -3 \\
s^2 & 0 \approx \epsilon & 2 \\
\\
s^1 & -3 - \dfrac{2}{\epsilon} \\
s^0 & 2
\end{array}
$$

One sign change:

There are two sign changes of the coefficients in the first column. So there are two roots in the right-half s plane. This agrees with the correct result indicated by the factored form of the polynomial equation.

If all the coefficients in any derived row are zero, it indicates that there are roots of equal magnitude lying radially opposite in the s plane—that is, two real roots with equal magnitudes and opposite signs and/or two conjugate imaginary roots. In such a case, the evaluation of the rest of the array can be continued by forming an auxiliary polynomial with the coefficients of the last row and by using the coefficients of the derivative of this polynomial in the next row. Such roots with equal magnitudes and lying radially opposite in the s plane can be found by solving the auxiliary polynomial, which is always even. For a $2n$-degree auxiliary polynomial, there are n pairs of equal and opposite roots. For example, consider the following equation:

$$s^5 + 2s^4 + 24s^3 + 48s^2 - 25s - 50 = 0$$

The array of coefficients is

$$
\begin{array}{llll}
s^5 & 1 & 24 & -25 \\
s^4 & 2 & 48 & -50 \quad \leftarrow \text{Auxiliary polynomial } P(s) \\
s^3 & 0 & 0 &
\end{array}
$$

The terms in the s^3 row are all zero. (Note that such a case occurs only in an odd-numbered row.) The auxiliary polynomial is then formed from the coefficients of the s^4 row. The auxiliary polynomial $P(s)$ is

$$P(s) = 2s^4 + 48s^2 - 50$$

which indicates that there are two pairs of roots of equal magnitude and opposite sign (that is, two real roots with the same magnitude but opposite signs or two complex-conjugate roots on the imaginary axis). These pairs are obtained by solving the auxiliary polynomial equation $P(s) = 0$. The derivative of $P(s)$ with respect to s is

$$\frac{dP(s)}{ds} = 8s^3 + 96s$$

The terms in the s^3 row are replaced by the coefficients of the last equation—that is, 8 and 96. The array of coefficients then becomes

$$
\begin{array}{lll}
s^5 & 1 & 24 & -25 \\
s^4 & 2 & 48 & -50 \\
s^3 & 8 & 96 & \quad \leftarrow \text{Coefficients of } dP(s)/ds \\
s^2 & 24 & -50 \\
s^1 & 112.7 & 0 \\
s^0 & -50 &
\end{array}
$$

We see that there is one change in sign in the first column of the new array. Thus, the original equation has one root with a positive real part. By solving for roots of the auxiliary polynomial equation,

$$2s^4 + 48s^2 - 50 = 0$$

we obtain

$$s^2 = 1, \qquad s^2 = -25$$

or

$$s = \pm 1, \qquad s = \pm j5$$

These two pairs of roots of $P(s)$ are a part of the roots of the original equation. As a matter of fact, the original equation can be written in factored form as follows:

$$(s + 1)(s - 1)(s + j5)(s - j5)(s + 2) = 0$$

Clearly, the original equation has one root with a positive real part.

Relative Stability Analysis. Routh's stability criterion provides the answer to the question of absolute stability. This, in many practical cases, is not sufficient. We usually require information about the relative stability of the system. A useful approach for examining relative stability is to shift the s-plane axis and apply Routh's stability criterion. That is, we substitute

$$s = \hat{s} - \sigma \qquad (\sigma = \text{constant})$$

into the characteristic equation of the system, write the polynomial in terms of $\hat{s}$; and apply Routh's stability criterion to the new polynomial in $\hat{s}$. The number of changes of sign in the first column of the array developed for the polynomial in $\hat{s}$ is equal to the number of roots that are located to the right of the vertical line $s = -\sigma$. Thus, this test reveals the number of roots that lie to the right of the vertical line $s = -\sigma$.

Application of Routh's Stability Criterion to Control-System Analysis. Routh's stability criterion is of limited usefulness in linear control-system analysis, mainly because it does not suggest how to improve relative stability or how to stabilize an unstable system. It is possible, however, to determine the effects of changing one or two parameters of a system by examining the values that cause instability. In the following, we shall consider the problem of determining the stability range of a parameter value.

Consider the system shown in Figure 5–35. Let us determine the range of K for stability. The closed-loop transfer function is

$$\frac{C(s)}{R(s)} = \frac{K}{s(s^2 + s + 1)(s + 2) + K}$$

The characteristic equation is

$$s^4 + 3s^3 + 3s^2 + 2s + K = 0$$

The array of coefficients becomes

$$
\begin{array}{c c c c}
s^4 & 1 & 3 & K \\
s^3 & 3 & 2 & 0 \\
s^2 & \frac{7}{3} & K & \\
s^1 & 2 - \frac{9}{7}K & & \\
s^0 & K & &
\end{array}
$$

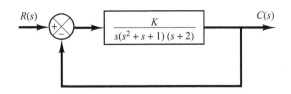

Figure 5–35
Control system.

For stability, K must be positive, and all coefficients in the first column must be positive. Therefore,

$$\frac{14}{9} > K > 0$$

When $K = \frac{14}{9}$, the system becomes oscillatory and, mathematically, the oscillation is sustained at constant amplitude.

Note that the ranges of design parameters that lead to stability may be determined by use of Routh's stability criterion.

5–7 EFFECTS OF INTEGRAL AND DERIVATIVE CONTROL ACTIONS ON SYSTEM PERFORMANCE

In this section, we shall investigate the effects of integral and derivative control actions on the system performance. Here we shall consider only simple systems, so that the effects of integral and derivative control actions on system performance can be clearly seen.

Integral Control Action. In the proportional control of a plant whose transfer function does not possess an integrator $1/s$, there is a steady-state error, or offset, in the response to a step input. Such an offset can be eliminated if the integral control action is included in the controller.

In the integral control of a plant, the control signal—the output signal from the controller—at any instant is the area under the actuating-error-signal curve up to that instant. The control signal $u(t)$ can have a nonzero value when the actuating error signal $e(t)$ is zero, as shown in Figure 5–36(a). This is impossible in the case of the proportional controller, since a nonzero control signal requires a nonzero actuating error signal. (A nonzero actuating error signal at steady state means that there is an offset.) Figure 5–36(b) shows the curve $e(t)$ versus t and the corresponding curve $u(t)$ versus t when the controller is of the proportional type.

Note that integral control action, while removing offset or steady-state error, may lead to oscillatory response of slowly decreasing amplitude or even increasing amplitude, both of which are usually undesirable.

Figure 5–36
(a) Plots of $e(t)$ and $u(t)$ curves showing nonzero control signal when the actuating error signal is zero (integral control); (b) plots of $e(t)$ and $u(t)$ curves showing zero control signal when the actuating error signal is zero (proportional control).

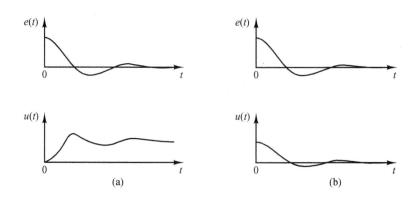

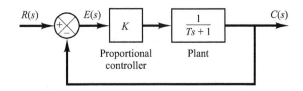

Figure 5–37
Proportional control
system.

Proportional Control of Systems. We shall show that the proportional control of a system without an integrator will result in a steady-state error with a step input. We shall then show that such an error can be eliminated if integral control action is included in the controller.

Consider the system shown in Figure 5–37. Let us obtain the steady-state error in the unit-step response of the system. Define

$$G(s) = \frac{K}{Ts + 1}$$

Since

$$\frac{E(s)}{R(s)} = \frac{R(s) - C(s)}{R(s)} = 1 - \frac{C(s)}{R(s)} = \frac{1}{1 + G(s)}$$

the error $E(s)$ is given by

$$E(s) = \frac{1}{1 + G(s)} R(s) = \frac{1}{1 + \dfrac{K}{Ts + 1}} R(s)$$

For the unit-step input $R(s) = 1/s$, we have

$$E(s) = \frac{Ts + 1}{Ts + 1 + K} \frac{1}{s}$$

The steady-state error is

$$e_{ss} = \lim_{t \to \infty} e(t) = \lim_{s \to 0} sE(s) = \lim_{s \to 0} \frac{Ts + 1}{Ts + 1 + K} = \frac{1}{K + 1}$$

Such a system without an integrator in the feedforward path always has a steady-state error in the step response. Such a steady-state error is called an offset. Figure 5–38 shows the unit-step response and the offset.

Figure 5–38
Unit-step response
and offset.

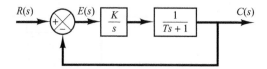

Figure 5–39
Integral control
system.

Integral Control of Systems.   Consider the system shown in Figure 5–39. The controller is an integral controller. The closed-loop transfer function of the system is

$$\frac{C(s)}{R(s)} = \frac{K}{s(Ts + 1) + K}$$

Hence

$$\frac{E(s)}{R(s)} = \frac{R(s) - C(s)}{R(s)} = \frac{s(Ts + 1)}{s(Ts + 1) + K}$$

Since the system is stable, the steady-state error for the unit-step response can be obtained by applying the final-value theorem, as follows:

$$e_{ss} = \lim_{s \to 0} sE(s)$$

$$= \lim_{s \to 0} \frac{s^2(Ts + 1)}{Ts^2 + s + K} \frac{1}{s}$$

$$= 0$$

Integral control of the system thus eliminates the steady-state error in the response to the step input. This is an important improvement over the proportional control alone, which gives offset.

Response to Torque Disturbances (Proportional Control).   Let us investigate the effect of a torque disturbance occurring at the load element. Consider the system shown in Figure 5–40. The proportional controller delivers torque T to position the load element, which consists of moment of inertia and viscous friction. Torque disturbance is denoted by D.

Assuming that the reference input is zero or $R(s) = 0$, the transfer function between $C(s)$ and $D(s)$ is given by

$$\frac{C(s)}{D(s)} = \frac{1}{Js^2 + bs + K_p}$$

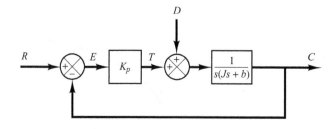

Figure 5–40
Control system with
a torque disturbance.

Hence

$$\frac{E(s)}{D(s)} = -\frac{C(s)}{D(s)} = -\frac{1}{Js^2 + bs + K_p}$$

The steady-state error due to a step disturbance torque of magnitude T_d is given by

$$e_{ss} = \lim_{s \to 0} sE(s)$$

$$= \lim_{s \to 0} \frac{-s}{Js^2 + bs + K_p} \frac{T_d}{s}$$

$$= -\frac{T_d}{K_p}$$

At steady state, the proportional controller provides the torque $-T_d$, which is equal in magnitude but opposite in sign to the disturbance torque T_d. The steady-state output due to the step disturbance torque is

$$c_{ss} = -e_{ss} = \frac{T_d}{K_p}$$

The steady-state error can be reduced by increasing the value of the gain K_p. Increasing this value, however, will cause the system response to be more oscillatory.

Response to Torque Disturbances (Proportional-Plus-Integral Control). To eliminate offset due to torque disturbance, the proportional controller may be replaced by a proportional-plus-integral controller.

If integral control action is added to the controller, then, as long as there is an error signal, a torque is developed by the controller to reduce this error, provided the control system is a stable one.

Figure 5–41 shows the proportional-plus-integral control of the load element, consisting of moment of inertia and viscous friction.

The closed-loop transfer function between $C(s)$ and $D(s)$ is

$$\frac{C(s)}{D(s)} = \frac{s}{Js^3 + bs^2 + K_ps + \dfrac{K_p}{T_i}}$$

In the absence of the reference input, or $r(t) = 0$, the error signal is obtained from

$$E(s) = -\frac{s}{Js^3 + bs^2 + K_ps + \dfrac{K_p}{T_i}} D(s)$$

Figure 5–41
Proportional-plus-
integral control of a
load element
consisting of moment
of inertia and viscous
friction.

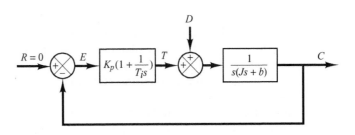

Figure 5-42
Integral control of a
load element
consisting of moment
of inertia and viscous
friction.

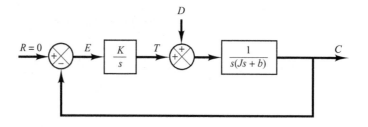

If this control system is stable—that is, if the roots of the characteristic equation

$$Js^3 + bs^2 + K_ps + \frac{K_p}{T_i} = 0$$

have negative real parts—then the steady-state error in the response to a unit-step disturbance torque can be obtained by applying the final-value theorem as follows:

$$e_{ss} = \lim_{s \to 0} sE(s)$$

$$= \lim_{s \to 0} \frac{-s^2}{Js^3 + bs^2 + K_ps + \dfrac{K_p}{T_i}} \frac{1}{s}$$

$$= 0$$

Thus steady-state error to the step disturbance torque can be eliminated if the controller is of the proportional-plus-integral type.

Note that the integral control action added to the proportional controller has converted the originally second-order system to a third-order one. Hence the control system may become unstable for a large value of K_p, since the roots of the characteristic equation may have positive real parts. (The second-order system is always stable if the coefficients in the system differential equation are all positive.)

It is important to point out that if the controller were an integral controller, as in Figure 5-42, then the system always becomes unstable, because the characteristic equation

$$Js^3 + bs^2 + K = 0$$

will have roots with positive real parts. Such an unstable system cannot be used in practice.

Note that in the system of Figure 5-41 the proportional control action tends to stabilize the system, while the integral control action tends to eliminate or reduce steady-state error in response to various inputs.

Derivative Control Action.   Derivative control action, when added to a proportional controller, provides a means of obtaining a controller with high sensitivity. An advantage of using derivative control action is that it responds to the rate of change of the actuating error and can produce a significant correction before the magnitude of the actuating error becomes too large. Derivative control thus anticipates the actuating error, initiates an early corrective action, and tends to increase the stability of the system.

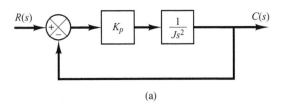

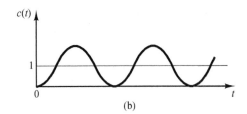

(a)

(b)

Figure 5–43
(a) Proportional
control of a system
with inertia load;
(b) response to a
unit-step input.

Although derivative control does not affect the steady-state error directly, it adds damping to the system and thus permits the use of a larger value of the gain K, which will result in an improvement in the steady-state accuracy.

Because derivative control operates on the rate of change of the actuating error and not the actuating error itself, this mode is never used alone. It is always used in combination with proportional or proportional-plus-integral control action.

Proportional Control of Systems with Inertia Load. Before we discuss further the effect of derivative control action on system performance, we shall consider the proportional control of an inertia load.

Consider the system shown in Figure 5–43(a). The closed-loop transfer function is obtained as

$$\frac{C(s)}{R(s)} = \frac{K_p}{Js^2 + K_p}$$

Since the roots of the characteristic equation

$$Js^2 + K_p = 0$$

are imaginary, the response to a unit-step input continues to oscillate indefinitely, as shown in Figure 5–43(b).

Control systems exhibiting such response characteristics are not desirable. We shall see that the addition of derivative control will stabilize the system.

Proportional-Plus-Derivative Control of a System with Inertia Load. Let us modify the proportional controller to a proportional-plus-derivative controller whose transfer function is $K_p(1 + T_d s)$. The torque developed by the controller is proportional to $K_p(e + T_d \dot{e})$. Derivative control is essentially anticipatory, measures the instantaneous error velocity, and predicts the large overshoot ahead of time and produces an appropriate counteraction before too large an overshoot occurs.

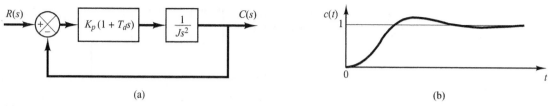

Figure 5–44
(a) Proportional-plus-derivative control of a system with inertia load; (b) response to a unit-step input.

Consider the system shown in Figure 5–44(a). The closed-loop transfer function is given by

$$\frac{C(s)}{R(s)} = \frac{K_p(1 + T_d s)}{Js^2 + K_p T_d s + K_p}$$

The characteristic equation

$$Js^2 + K_p T_d s + K_p = 0$$

now has two roots with negative real parts for positive values of J, K_p, and T_d. Thus derivative control introduces a damping effect. A typical response curve $c(t)$ to a unit-step input is shown in Figure 5–44(b). Clearly, the response curve shows a marked improvement over the original response curve shown in Figure 5–43(b).

Proportional-Plus-Derivative Control of Second-Order Systems.   A compromise between acceptable transient-response behavior and acceptable steady-state behavior may be achieved by use of proportional-plus-derivative control action.

Consider the system shown in Figure 5–45. The closed-loop transfer function is

$$\frac{C(s)}{R(s)} = \frac{K_p + K_d s}{Js^2 + (B + K_d)s + K_p}$$

The steady-state error for a unit-ramp input is

$$e_{ss} = \frac{B}{K_p}$$

The characteristic equation is

$$Js^2 + (B + K_d)s + K_p = 0$$

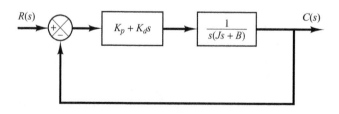

Figure 5–45
Control system.

The effective damping coefficient of this system is thus $B + K_d$ rather than B. Since the damping ratio ζ of this system is

$$\zeta = \frac{B + K_d}{2\sqrt{K_p J}}$$

it is possible to make both the steady-state error e_{ss} for a ramp input and the maximum overshoot for a step input small by making B small, K_p large, and K_d large enough so that ζ is between 0.4 and 0.7.

5–8 STEADY-STATE ERRORS IN UNITY-FEEDBACK CONTROL SYSTEMS

Errors in a control system can be attributed to many factors. Changes in the reference input will cause unavoidable errors during transient periods and may also cause steady-state errors. Imperfections in the system components, such as static friction, backlash, and amplifier drift, as well as aging or deterioration, will cause errors at steady state. In this section, however, we shall not discuss errors due to imperfections in the system components. Rather, we shall investigate a type of steady-state error that is caused by the incapability of a system to follow particular types of inputs.

Any physical control system inherently suffers steady-state error in response to certain types of inputs. A system may have no steady-state error to a step input, but the same system may exhibit nonzero steady-state error to a ramp input. (The only way we may be able to eliminate this error is to modify the system structure.) Whether a given system will exhibit steady-state error for a given type of input depends on the type of open-loop transfer function of the system, to be discussed in what follows.

Classification of Control Systems. Control systems may be classified according to their ability to follow step inputs, ramp inputs, parabolic inputs, and so on. This is a reasonable classification scheme, because actual inputs may frequently be considered combinations of such inputs. The magnitudes of the steady-state errors due to these individual inputs are indicative of the goodness of the system.

Consider the unity-feedback control system with the following open-loop transfer function $G(s)$:

$$G(s) = \frac{K(T_a s + 1)(T_b s + 1)\cdots(T_m s + 1)}{s^N(T_1 s + 1)(T_2 s + 1)\cdots(T_p s + 1)}$$

It involves the term s^N in the denominator, representing a pole of multiplicity N at the origin. The present classification scheme is based on the number of integrations indicated by the open-loop transfer function. A system is called type 0, type 1, type 2,..., if $N = 0$, $N = 1$, $N = 2,...$, respectively. Note that this classification is different from that of the order of a system. As the type number is increased, accuracy is improved; however, increasing the type number aggravates the stability problem. A compromise between steady-state accuracy and relative stability is always necessary.

We shall see later that, if $G(s)$ is written so that each term in the numerator and denominator, except the term s^N, approaches unity as s approaches zero, then the open-loop gain K is directly related to the steady-state error.

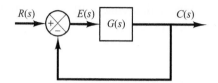

Figure 5–46
Control system.

Steady-State Errors. Consider the system shown in Figure 5–46. The closed-loop transfer function is

$$\frac{C(s)}{R(s)} = \frac{G(s)}{1 + G(s)}$$

The transfer function between the error signal $e(t)$ and the input signal $r(t)$ is

$$\frac{E(s)}{R(s)} = 1 - \frac{C(s)}{R(s)} = \frac{1}{1 + G(s)}$$

where the error $e(t)$ is the difference between the input signal and the output signal.

The final-value theorem provides a convenient way to find the steady-state performance of a stable system. Since $E(s)$ is

$$E(s) = \frac{1}{1 + G(s)} R(s)$$

the steady-state error is

$$e_{ss} = \lim_{t \to \infty} e(t) = \lim_{s \to 0} sE(s) = \lim_{s \to 0} \frac{sR(s)}{1 + G(s)}$$

The static error constants defined in the following are figures of merit of control systems. The higher the constants, the smaller the steady-state error. In a given system, the output may be the position, velocity, pressure, temperature, or the like. The physical form of the output, however, is immaterial to the present analysis. Therefore, in what follows, we shall call the output "position," the rate of change of the output "velocity," and so on. This means that in a temperature control system "position" represents the output temperature, "velocity" represents the rate of change of the output temperature, and so on.

Static Position Error Constant K_p. The steady-state error of the system for a unit-step input is

$$e_{ss} = \lim_{s \to 0} \frac{s}{1 + G(s)} \frac{1}{s}$$

$$= \frac{1}{1 + G(0)}$$

The static position error constant K_p is defined by

$$K_p = \lim_{s \to 0} G(s) = G(0)$$

Thus, the steady-state error in terms of the static position error constant K_p is given by

$$e_{ss} = \frac{1}{1 + K_p}$$

Chapter 5 / Transient and Steady-State Response Analyses

For a type 0 system,

$$K_p = \lim_{s \to 0} \frac{K(T_a s + 1)(T_b s + 1) \cdots}{(T_1 s + 1)(T_2 s + 1) \cdots} = K$$

For a type 1 or higher system,

$$K_p = \lim_{s \to 0} \frac{K(T_a s + 1)(T_b s + 1) \cdots}{s^N (T_1 s + 1)(T_2 s + 1) \cdots} = \infty, \qquad \text{for } N \geq 1$$

Hence, for a type 0 system, the static position error constant K_p is finite, while for a type 1 or higher system, K_p is infinite.

For a unit-step input, the steady-state error e_{ss} may be summarized as follows:

$$e_{ss} = \frac{1}{1 + K}, \qquad \text{for type 0 systems}$$

$$e_{ss} = 0, \qquad \text{for type 1 or higher systems}$$

From the foregoing analysis, it is seen that the response of a feedback control system to a step input involves a steady-state error if there is no integration in the feedforward path. (If small errors for step inputs can be tolerated, then a type 0 system may be permissible, provided that the gain K is sufficiently large. If the gain K is too large, however, it is difficult to obtain reasonable relative stability.) If zero steady-state error for a step input is desired, the type of the system must be one or higher.

Static Velocity Error Constant K_v. The steady-state error of the system with a unit-ramp input is given by

$$e_{ss} = \lim_{s \to 0} \frac{s}{1 + G(s)} \frac{1}{s^2}$$

$$= \lim_{s \to 0} \frac{1}{sG(s)}$$

The static velocity error constant K_v is defined by

$$K_v = \lim_{s \to 0} sG(s)$$

Thus, the steady-state error in terms of the static velocity error constant K_v is given by

$$e_{ss} = \frac{1}{K_v}$$

The term *velocity error* is used here to express the steady-state error for a ramp input. The dimension of the velocity error is the same as the system error. That is, velocity error is not an error in velocity, but it is an error in position due to a ramp input.

For a type 0 system,

$$K_v = \lim_{s \to 0} \frac{sK(T_a s + 1)(T_b s + 1) \cdots}{(T_1 s + 1)(T_2 s + 1) \cdots} = 0$$

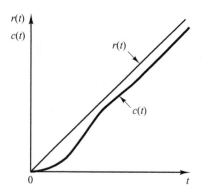

Figure 5–47
Response of a type 1
unity-feedback
system to a ramp
input.

For a type 1 system,

$$K_v = \lim_{s \to 0} \frac{sK(T_a s + 1)(T_b s + 1) \cdots}{s(T_1 s + 1)(T_2 s + 1) \cdots} = K$$

For a type 2 or higher system,

$$K_v = \lim_{s \to 0} \frac{sK(T_a s + 1)(T_b s + 1) \cdots}{s^N(T_1 s + 1)(T_2 s + 1) \cdots} = \infty, \qquad \text{for } N \geq 2$$

The steady-state error e_{ss} for the unit-ramp input can be summarized as follows:

$$e_{ss} = \frac{1}{K_v} = \infty, \qquad \text{for type 0 systems}$$

$$e_{ss} = \frac{1}{K_v} = \frac{1}{K}, \qquad \text{for type 1 systems}$$

$$e_{ss} = \frac{1}{K_v} = 0, \qquad \text{for type 2 or higher systems}$$

The foregoing analysis indicates that a type 0 system is incapable of following a ramp input in the steady state. The type 1 system with unity feedback can follow the ramp input with a finite error. In steady-state operation, the output velocity is exactly the same as the input velocity, but there is a positional error. This error is proportional to the velocity of the input and is inversely proportional to the gain K. Figure 5–47 shows an example of the response of a type 1 system with unity feedback to a ramp input. The type 2 or higher system can follow a ramp input with zero error at steady state.

Static Acceleration Error Constant K_a.   The steady-state error of the system with a unit-parabolic input (acceleration input), which is defined by

$$r(t) = \frac{t^2}{2}, \qquad \text{for } t \geq 0$$

$$= 0, \qquad \text{for } t < 0$$

is given by

$$e_{ss} = \lim_{s \to 0} \frac{s}{1 + G(s)} \frac{1}{s^3}$$

$$= \frac{1}{\lim_{s \to 0} s^2 G(s)}$$

The static acceleration error constant K_a is defined by the equation

$$K_a = \lim_{s \to 0} s^2 G(s)$$

The steady-state error is then

$$e_{ss} = \frac{1}{K_a}$$

Note that the acceleration error, the steady-state error due to a parabolic input, is an error in position.

The values of K_a are obtained as follows:
For a type 0 system,

$$K_a = \lim_{s \to 0} \frac{s^2 K(T_a s + 1)(T_b s + 1)\cdots}{(T_1 s + 1)(T_2 s + 1)\cdots} = 0$$

For a type 1 system,

$$K_a = \lim_{s \to 0} \frac{s^2 K(T_a s + 1)(T_b s + 1)\cdots}{s(T_1 s + 1)(T_2 s + 1)\cdots} = 0$$

For a type 2 system,

$$K_a = \lim_{s \to 0} \frac{s^2 K(T_a s + 1)(T_b s + 1)\cdots}{s^2(T_1 s + 1)(T_2 s + 1)\cdots} = K$$

For a type 3 or higher system,

$$K_a = \lim_{s \to 0} \frac{s^2 K(T_a s + 1)(T_b s + 1)\cdots}{s^N(T_1 s + 1)(T_2 s + 1)\cdots} = \infty, \qquad \text{for } N \geq 3$$

Thus, the steady-state error for the unit parabolic input is

$$e_{ss} = \infty, \quad \text{for type 0 and type 1 systems}$$

$$e_{ss} = \frac{1}{K}, \quad \text{for type 2 systems}$$

$$e_{ss} = 0, \quad \text{for type 3 or higher systems}$$

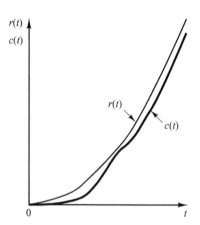

Figure 5–48
Response of a type 2
unity-feedback
system to a parabolic
input.

Note that both type 0 and type 1 systems are incapable of following a parabolic input in the steady state. The type 2 system with unity feedback can follow a parabolic input with a finite error signal. Figure 5–48 shows an example of the response of a type 2 system with unity feedback to a parabolic input. The type 3 or higher system with unity feedback follows a parabolic input with zero error at steady state.

Summary. Table 5–1 summarizes the steady-state errors for type 0, type 1, and type 2 systems when they are subjected to various inputs. The finite values for steady-state errors appear on the diagonal line. Above the diagonal, the steady-state errors are infinity; below the diagonal, they are zero.

Table 5–1 Steady-State Error in Terms of Gain K

	Step Input $r(t) = 1$	Ramp Input $r(t) = t$	Acceleration Input $r(t) = \frac{1}{2}t^2$
Type 0 system	$\dfrac{1}{1 + K}$	∞	∞
Type 1 system	0	$\dfrac{1}{K}$	∞
Type 2 system	0	0	$\dfrac{1}{K}$

Remember that the terms *position error, velocity error*, and *acceleration error* mean steady-state deviations in the output position. A finite velocity error implies that after transients have died out, the input and output move at the same velocity but have a finite position difference.

The error constants K_p, K_v, and K_a describe the ability of a unity-feedback system to reduce or eliminate steady-state error. Therefore, they are indicative of the steady-state performance. It is generally desirable to increase the error constants, while maintaining the transient response within an acceptable range. It is noted that to improve the steady-state performance we can increase the type of the system by adding an integrator or integrators to the feedforward path. This, however, introduces an additional stability problem. The design of a satisfactory system with more than two integrators in series in the feedforward path is generally not easy.

A–5–1. In the system of Figure 5–49, $x(t)$ is the input displacement and $\theta(t)$ is the output angular displacement. Assume that the masses involved are negligibly small and that all motions are restricted to be small; therefore, the system can be considered linear. The initial conditions for x and θ are zeros, or $x(0-) = 0$ and $\theta(0-) = 0$. Show that this system is a differentiating element. Then obtain the response $\theta(t)$ when $x(t)$ is a unit-step input.

Solution. The equation for the system is

$$b(\dot{x} - L\dot{\theta}) = kL\theta$$

or

$$L\dot{\theta} + \frac{k}{b}L\theta = \dot{x}$$

The Laplace transform of this last equation, using zero initial conditions, gives

$$\left(Ls + \frac{k}{b}L\right)\Theta(s) = sX(s)$$

And so

$$\frac{\Theta(s)}{X(s)} = \frac{1}{L}\frac{s}{s + (k/b)}$$

Thus the system is a differentiating system.

For the unit-step input $X(s) = 1/s$, the output $\Theta(s)$ becomes

$$\Theta(s) = \frac{1}{L}\frac{1}{s + (k/b)}$$

The inverse Laplace transform of $\Theta(s)$ gives

$$\theta(t) = \frac{1}{L}e^{-(k/b)t}$$

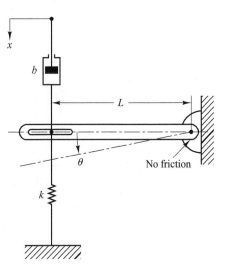

Figure 5–49
Mechanical system.

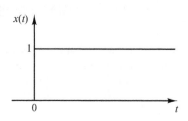

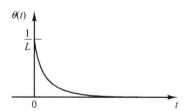

Figure 5–50
Unit-step input and
the response of the
mechanical system
shown in Figure
5–49.

Note that if the value of k/b is large, the response $\theta(t)$ approaches a pulse signal, as shown in Figure 5–50.

A–5–2. Gear trains are often used in servo systems to reduce speed, to magnify torque, or to obtain the most efficient power transfer by matching the driving member to the given load.

Consider the gear-train system shown in Figure 5–51. In this system, a load is driven by a motor through the gear train. Assuming that the stiffness of the shafts of the gear train is infinite (there is neither backlash nor elastic deformation) and that the number of teeth on each gear is proportional to the radius of the gear, obtain the equivalent moment of inertia and equivalent viscous-friction coefficient referred to the motor shaft and referred to the load shaft.

In Figure 5–51 the numbers of teeth on gears 1, 2, 3, and 4 are N_1, N_2, N_3, and N_4, respectively. The angular displacements of shafts, 1, 2, and 3 are θ_1, θ_2, and θ_3, respectively. Thus, $\theta_2/\theta_1 = N_1/N_2$ and $\theta_3/\theta_2 = N_3/N_4$. The moment of inertia and viscous-friction coefficient of each gear-train component are denoted by $J_1, b_1; J_2, b_2$; and J_3, b_3; respectively. (J_3 and b_3 include the moment of inertia and friction of the load.)

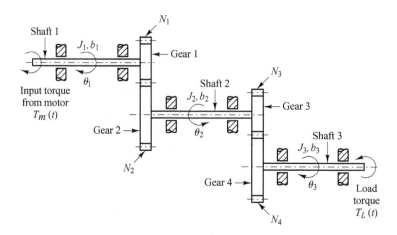

Figure 5–51
Gear-train system.

Chapter 5 / Transient and Steady-State Response Analyses

Solution. For this gear-train system, we can obtain the following equations: For shaft 1,

$$J_1\ddot{\theta}_1 + b_1\dot{\theta}_1 + T_1 = T_m \tag{5-63}$$

where T_m is the torque developed by the motor and T_1 is the load torque on gear 1 due to the rest of the gear train. For shaft 2,

$$J_2\ddot{\theta}_2 + b_2\dot{\theta}_2 + T_3 = T_2 \tag{5-64}$$

where T_2 is the torque transmitted to gear 2 and T_3 is the load torque on gear 3 due to the rest of the gear train. Since the work done by gear 1 is equal to that of gear 2,

$$T_1\theta_1 = T_2\theta_2 \quad \text{or} \quad T_2 = T_1\frac{N_2}{N_1}$$

If $N_1/N_2 < 1$, the gear ratio reduces the speed as well as magnifies the torque. For shaft 3,

$$J_3\ddot{\theta}_3 + b_3\dot{\theta}_3 + T_L = T_4 \tag{5-65}$$

where T_L is the load torque and T_4 is the torque transmitted to gear 4. T_3 and T_4 are related by

$$T_4 = T_3\frac{N_4}{N_3}$$

and θ_3 and θ_1 are related by

$$\theta_3 = \theta_2\frac{N_3}{N_4} = \theta_1\frac{N_1}{N_2}\frac{N_3}{N_4}$$

Eliminating T_1, T_2, T_3, and T_4 from Equations (5–63), (5–64), and (5–65) yields

$$J_1\ddot{\theta}_1 + b_1\dot{\theta}_1 + \frac{N_1}{N_2}(J_2\ddot{\theta}_2 + b_2\dot{\theta}_2) + \frac{N_1 N_3}{N_2 N_4}(J_3\ddot{\theta}_3 + b_3\dot{\theta}_3 + T_L) = T_m$$

Eliminating θ_2 and θ_3 from this last equation and writing the resulting equation in terms of θ_1 and its time derivatives, we obtain

$$\left[J_1 + \left(\frac{N_1}{N_2}\right)^2 J_2 + \left(\frac{N_1}{N_2}\right)^2\left(\frac{N_3}{N_4}\right)^2 J_3 \right]\ddot{\theta}_1$$

$$+ \left[b_1 + \left(\frac{N_1}{N_2}\right)^2 b_2 + \left(\frac{N_1}{N_2}\right)^2\left(\frac{N_3}{N_4}\right)^2 b_3 \right]\dot{\theta}_1 + \left(\frac{N_1}{N_2}\right)\left(\frac{N_3}{N_4}\right) T_L = T_m \tag{5-66}$$

Thus, the equivalent moment of inertia and viscous-friction coefficient of the gear train referred to shaft 1 are given, respectively, by

$$J_{1eq} = J_1 + \left(\frac{N_1}{N_2}\right)^2 J_2 + \left(\frac{N_1}{N_2}\right)^2\left(\frac{N_3}{N_4}\right)^2 J_3$$

$$b_{1eq} = b_1 + \left(\frac{N_1}{N_2}\right)^2 b_2 + \left(\frac{N_1}{N_2}\right)^2\left(\frac{N_3}{N_4}\right)^2 b_3$$

Similarly, the equivalent moment of inertia and viscous-friction coefficient of the gear train referred to the load shaft (shaft 3) are given, respectively, by

$$J_{3eq} = J_3 + \left(\frac{N_4}{N_3}\right)^2 J_2 + \left(\frac{N_2}{N_1}\right)^2\left(\frac{N_4}{N_3}\right)^2 J_1$$

$$b_{3eq} = b_3 + \left(\frac{N_4}{N_3}\right)^2 b_2 + \left(\frac{N_2}{N_1}\right)^2\left(\frac{N_4}{N_3}\right)^2 b_1$$

The relationship between J_{1eq} and J_{3eq} is thus

$$J_{1eq} = \left(\frac{N_1}{N_2}\right)^2 \left(\frac{N_3}{N_4}\right)^2 J_{3eq}$$

and that between b_{1eq} and b_{3eq} is

$$b_{1eq} = \left(\frac{N_1}{N_2}\right)^2 \left(\frac{N_3}{N_4}\right)^2 b_{3eq}$$

The effect of J_2 and J_3 on an equivalent moment of inertia is determined by the gear ratios N_1/N_2 and N_3/N_4. For speed-reducing gear trains, the ratios, N_1/N_2 and N_3/N_4 are usually less than unity. If $N_1/N_2 \ll 1$ and $N_3/N_4 \ll 1$, then the effect of J_2 and J_3 on the equivalent moment of inertia J_{1eq} is negligible. Similar comments apply to the equivalent viscous-friction coefficient b_{1eq} of the gear train. In terms of the equivalent moment of inertia J_{1eq} and equivalent viscous-friction coefficient b_{1eq}, Equation (5–66) can be simplified to give

$$J_{1eq}\ddot{\theta}_1 + b_{1eq}\dot{\theta}_1 + nT_L = T_m$$

where

$$n = \frac{N_1 N_3}{N_2 N_4}$$

A–5–3. When the system shown in Figure 5–52(a) is subjected to a unit-step input, the system output responds as shown in Figure 5–52(b). Determine the values of K and T from the response curve.

Solution. The maximum overshoot of 25.4% corresponds to $\zeta = 0.4$. From the response curve we have

$$t_p = 3$$

Consequently,

$$t_p = \frac{\pi}{\omega_d} = \frac{\pi}{\omega_n \sqrt{1 - \zeta^2}} = \frac{\pi}{\omega_n \sqrt{1 - 0.4^2}} = 3$$

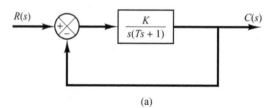

(a)

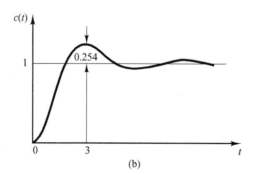

(b)

Figure 5–52
(a) Closed-loop system; (b) unit-step response curve.

Chapter 5 / Transient and Steady-State Response Analyses

It follows that

$$\omega_n = 1.14$$

From the block diagram we have

$$\frac{C(s)}{R(s)} = \frac{K}{Ts^2 + s + K}$$

from which

$$\omega_n = \sqrt{\frac{K}{T}}, \qquad 2\zeta\omega_n = \frac{1}{T}$$

Therefore, the values of T and K are determined as

$$T = \frac{1}{2\zeta\omega_n} = \frac{1}{2 \times 0.4 \times 1.14} = 1.09$$

$$K = \omega_n^2 T = 1.14^2 \times 1.09 = 1.42$$

A–5–4. Determine the values of K and k of the closed-loop system shown in Figure 5–53 so that the maximum overshoot in unit-step response is 25% and the peak time is 2 sec. Assume that $J = 1$ kg-m^2.

Solution. The closed-loop transfer function is

$$\frac{C(s)}{R(s)} = \frac{K}{Js^2 + Kks + K}$$

By substituting $J = 1$ kg-m^2 into this last equation, we have

$$\frac{C(s)}{R(s)} = \frac{K}{s^2 + Kks + K}$$

Note that in this problem

$$\omega_n = \sqrt{K}, \qquad 2\zeta\omega_n = Kk$$

The maximum overshoot M_p is

$$M_p = e^{-\zeta\pi/\sqrt{1-\zeta^2}}$$

which is specified as 25%. Hence

$$e^{-\zeta\pi/\sqrt{1-\zeta^2}} = 0.25$$

from which

$$\frac{\zeta\pi}{\sqrt{1-\zeta^2}} = 1.386$$

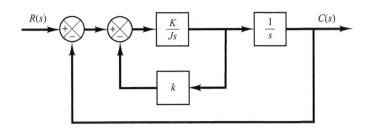

Figure 5–53
Closed-loop system.

or

$$\zeta = 0.404$$

The peak time t_p is specified as 2 sec. And so

$$t_p = \frac{\pi}{\omega_d} = 2$$

or

$$\omega_d = 1.57$$

Then the undamped natural frequency ω_n is

$$\omega_n = \frac{\omega_d}{\sqrt{1 - \zeta^2}} = \frac{1.57}{\sqrt{1 - 0.404^2}} = 1.72$$

Therefore, we obtain

$$K = \omega_n^2 = 1.72^2 = 2.95 \text{ N-m}$$

$$k = \frac{2\zeta\omega_n}{K} = \frac{2 \times 0.404 \times 1.72}{2.95} = 0.471 \text{ sec}$$

A–5–5. Figure 5–54(a) shows a mechanical vibratory system. When 2 lb of force (step input) is applied to the system, the mass oscillates, as shown in Figure 5–54(b). Determine m, b, and k of the system from this response curve. The displacement x is measured from the equilibrium position.

Solution. The transfer function of this system is

$$\frac{X(s)}{P(s)} = \frac{1}{ms^2 + bs + k}$$

Since

$$P(s) = \frac{2}{s}$$

we obtain

$$X(s) = \frac{2}{s(ms^2 + bs + k)}$$

It follows that the steady-state value of x is

$$x(\infty) = \lim_{s \to 0} sX(s) = \frac{2}{k} = 0.1 \text{ ft}$$

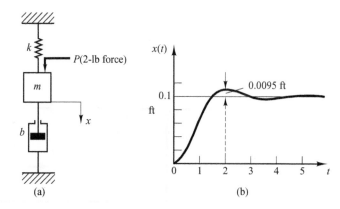

Figure 5–54
(a) Mechanical vibratory system; (b) step-response curve.

(a)

(b)

Chapter 5 / Transient and Steady-State Response Analyses

Hence

$$k = 20 \ lb_f/ft$$

Note that $M_p = 9.5\%$ corresponds to $\zeta = 0.6$. The peak time t_p is given by

$$t_p = \frac{\pi}{\omega_d} = \frac{\pi}{\omega_n \sqrt{1 - \zeta^2}} = \frac{\pi}{0.8\omega_n}$$

The experimental curve shows that $t_p = 2$ sec. Therefore,

$$\omega_n = \frac{3.14}{2 \times 0.8} = 1.96 \ rad/sec$$

Since $\omega_n^2 = k/m = 20/m$, we obtain

$$m = \frac{20}{\omega_n^2} = \frac{20}{1.96^2} = 5.2 \ slugs = 167 \ lb$$

(Note that 1 slug $= 1 \ lb_f\text{-}sec^2/ft$.) Then b is determined from

$$2\zeta\omega_n = \frac{b}{m}$$

or

$$b = 2\zeta\omega_n m = 2 \times 0.6 \times 1.96 \times 5.2 = 12.2 \ lb_f/ft/sec$$

A–5–6. Consider the unit-step response of the second-order system

$$\frac{C(s)}{R(s)} = \frac{\omega_n^2}{s^2 + 2\zeta\omega_n s + \omega_n^2}$$

The amplitude of the exponentially damped sinusoid changes as a geometric series. At time $t = t_p = \pi/\omega_d$, the amplitude is equal to $e^{-(\sigma/\omega_d)\pi}$. After one oscillation, or at $t = t_p + 2\pi/\omega_d = 3\pi/\omega_d$, the amplitude is equal to $e^{-(\sigma/\omega_d)3\pi}$; after another cycle of oscillation, the amplitude is $e^{-(\sigma/\omega_d)5\pi}$. The logarithm of the ratio of successive amplitudes is called the *logarithmic decrement*. Determine the logarithmic decrement for this second-order system. Describe a method for experimental determination of the damping ratio from the rate of decay of the oscillation.

Solution. Let us define the amplitude of the output oscillation at $t = t_i$ to be x_i, where $t_i = t_p + (i - 1)T$ ($T =$ period of oscillation). The amplitude ratio per one period of damped oscillation is

$$\frac{x_1}{x_2} = \frac{e^{-(\sigma/\omega_d)\pi}}{e^{-(\sigma/\omega_d)3\pi}} = e^{2(\sigma/\omega_d)\pi} = e^{2\zeta\pi/\sqrt{1-\zeta^2}}$$

Thus, the logarithmic decrement δ is

$$\delta = \ln\frac{x_1}{x_2} = \frac{2\zeta\pi}{\sqrt{1 - \zeta^2}}$$

It is a function only of the damping ratio ζ. Thus, the damping ratio ζ can be determined by use of the logarithmic decrement.

In the experimental determination of the damping ratio ζ from the rate of decay of the oscillation, we measure the amplitude x_1 at $t = t_p$ and amplitude x_n at $t = t_p + (n - 1)T$. Note that it is necessary to choose n large enough so that the ratio x_1/x_n is not near unity. Then

$$\frac{x_1}{x_n} = e^{(n-1)2\zeta\pi/\sqrt{1-\zeta^2}}$$

or

$$\ln\frac{x_1}{x_n} = (n-1)\frac{2\zeta\pi}{\sqrt{1-\zeta^2}}$$

Hence

$$\zeta = \frac{\dfrac{1}{n-1}\left(\ln\dfrac{x_1}{x_n}\right)}{\sqrt{4\pi^2 + \left[\dfrac{1}{n-1}\left(\ln\dfrac{x_1}{x_n}\right)\right]^2}}$$

A–5–7. In the system shown in Figure 5–55, the numerical values of m, b, and k are given as $m = 1$ kg, $b = 2$ N-sec/m, and $k = 100$ N/m. The mass is displaced 0.05 m and released without initial velocity. Find the frequency observed in the vibration. In addition, find the amplitude four cycles later. The displacement x is measured from the equilibrium position.

Solution. The equation of motion for the system is

$$m\ddot{x} + b\dot{x} + kx = 0$$

Substituting the numerical values for m, b, and k into this equation gives

$$\ddot{x} + 2\dot{x} + 100x = 0$$

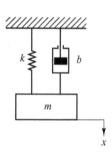

Figure 5–55
Spring-mass-damper system.

where the initial conditions are $x(0) = 0.05$ and $\dot{x}(0) = 0$. From this last equation the undamped natural frequency ω_n and the damping ratio ζ are found to be

$$\omega_n = 10, \qquad \zeta = 0.1$$

The frequency actually observed in the vibration is the damped natural frequency ω_d.

$$\omega_d = \omega_n\sqrt{1 - \zeta^2} = 10\sqrt{1 - 0.01} = 9.95 \text{ rad/sec}$$

In the present analysis, $\dot{x}(0)$ is given as zero. Thus, solution $x(t)$ can be written as

$$x(t) = x(0)e^{-\zeta\omega_n t}\left(\cos\omega_d t + \frac{\zeta}{\sqrt{1-\zeta^2}}\sin\omega_d t\right)$$

It follows that at $t = nT$, where $T = 2\pi/\omega_d$,

$$x(nT) = x(0)e^{-\zeta\omega_n nT}$$

Consequently, the amplitude four cycles later becomes

$$x(4T) = x(0)e^{-\zeta\omega_n 4T} = x(0)e^{-(0.1)(10)(4)(0.6315)}$$

$$= 0.05e^{-2.526} = 0.05 \times 0.07998 = 0.004 \text{ m}$$

A–5–8. Obtain both analytically and computationally the unit-step response of the following higher-order system:

$$\frac{C(s)}{R(s)} = \frac{3s^3 + 25s^2 + 72s + 80}{s^4 + 8s^3 + 40s^2 + 96s + 80}$$

[Obtain the partial-fraction expansion of $C(s)$ with MATLAB when $R(s)$ is a unit-step function.]

Solution. MATLAB Program 5–18 yields the unit-step response curve shown in Figure 5–56. It also yields the partial-fraction expansion of $C(s)$ as follows:

$$C(s) = \frac{3s^3 + 25s^2 + 72s + 80}{s^4 + 8s^3 + 40s^2 + 96s + 80} \frac{1}{s}$$

$$= \frac{-0.2813 - j0.1719}{s + 2 - j4} + \frac{-0.2813 + j0.1719}{s + 2 + j4}$$

$$+ \frac{-0.4375}{s + 2} + \frac{-0.375}{(s + 2)^2} + \frac{1}{s}$$

$$= \frac{-0.5626(s + 2)}{(s + 2)^2 + 4^2} + \frac{(0.3438) \times 4}{(s + 2)^2 + 4^2}$$

$$- \frac{0.4375}{s + 2} - \frac{0.375}{(s + 2)^2} + \frac{1}{s}$$

MATLAB Program 5–18

```
% ------- Unit-Step Response of C(s)/R(s) and Partial-Fraction Expansion of C(s) ------

num = [3  25  72  80];
den = [1  8  40  96  80];
step(num,den);
v = [0  3  0  1.2]; axis(v), grid

% To obtain the partial-fraction expansion of C(s), enter commands
%     num1 = [3  25  72  80];
%      den1 = [1  8  40  96  80  0];
%     [r,p,k] = residue(num1,den1)

num1 = [3  25  72  80];
den1 = [1  8  40  96  80  0];
[r,p,k] = residue(num1,den1)

r =

   -0.2813- 0.1719i
   -0.2813+ 0.1719i
   -0.4375
   -0.3750
    1.0000

p =

   -2.0000+ 4.0000i
   -2.0000- 4.0000i
   -2.0000
   -2.0000
    0

k =

   []
```

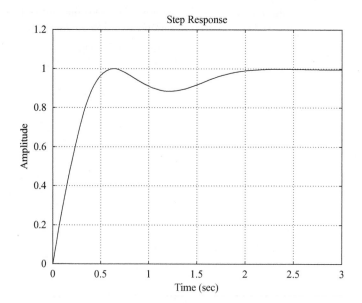

Figure 5–56
Unit-step response
curve.

Hence, the time response $c(t)$ can be given by

$$c(t) = -0.5626e^{-2t}\cos 4t + 0.3438e^{-2t}\sin 4t$$

$$- 0.4375e^{-2t} - 0.375te^{-2t} + 1$$

The fact that the response curve is an exponential curve superimposed by damped sinusoidal curves can be seen from Figure 5–56.

A–5–9. When the closed-loop system involves a numerator dynamics, the unit-step response curve may exhibit a large overshoot. Obtain the unit-step response of the following system with MATLAB:

$$\frac{C(s)}{R(s)} = \frac{10s + 4}{s^2 + 4s + 4}$$

Obtain also the unit-ramp response with MATLAB.

Solution. MATLAB Program 5–19 produces the unit-step response as well as the unit-ramp response of the system. The unit-step response curve and unit-ramp response curve, together with the unit-ramp input, are shown in Figures 5–57(a) and (b), respectively.

Notice that the unit-step response curve exhibits over 215% of overshoot. The unit-ramp response curve leads the input curve. These phenomena occurred because of the presence of a large derivative term in the numerator.

MATLAB Program 5–19

```
num = [10  4];
den = [1  4  4];
t = 0:0.02:10;
y = step(num,den,t);
plot(t,y)
grid
title('Unit-Step Response')
xlabel('t (sec)')
ylabel('Output')

num1 = [10  4];
den1 = [1  4  4  0];
y1 = step(num1,den1,t);
plot(t,t,'--',t,y1)
v = [0  10  0  10]; axis(v);
grid
title('Unit-Ramp Response')
xlabel('t (sec)')
ylabel('Unit-Ramp Input and Output')
text(6.1,5.0,'Unit-Ramp Input')
text(3.5,7.1,'Output')
```

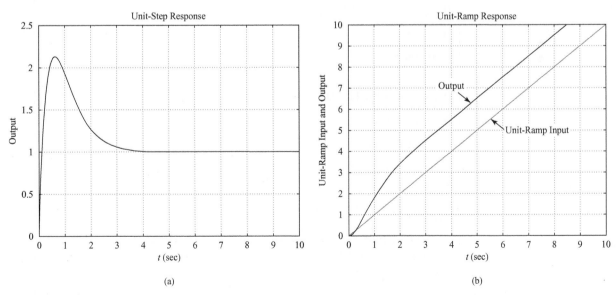

Figure 5–57

(a) Unit-step response curve; (b) unit-ramp response curve plotted with unit-ramp input.

A–5–10. Consider a higher-order system defined by

$$\frac{C(s)}{R(s)} = \frac{6.3223s^2 + 18s + 12.811}{s^4 + 6s^3 + 11.3223s^2 + 18s + 12.811}$$

Using MATLAB, plot the unit-step response curve of this system. Using MATLAB, obtain the rise time, peak time, maximum overshoot, and settling time.

Solution. MATLAB Program 5–20 plots the unit-step response curve as well as giving the rise time, peak time, maximum overshoot, and settling time. The unit-step response curve is shown in Figure 5–58.

MATLAB Program 5–20

```
% ------- This program is to plot the unit-step response curve, as well as to
% find the rise time, peak time, maximum overshoot, and settling time.
% In this program the rise time is calculated as the time required for the
% response to rise from 10% to 90% of its final value. -------

num = [6.3223 18 12.811];
den = [1  6  11.3223  18  12.811];
t = 0:0.02:20;
[y,x,t] = step(num,den,t);
plot(t,y)
grid
title('Unit-Step Response')
xlabel('t (sec)')
ylabel('Output y(t)')

r1 = 1; while y(r1) < 0.1, r1 = r1+1; end;
r2 = 1; while y(r2) < 0.9, r2 = r2+1; end;
rise_time = (r2-r1)*0. 02

rise_time =

   0.5800

[ymax,tp] = max(y);
peak_time = (tp-1)*0.02

peak_time =

   1.6600

max_overshoot = ymax-1

max_overshoot =

   0.6182

s = 1001; while y(s) > 0.98 & y(s) < 1.02; s = s-1; end;
settling_time = (s-1)*0.02

settling_time =

   10.0200
```

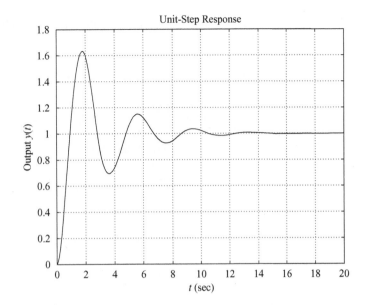

Unit-Step Response

Figure 5–58
Unit-step response curve.

A–5–11. Consider the closed-loop system defined by

$$\frac{C(s)}{R(s)} = \frac{\omega_n^2}{s^2 + 2\zeta\omega_n s + \omega_n^2}$$

Using a "for loop," write a MATLAB program to obtain unit-step response of this system for the following four cases:

Case 1: $\zeta = 0.3,$ $\omega_n = 1$

Case 2: $\zeta = 0.5,$ $\omega_n = 2$

Case 3: $\zeta = 0.7,$ $\omega_n = 4$

Case 4: $\zeta = 0.8,$ $\omega_n = 6$

Solution. Define $\omega_n^2 = a$ and $2\zeta\omega_n = b$. Then, a and b each have four elements as follows:

$$a = \begin{bmatrix} 1 & 4 & 16 & 36 \end{bmatrix}$$

$$b = \begin{bmatrix} 0.6 & 2 & 5.6 & 9.6 \end{bmatrix}$$

Using vectors a and b, MATLAB Program 5–21 will produce the unit-step response curves as shown in Figure 5–59.

MATLAB Program 5–21

```
a = [1  4  16  36];
b = [0.6  2  5.6  9.6];
t = 0:0.1:8;
y = zeros(81,4);
    for i = 1:4;
    num = [a(i)];
    den = [1  b(i)  a(i)];
    y(:,i) = step(num,den,t);
    end
plot(t,y(:,1),'o',t,y(:,2),'x',t,y(:,3),'-',t,y(:,4),'-.')
grid
title('Unit-Step Response Curves for Four Cases')
xlabel('t Sec')
ylabel('Outputs')
gtext('1')
gtext('2')
gtext('3')
gtext('4')
```

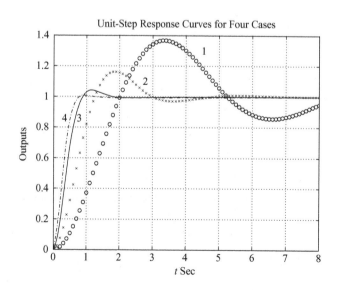

Figure 5–59
Unit-step response
curves for four cases.

A–5–12. Using MATLAB, obtain the unit-ramp response of the closed-loop control system whose closed-loop transfer function is

$$\frac{C(s)}{R(s)} = \frac{s + 10}{s^3 + 6s^2 + 9s + 10}$$

Also, obtain the response of this system when the input is given by

$$r = e^{-0.5t}$$

Solution. MATLAB Program 5–22 produces the unit-ramp response and the response to the exponential input $r = e^{-0.5t}$. The resulting response curves are shown in Figures 5–60(a) and (b), respectively.

MATLAB Program 5–22

```
% --------- Unit-Ramp Response ---------

num = [1  10];
den = [1  6  9  10];
t = 0:0.1:10;
r = t;
y = lsim(num,den,r,t);
plot(t,r,'-',t,y,'o')
grid
title('Unit-Ramp Response by Use of Command "lsim"')
xlabel('t Sec')
ylabel('Output')
text(3.2,6.5,'Unit-Ramp Input')
text(6.0,3.1,'Output')

% --------- Response to Input r1 = exp(-0.5t). ---------

num = [1  10];
den = [1  6  9  10];
t = 0:0.1:12;
r1 = exp(-0.5*t);
y1 = lsim(num,den,r1,t);
plot(t,r1,'-',t,y1,'o')
grid
title('Response to Input r1 = exp(-0.5t)')
xlabel('t Sec')
ylabel('Input and Output')
text(1.4,0.75,'Input r1 = exp(-0.5t)')
text(6.2,0.34,'Output')
```

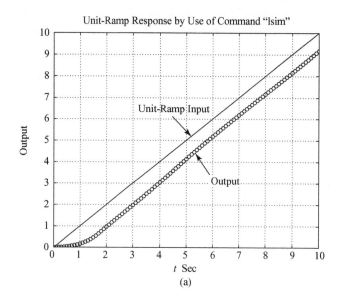

Unit-Ramp Response by Use of Command "lsim"

(a)

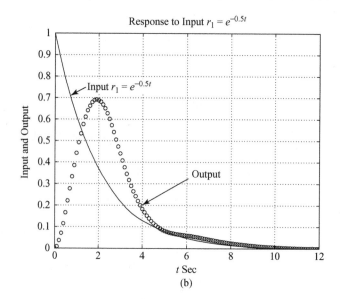

Response to Input $r_1 = e^{-0.5t}$

(b)

Figure 5–60
(a) Unit-ramp
response curve;
(b) response to
exponential input
$r_1 = e^{-0.5t}$.

A–5–13. Obtain the response of the closed-loop system defined by

$$\frac{C(s)}{R(s)} = \frac{5}{s^2 + s + 5}$$

when the input $r(t)$ is given by

$$r(t) = 2 + t$$

[The input $r(t)$ is a step input of magnitude 2 plus unit-ramp input.]

Chapter 5 / Transient and Steady-State Response Analyses

Solution. A possible MATLAB program is shown in MATLAB Program 5–23. The resulting response curve, together with a plot of the input function, is shown in Figure 5–61.

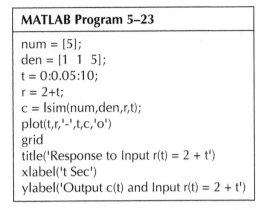

MATLAB Program 5–23

```
num = [5];
den = [1  1  5];
t = 0:0.05:10;
r = 2+t;
c = lsim(num,den,r,t);
plot(t,r,'-',t,c,'o')
grid
title('Response to Input r(t) = 2 + t')
xlabel('t Sec')
ylabel('Output c(t) and Input r(t) = 2 + t')
```

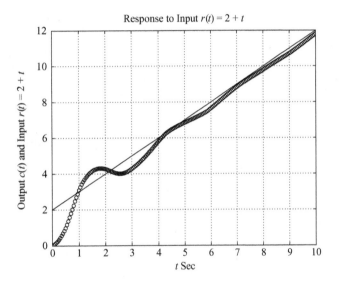

Figure 5–61
Response to input
$r(t) = 2 + t$.

A–5–14. Obtain the response of the system shown in Figure 5–62 when the input $r(t)$ is given by

$$r(t) = \frac{1}{2}t^2$$

[The input $r(t)$ is the unit-acceleration input.]

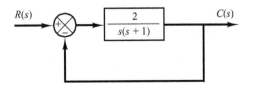

Figure 5–62
Control system.

Solution. The closed-loop transfer function is

$$\frac{C(s)}{R(s)} = \frac{2}{s^2 + s + 2}$$

MATLAB Program 5–24 produces the unit-acceleration response. The resulting response, together with the unit-acceleration input, is shown in Figure 5–63.

MATLAB Program 5–24

```
num = [2];
den = [1  1  2];
t = 0:0.2:10;
r = 0.5*t.^2;
y = lsim(num,den,r,t);
plot(t,r,'-',t,y,'o',t,y,'-')
grid
title('Unit-Acceleration Response')
xlabel('t Sec')
ylabel('Input and Output')
text(2.1,27.5,'Unit-Acceleration Input')
text(7.2,7.5,'Output')
```

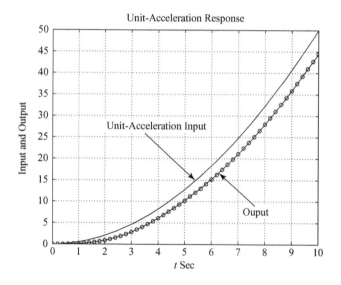

Figure 5–63
Response to unit-acceleration input.

A–5–15. Consider the system defined by

$$\frac{C(s)}{R(s)} = \frac{1}{s^2 + 2\zeta s + 1}$$

where $\zeta = 0,\ 0.2,\ 0.4,\ 0.6,\ 0.8$, and 1.0. Write a MATLAB program using a "for loop" to obtain the two-dimensional and three-dimensional plots of the system output. The input is the unit-step function.

Solution. MATLAB Program 5–25 is a possible program to obtain two-dimensional and three-dimensional plots. Figure 5–64(a) is the two-dimensional plot of the unit-step response curves for various values of ζ. Figure 5–64(b) is the three-dimensional plot obtained by use of the command "mesh(y)" and Figure 5–64(c) is obtained by use of the command "mesh(y')". (These two three-dimensional plots are basically the same. The only difference is that x axis and y axis are interchanged.)

MATLAB Program 5–25

```
t = 0:0.2:12;
    for n = 1:6;
    num = [1];
    den = [1  2*(n-1)*0.2  1];
    [y(1:61,n),x,t] = step(num,den,t);
    end
plot(t,y)
grid
title('Unit-Step Response Curves')
xlabel('t Sec')
ylabel('Outputs')
gtext('\zeta = 0'),
gtext('0.2')
gtext('0.4')
gtext('0.6')
gtext('0.8')
gtext('1.0')

% To draw a three-dimensional plot, enter the following command: mesh(y) or mesh(y').
% We shall show two three-dimensional plots, one using "mesh(y)" and the other using
% "mesh(y')". These two plots are the same, except that the x axis and y axis are
% interchanged.

mesh(y)
title('Three-Dimensional Plot of Unit-Step Response Curves using Command "mesh(y)"')
xlabel('n, where n = 1,2,3,4,5,6')
ylabel('Computation Time Points')
zlabel('Outputs')

mesh(y')
title('Three-Dimensional Plot of Unit-Step Response Curves using Command "mesh(y transpose)"')
xlabel('Computation Time Points')
ylabel('n, where n = 1,2,3,4,5,6')
zlabel('Outputs')
```

Figure 5–64
(a) Two-dimensional
plot of unit-step
response curves;
(b) three-dimensional
plot of unit-step
response curves
using command
"mesh(y)";
(c) three-dimensional
plot of unit-step
response curves
using command
"mesh(y')".

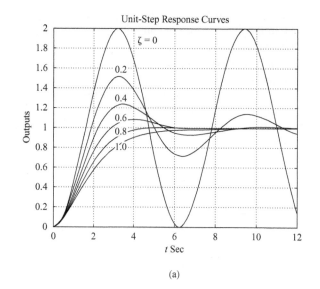

(a)

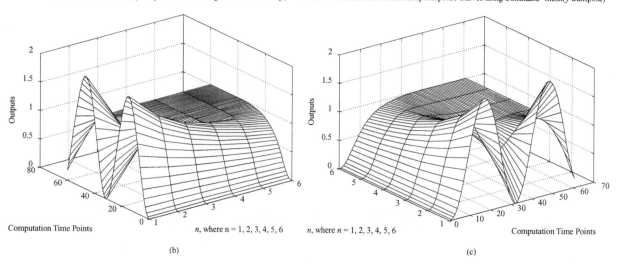

Three-Dimensional Plot of Unit-Step Response Curves using Command "mesh(y)"   Three-Dimensional Plot of Unit-Step Response Curves using Command "mesh(y transpose)"

(b)   (c)

A–5–16. Consider the system subjected to the initial condition as given below.

$$\begin{bmatrix} \dot{x}_1 \\ \dot{x}_2 \\ \dot{x}_3 \end{bmatrix} = \begin{bmatrix} 0 & 1 & 0 \\ 0 & 0 & 1 \\ -10 & -17 & -8 \end{bmatrix} \begin{bmatrix} x_1 \\ x_2 \\ x_3 \end{bmatrix}, \qquad \begin{bmatrix} x_1(0) \\ x_2(0) \\ x_3(0) \end{bmatrix} = \begin{bmatrix} 2 \\ 1 \\ 0.5 \end{bmatrix}$$

$$y = \begin{bmatrix} 1 & 0 & 0 \end{bmatrix} \begin{bmatrix} x_1 \\ x_2 \\ x_3 \end{bmatrix}$$

(There is no input or forcing function in this system.) Obtain the response y(t) versus t to the given initial condition by use of Equations (5–58) and (5–60).

Solution. A possible MATLAB program based on Equations (5–58) and (5–60) is given by MATLAB program 5–26. The response curve obtained here is shown in Figure 5–65. (Notice that this problem was solved by use of the command "initial" in Example 5–16. The response curve obtained here is exactly the same as that shown in Figure 5–34.)

MATLAB Program 5–26
t = 0:0.05:10; A = [0 1 0;0 0 1;-10 -17 -8]; B = [2;1;0.5]; C=[1 0 0]; [y,x,t] = step(A,B,C*A,C*B,1,t); plot(t,y) grid; title('Response to Initial Condition') xlabel('t (sec)') ylabel('Output y')

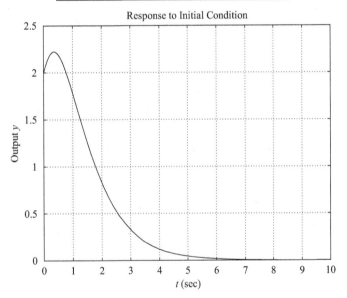

Figure 5–65
Response y(t) to the given initial condition.

A–5–17. Consider the following characteristic equation:

$$s^4 + Ks^3 + s^2 + s + 1 = 0$$

Determine the range of K for stability.

Solution. The Routh array of coefficients is

$$
\begin{array}{cccc}
s^4 & 1 & 1 & 1 \\
s^3 & K & 1 & 0 \\
s^2 & \dfrac{K-1}{K} & 1 & \\
s^1 & 1 - \dfrac{K^2}{K-1} & & \\
s^0 & 1 & &
\end{array}
$$

For stability, we require that

$$K > 0$$

$$\frac{K - 1}{K} > 0$$

$$1 - \frac{K^2}{K - 1} > 0$$

From the first and second conditions, K must be greater than 1. For $K > 1$, notice that the term $1 - [K^2/(K - 1)]$ is always negative, since

$$\frac{K - 1 - K^2}{K - 1} = \frac{-1 + K(1 - K)}{K - 1} < 0$$

Thus, the three conditions cannot be fulfilled simultaneously. Therefore, there is no value of K that allows stability of the system.

A–5–18. Consider the characteristic equation given by

$$a_0 s^n + a_1 s^{n-1} + a_2 s^{n-2} + \cdots + a_{n-1} s + a_n = 0 \qquad (5\text{--}67)$$

The Hurwitz stability criterion, given next, gives conditions for all the roots to have negative real parts in terms of the coefficients of the polynomial. As stated in the discussions of Routh's stability criterion in Section 5–6, for all the roots to have negative real parts, all the coefficients a's must be positive. This is a necessary condition but not a sufficient condition. If this condition is not satisfied, it indicates that some of the roots have positive real parts or are imaginary or zero. A sufficient condition for all the roots to have negative real parts is given in the following Hurwitz stability criterion: If all the coefficients of the polynomial are positive, arrange these coefficients in the following determinant:

$$\Delta_n = \begin{vmatrix} a_1 & a_3 & a_5 & \cdots & 0 & 0 & 0 \\ a_0 & a_2 & a_4 & \cdots & \cdot & \cdot & \cdot \\ 0 & a_1 & a_3 & \cdots & a_n & 0 & 0 \\ 0 & a_0 & a_2 & \cdots & a_{n-1} & 0 & 0 \\ \cdot & \cdot & \cdot & & a_{n-2} & a_n & 0 \\ \cdot & \cdot & \cdot & & a_{n-3} & a_{n-1} & 0 \\ 0 & 0 & 0 & \cdots & a_{n-4} & a_{n-2} & a_n \end{vmatrix}$$

where we substituted zero for a_s if $s > n$. For all the roots to have negative real parts, it is necessary and sufficient that successive principal minors of Δ_n be positive. The successive principal minors are the following determinants:

$$\Delta_i = \begin{vmatrix} a_1 & a_3 & \cdots & a_{2i-1} \\ a_0 & a_2 & \cdots & a_{2i-2} \\ 0 & a_1 & \cdots & a_{2i-3} \\ \cdot & \cdot & & \cdot \\ 0 & 0 & \cdots & a_i \end{vmatrix} \qquad (i = 1, 2, \ldots, n - 1)$$

where $a_s = 0$ if $s > n$. (It is noted that some of the conditions for the lower-order determinants are included in the conditions for the higher-order determinants.) If all these determinants are positive, and $a_0 > 0$ as already assumed, the equilibrium state of the system whose characteristic equation is given by Equation (5–67) is asymptotically stable. Note that exact values of determinants are not needed; instead, only signs of these determinants are needed for the stability criterion.

Now consider the following characteristic equation:

$$a_0 s^4 + a_1 s^3 + a_2 s^2 + a_3 s + a_4 = 0$$

Obtain the conditions for stability using the Hurwitz stability criterion.

Solution. The conditions for stability are that all the a's be positive and that

$$\Delta_2 = \begin{vmatrix} a_1 & a_3 \\ a_0 & a_2 \end{vmatrix} = a_1 a_2 - a_0 a_3 > 0$$

$$\Delta_3 = \begin{vmatrix} a_1 & a_3 & 0 \\ a_0 & a_2 & a_4 \\ 0 & a_1 & a_3 \end{vmatrix}$$

$$= a_1(a_2 a_3 - a_1 a_4) - a_0 a_3^2$$

$$= a_3(a_1 a_2 - a_0 a_3) - a_1^2 a_4 > 0$$

It is clear that, if all the a's are positive and if the condition $\Delta_3 > 0$ is satisfied, the condition $\Delta_2 > 0$ is also satisfied. Therefore, for all the roots of the given characteristic equation to have negative real parts, it is necessary and sufficient that all the coefficients a's are positive and $\Delta_3 > 0$.

A–5–19. Show that the first column of the Routh array of

$$s^n + a_1 s^{n-1} + a_2 s^{n-2} + \cdots + a_{n-1} s + a_n = 0$$

is given by

$$1, \quad \Delta_1, \quad \frac{\Delta_2}{\Delta_1}, \quad \frac{\Delta_3}{\Delta_2}, \dots, \quad \frac{\Delta_n}{\Delta_{n-1}}$$

where

$$\Delta_r = \begin{vmatrix} a_1 & 1 & 0 & 0 & \cdot & 0 \\ a_3 & a_2 & a_1 & 1 & \cdot & 0 \\ a_5 & a_4 & a_3 & a_2 & \cdot & 0 \\ \cdot & \cdot & \cdot & \cdot & & \cdot \\ \cdot & \cdot & \cdot & \cdot & & \cdot \\ \cdot & \cdot & \cdot & \cdot & & \cdot \\ a_{2r-1} & \cdot & \cdot & \cdot & \cdot & a_r \end{vmatrix}, \quad (n \geq r \geq 1)$$

$$a_k = 0 \quad \text{if } k > n$$

Solution. The Routh array of coefficients has the form

$$
\begin{array}{cccccc}
1 & a_2 & a_4 & a_6 & \cdots & a_n \\
a_1 & a_3 & a_5 & \cdots & & \\
b_1 & b_2 & b_3 & \cdots & & \\
c_1 & c_2 & \cdot & & & \\
\cdot & \cdot & \cdot & & & \\
\cdot & \cdot & \cdot & & & \\
\cdot & \cdot & \cdot & & &
\end{array}
$$

The first term in the first column of the Routh array is 1. The next term in the first column is a_1, which is equal to Δ_1. The next term is b_1, which is equal to

$$\frac{a_1 a_2 - a_3}{a_1} = \frac{\Delta_2}{\Delta_1}$$

The next term in the first column is c_1, which is equal to

$$\frac{b_1 a_3 - a_1 b_2}{b_1} = \frac{\left[\dfrac{a_1 a_2 - a_3}{a_1}\right] a_3 - a_1 \left[\dfrac{a_1 a_4 - a_5}{a_1}\right]}{\left[\dfrac{a_1 a_2 - a_3}{a_1}\right]}$$

$$= \frac{a_1 a_2 a_3 - a_3^2 - a_1^2 a_4 + a_1 a_5}{a_1 a_2 - a_3}$$

$$= \frac{\Delta_3}{\Delta_2}$$

In a similar manner the remaining terms in the first column of the Routh array can be found.

The Routh array has the property that the last nonzero terms of any columns are the same; that is, if the array is given by

$$
\begin{array}{cccc}
a_0 & a_2 & a_4 & a_6 \\
a_1 & a_3 & a_5 & a_7 \\
b_1 & b_2 & b_3 & \\
c_1 & c_2 & c_3 & \\
d_1 & d_2 & & \\
e_1 & e_2 & & \\
f_1 & & & \\
g_1 & & &
\end{array}
$$

then

$$a_7 = c_3 = e_2 = g_1$$

and if the array is given by

$$
\begin{array}{cccc}
a_0 & a_2 & a_4 & a_6 \\
a_1 & a_3 & a_5 & 0 \\
b_1 & b_2 & b_3 & \\
c_1 & c_2 & 0 & \\
d_1 & d_2 & & \\
e_1 & 0 & & \\
f_1 & & &
\end{array}
$$

then

$$a_6 = b_3 = d_2 = f_1$$

In any case, the last term of the first column is equal to a_n, or

$$a_n = \frac{\Delta_{n-1} a_n}{\Delta_{n-1}} = \frac{\Delta_n}{\Delta_{n-1}}$$

For example, if $n = 4$, then

$$\Delta_4 = \begin{vmatrix} a_1 & 1 & 0 & 0 \\ a_3 & a_2 & a_1 & 1 \\ a_5 & a_4 & a_3 & a_2 \\ a_7 & a_6 & a_5 & a_4 \end{vmatrix} = \begin{vmatrix} a_1 & 1 & 0 & 0 \\ a_3 & a_2 & a_1 & 1 \\ 0 & a_4 & a_3 & a_2 \\ 0 & 0 & 0 & a_4 \end{vmatrix} = \Delta_3 a_4$$

Thus it has been shown that the first column of the Routh array is given by

$$1, \quad \Delta_1, \quad \frac{\Delta_2}{\Delta_1}, \quad \frac{\Delta_3}{\Delta_2}, \quad \cdots, \quad \frac{\Delta_n}{\Delta_{n-1}}$$

A-5-20. Show that the Routh's stability criterion and Hurwitz stability criterion are equivalent.

Solution. If we write Hurwitz determinants in the triangular form

$$\Delta_i = \begin{vmatrix} a_{11} & & & * \\ & a_{22} & & \\ & & \ddots & \\ & & & \\ 0 & & & a_{ii} \end{vmatrix}, \quad (i = 1, 2, \ldots, n)$$

where the elements below the diagonal line are all zeros and the elements above the diagonal line any numbers, then the Hurwitz conditions for asymptotic stability become

$$\Delta_i = a_{11} a_{22} \cdots a_{ii} > 0, \quad (i = 1, 2, \ldots, n)$$

which are equivalent to the conditions

$$a_{11} > 0, \quad a_{22} > 0, \quad \ldots, \quad a_{nn} > 0$$

We shall show that these conditions are equivalent to

$$a_1 > 0, \quad b_1 > 0, \quad c_1 > 0, \quad \ldots$$

where $a_1, b_1, c_1, \ldots$, are the elements of the first column in the Routh array.

Consider, for example, the following Hurwitz determinant, which corresponds to $i = 4$:

$$\Delta_4 = \begin{vmatrix} a_1 & a_3 & a_5 & a_7 \\ a_0 & a_2 & a_4 & a_6 \\ 0 & a_1 & a_3 & a_5 \\ 0 & a_0 & a_2 & a_4 \end{vmatrix}$$

The determinant is unchanged if we subtract from the ith row k times the jth row. By subtracting from the second row a_0/a_1 times the first row, we obtain

$$\Delta_4 = \begin{vmatrix} a_{11} & a_3 & a_5 & a_7 \\ 0 & a_{22} & a_{23} & a_{24} \\ 0 & a_1 & a_3 & a_5 \\ 0 & a_0 & a_2 & a_4 \end{vmatrix}$$

Example Problems and Solutions

255

where

$$a_{11} = a_1$$

$$a_{22} = a_2 - \frac{a_0}{a_1} a_3$$

$$a_{23} = a_4 - \frac{a_0}{a_1} a_5$$

$$a_{24} = a_6 - \frac{a_0}{a_1} a_7$$

Similarly, subtracting from the fourth row a_0/a_1 times the third row yields

$$\Delta_4 = \begin{vmatrix} a_{11} & a_3 & a_5 & a_7 \\ 0 & a_{22} & a_{23} & a_{24} \\ 0 & a_1 & a_3 & a_5 \\ 0 & 0 & \hat{a}_{43} & \hat{a}_{44} \end{vmatrix}$$

where

$$\hat{a}_{43} = a_2 - \frac{a_0}{a_1} a_3$$

$$\hat{a}_{44} = a_4 - \frac{a_0}{a_1} a_5$$

Next, subtracting from the third row a_1/a_{22} times the second row yields

$$\Delta_4 = \begin{vmatrix} a_{11} & a_3 & a_5 & a_7 \\ 0 & a_{22} & a_{23} & a_{24} \\ 0 & 0 & a_{33} & a_{34} \\ 0 & 0 & \hat{a}_{43} & \hat{a}_{44} \end{vmatrix}$$

where

$$a_{33} = a_3 - \frac{a_1}{a_{22}} a_{23}$$

$$a_{34} = a_5 - \frac{a_1}{a_{22}} a_{24}$$

Finally, subtracting from the last row $\hat{a}_{43}/a_{33}$ times the third row yields

$$\Delta_4 = \begin{vmatrix} a_{11} & a_3 & a_5 & a_7 \\ 0 & a_{22} & a_{23} & a_{24} \\ 0 & 0 & a_{33} & a_{34} \\ 0 & 0 & 0 & a_{44} \end{vmatrix}$$

where

$$a_{44} = \hat{a}_{44} - \frac{\hat{a}_{43}}{a_{33}} a_{34}$$

From this analysis, we see that

$$\Delta_4 = a_{11}a_{22}a_{33}a_{44}$$

$$\Delta_3 = a_{11}a_{22}a_{33}$$

$$\Delta_2 = a_{11}a_{22}$$

$$\Delta_1 = a_{11}$$

The Hurwitz conditions for asymptotic stability

$$\Delta_1 > 0, \qquad \Delta_2 > 0, \qquad \Delta_3 > 0, \qquad \Delta_4 > 0, \qquad \cdots$$

reduce to the conditions

$$a_{11} > 0, \qquad a_{22} > 0, \qquad a_{33} > 0, \qquad a_{44} > 0, \qquad \cdots$$

The Routh array for the polynomial

$$a_0 s^4 + a_1 s^3 + a_2 s^2 + a_3 s + a_4 = 0$$

where $a_0 > 0$ and $n = 4$, is given by

$$
\begin{array}{ccc}
a_0 & a_2 & a_4 \\
a_1 & a_3 & \\
b_1 & b_2 & \\
c_1 & & \\
d_1 & &
\end{array}
$$

From this Routh array, we see that

$$a_{11} = a_1$$

$$a_{22} = a_2 - \frac{a_0}{a_1}a_3 = b_1$$

$$a_{33} = a_3 - \frac{a_1}{a_{22}}a_{23} = \frac{a_3 b_1 - a_1 b_2}{b_1} = c_1$$

$$a_{44} = \hat{a}_{44} - \frac{\hat{a}_{43}}{a_{33}}a_{34} = a_4 = d_1$$

(The last equation is obtained using the fact that $a_{34} = 0$, $\hat{a}_{44} = a_4$, and $a_4 = b_2 = d_1$.) Hence the Hurwitz conditions for asymptotic stability become

$$a_1 > 0, \qquad b_1 > 0, \qquad c_1 > 0, \qquad d_1 > 0$$

Thus we have demonstrated that Hurwitz conditions for asymptotic stability can be reduced to Routh's conditions for asymptotic stability. The same argument can be extended to Hurwitz determinants of any order, and the equivalence of Routh's stability criterion and Hurwitz stability criterion can be established.

A–5–21. Consider the characteristic equation

$$s^4 + 2s^3 + (4 + K)s^2 + 9s + 25 = 0$$

Using the Hurwitz stability criterion, determine the range of K for stability.

Solution. Comparing the given characteristic equation

$$s^4 + 2s^3 + (4 + K)s^2 + 9s + 25 = 0$$

with the following standard fourth-order characteristic equation:

$$a_0 s^4 + a_1 s^3 + a_2 s^2 + a_3 s + a_4 = 0$$

we find

$$a_0 = 1, \quad a_1 = 2, \quad a_2 = 4 + K, \quad a_3 = 9, \quad a_4 = 25$$

The Hurwitz stability criterion states that Δ_4 is given by

$$\Delta_4 = \begin{vmatrix} a_1 & a_3 & 0 & 0 \\ a_0 & a_2 & a_4 & 0 \\ 0 & a_1 & a_3 & 0 \\ 0 & a_0 & a_2 & a_4 \end{vmatrix}$$

For all the roots to have negative real parts, it is necessary and sufficient that succesive principal minors of Δ_4 be positive. The successive principal minors are

$$\Delta_1 = |a_1| = 2$$

$$\Delta_2 = \begin{vmatrix} a_1 & a_3 \\ a_0 & a_2 \end{vmatrix} = \begin{vmatrix} 2 & 9 \\ 1 & 4 + K \end{vmatrix} = 2K - 1$$

$$\Delta_3 = \begin{vmatrix} a_1 & a_3 & 0 \\ a_0 & a_2 & a_4 \\ 0 & a_1 & a_3 \end{vmatrix} = \begin{vmatrix} 2 & 9 & 0 \\ 1 & 4 + K & 25 \\ 0 & 2 & 9 \end{vmatrix} = 18K - 109$$

For all principal minors to be positive, we require that $\Delta_i (i = 1, 2, 3)$ be positive. Thus, we require

$$2K - 1 > 0$$

$$18K - 109 > 0$$

from which we obtain the region of K for stability to be

$$K > \frac{109}{18}$$

A–5–22. Explain why the proportional control of a plant that does not possess an integrating property (which means that the plant transfer function does not include the factor $1/s$) suffers offset in response to step inputs.

Solution. Consider, for example, the system shown in Figure 5–66. At steady state, if c were equal to a nonzero constant r, then $e = 0$ and $u = Ke = 0$, resulting in $c = 0$, which contradicts the assumption that $c = r = $ nonzero constant.

A nonzero offset must exist for proper operation of such a control system. In other words, at steady state, if e were equal to $r/(1 + K)$, then $u = Kr/(1 + K)$ and $c = Kr/(1 + K)$, which results in the assumed error signal $e = r/(1 + K)$. Thus the offset of $r/(1 + K)$ must exist in such a system.

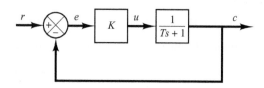

Figure 5–66
Control system.

A–5–23. The block diagram of Figure 5–67 shows a speed control system in which the output member of the system is subject to a torque disturbance. In the diagram, $\Omega_r(s)$, $\Omega(s)$, $T(s)$, and $D(s)$ are the Laplace transforms of the reference speed, output speed, driving torque, and disturbance torque, respectively. In the absence of a disturbance torque, the output speed is equal to the reference speed.

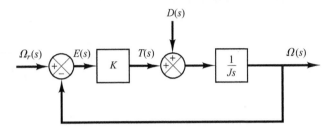

Figure 5–67
Block diagram of a
speed control system.

Investigate the response of this system to a unit-step disturbance torque. Assume that the reference input is zero, or $\Omega_r(s) = 0$.

Solution. Figure 5–68 is a modified block diagram convenient for the present analysis. The closed-loop transfer function is

$$\frac{\Omega_D(s)}{D(s)} = \frac{1}{Js + K}$$

where $\Omega_D(s)$ is the Laplace transform of the output speed due to the disturbance torque. For a unit-step disturbance torque, the steady-state output velocity is

$$\omega_D(\infty) = \lim_{s \to 0} s\Omega_D(s)$$

$$= \lim_{s \to 0} \frac{s}{Js + K} \frac{1}{s}$$

$$= \frac{1}{K}$$

From this analysis, we conclude that, if a step disturbance torque is applied to the output member of the system, an error speed will result so that the ensuing motor torque will exactly cancel the disturbance torque. To develop this motor torque, it is necessary that there be an error in speed so that nonzero torque will result. (Discussions continue to Problem **A–5–24**.)

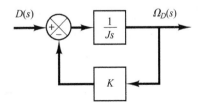

Figure 5–68
Block diagram of the
speed control system
of Figure 5–67 when
$\Omega_r(s) = 0$.

A-5-24. In the system considered in Problem **A-5-23**, it is desired to eliminate as much as possible the speed errors due to torque disturbances.

Is it possible to cancel the effect of a disturbance torque at steady state so that a constant disturbance torque applied to the output member will cause no speed change at steady state?

Solution. Suppose that we choose a suitable controller whose transfer function is $G_c(s)$, as shown in Figure 5–69. Then in the absence of the reference input the closed-loop transfer function between the output velocity $\Omega_D(s)$ and the disturbance torque $D(s)$ is

$$\frac{\Omega_D(s)}{D(s)} = \frac{\dfrac{1}{Js}}{1 + \dfrac{1}{Js}G_c(s)}$$

$$= \frac{1}{Js + G_c(s)}$$

The steady-state output speed due to a unit-step disturbance torque is

$$\omega_D(\infty) = \lim_{s \to 0} s\Omega_D(s)$$

$$= \lim_{s \to 0} \frac{s}{Js + G_c(s)} \frac{1}{s}$$

$$= \frac{1}{G_c(0)}$$

To satisfy the requirement that

$$\omega_D(\infty) = 0$$

we must choose $G_c(0) = \infty$. This can be realized if we choose

$$G_c(s) = \frac{K}{s}$$

Integral control action will continue to correct until the error is zero. This controller, however, presents a stability problem, because the characteristic equation will have two imaginary roots.

One method of stabilizing such a system is to add a proportional mode to the controller or choose

$$G_c(s) = K_p + \frac{K}{s}$$

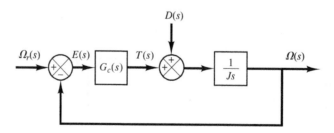

Figure 5–69
Block diagram of a speed control system.

Chapter 5 / Transient and Steady-State Response Analyses

Figure 5–70
Block diagram of the speed control system of Figure 5–69 when $G_c(s) = K_p + (K/s)$ and $\Omega_r(s) = 0$.

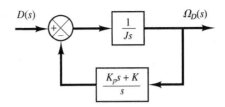

With this controller, the block diagram of Figure 5–69 in the absence of the reference input can be modified to that of Figure 5–70. The closed-loop transfer function $\Omega_D(s)/D(s)$ becomes

$$\frac{\Omega_D(s)}{D(s)} = \frac{s}{Js^2 + K_p s + K}$$

For a unit-step disturbance torque, the steady-state output speed is

$$\omega_D(\infty) = \lim_{s \to 0} s\Omega_D(s) = \lim_{s \to 0} \frac{s^2}{Js^2 + K_p s + K}\frac{1}{s} = 0$$

Thus, we see that the proportional-plus-integral controller eliminates speed error at steady state.

The use of integral control action has increased the order of the system by 1. (This tends to produce an oscillatory response.)

In the present system, a step disturbance torque will cause a transient error in the output speed, but the error will become zero at steady state. The integrator provides a nonzero output with zero error. (The nonzero output of the integrator produces a motor torque that exactly cancels the disturbance torque.)

Note that even if the system may have an integrator in the plant (such as an integrator in the transfer function of the plant), this does not eliminate the steady-state error due to a step disturbance torque. To eliminate this, we must have an integrator before the point where the disturbance torque enters.

A–5–25. Consider the system shown in Figure 5–71(a). The steady-state error to a unit-ramp input is $e_{ss} = 2\zeta/\omega_n$. Show that the steady-state error for following a ramp input may be eliminated if the input is introduced to the system through a proportional-plus-derivative filter, as shown in Figure 5–71(b), and the value of k is properly set. Note that the error $e(t)$ is given by $r(t) - c(t)$.

Solution. The closed-loop transfer function of the system shown in Figure 5–71(b) is

$$\frac{C(s)}{R(s)} = \frac{(1 + ks)\omega_n^2}{s^2 + 2\zeta\omega_n s + \omega_n^2}$$

Then

$$R(s) - C(s) = \left(\frac{s^2 + 2\zeta\omega_n s - \omega_n^2 ks}{s^2 + 2\zeta\omega_n s + \omega_n^2}\right)R(s)$$

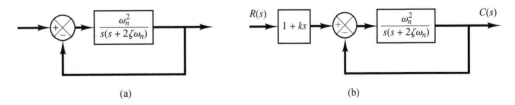

Figure 5–71
(a) Control system;
(b) control system with input filter.

(a)   (b)

If the input is a unit ramp, then the steady-state error is

$$e(\infty) = r(\infty) - c(\infty)$$

$$= \lim_{s \to 0} s \left(\frac{s^2 + 2\zeta\omega_n s - \omega_n^2 ks}{s^2 + 2\zeta\omega_n s + \omega_n^2} \right) \frac{1}{s^2}$$

$$= \frac{2\zeta\omega_n - \omega_n^2 k}{\omega_n^2}$$

Therefore, if k is chosen as

$$k = \frac{2\zeta}{\omega_n}$$

then the steady-state error for following a ramp input can be made equal to zero. Note that, if there are any variations in the values of ζ and/or ω_n due to environmental changes or aging, then a nonzero steady-state error for a ramp response may result.

A–5–26. Consider the stable unity-feedback control system with feedforward transfer function $G(s)$. Suppose that the closed-loop transfer function can be written

$$\frac{C(s)}{R(s)} = \frac{G(s)}{1 + G(s)} = \frac{(T_a s + 1)(T_b s + 1)\cdots(T_m s + 1)}{(T_1 s + 1)(T_2 s + 1)\cdots(T_n s + 1)} \qquad (m \le n)$$

Show that

$$\int_0^\infty e(t)\, dt = (T_1 + T_2 + \cdots + T_n) - (T_a + T_b + \cdots + T_m)$$

where $e(t) = r(t) - c(t)$ is the error in the unit-step response. Show also that

$$\frac{1}{K_v} = \frac{1}{\lim_{s \to 0} sG(s)} = (T_1 + T_2 + \cdots + T_n) - (T_a + T_b + \cdots + T_m)$$

Solution. Let us define

$$(T_a s + 1)(T_b s + 1)\cdots(T_m s + 1) = P(s)$$

and

$$(T_1 s + 1)(T_2 s + 1)\cdots(T_n s + 1) = Q(s)$$

Then

$$\frac{C(s)}{R(s)} = \frac{P(s)}{Q(s)}$$

and

$$E(s) = \frac{Q(s) - P(s)}{Q(s)} R(s)$$

For a unit-step input, $R(s) = 1/s$ and

$$E(s) = \frac{Q(s) - P(s)}{sQ(s)}$$

Since the system is stable, $\int_0^\infty e(t)\,dt$ converges to a constant value. Noting that

$$\int_0^\infty e(t)\,dt = \lim_{s \to 0} s\,\frac{E(s)}{s} = \lim_{s \to 0} E(s)$$

we have

$$\int_0^\infty e(t)\,dt = \lim_{s \to 0} \frac{Q(s) - P(s)}{sQ(s)}$$

$$= \lim_{s \to 0} \frac{Q'(s) - P'(s)}{Q(s) + sQ'(s)}$$

$$= \lim_{s \to 0} \left[Q'(s) - P'(s) \right]$$

Since

$$\lim_{s \to 0} P'(s) = T_a + T_b + \cdots + T_m$$

$$\lim_{s \to 0} Q'(s) = T_1 + T_2 + \cdots + T_n$$

we have

$$\int_0^\infty e(t)\,dt = \left(T_1 + T_2 + \cdots + T_n \right) - \left(T_a + T_b + \cdots + T_m \right)$$

For a unit-step input $r(t)$, since

$$\int_0^\infty e(t)\,dt = \lim_{s \to 0} E(s) = \lim_{s \to 0} \frac{1}{1 + G(s)} R(s) = \lim_{s \to 0} \frac{1}{1 + G(s)} \frac{1}{s} = \frac{1}{\lim\limits_{s \to 0} sG(s)} = \frac{1}{K_v}$$

we have

$$\frac{1}{K_v} = \frac{1}{\lim\limits_{s \to 0} sG(s)} = \left(T_1 + T_2 + \cdots + T_n \right) - \left(T_a + T_b + \cdots + T_m \right)$$

Note that zeros in the left half-plane (that is, positive $T_a, T_b, \ldots, T_m$) will improve K_v. Poles close to the origin cause low velocity-error constants unless there are zeros nearby.

PROBLEMS

B–5–1. A thermometer requires 1 min to indicate 98% of the response to a step input. Assuming the thermometer to be a first-order system, find the time constant.

If the thermometer is placed in a bath, the temperature of which is changing linearly at a rate of $10°/\text{min}$, how much error does the thermometer show?

B–5–2. Consider the unit-step response of a unity-feedback control system whose open-loop transfer function is

$$G(s) = \frac{1}{s(s + 1)}$$

Obtain the rise time, peak time, maximum overshoot, and settling time.

B–5–3. Consider the closed-loop system given by

$$\frac{C(s)}{R(s)} = \frac{\omega_n^2}{s^2 + 2\zeta\omega_n s + \omega_n^2}$$

Determine the values of ζ and ω_n so that the system responds to a step input with approximately 5% overshoot and with a settling time of 2 sec. (Use the 2% criterion.)

B–5–4. Consider the system shown in Figure 5–72. The system is initially at rest. Suppose that the cart is set into motion by an impulsive force whose strength is unity. Can it be stopped by another such impulsive force?

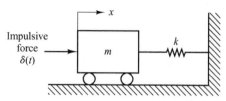

Figure 5–72
Mechanical system.

B–5–5. Obtain the unit-impulse response and the unit-step response of a unity-feedback system whose open-loop transfer function is

$$G(s) = \frac{2s + 1}{s^2}$$

B–5–6. An oscillatory system is known to have a transfer function of the following form:

$$G(s) = \frac{\omega_n^2}{s^2 + 2\zeta\omega_n s + \omega_n^2}$$

Assume that a record of a damped oscillation is available as shown in Figure 5–73. Determine the damping ratio ζ of the system from the graph.

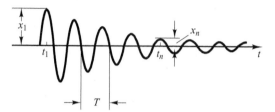

Figure 5–73
Decaying oscillation.

B–5–7. Consider the system shown in Figure 5–74(a). The damping ratio of this system is 0.158 and the undamped natural frequency is 3.16 rad/sec. To improve the relative stability, we employ tachometer feedback. Figure 5–74(b) shows such a tachometer-feedback system.

Determine the value of K_h so that the damping ratio of the system is 0.5. Draw unit-step response curves of both the original and tachometer-feedback systems. Also draw the error-versus-time curves for the unit-ramp response of both systems.

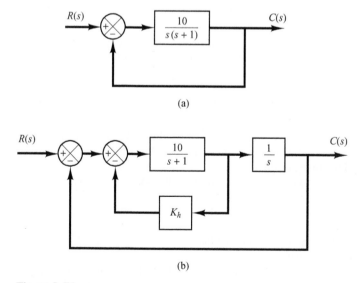

Figure 5–74
(a) Control system; (b) control system with tachometer feedback.

B–5–8. Referring to the system shown in Figure 5–75, determine the values of K and k such that the system has a damping ratio ζ of 0.7 and an undamped natural frequency ω_n of 4 rad/sec.

B–5–9. Consider the system shown in Figure 5–76. Determine the value of k such that the damping ratio ζ is 0.5. Then obtain the rise time t_r, peak time t_p, maximum overshoot M_p, and settling time t_s in the unit-step response.

B–5–10. Using MATLAB, obtain the unit-step response, unit-ramp response, and unit-impulse response of the following system:

$$\frac{C(s)}{R(s)} = \frac{10}{s^2 + 2s + 10}$$

where $R(s)$ and $C(s)$ are Laplace transforms of the input $r(t)$ and output $c(t)$, respectively.

B–5–11. Using MATLAB, obtain the unit-step response, unit-ramp response, and unit-impulse response of the following system:

$$\begin{bmatrix} \dot{x}_1 \\ \dot{x}_2 \end{bmatrix} = \begin{bmatrix} -1 & -0.5 \\ 1 & 0 \end{bmatrix} \begin{bmatrix} x_1 \\ x_2 \end{bmatrix} + \begin{bmatrix} 0.5 \\ 0 \end{bmatrix} u$$

$$y = \begin{bmatrix} 1 & 0 \end{bmatrix} \begin{bmatrix} x_1 \\ x_2 \end{bmatrix}$$

where u is the input and y is the output.

B–5–12. Obtain both analytically and computationally the rise time, peak time, maximum overshoot, and settling time in the unit-step response of a closed-loop system given by

$$\frac{C(s)}{R(s)} = \frac{36}{s^2 + 2s + 36}$$

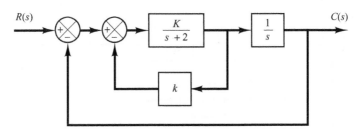

Figure 5–75
Closed-loop system.

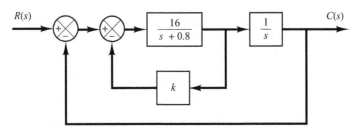

Figure 5–76
Block diagram of a system.

B–5–13. Figure 5–77 shows three systems. System I is a positional servo system. System II is a positional servo system with PD control action. System III is a positional servo system with velocity feedback. Compare the unit-step, unit-impulse, and unit-ramp responses of the three systems. Which system is best with respect to the speed of response and maximum overshoot in the step response?

B–5–14. Consider the position control system shown in Figure 5–78. Write a MATLAB program to obtain a unit-step response and a unit-ramp response of the system. Plot curves $x_1(t)$ versus t, $x_2(t)$ versus t, $x_3(t)$ versus t, and $e(t)$ versus t [where $e(t) = r(t) - x_1(t)$] for both the unit-step response and the unit-ramp response.

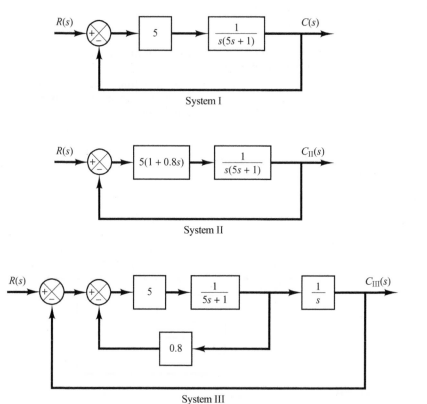

System I

System II

System III

Figure 5–77
Positional servo system (System I), positional servo system with PD control action (System II), and positional servo system with velocity feedback (System III).

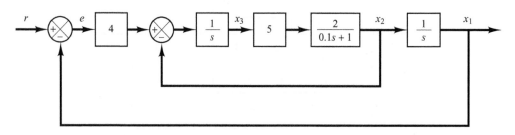

Figure 5–78
Position control system.

B–5–15. Using MATLAB, obtain the unit-step response curve for the unity-feedback control system whose open-loop transfer function is

$$G(s) = \frac{10}{s(s + 2)(s + 4)}$$

Using MATLAB, obtain also the rise time, peak time, maximum overshoot, and settling time in the unit-step response curve.

B–5–16. Consider the closed-loop system defined by

$$\frac{C(s)}{R(s)} = \frac{2\zeta s + 1}{s^2 + 2\zeta s + 1}$$

where $\zeta = 0.2, 0.4, 0.6, 0.8,$ and 1.0. Using MATLAB, plot a two-dimensional diagram of unit-impulse response curves. Also plot a three-dimensional plot of the response curves.

B–5–17. Consider the second-order system defined by

$$\frac{C(s)}{R(s)} = \frac{s + 1}{s^2 + 2\zeta s + 1}$$

where $\zeta = 0.2, 0.4, 0.6, 0.8, 1.0$. Plot a three-dimensional diagram of the unit-step response curves.

B–5–18. Obtain the unit-ramp response of the system defined by

$$\begin{bmatrix} \dot{x}_1 \\ \dot{x}_2 \end{bmatrix} = \begin{bmatrix} 0 & 1 \\ -1 & -1 \end{bmatrix} \begin{bmatrix} x_1 \\ x_2 \end{bmatrix} + \begin{bmatrix} 0 \\ 1 \end{bmatrix} u$$

$$y = \begin{bmatrix} 1 & 0 \end{bmatrix} \begin{bmatrix} x_1 \\ x_2 \end{bmatrix}$$

where u is the unit-ramp input. Use the lsim command to obtain the response.

B–5–19. Consider the differential equation system given by

$$\ddot{y} + 3\dot{y} + 2y = 0, \qquad y(0) = 0.1, \qquad \dot{y}(0) = 0.05$$

Using MATLAB, obtain the response $y(t)$, subject to the given initial condition.

B–5–20. Determine the range of K for stability of a unity-feedback control system whose open-loop transfer function is

$$G(s) = \frac{K}{s(s + 1)(s + 2)}$$

B–5–21. Consider the following characteristic equation:

$$s^4 + 2s^3 + (4 + K)s^2 + 9s + 25 = 0$$

Using the Routh stability criterion, determine the range of K for stability.

B–5–22. Consider the closed-loop system shown in Figure 5–79. Determine the range of K for stability. Assume that $K > 0$.

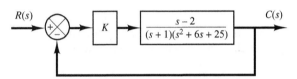

Figure 5–79 Closed-loop system.

B–5–23. Consider the satellite attitude control system shown in Figure 5–80(a). The output of this system exhibits continued oscillations and is not desirable. This system can be stabilized by use of tachometer feedback, as shown in Figure 5–80(b). If $K/J = 4$, what value of K_h will yield the damping ratio to be 0.6?

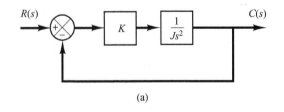

(a)

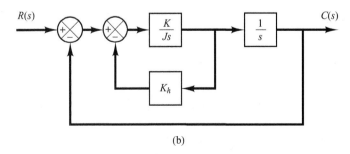

Figure 5–80
(a) Unstable satellite attitude control system;
(b) stabilized system.

(b)

B–5–24. Consider the servo system with tachometer feedback shown in Figure 5–81. Determine the ranges of stability for K and K_h. (Note that K_h must be positive.)

B–5–25. Consider the system

$$\dot{x} = Ax$$

where matrix A is given by

$$A = \begin{bmatrix} 0 & 1 & 0 \\ -b_3 & 0 & 1 \\ 0 & -b_2 & -b_1 \end{bmatrix}$$

(A is called Schwarz matrix.) Show that the first column of the Routh's array of the characteristic equation $|sI - A| = 0$ consists of $1, b_1, b_2,$ and $b_1 b_3$.

B–5–26. Consider a unity-feedback control system with the closed-loop transfer function

$$\frac{C(s)}{R(s)} = \frac{Ks + b}{s^2 + as + b}$$

Determine the open-loop transfer function $G(s)$.

Show that the steady-state error in the unit-ramp response is given by

$$e_{ss} = \frac{1}{K_v} = \frac{a - K}{b}$$

B–5–27. Consider a unity-feedback control system whose open-loop transfer function is

$$G(s) = \frac{K}{s(Js + B)}$$

Discuss the effects that varying the values of K and B has on the steady-state error in unit-ramp response. Sketch typical unit-ramp response curves for a small value, medium value, and large value of K, assuming that B is constant.

B–5–28. If the feedforward path of a control system contains at least one integrating element, then the output continues to change as long as an error is present. The output stops when the error is precisely zero. If an external disturbance enters the system, it is desirable to have an integrating element between the error-measuring element and the point where the disturbance enters, so that the effect of the external disturbance may be made zero at steady state.

Show that, if the disturbance is a ramp function, then the steady-state error due to this ramp disturbance may be eliminated only if two integrators precede the point where the disturbance enters.

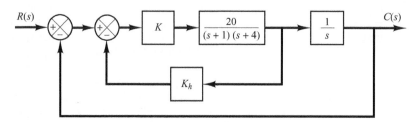

Figure 5–81
Servo system with tachometer feedback.

Control Systems Analysis and Design by the Root-Locus Method

6-1 INTRODUCTION

The basic characteristic of the transient response of a closed-loop system is closely related to the location of the closed-loop poles. If the system has a variable loop gain, then the location of the closed-loop poles depends on the value of the loop gain chosen. It is important, therefore, that the designer know how the closed-loop poles move in the s plane as the loop gain is varied.

From the design viewpoint, in some systems simple gain adjustment may move the closed-loop poles to desired locations. Then the design problem may become the selection of an appropriate gain value. If the gain adjustment alone does not yield a desired result, addition of a compensator to the system will become necessary. (This subject is discussed in detail in Sections 6–6 through 6–9.)

The closed-loop poles are the roots of the characteristic equation. Finding the roots of the characteristic equation of degree higher than 3 is laborious and will need computer solution. (MATLAB provides a simple solution to this problem.) However, just finding the roots of the characteristic equation may be of limited value, because as the gain of the open-loop transfer function varies, the characteristic equation changes and the computations must be repeated.

A simple method for finding the roots of the characteristic equation has been developed by W. R. Evans and used extensively in control engineering. This method, called the *root-locus method,* is one in which the roots of the characteristic equation

269

are plotted for all values of a system parameter. The roots corresponding to a particular value of this parameter can then be located on the resulting graph. Note that the parameter is usually the gain, but any other variable of the open-loop transfer function may be used. Unless otherwise stated, we shall assume that the gain of the open-loop transfer function is the parameter to be varied through all values, from zero to infinity.

By using the root-locus method the designer can predict the effects on the location of the closed-loop poles of varying the gain value or adding open-loop poles and/or open-loop zeros. Therefore, it is desired that the designer have a good understanding of the method for generating the root loci of the closed-loop system, both by hand and by use of a computer software program like MATLAB.

In designing a linear control system, we find that the root-locus method proves to be quite useful, since it indicates the manner in which the open-loop poles and zeros should be modified so that the response meets system performance specifications. This method is particularly suited to obtaining approximate results very quickly.

Because generating the root loci by use of MATLAB is very simple, one may think sketching the root loci by hand is a waste of time and effort. However, experience in sketching the root loci by hand is invaluable for interpreting computer-generated root loci, as well as for getting a rough idea of the root loci very quickly.

Outline of the Chapter. The outline of the chapter is as follows: Section 6–1 has presented an introduction to the root-locus method. Section 6–2 details the concepts underlying the root-locus method and presents the general procedure for sketching root loci using illustrative examples. Section 6–3 discusses generating root-locus plots with MATLAB. Section 6–4 treats a special case when the closed-loop system has positive feedback. Section 6–5 presents general aspects of the root-locus approach to the design of closed-loop systems. Section 6–6 discusses the control systems design by lead compensation. Section 6–7 treats the lag compensation technique. Section 6–8 deals with the lag–lead compensation technique. Finally, Section 6–9 discusses the parallel compensation technique.

6–2 ROOT-LOCUS PLOTS

Angle and Magnitude Conditions. Consider the negative feedback system shown in Figure 6–1. The closed-loop transfer function is

$$\frac{C(s)}{R(s)} = \frac{G(s)}{1 + G(s)H(s)} \tag{6–1}$$

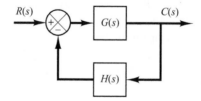

Figure 6–1
Control system.

The characteristic equation for this closed-loop system is obtained by setting the denominator of the right-hand side of Equation (6–1) equal to zero. That is,

$$1 + G(s)H(s) = 0$$

or

$$G(s)H(s) = -1 \qquad (6\text{–}2)$$

Here we assume that $G(s)H(s)$ is a ratio of polynomials in s. [It is noted that we can extend the analysis to the case when $G(s)H(s)$ involves the transport lag e^{-Ts}.] Since $G(s)H(s)$ is a complex quantity, Equation (6–2) can be split into two equations by equating the angles and magnitudes of both sides, respectively, to obtain the following:

Angle condition:

$$\underline{/G(s)H(s)} = \pm 180°(2k + 1) \qquad (k = 0, 1, 2, \dots) \qquad (6\text{–}3)$$

Magnitude condition:

$$|G(s)H(s)| = 1 \qquad (6\text{–}4)$$

The values of s that fulfill both the angle and magnitude conditions are the roots of the characteristic equation, or the closed-loop poles. A locus of the points in the complex plane satisfying the angle condition alone is the root locus. The roots of the characteristic equation (the closed-loop poles) corresponding to a given value of the gain can be determined from the magnitude condition. The details of applying the angle and magnitude conditions to obtain the closed-loop poles are presented later in this section.

In many cases, $G(s)H(s)$ involves a gain parameter K, and the characteristic equation may be written as

$$1 + \frac{K(s + z_1)(s + z_2)\cdots(s + z_m)}{(s + p_1)(s + p_2)\cdots(s + p_n)} = 0$$

Then the root loci for the system are the loci of the closed-loop poles as the gain K is varied from zero to infinity.

Note that to begin sketching the root loci of a system by the root-locus method we must know the location of the poles and zeros of $G(s)H(s)$. Remember that the angles of the complex quantities originating from the open-loop poles and open-loop zeros to the test point s are measured in the counterclockwise direction. For example, if $G(s)H(s)$ is given by

$$G(s)H(s) = \frac{K(s + z_1)}{(s + p_1)(s + p_2)(s + p_3)(s + p_4)}$$

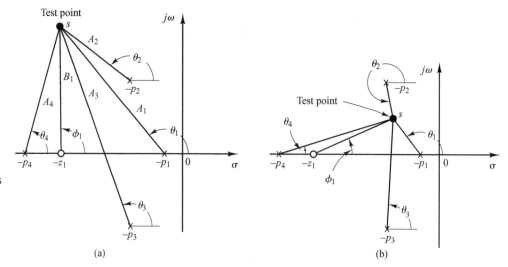

Figure 6–2
(a) and (b) Diagrams showing angle measurements from open-loop poles and open-loop zero to test point s.

(a)   (b)

where $-p_2$ and $-p_3$ are complex-conjugate poles, then the angle of $G(s)H(s)$ is

$$\underline{/G(s)H(s)} = \phi_1 - \theta_1 - \theta_2 - \theta_3 - \theta_4$$

where $\phi_1, \theta_1, \theta_2, \theta_3$, and θ_4 are measured counterclockwise as shown in Figures 6–2(a) and (b). The magnitude of $G(s)H(s)$ for this system is

$$|G(s)H(s)| = \frac{KB_1}{A_1 A_2 A_3 A_4}$$

where A_1, A_2, A_3, A_4, and B_1 are the magnitudes of the complex quantities $s + p_1$, $s + p_2$, $s + p_3$, $s + p_4$, and $s + z_1$, respectively, as shown in Figure 6–2(a).

Note that, because the open-loop complex-conjugate poles and complex-conjugate zeros, if any, are always located symmetrically about the real axis, the root loci are always symmetrical with respect to this axis. Therefore, we only need to construct the upper half of the root loci and draw the mirror image of the upper half in the lower-half s plane.

Illustrative Examples.   In what follows, two illustrative examples for constructing root-locus plots will be presented. Although computer approaches to the construction of the root loci are easily available, here we shall use graphical computation, combined with inspection, to determine the root loci upon which the roots of the characteristic equation of the closed-loop system must lie. Such a graphical approach will enhance understanding of how the closed-loop poles move in the complex plane as the open-loop poles and zeros are moved. Although we employ only simple systems for illustrative purposes, the procedure for finding the root loci is no more complicated for higher-order systems.

Because graphical measurements of angles and magnitudes are involved in the analysis, we find it necessary to use the same divisions on the abscissa as on the ordinate axis when sketching the root locus on graph paper.

Chapter 6 / Control Systems Analysis and Design by the Root-Locus Method

EXAMPLE 6–1   Consider the negative feedback system shown in Figure 6–3. (We assume that the value of gain K is nonnegative.) For this system,

$$G(s) = \frac{K}{s(s + 1)(s + 2)}, \qquad H(s) = 1$$

Let us sketch the root-locus plot and then determine the value of K such that the damping ratio ζ of a pair of dominant complex-conjugate closed-loop poles is 0.5.

For the given system, the angle condition becomes

$$\underline{/G(s)} = \underline{\left/\frac{K}{s(s + 1)(s + 2)}\right.}$$

$$= -\underline{/s} - \underline{/s + 1} - \underline{/s + 2}$$

$$= \pm 180°(2k + 1) \qquad (k = 0, 1, 2, \dots)$$

The magnitude condition is

$$|G(s)| = \left|\frac{K}{s(s + 1)(s + 2)}\right| = 1$$

A typical procedure for sketching the root-locus plot is as follows:

1. *Determine the root loci on the real axis.* The first step in constructing a root-locus plot is to locate the open-loop poles, $s = 0$, $s = -1$, and $s = -2$, in the complex plane. (There are no open-loop zeros in this system.) The locations of the open-loop poles are indicated by crosses. (The locations of the open-loop zeros in this book will be indicated by small circles.) Note that the starting points of the root loci (the points corresponding to $K = 0$) are open-loop poles. The number of individual root loci for this system is three, which is the same as the number of open-loop poles.

To determine the root loci on the real axis, we select a test point, s. If the test point is on the positive real axis, then

$$\underline{/s} = \underline{/s + 1} = \underline{/s + 2} = 0°$$

This shows that the angle condition cannot be satisfied. Hence, there is no root locus on the positive real axis. Next, select a test point on the negative real axis between 0 and -1. Then

$$\underline{/s} = 180°, \qquad \underline{/s + 1} = \underline{/s + 2} = 0°$$

Thus

$$-\underline{/s} - \underline{/s + 1} - \underline{/s + 2} = -180°$$

and the angle condition is satisfied. Therefore, the portion of the negative real axis between 0 and -1 forms a portion of the root locus. If a test point is selected between -1 and -2, then

$$\underline{/s} = \underline{/s + 1} = 180°, \qquad \underline{/s + 2} = 0°$$

and

$$-\underline{/s} - \underline{/s + 1} - \underline{/s + 2} = -360°$$

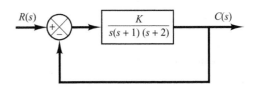

Figure 6–3
Control system.

It can be seen that the angle condition is not satisfied. Therefore, the negative real axis from -1 to -2 is not a part of the root locus. Similarly, if a test point is located on the negative real axis from -2 to $-\infty$, the angle condition is satisfied. Thus, root loci exist on the negative real axis between 0 and -1 and between -2 and $-\infty$.

2. *Determine the asymptotes of the root loci.* The asymptotes of the root loci as s approaches infinity can be determined as follows: If a test point s is selected very far from the origin, then

$$\lim_{s \to \infty} G(s) = \lim_{s \to \infty} \frac{K}{s(s+1)(s+2)} = \lim_{s \to \infty} \frac{K}{s^3}$$

and the angle condition becomes

$$-3\underline{/s} = \pm 180°(2k+1) \qquad (k = 0, 1, 2, \dots)$$

or

$$\text{Angles of asymptotes} = \frac{\pm 180°(2k+1)}{3} \qquad (k = 0, 1, 2, \dots)$$

Since the angle repeats itself as k is varied, the distinct angles for the asymptotes are determined as $60°$, $-60°$, and $180°$. Thus, there are three asymptotes. The one having the angle of $180°$ is the negative real axis.

Before we can draw these asymptotes in the complex plane, we must find the point where they intersect the real axis. Since

$$G(s) = \frac{K}{s(s+1)(s+2)}$$

if a test point is located very far from the origin, then $G(s)$ may be written as

$$G(s) = \frac{K}{s^3 + 3s^2 + \cdots}$$

For large values of s, this last equation may be approximated by

$$G(s) \doteq \frac{K}{(s+1)^3} \tag{6-5}$$

A root-locus diagram of $G(s)$ given by Equation (6–5) consists of three straight lines. This can be seen as follows: The equation of the root locus is

$$\underline{\bigg/ \frac{K}{(s+1)^3}} = \pm 180°(2k+1)$$

or

$$-3\underline{/s+1} = \pm 180°(2k+1)$$

which can be written as

$$\underline{/s+1} = \pm 60°(2k+1)$$

By substituting $s = \sigma + j\omega$ into this last equation, we obtain

$$\underline{/\sigma + j\omega + 1} = \pm60°(2k + 1)$$

or

$$\tan^{-1} \frac{\omega}{\sigma + 1} = 60°, \qquad -60°, \qquad 0°$$

Taking the tangent of both sides of this last equation,

$$\frac{\omega}{\sigma + 1} = \sqrt{3}, \qquad -\sqrt{3}, \qquad 0$$

which can be written as

$$\sigma + 1 - \frac{\omega}{\sqrt{3}} = 0, \qquad \sigma + 1 + \frac{\omega}{\sqrt{3}} = 0, \qquad \omega = 0$$

These three equations represent three straight lines, as shown in Figure 6–4. The three straight lines shown are the asymptotes. They meet at point $s = -1$. Thus, the abscissa of the intersection of the asymptotes and the real axis is obtained by setting the denominator of the right-hand side of Equation (6–5) equal to zero and solving for s. The asymptotes are almost parts of the root loci in regions very far from the origin.

3. *Determine the breakaway point.* To plot root loci accurately, we must find the breakaway point, where the root-locus branches originating from the poles at 0 and −1 break away (as K is increased) from the real axis and move into the complex plane. The breakaway point corresponds to a point in the s plane where multiple roots of the characteristic equation occur.

A simple method for finding the breakaway point is available. We shall present this method in the following: Let us write the characteristic equation as

$$f(s) = B(s) + KA(s) = 0 \tag{6–6}$$

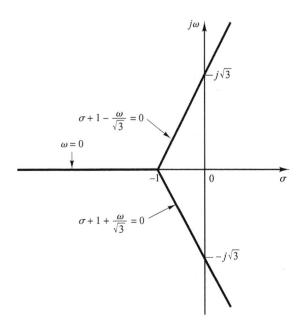

Figure 6–4
Three asymptotes.

where $A(s)$ and $B(s)$ do not contain K. Note that $f(s) = 0$ has multiple roots at points where

$$\frac{df(s)}{ds} = 0$$

This can be seen as follows: Suppose that $f(s)$ has multiple roots of order r, where $r \geq 2$. Then $f(s)$ may be written as

$$f(s) = (s - s_1)^r(s - s_2)\cdots(s - s_n)$$

Now we differentiate this equation with respect to s and evaluate $df(s)/ds$ at $s = s_1$. Then we get

$$\left.\frac{df(s)}{ds}\right|_{s=s_1} = 0 \tag{6-7}$$

This means that multiple roots of $f(s)$ will satisfy Equation (6–7). From Equation (6–6), we obtain

$$\frac{df(s)}{ds} = B'(s) + KA'(s) = 0 \tag{6-8}$$

where

$$A'(s) = \frac{dA(s)}{ds}, \qquad B'(s) = \frac{dB(s)}{ds}$$

The particular value of K that will yield multiple roots of the characteristic equation is obtained from Equation (6–8) as

$$K = -\frac{B'(s)}{A'(s)}$$

If we substitute this value of K into Equation (6–6), we get

$$f(s) = B(s) - \frac{B'(s)}{A'(s)}A(s) = 0$$

or

$$B(s)A'(s) - B'(s)A(s) = 0 \tag{6-9}$$

If Equation (6–9) is solved for s, the points where multiple roots occur can be obtained. On the other hand, from Equation (6–6) we obtain

$$K = -\frac{B(s)}{A(s)}$$

and

$$\frac{dK}{ds} = -\frac{B'(s)A(s) - B(s)A'(s)}{A^2(s)}$$

If dK/ds is set equal to zero, we get the same equation as Equation (6–9). Therefore, the breakaway points can be simply determined from the roots of

$$\frac{dK}{ds} = 0$$

It should be noted that not all the solutions of Equation (6–9) or of $dK/ds = 0$ correspond to actual breakaway points. If a point at which $dK/ds = 0$ is on a root locus, it is an actual breakaway or break-in point. Stated differently, if at a point at which $dK/ds = 0$ the value of K takes a real positive value, then that point is an actual breakaway or break-in point.

For the present example, the characteristic equation $G(s) + 1 = 0$ is given by

$$\frac{K}{s(s + 1)(s + 2)} + 1 = 0$$

or

$$K = -(s^3 + 3s^2 + 2s)$$

By setting $dK/ds = 0$, we obtain

$$\frac{dK}{ds} = -(3s^2 + 6s + 2) = 0$$

or

$$s = -0.4226, \qquad s = -1.5774$$

Since the breakaway point must lie on a root locus between 0 and -1, it is clear that $s = -0.4226$ corresponds to the actual breakaway point. Point $s = -1.5774$ is not on the root locus. Hence, this point is not an actual breakaway or break-in point. In fact, evaluation of the values of K corresponding to $s = -0.4226$ and $s = -1.5774$ yields

$$K = 0.3849, \qquad \text{for } s = -0.4226$$

$$K = -0.3849, \qquad \text{for } s = -1.5774$$

4. *Determine the points where the root loci cross the imaginary axis.* These points can be found by use of Routh's stability criterion as follows: Since the characteristic equation for the present system is

$$s^3 + 3s^2 + 2s + K = 0$$

the Routh array becomes

s^3	1	2
s^2	3	K
s^1	$\dfrac{6 - K}{3}$	
s^0	K	

The value of K that makes the s^1 term in the first column equal zero is $K = 6$. The crossing points on the imaginary axis can then be found by solving the auxiliary equation obtained from the s^2 row; that is,

$$3s^2 + K = 3s^2 + 6 = 0$$

which yields

$$s = \pm j\sqrt{2}$$

The frequencies at the crossing points on the imaginary axis are thus $\omega = \pm\sqrt{2}$. The gain value corresponding to the crossing points is $K = 6$.

An alternative approach is to let $s = j\omega$ in the characteristic equation, equate both the real part and the imaginary part to zero, and then solve for ω and K. For the present system, the characteristic equation, with $s = j\omega$, is

$$(j\omega)^3 + 3(j\omega)^2 + 2(j\omega) + K = 0$$

or

$$(K - 3\omega^2) + j(2\omega - \omega^3) = 0$$

Equating both the real and imaginary parts of this last equation to zero, respectively, we obtain

$$K - 3\omega^2 = 0, \qquad 2\omega - \omega^3 = 0$$

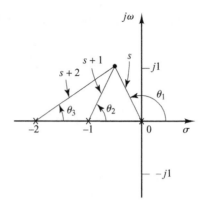

Figure 6–5
Construction of root
locus.

from which

$$\omega = \pm\sqrt{2}, \quad K = 6 \quad \text{or} \quad \omega = 0, \quad K = 0$$

Thus, root loci cross the imaginary axis at $\omega = \pm\sqrt{2}$, and the value of K at the crossing points is 6. Also, a root-locus branch on the real axis touches the imaginary axis at $\omega = 0$. The value of K is zero at this point.

5. *Choose a test point in the broad neighborhood of the $j\omega$ axis and the origin,* as shown in Figure 6–5, and apply the angle condition. If a test point is on the root loci, then the sum of the three angles, $\theta_1 + \theta_2 + \theta_3$, must be 180°. If the test point does not satisfy the angle condition, select another test point until it satisfies the condition. (The sum of the angles at the test point will indicate the direction in which the test point should be moved.) Continue this process and locate a sufficient number of points satisfying the angle condition.

6. *Draw the root loci,* based on the information obtained in the foregoing steps, as shown in Figure 6–6.

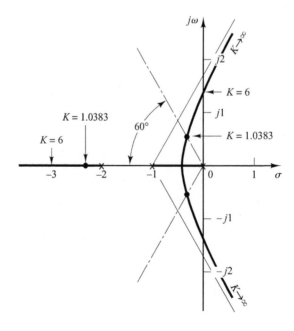

Figure 6–6
Root-locus plot.

Chapter 6 / Control Systems Analysis and Design by the Root-Locus Method

7. *Determine a pair of dominant complex-conjugate closed-loop poles such that the damping ratio ζ is 0.5.* Closed-loop poles with $\zeta = 0.5$ lie on lines passing through the origin and making the angles $\pm\cos^{-1}\zeta = \pm\cos^{-1}0.5 = \pm60°$ with the negative real axis. From Figure 6–6, such closed-loop poles having $\zeta = 0.5$ are obtained as follows:

$$s_1 = -0.3337 + j0.5780, \qquad s_2 = -0.3337 - j0.5780$$

The value of K that yields such poles is found from the magnitude condition as follows:

$$K = |s(s + 1)(s + 2)|_{s=-0.3337+j0.5780}$$

$$= 1.0383$$

Using this value of K, the third pole is found at $s = -2.3326$.

Note that, from step 4, it can be seen that for $K = 6$ the dominant closed-loop poles lie on the imaginary axis at $s = \pm j\sqrt{2}$. With this value of K, the system will exhibit sustained oscillations. For $K > 6$, the dominant closed-loop poles lie in the right-half s plane, resulting in an unstable system.

Finally, note that, if necessary, the root loci can be easily graduated in terms of K by use of the magnitude condition. We simply pick out a point on a root locus, measure the magnitudes of the three complex quantities s, $s + 1$, and $s + 2$, and multiply these magnitudes; the product is equal to the gain value K at that point, or

$$|s| \cdot |s + 1| \cdot |s + 2| = K$$

Graduation of the root loci can be done easily by use of MATLAB. (See Section 6–3.)

EXAMPLE 6–2 In this example, we shall sketch the root-locus plot of a system with complex-conjugate open-loop poles. Consider the negative feedback system shown in Figure 6–7. For this system,

$$G(s) = \frac{K(s + 2)}{s^2 + 2s + 3}, \qquad H(s) = 1$$

where $K \geq 0$. It is seen that $G(s)$ has a pair of complex-conjugate poles at

$$s = -1 + j\sqrt{2}, \qquad s = -1 - j\sqrt{2}$$

A typical procedure for sketching the root-locus plot is as follows:

1. *Determine the root loci on the real axis.* For any test point s on the real axis, the sum of the angular contributions of the complex-conjugate poles is 360°, as shown in Figure 6–8. Thus the net effect of the complex-conjugate poles is zero on the real axis. The location of the root locus on the real axis is determined from the open-loop zero on the negative real axis. A simple test reveals that a section of the negative real axis, that between -2 and $-\infty$, is a part of the root locus. It is noted that, since this locus lies between two zeros (at $s = -2$ and $s = -\infty$), it is actually a part of two root loci, each of which starts from one of the two complex-conjugate poles. In other words, two root loci break in the part of the negative real axis between -2 and $-\infty$.

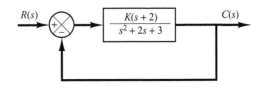

Figure 6–7
Control system.

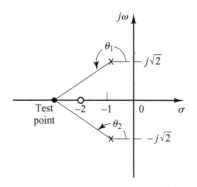

Figure 6–8
Determination of the
root locus on the real
axis.

Since there are two open-loop poles and one zero, there is one asymptote, which coincides with the negative real axis.

2. *Determine the angle of departure from the complex-conjugate open-loop poles.* The presence of a pair of complex-conjugate open-loop poles requires the determination of the angle of departure from these poles. Knowledge of this angle is important, since the root locus near a complex pole yields information as to whether the locus originating from the complex pole migrates toward the real axis or extends toward the asymptote.

Referring to Figure 6–9, if we choose a test point and move it in the very vicinity of the complex open-loop pole at $s = -p_1$, we find that the sum of the angular contributions from the pole at $s = p_2$ and zero at $s = -z_1$ to the test point can be considered remaining the same. If the test point is to be on the root locus, then the sum of ϕ_1', $-\theta_1$, and $-\theta_2'$ must be $\pm 180°(2k + 1)$, where $k = 0, 1, 2, \ldots$. Thus, in the example,

$$\phi_1' - (\theta_1 + \theta_2') = \pm 180°(2k + 1)$$

or

$$\theta_1 = 180° - \theta_2' + \phi_1' = 180° - \theta_2 + \phi_1$$

The angle of departure is then

$$\theta_1 = 180° - \theta_2 + \phi_1 = 180° - 90° + 55° = 145°$$

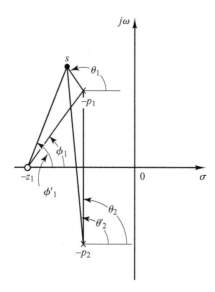

Figure 6–9
Determination of the
angle of departure.

Since the root locus is symmetric about the real axis, the angle of departure from the pole at $s = -p_2$ is $-145°$.

3. *Determine the break-in point.* A break-in point exists where a pair of root-locus branches coalesces as K is increased. For this problem, the break-in point can be found as follows: Since

$$K = -\frac{s^2 + 2s + 3}{s + 2}$$

we have

$$\frac{dK}{ds} = -\frac{(2s + 2)(s + 2) - (s^2 + 2s + 3)}{(s + 2)^2} = 0$$

which gives

$$s^2 + 4s + 1 = 0$$

or

$$s = -3.7320 \quad \text{or} \quad s = -0.2680$$

Notice that point $s = -3.7320$ is on the root locus. Hence this point is an actual break-in point. (Note that at point $s = -3.7320$ the corresponding gain value is $K = 5.4641$.) Since point $s = -0.2680$ is not on the root locus, it cannot be a break-in point. (For point $s = -0.2680$, the corresponding gain value is $K = -1.4641$.)

4. *Sketch a root-locus plot, based on the information obtained in the foregoing steps.* To determine accurate root loci, several points must be found by trial and error between the break-in point and the complex open-loop poles. (To facilitate sketching the root-locus plot, we should find the direction in which the test point should be moved by mentally summing up the changes on the angles of the poles and zeros.) Figure 6–10 shows a complete root-locus plot for the system considered.

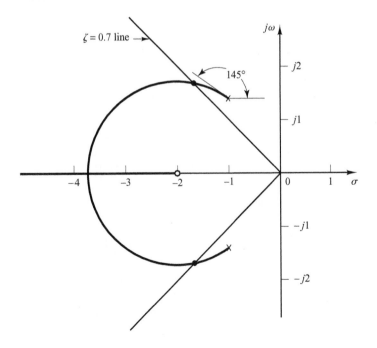

Figure 6–10
Root-locus plot.

The value of the gain K at any point on root locus can be found by applying the magnitude condition or by use of MATLAB (see Section 6–3). For example, the value of K at which the complex-conjugate closed-loop poles have the damping ratio $\zeta = 0.7$ can be found by locating the roots, as shown in Figure 6–10, and computing the value of K as follows:

$$K = \left| \frac{(s + 1 - j\sqrt{2})(s + 1 + j\sqrt{2})}{s + 2} \right|_{s=-1.67+j1.70} = 1.34$$

Or use MATLAB to find the value of K. (See Section 6–4.)

It is noted that in this system the root locus in the complex plane is a part of a circle. Such a circular root locus will not occur in most systems. Circular root loci may occur in systems that involve two poles and one zero, two poles and two zeros, or one pole and two zeros. Even in such systems, whether circular root loci occur depends on the locations of poles and zeros involved.

To show the occurrence of a circular root locus in the present system, we need to derive the equation for the root locus. For the present system, the angle condition is

$$\underline{/s + 2} - \underline{/s + 1 - j\sqrt{2}} - \underline{/s + 1 + j\sqrt{2}} = \pm 180°(2k + 1)$$

If $s = \sigma + j\omega$ is substituted into this last equation, we obtain

$$\underline{/\sigma + 2 + j\omega} - \underline{/\sigma + 1 + j\omega - j\sqrt{2}} - \underline{/\sigma + 1 + j\omega + j\sqrt{2}} = \pm 180°(2k + 1)$$

which can be written as

$$\tan^{-1}\left(\frac{\omega}{\sigma + 2}\right) - \tan^{-1}\left(\frac{\omega - \sqrt{2}}{\sigma + 1}\right) - \tan^{-1}\left(\frac{\omega + \sqrt{2}}{\sigma + 1}\right) = \pm 180°(2k + 1)$$

or

$$\tan^{-1}\left(\frac{\omega - \sqrt{2}}{\sigma + 1}\right) + \tan^{-1}\left(\frac{\omega + \sqrt{2}}{\sigma + 1}\right) = \tan^{-1}\left(\frac{\omega}{\sigma + 2}\right) \pm 180°(2k + 1)$$

Taking tangents of both sides of this last equation using the relationship

$$\tan(x \pm y) = \frac{\tan x \pm \tan y}{1 \mp \tan x \tan y} \tag{6–10}$$

we obtain

$$\tan\left[\tan^{-1}\left(\frac{\omega - \sqrt{2}}{\sigma + 1}\right) + \tan^{-1}\left(\frac{\omega + \sqrt{2}}{\sigma + 1}\right) \right] = \tan\left[\tan^{-1}\left(\frac{\omega}{\sigma + 2}\right) \pm 180°(2k + 1) \right]$$

or

$$\frac{\dfrac{\omega - \sqrt{2}}{\sigma + 1} + \dfrac{\omega + \sqrt{2}}{\sigma + 1}}{1 - \left(\dfrac{\omega - \sqrt{2}}{\sigma + 1}\right)\left(\dfrac{\omega + \sqrt{2}}{\sigma + 1}\right)} = \frac{\dfrac{\omega}{\sigma + 2} \pm 0}{1 \mp \dfrac{\omega}{\sigma + 2} \times 0}$$

which can be simplified to

$$\frac{2\omega(\sigma + 1)}{(\sigma + 1)^2 - (\omega^2 - 2)} = \frac{\omega}{\sigma + 2}$$

or

$$\omega\left[(\sigma + 2)^2 + \omega^2 - 3\right] = 0$$

This last equation is equivalent to

$$\omega = 0 \qquad \text{or} \qquad (\sigma + 2)^2 + \omega^2 = \left(\sqrt{3}\right)^2$$

These two equations are the equations for the root loci for the present system. Notice that the first equation, $\omega = 0$, is the equation for the real axis. The real axis from $s = -2$ to $s = -\infty$ corresponds to a root locus for $K \geq 0$. The remaining part of the real axis corresponds to a root locus when K is negative. (In the present system, K is nonnegative.) (Note that $K < 0$ corresponds to the positive-feedback case.) The second equation for the root locus is an equation of a circle with center at $\sigma = -2$, $\omega = 0$ and the radius equal to $\sqrt{3}$. That part of the circle to the left of the complex-conjugate poles corresponds to a root locus for $K \geq 0$. The remaining part of the circle corresponds to a root locus when K is negative.

It is important to note that easily interpretable equations for the root locus can be derived for simple systems only. For complicated systems having many poles and zeros, any attempt to derive equations for the root loci is discouraged. Such derived equations are very complicated and their configuration in the complex plane is difficult to visualize.

General Rules for Constructing Root Loci. For a complicated system with many open-loop poles and zeros, constructing a root-locus plot may seem complicated, but actually it is not difficult if the rules for constructing the root loci are applied. By locating particular points and asymptotes and by computing angles of departure from complex poles and angles of arrival at complex zeros, we can construct the general form of the root loci without difficulty.

We shall now summarize the general rules and procedure for constructing the root loci of the negative feedback control system shown in Figure 6–11.

First, obtain the characteristic equation

$$1 + G(s)H(s) = 0$$

Then rearrange this equation so that the parameter of interest appears as the multiplying factor in the form

$$1 + \frac{K(s + z_1)(s + z_2)\cdots(s + z_m)}{(s + p_1)(s + p_2)\cdots(s + p_n)} = 0 \qquad (6\text{–}11)$$

In the present discussions, we assume that the parameter of interest is the gain K, where $K > 0$. (If $K < 0$, which corresponds to the positive-feedback case, the angle condition must be modified. See Section 6–4.) Note, however, that the method is still applicable to systems with parameters of interest other than gain. (See Section 6–6.)

1. *Locate the poles and zeros of $G(s)H(s)$ on the s plane. The root-locus branches start from open-loop poles and terminate at zeros (finite zeros or zeros at infinity).* From the factored form of the open-loop transfer function, locate the open-loop poles and zeros in the s plane. [Note that the open-loop zeros are the zeros of $G(s)H(s)$, while the closed-loop zeros consist of the zeros of $G(s)$ and the poles of $H(s)$.]

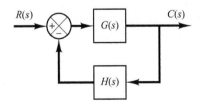

Figure 6–11
Control system.

Note that the root loci are symmetrical about the real axis of the s plane, because the complex poles and complex zeros occur only in conjugate pairs.

A root-locus plot will have just as many branches as there are roots of the characteristic equation. Since the number of open-loop poles generally exceeds that of zeros, the number of branches equals that of poles. If the number of closed-loop poles is the same as the number of open-loop poles, then the number of individual root-locus branches terminating at finite open-loop zeros is equal to the number m of the open-loop zeros. The remaining $n - m$ branches terminate at infinity ($n - m$ implicit zeros at infinity) along asymptotes.

If we include poles and zeros at infinity, the number of open-loop poles is equal to that of open-loop zeros. Hence we can always state that the root loci start at the poles of $G(s)H(s)$ and end at the zeros of $G(s)H(s)$, as K increases from zero to infinity, where the poles and zeros include both those in the finite s plane and those at infinity.

2. *Determine the root loci on the real axis.* Root loci on the real axis are determined by open-loop poles and zeros lying on it. The complex-conjugate poles and complex-conjugate zeros of the open-loop transfer function have no effect on the location of the root loci on the real axis because the angle contribution of a pair of complex-conjugate poles or complex-conjugate zeros is 360° on the real axis. Each portion of the root locus on the real axis extends over a range from a pole or zero to another pole or zero. In constructing the root loci on the real axis, choose a test point on it. If the total number of real poles and real zeros to the right of this test point is odd, then this point lies on a root locus. If the open-loop poles and open-loop zeros are simple poles and simple zeros, then the root locus and its complement form alternate segments along the real axis.

3. *Determine the asymptotes of root loci.* If the test point s is located far from the origin, then the angle of each complex quantity may be considered the same. One open-loop zero and one open-loop pole then cancel the effects of the other. Therefore, the root loci for very large values of s must be asymptotic to straight lines whose angles (slopes) are given by

$$\text{Angles of asymptotes} = \frac{\pm 180°(2k + 1)}{n - m} \qquad (k = 0, 1, 2, \dots)$$

where n = number of finite poles of $G(s)H(s)$

$\qquad m$ = number of finite zeros of $G(s)H(s)$

Here, $k = 0$ corresponds to the asymptotes with the smallest angle with the real axis. Although k assumes an infinite number of values, as k is increased the angle repeats itself, and the number of distinct asymptotes is $n - m$.

All the asymptotes intersect at a point on the real axis. The point at which they do so is obtained as follows: If both the numerator and denominator of the open-loop transfer function are expanded, the result is

$$G(s)H(s) = \frac{K[s^m + (z_1 + z_2 + \cdots + z_m)s^{m-1} + \cdots + z_1 z_2 \cdots z_m]}{s^n + (p_1 + p_2 + \cdots + p_n)s^{n-1} + \cdots + p_1 p_2 \cdots p_n}$$

If a test point is located very far from the origin, then by dividing the denominator by the numerator, it is possible to write $G(s)H(s)$ as

$$G(s)H(s) = \frac{K}{s^{n-m} + \left[(p_1 + p_2 + \cdots + p_n) - (z_1 + z_2 + \cdots + z_m)\right]s^{n-m-1} + \cdots}$$

or

$$G(s)H(s) = \frac{K}{\left[s + \dfrac{(p_1 + p_2 + \cdots + p_n) - (z_1 + z_2 + \cdots + z_m)}{n - m}\right]^{n-m}} \quad (6\text{--}12)$$

The abscissa of the intersection of the asymptotes and the real axis is then obtained by setting the denominator of the right-hand side of Equation (6–12) equal to zero and solving for s, or

$$s = -\frac{(p_1 + p_2 + \cdots + p_n) - (z_1 + z_2 + \cdots + z_m)}{n - m} \quad (6\text{--}13)$$

[Example 6–1 shows why Equation (6–13) gives the intersection.] Once this intersection is determined, the asymptotes can be readily drawn in the complex plane.

It is important to note that the asymptotes show the behavior of the root loci for $|s| \gg 1$. A root-locus branch may lie on one side of the corresponding asymptote or may cross the corresponding asymptote from one side to the other side.

4. *Find the breakaway and break-in points.* Because of the conjugate symmetry of the root loci, the breakaway points and break-in points either lie on the real axis or occur in complex-conjugate pairs.

If a root locus lies between two adjacent open-loop poles on the real axis, then there exists at least one breakaway point between the two poles. Similarly, if the root locus lies between two adjacent zeros (one zero may be located at $-\infty$) on the real axis, then there always exists at least one break-in point between the two zeros. If the root locus lies between an open-loop pole and a zero (finite or infinite) on the real axis, then there may exist no breakaway or break-in points or there may exist both breakaway and break-in points.

Suppose that the characteristic equation is given by

$$B(s) + KA(s) = 0$$

The breakaway points and break-in points correspond to multiple roots of the characteristic equation. Hence, as discussed in Example 6–1, the breakaway and break-in points can be determined from the roots of

$$\frac{dK}{ds} = -\frac{B'(s)A(s) - B(s)A'(s)}{A^2(s)} = 0 \quad (6\text{--}14)$$

where the prime indicates differentiation with respect to s. It is important to note that the breakaway points and break-in points must be the roots of Equation (6–14), but not all roots of Equation (6–14) are breakaway or break-in points. If a real root of Equation (6–14) lies on the root-locus portion of the real axis, then it is an actual breakaway or break-in point. If a real root of Equation (6–14) is not on the root-locus portion of the real axis, then this root corresponds to neither a breakaway point nor a break-in point.

If two roots $s = s_1$ and $s = -s_1$ of Equation (6–14) are a complex-conjugate pair and if it is not certain whether they are on root loci, then it is necessary to check the corresponding K value. If the value of K corresponding to a root $s = s_1$ of $dK/ds = 0$ is positive, point $s = s_1$ is an actual breakaway or break-in point. (Since K is assumed to be nonnegative, if the value of K thus obtained is negative, or a complex quantity, then point $s = s_1$ is neither a breakaway nor a break-in point.)

5. *Determine the angle of departure (angle of arrival) of the root locus from a complex pole (at a complex zero).* To sketch the root loci with reasonable accuracy, we must find the directions of the root loci near the complex poles and zeros. If a test point is chosen and moved in the very vicinity of a complex pole (or complex zero), the sum of the angular contributions from all other poles and zeros can be considered to remain the same. Therefore, the angle of departure (or angle of arrival) of the root locus from a complex pole (or at a complex zero) can be found by subtracting from 180° the sum of all the angles of vectors from all other poles and zeros to the complex pole (or complex zero) in question, with appropriate signs included.

Angle of departure from a complex pole = 180°
− (sum of the angles of vectors to a complex pole in question from other poles)
+ (sum of the angles of vectors to a complex pole in question from zeros)

Angle of arrival at a complex zero = 180°
− (sum of the angles of vectors to a complex zero in question from other zeros)
+ (sum of the angles of vectors to a complex zero in question from poles)

The angle of departure is shown in Figure 6–12.

6. *Find the points where the root loci may cross the imaginary axis.* The points where the root loci intersect the $j\omega$ axis can be found easily by (a) use of Routh's stability criterion or (b) letting $s = j\omega$ in the characteristic equation, equating both the real part and the imaginary part to zero, and solving for ω and K. The values of ω thus found give the frequencies at which root loci cross the imaginary axis. The K value corresponding to each crossing frequency gives the gain at the crossing point.

7. *Taking a series of test points in the broad neighborhood of the origin of the s plane, sketch the root loci.* Determine the root loci in the broad neighborhood of the $j\omega$ axis and the origin. The most important part of the root loci is on neither the real axis nor the asymptotes but is in the broad neighborhood of the $j\omega$ axis and the origin. The shape

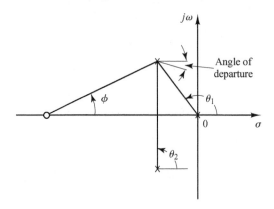

Figure 6–12
Construction of the root locus. [Angle of departure
= 180° −
$(\theta_1 + \theta_2) + \phi$.]

of the root loci in this important region in the s plane must be obtained with reasonable accuracy. (If accurate shape of the root loci is needed, MATLAB may be used rather than hand calculations of the exact shape of the root loci.)

8. *Determine closed-loop poles.* A particular point on each root-locus branch will be a closed-loop pole if the value of K at that point satisfies the magnitude condition. Conversely, the magnitude condition enables us to determine the value of the gain K at any specific root location on the locus. (If necessary, the root loci may be graduated in terms of K. The root loci are continuous with K.)

The value of K corresponding to any point s on a root locus can be obtained using the magnitude condition, or

$$K = \frac{\text{product of lengths between point } s \text{ to poles}}{\text{product of lengths between point } s \text{ to zeros}}$$

This value can be evaluated either graphically or analytically. (MATLAB can be used for graduating the root loci with K. See Section 6–3.)

If the gain K of the open-loop transfer function is given in the problem, then by applying the magnitude condition, we can find the correct locations of the closed-loop poles for a given K on each branch of the root loci by a trial-and-error approach or by use of MATLAB, which will be presented in Section 6–3.

Comments on the Root-Locus Plots. It is noted that the characteristic equation of the negative feedback control system whose open-loop transfer function is

$$G(s)H(s) = \frac{K(s^m + b_1 s^{m-1} + \cdots + b_m)}{s^n + a_1 s^{n-1} + \cdots + a_n} \qquad (n \geq m)$$

is an nth-degree algebraic equation in s. If the order of the numerator of $G(s)H(s)$ is lower than that of the denominator by two or more (which means that there are two or more zeros at infinity), then the coefficient a_1 is the negative sum of the roots of the equation and is independent of K. In such a case, if some of the roots move on the locus toward the left as K is increased, then the other roots must move toward the right as K is increased. This information is helpful in finding the general shape of the root loci.

It is also noted that a slight change in the pole–zero configuration may cause significant changes in the root-locus configurations. Figure 6–13 demonstrates the fact that a slight change in the location of a zero or pole will make the root-locus configuration look quite different.

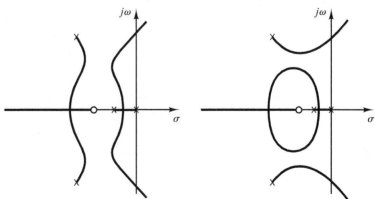

Figure 6–13
Root-locus plots.

Cancellation of Poles of G(s) with Zeros of H(s). It is important to note that if the denominator of $G(s)$ and the numerator of $H(s)$ involve common factors, then the corresponding open-loop poles and zeros will cancel each other, reducing the degree of the characteristic equation by one or more. For example, consider the system shown in Figure 6–14(a). (This system has velocity feedback.) By modifying the block diagram of Figure 6–14(a) to that shown in Figure 6–14(b), it is clearly seen that $G(s)$ and $H(s)$ have a common factor $s + 1$. The closed-loop transfer function $C(s)/R(s)$ is

$$\frac{C(s)}{R(s)} = \frac{K}{s(s+1)(s+2) + K(s+1)}$$

The characteristic equation is

$$[s(s+2) + K](s+1) = 0$$

Because of the cancellation of the terms $(s+1)$ appearing in $G(s)$ and $H(s)$, however, we have

$$1 + G(s)H(s) = 1 + \frac{K(s+1)}{s(s+1)(s+2)}$$
$$= \frac{s(s+2) + K}{s(s+2)}$$

The reduced characteristic equation is

$$s(s+2) + K = 0$$

The root-locus plot of $G(s)H(s)$ does not show all the roots of the characteristic equation, only the roots of the reduced equation.

To obtain the complete set of closed-loop poles, we must add the canceled pole of $G(s)H(s)$ to those closed-loop poles obtained from the root-locus plot of $G(s)H(s)$. The important thing to remember is that the canceled pole of $G(s)H(s)$ is a closed-loop pole of the system, as seen from Figure 6–14(c).

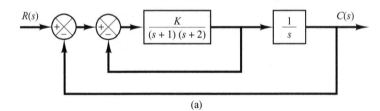

(a)

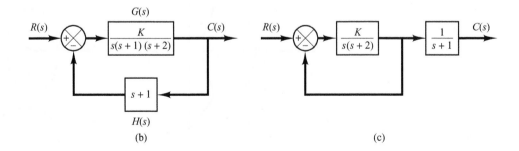

Figure 6–14
(a) Control system with velocity feedback; (b) and (c) modified block diagrams.

(b) (c)

Typical Pole–Zero Configurations and Corresponding Root Loci. In summarizing, we show several open-loop pole–zero configurations and their corresponding root loci in Table 6–1. The pattern of the root loci depends only on the relative separation of the open-loop poles and zeros. If the number of open-loop poles exceeds the number of finite zeros by three or more, there is a value of the gain K beyond which root loci enter the right-half s plane, and thus the system can become unstable. A stable system must have all its closed-loop poles in the left-half s plane.

Table 6–1   Open-Loop Pole–Zero Configurations and the Corresponding Root Loci

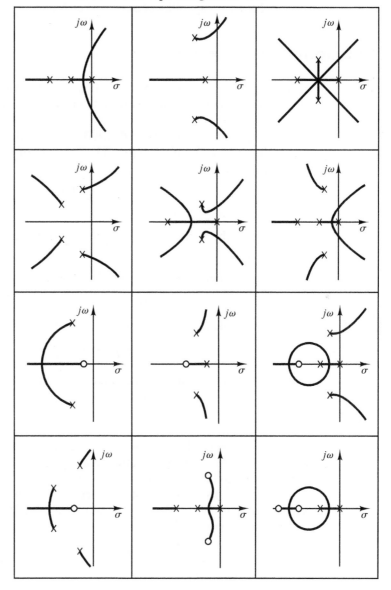

Note that once we have some experience with the method, we can easily evaluate the changes in the root loci due to the changes in the number and location of the open-loop poles and zeros by visualizing the root-locus plots resulting from various pole–zero configurations.

Summary. From the preceding discussions, it should be clear that it is possible to sketch a reasonably accurate root-locus diagram for a given system by following simple rules. (The reader should study the various root-locus diagrams shown in the solved problems at the end of the chapter.) At preliminary design stages, we may not need the precise locations of the closed-loop poles. Often their approximate locations are all that is needed to make an estimate of system performance. Thus, it is important that the designer have the capability of quickly sketching the root loci for a given system.

6–3 PLOTTING ROOT LOCI WITH MATLAB

In this section we present the MATLAB approach to the generation of root-locus plots and finding relevant information from the root-locus plots.

Plotting Root Loci with MATLAB. In plotting root loci with MATLAB we deal with the system equation given in the form of Equation (6–11), which may be written as

$$1 + K\frac{\text{num}}{\text{den}} = 0$$

where num is the numerator polynomial and den is the denominator polynomial. That is,

$$\text{num} = (s + z_1)(s + z_2)\cdots(s + z_m)$$
$$= s^m + (z_1 + z_2 + \cdots + z_m)s^{m-1} + \cdots + z_1 z_2 \cdots z_m$$
$$\text{den} = (s + p_1)(s + p_2)\cdots(s + p_n)$$
$$= s^n + (p_1 + p_2 + \cdots + p_n)s^{n-1} + \cdots + p_1 p_2 \cdots p_n$$

Note that both vectors num and den must be written in descending powers of s.

A MATLAB command commonly used for plotting root loci is

rlocus(num,den)

Using this command, the root-locus plot is drawn on the screen. The gain vector K is automatically determined. (The vector K contains all the gain values for which the closed-loop poles are to be computed.)

For the systems defined in state space, rlocus(A,B,C,D) plots the root locus of the system with the gain vector automatically determined.

Note that commands

rlocus(num,den,K)   and   rlocus(A,B,C,D,K)

use the user-supplied gain vector K.

If it is desired to plot the root loci with marks 'o' or 'x', it is necessary to use the following command:

$$r = \text{rlocus(num,den)}$$
$$\text{plot(r,'o')} \quad \text{or} \quad \text{plot(r,'x')}$$

Plotting root loci using marks o or x is instructive, since each calculated closed-loop pole is graphically shown; in some portion of the root loci those marks are densely placed and in another portion of the root loci they are sparsely placed. MATLAB supplies its own set of gain values used to calculate a root-locus plot. It does so by an internal adaptive step-size routine. Also, MATLAB uses the automatic axis-scaling feature of the plot command.

EXAMPLE 6–3   Consider the system shown in Figure 6–15. Plot root loci with a square aspect ratio so that a line with slope 1 is a true 45° line. Choose the region of root-locus plot to be

$$-6 \le x \le 6, \qquad -6 \le y \le 6$$

where x and y are the real-axis coordinate and imaginary-axis coordinate, respectively.

To set the given plot region on the screen to be square, enter the command

$$v = [\text{-6} \quad 6 \quad \text{-6} \quad 6]; \text{axis (v); axis('square')}$$

With this command, the region of the plot is as specified and a line with slope 1 is at a true 45°, not skewed by the irregular shape of the screen.

For this problem, the denominator is given as a product of first- and second-order terms. So we must multiply these terms to get a polynomial in s. The multiplication of these terms can be done easily by use of the convolution command, as shown next.

Define

$$a = s\,(s + 1): \qquad a = [1 \quad 1 \quad 0]$$
$$b = s^2 + 4s + 16: \qquad b = [1 \quad 4 \quad 16]$$

Then we use the following command:

$$c = \text{conv(a, b)}$$

Note that conv(a, b) gives the product of two polynomials a and b. See the following computer output:

```
a = [1  1  0];
b = [1  4  16];
c = conv (a,b)

c =

        1  5  20  16  0
```

The denominator polynomial is thus found to be

$$\text{den} = [1 \quad 5 \quad 20 \quad 16 \quad 0]$$

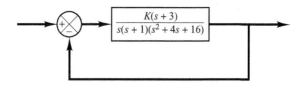

Figure 6–15
Control system.

To find the complex-conjugate open-loop poles (the roots of $s^2 + 4s + 16 = 0$), we may enter the roots command as follows:

```
r = roots(b)

r =

   -2.0000 + 3.464li
   -2.0000 - 3.464li
```

Thus, the system has the following open-loop zero and open-loop poles:

Open-loop zero: $s = -3$
Open-loop poles: $s = 0,\quad s = -1,\quad s = -2 \pm j3.4641$

MATLAB Program 6–1 will plot the root-locus diagram for this system. The plot is shown in Figure 6–16.

MATLAB Program 6–1

```
% --------- Root-locus plot ---------
num = [1   3];
den = [1   5  20  16   0];
rlocus(num,den)
v = [-6  6  -6  6];
axis(v); axis('square')
grid;
title ('Root-Locus Plot of G(s) = K(s + 3)/[s(s + 1)(s^2 + 4s + 16)]')
```

Note that in MATLAB Program 6–1, instead of

$$\text{den} = [1 \ 5 \ 20 \ 16 \ 0]$$

we may enter

$$\text{den} = \text{conv} ([1 \ 1 \ 0], [1 \ 4 \ 16])$$

The results are the same.

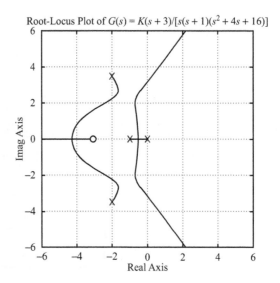

Root-Locus Plot of $G(s) = K(s + 3)/[s(s + 1)(s^2 + 4s + 16)]$

Figure 6–16
Root-locus plot.

EXAMPLE 6–4   Consider the negative feedback system whose open-loop transfer function $G(s)H(s)$ is

$$G(s)H(s) = \frac{K}{s(s + 0.5)(s^2 + 0.6s + 10)}$$

$$= \frac{K}{s^4 + 1.1s^3 + 10.3s^2 + 5s}$$

There are no open-loop zeros. Open-loop poles are located at $s = -0.3 + j3.1480$, $s = -0.3 - j3.1480$, $s = -0.5$, and $s = 0$.

Entering MATLAB Program 6–2 into the computer, we obtain the root-locus plot shown in Figure 6–17.

MATLAB Program 6–2

```
% --------- Root-locus plot ---------
num = [1];
den = [1  1.1  10.3  5  0];
r = rlocus(num,den);
plot(r,'o')
v = [-6  6  -6  6]; axis(v)
grid
title('Root-Locus Plot of G(s) = K/[s(s + 0.5)(s^2 + 0.6s + 10)]')
xlabel('Real Axis')
ylabel('Imag Axis')
```

Notice that in the regions near $x = -0.3$, $y = 2.3$ and $x = -0.3$, $y = -2.3$ two loci approach each other. We may wonder if these two branches should touch or not. To explore this situation, we may plot the root loci using smaller increments of K in the critical region.

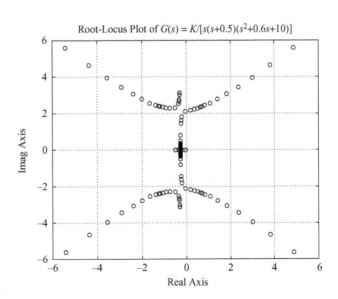

Figure 6–17
Root-locus plot.

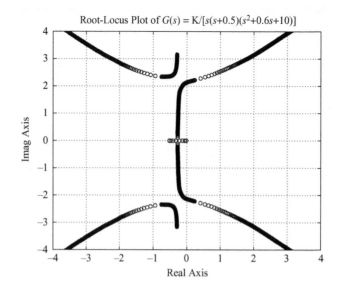

Figure 6–18
Root-locus plot.

By a conventional trial-and-error approach or using the command rlocfind to be presented later in this section, we find the particular region of interest to be $20 \le K \le 30$. By entering MATLAB Program 6–3, we obtain the root-locus plot shown in Figure 6–18. From this plot, it is clear that the two branches that approach in the upper half-plane (or in the lower half-plane) do not touch.

MATLAB Program 6–3

```
% --------- Root-locus plot ---------
num = [1];
den = [1  1.1  10.3  5  0];
K1 = 0:0.2:20;
K2 = 20:0.1:30;
K3 = 30:5:1000;
K = [K1  K2  K3];
r = rlocus(num,den,K);
plot(r, 'o')
v = [-4  4  -4  4]; axis(v)
grid
title('Root-Locus Plot of G(s) = K/[s(s + 0.5)(s^2 + 0.6s + 10)]')
xlabel('Real Axis')
ylabel('Imag Axis')
```

EXAMPLE 6–5   Consider the system shown in Figure 6–19. The system equations are

$$\dot{\mathbf{x}} = \mathbf{A}\mathbf{x} + \mathbf{B}u$$
$$y = \mathbf{C}\mathbf{x} + Du$$
$$u = r - y$$

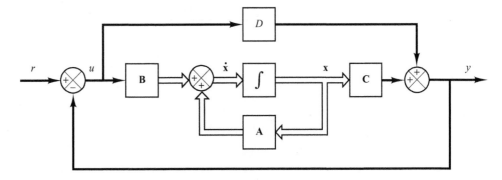

Figure 6–19
Closed-loop control
system.

In this example problem we shall obtain the root-locus diagram of the system defined in state space. As an example let us consider the case where matrices $\mathbf{A}$, $\mathbf{B}$, $\mathbf{C}$, and D are

$$\mathbf{A} = \begin{bmatrix} 0 & 1 & 0 \\ 0 & 0 & 1 \\ -160 & -56 & -14 \end{bmatrix}, \qquad \mathbf{B} = \begin{bmatrix} 0 \\ 1 \\ -14 \end{bmatrix} \qquad\qquad (6\text{--}15)$$

$$\mathbf{C} = \begin{bmatrix} 1 & 0 & 0 \end{bmatrix}, \qquad\qquad D = \begin{bmatrix} 0 \end{bmatrix}$$

The root-locus plot for this system can be obtained with MATLAB by use of the following command:

$$\text{rlocus(A,B,C,D)}$$

This command will produce the same root-locus plot as can be obtained by use of the rlocus (num,den) command, where num and den are obtained from

$$[\text{num,den}] = \text{ss2tf(A,B,C,D)}$$

as follows:

$$\text{num} = \begin{bmatrix} 0 & 0 & 1 & 0 \end{bmatrix}$$
$$\text{den} = \begin{bmatrix} 1 & 14 & 56 & 160 \end{bmatrix}$$

MATLAB Program 6–4 is a program that will generate the root-locus plot as shown in Figure 6–20.

MATLAB Program 6–4

```
% --------- Root-locus plot ---------

A = [0 1 0;0 0 1;-160 -56 -14];
B = [0;1;-14];
C = [1 0 0];
D = [0];
K = 0:0.1:400;
rlocus(A,B,C,D,K);
v = [-20 20 -20 20]; axis(v);
grid
title('Root-Locus Plot of System Defined in State Space')
```

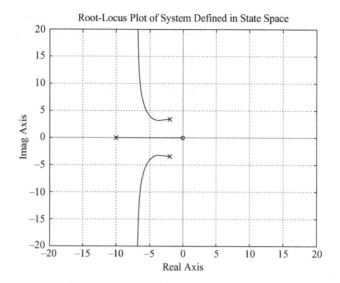

Root-Locus Plot of System Defined in State Space

Figure 6–20
Root-locus plot of
system defined in
state space, where **A**,
B, **C**, and D are as
given by Equation
(6–15).

Constant ζ Loci and Constant ω_n Loci.   Recall that in the complex plane the damping ratio ζ of a pair of complex-conjugate poles can be expressed in terms of the angle ϕ, which is measured from the negative real axis, as shown in Figure 6–21(a) with

$$\zeta = \cos\phi$$

In other words, lines of constant damping ratio ζ are radial lines passing through the origin as shown in Figure 6–21(b). For example, a damping ratio of 0.5 requires that the complex-conjugate poles lie on the lines drawn through the origin making angles of $\pm60°$ with the negative real axis. (If the real part of a pair of complex-conjugate poles is positive, which means that the system is unstable, the corresponding ζ is negative.) The damping ratio determines the angular location of the poles, while the

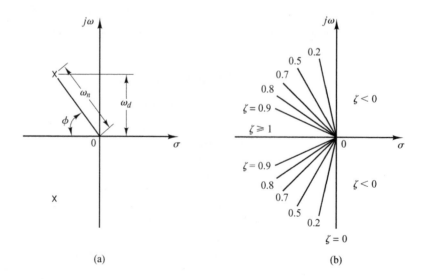

Figure 6–21
(a) Complex poles;
(b) lines of constant
damping ratio ζ.

(a)   (b)

distance of the pole from the origin is determined by the undamped natural frequency ω_n. The constant ω_n loci are circles.

To draw constant ζ lines and constant ω_n circles on the root-locus diagram with MATLAB, use the command sgrid.

Plotting Polar Grids in the Root-Locus Diagram.   The command

sgrid

overlays lines of constant damping ratio ($\zeta = 0 \sim 1$ with 0.1 increment) and circles of constant ω_n on the root-locus plot. See MATLAB Program 6–5 and the resulting diagram shown in Figure 6–22.

MATLAB Program 6–5

```
sgrid
v = [-3  3  -3  3]; axis(v); axis('square')
title('Constant \zeta Lines and Constant \omega_n Circles')
xlabel('Real Axis')
ylabel('Imag Axis')
```

If only particular constant ζ lines (such as the $\zeta = 0.5$ line and $\zeta = 0.707$ line) and particular constant ω_n circles (such as the $\omega_n = 0.5$ circle, $\omega_n = 1$ circle, and $\omega_n = 2$ circle) are desired, use the following command:

sgrid([0.5, 0.707], [0.5, 1, 2])

If we wish to overlay lines of constant ζ and circles of constant ω_n as given above to a root-locus plot of a negative feedback system with

num = [0 0 0 1]
den = [1 4 5 0]

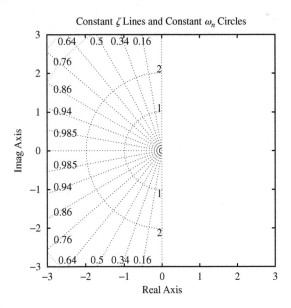

Figure 6–22
Constant ζ lines and constant ω_n circles.

then enter MATLAB Program 6–6 into the computer. The resulting root-locus plot is shown in Figure 6–23.

MATLAB Program 6–6

```
num = [1];
den = [1  4  5  0];
K = 0:0.01:1000;
r = rlocus(num,den,K);
plot(r,'-'); v = [-3  1  -2  2]; axis(v); axis('square')
sgrid([0.5,0.707], [0.5,1,2])
grid
title('Root-Locus Plot with \zeta = 0.5 and 0.707 Lines and \omega_n = 0.5,1, and 2 Circles')
xlabel('Real Axis'); ylabel('Imag Axis')
gtext('\omega_n = 2')
gtext('\omega_n = 1')
gtext('\omega_n = 0.5')
% Place 'x' mark at each of 3 open-loop poles.
gtext('x')
gtext('x')
gtext('x')
```

If we want to omit either the entire constant ζ lines or entire constant ω_n circles, we may use empty brackets [ ] in the arguments of the sgrid command. For example, if we want to overlay only the constant damping ratio line corresponding to $\zeta = 0.5$ and no constant ω_n circles on the root-locus plot, then we may use the command

$$\text{sgrid}(0.5, \ [])$$

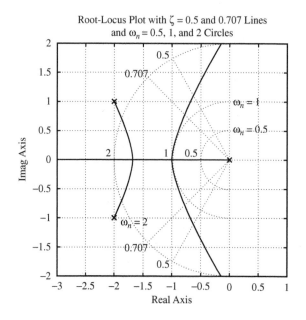

Figure 6–23
Constant ζ lines and constant ω_n circles superimposed on a root-locus plot.

Chapter 6 / Control Systems Analysis and Design by the Root-Locus Method

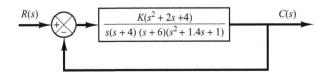

Figure 6–24
Control system.

Conditionally Stable Systems. Consider the negative feedback system shown in Figure 6–24. We can plot the root loci for this system by applying the general rules and procedure for constructing root loci, or use MATLAB to get root-locus plots. MATLAB Program 6–7 will plot the root-locus diagram for the system. The plot is shown in Figure 6–25.

MATLAB Program 6–7

```
num = [1  2  4];
den = conv(conv([1  4  0],[1  6]), [1  1.4  1]);
rlocus(num, den)
v = [-7  3  -5  5]; axis(v); axis('square')
grid
title('Root-Locus Plot of G(s) = K(s^2 + 2s + 4)/[s(s + 4)(s + 6)(s^2 + 1.4s + 1)]')
text(1.0, 0.55,'K = 12')
text(1.0,3.0,'K = 73')
text(1.0,4.15,'K = 154')
```

It can be seen from the root-locus plot of Figure 6–25 that this system is stable only for limited ranges of the value of K—that is, $0 < K < 12$ and $73 < K < 154$. The system becomes unstable for $12 < K < 73$ and $154 < K$. (If K assumes a value corresponding to unstable operation, the system may break down or may become nonlinear due to a saturation nonlinearity that may exist.) Such a system is called conditionally stable.

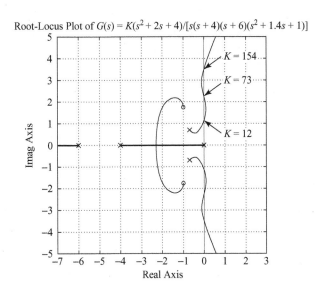

Figure 6–25
Root-locus plot of conditionally stable system.

In practice, conditionally stable systems are not desirable. Conditional stability is dangerous but does occur in certain systems—in particular, a system that has an unstable feedforward path. Such an unstable feedforward path may occur if the system has a minor loop. It is advisable to avoid such conditional stability since, if the gain drops beyond the critical value for any reason, the system becomes unstable. Note that the addition of a proper compensating network will eliminate conditional stability. [An addition of a zero will cause the root loci to bend to the left. (See Section 6–5.) Hence conditional stability may be eliminated by adding proper compensation.]

Nonminimum-Phase Systems. If all the poles and zeros of a system lie in the left-half s plane, then the system is called *minimum phase*. If a system has at least one pole or zero in the right-half s plane, then the system is called *nonminimum phase*. The term nonminimum phase comes from the phase-shift characteristics of such a system when subjected to sinusoidal inputs.

Consider the system shown in Figure 6–26(a). For this system

$$G(s) = \frac{K(1 - T_a s)}{s(Ts + 1)} \qquad (T_a > 0), \qquad H(s) = 1$$

This is a nonminimum-phase system, since there is one zero in the right-half s plane. For this system, the angle condition becomes

$$\underline{/G(s)} = \underline{/ -\frac{K(T_a s - 1)}{s(Ts + 1)}}$$

$$= \underline{/ \frac{K(T_a s - 1)}{s(Ts + 1)}} + 180°$$

$$= \pm 180°(2k + 1) \qquad (k = 0, 1, 2, \dots)$$

or

$$\underline{/ \frac{K(T_a s - 1)}{s(Ts + 1)}} = 0° \qquad (6\text{–}16)$$

The root loci can be obtained from Equation (6–16). Figure 6–26(b) shows a root-locus plot for this system. From the diagram, we see that the system is stable if the gain K is less than $1/T_a$.

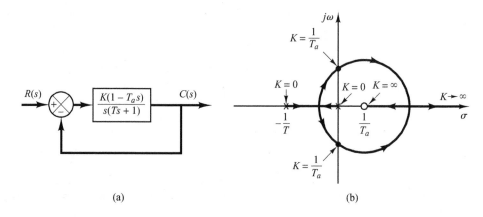

Figure 6–26
(a) Nonminimum-phase system;
(b) root-locus plot.

(a)

(b)

Chapter 6 / Control Systems Analysis and Design by the Root-Locus Method

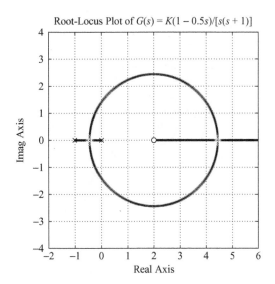

Figure 6–27
Root-locus plot of
$$G(s) = \frac{K(1 - 0.5s)}{s(s + 1)}.$$

To obtain a root-locus plot with MATLAB, enter the numerator and denominator as usual. For example, if $T = 1$ sec and $T_a = 0.5$ sec, enter the following num and den in the program:

$$\text{num} = [-0.5 \quad 1]$$
$$\text{den} = [1 \quad 1 \quad 0]$$

MATLAB Program 6–8 gives the plot of the root loci shown in Figure 6–27.

MATLAB Program 6–8

```
num = [-0.5  1];
den = [1  1  0];
k1 = 0:0.01:30;
k2 = 30:1:100;
K3 = 100:5:500;
K = [k1  k2  k3];
rlocus(num,den,K)
v = [-2  6  -4  4]; axis(v); axis('square')
grid
title('Root-Locus Plot of G(s) = K(1 - 0.5s)/[s(s + 1)]')
% Place 'x' mark at each of 2 open-loop poles.
% Place 'o' mark at open-loop zero.
gtext('x')
gtext('x')
gtext('o')
```

Orthogonality of Root Loci and Constant-Gain Loci. Consider the negative feedback system whose open-loop transfer function is $G(s)H(s)$. In the $G(s)H(s)$ plane, the loci of $|G(s)H(s)|$ = constant are circles centered at the origin, and the loci corresponding to $\underline{/G(s)H(s)} = \pm180°(2k + 1)$ $(k = 0, 1, 2, \ldots)$ lie on the negative real axis

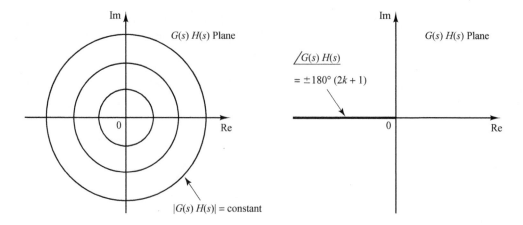

Figure 6–28
Plots of constant-gain and constant-phase loci in the $G(s)H(s)$ plane.

of the $G(s)H(s)$ plane, as shown in Figure 6–28. [Note that the complex plane employed here is not the s plane, but the $G(s)H(s)$ plane.]

The root loci and constant-gain loci in the s plane are conformal mappings of the loci of $\underline{/G(s)H(s)} = \pm180°(2k + 1)$ and of $|G(s)H(s)| = $ constant in the $G(s)H(s)$ plane.

Since the constant-phase and constant-gain loci in the $G(s)H(s)$ plane are orthogonal, the root loci and constant-gain loci in the s plane are orthogonal. Figure 6–29(a) shows the root loci and constant-gain loci for the following system:

$$G(s) = \frac{K(s + 2)}{s^2 + 2s + 3}, \qquad H(s) = 1$$

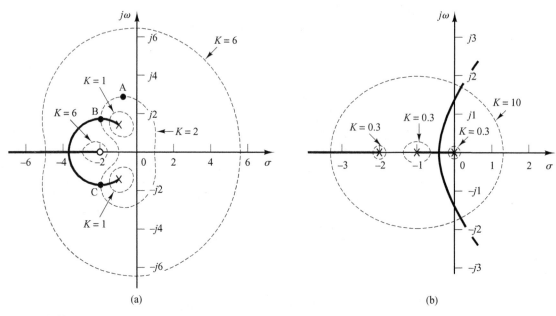

(a)   (b)

Figure 6–29
Plots of root loci and constant-gain loci. (a) System with $G(s) = K(s + 2)/(s^2 + 2s + 3)$, $H(s) = 1$; (b) system with $G(s) = K/[s(s + 1)(s + 2)], H(s) = 1$.

Notice that since the pole–zero configuration is symmetrical about the real axis, the constant-gain loci are also symmetrical about the real axis.

Figure 6–29(b) shows the root loci and constant-gain loci for the system:

$$G(s) = \frac{K}{s(s+1)(s+2)}, \qquad H(s) = 1$$

Notice that since the configuration of the poles in the s plane is symmetrical about the real axis and the line parallel to the imaginary axis passing through point $(\sigma = -1, \omega = 0)$, the constant-gain loci are symmetrical about the $\omega = 0$ line (real axis) and the $\sigma = -1$ line.

From Figures 6–29(a) and (b), notice that every point in the s plane has the corresponding K value. If we use a command rlocfind (presented next), MATLAB will give the K value of the specified point as well as the nearest closed-loop poles corresponding to this K value.

Finding the Gain Value K at an Arbitrary Point on the Root Loci. In MATLAB analysis of closed-loop systems, it is frequently desired to find the gain value K at an arbitrary point on the root locus. This can be accomplished by using the following rlocfind command:

$$[K, r] = \text{rlocfind(num, den)}$$

The rlocfind command, which must follow an rlocus command, overlays movable x-y coordinates on the screen. Using the mouse, we position the origin of the x-y coordinates over the desired point on the root locus and press the mouse button. Then MATLAB displays on the screen the coordinates of that point, the gain value at that point, and the closed-loop poles corresponding to this gain value.

If the selected point is not on the root locus, such as point A in Figure 6–29(a), the rlocfind command gives the coordinates of this selected point, the gain value of this point, such as $K = 2$, and the locations of the closed-loop poles, such as points B and C corresponding to this K value. [Note that every point on the s plane has a gain value. See, for example, Figures 6–29 (a) and (b).]

6–4 ROOT-LOCUS PLOTS OF POSITIVE FEEDBACK SYSTEMS

Root Loci for Positive-Feedback Systems.* In a complex control system, there may be a positive-feedback inner loop as shown in Figure 6–30. Such a loop is usually stabilized by the outer loop. In what follows, we shall be concerned only with the positive-feedback inner loop. The closed-loop transfer function of the inner loop is

$$\frac{C(s)}{R(s)} = \frac{G(s)}{1 - G(s)H(s)}$$

The characteristic equation is

$$1 - G(s)H(s) = 0 \qquad\qquad (6\text{–}17)$$

---

* Reference W-4

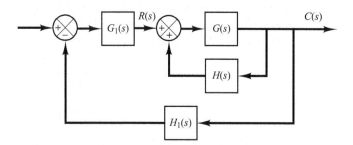

Figure 6–30
Control system.

This equation can be solved in a manner similar to the development of the root-locus method for negative-feedback systems presented in Section 6–2. The angle condition, however, must be altered.

Equation (6–17) can be rewritten as

$$G(s)H(s) = 1$$

which is equivalent to the following two equations:

$$\underline{/G(s)H(s)} = 0° \pm k360° \qquad (k = 0, 1, 2, \dots)$$

$$|G(s)H(s)| = 1$$

For the positive-feedback case, the total sum of all angles from the open-loop poles and zeros must be equal to $0° \pm k360°$. Thus the root locus follows a $0°$ locus in contrast to the $180°$ locus considered previously. The magnitude condition remains unaltered.

To illustrate the root-locus plot for the positive-feedback system, we shall use the following transfer functions $G(s)$ and $H(s)$ as an example.

$$G(s) = \frac{K(s + 2)}{(s + 3)(s^2 + 2s + 2)}, \qquad H(s) = 1$$

The gain K is assumed to be positive.

The general rules for constructing root loci for negative-feedback systems given in Section 6–2 must be modified in the following way:

Rule 2 is Modified as Follows: If the total number of real poles and real zeros to the right of a test point on the real axis is even, then this test point lies on the root locus.

Rule 3 is Modified as Follows:

$$\text{Angles of asymptotes} = \frac{\pm k360°}{n - m} \qquad (k = 0, 1, 2, \dots)$$

where n = number of finite poles of $G(s)H(s)$
m = number of finite zeros of $G(s)H(s)$

Rule 5 is Modified as Follows: When calculating the angle of departure (or angle of arrival) from a complex open-loop pole (or at a complex zero), subtract from $0°$ the sum of all angles of the vectors from all the other poles and zeros to the complex pole (or complex zero) in question, with appropriate signs included.

Other rules for constructing the root-locus plot remain the same. We shall now apply the modified rules to construct the root-locus plot.

1. Plot the open-loop poles ($s = -1 + j$, $s = -1 - j$, $s = -3$) and zero ($s = -2$) in the complex plane. As K is increased from 0 to ∞, the closed-loop poles start at the open-loop poles and terminate at the open-loop zeros (finite or infinite), just as in the case of negative-feedback systems.

2. Determine the root loci on the real axis. Root loci exist on the real axis between -2 and $+\infty$ and between -3 and $-\infty$.

3. Determine the asymptotes of the root loci. For the present system,

$$\text{Angles of asymptote} = \frac{\pm k360°}{3 - 1} = \pm 180°$$

This simply means that asymptotes are on the real axis.

4. Determine the breakaway and break-in points. Since the characteristic equation is

$$(s + 3)(s^2 + 2s + 2) - K(s + 2) = 0$$

we obtain

$$K = \frac{(s + 3)(s^2 + 2s + 2)}{s + 2}$$

By differentiating K with respect to s, we obtain

$$\frac{dK}{ds} = \frac{2s^3 + 11s^2 + 20s + 10}{(s + 2)^2}$$

Note that

$$2s^3 + 11s^2 + 20s + 10 = 2(s + 0.8)(s^2 + 4.7s + 6.24)$$
$$= 2(s + 0.8)(s + 2.35 + j0.77)(s + 2.35 - j0.77)$$

Point $s = -0.8$ is on the root locus. Since this point lies between two zeros (a finite zero and an infinite zero), it is an actual break-in point. Points $s = -2.35 \pm j0.77$ do not satisfy the angle condition and, therefore, they are neither breakaway nor break-in points.

5. Find the angle of departure of the root locus from a complex pole. For the complex pole at $s = -1 + j$, the angle of departure θ is

$$\theta = 0° - 27° - 90° + 45°$$

or

$$\theta = -72°$$

(The angle of departure from the complex pole at $s = -1 - j$ is 72°.)

6. Choose a test point in the broad neighborhood of the $j\omega$ axis and the origin and apply the angle condition. Locate a sufficient number of points that satisfy the angle condition.

Figure 6–31 shows the root loci for the given positive-feedback system. The root loci are shown with dashed lines and a curve.

Note that if

$$K > \left. \frac{(s + 3)(s^2 + 2s + 2)}{s + 2} \right|_{s=0} = 3$$

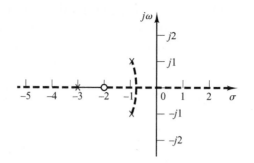

Figure 6–31
Root-locus plot for the positive-feedback system with
$G(s) = K(s + 2)/$
$[(s + 3)(s^2 + 2s + 2)]$,
$H(s) = 1$.

one real root enters the right-half s plane. Hence, for values of K greater than 3, the system becomes unstable. (For $K > 3$, the system must be stabilized with an outer loop.)

Note that the closed-loop transfer function for the positive-feedback system is given by

$$\frac{C(s)}{R(s)} = \frac{G(s)}{1 - G(s)H(s)}$$

$$= \frac{K(s + 2)}{(s + 3)(s^2 + 2s + 2) - K(s + 2)}$$

To compare this root-locus plot with that of the corresponding negative-feedback system, we show in Figure 6–32 the root loci for the negative-feedback system whose closed-loop transfer function is

$$\frac{C(s)}{R(s)} = \frac{K(s + 2)}{(s + 3)(s^2 + 2s + 2) + K(s + 2)}$$

Table 6–2 shows various root-locus plots of negative-feedback and positive-feedback systems. The closed-loop transfer functions are given by

$$\frac{C}{R} = \frac{G}{1 + GH}, \qquad \text{for negative-feedback systems}$$

$$\frac{C}{R} = \frac{G}{1 - GH}, \qquad \text{for positive-feedback systems}$$

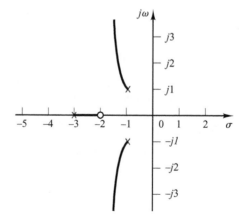

Figure 6–32
Root-locus plot for the negative-feedback system with
$G(s) = K(s + 2)/$
$[(s + 3)(s^2 + 2s + 2)]$,
$H(s) = 1$.

where *GH* is the open-loop transfer function. In Table 6–2, the root loci for negative-feedback systems are drawn with heavy lines and curves, and those for positive-feedback systems are drawn with dashed lines and curves.

Table 6–2 Root-Locus Plots of Negative-Feedback and Positive-Feedback Systems

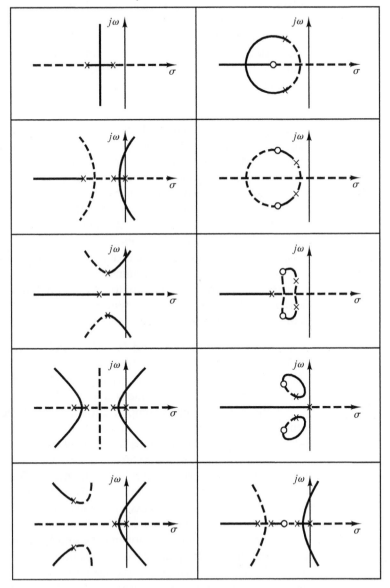

Heavy lines and curves correspond to negative-feedback systems; dashed lines and curves correspond to positive-feedback systems.

Preliminary Design Consideration.   In building a control system, we know that proper modification of the plant dynamics may be a simple way to meet the performance specifications. This, however, may not be possible in many practical situations because the plant may be fixed and not modifiable. Then we must adjust parameters other than those in the fixed plant. In this book, we assume that the plant is given and unalterable.

In practice, the root-locus plot of a system may indicate that the desired performance cannot be achieved just by the adjustment of gain (or some other adjustable parameter). In fact, in some cases, the system may not be stable for all values of gain (or other adjustable parameter). Then it is necessary to reshape the root loci to meet the performance specifications.

The design problems, therefore, become those of improving system performance by insertion of a compensator. Compensation of a control system is reduced to the design of a filter whose characteristics tend to compensate for the undesirable and unalterable characteristics of the plant.

Design by Root-Locus Method.   The design by the root-locus method is based on reshaping the root locus of the system by adding poles and zeros to the system's open-loop transfer function and forcing the root loci to pass through desired closed-loop poles in the s plane. The characteristic of the root-locus design is its being based on the assumption that the closed-loop system has a pair of dominant closed-loop poles. This means that the effects of zeros and additional poles do not affect the response characteristics very much.

In designing a control system, if other than a gain adjustment (or other parameter adjustment) is required, we must modify the original root loci by inserting a suitable compensator. Once the effects on the root locus of the addition of poles and/or zeros are fully understood, we can readily determine the locations of the pole(s) and zero(s) of the compensator that will reshape the root locus as desired. In essence, in the design by the root-locus method, the root loci of the system are reshaped through the use of a compensator so that a pair of dominant closed-loop poles can be placed at the desired location.

Series Compensation and Parallel (or Feedback) Compensation.   Figures 6–33(a) and (b) show compensation schemes commonly used for feedback control systems. Figure 6–33(a) shows the configuration where the compensator $G_c(s)$ is placed in series with the plant. This scheme is called *series compensation.*

An alternative to series compensation is to feed back the signal(s) from some element(s) and place a compensator in the resulting inner feedback path, as shown in Figure 6–33(b). Such compensation is called *parallel compensation* or *feedback compensation.*

In compensating control systems, we see that the problem usually boils down to a suitable design of a series or parallel compensator. The choice between series compensation and parallel compensation depends on the nature of the signals in the system, the power levels at various points, available components, the designer's experience, economic considerations, and so on.

In general, series compensation may be simpler than parallel compensation; however, series compensation frequently requires additional amplifiers to increase the gain and/or to provide isolation. (To avoid power dissipation, the series compensator is inserted at the lowest energy point in the feedforward path.) Note that, in general, the number of components required in parallel compensation will be less than the number of components

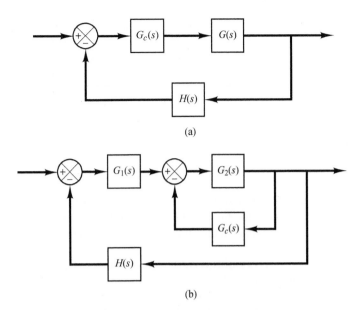

Figure 6–33
(a) Series compensation;
(b) parallel or feed-back compensation.

(a)

(b)

in series compensation, provided a suitable signal is available, because the energy transfer is from a higher power level to a lower level. (This means that additional amplifiers may not be necessary.)

In Sections 6–6 through 6–9 we first discuss series compensation techniques and then present a parallel compensation technique using a design of a velocity-feedback control system.

Commonly Used Compensators. If a compensator is needed to meet the performance specifications, the designer must realize a physical device that has the prescribed transfer function of the compensator.

Numerous physical devices have been used for such purposes. In fact, many noble and useful ideas for physically constructing compensators may be found in the literature.

If a sinusoidal input is applied to the input of a network, and the steady-state output (which is also sinusoidal) has a phase lead, then the network is called a lead network. (The amount of phase lead angle is a function of the input frequency.) If the steady-state output has a phase lag, then the network is called a lag network. In a lag–lead network, both phase lag and phase lead occur in the output but in different frequency regions; phase lag occurs in the low-frequency region and phase lead occurs in the high-frequency region. A compensator having a characteristic of a lead network, lag network, or lag–lead network is called a lead compensator, lag compensator, or lag–lead compensator.

Among the many kinds of compensators, widely employed compensators are the lead compensators, lag compensators, lag–lead compensators, and velocity-feedback (tachometer) compensators. In this chapter we shall limit our discussions mostly to these types. Lead, lag, and lag–lead compensators may be electronic devices (such as circuits using operational amplifiers) or RC networks (electrical, mechanical, pneumatic, hydraulic, or combinations thereof) and amplifiers.

Frequently used series compensators in control systems are lead, lag, and lag–lead compensators. PID controllers which are frequently used in industrial control systems are discussed in Chapter 8.

Figure 6–34
(a) Root-locus plot
of a single-pole
system;
(b) root-locus plot of
a two-pole system;
(c) root-locus plot of
a three-pole system.

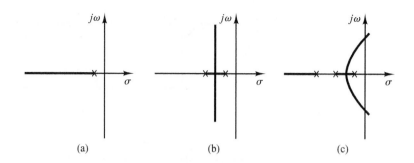

(a)   (b)   (c)

It is noted that in designing control systems by the root-locus or frequency-response methods the final result is not unique, because the best or optimal solution may not be precisely defined if the time-domain specifications or frequency-domain specifications are given.

Effects of the Addition of Poles.   The addition of a pole to the open-loop transfer function has the effect of pulling the root locus to the right, tending to lower the system's relative stability and to slow down the settling of the response. (Remember that the addition of integral control adds a pole at the origin, thus making the system less stable.) Figure 6–34 shows examples of root loci illustrating the effects of the addition of a pole to a single-pole system and the addition of two poles to a single-pole system.

Effects of the Addition of Zeros.   The addition of a zero to the open-loop transfer function has the effect of pulling the root locus to the left, tending to make the system more stable and to speed up the settling of the response. (Physically, the addition of a zero in the feedforward transfer function means the addition of derivative control to the system. The effect of such control is to introduce a degree of anticipation into the system and speed up the transient response.) Figure 6–35(a) shows the root loci for a system

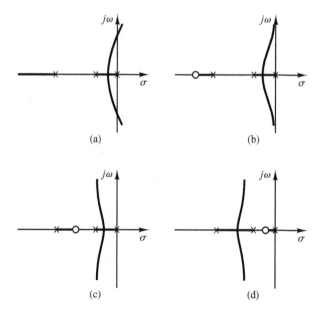

Figure 6–35
(a) Root-locus plot
of a three-pole
system; (b), (c), and
(d) root-locus plots
showing effects of
addition of a zero to
the three-pole
system.

(a)   (b)

(c)   (d)

that is stable for small gain but unstable for large gain. Figures 6–35(b), (c), and (d) show root-locus plots for the system when a zero is added to the open-loop transfer function. Notice that when a zero is added to the system of Figure 6–35(a), it becomes stable for all values of gain.

6–6 LEAD COMPENSATION

In Section 6–5 we presented an introduction to compensation of control systems and discussed preliminary materials for the root-locus approach to control-systems design and compensation. In this section we shall present control-systems design by use of the lead compensation technique. In carrying out a control-system design, we place a compensator in series with the unalterable transfer function $G(s)$ to obtain desirable behavior. The main problem then involves the judicious choice of the pole(s) and zero(s) of the compensator $G_c(s)$ to have the dominant closed-loop poles at the desired location in the s plane so that the performance specifications will be met.

Lead Compensators and Lag Compensators. There are many ways to realize lead compensators and lag compensators, such as electronic networks using operational amplifiers, electrical RC networks, and mechanical spring-dashpot systems.

Figure 6–36 shows an electronic circuit using operational amplifiers. The transfer function for this circuit was obtained in Chapter 3 as follows [see Equation (3–36)]:

$$\frac{E_o(s)}{E_i(s)} = \frac{R_2 R_4}{R_1 R_3} \frac{R_1 C_1 s + 1}{R_2 C_2 s + 1} = \frac{R_4 C_1}{R_3 C_2} \frac{s + \dfrac{1}{R_1 C_1}}{s + \dfrac{1}{R_2 C_2}}$$

$$= K_c \alpha \frac{Ts + 1}{\alpha Ts + 1} = K_c \frac{s + \dfrac{1}{T}}{s + \dfrac{1}{\alpha T}} \tag{6–18}$$

where

$$T = R_1 C_1, \qquad \alpha T = R_2 C_2, \qquad K_c = \frac{R_4 C_1}{R_3 C_2}$$

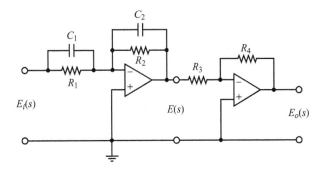

Figure 6–36
Electronic circuit that is a lead network if $R_1 C_1 > R_2 C_2$ and a lag network if $R_1 C_1 < R_2 C_2$.

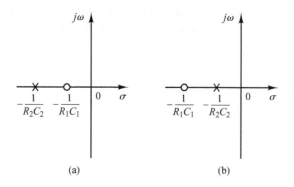

Figure 6–37
Pole-zero
configurations:
(a) lead network;
(b) lag network.

(a)   (b)

Notice that

$$K_c\alpha = \frac{R_4 C_1}{R_3 C_2}\frac{R_2 C_2}{R_1 C_1} = \frac{R_2 R_4}{R_1 R_3}, \qquad \alpha = \frac{R_2 C_2}{R_1 C_1}$$

This network has a dc gain of $K_c\alpha = R_2 R_4/(R_1 R_3)$.

From Equation (6–18), we see that this network is a lead network if $R_1 C_1 > R_2 C_2$, or $\alpha < 1$. It is a lag network if $R_1 C_1 < R_2 C_2$. The pole-zero configurations of this network when $R_1 C_1 > R_2 C_2$ and $R_1 C_1 < R_2 C_2$ are shown in Figure 6–37(a) and (b), respectively.

Lead Compensation Techniques Based on the Root-Locus Approach.   The root-locus approach to design is very powerful when the specifications are given in terms of time-domain quantities, such as the damping ratio and undamped natural frequency of the desired dominant closed-loop poles, maximum overshoot, rise time, and settling time.

Consider a design problem in which the original system either is unstable for all values of gain or is stable but has undesirable transient-response characteristics. In such a case, the reshaping of the root locus is necessary in the broad neighborhood of the $j\omega$ axis and the origin in order that the dominant closed-loop poles be at desired locations in the complex plane. This problem may be solved by inserting an appropriate lead compensator in cascade with the feedforward transfer function.

The procedures for designing a lead compensator for the system shown in Figure 6–38 by the root-locus method may be stated as follows:

1. From the performance specifications, determine the desired location for the dominant closed-loop poles.

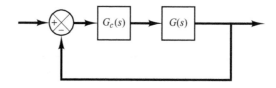

Figure 6–38
Control system.

2. By drawing the root-locus plot of the uncompensated system (original system), ascertain whether or not the gain adjustment alone can yield the desired closed-loop poles. If not, calculate the angle deficiency ϕ. This angle must be contributed by the lead compensator if the new root locus is to pass through the desired locations for the dominant closed-loop poles.

3. Assume the lead compensator $G_c(s)$ to be

$$G_c(s) = K_c \alpha \frac{Ts + 1}{\alpha Ts + 1} = K_c \frac{s + \dfrac{1}{T}}{s + \dfrac{1}{\alpha T}}, \qquad (0 < \alpha < 1)$$

where α and T are determined from the angle deficiency. K_c is determined from the requirement of the open-loop gain.

4. If static error constants are not specified, determine the location of the pole and zero of the lead compensator so that the lead compensator will contribute the necessary angle ϕ. If no other requirements are imposed on the system, try to make the value of α as large as possible. A larger value of α generally results in a larger value of K_v, which is desirable. Note that

$$K_v = \lim_{s \to 0} sG_c(s)G(s) = K_c \alpha \lim_{s \to 0} sG(s)$$

5. Determine the value of K_c of the lead compensator from the magnitude condition.

Once a compensator has been designed, check to see whether all performance specifications have been met. If the compensated system does not meet the performance specifications, then repeat the design procedure by adjusting the compensator pole and zero until all such specifications are met. If a large static error constant is required, cascade a lag network or alter the lead compensator to a lag–lead compensator.

Note that if the selected dominant closed-loop poles are not really dominant, or if the selected dominant closed-loop poles do not yield the desired result, it will be necessary to modify the location of the pair of such selected dominant closed-loop poles. (The closed-loop poles other than dominant ones modify the response obtained from the dominant closed-loop poles alone. The amount of modification depends on the location of these remaining closed-loop poles.) Also, the closed-loop zeros affect the response if they are located near the origin.

EXAMPLE 6-6 Consider the position control system shown in Figure 6–39(a). The feedforward transfer function is

$$G(s) = \frac{10}{s(s + 1)}$$

The root-locus plot for this system is shown in Figure 6–39(b). The closed-loop transfer function for the system is

$$\frac{C(s)}{R(s)} = \frac{10}{s^2 + s + 10}$$

$$= \frac{10}{(s + 0.5 + j3.1225)(s + 0.5 - j3.1225)}$$

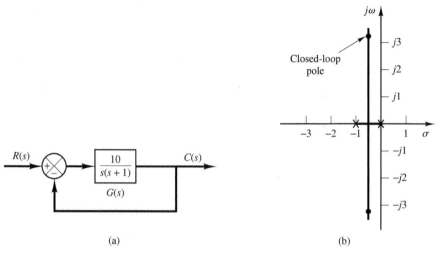

R(s) — G(s) = 10/(s(s+1)) — C(s)

Figure 6–39
(a) Control system;
(b) root-locus plot.

(a)　　　　　　　　　　　　　　(b)

The closed-loop poles are located at

$$s = -0.5 \pm j3.1225$$

The damping ratio of the closed-loop poles is $\zeta = (1/2)/\sqrt{10} = 0.1581$. The undamped natural frequency of the closed-loop poles is $\omega_n = \sqrt{10} = 3.1623$ rad/sec. Because the damping ratio is small, this system will have a large overshoot in the step response and is not desirable.

It is desired to design a lead compensator $G_c(s)$ as shown in Figure 6–40(a) so that the dominant closed-loop poles have the damping ratio $\zeta = 0.5$ and the undamped natural frequency $\omega_n = 3$ rad/sec. The desired location of the dominant closed-loop poles can be determined from

$$s^2 + 2\zeta \omega_n s + \omega_n^2 = s^2 + 3s + 9$$

$$= (s + 1.5 + j2.5981)(s + 1.5 - j2.5981)$$

as follows:

$$s = -1.5 \pm j2.5981$$

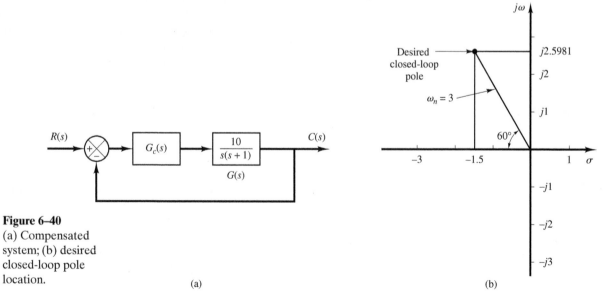

Figure 6–40
(a) Compensated
system; (b) desired
closed-loop pole
location.

(a)　　　　　　　　　　　　　　(b)

[See Figure 6–40 (b).] In some cases, after the root loci of the original system have been obtained, the dominant closed-loop poles may be moved to the desired location by simple gain adjustment. This is, however, not the case for the present system. Therefore, we shall insert a lead compensator in the feedforward path.

A general procedure for determining the lead compensator is as follows: First, find the sum of the angles at the desired location of one of the dominant closed-loop poles with the open-loop poles and zeros of the original system, and determine the necessary angle ϕ to be added so that the total sum of the angles is equal to $\pm180°(2k + 1)$. The lead compensator must contribute this angle ϕ. (If the angle ϕ is quite large, then two or more lead networks may be needed rather than a single one.)

Assume that the lead compensator $G_c(s)$ has the transfer function as follows:

$$G_c(s) = K_c\alpha\frac{Ts + 1}{\alpha Ts + 1} = K_c\frac{s + \dfrac{1}{T}}{s + \dfrac{1}{\alpha T}}, \qquad (0 < \alpha < 1)$$

The angle from the pole at the origin to the desired dominant closed-loop pole at $s = -1.5 + j2.5981$ is $120°$. The angle from the pole at $s = -1$ to the desired closed-loop pole is $100.894°$. Hence, the angle deficiency is

$$\text{Angle deficiency} = 180° - 120° - 100.894° = -40.894°$$

Deficit angle $40.894°$ must be contributed by a lead compensator.

Note that the solution to such a problem is not unique. There are infinitely many solutions. We shall present two solutions to the problem in what follows.

Method 1. There are many ways to determine the locations of the zero and pole of the lead compensator. In what follows we shall introduce a procedure to obtain a largest possible value for α. (Note that a larger value of α will produce a larger value of K_v. In most cases, the larger the K_v is, the better the system performance.) First, draw a horizontal line passing through point P, the desired location for one of the dominant closed-loop poles. This is shown as line PA in Figure 6–41. Draw also a line connecting point P and the origin. Bisect the angle between the lines PA and PO, as shown in Figure 6–41. Draw two lines PC and PD that make angles $\pm\phi/2$ with the bisector PB. The intersections of PC and PD with the negative real axis give the necessary locations for the pole and zero of the lead network. The compensator thus designed will make point P a point on the root locus of the compensated system. The open-loop gain is determined by use of the magnitude condition.

In the present system, the angle of $G(s)$ at the desired closed-loop pole is

$$\left.\underline{/\frac{10}{s(s + 1)}}\right|_{s=-1.5+j2.5981} = -220.894°$$

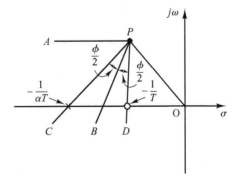

Figure 6–41
Determination of the pole and zero of a lead network.

Thus, if we need to force the root locus to go through the desired closed-loop pole, the lead compensator must contribute $\phi = 40.894°$ at this point. By following the foregoing design procedure, we can determine the zero and pole of the lead compensator.

Referring to Figure 6–42, if we bisect angle APO and take 40.894°/2 each side, then the locations of the zero and pole are found as follows:

$$\text{zero at } s = -1.9432$$
$$\text{pole at } s = -4.6458$$

Thus, $G_c(s)$ can be given as

$$G_c(s) = K_c \frac{s + \dfrac{1}{T}}{s + \dfrac{1}{\alpha T}} = K_c \frac{s + 1.9432}{s + 4.6458}$$

(For this compensator the value of α is $\alpha = 1.9432/4.6458 = 0.418$.)

The value of K_c can be determined by use of the magnitude condition.

$$\left| K_c \frac{s + 1.9432}{s + 4.6458} \frac{10}{s(s + 1)} \right|_{s=-1.5+j2.5981} = 1$$

or

$$K_c = \left| \frac{(s + 4.6458)s(s + 1)}{10(s + 1.9432)} \right|_{s=-1.5+j2.5981} = 1.2287$$

Hence, the lead compensator $G_c(s)$ just designed is given by

$$G_c(s) = 1.2287 \frac{s + 1.9432}{s + 4.6458}$$

Then, the open-loop transfer function of the designed system becomes

$$G_c(s)G(s) = 1.2287 \left(\frac{s + 1.9432}{s + 4.6458} \right) \frac{10}{s(s + 1)}$$

and the closed-loop transfer function becomes

$$\frac{C(s)}{R(s)} = \frac{12.287(s + 1.9432)}{s(s + 1)(s + 4.6458) + 12.287(s + 1.9432)}$$

$$= \frac{12.287s + 23.876}{s^3 + 5.646s^2 + 16.933s + 23.876}$$

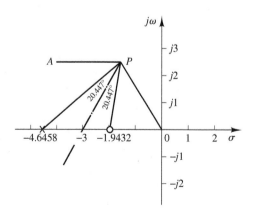

Figure 6–42
Determination of the pole and zero of the lead compensator.

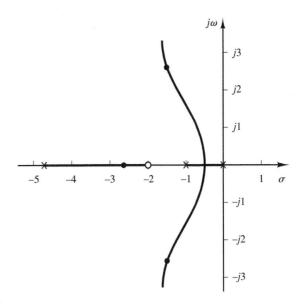

Figure 6–43
Root-locus plot
of the designed
system.

Figure 6–43 shows the root-locus plot for the designed system.

It is worthwhile to check the static velocity error constant K_v for the system just designed.

$$K_v = \lim_{s \to 0} sG_c(s)G(s)$$

$$= \lim_{s \to 0} s\left[1.2287 \frac{s + 1.9432}{s + 4.6458} \frac{10}{s(s + 1)} \right]$$

$$= 5.139$$

Note that the third closed-loop pole of the designed system is found by dividing the characteristic equation by the known factors as follows:

$$s^3 + 5.646s^2 + 16.933s + 23.875 = (s + 1.5 + j2.5981)(s + 1.5 - j2.5981)(s + 2.65)$$

The foregoing compensation method enables us to place the dominant closed-loop poles at the desired points in the complex plane. The third pole at $s = -2.65$ is fairly close to the added zero at -1.9432. Therefore, the effect of this pole on the transient response is relatively small. Since no restriction has been imposed on the nondominant pole and no specification has been given concerning the value of the static velocity error coefficient, we conclude that the present design is satisfactory.

Method 2. If we choose the zero of the lead compensator at $s = -1$ so that it will cancel the plant pole at $s = -1$, then the compensator pole must be located at $s = -3$. (See Figure 6–44.) Hence the lead compensator becomes

$$G_c(s) = K_c \frac{s + 1}{s + 3}$$

The value of K_c can be determined by use of the magnitude condition.

$$\left| K_c \frac{s + 1}{s + 3} \frac{10}{s(s + 1)} \right|_{s = -1.5 + j2.5981} = 1$$

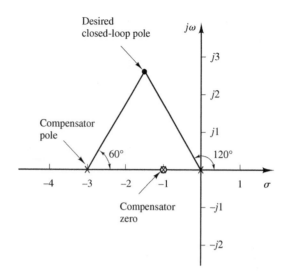

Desired
closed-loop pole

Compensator
pole

60°

120°

Compensator
zero

Figure 6–44
Compensator pole
and zero.

or

$$K_c = \left| \frac{s(s+3)}{10} \right|_{s=-1.5+j2.5981} = 0.9$$

Hence

$$G_c(s) = 0.9\frac{s+1}{s+3}$$

The open-loop transfer function of the designed system then becomes

$$G_c(s)G(s) = 0.9\frac{s+1}{s+3}\frac{10}{s(s+1)} = \frac{9}{s(s+3)}$$

The closed-loop transfer function of the compensated system becomes

$$\frac{C(s)}{R(s)} = \frac{9}{s^2+3s+9}$$

Note that in the present case the zero of the lead compensator will cancel a pole of the plant, resulting in the second-order system, rather than the third-order system as we designed using Method 1.

The static velocity error constant for the present case is obtained as follows:

$$K_v = \lim_{s\to 0} sG_c(s)G(s)$$

$$= \lim_{s\to 0} s\left[\frac{9}{s(s+3)}\right] = 3$$

Notice that the system designed by Method 1 gives a larger value of the static velocity error constant. This means that the system designed by Method 1 will give smaller steady-state errors in following ramp inputs than the system designed by Method 2.

For different combinations of a zero and pole of the compensator that contributes 40.894°, the value of K_v will be different. Although a certain change in the value of K_v can be made by altering the pole-zero location of the lead compensator, if a large increase in the value of K_v is desired, then we must alter the lead compensator to a lag–lead compensator.

Comparison of step and ramp responses of the compensated and uncompensated systems. In what follows we shall compare the unit-step and unit-ramp responses of the three systems: the original uncompensated system, the system designed by Method 1, and the system designed by Method 2. The MATLAB program used to obtain unit-step response curves is given in

MATLAB Program 6–9, where num1 and den1 denote the numerator and denominator of the system designed by Method 1 and num2 and den2 denote that designed by Method 2. Also, num and den are used for the original uncompensated system. The resulting unit-step response curves are shown in Figure 6–45. The MATLAB program to obtain the unit-ramp response curves of the

---

MATLAB Program 6–9

```
% ***** Unit-Step Response of Compensated and Uncompensated Systems *****

num1 = [12.287  23.876];
den1 = [1  5.646  16.933  23.876];
num2 = [9];
den2 = [1  3  9];
num = [10];
den = [1  1  10];
t = 0:0.05:5;
c1 = step(num1,den1,t);
c2 = step(num2,den2,t);
c = step(num,den,t);
plot(t,c1,'-',t,c2,'.',t,c,'x')
grid
title('Unit-Step Responses of Compensated Systems and Uncompensated System')
xlabel('t Sec')
ylabel('Outputs c1, c2, and c')
text(1.51,1.48,'Compensated System (Method 1)')
text(0.9,0.48,'Compensated System (Method 2)')
text(2.51,0.67,'Uncompensated System')
```

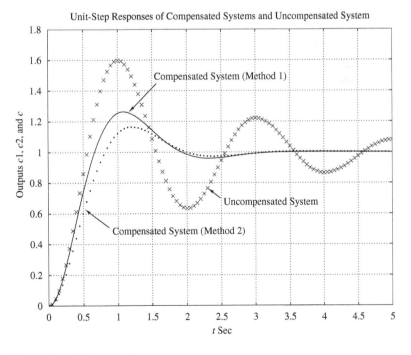

Figure 6–45
Unit-step response curves of designed systems and original uncompensated system.

designed systems is given in MATLAB Program 6–10, where we used the step command to obtain unit-ramp responses by using the numerators and denominators for the systems designed by Method 1 and Method 2 as follows:

$$num1 = [12.287 \ 23.876]$$
$$den1 = [1 \ 5.646 \ 16.933 \ 23.876 \ 0]$$
$$num2 = [9]$$
$$den2 = [1 \ 3 \ 9 \ 0]$$

The resulting unit-ramp response curves are shown in Figure 6–46.

MATLAB Program 6–10

```
% ***** Unit-Ramp Responses of Compensated Systems *****
num1 = [12.287  23.876];
den1 = [1  5.646  16.933  23.876  0];
num2 = [9];
den2 = [1  3  9  0];
t = 0:0.05:5;
c1 = step(num1,den1,t);
c2 = step(num2,den2,t);
plot(t,c1,'-',t,c2,'.',t,t,'-')
grid
title('Unit-Ramp Responses of Compensated Systems')
xlabel('t Sec')
ylabel('Unit-Ramp Input and Outputs c1 and c2')
text(2.55,3.8,'Input')
text(0.55,2.8,'Compensated System (Method 1)')
text(2.35,1.75,'Compensated System (Method 2)')
```

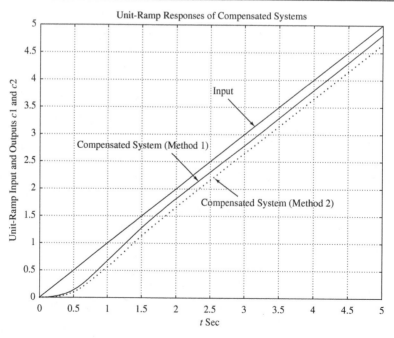

Figure 6–46
Unit-ramp response curves of designed systems.

Chapter 6 / Control Systems Analysis and Design by the Root-Locus Method

In examining these response curves notice that the compensated system designed by Method 1 exhibits a little bit larger overshoot in the step response than the compensated system designed by Method 2. However, the former has better response characteristics for the ramp input than the latter. So it is difficult to say which one is better. The decision on which one to choose should be made by the response requirements (such as smaller overshoots for step type inputs or smaller steady-state errors in following ramp or changing inputs) expected in the designed system. If both smaller overshoots in step inputs and smaller steady-state errors in following changing inputs are required, then we might use a lag–lead compensator. (See Section 6–8 for the lag–lead compensation techniques.)

6–7 LAG COMPENSATION

Electronic Lag Compensator Using Operational Amplifiers. The configuration of the electronic lag compensator using operational amplifiers is the same as that for the lead compensator shown in Figure 6–36. If we choose $R_2C_2 > R_1C_1$ in the circuit shown in Figure 6–36, it becomes a lag compensator. Referring to Figure 6–36, the transfer function of the lag compensator is given by

$$\frac{E_o(s)}{E_i(s)} = \hat{K}_c\beta\frac{Ts + 1}{\beta Ts + 1} = \hat{K}_c\frac{s + \dfrac{1}{T}}{s + \dfrac{1}{\beta T}}$$

where

$$T = R_1C_1, \qquad \beta T = R_2C_2, \qquad \beta = \frac{R_2C_2}{R_1C_1} > 1, \qquad \hat{K}_c = \frac{R_4C_1}{R_3C_2}$$

Note that we use β instead of α in the above expressions. [In the lead compensator we used α to indicate the ratio $R_2C_2/(R_1C_1)$, which was less than 1, or $0 < \alpha < 1$.] In this book we always assume that $0 < \alpha < 1$ and $\beta > 1$.

Lag Compensation Techniques Based on the Root-Locus Approach. Consider the problem of finding a suitable compensation network for the case where the system exhibits satisfactory transient-response characteristics but unsatisfactory steady-state characteristics. Compensation in this case essentially consists of increasing the open-loop gain without appreciably changing the transient-response characteristics. This means that the root locus in the neighborhood of the dominant closed-loop poles should not be changed appreciably, but the open-loop gain should be increased as much as needed. This can be accomplished if a lag compensator is put in cascade with the given feedforward transfer function.

To avoid an appreciable change in the root loci, the angle contribution of the lag network should be limited to a small amount, say less than 5°. To assure this, we place the pole and zero of the lag network relatively close together and near the origin of the s plane. Then the closed-loop poles of the compensated system will be shifted only slightly from their original locations. Hence, the transient-response characteristics will be changed only slightly.

Consider a lag compensator $G_c(s)$, where

$$G_c(s) = \hat{K}_c \beta \frac{Ts + 1}{\beta Ts + 1} = \hat{K}_c \frac{s + \dfrac{1}{T}}{s + \dfrac{1}{\beta T}} \tag{6–19}$$

If we place the zero and pole of the lag compensator very close to each other, then at $s = s_1$, where s_1 is one of the dominant closed-loop poles, the magnitudes $s_1 + (1/T)$ and $s_1 + \left[1/(\beta T)\right]$ are almost equal, or

$$|G_c(s_1)| = \left| \hat{K}_c \frac{s_1 + \dfrac{1}{T}}{s_1 + \dfrac{1}{\beta T}} \right| \doteq \hat{K}_c$$

To make the angle contribution of the lag portion of the compensator small, we require

$$-5° < \left| \frac{s_1 + \dfrac{1}{T}}{s_1 + \dfrac{1}{\beta T}} \right. < 0°$$

This implies that if gain $\hat{K}_c$ of the lag compensator is set equal to 1, the alteration in the transient-response characteristics will be very small, despite the fact that the overall gain of the open-loop transfer function is increased by a factor of β, where $\beta > 1$. If the pole and zero are placed very close to the origin, then the value of β can be made large. (A large value of β may be used, provided physical realization of the lag compensator is possible.) It is noted that the value of T must be large, but its exact value is not critical. However, it should not be too large in order to avoid difficulties in realizing the phase-lag compensator by physical components.

An increase in the gain means an increase in the static error constants. If the open-loop transfer function of the uncompensated system is $G(s)$, then the static velocity error constant K_v of the uncompensated system is

$$K_v = \lim_{s \to 0} sG(s)$$

If the compensator is chosen as given by Equation (6–19), then for the compensated system with the open-loop transfer function $G_c(s)G(s)$ the static velocity error constant $\hat{K}_v$ becomes

$$\hat{K}_v = \lim_{s \to 0} sG_c(s)G(s) = \lim_{s \to 0} G_c(s)K_v = \hat{K}_c \beta K_v$$

where K_v is the static velocity error constant of the uncompensated system.

Thus if the compensator is given by Equation (6–19), then the static velocity error constant is increased by a factor of $\hat{K}_c \beta$, where $\hat{K}_c$ is approximately unity.

The main negative effect of the lag compensation is that the compensator zero that will be generated near the origin creates a closed-loop pole near the origin. This closed-loop pole and compensator zero will generate a long tail of small amplitude in the step response, thus increasing the settling time.

Design Procedures for Lag Compensation by the Root-Locus Method. The procedure for designing lag compensators for the system shown in Figure 6–47 by the root-locus method may be stated as follows (we assume that the uncompensated system meets the transient-response specifications by simple gain adjustment; if this is not the case, refer to Section 6–8):

1. Draw the root-locus plot for the uncompensated system whose open-loop transfer function is $G(s)$. Based on the transient-response specifications, locate the dominant closed-loop poles on the root locus.

2. Assume the transfer function of the lag compensator to be given by Equation (6–19):

$$G_c(s) = \hat{K}_c \beta \frac{Ts + 1}{\beta Ts + 1} = \hat{K}_c \frac{s + \dfrac{1}{T}}{s + \dfrac{1}{\beta T}}$$

Then the open-loop transfer function of the compensated system becomes $G_c(s)G(s)$.

3. Evaluate the particular static error constant specified in the problem.

4. Determine the amount of increase in the static error constant necessary to satisfy the specifications.

5. Determine the pole and zero of the lag compensator that produce the necessary increase in the particular static error constant without appreciably altering the original root loci. (Note that the ratio of the value of gain required in the specifications and the gain found in the uncompensated system is the required ratio between the distance of the zero from the origin and that of the pole from the origin.)

6. Draw a new root-locus plot for the compensated system. Locate the desired dominant closed-loop poles on the root locus. (If the angle contribution of the lag network is very small—that is, a few degrees—then the original and new root loci are almost identical. Otherwise, there will be a slight discrepancy between them. Then locate, on the new root locus, the desired dominant closed-loop poles based on the transient-response specifications.)

7. Adjust gain $\hat{K}_c$ of the compensator from the magnitude condition so that the dominant closed-loop poles lie at the desired location. ($\hat{K}_c$ will be approximately 1.)

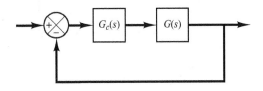

Figure 6–47
Control system.

EXAMPLE 6–7 Consider the system shown in Figure 6–48(a). The feedforward transfer function is

$$G(s) = \frac{1.06}{s(s + 1)(s + 2)}$$

The root-locus plot for the system is shown in Figure 6–48(b). The closed-loop transfer function becomes

$$\frac{C(s)}{R(s)} = \frac{1.06}{s(s + 1)(s + 2) + 1.06}$$

$$= \frac{1.06}{(s + 0.3307 - j0.5864)(s + 0.3307 + j0.5864)(s + 2.3386)}$$

The dominant closed-loop poles are

$$s = -0.3307 \pm j0.5864$$

The damping ratio of the dominant closed-loop poles is $\zeta = 0.491$. The undamped natural frequency of the dominant closed-loop poles is 0.673 rad/sec. The static velocity error constant is 0.53 sec^{-1}.

It is desired to increase the static velocity error constant K_v to about 5 sec^{-1} without appreciably changing the location of the dominant closed-loop poles.

To meet this specification, let us insert a lag compensator as given by Equation (6–19) in cascade with the given feedforward transfer function. To increase the static velocity error constant by a factor of about 10, let us choose $\beta = 10$ and place the zero and pole of the lag compensator at $s = -0.05$ and $s = -0.005$, respectively. The transfer function of the lag compensator becomes

$$G_c(s) = \hat{K}_c \frac{s + 0.05}{s + 0.005}$$

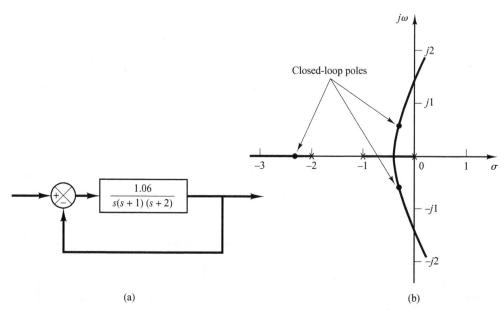

Figure 6–48
(a) Control system;
(b) root-locus plot.

(a)

(b)

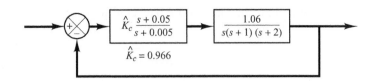

Figure 6–49
Compensated
system.

The angle contribution of this lag network near a dominant closed-loop pole is about 4°. Because this angle contribution is not very small, there is a small change in the new root locus near the desired dominant closed-loop poles.

The open-loop transfer function of the compensated system then becomes

$$G_c(s)G(s) = \hat{K}_c \frac{s + 0.05}{s + 0.005} \frac{1.06}{s(s + 1)(s + 2)}$$

$$= \frac{K(s + 0.05)}{s(s + 0.005)(s + 1)(s + 2)}$$

where

$$K = 1.06\hat{K}_c$$

The block diagram of the compensated system is shown in Figure 6–49. The root-locus plot for the compensated system near the dominant closed-loop poles is shown in Figure 6–50(a), together with the original root-locus plot. Figure 6–50(b) shows the root-locus plot of the compensated system

Root-Locus Plots of Compensated and Uncompensated Systems

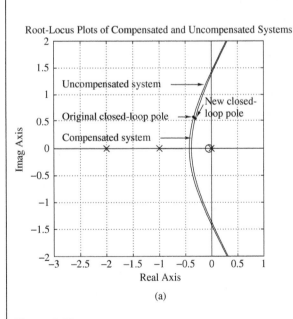

(a)

Root-Locus Plot of Compensated System near the Origin

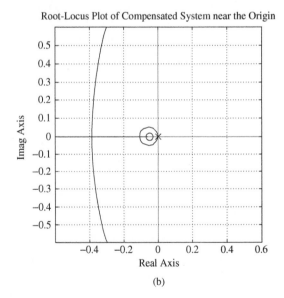

(b)

Figure 6–50
(a) Root-locus plots of the compensated system and uncompensated system; (b) root-locus plot of compensated system near the origin.

near the origin. The MATLAB program to generate the root-locus plots shown in Figures 6–50(a) and (b) is given in MATLAB Program 6–11.

MATLAB Program 6–11

```
% ***** Root-locus plots of the compensated system and
% uncompensated system *****

% ***** Enter the numerators and denominators of the
% compensated and uncompensated systems *****

numc = [1  0.05];
denc = [1  3.005  2.015  0.01  0];
num = [1.06];
den = [1  3  2  0];

% ***** Enter rlocus command. Plot the root loci of both
% systems *****

rlocus(numc,denc)
hold
Current plot held
rlocus(num,den)
v = [-3  1  -2  2]; axis(v); axis('square')
grid
text(-2.8,0.2,'Compensated system')
text(-2.8,1.2,'Uncompensated system')
text(-2.8,0.58,'Original closed-loop pole')
text(-0.1,0.85,'New closed-')
text(-0.1,0.62,'loop pole')
title('Root-Locus Plots of Compensated and Uncompensated Systems')

hold
Current plot released

% ***** Plot root loci of the compensated system near the origin *****

rlocus(numc,denc)
v = [-0.6  0.6  -0.6  0.6]; axis(v); axis('square')
grid
title('Root-Locus Plot of Compensated System near the Origin')
```

If the damping ratio of the new dominant closed-loop poles is kept the same, then these poles are obtained from the new root-locus plot as follows:

$$s_1 = -0.31 + j0.55, \qquad s_2 = -0.31 - j0.55$$

The open-loop gain K is determined from the magnitude condition as follows:

$$K = \left| \frac{s(s + 0.005)(s + 1)(s + 2)}{s + 0.05} \right|_{s=-0.31+j0.55}$$

$$= 1.0235$$

Then the lag compensator gain $\hat{K}_c$ is determined as

$$\hat{K}_c = \frac{K}{1.06} = \frac{1.0235}{1.06} = 0.9656$$

Thus the transfer function of the lag compensator designed is

$$G_c(s) = 0.9656 \frac{s + 0.05}{s + 0.005} = 9.656 \frac{20s + 1}{200s + 1} \qquad (6-20)$$

Then the compensated system has the following open-loop transfer function:

$$G_1(s) = \frac{1.0235(s + 0.05)}{s(s + 0.005)(s + 1)(s + 2)}$$

$$= \frac{5.12(20s + 1)}{s(200s + 1)(s + 1)(0.5s + 1)}$$

The static velocity error constant K_v is

$$K_v = \lim_{s \to 0} sG_1(s) = 5.12 \text{ sec}^{-1}$$

In the compensated system, the static velocity error constant has increased to 5.12 sec^{-1}, or $5.12/0.53 = 9.66$ times the original value. (The steady-state error with ramp inputs has decreased to about 10% of that of the original system.) We have essentially accomplished the design objective of increasing the static velocity error constant to 5 sec^{-1}.

Note that, since the pole and zero of the lag compensator are placed close together and are located very near the origin, their effect on the shape of the original root loci has been small. Except for the presence of a small closed root locus near the origin, the root loci of the compensated and the uncompensated systems are very similar to each other. However, the value of the static velocity error constant of the compensated system is 9.66 times greater than that of the uncompensated system.

The two other closed-loop poles for the compensated system are found as follows:

$$s_3 = -2.326, \qquad s_4 = -0.0549$$

The addition of the lag compensator increases the order of the system from 3 to 4, adding one additional closed-loop pole close to the zero of the lag compensator. (The added closed-loop pole at $s = -0.0549$ is close to the zero at $s = -0.05$.) Such a pair of a zero and pole creates a long tail of small amplitude in the transient response, as we will see later in the unit-step response. Since the pole at $s = -2.326$ is very far from the $j\omega$ axis compared with the dominant closed-loop poles, the effect of this pole on the transient response is also small. Therefore, we may consider the closed-loop poles at $s = -0.31 \pm j0.55$ to be the dominant closed-loop poles.

The undamped natural frequency of the dominant closed-loop poles of the compensated system is 0.631 rad/sec. This value is about 6% less than the original value, 0.673 rad/sec. This implies that the transient response of the compensated system is slower than that of the original system. The response will take a longer time to settle down. The maximum overshoot in the step response will increase in the compensated system. If such adverse effects can be tolerated, the lag compensation as discussed here presents a satisfactory solution to the given design problem.

Next, we shall compare the unit-ramp responses of the compensated system against the uncompensated system and verify that the steady-state performance is much better in the compensated system than the uncompensated system.

To obtain the unit-ramp response with MATLAB, we use the step command for the system $C(s)/[sR(s)]$. Since $C(s)/[sR(s)]$ for the compensated system is

$$\frac{C(s)}{sR(s)} = \frac{1.0235(s + 0.05)}{s[s(s + 0.005)(s + 1)(s + 2) + 1.0235(s + 0.05)]}$$

$$= \frac{1.0235s + 0.0512}{s^5 + 3.005s^4 + 2.015s^3 + 1.0335s^2 + 0.0512s}$$

we have

$$\text{numc} = [1.0235 \ \ 0.0512]$$
$$\text{denc} = [1 \ \ 3.005 \ \ 2.015 \ \ 1.0335 \ \ 0.0512 \ \ 0]$$

Also, $C(s)/[sR(s)]$ for the uncompensated system is

$$\frac{C(s)}{sR(s)} = \frac{1.06}{s[s(s+1)(s+2) + 1.06]}$$

$$= \frac{1.06}{s^4 + 3s^3 + 2s^2 + 1.06s}$$

Hence,

$$\text{num} = [1.06]$$
$$\text{den} = [1 \ \ 3 \ \ 2 \ \ 1.06 \ \ 0]$$

MATLAB Program 6–12 produces the plot of the unit-ramp response curves. Figure 6–51 shows the result. Clearly, the compensated system shows much smaller steady-state error (one-tenth of the original steady-state error) in following the unit-ramp input.

MATLAB Program 6–12

```
% ***** Unit-ramp responses of compensated system and
% uncompensated system *****

% ***** Unit-ramp response will be obtained as the unit-step
% response of C(s)/[sR(s)] *****
% ***** Enter the numerators and denominators of C1(s)/[sR(s)]
% and C2(s)/[sR(s)], where C1(s) and C2(s) are Laplace
% transforms of the outputs of the compensated and un-
% compensated systems, respectively. *****

numc = [1.0235  0.0512];
denc = [1  3.005  2.015  1.0335  0.0512  0];
num = [1.06];
den = [1  3  2  1.06  0];

% ***** Specify the time range (such as t= 0:0.1:50) and enter
% step command and plot command. *****

t = 0:0.1:50;
c1 = step(numc,denc,t);
c2 = step(num,den,t);
plot(t,c1,'-',t,c2,'.',t,t,'--')
grid
text(2.2,27,'Compensated system');
text(26,21.3,'Uncompensated system');
title('Unit-Ramp Responses of Compensated and Uncompensated Systems')
xlabel('t Sec');
ylabel('Outputs c1 and c2')
```

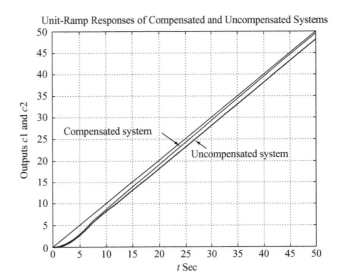

Figure 6–51
Unit-ramp responses of compensated and uncompensated systems. [The compensator is given by Equation (6–20).]

MATLAB Program 6–13 gives the unit-step response curves of the compensated and uncompensated systems. The unit-step response curves are shown in Figure 6–52. Notice that the lag-compensated system exhibits a larger maximum overshoot and slower response than the original uncompensated system. Notice that a pair of the pole at $s = -0.0549$ and zero at

---

MATLAB Program 6–13

```
% ***** Unit-step responses of compensated system and
% uncompensated system *****

% ***** Enter the numerators and denominators of the
% compensated and uncompensated systems *****

numc = [1.0235  0.0512];
denc = [1  3.005  2.015  1.0335  0.0512];
num = [1.06];
den = [1  3  2  1.06];

% ***** Specify the time range (such as t = 0:0.1:40) and enter
% step command and plot command. *****

t = 0:0.1:40;
c1 = step(numc,denc,t);
c2 = step(num,den,t);
plot(t,c1,'-',t,c2,'.')
grid
text(13,1.12,'Compensated system')
text(13.6,0.88,'Uncompensated system')
title('Unit-Step Responses of Compensated and Uncompensated Systems')
xlabel('t Sec')
ylabel('Outputs c1 and c2')
```

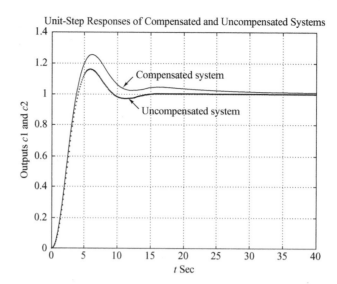

Figure 6–52
Unit-step responses
of compensated and
uncompensated
systems. [The
compensator is given
by Equation (6–20).]

$s = -0.05$ generates a long tail of small amplitude in the transient response. If a larger maximum overshoot and a slower response are not desired, we need to use a lag–lead compensator as presented in Section 6–8.

Comments. It is noted that under certain circumstances, however, both lead compensator and lag compensator may satisfy the given specifications (both transient-response specifications and steady-state specifications.) Then either compensation may be used.

6–8 LAG–LEAD COMPENSATION

Lead compensation basically speeds up the response and increases the stability of the system. Lag compensation improves the steady-state accuracy of the system, but reduces the speed of the response.

If improvements in both transient response and steady-state response are desired, then both a lead compensator and a lag compensator may be used simultaneously. Rather than introducing both a lead compensator and a lag compensator as separate units, however, it is economical to use a single lag–lead compensator.

Lag–lead compensation combines the advantages of lag and lead compensations. Since the lag–lead compensator possesses two poles and two zeros, such a compensation increases the order of the system by 2, unless cancellation of pole(s) and zero(s) occurs in the compensated system.

Electronic Lag–Lead Compensator Using Operational Amplifiers. Figure 6–53 shows an electronic lag–lead compensator using operational amplifiers. The transfer

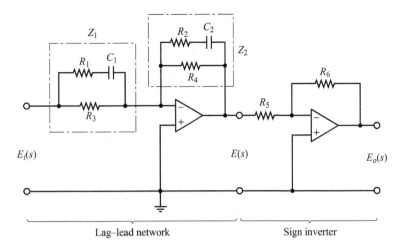

Figure 6–53
Lag–lead
compensator.

Lag–lead network   Sign inverter

function for this compensator may be obtained as follows: The complex impedance Z_1 is given by

$$\frac{1}{Z_1} = \frac{1}{R_1 + \dfrac{1}{C_1 s}} + \frac{1}{R_3}$$

or

$$Z_1 = \frac{(R_1 C_1 s + 1)R_3}{(R_1 + R_3)C_1 s + 1}$$

Similarly, complex impedance Z_2 is given by

$$Z_2 = \frac{(R_2 C_2 s + 1)R_4}{(R_2 + R_4)C_2 s + 1}$$

Hence, we have

$$\frac{E(s)}{E_i(s)} = -\frac{Z_2}{Z_1} = -\frac{R_4}{R_3}\frac{(R_1 + R_3)C_1 s + 1}{R_1 C_1 s + 1} \cdot \frac{R_2 C_2 s + 1}{(R_2 + R_4)C_2 s + 1}$$

The sign inverter has the transfer function

$$\frac{E_o(s)}{E(s)} = -\frac{R_6}{R_5}$$

Thus the transfer function of the compensator shown in Figure 6–53 is

$$\frac{E_o(s)}{E_i(s)} = \frac{E_o(s)}{E(s)}\frac{E(s)}{E_i(s)} = \frac{R_4 R_6}{R_3 R_5}\left[\frac{(R_1 + R_3)C_1 s + 1}{R_1 C_1 s + 1}\right]\left[\frac{R_2 C_2 s + 1}{(R_2 + R_4)C_2 s + 1}\right] \quad (6\text{–}21)$$

Let us define

$$T_1 = (R_1 + R_3)C_1, \qquad \frac{T_1}{\gamma} = R_1 C_1, \qquad T_2 = R_2 C_2, \qquad \beta T_2 = (R_2 + R_4)C_2$$

Then Equation (6–21) becomes

$$\frac{E_o(s)}{E_i(s)} = K_c \frac{\beta}{\gamma} \left(\frac{T_1 s + 1}{\frac{T_1}{\gamma} s + 1} \right) \left(\frac{T_2 s + 1}{\beta T_2 s + 1} \right) = K_c \frac{\left(s + \dfrac{1}{T_1} \right)\left(s + \dfrac{1}{T_2} \right)}{\left(s + \dfrac{\gamma}{T_1} \right)\left(s + \dfrac{1}{\beta T_2} \right)} \quad (6\text{--}22)$$

where

$$\gamma = \frac{R_1 + R_3}{R_1} > 1, \qquad \beta = \frac{R_2 + R_4}{R_2} > 1, \qquad K_c = \frac{R_2 R_4 R_6}{R_1 R_3 R_5} \frac{R_1 + R_3}{R_2 + R_4}$$

Note that γ is often chosen to be equal to β.

Lag–lead Compensation Techniques Based on the Root-Locus Approach.
Consider the system shown in Figure 6–54. Assume that we use the lag–lead compensator:

$$G_c(s) = K_c \frac{\beta}{\gamma} \frac{(T_1 s + 1)(T_2 s + 1)}{\left(\dfrac{T_1}{\gamma} s + 1 \right)(\beta T_2 s + 1)} = K_c \frac{\left(s + \dfrac{1}{T_1} \right)\left(s + \dfrac{1}{T_2} \right)}{\left(s + \dfrac{\gamma}{T_1} \right)\left(s + \dfrac{1}{\beta T_2} \right)} \quad (6\text{--}23)$$

where $\beta > 1$ and $\gamma > 1$. (Consider K_c to belong to the lead portion of the lag–lead compensator.)

In designing lag–lead compensators, we consider two cases where $\gamma \neq \beta$ and $\gamma = \beta$.

Case 1. $\gamma \neq \beta$.   In this case, the design process is a combination of the design of the lead compensator and that of the lag compensator. The design procedure for the lag–lead compensator follows:

1. From the given performance specifications, determine the desired location for the dominant closed-loop poles.
2. Using the uncompensated open-loop transfer function $G(s)$, determine the angle deficiency ϕ if the dominant closed-loop poles are to be at the desired location. The phase-lead portion of the lag–lead compensator must contribute this angle ϕ.
3. Assuming that we later choose T_2 sufficiently large so that the magnitude of the lag portion

$$\left| \frac{s_1 + \dfrac{1}{T_2}}{s_1 + \dfrac{1}{\beta T_2}} \right|$$

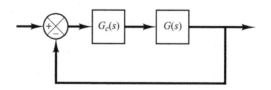

Figure 6–54
Control system.

332

is approximately unity, where $s = s_1$ is one of the dominant closed-loop poles, choose the values of T_1 and γ from the requirement that

$$\left| \frac{s_1 + \dfrac{1}{T_1}}{s_1 + \dfrac{\gamma}{T_1}} \right| = \phi$$

The choice of T_1 and γ is not unique. (Infinitely many sets of T_1 and γ are possible.) Then determine the value of K_c from the magnitude condition:

$$\left| K_c \frac{s_1 + \dfrac{1}{T_1}}{s_1 + \dfrac{\gamma}{T_1}} G(s_1) \right| = 1$$

4. If the static velocity error constant K_v is specified, determine the value of β to satisfy the requirement for K_v. The static velocity error constant K_v is given by

$$K_v = \lim_{s \to 0} s G_c(s) G(s)$$

$$= \lim_{s \to 0} s K_c \left(\frac{s + \dfrac{1}{T_1}}{s + \dfrac{\gamma}{T_1}} \right)\left(\frac{s + \dfrac{1}{T_2}}{s + \dfrac{1}{\beta T_2}} \right) G(s)$$

$$= \lim_{s \to 0} s K_c \frac{\beta}{\gamma} G(s)$$

where K_c and γ are already determined in step 3. Hence, given the value of K_v, the value of β can be determined from this last equation. Then, using the value of β thus determined, choose the value of T_2 such that

$$\left| \frac{s_1 + \dfrac{1}{T_2}}{s_1 + \dfrac{1}{\beta T_2}} \right| \doteq 1$$

$$-5° < \left| \frac{s_1 + \dfrac{1}{T_2}}{s_1 + \dfrac{1}{\beta T_2}} \right| < 0°$$

(The preceding design procedure is illustrated in Example 6–8.)

Case 2. $\gamma = \beta$. If $\gamma = \beta$ is required in Equation (6–23), then the preceeding design procedure for the lag–lead compensator may be modified as follows:

1. From the given performance specifications, determine the desired location for the dominant closed-loop poles.

2. The lag–lead compensator given by Equation (6–23) is modified to

$$G_c(s) = K_c \frac{(T_1 s + 1)(T_2 s + 1)}{\left(\dfrac{T_1}{\beta} s + 1\right)(\beta T_2 s + 1)} = K_c \frac{\left(s + \dfrac{1}{T_1}\right)\left(s + \dfrac{1}{T_2}\right)}{\left(s + \dfrac{\beta}{T_1}\right)\left(s + \dfrac{1}{\beta T_2}\right)} \qquad (6\text{--}24)$$

where $\beta > 1$. The open-loop transfer function of the compensated system is $G_c(s)G(s)$. If the static velocity error constant K_v is specified, determine the value of constant K_c from the following equation:

$$K_v = \lim_{s \to 0} s G_c(s) G(s)$$
$$= \lim_{s \to 0} s K_c G(s)$$

3. To have the dominant closed-loop poles at the desired location, calculate the angle contribution ϕ needed from the phase-lead portion of the lag–lead compensator.

4. For the lag–lead compensator, we later choose T_2 sufficiently large so that

$$\left| \frac{s_1 + \dfrac{1}{T_2}}{s_1 + \dfrac{1}{\beta T_2}} \right|$$

is approximately unity, where $s = s_1$ is one of the dominant closed-loop poles. Determine the values of T_1 and β from the magnitude and angle conditions:

$$\left| K_c \left(\frac{s_1 + \dfrac{1}{T_1}}{s_1 + \dfrac{\beta}{T_1}} \right) G(s_1) \right| = 1$$

$$\left| \frac{s_1 + \dfrac{1}{T_1}}{s_1 + \dfrac{\beta}{T_1}} \right. = \phi$$

5. Using the value of β just determined, choose T_2 so that

$$\left| \frac{s_1 + \dfrac{1}{T_2}}{s_1 + \dfrac{1}{\beta T_2}} \right| \doteq 1$$

$$-5° < \left| \frac{s_1 + \dfrac{1}{T_2}}{s_1 + \dfrac{1}{\beta T_2}} \right. < 0°$$

The value of βT_2, the largest time constant of the lag–lead compensator, should not be too large to be physically realized. (An example of the design of the lag–lead compensator when $\gamma = \beta$ is given in Example 6–9.)

EXAMPLE 6–8   Consider the control system shown in Figure 6–55. The feedforward transfer function is

$$G(s) = \frac{4}{s(s + 0.5)}$$

This system has closed-loop poles at

$$s = -0.2500 \pm j1.9843$$

The damping ratio is 0.125, the undamped natural frequency is 2 rad/sec, and the static velocity error constant is 8 sec^{-1}.

It is desired to make the damping ratio of the dominant closed-loop poles equal to 0.5 and to increase the undamped natural frequency to 5 rad/sec and the static velocity error constant to 80 sec^{-1}. Design an appropriate compensator to meet all the performance specifications.

Let us assume that we use a lag–lead compensator having the transfer function

$$G_c(s) = K_c \left(\frac{s + \dfrac{1}{T_1}}{s + \dfrac{\gamma}{T_1}} \right) \left(\frac{s + \dfrac{1}{T_2}}{s + \dfrac{1}{\beta T_2}} \right) \qquad (\gamma > 1, \beta > 1)$$

where γ is not equal to β. Then the compensated system will have the open-loop transfer function

$$G_c(s)G(s) = K_c \left(\frac{s + \dfrac{1}{T_1}}{s + \dfrac{\gamma}{T_1}} \right) \left(\frac{s + \dfrac{1}{T_2}}{s + \dfrac{1}{\beta T_2}} \right) G(s)$$

From the performance specifications, the dominant closed-loop poles must be at

$$s = -2.50 \pm j4.33$$

Since

$$\left/ \frac{4}{s(s + 0.5)} \right|_{s=-2.50+j4.33} = -235°$$

the phase-lead portion of the lag–lead compensator must contribute 55° so that the root locus passes through the desired location of the dominant closed-loop poles.

To design the phase-lead portion of the compensator, we first determine the location of the zero and pole that will give 55° contribution. There are many possible choices, but we shall here choose the zero at $s = -0.5$ so that this zero will cancel the pole at $s = -0.5$ of the plant. Once the zero is chosen, the pole can be located such that the angle contribution is 55°. By simple calculation or graphical analysis, the pole must be located at $s = -5.02$. Thus, the phase-lead portion of the lag–lead compensator becomes

$$K_c \frac{s + \dfrac{1}{T_1}}{s + \dfrac{\gamma}{T_1}} = K_c \frac{s + 0.5}{s + 5.02}$$

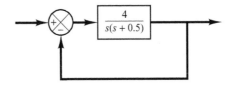

Figure 6–55
Control system.

Thus

$$T_1 = 2, \qquad \gamma = \frac{5.02}{0.5} = 10.04$$

Next we determine the value of K_c from the magnitude condition:

$$\left| K_c \frac{s + 0.5}{s + 5.02} \frac{4}{s(s + 0.5)} \right|_{s=-2.5+j4.33} = 1$$

Hence,

$$K_c = \left| \frac{(s + 5.02)s}{4} \right|_{s=-2.5+j4.33} = 6.26$$

The phase-lag portion of the compensator can be designed as follows: First the value of β is determined to satisfy the requirement on the static velocity error constant:

$$K_v = \lim_{s \to 0} sG_c(s)G(s) = \lim_{s \to 0} sK_c \frac{\beta}{\gamma} G(s)$$

$$= \lim_{s \to 0} s(6.26) \frac{\beta}{10.04} \frac{4}{s(s + 0.5)} = 4.988\beta = 80$$

Hence, β is determined as

$$\beta = 16.04$$

Finally, we choose the value T_2 such that the following two conditions are satisfied:

$$\left| \frac{s + \dfrac{1}{T_2}}{s + \dfrac{1}{16.04T_2}} \right|_{s=-2.5+j4.33} \doteq 1, \qquad -5° < \left/ \frac{s + \dfrac{1}{T_2}}{s + \dfrac{1}{16.04T_2}} \right._{s=-2.5+j4.33} < 0°$$

We may choose several values for T_2 and check if the magnitude and angle conditions are satisfied. After simple calculations we find for $T_2 = 5$

$$1 > \text{magnitude} > 0.98, \qquad -2.10° < \text{angle} < 0°$$

Since $T_2 = 5$ satisfies the two conditions, we may choose

$$T_2 = 5$$

Now the transfer function of the designed lag–lead compensator is given by

$$G_c(s) = (6.26) \left(\frac{s + \dfrac{1}{2}}{s + \dfrac{10.04}{2}} \right) \left(\frac{s + \dfrac{1}{5}}{s + \dfrac{1}{16.04 \times 5}} \right)$$

$$= 6.26 \left(\frac{s + 0.5}{s + 5.02} \right) \left(\frac{s + 0.2}{s + 0.01247} \right)$$

$$= \frac{10(2s + 1)(5s + 1)}{(0.1992s + 1)(80.19s + 1)}$$

The compensated system will have the open-loop transfer function

$$G_c(s)G(s) = \frac{25.04(s + 0.2)}{s(s + 5.02)(s + 0.01247)}$$

Because of the cancellation of the $(s + 0.5)$ terms, the compensated system is a third-order system. (Mathematically, this cancellation is exact, but practically such cancellation will not be exact because some approximations are usually involved in deriving the mathematical model of the system and, as a result, the time constants are not precise.) The root-locus plot of the compensated system is shown in Figure 6–56(a). An enlarged view of the root-locus plot near the origin is shown in Figure 6–56(b). Because the angle contribution of the phase lag portion of the lag–lead compensator is quite small, there is only a small change in the location of the dominant closed-loop poles from the desired location, $s = -2.5 \pm j4.33$. The characteristic equation for the compensated system is

$$s(s + 5.02)(s + 0.01247) + 25.04(s + 0.2) = 0$$

or

$$s^3 + 5.0325s^2 + 25.1026s + 5.008$$

$$= (s + 2.4123 + j4.2756)(s + 2.4123 - j4.2756)(s + 0.2078) = 0$$

Hence the new closed-loop poles are located at

$$s = -2.4123 \pm j4.2756$$

The new damping ratio is $\zeta = 0.491$. Therefore the compensated system meets all the required performance specifications. The third closed-loop pole of the compensated system is located at $s = -0.2078$. Since this closed-loop pole is very close to the zero at $s = -0.2$, the effect of this pole on the response is small. (Note that, in general, if a pole and a zero lie close to each other on the negative real axis near the origin, then such a pole-zero combination will yield a long tail of small amplitude in the transient response.)

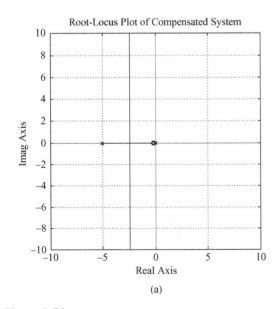

(a)

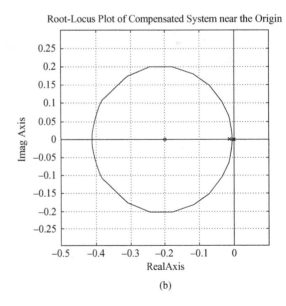

(b)

Figure 6–56
(a) Root-locus plot of the compensated system; (b) root-locus plot near the origin.

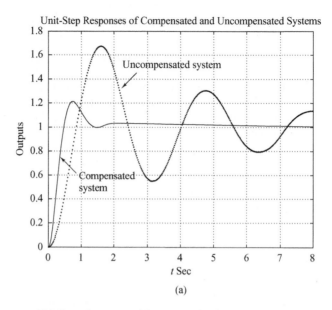

Unit-Step Responses of Compensated and Uncompensated Systems

(a)

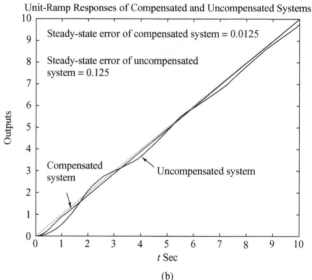

Unit-Ramp Responses of Compensated and Uncompensated Systems

Steady-state error of compensated system = 0.0125

Steady-state error of uncompensated system = 0.125

(b)

Figure 6–57
Transient-response
curves for the
compensated system
and uncompensated
system. (a) Unit-step
response curves;
(b) unit-ramp
response curves.

The unit-step response curves and unit-ramp response curves before and after compensation are shown in Figure 6–57. (Notice a long tail of a small amplitude in the unit-step response of the compensated system.)

EXAMPLE 6–9 Consider the control system of Example 6–8 again. Suppose that we use a lag–lead compensator of the form given by Equation (6–24), or

$$G_c(s) = K_c \frac{\left(s + \dfrac{1}{T_1}\right)\left(s + \dfrac{1}{T_2}\right)}{\left(s + \dfrac{\beta}{T_1}\right)\left(s + \dfrac{1}{\beta T_2}\right)} \quad (\beta > 1)$$

Assuming the specifications are the same as those given in Example 6–8, design a compensator $G_c(s)$.

The desired locations for the dominant closed-loop poles are at

$$s = -2.50 \pm j4.33$$

The open-loop transfer function of the compensated system is

$$G_c(s)G(s) = K_c \frac{\left(s + \dfrac{1}{T_1}\right)\left(s + \dfrac{1}{T_2}\right)}{\left(s + \dfrac{\beta}{T_1}\right)\left(s + \dfrac{1}{\beta T_2}\right)} \cdot \frac{4}{s(s + 0.5)}$$

Since the requirement on the static velocity error constant K_v is 80 sec^{-1}, we have

$$K_v = \lim_{s \to 0} sG_c(s)G(s) = \lim_{s \to 0} K_c \frac{4}{0.5} = 8K_c = 80$$

Thus

$$K_c = 10$$

The time constant T_1 and the value of β are determined from

$$\left|\frac{s + \dfrac{1}{T_1}}{s + \dfrac{\beta}{T_1}}\right| \left|\frac{40}{s(s + 0.5)}\right|_{s=-2.5+j4.33} = \left|\frac{s + \dfrac{1}{T_1}}{s + \dfrac{\beta}{T_1}}\right| \frac{8}{4.77} = 1$$

$$\left| \frac{s + \dfrac{1}{T_1}}{s + \dfrac{\beta}{T_1}} \right|_{s=-2.5+j4.33} = 55°$$

(The angle deficiency of 55° was obtained in Example 6–8.) Referring to Figure 6–58, we can easily locate points A and B such that

$$\underline{/APB} = 55°, \qquad \frac{\overline{PA}}{\overline{PB}} = \frac{4.77}{8}$$

(Use a graphical approach or a trigonometric approach.) The result is

$$\overline{AO} = 2.38, \qquad \overline{BO} = 8.34$$

or

$$T_1 = \frac{1}{2.38} = 0.420, \qquad \beta = 8.34T_1 = 3.503$$

The phase-lead portion of the lag–lead network thus becomes

$$10\left(\frac{s + 2.38}{s + 8.34}\right)$$

For the phase-lag portion, we choose T_2 such that it satisfies the conditions

$$\left| \frac{s + \dfrac{1}{T_2}}{s + \dfrac{1}{3.503T_2}} \right|_{s=-2.50+j4.33} \doteq 1, \qquad -5° < \underline{/\frac{s + \dfrac{1}{T_2}}{s + \dfrac{1}{3.503T_2}}}\Bigg|_{s=-2.50+j4.33} < 0°$$

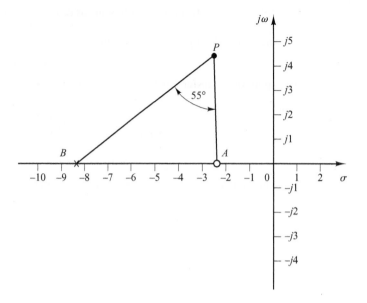

Figure 6–58
Determination of the desired pole-zero location.

By simple calculations, we find that if we choose $T_2 = 5$, then

$$1 > \text{magnitude} > 0.98, \qquad -1.5° < \text{angle} < 0°$$

and if we choose $T_2 = 10$, then

$$1 > \text{magnitude} > 0.99, \qquad -1° < \text{angle} < 0°$$

Since T_2 is one of the time constants of the lag–lead compensator, it should not be too large. If $T_2 = 10$ can be acceptable from practical viewpoint, then we may choose $T_2 = 10$. Then

$$\frac{1}{\beta T_2} = \frac{1}{3.503 \times 10} = 0.0285$$

Thus, the lag–lead compensator becomes

$$G_c(s) = (10)\left(\frac{s + 2.38}{s + 8.34}\right)\left(\frac{s + 0.1}{s + 0.0285}\right)$$

The compensated system will have the open-loop transfer function

$$G_c(s)G(s) = \frac{40(s + 2.38)(s + 0.1)}{(s + 8.34)(s + 0.0285)s(s + 0.5)}$$

No cancellation occurs in this case, and the compensated system is of fourth order. Because the angle contribution of the phase lag portion of the lag–lead network is quite small, the dominant closed-loop poles are located very near the desired location. In fact, the location of the dominant closed-loop poles can be found from the characteristic equation as follows: The characteristic equation of the compensated system is

$$(s + 8.34)(s + 0.0285)s(s + 0.5) + 40(s + 2.38)(s + 0.1) = 0$$

which can be simplified to

$$s^4 + 8.8685s^3 + 44.4219s^2 + 99.3188s + 9.52$$

$$= (s + 2.4539 + j4.3099)(s + 2.4539 - j4.3099)(s + 0.1003)(s + 3.8604) = 0$$

The dominant closed-loop poles are located at

$$s = -2.4539 \pm j4.3099$$

The other closed-loop poles are located at

$$s = -0.1003; \qquad s = -3.8604$$

Since the closed-loop pole at $s = -0.1003$ is very close to a zero at $s = -0.1$, they almost cancel each other. Thus, the effect of this closed-loop pole is very small. The remaining closed-loop pole ($s = -3.8604$) does not quite cancel the zero at $s = -2.38$. The effect of this zero is to cause a larger overshoot in the step response than a similar system without such a zero. The unit-step response curves of the compensated and uncompensated systems are shown in Figure 6–59(a). The unit-ramp response curves for both systems are depicted in Figure 6–59(b).

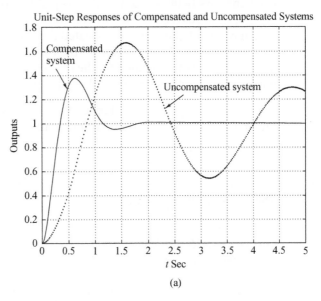

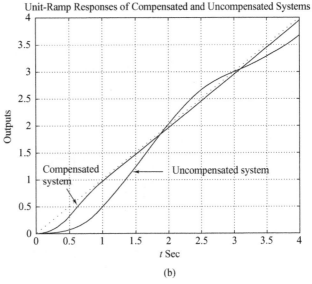

Figure 6–59
(a) Unit-step response curves for the compensated and uncompensated systems; (b) unit-ramp response curves for both systems.

The maximum overshoot in the step response of the compensated system is approximately 38%. (This is much larger than the maximum overshoot of 21% in the design presented in Example 6–8.) It is possible to decrease the maximum overshoot by a small amount from 38%, but not to 20% if $\gamma = \beta$ is required, as in this example. Note that by not requiring $\gamma = \beta$, we have an additional parameter to play with and thus can reduce the maximum overshoot.

6–9 PARALLEL COMPENSATION

Thus far we have presented series compensation techniques using lead, lag, or lag–lead compensators. In this section we discuss parallel compensation technique. Because in the parallel compensation design the controller (or compensator) is in a minor loop, the design may seem to be more complicated than in the series compensation case. It is, however, not complicated if we rewrite the characteristic equation to be of the same form as the characteristic equation for the series compensated system. In this section we present a simple design problem involving parallel compensation.

Basic Principle for Designing Parallel Compensated System. Referring to Figure 6–60(a), the closed-loop transfer function for the system with series compensation is

$$\frac{C}{R} = \frac{G_c G}{1 + G_c G H}$$

The characteristic equation is

$$1 + G_c G H = 0$$

Given G and H, the design problem becomes that of determining the compensator G_c that satisfies the given specification.

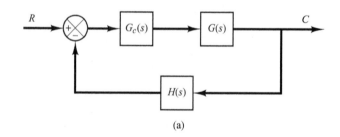

(a)

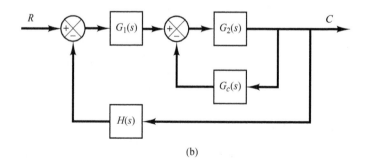

(b)

Figure 6–60
(a) Series compensation;
(b) parallel or feedback compensation.

Chapter 6 / Control Systems Analysis and Design by the Root-Locus Method

The closed-loop transfer function for the system with parallel compensation [Figure 6–60(b)] is

$$\frac{C}{R} = \frac{G_1 G_2}{1 + G_2 G_c + G_1 G_2 H}$$

The characteristic equation is

$$1 + G_1 G_2 H + G_2 G_c = 0$$

By dividing this characteristic equation by the sum of the terms that do not involve G_c, we obtain

$$1 + \frac{G_c G_2}{1 + G_1 G_2 H} = 0 \qquad (6\text{--}25)$$

If we define

$$G_f = \frac{G_2}{1 + G_1 G_2 H}$$

then Equation (6–25) becomes

$$1 + G_c G_f = 0$$

Since G_f is a fixed transfer function, the design of G_c becomes the same as the case of series compensation. Hence the same design approach applies to the parallel compensated system.

Velocity Feedback Systems. A velocity feedback system (tachometer feedback system) is an example of parallel compensated systems. The controller (or compensator) in such a system is a gain element. The gain of the feedback element in a minor loop must be determined properly so that the entire system satisfies the given design specifications. The characteristic of such a velocity feedback system is that the variable parameter does not appear as a multiplying factor in the open-loop transfer function, so that direct application of the root-locus design technique is not possible. However, by rewriting the characteristic equation such that the variable parameter appears as a multiplying factor, then the root-locus approach to the design is possible.

An example of control system design using parallel compensation technique is presented in Example 6–10.

EXAMPLE 6–10   Consider the system shown in Figure 6–61. Draw a root-locus diagram. Then determine the value of k such that the damping ratio of the dominant closed-loop poles is 0.4.

Here the system involves velocity feedback. The open-loop transfer function is

$$\text{Open-loop transfer function} = \frac{20}{s(s+1)(s+4) + 20ks}$$

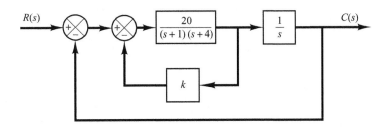

Figure 6–61
Control system.

Notice that the adjustable variable k does not appear as a multiplying factor. The characteristic equation for the system is

$$s^3 + 5s^2 + 4s + 20ks + 20 = 0 \qquad (6\text{--}26)$$

Define

$$20k = K$$

Then Equation (6–26) becomes

$$s^3 + 5s^2 + 4s + Ks + 20 = 0 \qquad (6\text{--}27)$$

Dividing both sides of Equation (6–27) by the sum of the terms that do not contain K, we get

$$1 + \frac{Ks}{s^3 + 5s^2 + 4s + 20} = 0$$

or

$$1 + \frac{Ks}{(s + j2)(s - j2)(s + 5)} = 0 \qquad (6\text{--}28)$$

Equation (6–28) is of the form of Equation (6–11).

We shall now sketch the root loci of the system given by Equation (6–28). Notice that the open-loop poles are located at $s = j2$, $s = -j2$, $s = -5$, and the open-loop zero is located at $s = 0$. The root locus exists on the real axis between 0 and -5. Since

$$\lim_{s \to \infty} \frac{Ks}{(s + j2)(s - j2)(s + 5)} = \lim_{s \to \infty} \frac{K}{s^2}$$

we have

$$\text{Angles of asymptote} = \frac{\pm 180°(2k + 1)}{2} = \pm 90°$$

The intersection of the asymptotes with the real axis can be found from

$$\lim_{s \to \infty} \frac{Ks}{s^3 + 5s^2 + 4s + 20} = \lim_{s \to \infty} \frac{K}{s^2 + 5s + \cdots} = \lim_{s \to \infty} \frac{K}{(s + 2.5)^2}$$

as

$$s = -2.5$$

The angle of departure (angle θ) from the pole at $s = j2$ is obtained as follows:

$$\theta = 180° - 90° - 21.8° + 90° = 158.2°$$

Thus, the angle of departure from the pole $s = j2$ is 158.2°. Figure 6–62 shows a root-locus plot for the system. Notice that two branches of the root locus originate from the poles at $s = \pm j2$ and terminate on the zeros at infinity. The remaining one branch originates from the pole at $s = -5$ and terminates on the zero at $s = 0$.

Note that the closed-loop poles with $\zeta = 0.4$ must lie on straight lines passing through the origin and making the angles $\pm 66.42°$ with the negative real axis. In the present case, there are two intersections of the root-locus branch in the upper half s plane and the straight line of angle 66.42°. Thus, two values of K will give the damping ratio ζ of the closed-loop poles equal to 0.4. At point P, the value of K is

$$K = \left| \frac{(s + j2)(s - j2)(s + 5)}{s} \right|_{s=-1.0490+j2.4065} = 8.9801$$

Hence

$$k = \frac{K}{20} = 0.4490 \qquad \text{at point } P$$

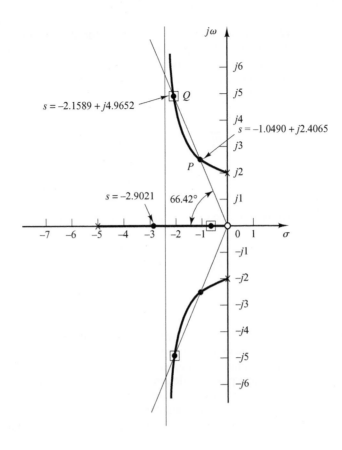

Figure 6–62
Root-locus plot for
the system shown in
Figure 6–61.

At point Q, the value of K is

$$K = \left| \frac{(s + j2)(s - j2)(s + 5)}{s} \right|_{s=-2.1589+j4.9652} = 28.260$$

Hence

$$k = \frac{K}{20} = 1.4130 \qquad \text{at point } Q$$

Thus, we have two solutions for this problem. For $k = 0.4490$, the three closed-loop poles are located at

$$s = -1.0490 + j2.4065, \qquad s = -1.0490 - j2.4065, \qquad s = -2.9021$$

For $k = 1.4130$, the three closed-loop poles are located at

$$s = -2.1589 + j4.9652, \qquad s = -2.1589 - j4.9652, \qquad s = -0.6823$$

It is important to point out that the zero at the origin is the open-loop zero, but not the closed-loop zero. This is evident, because the original system shown in Figure 6–61 does not have a closed-loop zero, since

$$\frac{G(s)}{R(s)} = \frac{20}{s(s + 1)(s + 4) + 20(1 + ks)}$$

The open-loop zero at $s = 0$ was introduced in the process of modifying the characteristic equation such that the adjustable variable $K = 20k$ was to appear as a multiplying factor.

We have obtained two different values of k to satisfy the requirement that the damping ratio of the dominant closed-loop poles be equal to 0.4. The closed-loop transfer function with $k = 0.4490$ is given by

$$\frac{C(s)}{R(s)} = \frac{20}{s^3 + 5s^2 + 12.98s + 20}$$

$$= \frac{20}{(s + 1.0490 + j2.4065)(s + 1.0490 - j2.4065)(s + 2.9021)}$$

The closed-loop transfer function with $k = 1.4130$ is given by

$$\frac{C(s)}{R(s)} = \frac{20}{s^3 + 5s^2 + 32.26s + 20}$$

$$= \frac{20}{(s + 2.1589 + j4.9652)(s + 2.1589 - j4.9652)(s + 0.6823)}$$

Notice that the system with $k = 0.4490$ has a pair of dominant complex-conjugate closed-loop poles, while in the system with $k = 1.4130$ the real closed-loop pole at $s = -0.6823$ is dominant, and the complex-conjugate closed-loop poles are not dominant. In this case, the response characteristic is primarily determined by the real closed-loop pole.

Let us compare the unit-step responses of both systems. MATLAB Program 6–14 may be used for plotting the unit-step response curves in one diagram. The resulting unit-step response curves $[c_1(t)$ for $k = 0.4490$ and $c_2(t)$ for $k = 1.4130]$ are shown in Figure 6–63.

MATLAB Program 6–14

```
% ---------- Unit-step response ----------

% ***** Enter numerators and denominators of systems with
% k = 0.4490 and k = 1.4130, respectively. *****

num1 = [20];
den1 = [1  5  12.98  20];
num2 = [20];
den2 = [1  5  32.26  20];
t = 0:0.1:10;
c1 = step(num1,den1,t);
c2 = step(num2,den2,t);
plot(t,c1,t,c2)
text(2.5,1.12,'k = 0.4490')
text(3.7,0.85,'k = 1.4130')
grid
title('Unit-step Responses of Two Systems')
xlabel('t Sec')
ylabel('Outputs c1 and c2')
```

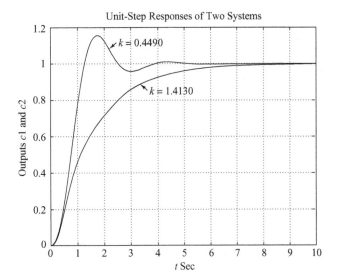

Figure 6-63
Unit-step response curves for the system shown in Figure 6-61 when the damping ratio ζ of the dominant closed-loop poles is set equal to 0.4. (Two possible values of k give the damping ratio ζ equal to 0.4.)

From Figure 6-63 we notice that the response of the system with $k = 0.4490$ is oscillatory. (The effect of the closed-loop pole at $s = -2.9021$ on the unit-step response is small.) For the system with $k = 1.4130$, the oscillations due to the closed-loop poles at $s = -2.1589 \pm j4.9652$ damp out much faster than purely exponential response due to the closed-loop pole at $s = -0.6823$.

The system with $k = 0.4490$ (which exhibits a faster response with relatively small overshoot) has a much better response characteristic than the system with $k = 1.4130$ (which exhibits a slow overdamped response). Therefore, we should choose $k = 0.4490$ for the present system.

EXAMPLE PROBLEMS AND SOLUTIONS

A-6-1. Sketch the root loci for the system shown in Figure 6-64(a). (The gain K is assumed to be positive.) Observe that for small or large values of K the system is overdamped and for medium values of K it is underdamped.

Solution. The procedure for plotting the root loci is as follows:

1. Locate the open-loop poles and zeros on the complex plane. Root loci exist on the negative real axis between 0 and -1 and between -2 and -3.

2. The number of open-loop poles and that of finite zeros are the same. This means that there are no asymptotes in the complex region of the s plane.

3. Determine the breakaway and break-in points. The characteristic equation for the system is

$$1 + \frac{K(s + 2)(s + 3)}{s(s + 1)} = 0$$

or

$$K = -\frac{s(s + 1)}{(s + 2)(s + 3)}$$

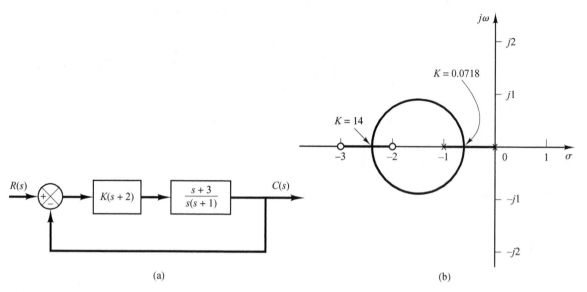

$R(s)$ → $K(s + 2)$ → $\dfrac{s + 3}{s(s + 1)}$ → $C(s)$

(a)

(b)

Figure 6–64
(a) Control system; (b) root-locus plot.

The breakaway and break-in points are determined from

$$\frac{dK}{ds} = -\frac{(2s + 1)(s + 2)(s + 3) - s(s + 1)(2s + 5)}{\left[(s + 2)(s + 3)\right]^2}$$

$$= -\frac{4(s + 0.634)(s + 2.366)}{\left[(s + 2)(s + 3)\right]^2}$$

$$= 0$$

as follows:

$$s = -0.634, \qquad s = -2.366$$

Notice that both points are on root loci. Therefore, they are actual breakaway or break-in points. At point $s = -0.634$, the value of K is

$$K = -\frac{(-0.634)(0.366)}{(1.366)(2.366)} = 0.0718$$

Similarly, at $s = -2.366$,

$$K = -\frac{(-2.366)(-1.366)}{(-0.366)(0.634)} = 14$$

(Because point $s = -0.634$ lies between two poles, it is a breakaway point, and because point $s = -2.366$ lies between two zeros, it is a break-in point.)

Chapter 6 / **Control Systems Analysis and Design by the Root-Locus Method**

4. Determine a sufficient number of points that satisfy the angle condition. (It can be found that the root loci involve a circle with center at -1.5 that passes through the breakaway and break-in points.) The root-locus plot for this system is shown in Figure 6–64(b).

Note that this system is stable for any positive value of K since all the root loci lie in the left-half s plane.

Small values of K $(0 < K < 0.0718)$ correspond to an overdamped system. Medium values of K $(0.0718 < K < 14)$ correspond to an underdamped system. Finally, large values of K $(14 < K)$ correspond to an overdamped system. With a large value of K, the steady state can be reached in much shorter time than with a small value of K.

The value of K should be adjusted so that system performance is optimum according to a given performance index.

A–6–2. Sketch the root loci for the system shown in Figure 6–65(a).

Solution. A root locus exists on the real axis between points $s = -1$ and $s = -3.6$. The asymptotes can be determined as follows:

$$\text{Angles of asymptotes} = \frac{\pm 180°(2k + 1)}{3 - 1} = 90°, -90°$$

The intersection of the asymptotes and the real axis is found from

$$s = -\frac{0 + 0 + 3.6 - 1}{3 - 1} = -1.3$$

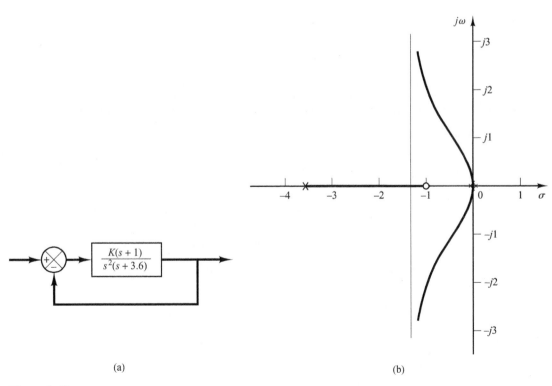

(a)   (b)

Figure 6–65
(a) Control system; (b) root-locus plot.

Since the characteristic equation is

$$s^3 + 3.6s^2 + K(s + 1) = 0$$

we have

$$K = -\frac{s^3 + 3.6s^2}{s + 1}$$

The breakaway and break-in points are found from

$$\frac{dK}{ds} = -\frac{(3s^2 + 7.2s)(s + 1) - (s^3 + 3.6s^2)}{(s + 1)^2} = 0$$

or

$$s^3 + 3.3s^2 + 3.6s = 0$$

from which we get

$$s = 0, \qquad s = -1.65 + j0.9367, \qquad s = -1.65 - j0.9367$$

Point $s = 0$ corresponds to the actual breakaway point. But points $s = 1.65 \pm j0.9367$ are neither breakaway nor break-in points, because the corresponding gain values K become complex quantities.

To check the points where root-locus branches may cross the imaginary axis, substitute $s = j\omega$ into the characteristic equation, yielding.

$$(j\omega)^3 + 3.6(j\omega)^2 + Kj\omega + K = 0$$

or

$$(K - 3.6\omega^2) + j\omega(K - \omega^2) = 0$$

Notice that this equation can be satisfied only if $\omega = 0$, $K = 0$. Because of the presence of a double pole at the origin, the root locus is tangent to the $j\omega$ axis at $\omega = 0$. The root-locus branches do not cross the $j\omega$ axis. Figure 6–65(b) is a sketch of the root loci for this system.

A–6–3. Sketch the root loci for the system shown in Figure 6–66(a).

Solution. A root locus exists on the real axis between point $s = -0.4$ and $s = -3.6$. The angles of asymptotes can be found as follows:

$$\text{Angles of asymptotes} = \frac{\pm 180°(2k + 1)}{3 - 1} = 90°, -90°$$

The intersection of the asymptotes and the real axis is obtained from

$$s = -\frac{0 + 0 + 3.6 - 0.4}{3 - 1} = -1.6$$

Next we shall find the breakaway points. Since the characteristic equation is

$$s^3 + 3.6s^2 + Ks + 0.4K = 0$$

we have

$$K = -\frac{s^3 + 3.6s^2}{s + 0.4}$$

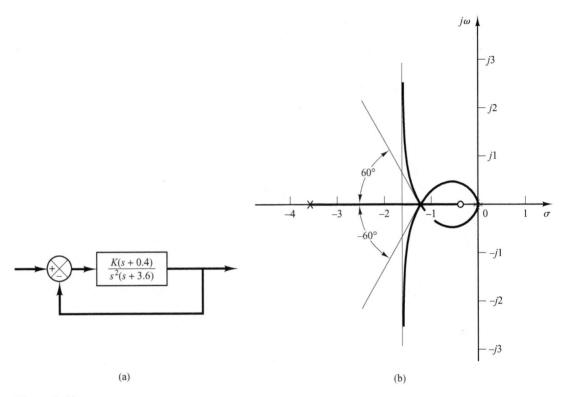

(a)   (b)

Figure 6–66
(a) Control system; (b) root-locus plot.

The breakaway and break-in points are found from

$$\frac{dK}{ds} = -\frac{(3s^2 + 7.2s)(s + 0.4) - (s^3 + 3.6s^2)}{(s + 0.4)^2} = 0$$

from which we get

$$s^3 + 2.4s^2 + 1.44s = 0$$

or

$$s(s + 1.2)^2 = 0$$

Thus, the breakaway or break-in points are at $s = 0$ and $s = -1.2$. Note that $s = -1.2$ is a double root. When a double root occurs in $dK/ds = 0$ at point $s = -1.2$, $d^2K/(ds^2) = 0$ at this point. The value of gain K at point $s = -1.2$ is

$$K = -\frac{s^3 + 3.6s^2}{s + 4}\bigg|_{s=-1.2} = 4.32$$

This means that with $K = 4.32$ the characteristic equation has a triple root at point $s = -1.2$. This can be easily verified as follows:

$$s^3 + 3.6s^2 + 4.32s + 1.728 = (s + 1.2)^3 = 0$$

Example Problems and Solutions   **351**

Hence, three root-locus branches meet at point $s = -1.2$. The angles of departures at point $s = -1.2$ of the root locus branches that approach the asymptotes are $\pm 180°/3$, that is, $60°$ and $-60°$. (See Problem **A–6–4.**)

Finally, we shall examine if root-locus branches cross the imaginary axis. By substituting $s = j\omega$ into the characteristic equation, we have

$$(j\omega)^3 + 3.6(j\omega)^2 + K(j\omega) + 0.4K = 0$$

or

$$(0.4K - 3.6\omega^2) + j\omega(K - \omega^2) = 0$$

This equation can be satisfied only if $\omega = 0$, $K = 0$. At point $\omega = 0$, the root locus is tangent to the $j\omega$ axis because of the presence of a double pole at the origin. There are no points where root-locus branches cross the imaginary axis.

A sketch of the root loci for this system is shown in Figure 6–66(b).

A–6–4. Referring to Problem **A–6–3**, obtain the equations for the root-locus branches for the system shown in Figure 6–66(a). Show that the root-locus branches cross the real axis at the breakaway point at angles $\pm 60°$.

Solution. The equations for the root-locus branches can be obtained from the angle condition

$$\underline{\bigg/ \dfrac{K(s + 0.4)}{s^2(s + 3.6)}} = \pm 180°(2k + 1)$$

which can be rewritten as

$$\underline{\big/ s + 0.4} - 2\underline{\big/ s} - \underline{\big/ s + 3.6} = \pm 180°(2k + 1)$$

By substituting $s = \sigma + j\omega$, we obtain

$$\underline{\big/ \sigma + j\omega + 0.4} - 2\underline{\big/ \sigma + j\omega} - \underline{\big/ \sigma + j\omega + 3.6} = \pm 180°(2k + 1)$$

or

$$\tan^{-1}\left(\dfrac{\omega}{\sigma + 0.4}\right) - 2\tan^{-1}\left(\dfrac{\omega}{\sigma}\right) - \tan^{-1}\left(\dfrac{\omega}{\sigma + 3.6}\right) = \pm 180°(2k + 1)$$

By rearranging, we have

$$\tan^{-1}\left(\dfrac{\omega}{\sigma + 0.4}\right) - \tan^{-1}\left(\dfrac{\omega}{\sigma}\right) = \tan^{-1}\left(\dfrac{\omega}{\sigma}\right) + \tan^{-1}\left(\dfrac{\omega}{\sigma + 3.6}\right) \pm 180°(2k + 1)$$

Taking tangents of both sides of this last equation, and noting that

$$\tan\left[\tan^{-1}\left(\dfrac{\omega}{\sigma + 3.6}\right) \pm 180°(2k + 1)\right] = \dfrac{\omega}{\sigma + 3.6}$$

we obtain

$$\dfrac{\dfrac{\omega}{\sigma + 0.4} - \dfrac{\omega}{\sigma}}{1 + \dfrac{\omega}{\sigma + 0.4}\dfrac{\omega}{\sigma}} = \dfrac{\dfrac{\omega}{\sigma} + \dfrac{\omega}{\sigma + 3.6}}{1 - \dfrac{\omega}{\sigma}\dfrac{\omega}{\sigma + 3.6}}$$

which can be simplified to

$$\dfrac{\omega\sigma - \omega(\sigma + 0.4)}{(\sigma + 0.4)\sigma + \omega^2} = \dfrac{\omega(\sigma + 3.6) + \omega\sigma}{\sigma(\sigma + 3.6) - \omega^2}$$

or

$$\omega(\sigma^3 + 2.4\sigma^2 + 1.44\sigma + 1.6\omega^2 + \sigma\omega^2) = 0$$

which can be further simplified to

$$\omega\left[\sigma(\sigma + 1.2)^2 + (\sigma + 1.6)\omega^2\right] = 0$$

For $\sigma \neq -1.6$, we may write this last equation as

$$\omega\left[\omega - (\sigma + 1.2)\sqrt{\frac{-\sigma}{\sigma + 1.6}}\right]\left[\omega + (\sigma + 1.2)\sqrt{\frac{-\sigma}{\sigma + 1.6}}\right] = 0$$

which gives the equations for the root locus as follows:

$$\omega = 0$$

$$\omega = (\sigma + 1.2)\sqrt{\frac{-\sigma}{\sigma + 1.6}}$$

$$\omega = -(\sigma + 1.2)\sqrt{\frac{-\sigma}{\sigma + 1.6}}$$

The equation $\omega = 0$ represents the real axis. The root locus for $0 \leq K \leq \infty$ is between points $s = -0.4$ and $s = -3.6$. (The real axis other than this line segment and the origin $s = 0$ corresponds to the root locus for $-\infty \leq K < 0$.)

The equations

$$\omega = \pm(\sigma + 1.2)\sqrt{\frac{-\sigma}{\sigma + 1.6}} \qquad (6\text{--}29)$$

represent the complex branches for $0 \leq K \leq \infty$. These two branches lie between $\sigma = -1.6$ and $\sigma = 0$. [See Figure 6–66(b).] The slopes of the complex root-locus branches at the breakaway point ($\sigma = -1.2$) can be found by evaluating $d\omega/d\sigma$ of Equation (6–29) at point $\sigma = -1.2$.

$$\left.\frac{d\omega}{d\sigma}\right|_{\sigma=-1.2} = \pm\sqrt{\frac{-\sigma}{\sigma + 1.6}}\,\bigg|_{\sigma=-1.2} = \pm\sqrt{\frac{1.2}{0.4}} = \pm\sqrt{3}$$

Since $\tan^{-1}\sqrt{3} = 60°$, the root-locus branches intersect the real axis with angles $\pm 60°$.

A–6–5. Consider the system shown in Figure 6–67(a). Sketch the root loci for the system. Observe that for small or large values of K the system is underdamped and for medium values of K it is overdamped.

Solution. A root locus exists on the real axis between the origin and $-\infty$. The angles of asymptotes of the root-locus branches are obtained as

$$\text{Angles of asymptotes} = \frac{\pm 180°(2k + 1)}{3} = 60°, -60°, -180°$$

The intersection of the asymptotes and the real axis is located on the real axis at

$$s = -\frac{0 + 2 + 2}{3} = -1.3333$$

The breakaway and break-in points are found from $dK/ds = 0$. Since the characteristic equation is

$$s^3 + 4s^2 + 5s + K = 0$$

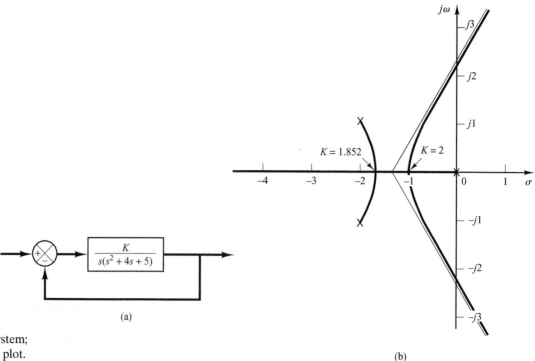

Figure 6–67
(a) Control system;
(b) root-locus plot.

(a)

(b)

we have

$$K = -(s^3 + 4s^2 + 5s)$$

Now we set

$$\frac{dK}{ds} = -(3s^2 + 8s + 5) = 0$$

which yields

$$s = -1, \qquad s = -1.6667$$

Since these points are on root loci, they are actual breakaway or break-in points. (At point $s = -1$, the value of K is 2, and at point $s = -1.6667$, the value of K is 1.852.)

The angle of departure from a complex pole in the upper-half s plane is obtained from

$$\theta = 180° - 153.43° - 90°$$

or

$$\theta = -63.43°$$

The root-locus branch from the complex pole in the upper-half s plane breaks into the real axis at $s = -1.6667$.

Next we determine the points where root-locus branches cross the imaginary axis. By substituting $s = j\omega$ into the characteristic equation, we have

$$(j\omega)^3 + 4(j\omega)^2 + 5(j\omega) + K = 0$$

or

$$(K - 4\omega^2) + j\omega(5 - \omega^2) = 0$$

from which we obtain

$$\omega = \pm \sqrt{5}, \qquad K = 20 \qquad \text{or} \qquad \omega = 0, \qquad K = 0$$

Root-locus branches cross the imaginary axis at $\omega = \sqrt{5}$ and $\omega = -\sqrt{5}$. The root-locus branch on the real axis touches the $j\omega$ axis at $\omega = 0$. A sketch of the root loci for the system is shown in Figure 6–67(b).

Note that since this system is of third order, there are three closed-loop poles. The nature of the system response to a given input depends on the locations of the closed-loop poles.

For $0 < K < 1.852$, there are a set of complex-conjugate closed-loop poles and a real closed-loop pole. For $1.852 \le K \le 2$, there are three real closed-loop poles. For example, the closed-loop poles are located at

$$s = -1.667, \quad s = -1.667, \quad s = -0.667, \quad \text{for } K = 1.852$$

$$s = -1, \quad s = -1, \quad s = -2, \quad \text{for } K = 2$$

For $2 < K$, there are a set of complex-conjugate closed-loop poles and a real closed-loop pole. Thus, small values of K $(0 < K < 1.852)$ correspond to an underdamped system. (Since the real closed-loop pole dominates, only a small ripple may show up in the transient response.) Medium values of K $(1.852 \le K \le 2)$ correspond to an overdamped system. Large values of K $(2 < K)$ correspond to an underdamped system. With a large value of K, the system responds much faster than with a smaller value of K.

A–6–6. Sketch the root loci for the system shown in Figure 6–68(a).

Solution. The open-loop poles are located at $s = 0$, $s = -1$, $s = -2 + j3$, and $s = -2 - j3$. A root locus exists on the real axis between points $s = 0$ and $s = -1$. The angles of the asymptotes are found as follows:

$$\text{Angles of asymptotes} = \frac{\pm 180°(2k + 1)}{4} = 45°, -45°, 135°, -135°$$

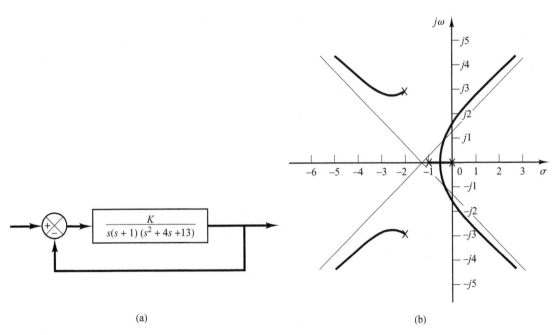

(a)   (b)

Figure 6–68
(a) Control system; (b) root-locus plot.

The intersection of the asymptotes and the real axis is found from

$$s = -\frac{0 + 1 + 2 + 2}{4} = -1.25$$

The breakaway and break-in points are found from $dK/ds = 0$. Noting that

$$K = -s(s + 1)(s^2 + 4s + 13) = -(s^4 + 5s^3 + 17s^2 + 13s)$$

we have

$$\frac{dK}{ds} = -(4s^3 + 15s^2 + 34s + 13) = 0$$

from which we get

$$s = -0.467, \quad s = -1.642 + j2.067, \quad s = -1.642 - j2.067$$

Point $s = -0.467$ is on a root locus. Therefore, it is an actual breakaway point. The gain values K corresponding to points $s = -1.642 \pm j2.067$ are complex quantities. Since the gain values are not real positive, these points are neither breakaway nor break-in points.

The angle of departure from the complex pole in the upper-half s plane is

$$\theta = 180° - 123.69° - 108.44° - 90°$$

or

$$\theta = -142.13°$$

Next we shall find the points where root loci may cross the $j\omega$ axis. Since the characteristic equation is

$$s^4 + 5s^3 + 17s^2 + 13s + K = 0$$

by substituting $s = j\omega$ into it we obtain

$$(j\omega)^4 + 5(j\omega)^3 + 17(j\omega)^2 + 13(j\omega) + K = 0$$

or

$$(K + \omega^4 - 17\omega^2) + j\omega(13 - 5\omega^2) = 0$$

from which we obtain

$$\omega = \pm 1.6125, \quad K = 37.44 \quad \text{or} \quad \omega = 0, \quad K = 0$$

The root-locus branches that extend to the right-half s plane cross the imaginary axis at $\omega = \pm1.6125$. Also, the root-locus branch on the real axis touches the imaginary axis at $\omega = 0$. Figure 6–68(b) shows a sketch of the root loci for the system. Notice that each root-locus branch that extends to the right-half s plane crosses its own asymptote.

A–6–7. Sketch the root loci of the control system shown in Figure 6–69(a). Determine the range of gain K for stability.

Solution. Open-loop poles are located at $s = 1$, $s = -2 + j\sqrt{3}$, and $s = -2 - j\sqrt{3}$. A root locus exists on the real axis between points $s = 1$ and $s = -\infty$. The asymptotes of the root-locus branches are found as follows:

$$\text{Angles of asymptotes} = \frac{\pm 180°(2k + 1)}{3} = 60°, -60°, 180°$$

The intersection of the asymptotes and the real axis is obtained as

$$s = -\frac{-1 + 2 + 2}{3} = -1$$

The breakaway and break-in points can be located from $dK/ds = 0$. Since

$$K = -(s - 1)(s^2 + 4s + 7) = -(s^3 + 3s^2 + 3s - 7)$$

we have

$$\frac{dK}{ds} = -(3s^2 + 6s + 3) = 0$$

which yields

$$(s + 1)^2 = 0$$

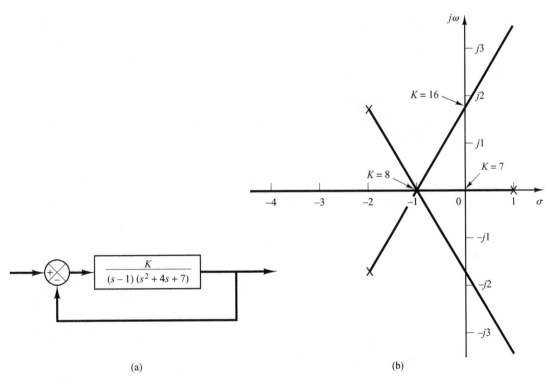

(a)

(b)

Figure 6–69
(a) Control system; (b) root-locus plot.

Thus the equation $dK/ds = 0$ has a double root at $s = -1$. (This means that the characteristic equation has a triple root at $s = -1$.) The breakaway point is located at $s = -1$. Three root-locus branches meet at this breakaway point. The angles of departure of the branches at the breakaway point are $\pm180°/3$—that is, $60°$ and $-60°$.

We shall next determine the points where root-locus branches may cross the imaginary axis. Noting that the characteristic equation is

$$(s - 1)(s^2 + 4s + 7) + K = 0$$

or

$$s^3 + 3s^2 + 3s - 7 + K = 0$$

we substitute $s = j\omega$ into it and obtain

$$(j\omega)^3 + 3(j\omega)^2 + 3(j\omega) - 7 + K = 0$$

By rewriting this last equation, we have

$$(K - 7 - 3\omega^2) + j\omega(3 - \omega^2) = 0$$

This equation is satisfied when

$$\omega = \pm\sqrt{3}, \quad K = 7 + 3\omega^2 = 16 \quad \text{or} \quad \omega = 0, \quad K = 7$$

The root-locus branches cross the imaginary axis at $\omega = \pm\sqrt{3}$ (where $K = 16$) and $\omega = 0$ (where $K = 7$). Since the value of gain K at the origin is 7, the range of gain value K for stability is

$$7 < K < 16$$

Figure 6–69(b) shows a sketch of the root loci for the system. Notice that all branches consist of parts of straight lines.

The fact that the root-locus branches consist of straight lines can be verified as follows: Since the angle condition is

$$\left/ \frac{K}{(s - 1)(s + 2 + j\sqrt{3})(s + 2 - j\sqrt{3})} \right. = \pm180°(2k + 1)$$

we have

$$-\left/ s - 1 \right. - \left/ s + 2 + j\sqrt{3} \right. - \left/ s + 2 - j\sqrt{3} \right. = \pm180°(2k + 1)$$

By substituting $s = \sigma + j\omega$ into this last equation,

$$\left/ \sigma - 1 + j\omega \right. + \left/ \sigma + 2 + j\omega + j\sqrt{3} \right. + \left/ \sigma + 2 + j\omega - j\sqrt{3} \right. = \pm180°(2k + 1)$$

or

$$\left/ \sigma + 2 + j(\omega + \sqrt{3}) \right. + \left/ \sigma + 2 + j(\omega - \sqrt{3}) \right. = -\left/ \sigma - 1 + j\omega \right. \pm 180°(2k + 1)$$

which can be rewritten as

$$\tan^{-1}\left(\frac{\omega + \sqrt{3}}{\sigma + 2}\right) + \tan^{-1}\left(\frac{\omega - \sqrt{3}}{\sigma + 2}\right) = -\tan^{-1}\left(\frac{\omega}{\sigma - 1}\right) \pm 180°(2k + 1)$$

Taking tangents of both sides of this last equation, we obtain

$$\frac{\dfrac{\omega + \sqrt{3}}{\sigma + 2} + \dfrac{\omega - \sqrt{3}}{\sigma + 2}}{1 - \left(\dfrac{\omega + \sqrt{3}}{\sigma + 2}\right)\left(\dfrac{\omega - \sqrt{3}}{\sigma + 2}\right)} = -\frac{\omega}{\sigma - 1}$$

or

$$\frac{2\omega(\sigma + 2)}{\sigma^2 + 4\sigma + 4 - \omega^2 + 3} = -\frac{\omega}{\sigma - 1}$$

which can be simplified to

$$2\omega(\sigma + 2)(\sigma - 1) = -\omega(\sigma^2 + 4\sigma + 7 - \omega^2)$$

or

$$\omega(3\sigma^2 + 6\sigma + 3 - \omega^2) = 0$$

Further simplification of this last equation yields

$$\omega\left(\sigma + 1 + \frac{1}{\sqrt{3}}\omega\right)\left(\sigma + 1 - \frac{1}{\sqrt{3}}\omega\right) = 0$$

which defines three lines:

$$\omega = 0, \qquad \sigma + 1 + \frac{1}{\sqrt{3}}\omega = 0, \qquad \sigma + 1 - \frac{1}{\sqrt{3}}\omega = 0$$

Thus the root-locus branches consist of three lines. Note that the root loci for $K > 0$ consist of portions of the straight lines as shown in Figure 6–69(b). (Note that each straight line starts from an open-loop pole and extends to infinity in the direction of $180°, 60°,$ or $-60°$ measured from the real axis.) The remaining portion of each straight line corresponds to $K < 0$.

A–6–8. Consider a unity-feedback control system with the following feedforward transfer function:

$$G(s) = \frac{K}{s(s + 1)(s + 2)}$$

Using MATLAB, plot the root loci and their asymptotes.

Solution. We shall plot the root loci and asymptotes on one diagram. Since the feedforward transfer function is given by

$$G(s) = \frac{K}{s(s + 1)(s + 2)}$$

$$= \frac{K}{s^3 + 3s^2 + 2s}$$

the equation for the asymptotes may be obtained as follows: Noting that

$$\lim_{s \to \infty} \frac{K}{s^3 + 3s^2 + 2s} \doteq \lim_{s \to \infty} \frac{K}{s^3 + 3s^2 + 3s + 1} = \frac{K}{(s + 1)^3}$$

the equation for the asymptotes may be given by

$$G_a(s) = \frac{K}{(s + 1)^3}$$

Hence, for the system we have

$$\text{num} = [1]$$
$$\text{den} = [1 \ \ 3 \ \ 2 \ \ 0]$$

and for the asymptotes,

$$\text{numa} = [1]$$
$$\text{dena} = [1 \ \ 3 \ \ 3 \ \ 1]$$

In using the following root-locus and plot commands

```
r = rlocus(num,den)
a = rlocus(numa,dena)
plot([r  a])
```

the number of rows of r and that of a must be the same. To ensure this, we include the gain constant K in the commands. For example,

```
K1 = 0:0.1:0.3;
K2 = 0.3:0.005:0.5:
K3 = 0.5:0.5:10;
K4 = 10:5:100;
K = [K1  K2  K3  K4]
r = rlocus(num,den,K)
a = rlocus(numa,dena,K)
y = [r    a]
plot(y, '-')
```

MATLAB Program 6–15 will generate a plot of root loci and their asymptotes as shown in Figure 6–70.

MATLAB Program 6–15

```
% ---------- Root-Locus Plots ----------
num = [1];
den = [1  3  2  0];
numa = [1];
dena = [1  3  3  1];
K1 = 0:0.1:0.3;
K2 = 0.3:0.005:0.5;
K3 = 0.5:0.5:10;
K4 = 10:5:100;
K = [K1  K2  K3  K4];
r = rlocus(num,den,K);
a = rlocus(numa,dena,K);
y = [r    a];
plot(y,'-')
v = [-4  4  -4  4]; axis(v)
grid
title('Root-Locus Plot of G(s) = K/[s(s + 1)(s + 2)] and Asymptotes')
xlabel('Real Axis')
ylabel('Imag Axis')
% ***** Manually draw open-loop poles in the hard copy *****
```

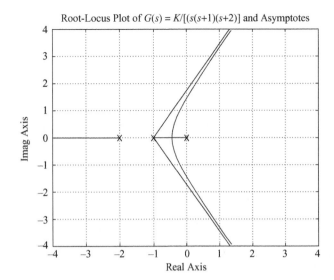

Figure 6–70
Root-locus plot.

Drawing two or more plots in one diagram can also be accomplished by using the hold command. MATLAB Program 6–16 uses the hold command. The resulting root-locus plot is shown in Figure 6–71.

MATLAB Program 6–16

```
% ------------ Root-Locus Plots ------------
num = [1];
den = [1  3  2  0];
numa = [1];
dena = [1  3  3  1];
K1 = 0:0.1:0.3;
K2 = 0.3:0.005:0.5;
K3 = 0.5:0.5:10;
K4 = 10:5:100;
K = [K1  K2  K3  K4];
r = rlocus(num,den,K);
a = rlocus(numa,dena,K);
plot(r,'o')
hold
Current plot held
plot(a,'-')
v = [-4  4  -4  4]; axis(v)
grid
title('Root-Locus Plot of G(s) = K/[s(s+1)(s+2)] and Asymptotes')
xlabel('Real Axis')
ylabel('Imag Axis')
```

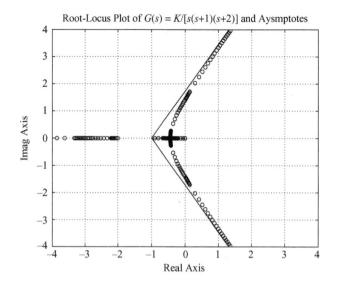

Root-Locus Plot of $G(s) = K/[s(s+1)(s+2)]$ and Aysmptotes

Figure 6–71
Root-locus plot.

A–6–9. Plot the root loci and asymptotes for a unity-feedback system with the following feedforward transfer function:

$$G(s) = \frac{K}{(s^2 + 2s + 2)(s^2 + 2s + 5)}$$

Determine the exact points where the root loci cross the $j\omega$ axis

Solution. The feedforward transfer function $G(s)$ can be written as

$$G(s) = \frac{K}{s^4 + 4s^3 + 11s^2 + 14s + 10}$$

Note that as s approaches infinity, $\lim_{s\to\infty} G(s)$ can be written as

$$\lim_{s\to\infty} G(s) = \lim_{s\to\infty} \frac{K}{s^4 + 4s^3 + 11s^2 + 14s + 10}$$

$$\doteq \lim_{s\to\infty} \frac{K}{s^4 + 4s^3 + 6s^2 + 4s + 1}$$

$$= \lim_{s\to\infty} \frac{K}{(s + 1)^4}$$

where we used the following formula:

$$(s + a)^4 = s^4 + 4as^3 + 6a^2s^2 + 4a^3s + a^4$$

The expression

$$\lim_{s\to\infty} G(s) = \lim_{s\to\infty} \frac{K}{(s + 1)^4}$$

gives the equation for the asymptotes.

Chapter 6 / Control Systems Analysis and Design by the Root-Locus Method

The MATLAB program to plot the root loci of $G(s)$ and the asymptotes is given in MATLAB Program 6–17. Note that the numerator and denominator for $G(s)$ are

$$\text{num} = [1]$$
$$\text{den} = [1 \quad 4 \quad 11 \quad 14 \quad 10]$$

For the numerator and denominator of the asymptotes $\lim\limits_{s \to \infty} G(s)$ we used

$$\text{numa} = [1]$$
$$\text{dena} = [1 \quad 4 \quad 6 \quad 4 \quad 1]$$

Figure 6–72 shows the plot of the root loci and asymptotes.

Since the characteristic equation for the system is

$$(s^2 + 2s + 2)(s^2 + 2s + 5) + K = 0$$

MATLAB Program 6–17

```
% ***** Root-locus plot *****
num = [1];
den = [1   4  11  14   10];
numa = [1];
dena = [1   4  6  4   1];
r = rlocus(num,den);
plot(r,'-')
hold
Current plot held
plot(r,'o')
rlocus(numa,dena);
v = [-6  4  -5  5]; axis(v); axis('square')
grid
title('Plot of Root Loci and Asymptotes')
```

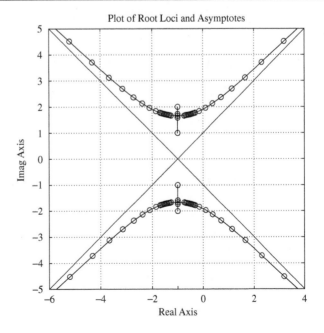

Figure 6–72
Plot of root loci and asymptotes.

the points where the root loci cross the imaginary axis can be found by substituting $s = j\omega$ with the characteristic equation as follows:

$$[(j\omega)^2 + 2j\omega + 2][(j\omega)^2 + 2j\omega + 5] + K$$

$$= (\omega^4 - 11\omega^2 + 10 + K) + j(-4\omega^3 + 14\omega) = 0$$

and equating the imaginary part to zero. The result is

$$\omega = \pm 1.8708$$

Thus the exact points where the root loci cross the $j\omega$ axis are $\omega = \pm 1.8708$. By equating the real part to zero, we get the gain value K at the crossing points to be 16.25.

A–6–10. Consider a unity-feedback control system with the feed-forward transfer function $G(s)$ given by

$$G(s) = \frac{K(s + 1)}{(s^2 + 2s + 2)(s^2 + 2s + 5)}$$

Plot a root-locus diagram with MATLAB.

Solution. The feedforward transfer function $G(s)$ can be written as

$$G(s) = \frac{K(s + 1)}{s^4 + 4s^3 + 11s^2 + 14s + 10}$$

A possible MATLAB program to plot a root-locus diagram is shown in MATLAB Program 6–18. The resulting root-locus plot is shown in Figure 6–73.

MATLAB Program 6–18

```
num = [1  1];
den = [1  4  11  14  10];
K1 = 0:0.1:2;
K2 = 2:0.0.2:2.5;
K3 = 2.5:0.5:10;
K4 = 10:1:50;
K = [K1  K2  K3  K4]
r = rlocus(num,den,K);
plot(r, 'o')
v = [-8  2  -5  5]; axis(v);
grid
title('Root-Locus Plot of G(s) = K(s+1)/[(s^2+2s+2)(s^2+2s+5)]')
xlabel('Real Axis')
ylabel('Imag Axis')
```

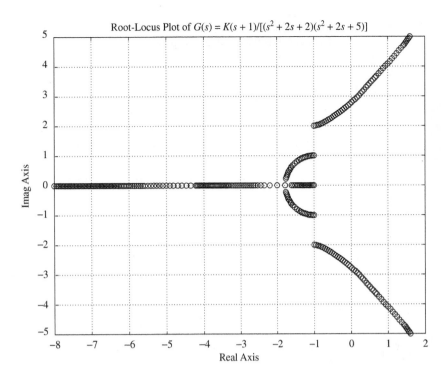

Figure 6–73
Plot of root loci.

A–6–11. Obtain the transfer function of the mechanical system shown in Figure 6–74. Assume that the displacement x_i is the input and displacement x_o is the output of the system.

Solution. From the diagram we obtain the following equations of motion:

$$b_2(\dot{x}_i - \dot{x}_o) = b_1(\dot{x}_o - \dot{y})$$
$$b_1(\dot{x}_o - \dot{y}) = ky$$

Taking the Laplace transforms of these two equations, assuming zero initial conditions, and then eliminating $Y(s)$, we obtain

$$\frac{X_o(s)}{X_i(s)} = \frac{b_2}{b_1 + b_2} \cdot \frac{\dfrac{b_1}{k}s + 1}{\dfrac{b_2}{b_1 + b_2}\dfrac{b_1}{k}s + 1}$$

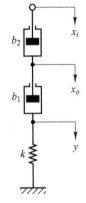

Figure 6–74
Mechanical system.

This is the transfer function between $X_o(s)$ and $X_i(s)$. By defining

$$\frac{b_1}{k} = T, \qquad \frac{b_2}{b_1 + b_2} = \alpha < 1$$

we obtain

$$\frac{X_o(s)}{X_i(s)} = \alpha \frac{Ts + 1}{\alpha Ts + 1} = \frac{s + \dfrac{1}{T}}{s + \dfrac{1}{\alpha T}}$$

This mechanical system is a mechanical lead network.

Example Problems and Solutions

Obtain the transfer function of the mechanical system shown in Figure 6–75. Assume that the displacement x_i is the input and displacement x_o is the output.

Solution. The equations of motion for this system are

$$b_2(\dot{x}_i - \dot{x}_o) + k_2(x_i - x_o) = b_1(\dot{x}_o - \dot{y})$$

$$b_1(\dot{x}_o - \dot{y}) = k_1 y$$

By taking the Laplace transforms of these two equations, assuming zero initial conditions, we obtain

$$b_2[sX_i(s) - sX_o(s)] + k_2[X_i(s) - X_o(s)] = b_1[sX_o(s) - sY(s)]$$

$$b_1[sX_o(s) - sY(s)] = k_1 Y(s)$$

If we eliminate $Y(s)$ from the last two equations, the transfer function $X_o(s)/X_i(s)$ can be obtained as

$$\frac{X_o(s)}{X_i(s)} = \frac{\left(\dfrac{b_1}{k_1}s + 1\right)\left(\dfrac{b_2}{k_2}s + 1\right)}{\left(\dfrac{b_1}{k_1}s + 1\right)\left(\dfrac{b_2}{k_2}s + 1\right) + \dfrac{b_1}{k_2}s}$$

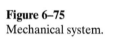

Figure 6–75
Mechanical system.

Define

$$T_1 = \frac{b_1}{k_1}, \qquad T_2 = \frac{b_2}{k_2},$$

If k_1, k_2, b_1, and b_2 are chosen such that there exists a β that satisfies the following equation:

$$\frac{b_1}{k_1} + \frac{b_2}{k_2} + \frac{b_1}{k_2} = \frac{T_1}{\beta} + \beta T_2 \qquad (\beta > 1) \tag{6-30}$$

then $X_o(s)/X_i(s)$ can be obtained as

$$\frac{X_o(s)}{X_i(s)} = \frac{(T_1 s + 1)(T_2 s + 1)}{\left(\dfrac{T_1}{\beta}s + 1\right)(\beta T_2 s + 1)} = \frac{\left(s + \dfrac{1}{T_1}\right)\left(s + \dfrac{1}{T_2}\right)}{\left(s + \dfrac{\beta}{T_1}\right)\left(s + \dfrac{1}{\beta T_2}\right)}$$

[Note that depending on the choice of k_1, k_2, b_1, and b_2, there does not exist a β that satisfies Equation (6–30).]

If such a β exists and if for a given s_1 (where $s = s_1$ is one of the dominant closed-loop poles of the control system to which we wish to use this mechanical device) the following conditions are satisfied:

$$\left| \frac{s_1 + \dfrac{1}{T_2}}{s_1 + \dfrac{1}{\beta T_2}} \right| \doteq 1, \qquad -5° < \left/ \frac{s_1 + \dfrac{1}{T_2}}{s_1 + \dfrac{1}{\beta T_2}} \right. < 0°$$

then the mechanical system shown in Figure 6–75 acts as a lag–lead compensator.

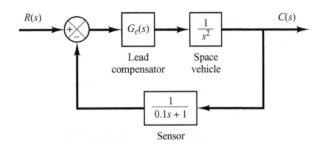

Figure 6–76
Space-vehicle control
system.

A–6–13. Consider the model for a space-vehicle control system shown in Figure 6–76. Design a lead compensator $G_c(s)$ such that the damping ratio ζ and the undamped natural frequency ω_n of the dominant closed-loop poles are 0.5 and 2 rad/sec, respectively.

Solution.
First Attempt: Assume the lead compensator $G_c(s)$ to be

$$G_c(s) = K_c \left(\frac{s + \dfrac{1}{T}}{s + \dfrac{1}{\alpha T}} \right) \qquad (0 < \alpha < 1)$$

From the given specifications, $\zeta = 0.5$ and $\omega_n = 2$ rad/sec, the dominant closed-loop poles must be located at

$$s = -1 \pm j\sqrt{3}$$

We first calculate the angle deficiency at this closed-loop pole.

$$\text{Angle deficiency} = -120° - 120° - 10.8934° + 180°$$

$$= -70.8934°$$

This angle deficiency must be compensated by the lead compensator. There are many ways to determine the locations of the pole and zero of the lead network. Let us choose the zero of the compensator at $s = -1$. Then, referring to Figure 6–77, we have the following equation:

$$\frac{1.73205}{x - 1} = \tan(90° - 70.8934°) = 0.34641$$

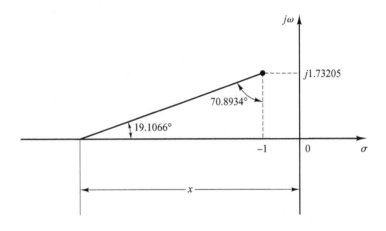

Figure 6–77
Determination of the
pole of the lead
network.

or

$$x = 1 + \frac{1.73205}{0.34641} = 6$$

Hence,

$$G_c(s) = K_c \frac{s + 1}{s + 6}$$

The value of K_c can be determined from the magnitude condition

$$K_c \left| \frac{s + 1}{s + 6} \frac{1}{s^2} \frac{1}{0.1s + 1} \right|_{s=-1+j\sqrt{3}} = 1$$

as follows:

$$K_c = \left| \frac{(s + 6)s^2(0.1s + 1)}{s + 1} \right|_{s=-1+j\sqrt{3}} = 11.2000$$

Thus

$$G_c(s) = 11.2 \frac{s + 1}{s + 6}$$

Since the open-loop transfer function becomes

$$G_c(s)G(s)H(s) = 11.2 \frac{s + 1}{(s + 6)s^2(0.1s + 1)}$$

$$= \frac{11.2(s + 1)}{0.1s^4 + 1.6s^3 + 6s^2}$$

a root-locus plot of the compensated system can be obtained easily with MATLAB by entering num and den and using rlocus command. The result is shown in Figure 6–78.

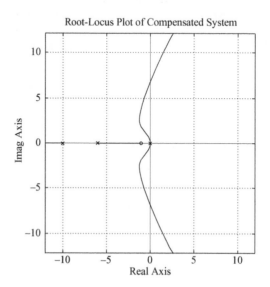

Figure 6–78
Root-locus plot of the compensated system.

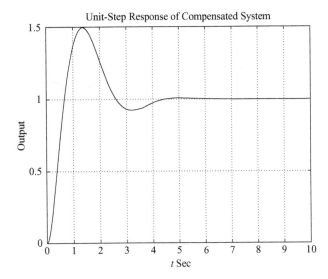

Figure 6–79
Unit-step response of
the compensated
system.

The closed-loop transfer function for the compensated system becomes

$$\frac{C(s)}{R(s)} = \frac{11.2(s + 1)(0.1s + 1)}{(s + 6)s^2(0.1s + 1) + 11.2(s + 1)}$$

Figure 6–79 shows the unit-step response curve. Even though the damping ratio of the dominant closed-loop poles is 0.5, the amount of overshoot is very much higher than expected. A closer look at the root-locus plot reveals that the presence of the zero at $s = -1$ is increasing the amount of the maximum overshoot. [In general, if a closed-loop zero or zeros (compensator zero or zeros) lie to the right of the dominant pair of the complex poles, then the dominant poles are no longer dominant.] If large maximum overshoot cannot be tolerated, the compensator zero(s) should be shifted sufficiently to the left.

In the current design, it is desirable to modify the compensator and make the maximum overshoot smaller. This can be done by modifying the lead compensator, as presented in the following second attempt.

Second Attempt: To modify the shape of the root loci, we may use two lead networks, each contributing half the necessary lead angle, which is 70.8934°/2 = 35.4467°. Let us choose the location of the zeros at $s = -3$. (This is an arbitrary choice. Other choices such as $s = -2.5$ and $s = -4$ may be made.)

Once we choose two zeros at $s = -3$, the necessary location of the poles can be determined as shown in Figure 6–80, or

$$\frac{1.73205}{y - 1} = \tan(40.89334° - 35.4467°)$$

$$= \tan 5.4466° = 0.09535$$

which yields

$$y = 1 + \frac{1.73205}{0.09535} = 19.1652$$

Example Problems and Solutions

369

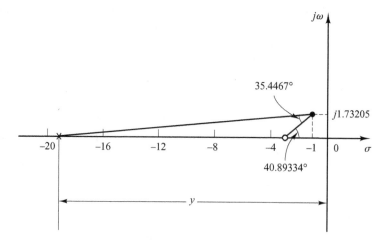

Figure 6–80
Determination of the pole of the lead network.

Hence, the lead compensator will have the following transfer function:

$$G_c(s) = K_c \left(\frac{s + 3}{s + 19.1652} \right)^2$$

The value of K_c can be determined from the magnitude condition as follows:

$$\left| K_c \left(\frac{s + 3}{s + 19.1652} \right)^2 \frac{1}{s^2} \frac{1}{0.1s + 1} \right|_{s=-1+j\sqrt{3}} = 1$$

or

$$K_c = 174.3864$$

Then the lead compensator just designed is

$$G_c(s) = 174.3864 \left(\frac{s + 3}{s + 19.1652} \right)^2$$

Then the open-loop transfer function becomes

$$G_c(s)G(s)H(s) = 174.3864 \left(\frac{s + 3}{s + 19.1652} \right)^2 \frac{1}{s^2} \frac{1}{0.1s + 1}$$

A root-locus plot for the compensated system is shown in Figure 6–81(a). Notice that there is no closed-loop zero near the origin. An expanded view of the root-locus plot near the origin is shown in Figure 6–81(b).

The closed-loop transfer function becomes

$$\frac{C(s)}{R(s)} = \frac{174.3864(s + 3)^2(0.1s + 1)}{(s + 19.1652)^2 s^2(0.1s + 1) + 174.3864(s + 3)^2}$$

The closed-loop poles are found as follows:

$$s = -1 \pm j1.73205$$

$$s = -9.1847 \pm j7.4814$$

$$s = -27.9606$$

Root-Locus Plot of Compensated System

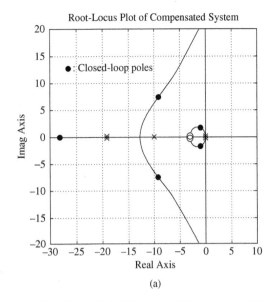

(a)

Root-Locus Plot of Compensated System near Origin

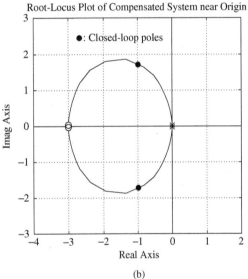

(b)

Figure 6–81
(a) Root-locus plot of compensated system; (b) root-locus plot near the origin.

Figures 6–82(a) and (b) show the unit-step response and unit-ramp response of the compensated system. The unit-step response curve is reasonable and the unit-ramp response looks acceptable. Notice that in the unit-ramp response the output leads the input by a small amount. This is because the system has a feedback transfer function $1/(0.1s + 1)$. If the feedback signal versus t is plotted, together with the unit-ramp input, the former will not lead the input ramp at steady state. See Figure 6–82(c).

Example Problems and Solutions

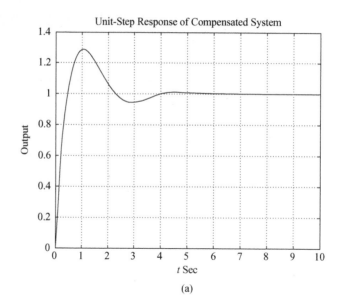

Figure 6–82
(a) unit-step
response of the
compensated system;
(b) unit-ramp
response of the
compensated system;
(c) a plot of feedback
signal versus t in the
unit-ramp response.

(a)

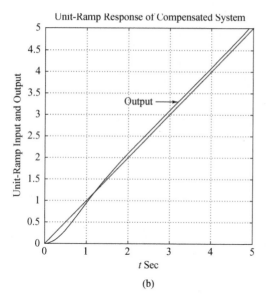

(b)

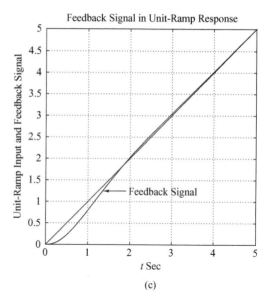

(c)

A–6–14. Consider a system with an unstable plant as shown in Figure 6–83(a). Using the root-locus approach, design a proportional-plus-derivative controller (that is, determine the values of K_p and T_d) such that the damping ratio ζ of the closed-loop system is 0.7 and the undamped natural frequency ω_n is 0.5 rad/sec.

Solution. Note that the open-loop transfer function involves two poles at $s = 1.085$ and $s = -1.085$ and one zero at $s = -1/T_d$, which is unknown at this point.

Since the desired closed-loop poles must have $\omega_n = 0.5$ rad/sec and $\zeta = 0.7$, they must be located at

$$s = 0.5 \, \underline{/180° \pm 45.573°}$$

Chapter 6 / **Control Systems Analysis and Design by the Root-Locus Method**

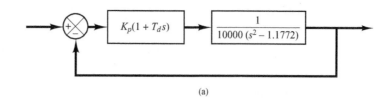

(a)

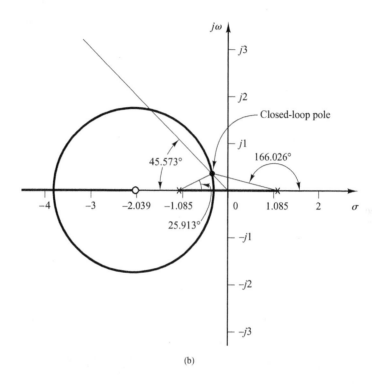

Figure 6–83
(a) PD control of an
unstable plant;
(b) root-locus
diagram for the
system.

(b)

($\zeta = 0.7$ corresponds to a line having an angle of 45.573° with the negative real axis.) Hence, the desired closed-loop poles are at

$$s = -0.35 \pm j0.357$$

The open-loop poles and the desired closed-loop pole in the upper half-plane are located in the diagram shown in Figure 6–83(b). The angle deficiency at point $s = -0.35 + j0.357$ is

$$-166.026° - 25.913° + 180° = -11.939°$$

This means that the zero at $s = -1/T_d$ must contribute 11.939°, which, in turn, determines the location of the zero as follows:

$$s = -\frac{1}{T_d} = -2.039$$

Example Problems and Solutions

373

Hence, we have

$$K_p(1 + T_d s) = K_p T_d \left(\frac{1}{T_d} + s \right) = K_p T_d (s + 2.039) \tag{6-31}$$

The value of T_d is

$$T_d = \frac{1}{2.039} = 0.4904$$

The value of gain K_p can be determined from the magnitude condition as follows:

$$\left| K_p T_d \frac{s + 2.039}{10000(s^2 - 1.1772)} \right|_{s=-0.35+j0.357} = 1$$

or

$$K_p T_d = 6999.5$$

Hence,

$$K_p = \frac{6999.5}{0.4904} = 14{,}273$$

By substituting the numerical values of T_d and K_p into Equation (6–31), we obtain

$$K_p(1 + T_d s) = 14{,}273(1 + 0.4904s) = 6999.5(s + 2.039)$$

which gives the transfer function of the desired proportional-plus-derivative controller.

A–6–15. Consider the control system shown in Figure 6–84. Design a lag compensator $G_c(s)$ such that the static velocity error constant K_v is 50 sec^{-1} without appreciably changing the location of the original closed-loop poles, which are at $s = -2 \pm j\sqrt{6}$.

Solution. Assume that the transfer function of the lag compensator is

$$G_c(s) = \hat{K}_c \frac{s + \dfrac{1}{T}}{s + \dfrac{1}{\beta T}} \qquad (\beta > 1)$$

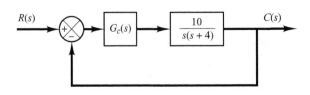

Figure 6–84
Control system.

Since K_v is specified as 50 sec^{-1}, we have

$$K_v = \lim_{s \to 0} sG_c(s) \frac{10}{s(s+4)} = \hat{K}_c \beta 2.5 = 50$$

Thus

$$\hat{K}_c \beta = 20$$

Now choose $\hat{K}_c = 1$. Then

$$\beta = 20$$

Choose $T = 10$. Then the lag compensator can be given by

$$G_c(s) = \frac{s + 0.1}{s + 0.005}$$

The angle contribution of the lag compensator at the closed-loop pole $s = -2 + j\sqrt{6}$ is

$$\underline{/G_c(s)}\bigg|_{s=-2+j\sqrt{6}} = \tan^{-1}\frac{\sqrt{6}}{-1.9} - \tan^{-1}\frac{\sqrt{6}}{-1.995}$$

$$= -1.3616°$$

which is small. The magnitude of $G_c(s)$ at $s = -2 + j6$ is 0.981. Hence the change in the location of the dominant closed-loop poles is very small.

The open-loop transfer function of the system becomes

$$G_c(s)G(s) = \frac{s + 0.1}{s + 0.005} \frac{10}{s(s+4)}$$

The closed-loop transfer function is

$$\frac{C(s)}{R(s)} = \frac{10s + 1}{s^3 + 4.005s^2 + 10.02s + 1}$$

To compare the transient-response characteristics before and after the compensation, the unit-step and unit-ramp responses of the compensated and uncompensated systems are shown in Figures 6–85(a) and (b), respectively. The steady-state error in the unit-ramp response is shown in Figure 6–85(c). The designed lag compensator is acceptable.

Example Problems and Solutions

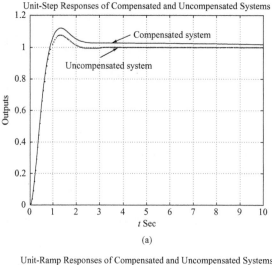

(a)

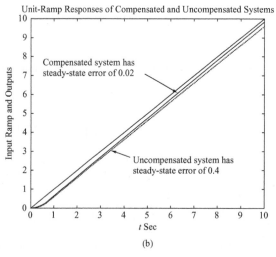

(b)

Figure 6–85
(a) Unit-step responses of the compensated and uncompensated systems; (b) unit-ramp responses of both systems; (c) unit-ramp responses showing steady-state errors.

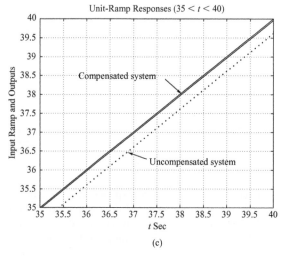

(c)

A–6–16. Consider a unity-feedback control system whose feedforward transfer function is given by

$$G(s) = \frac{10}{s(s + 2)(s + 8)}$$

Design a compensator such that the dominant closed-loop poles are located at $s = -2 \pm j2\sqrt{3}$ and the static velocity error constant K_v is equal to 80 sec^{-1}.

Solution. The static velocity error constant of the uncompensated system is $K_v = \frac{10}{16} = 0.625$. Since $K_v = 80$ is required, we need to increase the open-loop gain by 128. (This implies that we need a lag compensator.) The root-locus plot of the uncompensated system reveals that it is not possible to bring the dominant closed-loop poles to $-2 \pm j2\sqrt{3}$ by just a gain adjustment alone. See Figure 6–86. (This means that we also need a lead compensator.) Therefore, we shall employ a lag–lead compensator.

Let us assume the transfer function of the lag–lead compensator to be

$$G_c(s) = K_c \left(\frac{s + \dfrac{1}{T_1}}{s + \dfrac{\beta}{T_1}} \right) \left(\frac{s + \dfrac{1}{T_2}}{s + \dfrac{1}{\beta T_2}} \right)$$

where $K_c = 128$. This is because

$$K_v = \lim_{s \to 0} sG_c(s)G(s) = \lim_{s \to 0} sK_cG(s) = K_c \frac{10}{16} = 80$$

and we obtain $K_c = 128$. The angle deficiency at the desired closed-loop pole $s = -2 + j2\sqrt{3}$ is

$$\text{Angle deficiency} = -120° - 90° - 30° + 180° = -60°$$

The lead portion of the lag–lead compensator must contribute 60°. To choose T_1 we may use the graphical method presented in Section 6–8.

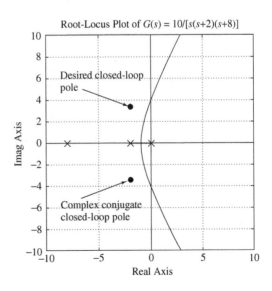

Figure 6–86
Root-locus plot of $G(s) = 10/ [s(s + 2)(s + 8)]$.

The lead portion must satisfy the following conditions:

$$\left| 128 \left(\frac{s_1 + \dfrac{1}{T_1}}{s_1 + \dfrac{\beta}{T_1}} \right) G(s_1) \right|_{s_1 = -2 + j2\sqrt{3}} = 1$$

and

$$\left/ \frac{s_1 + \dfrac{1}{T_1}}{s_1 + \dfrac{\beta}{T_1}} \right._{s_1 = -2 + j2\sqrt{3}} = 60°$$

The first condition can be simplified as

$$\left| \frac{s_1 + \dfrac{1}{T_1}}{s_1 + \dfrac{\beta}{T_1}} \right|_{s_1 = -2 + j2\sqrt{3}} = \frac{1}{13.3333}$$

By using the same approach as used in Section 6–8, the zero $(s = 1/T_1)$ and pole $(s = \beta/T_1)$ can be determined as follows:

$$\frac{1}{T_1} = 3.70, \qquad \frac{\beta}{T_1} = 53.35$$

See Figure 6–87. The value of β is thus determined as

$$\beta = 14.419$$

For the lag portion of the compensator, we may choose

$$\frac{1}{\beta T_2} = 0.01$$

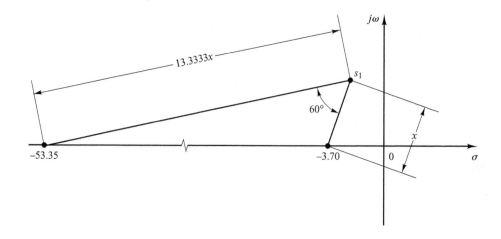

Figure 6–87
Graphical
determination of the
zero and pole of
the lead portion
of the compensator.

Then

$$\frac{1}{T_2} = 0.1442$$

Noting that

$$\left|\frac{s_1 + 0.1442}{s_1 + 0.01}\right|_{s_1=-2+j2\sqrt{3}} = 0.9837$$

$$\left/\frac{s_1 + 0.1442}{s_1 + 0.01}\right._{s_1=-2+j2\sqrt{3}} = -1.697°$$

the angle contribution of the lag portion is $-1.697°$ and the magnitude contribution is 0.9837. This means that the dominant closed-loop poles lie close to the desired location $s = -2 \pm j2\sqrt{3}$. Thus the compensator designed,

$$G_c(s) = 128\left(\frac{s + 3.70}{s + 53.35}\right)\left(\frac{s + 0.1442}{s + 0.01}\right)$$

is acceptable. The feedforward transfer function of the compensated system becomes

$$G_c(s)G(s) = \frac{1280(s + 3.7)(s + 0.1442)}{s(s + 53.35)(s + 0.01)(s + 2)(s + 8)}$$

A root-locus plot of the compensated system is shown in Figure 6–88(a). An enlarged root-locus plot near the origin is shown in Figure 6–88(b).

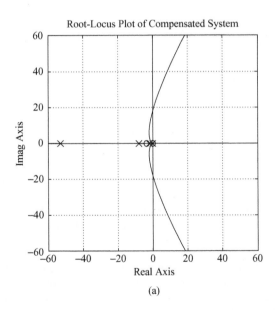

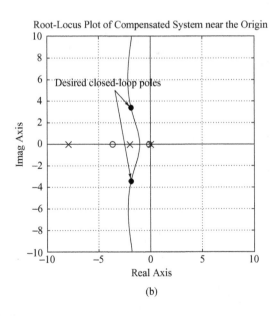

(a)   (b)

Figure 6–88
(a) Root-locus plot of compensated system; (b) root-locus plot near the origin.

Example Problems and Solutions

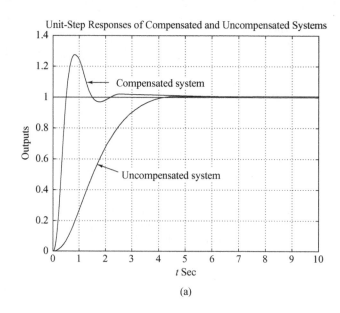

Unit-Step Responses of Compensated and Uncompensated Systems

Compensated system

Uncompensated system

t Sec

(a)

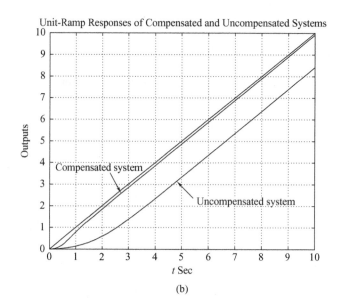

Unit-Ramp Responses of Compensated and Uncompensated Systems

Compensated system

Uncompensated system

t Sec

(b)

Figure 6–89
(a) Unit-step responses of compensated and uncompensated systems; (b) unit-ramp responses of both systems.

To verify the improved system performance of the compensated system, see the unit-step responses and unit-ramp responses of the compensated and uncompensated systems shown in Figures 6–89 (a) and (b), respectively.

A–6–17. Consider the system shown in Figure 6–90. Design a lag–lead compensator such that the static velocity error constant K_v is 50 sec^{-1} and the damping ratio ζ of the dominant closed-loop poles is 0.5. (Choose the zero of the lead portion of the lag–lead compensator to cancel the pole at $s = -1$ of the plant.) Determine all closed-loop poles of the compensated system.

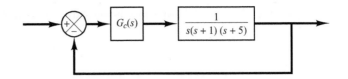

Figure 6–90
Control system.

Solution. Let us employ the lag–lead compensator given by

$$G_c(s) = K_c \left(\frac{s + \dfrac{1}{T_1}}{s + \dfrac{\beta}{T_1}} \right) \left(\frac{s + \dfrac{1}{T_2}}{s + \dfrac{1}{\beta T_2}} \right) = K_c \frac{(T_1 s + 1)(T_2 s + 1)}{\left(\dfrac{T_1}{\beta} s + 1 \right)(\beta T_2 s + 1)}$$

where $\beta > 1$. Then

$$K_v = \lim_{s \to 0} s G_c(s) G(s)$$

$$= \lim_{s \to 0} s \frac{K_c(T_1 s + 1)(T_2 s + 1)}{\left(\dfrac{T_1}{\beta} s + 1 \right)(\beta T_2 s + 1)} \frac{1}{s(s + 1)(s + 5)}$$

$$= \frac{K_c}{5}$$

The specification that $K_v = 50 \text{ sec}^{-1}$ determines the value of K_c, or

$$K_c = 250$$

We now choose $T_1 = 1$ so that $s + (1/T_1)$ will cancel the $(s + 1)$ term of the plant. The lead portion then becomes

$$\frac{s + 1}{s + \beta}$$

For the lag portion of the lag–lead compensator we require

$$\left| \frac{s_1 + \dfrac{1}{T_2}}{s_1 + \dfrac{1}{\beta T_2}} \right| \doteq 1, \qquad -5° < \left/ \frac{s_1 + \dfrac{1}{T_2}}{s_1 + \dfrac{1}{\beta T_2}} \right. < 0°$$

where $s = s_1$ is one of the dominant closed-loop poles. Noting these requirements for the lag portion of the compensator, at $s = s_1$, the open-loop transfer function becomes

$$G_c(s_1)G(s_1) \doteq K_c \left(\frac{s_1 + 1}{s_1 + \beta} \right) \frac{1}{s_1(s_1 + 1)(s_1 + 5)} = K_c \frac{1}{s_1(s_1 + \beta)(s_1 + 5)}$$

Example Problems and Solutions

Then at $s = s_1$, the following magnitude and angle conditions must be satisfied:

$$\left| K_c \frac{1}{s_1(s_1 + \beta)(s_1 + 5)} \right| = 1 \qquad (6\text{–}32)$$

$$\underline{/K_c \frac{1}{s_1(s_1 + \beta)(s_1 + 5)}} = \pm 180°(2k + 1) \qquad (6\text{–}33)$$

where $k = 0, 1, 2, \ldots$. In Equations (6–32) and (6–33), β and s_1 are unknowns. Since the damping ratio ζ of the dominant closed-loop poles is specified as 0.5, the closed-loop pole $s = s_1$ can be written as

$$s_1 = -x + j\sqrt{3}x$$

where x is as yet undetermined.

Notice that the magnitude condition, Equation (6–32), can be rewritten as

$$\left| \frac{K_c}{(-x + j\sqrt{3}x)(-x + \beta + j\sqrt{3}x)(-x + 5 + j\sqrt{3}x)} \right| = 1$$

Noting that $K_c = 250$, we have

$$x\sqrt{(\beta - x)^2 + 3x^2} \ \sqrt{(5 - x)^2 + 3x^2} = 125 \qquad (6\text{–}34)$$

The angle condition, Equation (6–33), can be rewritten as

$$\underline{/K_c \frac{1}{(-x + j\sqrt{3}x)(-x + \beta + j\sqrt{3}x)(-x + 5 + j\sqrt{3}x)}}$$

$$= -120° - \tan^{-1}\left(\frac{\sqrt{3}x}{-x + \beta} \right) - \tan^{-1}\left(\frac{\sqrt{3}x}{-x + 5} \right) = -180°$$

or

$$\tan^{-1}\left(\frac{\sqrt{3}x}{-x + \beta} \right) + \tan^{-1}\left(\frac{\sqrt{3}x}{-x + 5} \right) = 60° \qquad (6\text{–}35)$$

We need to solve Equations (6–34) and (6–35) for β and x. By several trial-and-error calculations, it can be found that

$$\beta = 16.025, \qquad x = 1.9054$$

Thus

$$s_1 = -1.9054 + j\sqrt{3}\,(1.9054) = -1.9054 + j3.3002$$

The lag portion of the lag–lead compensator can be determined as follows: Noting that the pole and zero of the lag portion of the compensator must be located near the origin, we may choose

$$\frac{1}{\beta T_2} = 0.01$$

That is,

$$\frac{1}{T_2} = 0.16025 \qquad \text{or} \qquad T_2 = 6.24$$

With the choice of $T_2 = 6.24$, we find

$$\left| \frac{s_1 + \dfrac{1}{T_2}}{s_1 + \dfrac{1}{\beta T_2}} \right| = \left| \frac{-1.9054 + j3.3002 + 0.16025}{-1.9054 + j3.3002 + 0.01} \right|$$

$$= \left| \frac{-1.74515 + j3.3002}{-1.89054 + j3.3002} \right| = 0.98 \doteq 1 \qquad (6\text{--}36)$$

and

$$\underline{\left/ \frac{s_1 + \dfrac{1}{T_2}}{s_1 + \dfrac{1}{\beta T_2}} \right.} = \underline{\left/ \frac{-1.9054 + j3.3002 + 0.16025}{-1.9054 + j3.3002 + 0.01} \right.}$$

$$= \tan^{-1}\left(\frac{3.3002}{-1.74515} \right) - \tan^{-1}\left(\frac{3.3002}{-1.89054} \right) = -1.937° \qquad (6\text{--}37)$$

Since

$$-5° < -1.937° < 0°$$

our choice of $T_2 = 6.24$ is acceptable. Then the lag–lead compensator just designed can be written as

$$G_c(s) = 250\left(\frac{s + 1}{s + 16.025} \right)\left(\frac{s + 0.16025}{s + 0.01} \right)$$

Therefore, the compensated system has the following open-loop transfer function:

$$G_c(s)G(s) = \frac{250(s + 0.16025)}{s(s + 0.01)(s + 5)(s + 16.025)}$$

A root-locus plot of the compensated system is shown in Figure 6–91(a). An enlarged root-locus plot near the origin is shown in Figure 6–91(b).

The closed loop transfer function becomes

$$\frac{C(s)}{R(s)} = \frac{250(s + 0.16025)}{s(s + 0.01)(s + 5)(s + 16.025) + 250(s + 0.16025)}$$

The closed-loop poles are located at

$$s = -1.8308 \pm j3.2359$$
$$s = -0.1684$$
$$s = -17.205$$

Notice that the dominant closed-loop poles $s = -1.8308 \pm j3.2359$ differ from the dominant closed-loop poles $s = \pm s_1$ assumed in the computation of β and T_2. Small deviations of the dominant closed-loop poles $s = -1.8308 \pm j3.2359$ from $s = \pm s_1 = -1.9054 \pm j3.3002$ are due to the approximations involved in determining the lag portion of the compensator. [See Equations (6–36) and (6–37).]

Example Problems and Solutions   **383**

Root-Locus Plot of Compensated System

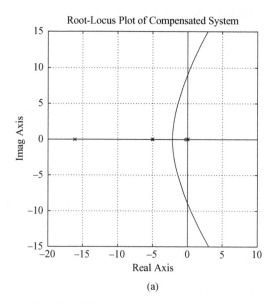

(a)

Root-Locus Plot of Compensated System near the Origin

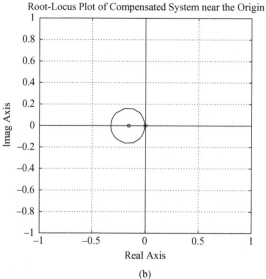

(b)

Figure 6–91
(a) Root-locus plot of compensated system; (b) root-locus plot near the origin.

Figures 6–92(a) and (b) show the unit-step response and unit-ramp response of the designed system, respectively. Note that the closed-loop pole at $s = -0.1684$ almost cancels the zero at $s = -0.16025$. However, this pair of closed-loop pole and zero located near the origin produces a long tail of small amplitude. Since the closed-loop pole at $s = -17.205$ is located very much farther to the left compared to the closed-loop poles at $s = -1.8308 \pm j3.2359$, the effect of this real pole on the system response is very small. Therefore, the closed-loop poles at $s = -1.8308 \pm j3.2359$ are indeed dominant closed-loop poles that determine the response characteristics of the closed-loop system. In the unit-ramp response, the steady-state error in following the unit-ramp input eventually becomes $1/K_v = \frac{1}{50} = 0.02$.

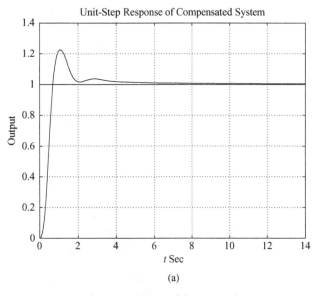

(a)

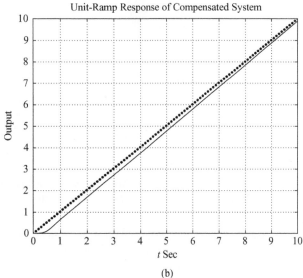

Figure 6–92
(a) Unit-step response of the compensated system; (b) unit-ramp response of the compensated system.

(b)

A–6–18. Figure 6–93(a) is a block diagram of a model for an attitude-rate control system. The closed-loop transfer function for this system is

$$\frac{C(s)}{R(s)} = \frac{2s + 0.1}{s^3 + 0.1s^2 + 6s + 0.1}$$

$$= \frac{2(s + 0.05)}{(s + 0.0417 + j2.4489)(s + 0.0417 - j2.4489)(s + 0.0167)}$$

The unit-step response of this system is shown in Figure 6–93(b). The response shows high-frequency oscillations at the beginning of the response due to the poles at $s = -0.0417 \pm j2.4489$. The response is dominated by the pole at $s = -0.0167$. The settling time is approximately 240 sec.

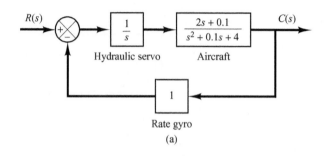

R(s) $\xrightarrow{}$ $\boxed{+ \bigotimes -}$ $\xrightarrow{}$ $\boxed{\dfrac{1}{s}}$ $\xrightarrow{}$ $\boxed{\dfrac{2s + 0.1}{s^2 + 0.1s + 4}}$ $\xrightarrow{}$ C(s)

Hydraulic servo   Aircraft

$\boxed{1}$

Rate gyro

(a)

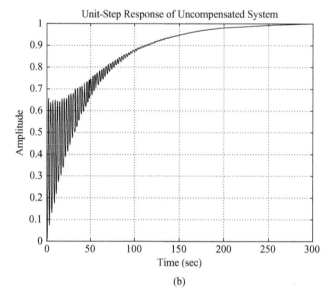

(b)

Figure 6–93
(a) Attitude-rate control system; (b) unit-step response.

It is desired to speed up the response and also eliminate the oscillatory mode at the beginning of the response. Design a suitable compensator such that the dominant closed-loop poles are at $s = -2 \pm j2\sqrt{3}$.

Solution. Figure 6–94 shows a block diagram for the compensated system. Note that the open-loop zero at $s = -0.05$ and the open-loop pole at $s = 0$ generate a closed-loop pole between $s = 0$ and $s = -0.05$. Such a closed-loop pole becomes a dominant closed-loop pole and makes the response quite slow. Hence, it is necessary to replace this zero by a zero that is located far away from the $j\omega$ axis—for example, a zero at $s = -4$.

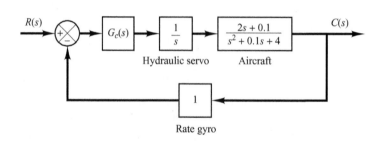

R(s) $\xrightarrow{}$ $\boxed{+ \bigotimes -}$ $\xrightarrow{}$ $\boxed{G_c(s)}$ $\xrightarrow{}$ $\boxed{\dfrac{1}{s}}$ $\xrightarrow{}$ $\boxed{\dfrac{2s + 0.1}{s^2 + 0.1s + 4}}$ $\xrightarrow{}$ C(s)

Hydraulic servo   Aircraft

$\boxed{1}$

Rate gyro

Figure 6–94
Compensated attitude-rate control system.

We now choose the compensator in the following form:

$$G_c(s) = \hat{G}_c(s) \frac{s + 4}{2s + 0.1}$$

Then the open-loop transfer function of the compensated system becomes

$$G_c(s)G(s) = \hat{G}_c(s) \frac{s + 4}{2s + 0.1} \frac{1}{s} \frac{2s + 0.1}{s^2 + 0.1s + 4}$$

$$= \hat{G}_c(s) \frac{s + 4}{s(s^2 + 0.1s + 4)}$$

To determine $\hat{G}_c(s)$ by the root-locus method, we need to find the angle deficiency at the desired closed-loop pole $s = -2 + j2\sqrt{3}$. The angle deficiency can be found as follows:

$$\text{Angle deficiency} = -143.088° - 120° - 109.642° + 60° + 180°$$

$$= -132.73°$$

Hence, the lead compensator $\hat{G}_c(s)$ must provide 132.73°. Since the angle deficiency is –132.73°, we need two lead compensators, each providing 66.365°. Thus $\hat{G}_c(s)$ will have the following form:

$$\hat{G}_c(s) = K_c \left(\frac{s + s_z}{s + s_p} \right)^2$$

Suppose that we choose two zeros at $s = -2$. Then the two poles of the lead compensators can be obtained from

$$\frac{3.4641}{s_p - 2} = \tan(90° - 66.365°) = 0.4376169$$

or

$$s_p = 2 + \frac{3.4641}{0.4376169}$$

$$= 9.9158$$

(See Figure 6–95.) Hence,

$$\hat{G}_c(s) = K_c \left(\frac{s + 2}{s + 9.9158} \right)^2$$

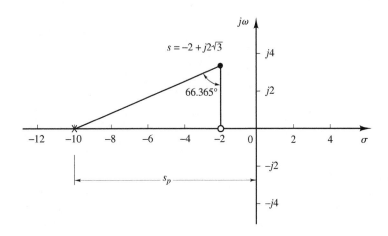

Figure 6–95
Pole and zero of
$\hat{G}_c(s)$.

The entire compensator $G_c(s)$ for the system becomes

$$G_c(s) = \hat{G}_c(s)\,\frac{s+4}{2s+0.1} = K_c\,\frac{(s+2)^2}{(s+9.9158)^2}\,\frac{s+4}{2s+0.1}$$

The value of K_c can be determined from the magnitude condition. Since the open-loop transfer function is

$$G_c(s)G(s) = K_c\,\frac{(s+2)^2(s+4)}{(s+9.9158)^2 s(s^2+0.1s+4)}$$

the magnitude condition becomes

$$\left|K_c\,\frac{(s+2)^2(s+4)}{(s+9.9158)^2 s(s^2+0.1s+4)}\right|_{s=-2+j2\sqrt{3}} = 1$$

Hence,

$$K_c = \left|\frac{(s+9.9158)^2 s(s^2+0.1s+4)}{(s+2)^2(s+4)}\right|_{s=-2+j2\sqrt{3}}$$

$$= 88.0227$$

Thus the compensator $G_c(s)$ becomes

$$G_c(s) = 88.0227\,\frac{(s+2)^2(s+4)}{(s+9.9158)^2(2s+0.1)}$$

The open-loop transfer function is given by

$$G_c(s)G(s) = \frac{88.0227(s+2)^2(s+4)}{(s+9.9158)^2 s(s^2+0.1s+4)}$$

A root-locus plot for the compensated system is shown in Figure 6–96. The closed-loop poles for the compensated system are indicated in the plot. The closed-loop poles, the roots of the characteristic equation

$$(s+9.9158)^2 s(s^2+0.1s+4) + 88.0227(s+2)^2(s+4) = 0$$

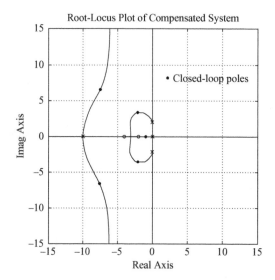

Root-Locus Plot of Compensated System

Figure 6–96
Root-locus plot of
the compensated
system.

are as follows:

$$s = -2.0000 \pm j3.4641$$
$$s = -7.5224 \pm j6.5326$$
$$s = -0.8868$$

Now that the compensator has been designed, we shall examine the transient-response characteristics with MATLAB. The closed-loop transfer function is given by

$$\frac{C(s)}{R(s)} = \frac{88.0227(s + 2)^2(s + 4)}{(s + 9.9158)^2 s(s^2 + 0.1s + 4) + 88.0227(s + 2)^2(s + 4)}$$

Figures 6–97(a) and (b) show the plots of the unit-step response and unit-ramp response of the compensated system. These response curves show that the designed system is acceptable.

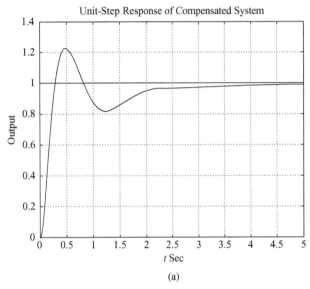

(a)

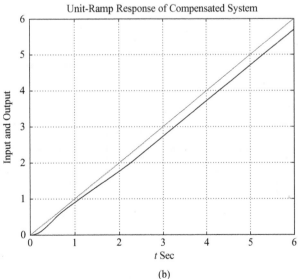

Figure 6–97
(a) Unit-step
response of the
compensated system;
(b) unit-ramp
response of the
compensated system.

(b)

A–6–19. Consider the system shown in Figure 6–98(a). Determine the value of a such that the damping ratio ζ of the dominant closed poles is 0.5.

Solution. The characteristic equation is

$$1 + \frac{10(s + a)}{s(s + 1)(s + 8)} = 0$$

The variable a is not a multiplying factor. Hence, we need to modify the characteristic equation. Since the characteristic equation can be written as

$$s^3 + 9s^2 + 18s + 10a = 0$$

we rewrite this equation such that a appears as a multiplying factor as follows:

$$1 + \frac{10a}{s(s^2 + 9s + 18)} = 0$$

Define

$$10a = K$$

Then the characteristic equation becomes

$$1 + \frac{K}{s(s^2 + 9s + 18)} = 0$$

Notice that the characteristic equation is in a suitable form for the construction of the root loci.

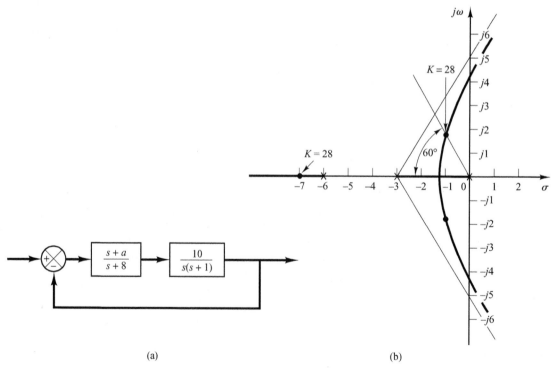

(a)   (b)

Figure 6–98
(a) Control system; (b) root-locus plot, where $K = 10a$.

Chapter 6 / Control Systems Analysis and Design by the Root-Locus Method

This system involves three poles and no zero. The three poles are at $s = 0$, $s = -3$, and $s = -6$. A root-locus branch exists on the real axis between points $s = 0$ and $s = -3$. Also, another branch exists between points $s = -6$ and $s = -\infty$.

The asymptotes for the root loci are found as follows:

$$\text{Angles of asymptotes} = \frac{\pm 180°(2k + 1)}{3} = 60°, -60°, 180°$$

The intersection of the asymptotes and the real axis is obtained from

$$s = -\frac{0 + 3 + 6}{3} = -3$$

The breakaway and break-in points can be determined from $dK/ds = 0$, where

$$K = -\left(s^3 + 9s^2 + 18s\right)$$

Now we set

$$\frac{dK}{ds} = -\left(3s^2 + 18s + 18\right) = 0$$

which yields

$$s^2 + 6s + 6 = 0$$

or

$$s = -1.268, \qquad s = -4.732$$

Point $s = -1.268$ is on a root-locus branch. Hence, point $s = -1.268$ is an actual breakaway point. But point $s = -4.732$ is not on the root locus and therefore is neither a breakaway nor break-in point.

Next we shall find points where root-locus branches cross the imaginary axis. We substitute $s = j\omega$ in the characteristic equation, which is

$$s^3 + 9s^2 + 18s + K = 0$$

as follows:

$$(j\omega)^3 + 9(j\omega)^2 + 18(j\omega) + K = 0$$

or

$$\left(K - 9\omega^2\right) + j\omega\left(18 - \omega^2\right) = 0$$

from which we get

$$\omega = \pm 3\sqrt{2}, \qquad K = 9\omega^2 = 162 \qquad \text{or} \qquad \omega = 0, \qquad K = 0$$

The crossing points are at $\omega = \pm 3\sqrt{2}$ and the corresponding value of gain K is 162. Also, a root-locus branch touches the imaginary axis at $\omega = 0$. Figure 6–98(b) shows a sketch of the root loci for the system.

Since the damping ratio of the dominant closed-loop poles is specified as 0.5, the desired closed-loop pole in the upper-half s plane is located at the intersection of the root-locus branch in the upper-half s plane and a straight line having an angle of 60° with the negative real axis. The desired dominant closed-loop poles are located at

$$s = -1 + j1.732, \qquad s = -1 - j1.732$$

At these points, the value of gain K is 28. Hence,

$$a = \frac{K}{10} = 2.8$$

Example Problems and Solutions

Since the system involves two or more poles than zeros (in fact, three poles and no zero), the third pole can be located on the negative real axis from the fact that the sum of the three closed-loop poles is −9. Hence, the third pole is found to be at

$$s = -9 - (-1 + j1.732) - (-1 - j1.732)$$

or

$$s = -7$$

A–6–20. Consider the system shown in Figure 6–99(a). Sketch the root loci of the system as the velocity feedback gain k varies from zero to infinity. Determine the value of k such that the closed-loop poles have the damping ratio ζ of 0.7.

Solution. The open-loop transfer function is

$$\text{Open-loop transfer function} = \frac{10}{(s + 1 + 10k)s}$$

Since k is not a multilying factor, we modify the equation such that k appears as a multiplying factor. Since the characteristic equation is

$$s^2 + s + 10ks + 10 = 0$$

we rewrite this equation as follows:

$$1 + \frac{10ks}{s^2 + s + 10} = 0 \tag{6–38}$$

Define

$$10k = K$$

Then Equation (6–38) becomes

$$1 + \frac{Ks}{s^2 + s + 10} = 0$$

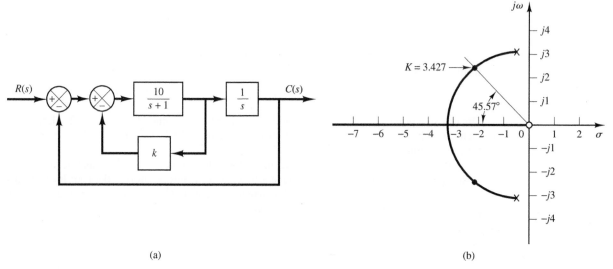

(a) (b)

Figure 6–99
(a) Control system; (b) root-locus plot, where $K = 10k$.

Notice that the system has a zero at $s = 0$ and two poles at $s = -0.5 \pm j3.1225$. Since this system involves two poles and one zero, there is a possibility that a circular root locus exists. In fact, this system has a circular root locus, as will be shown. Since the angle condition is

$$\Bigg/ \frac{Ks}{s^2 + s + 10} = \pm 180°(2k + 1)$$

we have

$$\underline{/s} - \underline{/s + 0.5 + j3.1225} - \underline{/s + 0.5 - j3.1225} = \pm 180°(2k + 1)$$

By substituting $s = \sigma + j\omega$ into this last equation and rearranging, we obtain

$$\underline{/\sigma + 0.5 + j(\omega + 3.1225)} + \underline{/\sigma + 0.5 + j(\omega - 3.1225)} = \underline{/\sigma + j\omega} \pm 180°(2k + 1)$$

which can be rewritten as

$$\tan^{-1}\left(\frac{\omega + 3.1225}{\sigma + 0.5}\right) + \tan^{-1}\left(\frac{\omega - 3.1225}{\sigma + 0.5}\right) = \tan^{-1}\left(\frac{\omega}{\sigma}\right) \pm 180°(2k + 1)$$

Taking tangents of both sides of this last equation, we obtain

$$\frac{\dfrac{\omega + 3.1225}{\sigma + 0.5} + \dfrac{\omega - 3.1225}{\sigma + 0.5}}{1 - \left(\dfrac{\omega + 3.1225}{\sigma + 0.5}\right)\left(\dfrac{\omega - 3.1225}{\sigma + 0.5}\right)} = \frac{\omega}{\sigma}$$

which can be simplified to

$$\frac{2\omega(\sigma + 0.5)}{(\sigma + 0.5)^2 - (\omega^2 - 3.1225^2)} = \frac{\omega}{\sigma}$$

or

$$\omega(\sigma^2 - 10 + \omega^2) = 0$$

which yields

$$\omega = 0 \qquad \text{or} \qquad \sigma^2 + \omega^2 = 10$$

Notice that $\omega = 0$ corresponds to the real axis. The negative real axis (between $s = 0$ and $s = -\infty$) corresponds to $K \geq 0$, and the positive real axis corresponds to $K < 0$. The equation

$$\sigma^2 + \omega^2 = 10$$

is an equation of a circle with center at $\sigma = 0$, $\omega = 0$ with the radius equal to $\sqrt{10}$. A portion of this circle that lies to the left of the complex poles corresponds to the root locus for $K > 0$. (The portion of the circle which lies to the right of the complex poles corresponds to the root locus for $K < 0$.) Figure 6–99(b) shows a sketch of the root loci for $K > 0$.

Since we require $\zeta = 0.7$ for the closed-loop poles, we find the intersection of the circular root locus and a line having an angle of $45.57°$ (note that $\cos 45.57° = 0.7$) with the negative real axis. The intersection is at $s = -2.214 + j2.258$. The gain K corresponding to this point is 3.427. Hence, the desired value of the velocity feedback gain k is

$$k = \frac{K}{10} = 0.3427$$

PROBLEMS

B–6–1. Plot the root loci for the closed-loop control system with

$$G(s) = \frac{K(s + 1)}{s^2}, \qquad H(s) = 1$$

B–6–2. Plot the root loci for the closed-loop control system with

$$G(s) = \frac{K}{s(s + 1)(s^2 + 4s + 5)}, \qquad H(s) = 1$$

B–6–3. Plot the root loci for the system with

$$G(s) = \frac{K}{s(s + 0.5)(s^2 + 0.6s + 10)}, \qquad H(s) = 1$$

B–6–4. Show that the root loci for a control system with

$$G(s) = \frac{K(s^2 + 6s + 10)}{s^2 + 2s + 10}, \qquad H(s) = 1$$

are arcs of the circle centered at the origin with radius equal to $\sqrt{10}$.

B–6–5. Plot the root loci for a closed-loop control system with

$$G(s) = \frac{K(s + 0.2)}{s^2(s + 3.6)}, \qquad H(s) = 1$$

B–6–6. Plot the root loci for a closed-loop control system with

$$G(s) = \frac{K(s + 9)}{s(s^2 + 4s + 11)}, \qquad H(s) = 1$$

Locate the closed-loop poles on the root loci such that the dominant closed-loop poles have a damping ratio equal to 0.5. Determine the corresponding value of gain K.

B–6–7. Plot the root loci for the system shown in Figure 6–100. Determine the range of gain K for stability.

B–6–8. Consider a unity-feedback control system with the following feedforward transfer function:

$$G(s) = \frac{K}{s(s^2 + 4s + 8)}$$

Plot the root loci for the system. If the value of gain K is set equal to 2, where are the closed-loop poles located?

B–6–9. Consider the system whose open-loop transfer function is given by

$$G(s)H(s) = \frac{K(s - 0.6667)}{s^4 + 3.3401s^3 + 7.0325s^2}$$

Show that the equation for the asymptotes is given by

$$G_a(s)H_a(s) = \frac{K}{s^3 + 4.0068s^2 + 5.3515s + 2.3825}$$

Using MATLAB, plot the root loci and asymptotes for the system.

B–6–10. Consider the unity-feedback system whose feedforward transfer function is

$$G(s) = \frac{K}{s(s + 1)}$$

The constant-gain locus for the system for a given value of K is defined by the following equation:

$$\left| \frac{K}{s(s + 1)} \right| = 1$$

Show that the constant-gain loci for $0 \le K \le \infty$ may be given by

$$[\sigma(\sigma + 1) + \omega^2]^2 + \omega^2 = K^2$$

Sketch the constant-gain loci for $K = 1, 2, 5, 10,$ and 20 on the s plane.

B–6–11. Consider the system shown in Figure 6–101. Plot the root loci with MATLAB. Locate the closed-loop poles when the gain K is set equal to 2.

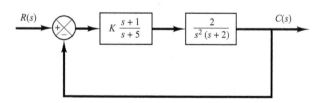

Figure 6–100
Control system.

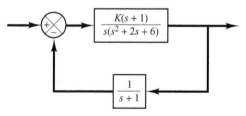

Figure 6–101
Control system.

B–6–12. Plot root-locus diagrams for the nonminimum-phase systems shown in Figures 6–102(a) and (b), respectively.

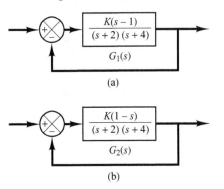

(a)

(b)

Figure 6–102 (a) and (b) Nonminimum-phase systems.

B–6–13. Consider the mechanical system shown in Figure 6–103. It consists of a spring and two dashpots. Obtain the transfer function of the system. The displacement x_i is the input and displacement x_o is the output. Is this system a mechanical lead network or lag network?

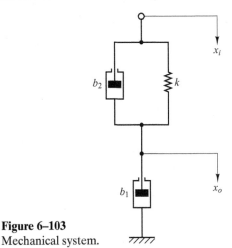

Figure 6–103
Mechanical system.

B–6–14. Consider the system shown in Figure 6–104. Plot the root loci for the system. Determine the value of K such that the damping ratio ζ of the dominant closed-loop poles is 0.5. Then determine all closed-loop poles. Plot the unit-step response curve with MATLAB.

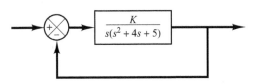

Figure 6–104 Control system.

B–6–15. Determine the values of K, T_1, and T_2 of the system shown in Figure 6–105 so that the dominant closed-loop poles have the damping ratio $\zeta = 0.5$ and the undamped natural frequency $\omega_n = 3$ rad/sec.

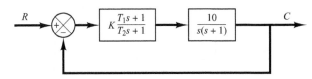

Figure 6–105 Control system.

B–6–16. Consider the control system shown in Figure 6–106. Determine the gain K and time constant T of the controller $G_c(s)$ such that the closed-loop poles are located at $s = -2 \pm j2$.

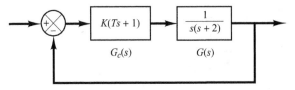

Figure 6–106 Control system.

B–6–17. Consider the system shown in Figure 6–107. Design a lead compensator such that the dominant closed-loop poles are located at $s = -2 \pm j2\sqrt{3}$. Plot the unit-step response curve of the designed system with MATLAB.

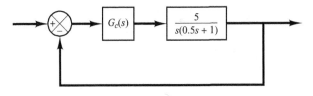

Figure 6–107 Control system.

B–6–18. Consider the system shown in Figure 6–108. Design a compensator such that the dominant closed-loop poles are located at $s = -1 \pm j1$.

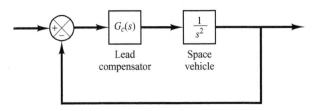

Figure 6–108 Control system.

B–6–19. Referring to the system shown in Figure 6–109, design a compensator such that the static velocity error constant K_v is 20 sec^{-1} without appreciably changing the original location $(s = -2 \pm j2\sqrt{3})$ of a pair of the complex-conjugate closed-loop poles.

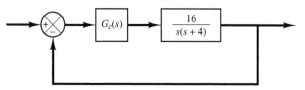

Figure 6–109
Control system.

B–6–20. Consider the angular-positional system shown in Figure 6–110. The dominant closed-loop poles are located at $s = -3.60 \pm j4.80$. The damping ratio ζ of the dominant closed-loop poles is 0.6. The static velocity error constant K_v is 4.1 sec^{-1}, which means that for a ramp input of 360°/sec the steady-state error in following the ramp input is

$$e_v = \frac{\theta_i}{K_v} = \frac{360°/\text{sec}}{4.1 \text{ sec}^{-1}} = 87.8°$$

It is desired to decrease e_v to one-tenth of the present value, or to increase the value of the static velocity error constant K_v to 41 sec^{-1}. It is also desired to keep the damping ratio ζ of the dominant closed-loop poles at 0.6. A small change in the undamped natural frequency ω_n of the dominant closed-loop poles is permissible. Design a suitable lag compensator to increase the static velocity error constant as desired.

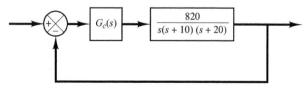

Figure 6–110
Angular-positional system.

B–6–21. Consider the control system shown in Figure 6–111. Design a compensator such that the dominant closed-loop poles are located at $s = -2 \pm j2\sqrt{3}$ and the static velocity error constant K_v is 50 sec^{-1}.

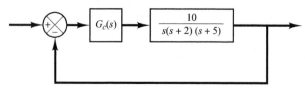

Figure 6–111
Control system.

B–6–22. Consider the control system shown in Figure 6–112. Design a compensator such that the unit-step response curve will exhibit maximum overshoot of 30% or less and settling time of 3 sec or less.

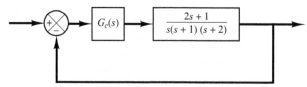

Figure 6–112
Control system.

B–6–23. Consider the control system shown in Figure 6–113. Design a compensator such that the unit-step response curve will exhibit maximum overshoot of 25% or less and settling time of 5 sec or less.

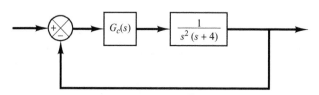

Figure 6–113
Control system.

B–6–24. Consider the system shown in Figure 6–114, which involves velocity feedback. Determine the values of the amplifier gain K and the velocity feedback gain K_h so that the following specifications are satisfied:

1. Damping ratio of the closed-loop poles is 0.5
2. Settling time $\leq$ 2 sec
3. Static velocity error constant $K_v \geq 50$ sec^{-1}
4. $0 < K_h < 1$

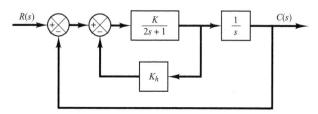

Figure 6–114
Control system.

B–6–25. Consider the system shown in Figure 6–115. The system involves velocity feedback. Determine the value of gain K such that the dominant closed-loop poles have a damping ratio of 0.5. Using the gain K thus determined, obtain the unit-step response of the system.

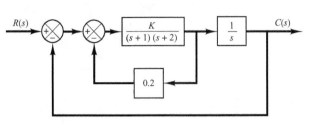

Figure 6–115
Control system.

B–6–26. Consider the system shown in Figure 6–116. Plot the root loci as a varies from 0 to ∞. Determine the value of a such that the damping ratio of the dominant closed-loop poles is 0.5.

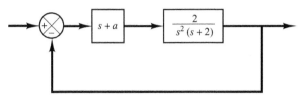

Figure 6–116
Control system.

B–6–27. Consider the system shown in Figure 6–117. Plot the root loci as the value of k varies from 0 to ∞. What value of k will give a damping ratio of the dominant closed-loop poles equal to 0.5? Find the static velocity error constant of the system with this value of k.

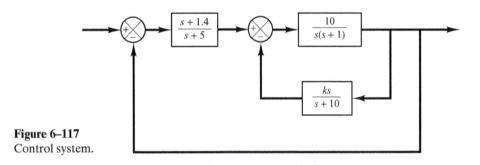

Figure 6–117
Control system.

B–6–28. Consider the system shown in Figure 6–118. Assuming that the value of gain K varies from 0 to ∞, plot the root loci when $K_h = 0.1, 0.3,$ and 0.5.

Compare unit-step responses of the system for the following three cases:

(1) $K = 10,$ $K_h = 0.1$

(2) $K = 10,$ $K_h = 0.3$

(3) $K = 10,$ $K_h = 0.5$

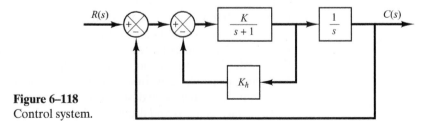

Figure 6–118
Control system.

Control Systems Analysis and Design by the Frequency-Response Method

7–1 INTRODUCTION

By the term *frequency response,* we mean the steady-state response of a system to a sinusoidal input. In frequency-response methods, we vary the frequency of the input signal over a certain range and study the resulting response.

In this chapter we present frequency-response approaches to the analysis and design of control systems. The information we get from such analysis is different from what we get from root-locus analysis. In fact, the frequency response and root-locus approaches complement each other. One advantage of the frequency-response approach is that we can use the data obtained from measurements on the physical system without deriving its mathematical model. In many practical designs of control systems both approaches are employed. Control engineers must be familiar with both.

Frequency-response methods were developed in 1930s and 1940s by Nyquist, Bode, Nichols, and many others. The frequency-response methods are most powerful in conventional control theory. They are also indispensable to robust control theory.

The Nyquist stability criterion enables us to investigate both the absolute and relative stabilities of linear closed-loop systems from a knowledge of their open-loop frequency-response characteristics. An advantage of the frequency-response approach is that frequency-response tests are, in general, simple and can be made accurately by use of readily available sinusoidal signal generators and precise measurement equipment. Often the transfer functions of complicated components can be determined experimentally by frequency-response tests. In addition, the frequency-response approach has the advantages that a system may be designed so that the effects of undesirable noise are negligible and that such analysis and design can be extended to certain nonlinear control systems.

Although the frequency response of a control system presents a qualitative picture of the transient response, the correlation between frequency and transient responses is indirect, except for the case of second-order systems. In designing a closed-loop system, we adjust the frequency-response characteristic of the open-loop transfer function by using several design criteria in order to obtain acceptable transient-response characteristics for the system.

Obtaining Steady-State Outputs to Sinusoidal Inputs. We shall show that the steady-state output of a transfer function system can be obtained directly from the sinusoidal transfer function—that is, the transfer function in which s is replaced by $j\omega$, where ω is frequency.

Consider the stable, linear, time-invariant system shown in Figure 7–1. The input and output of the system, whose transfer function is $G(s)$, are denoted by $x(t)$ and $y(t)$, respectively. If the input $x(t)$ is a sinusoidal signal, the steady-state output will also be a sinusoidal signal of the same frequency, but with possibly different magnitude and phase angle.

Let us assume that the input signal to the system is given by

$$x(t) = X \sin \omega t$$

[In this book "ω" is always measured in rad/sec. When the frequency is measured in cycle/sec, we use notation "f". That is, $\omega = 2\pi f$.]

Suppose that the transfer function $G(s)$ of the system can be written as a ratio of two polynomials in s; that is,

$$G(s) = \frac{p(s)}{q(s)} = \frac{p(s)}{(s + s_1)(s + s_2)\cdots(s + s_n)}$$

The Laplace-transformed output $Y(s)$ of the system is then

$$Y(s) = G(s)X(s) = \frac{p(s)}{q(s)} X(s) \tag{7–1}$$

where $X(s)$ is the Laplace transform of the input $x(t)$.

It will be shown that, after waiting until steady-state conditions are reached, the frequency response can be calculated by replacing s in the transfer function by $j\omega$. It will also be shown that the steady-state response can be given by

$$G(j\omega) = Me^{j\phi} = M \underline{/\phi}$$

where M is the amplitude ratio of the output and input sinusoids and ϕ is the phase shift between the input sinusoid and the output sinusoid. In the frequency-response test, the input frequency ω is varied until the entire frequency range of interest is covered.

The steady-state response of a stable, linear, time-invariant system to a sinusoidal input does not depend on the initial conditions. (Thus, we can assume the zero initial condition.) If $Y(s)$ has only distinct poles, then the partial fraction expansion of Equation (7–1) when $x(t) = X \sin \omega t$ yields

$$Y(s) = G(s)X(s) = G(s)\frac{\omega X}{s^2 + \omega^2}$$

$$= \frac{a}{s + j\omega} + \frac{\bar{a}}{s - j\omega} + \frac{b_1}{s + s_1} + \frac{b_2}{s + s_2} + \cdots + \frac{b_n}{s + s_n} \tag{7–2}$$

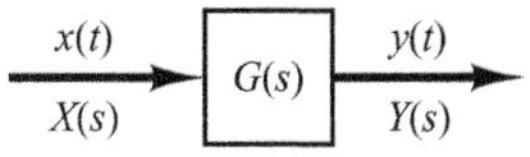

Figure 7–1
Stable, linear, time-invariant system.

where a and the b_i (where $i = 1, 2, \ldots, n$) are constants and $\bar{a}$ is the complex conjugate of a. The inverse Laplace transform of Equation (7–2) gives

$$y(t) = ae^{-j\omega t} + \bar{a}e^{j\omega t} + b_1e^{-s_1 t} + b_2e^{-s_2 t} + \cdots + b_ne^{-s_n t} \qquad (t \geq 0) \qquad (7\text{–}3)$$

For a stable system, $-s_1, -s_2, \ldots, -s_n$ have negative real parts. Therefore, as t approaches infinity, the terms $e^{-s_1 t}$, $e^{-s_2 t}$, $\ldots$, and $e^{-s_n t}$ approach zero. Thus, all the terms on the right-hand side of Equation (7–3), except the first two, drop out at steady state.

If $Y(s)$ involves multiple poles s_j of multiplicity m_j, then $y(t)$ will involve terms such as $t^{h_j}e^{-s_j t}$ $(h_j = 0, 1, 2, \ldots, m_j - 1)$. For a stable system, the terms $t^{h_j}e^{-s_j t}$ approach zero as t approaches infinity.

Thus, regardless of whether the system is of the distinct-pole type or multiple-pole type, the steady-state response becomes

$$y_{ss}(t) = ae^{-j\omega t} + \bar{a}e^{j\omega t} \qquad (7\text{–}4)$$

where the constant a can be evaluated from Equation (7–2) as follows:

$$a = G(s)\frac{\omega X}{s^2 + \omega^2}(s + j\omega)\bigg|_{s=-j\omega} = -\frac{XG(-j\omega)}{2j}$$

Note that

$$\bar{a} = G(s)\frac{\omega X}{s^2 + \omega^2}(s - j\omega)\bigg|_{s=j\omega} = \frac{XG(j\omega)}{2j}$$

Since $G(j\omega)$ is a complex quantity, it can be written in the following form:

$$G(j\omega) = |G(j\omega)|e^{j\phi}$$

where $|G(j\omega)|$ represents the magnitude and ϕ represents the angle of $G(j\omega)$; that is,

$$\phi = \underline{/G(j\omega)} = \tan^{-1}\left[\frac{\text{imaginary part of } G(j\omega)}{\text{real part of } G(j\omega)}\right]$$

The angle ϕ may be negative, positive, or zero. Similarly, we obtain the following expression for $G(-j\omega)$:

$$G(-j\omega) = |G(-j\omega)|e^{-j\phi} = |G(j\omega)|e^{-j\phi}$$

Then, noting that

$$a = -\frac{X|G(j\omega)|e^{-j\phi}}{2j}, \qquad \bar{a} = \frac{X|G(j\omega)|e^{j\phi}}{2j}$$

Equation (7–4) can be written

$$y_{ss}(t) = X|G(j\omega)|\frac{e^{j(\omega t+\phi)} - e^{-j(\omega t+\phi)}}{2j}$$

$$= X|G(j\omega)|\sin(\omega t + \phi)$$

$$= Y\sin(\omega t + \phi) \qquad (7\text{–}5)$$

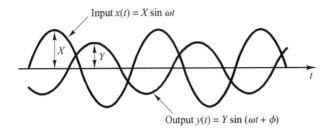

Figure 7–2
Input and output
sinusoidal signals.

Input $x(t) = X \sin \omega t$

Output $y(t) = Y \sin (\omega t + \phi)$

where $Y = X|G(j\omega)|$. We see that a stable, linear, time-invariant system subjected to a sinusoidal input will, at steady state, have a sinusoidal output of the same frequency as the input. But the amplitude and phase of the output will, in general, be different from those of the input. In fact, the amplitude of the output is given by the product of that of the input and $|G(j\omega)|$, while the phase angle differs from that of the input by the amount $\phi = \underline{/G(j\omega)}$. An example of input and output sinusoidal signals is shown in Figure 7–2.

On the basis of this, we obtain this important result: For sinusoidal inputs,

$$|G(j\omega)| = \left|\frac{Y(j\omega)}{X(j\omega)}\right| = \begin{array}{l}\text{amplitude ratio of the output sinuisoid to the}\\ \text{input sinusoid}\end{array}$$

$$\underline{/G(j\omega)} = \underline{\bigg/\frac{Y(j\omega)}{X(j\omega)}} = \begin{array}{l}\text{phase shift of the output sinusoid with respect}\\ \text{to the input sinusoid}\end{array}$$

Hence, the steady-state response characteristics of a system to a sinusoidal input can be obtained directly from

$$\frac{Y(j\omega)}{X(j\omega)} = G(j\omega)$$

The function $G(j\omega)$ is called the sinusoidal transfer function. It is the ratio of $Y(j\omega)$ to $X(j\omega)$, is a complex quantity, and can be represented by the magnitude and phase angle with frequency as a parameter. The sinusoidal transfer function of any linear system is obtained by substituting $j\omega$ for s in the transfer function of the system.

As already mentioned in Chapter 6, a positive phase angle is called phase lead, and a negative phase angle is called phase lag. A network that has phase-lead characteristics is called a lead network, while a network that has phase-lag characteristics is called a lag network.

EXAMPLE 7–1   Consider the system shown in Figure 7–3. The transfer function $G(s)$ is

$$G(s) = \frac{K}{Ts + 1}$$

For the sinusoidal input $x(t) = X \sin \omega t$, the steady-state output $y_{ss}(t)$ can be found as follows: Substituting $j\omega$ for s in $G(s)$ yields

$$G(j\omega) = \frac{K}{jT\omega + 1}$$

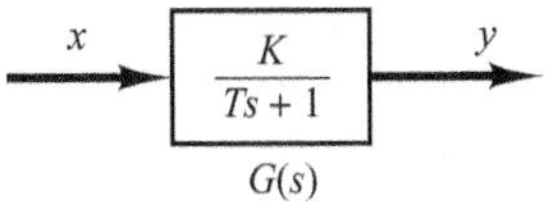

$x \longrightarrow \boxed{\dfrac{K}{Ts + 1}} \longrightarrow y$

$G(s)$

Figure 7–3
First-order system.

The amplitude ratio of the output to the input is

$$|G(j\omega)| = \frac{K}{\sqrt{1 + T^2\omega^2}}$$

while the phase angle ϕ is

$$\phi = \underline{/G(j\omega)} = -\tan^{-1}T\omega$$

Thus, for the input $x(t) = X \sin \omega t$, the steady-state output $y_{ss}(t)$ can be obtained from Equation (7–5) as follows:

$$y_{ss}(t) = \frac{XK}{\sqrt{1 + T^2\omega^2}} \sin(\omega t - \tan^{-1}T\omega) \tag{7–6}$$

From Equation (7–6), it can be seen that for small ω, the amplitude of the steady-state output $y_{ss}(t)$ is almost equal to K times the amplitude of the input. The phase shift of the output is small for small ω. For large ω, the amplitude of the output is small and almost inversely proportional to ω. The phase shift approaches $-90°$ as ω approaches infinity. This is a phase-lag network.

EXAMPLE 7–2   Consider the network given by

$$G(s) = \frac{s + \dfrac{1}{T_1}}{s + \dfrac{1}{T_2}}$$

Determine whether this network is a lead network or lag network.

For the sinusoidal input $x(t) = X \sin \omega t$, the steady-state output $y_{ss}(t)$ can be found as follows: Since

$$G(j\omega) = \frac{j\omega + \dfrac{1}{T_1}}{j\omega + \dfrac{1}{T_2}} = \frac{T_2(1 + T_1 j\omega)}{T_1(1 + T_2 j\omega)}$$

we have

$$|G(j\omega)| = \frac{T_2\sqrt{1 + T_1^2\omega^2}}{T_1\sqrt{1 + T_2^2\omega^2}}$$

and

$$\phi = \underline{/G(j\omega)} = \tan^{-1}T_1\omega - \tan^{-1}T_2\omega$$

Thus the steady-state output is

$$y_{ss}(t) = \frac{XT_2\sqrt{1 + T_1^2\omega^2}}{T_1\sqrt{1 + T_2^2\omega^2}} \sin(\omega t + \tan^{-1}T_1\omega - \tan^{-1}T_2\omega)$$

From this expression, we find that if $T_1 > T_2$, then $\tan^{-1}T_1\omega - \tan^{-1}T_2\omega > 0$. Thus, if $T_1 > T_2$, then the network is a lead network. If $T_1 < T_2$, then the network is a lag network.

Presenting Frequency-Response Characteristics in Graphical Forms.   The sinusoidal transfer function, a complex function of the frequency ω, is characterized by its magnitude and phase angle, with frequency as the parameter. There are three commonly used representations of sinusoidal transfer functions:

Chapter 7 / Control Systems Analysis and Design by the Frequency-Response Method

1. Bode diagram or logarithmic plot

2. Nyquist plot or polar plot

3. Log-magnitude-versus-phase plot (Nichols plots)

We shall discuss these representations in detail in this chapter. We shall include the MATLAB approach to obtain Bode diagrams, Nyquist plots, and Nichols plots.

Outline of the Chapter. Section 7–1 has presented introductory material on the frequency response. Section 7–2 presents Bode diagrams of various transfer-function systems. Section 7–3 treats polar plots of transfer functions. Section 7–4 discusses log-magnitude-versus-phase plots. Section 7–5 gives a detailed account of the Nyquist stability criterion. Section 7–6 discusses the stability analysis based on the Nyquist stability criterion. Section 7–7 introduces measures of relative stability analysis. Section 7–8 presents a method for obtaining the closed-loop frequency response from the open-loop frequency response by use of the M and N circles. The Nichols chart is introduced here. Section 7–9 treats experimental determination of transfer functions. Section 7–10 presents introductory aspects of control systems design by the frequency-response approach. Sections 7–11, 7–12, and 7–13 give detailed accounts of lead compensation, lag compensation, and lag–lead compensation techniques, respectively.

7–2 BODE DIAGRAMS

Bode Diagrams or Logarithmic Plots. A Bode diagram consists of two graphs: One is a plot of the logarithm of the magnitude of a sinusoidal transfer function; the other is a plot of the phase angle; both are plotted against the frequency on a logarithmic scale.

The standard representation of the logarithmic magnitude of $G(j\omega)$ is $20\log|G(j\omega)|$, where the base of the logarithm is 10. The unit used in this representation of the magnitude is the decibel, usually abbreviated dB. In the logarithmic representation, the curves are drawn on semilog paper, using the log scale for frequency and the linear scale for either magnitude (but in decibels) or phase angle (in degrees). (The frequency range of interest determines the number of logarithmic cycles required on the abscissa.)

The main advantage of using the Bode diagram is that multiplication of magnitudes can be converted into addition. Furthermore, a simple method for sketching an approximate log-magnitude curve is available. It is based on asymptotic approximations. Such approximation by straight-line asymptotes is sufficient if only rough information on the frequency-response characteristics is needed. Should the exact curve be desired, corrections can be made easily to these basic asymptotic plots. Expanding the low-frequency range by use of a logarithmic scale for the frequency is highly advantageous, since characteristics at low frequencies are most important in practical systems. Although it is not possible to plot the curves right down to zero frequency because of the logarithmic frequency ($\log 0 = -\infty$), this does not create a serious problem.

Note that the experimental determination of a transfer function can be made simple if frequency-response data are presented in the form of a Bode diagram.

Basic Factors of $G(j\omega)H(j\omega)$.   As stated earlier, the main advantage in using the logarithmic plot is the relative ease of plotting frequency-response curves. The basic factors that very frequently occur in an arbitrary transfer function $G(j\omega)H(j\omega)$ are

1. Gain K
2. Integral and derivative factors $(j\omega)^{\mp 1}$
3. First-order factors $(1 + j\omega T)^{\mp 1}$
4. Quadratic factors $\left[1 + 2\zeta(j\omega/\omega_n) + (j\omega/\omega_n)^2\right]^{\mp 1}$

Once we become familiar with the logarithmic plots of these basic factors, it is possible to utilize them in constructing a composite logarithmic plot for any general form of $G(j\omega)H(j\omega)$ by sketching the curves for each factor and adding individual curves graphically, because adding the logarithms of the gains corresponds to multiplying them together.

The Gain K.   A number greater than unity has a positive value in decibels, while a number smaller than unity has a negative value. The log-magnitude curve for a constant gain K is a horizontal straight line at the magnitude of $20 \log K$ decibels. The phase angle of the gain K is zero. The effect of varying the gain K in the transfer function is that it raises or lowers the log-magnitude curve of the transfer function by the corresponding constant amount, but it has no effect on the phase curve.

A number–decibel conversion line is given in Figure 7–4. The decibel value of any number can be obtained from this line. As a number increases by a factor of 10, the corresponding decibel value increases by a factor of 20. This may be seen from the following:

$$20 \log(K \times 10) = 20 \log K + 20$$

Similarly,

$$20 \log(K \times 10^n) = 20 \log K + 20n$$

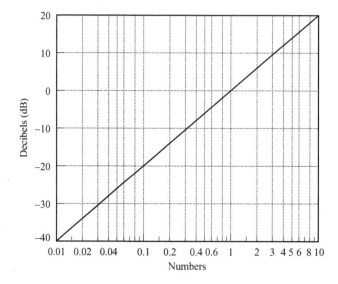

Figure 7–4
Number–decibel
conversion line.

Note that, when expressed in decibels, the reciprocal of a number differs from its value only in sign; that is, for the number K,

$$20 \log K = -20 \log \frac{1}{K}$$

Integral and Derivative Factors $(j\omega)^{\mp 1}$.   The logarithmic magnitude of $1/j\omega$ in decibels is

$$20 \log \left| \frac{1}{j\omega} \right| = -20 \log \omega \text{ dB}$$

The phase angle of $1/j\omega$ is constant and equal to $-90°$.

In Bode diagrams, frequency ratios are expressed in terms of octaves or decades. An octave is a frequency band from ω_1 to $2\omega_1$, where ω_1 is any frequency value. A decade is a frequency band from ω_1 to $10\omega_1$, where again ω_1 is any frequency. (On the logarithmic scale of semilog paper, any given frequency ratio can be represented by the same horizontal distance. For example, the horizontal distance from $\omega = 1$ to $\omega = 10$ is equal to that from $\omega = 3$ to $\omega = 30$.)

If the log magnitude $-20 \log \omega$ dB is plotted against ω on a logarithmic scale, it is a straight line. To draw this straight line, we need to locate one point (0 dB, $\omega = 1$) on it. Since

$$(-20 \log 10\omega) \text{ dB} = (-20 \log \omega - 20) \text{ dB}$$

the slope of the line is -20 dB/decade (or -6 dB/octave).

Similarly, the log magnitude of $j\omega$ in decibels is

$$20 \log |j\omega| = 20 \log \omega \text{ dB}$$

The phase angle of $j\omega$ is constant and equal to $90°$. The log-magnitude curve is a straight line with a slope of 20 dB/decade. Figures 7–5(a) and (b) show frequency-response curves for $1/j\omega$ and $j\omega$, respectively. We can clearly see that the differences in the frequency responses of the factors $1/j\omega$ and $j\omega$ lie in the signs of the slopes of the log-magnitude curves and in the signs of the phase angles. Both log magnitudes become equal to 0 dB at $\omega = 1$.

If the transfer function contains the factor $(1/j\omega)^n$ or $(j\omega)^n$, the log magnitude becomes, respectively,

$$20 \log \left| \frac{1}{(j\omega)^n} \right| = -n \times 20 \log |j\omega| = -20n \log \omega \text{ dB}$$

or

$$20 \log |(j\omega)^n| = n \times 20 \log |j\omega| = 20n \log \omega \text{ dB}$$

The slopes of the log-magnitude curves for the factors $(1/j\omega)^n$ and $(j\omega)^n$ are thus $-20n$ dB/decade and $20n$ dB/decade, respectively. The phase angle of $(1/j\omega)^n$ is equal to $-90° \times n$ over the entire frequency range, while that of $(j\omega)^n$ is equal to $90° \times n$ over the entire frequency range. The magnitude curves will pass through the point (0 dB, $\omega = 1$).

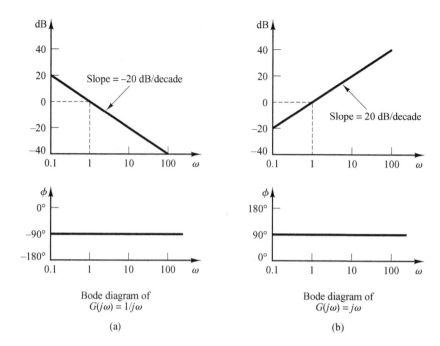

Figure 7–5
(a) Bode diagram of
$G(j\omega) = 1/j\omega$;
(b) Bode diagram of
$G(j\omega) = j\omega$.

Bode diagram of
$G(j\omega) = 1/j\omega$

(a)

Bode diagram of
$G(j\omega) = j\omega$

(b)

First-Order Factors $(1 + j\omega T)^{\mp 1}$. The log magnitude of the first-order factor $1/(1 + j\omega T)$ is

$$20 \log \left| \frac{1}{1 + j\omega T} \right| = -20 \log \sqrt{1 + \omega^2 T^2} \text{ dB}$$

For low frequencies, such that $\omega \ll 1/T$, the log magnitude may be approximated by

$$-20 \log \sqrt{1 + \omega^2 T^2} \doteq -20 \log 1 = 0 \text{ dB}$$

Thus, the log-magnitude curve at low frequencies is the constant 0-dB line. For high frequencies, such that $\omega \gg 1/T$,

$$-20 \log \sqrt{1 + \omega^2 T^2} \doteq -20 \log \omega T \text{ dB}$$

This is an approximate expression for the high-frequency range. At $\omega = 1/T$, the log magnitude equals 0 dB; at $\omega = 10/T$, the log magnitude is -20 dB. Thus, the value of $-20 \log \omega T$ dB decreases by 20 dB for every decade of ω. For $\omega \gg 1/T$, the log-magnitude curve is thus a straight line with a slope of -20 dB/decade (or -6 dB/octave).

Our analysis shows that the logarithmic representation of the frequency-response curve of the factor $1/(1 + j\omega T)$ can be approximated by two straight-line asymptotes, one a straight line at 0 dB for the frequency range $0 < \omega < 1/T$ and the other a straight line with slope -20 dB/decade (or -6 dB/octave) for the frequency range $1/T < \omega < \infty$. The exact log-magnitude curve, the asymptotes, and the exact phase-angle curve are shown in Figure 7–6.

The frequency at which the two asymptotes meet is called the *corner* frequency or *break* frequency. For the factor $1/(1 + j\omega T)$, the frequency $\omega = 1/T$ is the corner frequency, since at $\omega = 1/T$ the two asymptotes have the same value. (The low-frequency asymptotic expression at $\omega = 1/T$ is 20 log 1 dB = 0 dB, and the high-frequency

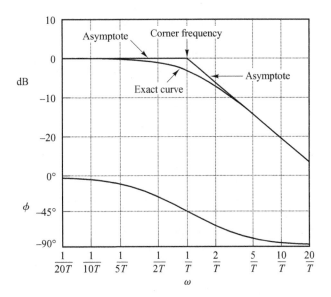

Figure 7–6
Log-magnitude curve, together with the asymptotes, and phase-angle curve of $1/(1 + j\omega T)$.

asymptotic expression at $\omega = 1/T$ is also 20 log 1 dB = 0 dB.) The corner frequency divides the frequency-response curve into two regions: a curve for the low-frequency region and a curve for the high-frequency region. The corner frequency is very important in sketching logarithmic frequency-response curves.

The exact phase angle ϕ of the factor $1/(1 + j\omega T)$ is

$$\phi = -\tan^{-1}\omega T$$

At zero frequency, the phase angle is 0°. At the corner frequency, the phase angle is

$$\phi = -\tan^{-1}\frac{T}{T} = -\tan^{-1}1 = -45°$$

At infinity, the phase angle becomes −90°. Since the phase angle is given by an inverse-tangent function, the phase angle is skew symmetric about the inflection point at $\phi = -45°$.

The error in the magnitude curve caused by the use of asymptotes can be calculated. The maximum error occurs at the corner frequency and is approximately equal to −3 dB, since

$$-20\log\sqrt{1+1} + 20\log 1 = -10\log 2 = -3.03 \text{ dB}$$

The error at the frequency one octave below the corner frequency—that is, at $\omega = 1/(2T)$—is

$$-20\log\sqrt{\frac{1}{4}+1} + 20\log 1 = -20\log\frac{\sqrt{5}}{2} = -0.97 \text{ dB}$$

The error at the frequency one octave above the corner frequency—that is, at $\omega = 2/T$—is

$$-20\log\sqrt{2^2+1} + 20\log 2 = -20\log\frac{\sqrt{5}}{2} = -0.97 \text{ dB}$$

Thus, the error at one octave below or above the corner frequency is approximately equal to −1 dB. Similarly, the error at one decade below or above the corner frequency is approximately −0.04 dB. The error in decibels involved in using the asymptotic expression for the frequency-response curve of $1/(1 + j\omega T)$ is shown in Figure 7–7. The error is symmetric with respect to the corner frequency.

Since the asymptotes are quite easy to draw and are sufficiently close to the exact curve, the use of such approximations in drawing Bode diagrams is convenient in establishing the general nature of the frequency-response characteristics quickly with a minimum amount of calculation and may be used for most preliminary design work. If accurate frequency-response curves are desired, corrections may easily be made by referring to the curve given in Figure 7–7. In practice, an accurate frequency-response curve can be drawn by introducing a correction of 3 dB at the corner frequency and a correction of 1 dB at points one octave below and above the corner frequency and then connecting these points by a smooth curve.

Note that varying the time constant T shifts the corner frequency to the left or to the right, but the shapes of the log-magnitude and the phase-angle curves remain the same.

The transfer function $1/(1 + j\omega T)$ has the characteristics of a low-pass filter. For frequencies above $\omega = 1/T$, the log magnitude falls off rapidly toward $-\infty$. This is essentially due to the presence of the time constant. In the low-pass filter, the output can follow a sinusoidal input faithfully at low frequencies. But as the input frequency is increased, the output cannot follow the input because a certain amount of time is required for the system to build up in magnitude. Thus, at high frequencies, the amplitude of the output approaches zero and the phase angle of the output approaches $-90°$. Therefore, if the input function contains many harmonics, then the low-frequency components are reproduced faithfully at the output, while the high-frequency components are attenuated in amplitude and shifted in phase. Thus, a first-order element yields exact, or almost exact, duplication only for constant or slowly varying phenomena.

An advantage of the Bode diagram is that for reciprocal factors—for example, the factor $1 + j\omega T$—the log-magnitude and the phase-angle curves need only be changed in sign, since

$$20 \log|1 + j\omega T| = -20 \log\left|\frac{1}{1 + j\omega T}\right|$$

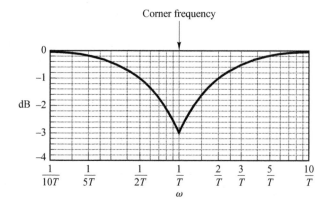

Figure 7–7
Log-magnitude error in the asymptotic expression of the frequency-response curve of $1/(1 + j\omega T)$.

Chapter 7 / Control Systems Analysis and Design by the Frequency-Response Method

and

$$\underline{/1 + j\omega T} = \tan^{-1}\omega T = -\left/\underline{\frac{1}{1 + j\omega T}}\right.$$

The corner frequency is the same for both cases. The slope of the high-frequency asymptote of $1 + j\omega T$ is 20 dB/decade, and the phase angle varies from 0° to 90° as the frequency ω is increased from zero to infinity. The log-magnitude curve, together with the asymptotes, and the phase-angle curve for the factor $1 + j\omega T$ are shown in Figure 7–8.

To draw a phase curve accurately, we have to locate several points on the curve. The phase angles of $(1 + j\omega T)^{\mp 1}$ are

$$\mp 45° \qquad \text{at} \qquad \omega = \frac{1}{T}$$

$$\mp 26.6° \qquad \text{at} \qquad \omega = \frac{1}{2T}$$

$$\mp 5.7° \qquad \text{at} \qquad \omega = \frac{1}{10T}$$

$$\mp 63.4° \qquad \text{at} \qquad \omega = \frac{2}{T}$$

$$\mp 84.3° \qquad \text{at} \qquad \omega = \frac{10}{T}$$

For the case where a given transfer function involves terms like $(1 + j\omega T)^{\mp n}$, a similar asymptotic construction may be made. The corner frequency is still at $\omega = 1/T$, and the asymptotes are straight lines. The low-frequency asymptote is a horizontal straight line

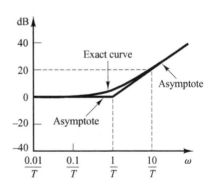

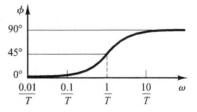

Figure 7–8
Log-magnitude curve, together with the asymptotes, and phase-angle curve for $1 + j\omega T$.

at 0 dB, while the high-frequency asymptote has the slope of $-20n$ dB/decade or $20n$ dB/decade. The error involved in the asymptotic expressions is n times that for $(1 + j\omega T)^{\mp 1}$. The phase angle is n times that of $(1 + j\omega T)^{\mp 1}$ at each frequency point.

Quadratic Factors $\left[1 + 2\zeta(j\omega/\omega_n) + (j\omega/\omega_n)^2\right]^{\mp 1}$. Control systems often possess quadratic factors of the form

$$G(j\omega) = \frac{1}{1 + 2\zeta\left(j\dfrac{\omega}{\omega_n}\right) + \left(j\dfrac{\omega}{\omega_n}\right)^2} \tag{7-7}$$

If $\zeta > 1$, this quadratic factor can be expressed as a product of two first-order factors with real poles. If $0 < \zeta < 1$, this quadratic factor is the product of two complex-conjugate factors. Asymptotic approximations to the frequency-response curves are not accurate for a factor with low values of ζ. This is because the magnitude and phase of the quadratic factor depend on both the corner frequency and the damping ratio ζ.

The asymptotic frequency-response curve may be obtained as follows: Since

$$20 \log \left| \frac{1}{1 + 2\zeta\left(j\dfrac{\omega}{\omega_n}\right) + \left(j\dfrac{\omega}{\omega_n}\right)^2} \right| = -20 \log \sqrt{\left(1 - \frac{\omega^2}{\omega_n^2}\right)^2 + \left(2\zeta\frac{\omega}{\omega_n}\right)^2}$$

for low frequencies such that $\omega \ll \omega_n$, the log magnitude becomes

$$-20 \log 1 = 0 \text{ dB}$$

The low-frequency asymptote is thus a horizontal line at 0 dB. For high frequencies such that $\omega \gg \omega_n$, the log magnitude becomes

$$-20 \log \frac{\omega^2}{\omega_n^2} = -40 \log \frac{\omega}{\omega_n} \text{ dB}$$

The equation for the high-frequency asymptote is a straight line having the slope -40 dB/decade, since

$$-40 \log \frac{10\omega}{\omega_n} = -40 - 40 \log \frac{\omega}{\omega_n}$$

The high-frequency asymptote intersects the low-frequency one at $\omega = \omega_n$, since at this frequency

$$-40 \log \frac{\omega_n}{\omega_n} = -40 \log 1 = 0 \text{ dB}$$

This frequency, ω_n, is the corner frequency for the quadratic factor considered.

The two asymptotes just derived are independent of the value of ζ. Near the frequency $\omega = \omega_n$, a resonant peak occurs, as may be expected from Equation (7–7). The damping ratio ζ determines the magnitude of this resonant peak. Errors obviously exist in the approximation by straight-line asymptotes. The magnitude of the error depends on the value of ζ. It is large for small values of ζ. Figure 7–9 shows the exact log-magnitude curves, together with the straight-line asymptotes and the exact

Chapter 7 / Control Systems Analysis and Design by the Frequency-Response Method

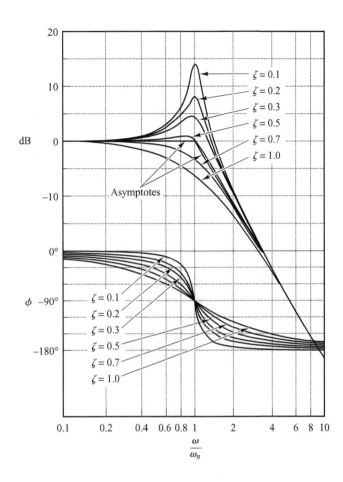

Figure 7–9
Log-magnitude
curves, together with
the asymptotes, and
phase-angle curves
of the quadratic
transfer function
given by
Equation (7–7).

phase-angle curves for the quadratic factor given by Equation (7–7) with several values of ζ. If corrections are desired in the asymptotic curves, the necessary amounts of correction at a sufficient number of frequency points may be obtained from Figure 7–9.

The phase angle of the quadratic factor $\left[1 + 2\zeta(j\omega/\omega_n) + (j\omega/\omega_n)^2\right]^{-1}$ is

$$\phi = \left| \frac{1}{1 + 2\zeta\left(j\dfrac{\omega}{\omega_n}\right) + \left(j\dfrac{\omega}{\omega_n}\right)^2} = -\tan^{-1}\left[\frac{2\zeta\dfrac{\omega}{\omega_n}}{1 - \left(\dfrac{\omega}{\omega_n}\right)^2}\right] \right. \tag{7-8}$$

The phase angle is a function of both ω and ζ. At $\omega = 0$, the phase angle equals $0°$. At the corner frequency $\omega = \omega_n$, the phase angle is $-90°$ regardless of ζ, since

$$\phi = -\tan^{-1}\left(\frac{2\zeta}{0}\right) = -\tan^{-1}\infty = -90°$$

At $\omega = \infty$, the phase angle becomes $-180°$. The phase-angle curve is skew symmetric about the inflection point—the point where $\phi = -90°$. There are no simple ways to sketch such phase curves. We need to refer to the phase-angle curves shown in Figure 7–9.

The frequency-response curves for the factor

$$1 + 2\zeta\left(j\frac{\omega}{\omega_n}\right) + \left(j\frac{\omega}{\omega_n}\right)^2$$

can be obtained by merely reversing the sign of the log magnitude and that of the phase angle of the factor

$$\frac{1}{1 + 2\zeta\left(j\frac{\omega}{\omega_n}\right) + \left(j\frac{\omega}{\omega_n}\right)^2}$$

To obtain the frequency-response curves of a given quadratic transfer function, we must first determine the value of the corner frequency ω_n and that of the damping ratio ζ. Then, by using the family of curves given in Figure 7–9, the frequency-response curves can be plotted.

The Resonant Frequency ω_r and the Resonant Peak Value M_r. The magnitude of

$$G(j\omega) = \frac{1}{1 + 2\zeta\left(j\frac{\omega}{\omega_n}\right) + \left(j\frac{\omega}{\omega_n}\right)^2}$$

is

$$|G(j\omega)| = \frac{1}{\sqrt{\left(1 - \frac{\omega^2}{\omega_n^2}\right)^2 + \left(2\zeta\frac{\omega}{\omega_n}\right)^2}} \tag{7–9}$$

If $|G(j\omega)|$ has a peak value at some frequency, this frequency is called the *resonant frequency*. Since the numerator of $|G(j\omega)|$ is constant, a peak value of $|G(j\omega)|$ will occur when

$$g(\omega) = \left(1 - \frac{\omega^2}{\omega_n^2}\right)^2 + \left(2\zeta\frac{\omega}{\omega_n}\right)^2 \tag{7–10}$$

is a minimum. Since Equation (7–10) can be written

$$g(\omega) = \left[\frac{\omega^2 - \omega_n^2(1 - 2\zeta^2)}{\omega_n^2}\right]^2 + 4\zeta^2(1 - \zeta^2) \tag{7–11}$$

the minimum value of $g(\omega)$ occurs at $\omega = \omega_n\sqrt{1 - 2\zeta^2}$. Thus the resonant frequency ω_r is

$$\omega_r = \omega_n\sqrt{1 - 2\zeta^2}, \qquad \text{for } 0 \le \zeta \le 0.707 \tag{7–12}$$

As the damping ratio ζ approaches zero, the resonant frequency approaches ω_n. For $0 < \zeta \le 0.707$, the resonant frequency ω_r is less than the damped natural frequency $\omega_d = \omega_n\sqrt{1 - \zeta^2}$, which is exhibited in the transient response. From Equation (7–12), it can be seen that for $\zeta > 0.707$, there is no resonant peak. The magnitude $|G(j\omega)|$ decreases monotonically with increasing frequency ω. (The magnitude is less than 0 dB for all values of $\omega > 0$. Recall that, for $0.7 < \zeta < 1$, the step response is oscillatory, but the oscillations are well damped and are hardly perceptible.)

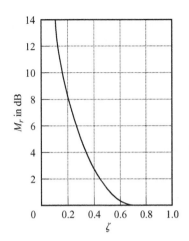

Figure 7–10
M_r-versus-ζ curve for the second-order system $1/[1 + 2\zeta(j\omega/\omega_n) + (j\omega/\omega_n)^2]$.

For $0 \le \zeta \le 0.707$, the magnitude of the resonant peak, $M_r = |G(j\omega_r)|$, can be found from Equations (7–12) and (7–9). For $0 \le \zeta \le 0.707$,

$$M_r = |G(j\omega)|_{\max} = |G(j\omega_r)| = \frac{1}{2\zeta\sqrt{1 - \zeta^2}} \qquad (7\text{–}13)$$

For $\zeta > 0.707$,

$$M_r = 1 \qquad (7\text{–}14)$$

As ζ approaches zero, M_r approaches infinity. This means that if the undamped system is excited at its natural frequency, the magnitude of $G(j\omega)$ becomes infinity. The relationship between M_r and ζ is shown in Figure 7–10.

The phase angle of $G(j\omega)$ at the frequency where the resonant peak occurs can be obtained by substituting Equation (7–12) into Equation (7–8). Thus, at the resonant frequency ω_r,

$$\underline{/G(j\omega_r)} = -\tan^{-1}\frac{\sqrt{1 - 2\zeta^2}}{\zeta} = -90° + \sin^{-1}\frac{\zeta}{\sqrt{1 - \zeta^2}}$$

General Procedure for Plotting Bode Diagrams. MATLAB provides an easy way to plot Bode diagrams. (The MATLAB approach is presented later in this section.) Here, however, we consider the case where we want to draw Bode diagrams manually without using MATLAB.

First rewrite the sinusoidal transfer function $G(j\omega)H(j\omega)$ as a product of basic factors discussed above. Then identify the corner frequencies associated with these basic factors. Finally, draw the asymptotic log-magnitude curves with proper slopes between the corner frequencies. The exact curve, which lies close to the asymptotic curve, can be obtained by adding proper corrections.

The phase-angle curve of $G(j\omega)H(j\omega)$ can be drawn by adding the phase-angle curves of individual factors.

The use of Bode diagrams employing asymptotic approximations requires much less time than other methods that may be used for computing the frequency response of a transfer function. The ease of plotting the frequency-response curves for a given transfer function and the ease of modification of the frequency-response curve as compensation is added are the main reasons why Bode diagrams are very frequently used in practice.

EXAMPLE 7–3 Draw the Bode diagram for the following transfer function:

$$G(j\omega) = \frac{10(j\omega + 3)}{(j\omega)(j\omega + 2)[(j\omega)^2 + j\omega + 2]}$$

Make corrections so that the log-magnitude curve is accurate.

To avoid any possible mistakes in drawing the log-magnitude curve, it is desirable to put $G(j\omega)$ in the following normalized form, where the low-frequency asymptotes for the first-order factors and the second-order factor are the 0-dB line:

$$G(j\omega) = \frac{7.5\left(\dfrac{j\omega}{3} + 1\right)}{(j\omega)\left(\dfrac{j\omega}{2} + 1\right)\left[\dfrac{(j\omega)^2}{2} + \dfrac{j\omega}{2} + 1\right]}$$

This function is composed of the following factors:

$$7.5, \quad (j\omega)^{-1}, \quad 1 + j\frac{\omega}{3}, \quad \left(1 + j\frac{\omega}{2}\right)^{-1}, \quad \left[1 + j\frac{\omega}{2} + \frac{(j\omega)^2}{2}\right]^{-1}$$

The corner frequencies of the third, fourth, and fifth terms are $\omega = 3$, $\omega = 2$, and $\omega = \sqrt{2}$, respectively. Note that the last term has the damping ratio of 0.3536.

To plot the Bode diagram, the separate asymptotic curves for each of the factors are shown in Figure 7–11. The composite curve is then obtained by algebraically adding the individual curves, also shown in Figure 7–11. Note that when the individual asymptotic curves are added at each frequency, the slope of the composite curve is cumulative. Below $\omega = \sqrt{2}$, the plot has the slope of -20 dB/decade. At the first corner frequency $\omega = \sqrt{2}$, the slope changes to -60 dB/decade and continues to the next corner frequency $\omega = 2$, where the slope becomes -80 dB/decade. At the last corner frequency $\omega = 3$, the slope changes to -60 dB/decade.

Once such an approximate log-magnitude curve has been drawn, the actual curve can be obtained by adding corrections at each corner frequency and at frequencies one octave below and above the corner frequencies. For first-order factors $(1 + j\omega T)^{\mp1}$, the corrections are ±3 dB at the corner frequency and ±1 dB at the frequencies one octave below and above the corner frequency. Corrections necessary for the quadratic factor are obtained from Figure 7–9. The exact log-magnitude curve for $G(j\omega)$ is shown by a dashed curve in Figure 7–11.

Note that any change in the slope of the magnitude curve is made only at the corner frequencies of the transfer function $G(j\omega)$. Therefore, instead of drawing individual magnitude curves and adding them up, as shown, we may sketch the magnitude curve without sketching individual curves. We may start drawing the lowest-frequency portion of the straight line (that is, the straight line with the slope -20 dB/decade for $\omega < \sqrt{2}$). As the frequency is increased, we get the effect of the complex-conjugate poles (quadratic term) at the corner frequency $\omega = \sqrt{2}$. The complex-conjugate poles cause the slopes of the magnitude curve to change from -20 to -60 dB/decade. At the next corner frequency, $\omega = 2$, the effect of the pole is to change the slope to -80 dB/decade. Finally, at the corner frequency $\omega = 3$, the effect of the zero is to change the slope from -80 to -60 dB/decade.

For plotting the complete phase-angle curve, the phase-angle curves for all factors have to be sketched. The algebraic sum of all phase-angle curves provides the complete phase-angle curve, as shown in Figure 7–11.

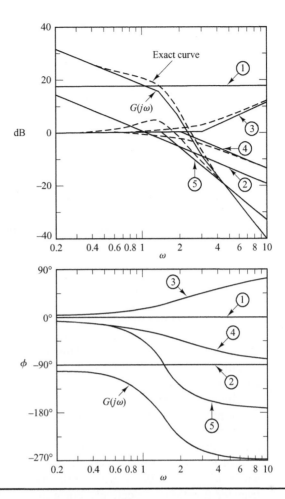

Figure 7–11
Bode diagram of the
system considered in
Example 7–3.

Minimum-Phase Systems and Nonminimum-Phase Systems. Transfer functions having neither poles nor zeros in the right-half s plane are minimum-phase transfer functions, whereas those having poles and/or zeros in the right-half s plane are nonminimum-phase transfer functions. Systems with minimum-phase transfer functions are called *minimum-phase* systems, whereas those with nonminimum-phase transfer functions are called *nonminimum-phase* systems.

For systems with the same magnitude characteristic, the range in phase angle of the minimum-phase transfer function is minimum among all such systems, while the range in phase angle of any nonminimum-phase transfer function is greater than this minimum.

It is noted that for a minimum-phase system, the transfer function can be uniquely determined from the magnitude curve alone. For a nonminimum-phase system, this is not the case. Multiplying any transfer function by all-pass filters does not alter the magnitude curve, but the phase curve is changed.

Consider as an example the two systems whose sinusoidal transfer functions are, respectively,

$$G_1(j\omega) = \frac{1 + j\omega T}{1 + j\omega T_1}, \qquad G_2(j\omega) = \frac{1 - j\omega T}{1 + j\omega T_1}, \qquad 0 < T < T_1$$

Figure 7–12
Pole–zero
configurations of a
minimum-phase
system $G_1(s)$ and
nonminimum-phase
system $G_2(s)$.

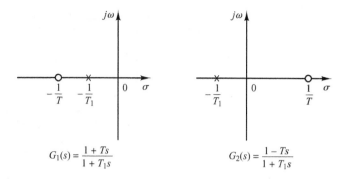

$$G_1(s) = \frac{1 + Ts}{1 + T_1 s} \qquad G_2(s) = \frac{1 - Ts}{1 + T_1 s}$$

The pole–zero configurations of these systems are shown in Figure 7–12. The two sinu-soidal transfer functions have the same magnitude characteristics, but they have differ-ent phase-angle characteristics, as shown in Figure 7–13. These two systems differ from each other by the factor

$$G(j\omega) = \frac{1 - j\omega T}{1 + j\omega T}$$

The magnitude of the factor $(1 - j\omega T)/(1 + j\omega T)$ is always unity. But the phase angle equals $-2 \tan^{-1} \omega T$ and varies from $0°$ to $-180°$ as ω is increased from zero to infinity.

As stated earlier, for a minimum-phase system, the magnitude and phase-angle char-acteristics are uniquely related. This means that if the magnitude curve of a system is specified over the entire frequency range from zero to infinity, then the phase-angle curve is uniquely determined, and vice versa. This, however, does not hold for a non-minimum-phase system.

Nonminimum-phase situations may arise in two different ways. One is simply when a system includes a nonminimum-phase element or elements. The other situation may arise in the case where a minor loop is unstable.

For a minimum-phase system, the phase angle at $\omega = \infty$ becomes $-90°(q - p)$, where p and q are the degrees of the numerator and denominator polynomials of the transfer function, respectively. For a nonminimum-phase system, the phase angle at $\omega = \infty$ differs from $-90°(q - p)$. In either system, the slope of the log-magnitude curve at $\omega = \infty$ is equal to $-20(q - p)$ dB/decade. It is therefore possible to detect whether the system is minimum phase by examining both the slope of the high-frequency asymptote of the log-magnitude curve and the phase angle at $\omega = \infty$. If the slope of the log-magnitude curve as ω approaches infinity is $-20(q - p)$ dB/decade and the phase angle at $\omega = \infty$ is equal to $-90°(q - p)$, then the system is minimum phase.

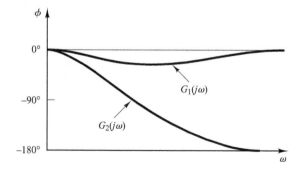

Figure 7–13
Phase-angle
characteristics of the
systems $G_1(s)$ and
$G_2(s)$ shown in
Figure 7–12.

Nonminimum-phase systems are slow in responding because of their faulty behavior at the start of a response. In most practical control systems, excessive phase lag should be carefully avoided. In designing a system, if fast speed of response is of primary importance, we should not use nonminimum-phase components. (A common example of nonminimum-phase elements that may be present in control systems is transport lag or dead time.)

It is noted that the techniques of frequency-response analysis and design to be presented in this and the next chapter are valid for both minimum-phase and nonminimum-phase systems.

Transport Lag. Transport lag, which is also called dead time, is of nonminimum-phase behavior and has an excessive phase lag with no attenuation at high frequencies. Such transport lags normally exist in thermal, hydraulic, and pneumatic systems.

Consider the transport lag given by

$$G(j\omega) = e^{-j\omega T}$$

The magnitude is always equal to unity, since

$$|G(j\omega)| = |\cos \omega T - j \sin \omega T| = 1$$

Therefore, the log magnitude of the transport lag $e^{-j\omega T}$ is equal to 0 dB. The phase angle of the transport lag is

$$\underline{/G(j\omega)} = -\omega T \quad \text{(radians)}$$
$$= -57.3\,\omega T \quad \text{(degrees)}$$

The phase angle varies linearly with the frequency ω. The phase-angle characteristic of transport lag is shown in Figure 7–14.

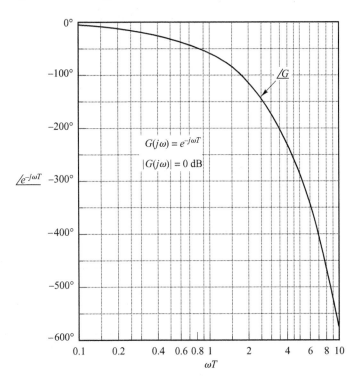

Figure 7–14
Phase-angle characteristic of transport lag.

EXAMPLE 7–4   Draw the Bode diagram of the following transfer function:

$$G(j\omega) = \frac{e^{-j\omega L}}{1 + j\omega T}$$

The log magnitude is

$$20 \log |G(j\omega)| = 20 \log |e^{-j\omega L}| + 20 \log \left| \frac{1}{1 + j\omega T} \right|$$

$$= 0 + 20 \log \left| \frac{1}{1 + j\omega T} \right|$$

The phase angle of $G(j\omega)$ is

$$\underline{/G(j\omega)} = \underline{/e^{-j\omega L}} + \underline{\Big/ \frac{1}{1 + j\omega T}}$$

$$= -\omega L - \tan^{-1} \omega T$$

The log-magnitude and phase-angle curves for this transfer function with $L = 0.5$ and $T = 1$ are shown in Figure 7–15.

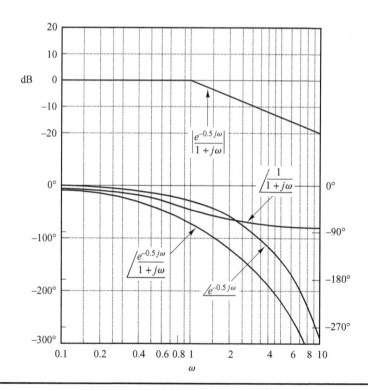

Figure 7–15
Bode diagram for the system $e^{-j\omega L}/(1 + j\omega T)$ with $L = 0.5$ and $T = 1$.

Relationship between System Type and Log-Magnitude Curve. Consider the unity-feedback control system. The static position, velocity, and acceleration error constants describe the low-frequency behavior of type 0, type 1, and type 2 systems, respectively. For a given system, only one of the static error constants is finite and significant. (The larger the value of the finite static error constant, the higher the loop gain is as ω approaches zero.)

The type of the system determines the slope of the log-magnitude curve at low frequencies. Thus, information concerning the existence and magnitude of the steady-state error of a control system to a given input can be determined from the observation of the low-frequency region of the log-magnitude curve.

Determination of Static Position Error Constants. Consider the unity-feedback control system shown in Figure 7–16. Assume that the open-loop transfer function is given by

$$G(s) = \frac{K(T_a s + 1)(T_b s + 1)\cdots(T_m s + 1)}{s^N(T_1 s + 1)(T_2 s + 1)\cdots(T_p s + 1)}$$

or

$$G(j\omega) = \frac{K(T_a j\omega + 1)(T_b j\omega + 1)\cdots(T_m j\omega + 1)}{(j\omega)^N(T_1 j\omega + 1)(T_2 j\omega + 1)\cdots(T_p j\omega + 1)}$$

Figure 7–17 shows an example of the log-magnitude plot of a type 0 system. In such a system, the magnitude of $G(j\omega)$ equals K_p at low frequencies, or

$$\lim_{\omega \to 0} G(j\omega) = K = K_p$$

It follows that the low-frequency asymptote is a horizontal line at $20 \log K_p$ dB.

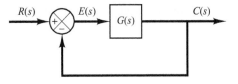

Figure 7–16
Unity-feedback
control system.

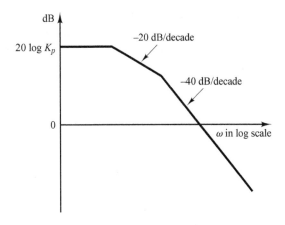

Figure 7–17
Log-magnitude curve
of a type 0 system.

Determination of Static Velocity Error Constants. Consider the unity-feedback control system shown in Figure 7–16. Figure 7–18 shows an example of the log-magnitude plot of a type 1 system. The intersection of the initial −20-dB/decade segment (or its extension) with the line $\omega = 1$ has the magnitude $20 \log K_v$. This may be seen as follows: In a type 1 system

$$G(j\omega) = \frac{K_v}{j\omega}, \qquad \text{for } \omega \ll 1$$

Thus,

$$20 \log \left| \frac{K_v}{j\omega} \right|_{\omega=1} = 20 \log K_v$$

The intersection of the initial −20-dB/decade segment (or its extension) with the 0-dB line has a frequency numerically equal to K_v. To see this, define the frequency at this intersection to be ω_1; then

$$\left| \frac{K_v}{j\omega_1} \right| = 1$$

or

$$K_v = \omega_1$$

As an example, consider the type 1 system with unity feedback whose open-loop transfer function is

$$G(s) = \frac{K}{s(Js + F)}$$

If we define the corner frequency to be ω_2 and the frequency at the intersection of the −40-dB/decade segment (or its extension) with 0-dB line to be ω_3, then

$$\omega_2 = \frac{F}{J}, \qquad \omega_3^2 = \frac{K}{J}$$

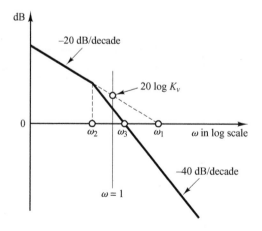

Figure 7–18
Log-magnitude curve
of a type 1 system.

Since

$$\omega_1 = K_v = \frac{K}{F}$$

it follows that

$$\omega_1 \omega_2 = \omega_3^2$$

or

$$\frac{\omega_1}{\omega_3} = \frac{\omega_3}{\omega_2}$$

On the Bode diagram,

$$\log \omega_1 - \log \omega_3 = \log \omega_3 - \log \omega_2$$

Thus, the ω_3 point is just midway between the ω_2 and ω_1 points. The damping ratio ζ of the system is then

$$\zeta = \frac{F}{2\sqrt{KJ}} = \frac{\omega_2}{2\omega_3}$$

Determination of Static Acceleration Error Constants. Consider the unity-feedback control system shown in Figure 7–16. Figure 7–19 shows an example of the log-magnitude plot of a type 2 system. The intersection of the initial −40-dB/decade segment (or its extension) with the $\omega = 1$ line has the magnitude of $20 \log K_a$. Since at low frequencies

$$G(j\omega) = \frac{K_a}{(j\omega)^2}, \qquad \text{for } \omega \ll 1$$

it follows that

$$20 \log \left| \frac{K_a}{(j\omega)^2} \right|_{\omega=1} = 20 \log K_a$$

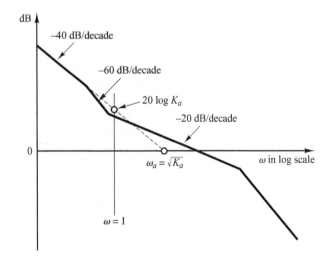

Figure 7–19
Log-magnitude curve
of a type 2 system.

The frequency ω_a at the intersection of the initial -40-dB/decade segment (or its extension) with the 0-dB line gives the square root of K_a numerically. This can be seen from the following:

$$20 \log \left| \frac{K_a}{(j\omega_a)^2} \right| = 20 \log 1 = 0$$

which yields

$$\omega_a = \sqrt{K_a}$$

Plotting Bode Diagrams with MATLAB. The command bode computes magnitudes and phase angles of the frequency response of continuous-time, linear, time-invariant systems.

When the command bode (without left-hand arguments) is entered in the computer, MATLAB produces a Bode plot on the screen. Most commonly used bode commands are

```
bode(num,den)
bode(num,den,w)
bode(A,B,C,D)
bode(A,B,C,D,w)
bode(A,B,C,D,iu,w)
bode(sys)
```

When invoked with left-hand arguments, such as

```
[mag,phase,w] = bode(num,den,w)
```

bode returns the frequency response of the system in matrices mag, phase, and w. No plot is drawn on the screen. The matrices mag and phase contain magnitudes and phase angles of the frequency response of the system, evaluated at user-specified frequency points. The phase angle is returned in degrees. The magnitude can be converted to decibels with the statement

```
magdB = 20*log10(mag)
```

Other Bode commands with left-hand arguments are

```
[mag,phase,w] = bode(num,den)
[mag,phase,w] = bode(num,den,w)
[mag,phase,w] = bode(A,B,C,D)
[mag,phase,w] = bode(A,B.C,D,w)
[mag,phase,w] = bode(A,B,C,D,iu,w)
[mag,phase,w] = bode(sys)
```

To specify the frequency range, use the command logspace(d1,d2) or logspace (d1,d2,n). logspace(d1,d2) generates a vector of 50 points logarithmically equally spaced between decades 10^{d1} and 10^{d2}. (50 points include both endpoints. There are 48 points between the endpoints.) To generate 50 points between 0.1 rad/sec and 100 rad/sec, enter the command

```
w = logspace(-1,2)
```

logspace(dl,d2,n) generates n points logarithmically equally spaced between decades 10^{d1} and 10^{d2}. (n points include both endpoints.) For example, to generate 100 points including both endpoints between 1 rad/sec and 1000 rad/sec, enter the following command:

$$w = logspace(0,3,100)$$

To incorporate the user-specified frequency points when plotting Bode diagrams, the bode command must include the frequency vector w, such as bode(num,den,w) and [mag,phase,w] = bode(A,B,C,D,w).

EXAMPLE 7–5 Consider the following transfer function:

$$G(s) = \frac{25}{s^2 + 4s + 25}$$

Plot a Bode diagram for this transfer function.
When the system is defined in the form

$$G(s) = \frac{num(s)}{den(s)}$$

use the command bode(num,den) to draw the Bode diagram. [When the numerator and denominator contain the polynomial coefficients in descending powers of s, bode(num,den) draws the Bode diagram.] MATLAB Program 7–1 shows a program to plot the Bode diagram for this system. The resulting Bode diagram is shown in Figure 7–20.

MATLAB Program 7–1

```
num = [25];
den = [1  4  25];
bode(num,den)
title('Bode Diagram of G(s) = 25/(s^2 + 4s + 25)')
```

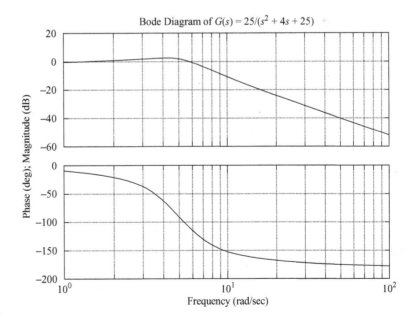

Figure 7–20
Bode diagram of
$G(s) = \dfrac{25}{s^2 + 4s + 25}$.

EXAMPLE 7–6   Consider the system shown in Figure 7–21. The open-loop transfer function is

$$G(s) = \frac{9(s^2 + 0.2s + 1)}{s(s^2 + 1.2s + 9)}$$

Plot a bode diagram.

MATLAB Program 7–2 plots a Bode diagram for the system. The resulting plot is shown in Figure 7–22. The frequency range in this case is automatically determined to be from 0.01 to 10 rad/sec.

MATLAB Program 7–2

```
num = [9  1.8  9];
den = [1  1.2  9  0];
bode(num,den)
title('Bode Diagram of G(s) = 9(s^2 + 0.2s + 1)/[s(s^2 + 1.2s + 9)]')
```

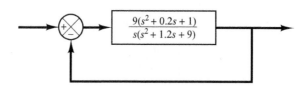

Figure 7–21
Control system.

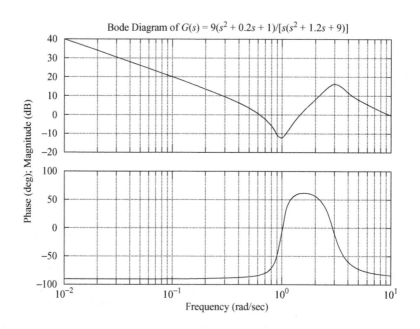

Figure 7–22
Bode diagram of
$$G(s) = \frac{9(s^2 + 0.2s + 1)}{s(s^2 + 1.2s + 9)}.$$

If it is desired to plot the Bode diagram from 0.01 to 1000 rad/sec, enter the following command:

$$w = \text{logspace}(-2,3,100)$$

This command generates 100 points logarithmically equally spaced between 0.01 and 1000 rad/sec. (Note that such a vector w specifies the frequencies in radians per second at which the frequency response will be calculated.)

If we use the command

$$\text{bode(num,den,w)}$$

then the frequency range is as the user specified, but the magnitude range and phase-angle range will be automatically determined. See MATLAB Program 7–3 and the resulting plot in Figure 7–23.

MATLAB Program 7–3

```
num = [9  1.8  9];
den = [1  1.2  9  0];
w = logspace(-2,3,100);
bode(num,den,w)
title('Bode Diagram of G(s) = 9(s^2 + 0.2s + 1)/[s(s^2 + 1.2s + 9)]')
```

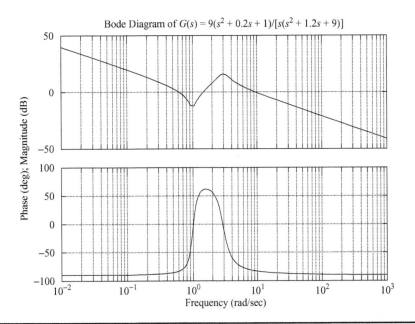

Bode Diagram of $G(s) = 9(s^2 + 0.2s + 1)/[s(s^2 + 1.2s + 9)]$

Figure 7–23
Bode diagram of
$$G(s) = \frac{9(s^2 + 0.2s + 1)}{s(s^2 + 1.2s + 9)}.$$

Obtaining Bode Diagrams of Systems Defined in State Space. Consider the system defined by

$$\dot{\mathbf{x}} = \mathbf{A}\mathbf{x} + \mathbf{B}\mathbf{u}$$
$$\mathbf{y} = \mathbf{C}\mathbf{x} + \mathbf{D}\mathbf{u}$$

where $\mathbf{x}$ = state vector (n-vector)
  $\mathbf{y}$ = output vector (m-vector)
  $\mathbf{u}$ = control vector (r-vector)
  $\mathbf{A}$ = state matrix ($n \times n$ matrix)
  $\mathbf{B}$ = control matrix ($n \times r$ matrix)
  $\mathbf{C}$ = output matrix ($m \times n$ matrix)
  $\mathbf{D}$ = direct transmission matrix ($m \times r$ matrix)

A Bode diagram for this system may be obtained by entering the command

$$\text{bode(A,B,C,D)}$$

or others listed earlier in this section.

The command bode(A,B,C,D) produces a series of Bode plots, one for each input of the system, with the frequency range automatically determined. (More points are used when the response is changing rapidly.)

The command bode(A,B,C,D,iu), where iu is the ith input of the system, produces the Bode diagrams from the input iu to all the outputs $(y_1, y_2, \ldots, y_m)$ of the system, with a frequency range automatically determined. (The scalar iu is an index into the inputs of the system and specifies which input is to be used for plotting Bode diagrams). If the control vector $\mathbf{u}$ has three inputs such that

$$\mathbf{u} = \begin{bmatrix} u_1 \\ u_2 \\ u_3 \end{bmatrix}$$

then iu must be set to either 1, 2, or 3.

If the system has only one input u, then either of the following commands may be used:

$$\text{bode(A,B,C,D)}$$

or

$$\text{bode(A,B,C,D,1)}$$

EXAMPLE 7–7 Consider the following system:

$$\begin{bmatrix} \dot{x}_1 \\ \dot{x}_2 \end{bmatrix} = \begin{bmatrix} 0 & 1 \\ -25 & -4 \end{bmatrix} \begin{bmatrix} x_1 \\ x_2 \end{bmatrix} + \begin{bmatrix} 0 \\ 25 \end{bmatrix} u$$

$$y = \begin{bmatrix} 1 & 0 \end{bmatrix} \begin{bmatrix} x_1 \\ x_2 \end{bmatrix}$$

This system has one input u and one output y. By using the command

$$\text{bode(A,B,C,D)}$$

and entering MATLAB Program 7–4 into the computer, we obtain the Bode diagram shown in Figure 7–24.

MATLAB Program 7–4
A = [0 1;-25 -4]; B = [0;25]; C = [1 0]; D = [0]; bode(A,B,C,D) title('Bode Diagram')

If we replace the command bode(A,B,C,D) in MATLAB Program 7–4 with

$$bode(A,B,C,D,1)$$

then MATLAB will produce the Bode diagram identical to that shown in Figure 7–24.

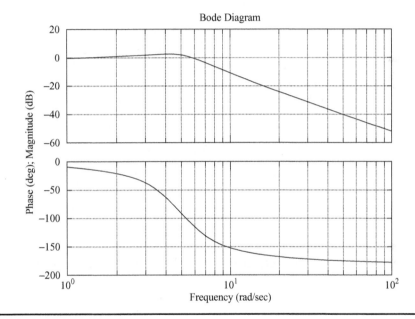

Figure 7–24
Bode diagram of the system considered in Example 7–7.

7–3 POLAR PLOTS

The polar plot of a sinusoidal transfer function $G(j\omega)$ is a plot of the magnitude of $G(j\omega)$ versus the phase angle of $G(j\omega)$ on polar coordinates as ω is varied from zero to infinity. Thus, the polar plot is the locus of vectors $|G(j\omega)| \underline{/G(j\omega)}$ as ω is varied from zero to infinity. Note that in polar plots a positive (negative) phase angle is measured counterclockwise (clockwise) from the positive real axis. The polar plot is often called the Nyquist plot. An example of such a plot is shown in Figure 7–25. Each point on the polar plot of $G(j\omega)$ represents the terminal point of a vector at a particular value of ω. In the polar plot, it is important to show the frequency graduation of the locus. The projections of $G(j\omega)$ on the real and imaginary axes are its real and imaginary components.

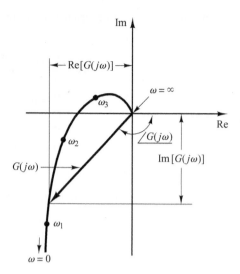

Figure 7–25
Polar plot.

MATLAB may be used to obtain a polar plot $G(j\omega)$ or to obtain $|G(j\omega)|$ and $\underline{/G(j\omega)}$ accurately for various values of ω in the frequency range of interest.

An advantage in using a polar plot is that it depicts the frequency-response characteristics of a system over the entire frequency range in a single plot. One disadvantage is that the plot does not clearly indicate the contributions of each individual factor of the open-loop transfer function.

Integral and Derivative Factors $(j\omega)^{\mp 1}$. The polar plot of $G(j\omega) = 1/j\omega$ is the negative imaginary axis, since

$$G(j\omega) = \frac{1}{j\omega} = -j\frac{1}{\omega} = \frac{1}{\omega}\,\underline{/-90°}$$

The polar plot of $G(j\omega) = j\omega$ is the positive imaginary axis.

First-Order Factors $(1 + j\omega T)^{\mp 1}$. For the sinusoidal transfer function

$$G(j\omega) = \frac{1}{1 + j\omega T} = \frac{1}{\sqrt{1 + \omega^2 T^2}}\,\underline{/-\tan^{-1}\omega T}$$

the values of $G(j\omega)$ at $\omega = 0$ and $\omega = 1/T$ are, respectively,

$$G(j0) = 1\,\underline{/0°} \quad \text{and} \quad G\!\left(j\frac{1}{T}\right) = \frac{1}{\sqrt{2}}\,\underline{/-45°}$$

If ω approaches infinity, the magnitude of $G(j\omega)$ approaches zero and the phase angle approaches $-90°$. The polar plot of this transfer function is a semicircle as the frequency ω is varied from zero to infinity, as shown in Figure 7–26(a). The center is located at 0.5 on the real axis, and the radius is equal to 0.5.

To prove that the polar plot of the first-order factor $G(j\omega) = 1/(1 + j\omega T)$ is a semicircle, define

$$G(j\omega) = X + jY$$

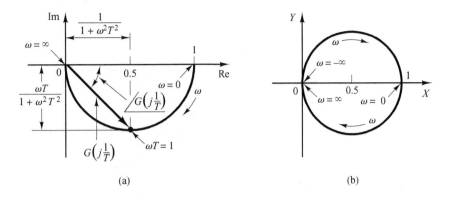

Figure 7–26
(a) Polar plot of $1/(1 + j\omega T)$; (b) plot of $G(j\omega)$ in X-Y plane.

where

$$X = \frac{1}{1 + \omega^2 T^2} = \text{real part of } G(j\omega)$$

$$Y = \frac{-\omega T}{1 + \omega^2 T^2} = \text{imaginary part of } G(j\omega)$$

Then we obtain

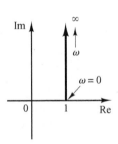

$$\left(X - \frac{1}{2}\right)^2 + Y^2 = \left(\frac{1}{2}\frac{1 - \omega^2 T^2}{1 + \omega^2 T^2}\right)^2 + \left(\frac{-\omega T}{1 + \omega^2 T^2}\right)^2 = \left(\frac{1}{2}\right)^2$$

Thus, in the X-Y plane $G(j\omega)$ is a circle with center at $X = \frac{1}{2}, Y = 0$ and with radius $\frac{1}{2}$, as shown in Figure 7–26(b). The lower semicircle corresponds to $0 \leq \omega \leq \infty$, and the upper semicircle corresponds to $-\infty \leq \omega \leq 0$.

The polar plot of the transfer function $1 + j\omega T$ is simply the upper half of the straight line passing through point $(1,0)$ in the complex plane and parallel to the imaginary axis, as shown in Figure 7–27. The polar plot of $1 + j\omega T$ has an appearance completely different from that of $1/(1 + j\omega T)$.

Figure 7–27
Polar plot of $1 + j\omega T$.

Quadratic Factors $[1 + 2\zeta(j\omega/\omega_n) + (j\omega/\omega_n)^2]^{\mp 1}$. The low- and high-frequency portions of the polar plot of the following sinusoidal transfer function

$$G(j\omega) = \frac{1}{1 + 2\zeta\left(j\dfrac{\omega}{\omega_n}\right) + \left(j\dfrac{\omega}{\omega_n}\right)^2}, \qquad \text{for } \zeta > 0$$

are given, respectively, by

$$\lim_{\omega \to 0} G(j\omega) = 1\,\underline{/0^\circ} \quad \text{and} \quad \lim_{\omega \to \infty} G(j\omega) = 0\,\underline{/-180^\circ}$$

The polar plot of this sinusoidal transfer function starts at $1\,\underline{/0^\circ}$ and ends at $0\,\underline{/-180^\circ}$ as ω increases from zero to infinity. Thus, the high-frequency portion of $G(j\omega)$ is tangent to the negative real axis.

Section 7–3 / Polar Plots

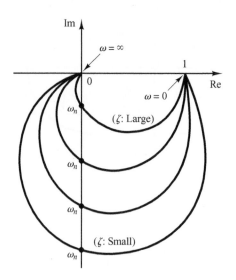

Figure 7–28
Polar plots of
$$\frac{1}{1 + 2\zeta\left(j\dfrac{\omega}{\omega_n}\right) + \left(j\dfrac{\omega}{\omega_n}\right)^2} \quad \text{for } \zeta > 0.$$

Examples of polar plots of the transfer function just considered are shown in Figure 7–28. The exact shape of a polar plot depends on the value of the damping ratio ζ, but the general shape of the plot is the same for both the underdamped case $(1 > \zeta > 0)$ and overdamped case $(\zeta > 1)$.

For the underdamped case at $\omega = \omega_n$, we have $G(j\omega_n) = 1/(j2\zeta)$, and the phase angle at $\omega = \omega_n$ is $-90°$. Therefore, it can be seen that the frequency at which the $G(j\omega)$ locus intersects the imaginary axis is the undamped natural frequency ω_n. In the polar plot, the frequency point whose distance from the origin is maximum corresponds to the resonant frequency ω_r. The peak value of $G(j\omega)$ is obtained as the ratio of the magnitude of the vector at the resonant frequency ω_r to the magnitude of the vector at $\omega = 0$. The resonant frequency ω_r is indicated in the polar plot shown in Figure 7–29.

For the overdamped case, as ζ increases well beyond unity, the $G(j\omega)$ locus approaches a semicircle. This may be seen from the fact that, for a heavily damped system, the characteristic roots are real, and one is much smaller than the other. Since, for sufficiently large ζ, the effect of the larger root (larger in the absolute value) on the response becomes very small, the system behaves like a first-order one.

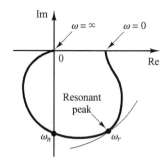

Figure 7–29
Polar plot showing the resonant peak and resonant frequency ω_r.

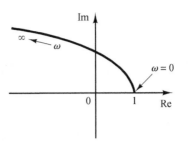

Figure 7–30
Polar plot of
$1 + 2\zeta\left(j\dfrac{\omega}{\omega_n}\right) + \left(j\dfrac{\omega}{\omega_n}\right)^2$ for $\zeta > 0$.

Next, consider the following sinusoidal transfer function:

$$G(j\omega) = 1 + 2\zeta\left(j\frac{\omega}{\omega_n}\right) + \left(j\frac{\omega}{\omega_n}\right)^2$$

$$= \left(1 - \frac{\omega^2}{\omega_n^2}\right) + j\left(\frac{2\zeta\omega}{\omega_n}\right)$$

The low-frequency portion of the curve is

$$\lim_{\omega \to 0} G(j\omega) = 1\,\underline{/0^\circ}$$

and the high-frequency portion is

$$\lim_{\omega \to \infty} G(j\omega) = \infty\,\underline{/180^\circ}$$

Since the imaginary part of $G(j\omega)$ is positive for $\omega > 0$ and is monotonically increasing, and the real part of $G(j\omega)$ is monotonically decreasing from unity, the general shape of the polar plot of $G(j\omega)$ is as shown in Figure 7–30. The phase angle is between 0° and 180°.

EXAMPLE 7–8 Consider the following second-order transfer function:

$$G(s) = \frac{1}{s(Ts + 1)}$$

Sketch a polar plot of this transfer function.
Since the sinusoidal transfer function can be written

$$G(j\omega) = \frac{1}{j\omega(1 + j\omega T)} = -\frac{T}{1 + \omega^2 T^2} - j\frac{1}{\omega(1 + \omega^2 T^2)}$$

the low-frequency portion of the polar plot becomes

$$\lim_{\omega \to 0} G(j\omega) = -T - j\infty$$

and the high-frequency portion becomes

$$\lim_{\omega \to \infty} G(j\omega) = 0 - j0$$

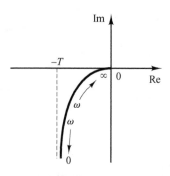

Figure 7–31
Polar plot of
$1/[j\omega(1 + j\omega T)]$.

The general shape of the polar plot of $G(j\omega)$ is shown in Figure 7–31. The $G(j\omega)$ plot is asymptotic to the vertical line passing through the point $(-T, 0)$. Since this transfer function involves an integrator $(1/s)$, the general shape of the polar plot differs substantially from those of second-order transfer functions that do not have an integrator.

EXAMPLE 7–9   Obtain the polar plot of the following transfer function:

$$G(j\omega) = \frac{e^{-j\omega L}}{1 + j\omega T}$$

Since $G(j\omega)$ can be written

$$G(j\omega) = \left(e^{-j\omega L}\right)\left(\frac{1}{1 + j\omega T}\right)$$

the magnitude and phase angle are, respectively,

$$|G(j\omega)| = |e^{-j\omega L}| \cdot \left|\frac{1}{1 + j\omega T}\right| = \frac{1}{\sqrt{1 + \omega^2 T^2}}$$

and

$$\underline{/G(j\omega)} = \underline{/e^{-j\omega L}} + \underline{/\frac{1}{1 + j\omega T}} = -\omega L - \tan^{-1}\omega T$$

Since the magnitude decreases from unity monotonically and the phase angle also decreases monotonically and indefinitely, the polar plot of the given transfer function is a spiral, as shown in Figure 7–32.

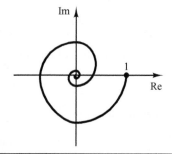

Figure 7–32
Polar plot of
$e^{-j\omega L}/(1 + j\omega T)$.

General Shapes of Polar Plots. The polar plots of a transfer function of the form

$$G(j\omega) = \frac{K(1 + j\omega T_a)(1 + j\omega T_b)\cdots}{(j\omega)^\lambda(1 + j\omega T_1)(1 + j\omega T_2)\cdots}$$

$$= \frac{b_0(j\omega)^m + b_1(j\omega)^{m-1} + \cdots}{a_0(j\omega)^n + a_1(j\omega)^{n-1} + \cdots}$$

where $n > m$ or the degree of the denominator polynomial is greater than that of the numerator, will have the following general shapes:

1. *For $\lambda = 0$ or type 0 systems:* The starting point of the polar plot (which corresponds to $\omega = 0$) is finite and is on the positive real axis. The tangent to the polar plot at $\omega = 0$ is perpendicular to the real axis. The terminal point, which corresponds to $\omega = \infty$, is at the origin, and the curve is tangent to one of the axes.

2. *For $\lambda = 1$ or type 1 systems:* The $j\omega$ term in the denominator contributes $-90°$ to the total phase angle of $G(j\omega)$ for $0 \le \omega \le \infty$. At $\omega = 0$, the magnitude of $G(j\omega)$ is infinity, and the phase angle becomes $-90°$. At low frequencies, the polar plot is asymptotic to a line parallel to the negative imaginary axis. At $\omega = \infty$, the magnitude becomes zero, and the curve converges to the origin and is tangent to one of the axes.

3. *For $\lambda = 2$ or type 2 systems:* The $(j\omega)^2$ term in the denominator contributes $-180°$ to the total phase angle of $G(j\omega)$ for $0 \le \omega \le \infty$. At $\omega = 0$, the magnitude of $G(j\omega)$ is infinity, and the phase angle is equal to $-180°$. At low frequencies, the polar plot may be asymptotic to the negative real axis. At $\omega = \infty$, the magnitude becomes zero, and the curve is tangent to one of the axes.

The general shapes of the low-frequency portions of the polar plots of type 0, type 1, and type 2 systems are shown in Figure 7–33. It can be seen that, if the degree of the

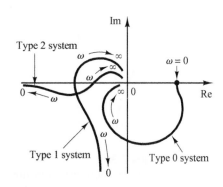

Figure 7–33
Polar plots of type 0,
type 1, and type 2
systems.

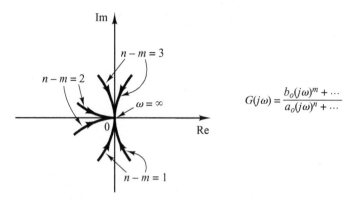

$$G(j\omega) = \frac{b_o(j\omega)^m + \cdots}{a_o(j\omega)^n + \cdots}$$

Figure 7–34
Polar plots in the high-frequency range.

denominator polynomial of $G(j\omega)$ is greater than that of the numerator, then the $G(j\omega)$ loci converge to the origin clockwise. At $\omega = \infty$, the loci are tangent to one or the other axes, as shown in Figure 7–34.

Note that any complicated shapes in the polar plot curves are caused by the numerator dynamics—that is, by the time constants in the numerator of the transfer function. Figure 7–35 shows examples of polar plots of transfer functions with numerator dynamics. In analyzing control systems, the polar plot of $G(j\omega)$ in the frequency range of interest must be accurately determined.

Table 7–1 shows sketches of polar plots of several transfer functions.

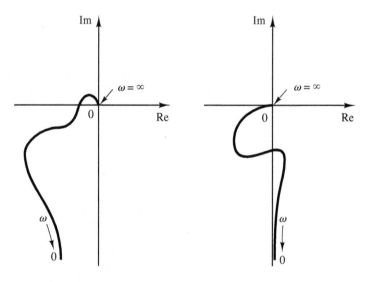

Figure 7–35
Polar plots of transfer functions with numerator dynamics.

Table 7–1 Polar Plots of Simple Transfer Functions

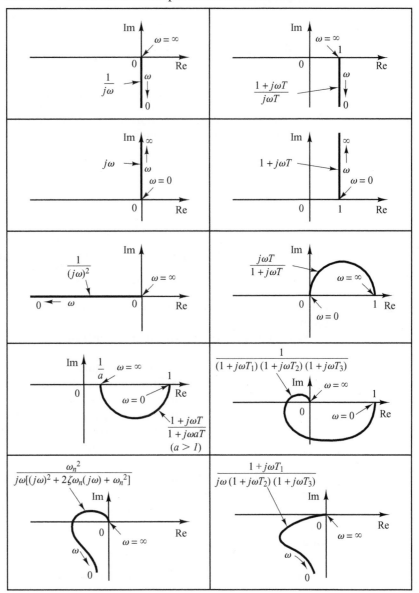

Drawing Nyquist Plots with MATLAB.   Nyquist plots, just like Bode diagrams, are commonly used in the frequency-response representation of linear, time-invariant, feedback control systems. Nyquist plots are polar plots, while Bode diagrams are rectangular plots. One plot or the other may be more convenient for a particular operation, but a given operation can always be carried out in either plot.

The MATLAB command nyquist computes the frequency response for continuous-time, linear, time-invariant systems. When invoked without left-hand arguments, nyquist produces a Nyquist plot on the screen.

The command

$$\text{nyquist(num,den)}$$

draws the Nyquist plot of the transfer function

$$G(s) = \frac{\text{num}(s)}{\text{den}(s)}$$

where num and den contain the polynomial coefficients in descending powers of s. Other commonly used nyquist commands are

nyquist(num,den,w)
nyquist(A,B,C,D)
nyquist(A,B,C,D,w)
nyquist(A,B,C,D,iu,w)
nyquist(sys)

The command involving the user-specified frequency vector w, such as

nyquist(num,den,w)

calculates the frequency response at the specified frequency points in radians per second.

When invoked with left-hand arguments such as

[re,im,w] = nyquist(num,den)
[re,im,w] = nyquist(num,den,w)
[re,im,w] = nyquist(A,B,C,D)
[re,im,w] = nyquist(A,B,C,D,w)
[re,im,w] = nyquist(A,B,C,D,iu,w)
[re,im,w] = nyquist(sys)

MATLAB returns the frequency response of the system in the matrices re, im, and w. No plot is drawn on the screen. The matrices re and im contain the real and imaginary parts of the frequency response of the system, evaluated at the frequency points specified in the vector w. Note that re and im have as many columns as outputs and one row for each element in w.

EXAMPLE 7–10   Consider the following open-loop transfer function:

$$G(s) = \frac{1}{s^2 + 0.8s + 1}$$

Draw a Nyquist plot with MATLAB.

Since the system is given in the form of the transfer function, the command

nyquist(num,den)

may be used to draw a Nyquist plot. MATLAB Program 7–5 produces the Nyquist plot shown in Figure 7–36. In this plot, the ranges for the real axis and imaginary axis are automatically determined.

MATLAB Program 7–5

```
num = [1];
den = [1  0.8  1];
nyquist(num,den)
grid
title('Nyquist Plot of G(s) = 1/(s^2 + 0.8s + 1)')
```

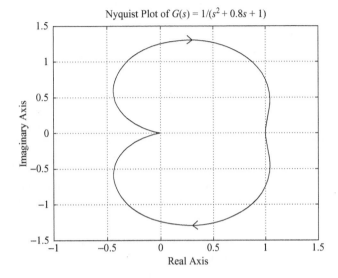

Figure 7–36
Nyquist plot of
$$G(s) = \frac{1}{s^2 + 0.8s + 1}.$$

If we wish to draw the Nyquist plot using manually determined ranges—for example, from -2 to 2 on the real axis and from -2 to 2 on the imaginary axis—enter the following command into the computer:

$$v = [-2 \ 2 \ -2 \ 2];$$
$$axis(v);$$

or, combining these two lines into one,

$$axis([-2 \ 2 \ -2 \ 2]);$$

See MATLAB Program 7–6 and the resulting Nyquist plot shown in Figure 7–37.

MATLAB Program 7–6

```
% ---------- Nyquist plot ----------

num = [1];
den = [1  0.8  1];
nyquist(num,den)
v = [-2  2  -2  2]; axis(v)
grid
title('Nyquist Plot of G(s) = 1/(s^2 + 0.8s + 1)')
```

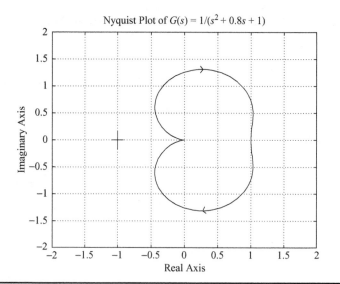

Figure 7–37
Nyquist plot of
$$G(s) = \frac{1}{s^2 + 0.8s + 1}.$$

Caution. In drawing a Nyquist plot, where a MATLAB operation involves "Divide by zero," the resulting Nyquist plot may have an erroneous or undesirable appearance. For example, if the transfer function $G(s)$ is given by

$$G(s) = \frac{1}{s(s + 1)}$$

then the MATLAB command

```
num = [1];
den = [1  1  0];
nyquist(num,den)
```

produces an undesirable Nyquist plot. An example of an undesirable Nyquist plot is shown in Figure 7–38. If such an undesirable Nyquist plot appears on the computer,

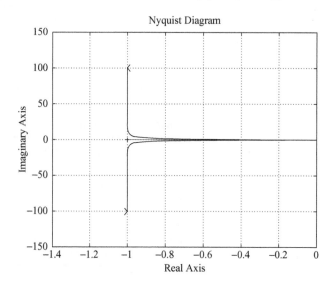

Figure 7–38
Undesirable Nyquist plot.

then it can be corrected if we specify the axis(v). For example, if we enter the axis command

$$v = [-2 \quad 2 \quad -5 \quad 5]; \; \text{axis}(v)$$

in the computer, then a desirable form of Nyquist plot can be obtained. See Example 7–11.

EXAMPLE 7–11 Draw a Nyquist plot for the following $G(s)$:

$$G(s) = \frac{1}{s(s + 1)}$$

MATLAB Program 7–7 will produce a desirable form of Nyquist plot on the computer, even though a warning message "Divide by zero" may appear on the screen. The resulting Nyquist plot is shown in Figure 7–39.

MATLAB Program 7–7

```
% ---------- Nyquist plot----------

num = [1];
den = [1  1  0];
nyquist(num,den)
v = [-2  2  -5  5]; axis(v)
grid
title('Nyquist Plot of G(s) = 1/[s(s + 1)]')
```

Notice that the Nyquist plot shown in Figure 7–39 includes the loci for both $\omega > 0$ and $\omega < 0$. If we wish to draw the Nyquist plot for only the positive frequency region ($\omega > 0$), then we need to use the command

$$[re,im,w] = nyquist(num,den,w)$$

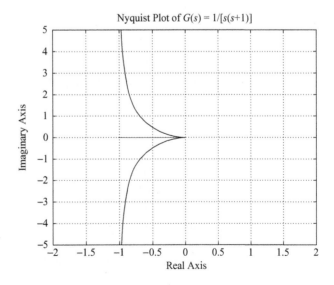

Figure 7–39
Nyquist plot of
$G(s) = \dfrac{1}{s(s + 1)}$.

A MATLAB program using this nyquist command is shown in MATLAB Program 7–8. The resulting Nyquist plot is presented in Figure 7–40.

MATLAB Program 7–8

```
% ---------- Nyquist plot----------

num = [1];
den = [1  1  0];
w = 0.1:0.1:100;
[re,im,w] = nyquist(num,den,w);
plot(re,im)
v = [-2  2  -5  5]; axis(v);
grid
title('Nyquist Plot of G(s) = 1/[s(s + 1)]')
xlabel('Real Axis')
ylabel('Imag Axis')
```

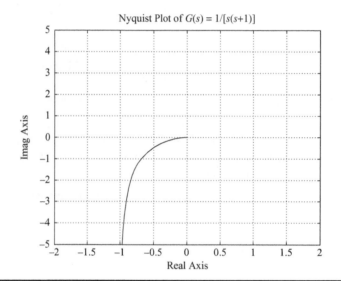

Figure 7–40
Nyquist plot of
$$G(s) = \frac{1}{s(s + 1)}$$
for $\omega > 0$.

Drawing Nyquist Plots of a System Defined in State Space. Consider the system defined by

$$\dot{\mathbf{x}} = \mathbf{Ax} + \mathbf{Bu}$$
$$\mathbf{y} = \mathbf{Cx} + \mathbf{Du}$$

where $\mathbf{x}$ = state vector (n-vector)
$\mathbf{y}$ = output vector (m-vector)
$\mathbf{u}$ = control vector (r-vector)
$\mathbf{A}$ = state matrix ($n \times n$ matrix)
$\mathbf{B}$ = control matrix ($n \times r$ matrix)
$\mathbf{C}$ = output matrix ($m \times n$ matrix)
$\mathbf{D}$ = direct transmission matrix ($m \times r$ matrix)

Nyquist plots for this system may be obtained by the use of the command

$$\text{nyquist(A,B,C,D)}$$

This command produces a series of Nyquist plots, one for each input and output combination of the system. The frequency range is automatically determined.

The command

$$\text{nyquist(A,B,C,D,iu)}$$

produces Nyquist plots from the single input iu to all the outputs of the system, with the frequency range determined automatically. The scalar iu is an index into the inputs of the system and specifies which input to use for the frequency response.

The command

$$\text{nyquist(A,B,C,D,iu,w)}$$

uses the user-supplied frequency vector w. The vector w specifies the frequencies in radians per second at which the frequency response should be calculated.

EXAMPLE 7–12   Consider the system defined by

$$\begin{bmatrix} \dot{x}_1 \\ \dot{x}_2 \end{bmatrix} = \begin{bmatrix} 0 & 1 \\ -25 & -4 \end{bmatrix} \begin{bmatrix} x_1 \\ x_2 \end{bmatrix} + \begin{bmatrix} 0 \\ 25 \end{bmatrix} u$$

$$y = \begin{bmatrix} 1 & 0 \end{bmatrix} \begin{bmatrix} x_1 \\ x_2 \end{bmatrix} + [0]u$$

Draw a Nyquist plot.

This system has a single input u and a single output y. A Nyquist plot may be obtained by entering the command

$$\text{nyquist(A,B,C,D)}$$

or

$$\text{nyquist(A,B,C,D,1)}$$

MATLAB Program 7–9 will provide the Nyquist plot. (Note that we obtain the identical result by using either of these two commands.) Figure 7–41 shows the Nyquist plot produced by MATLAB Program 7–9.

MATLAB Program 7–9
A = [0 1;-25 -4]; B = [0;25]; C = [1 0]; D = [0]; nyquist(A,B,C,D) grid title('Nyquist Plot')

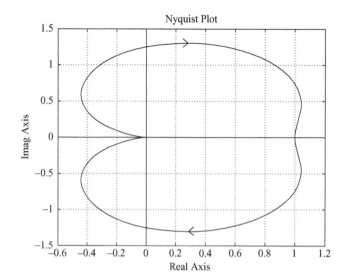

Figure 7–41
Nyquist plot of
system considered in
Example 7–12.

EXAMPLE 7–13   Consider the system defined by

$$\begin{bmatrix} \dot{x}_1 \\ \dot{x}_2 \end{bmatrix} = \begin{bmatrix} -1 & -1 \\ 6.5 & 0 \end{bmatrix} \begin{bmatrix} x_1 \\ x_2 \end{bmatrix} + \begin{bmatrix} 1 & 1 \\ 1 & 0 \end{bmatrix} \begin{bmatrix} u_1 \\ u_2 \end{bmatrix}$$

$$\begin{bmatrix} y_1 \\ y_2 \end{bmatrix} = \begin{bmatrix} 1 & 0 \\ 0 & 1 \end{bmatrix} \begin{bmatrix} x_1 \\ x_2 \end{bmatrix} + \begin{bmatrix} 0 & 0 \\ 0 & 0 \end{bmatrix} \begin{bmatrix} u_1 \\ u_2 \end{bmatrix}$$

This system involves two inputs and two outputs. There are four sinusoidal output–input re-
lationships: $Y_1(j\omega)/U_1(j\omega)$, $Y_2(j\omega)/U_1(j\omega)$, $Y_1(j\omega)/U_2(j\omega)$, and $Y_2(j\omega)/U_2(j\omega)$. Draw Nyquist
plots for the system. (When considering input u_1, we assume that input u_2 is zero, and vice
versa.)
    The four individual Nyquist plots can be obtained by the use of the command

$$\text{nyquist(A,B,C,D)}$$

MATLAB Program 7–10 produces the four Nyquist plots. They are shown in Figure 7–42.

MATLAB Program 7–10
A = [-1 -1;6.5 0];
B = [1 1;1 0];
C = [1 0;0 1];
D = [0 0;0 0];
nyquist(A,B,C,D)

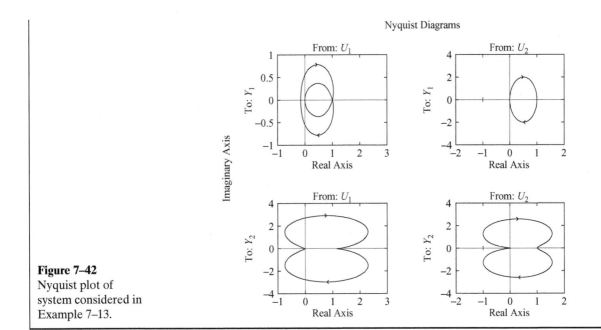

Figure 7–42
Nyquist plot of
system considered in
Example 7–13.

7–4 LOG-MAGNITUDE-VERSUS-PHASE PLOTS

Another approach to graphically portraying the frequency-response characteristics is to use the log-magnitude-versus-phase plot, which is a plot of the logarithmic magnitude in decibels versus the phase angle or phase margin for a frequency range of interest. [The phase margin is the difference between the actual phase angle ϕ and $-180°$; that is, $\phi - (-180°) = 180° + \phi$.] The curve is graduated in terms of the frequency ω. Such log-magnitude-versus-phase plots are commonly called Nichols plots.

In the Bode diagram, the frequency-response characteristics of $G(j\omega)$ are shown on semilog paper by two separate curves, the log-magnitude curve and the phase-angle curve, while in the log-magnitude-versus-phase plot, the two curves in the Bode diagram are combined into one. In the manual approach the log-magnitude-versus-phase plot can easily be constructed by reading values of the log magnitude and phase angle from the Bode diagram. Notice that in the log-magnitude-versus-phase plot a change in the gain constant of $G(j\omega)$ merely shifts the curve up (for increasing gain) or down (for decreasing gain), but the shape of the curve remains the same.

Advantages of the log-magnitude-versus-phase plot are that the relative stability of the closed-loop system can be determined quickly and that compensation can be worked out easily.

The log-magnitude-versus-phase plot for the sinusoidal transfer function $G(j\omega)$ and that for $1/G(j\omega)$ are skew symmetrical about the origin, since

$$\left| \frac{1}{G(j\omega)} \right| \text{ in dB} = -\left| G(j\omega) \right| \text{ in dB}$$

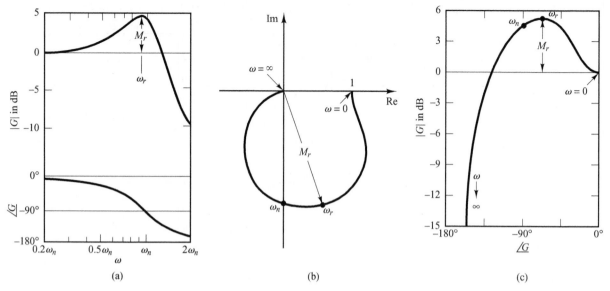

(a) (b) (c)

Figure 7–43

Three representations of the frequency response of $\dfrac{1}{1 + 2\zeta\left(j\dfrac{\omega}{\omega_n}\right) + \left(j\dfrac{\omega}{\omega_n}\right)^2}$, for $\zeta > 0$.

(a) Bode diagram; (b) polar plot; (c) log-magnitude-versus-phase plot.

and

$$\left/\frac{1}{G(j\omega)}\right. = - \left/G(j\omega)\right.$$

Figure 7–43 compares frequency-response curves of

$$G(j\omega) = \frac{1}{1 + 2\zeta\left(j\dfrac{\omega}{\omega_n}\right) + \left(j\dfrac{\omega}{\omega_n}\right)^2}$$

in three different representations. In the log-magnitude-versus-phase plot, the vertical distance between the points $\omega = 0$ and $\omega = \omega_r$, where ω_r is the resonant frequency, is the peak value of $G(j\omega)$ in decibels.

Since log-magnitude and phase-angle characteristics of basic transfer functions have been discussed in detail in Sections 7–2 and 7–3, it will be sufficient here to give examples of some log-magnitude-versus-phase plots. Table 7–2 shows such examples. (However, more on Nichols charts will be discussed in Section 7–6.)

Table 7–2   Log-Magnitude-versus-Phase Plots of Simple Transfer Functions

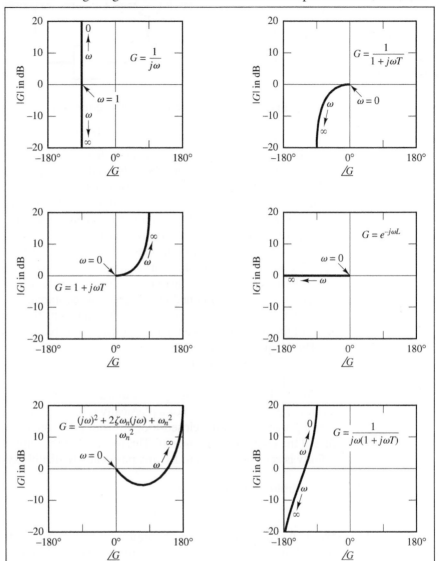

7–5 NYQUIST STABILITY CRITERION

The Nyquist stability criterion determines the stability of a closed-loop system from its open-loop frequency response and open-loop poles.

This section presents mathematical background for understanding the Nyquist stability criterion. Consider the closed-loop system shown in Figure 7–44. The closed-loop transfer function is

$$\frac{C(s)}{R(s)} = \frac{G(s)}{1 + G(s)H(s)}$$

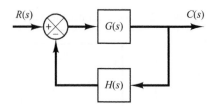

Figure 7–44
Closed-loop system.

For stability, all roots of the characteristic equation

$$1 + G(s)H(s) = 0$$

must lie in the left-half s plane. [It is noted that, although poles and zeros of the open-loop transfer function $G(s)H(s)$ may be in the right-half s plane, the system is stable if all the poles of the closed-loop transfer function (that is, the roots of the characteristic equation) are in the left-half s plane.] The Nyquist stability criterion relates the open-loop frequency response $G(j\omega)H(j\omega)$ to the number of zeros and poles of $1 + G(s)H(s)$ that lie in the right-half s plane. This criterion, derived by H. Nyquist, is useful in control engineering because the absolute stability of the closed-loop system can be determined graphically from open-loop frequency-response curves, and there is no need for actually determining the closed-loop poles. Analytically obtained open-loop frequency-response curves, as well as those experimentally obtained, can be used for the stability analysis. This is convenient because, in designing a control system, it often happens that mathematical expressions for some of the components are not known; only their frequency-response data are available.

The Nyquist stability criterion is based on a theorem from the theory of complex variables. To understand the criterion, we shall first discuss mappings of contours in the complex plane.

We shall assume that the open-loop transfer function $G(s)H(s)$ is representable as a ratio of polynomials in s. For a physically realizable system, the degree of the denominator polynomial of the closed-loop transfer function must be greater than or equal to that of the numerator polynomial. This means that the limit of $G(s)H(s)$ as s approaches infinity is zero or a constant for any physically realizable system.

Preliminary Study.   The characteristic equation of the system shown in Figure 7–44 is

$$F(s) = 1 + G(s)H(s) = 0$$

We shall show that, for a given continuous closed path in the s plane that does not go through any singular points, there corresponds a closed curve in the $F(s)$ plane. The number and direction of encirclements of the origin of the $F(s)$ plane by the closed curve play a particularly important role in what follows, for later we shall correlate the number and direction of encirclements with the stability of the system.

Consider, for example, the following open-loop transfer function:

$$G(s)H(s) = \frac{2}{s - 1}$$

The characteristic equation is

$$F(s) = 1 + G(s)H(s)$$

$$= 1 + \frac{2}{s - 1} = \frac{s + 1}{s - 1} = 0 \tag{7–15}$$

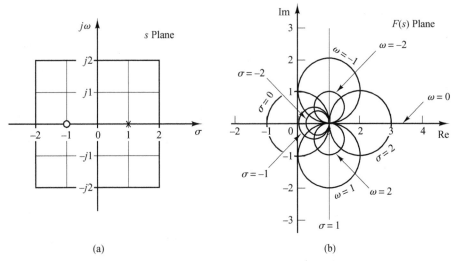

Figure 7–45
Conformal mapping of the
s-plane grids into the $F(s)$
plane, where
$F(s) = (s + 1)/(s - 1)$.

(a)

(b)

The function $F(s)$ is analytic[#] everywhere in the s plane except at its singular points. For each point of analyticity in the s plane, there corresponds a point in the $F(s)$ plane. For example, if $s = 2 + j1$, then $F(s)$ becomes

$$F(2 + j1) = \frac{2 + j1 + 1}{2 + j1 - 1} = 2 - j1$$

Thus, point $s = 2 + j1$ in the s plane maps into point $2 - j1$ in the $F(s)$ plane.

Thus, as stated previously, for a given continuous closed path in the s plane, which does not go through any singular points, there corresponds a closed curve in the $F(s)$ plane.

For the characteristic equation $F(s)$ given by Equation (7–15), the conformal mapping of the lines $\omega = 0, \pm 1, \pm 2$ and the lines $\sigma = 0, \pm 1, \pm 2$ [see Figure 7–45(a)] yield circles in the $F(s)$ plane, as shown in Figure 7–45(b). Suppose that representative point s traces out a contour in the s plane in the clockwise direction. If the contour in the s plane encloses the pole of $F(s)$, there is one encirclement of the origin of the $F(s)$ plane by the locus of $F(s)$ in the counterclockwise direction. [See Figure 7–46(a).] If the contour in the s plane encloses the zero of $F(s)$, there is one encirclement of the origin of the $F(s)$ plane by the locus of $F(s)$ in the clockwise direction. [See Figure 7–46(b).] If the contour in the s plane encloses both the zero and the pole or if the contour encloses neither the zero nor the pole, then there is no encirclement of the origin of the $F(s)$ plane by the locus of $F(s)$. [See Figures 7–46(c) and (d).]

From the foregoing analysis, we can say that the direction of encirclement of the origin of the $F(s)$ plane by the locus of $F(s)$ depends on whether the contour in the s plane encloses a pole or a zero. Note that the location of a pole or zero in the s plane, whether in the right-half or left-half s plane, does not make any difference, but the enclosure of a pole or zero does. If the contour in the s plane encloses equal numbers of poles and zeros, then the corresponding closed curve in the $F(s)$ plane does not encircle the origin of the $F(s)$ plane. The foregoing discussion is a graphical explanation of the mapping theorem, which is the basis for the Nyquist stability criterion.

[#]A complex function $F(s)$ is said to be analytic in a region if $F(s)$ and all its derivatives exist in that region.

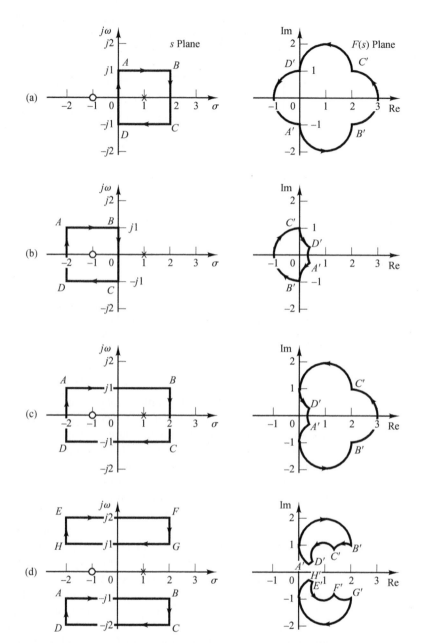

Figure 7–46
Closed contours in the s plane and their corresponding closed curves in the $F(s)$ plane, where $F(s) = (s + 1)/(s - 1)$.

Mapping Theorem. Let $F(s)$ be a ratio of two polynomials in s. Let P be the number of poles and Z be the number of zeros of $F(s)$ that lie inside some closed contour in the s plane, with multiplicity of poles and zeros accounted for. Let the contour be such that it does not pass through any poles or zeros of $F(s)$. This closed contour in the s plane is then mapped into the $F(s)$ plane as a closed curve. The total number N of clockwise encirclements of the origin of the $F(s)$ plane, as a representative point s traces out the entire contour in the clockwise direction, is equal to $Z - P$. (Note that by this mapping theorem, the numbers of zeros and of poles cannot be found—only their difference.)

We shall not present a formal proof of this theorem here, but leave the proof to Problem **A–7–6**. Note that a positive number N indicates an excess of zeros over poles of the function $F(s)$ and a negative N indicates an excess of poles over zeros. In control system applications, the number P can be readily determined for $F(s) = 1 + G(s)H(s)$ from the function $G(s)H(s)$. Therefore, if N is determined from the plot of $F(s)$, the number of zeros in the closed contour in the s plane can be determined readily. Note that the exact shapes of the s-plane contour and $F(s)$ locus are immaterial so far as encirclements of the origin are concerned, since encirclements depend only on the enclosure of poles and/or zeros of $F(s)$ by the s-plane contour.

Application of the Mapping Theorem to the Stability Analysis of Closed-Loop Systems. For analyzing the stability of linear control systems, we let the closed contour in the s plane enclose the entire right-half s plane. The contour consists of the entire $j\omega$ axis from $\omega = -\infty$ to $+\infty$ and a semicircular path of infinite radius in the right-half s plane. Such a contour is called the Nyquist path. (The direction of the path is clockwise.) The Nyquist path encloses the entire right-half s plane and encloses all the zeros and poles of $1 + G(s)H(s)$ that have positive real parts. [If there are no zeros of $1 + G(s)H(s)$ in the right-half s plane, then there are no closed-loop poles there, and the system is stable.] It is necessary that the closed contour, or the Nyquist path, not pass through any zeros and poles of $1 + G(s)H(s)$. If $G(s)H(s)$ has a pole or poles at the origin of the s plane, mapping of the point $s = 0$ becomes indeterminate. In such cases, the origin is avoided by taking a detour around it. (A detailed discussion of this special case is given later.)

If the mapping theorem is applied to the special case in which $F(s)$ is equal to $1 + G(s)H(s)$, then we can make the following statement: If the closed contour in the s plane encloses the entire right-half s plane, as shown in Figure 7–47, then the number of right-half plane zeros of the function $F(s) = 1 + G(s)H(s)$ is equal to the number of poles of the function $F(s) = 1 + G(s)H(s)$ in the right-half s plane plus the number of clockwise encirclements of the origin of the $1 + G(s)H(s)$ plane by the corresponding closed curve in this latter plane.

Because of the assumed condition that

$$\lim_{s \to \infty} \left[1 + G(s)H(s)\right] = \text{constant}$$

the function of $1 + G(s)H(s)$ remains constant as s traverses the semicircle of infinite radius. Because of this, whether the locus of $1 + G(s)H(s)$ encircles the origin of the $1 + G(s)H(s)$ plane can be determined by considering only a part of the closed contour in the s plane—that is, the $j\omega$ axis. Encirclements of the origin, if there are any, occur only

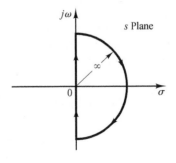

Figure 7–47
Closed contour in the s plane.

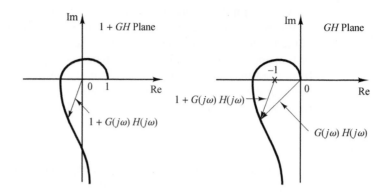

Figure 7–48
Plots of
$1 + G(j\omega)H(j\omega)$ in
the $1 + GH$ plane
and GH plane.

while a representative point moves from $-j\infty$ to $+j\infty$ along the $j\omega$ axis, provided that no zeros or poles lie on the $j\omega$ axis.

Note that the portion of the $1 + G(s)H(s)$ contour from $\omega = -\infty$ to $\omega = \infty$ is simply $1 + G(j\omega)H(j\omega)$. Since $1 + G(j\omega)H(j\omega)$ is the vector sum of the unit vector and the vector $G(j\omega)H(j\omega)$, $1 + G(j\omega)H(j\omega)$ is identical to the vector drawn from the $-1 + j0$ point to the terminal point of the vector $G(j\omega)H(j\omega)$, as shown in Figure 7–48. Encirclement of the origin by the graph of $1 + G(j\omega)H(j\omega)$ is equivalent to encirclement of the $-1 + j0$ point by just the $G(j\omega)H(j\omega)$ locus. Thus, the stability of a closed-loop system can be investigated by examining encirclements of the $-1 + j0$ point by the locus of $G(j\omega)H(j\omega)$. The number of clockwise encirclements of the $-1 + j0$ point can be found by drawing a vector from the $-1 + j0$ point to the $G(j\omega)H(j\omega)$ locus, starting from $\omega = -\infty$, going through $\omega = 0$, and ending at $\omega = +\infty$, and by counting the number of clockwise rotations of the vector.

Plotting $G(j\omega)H(j\omega)$ for the Nyquist path is straightforward. The map of the negative $j\omega$ axis is the mirror image about the real axis of the map of the positive $j\omega$ axis. That is, the plot of $G(j\omega)H(j\omega)$ and the plot of $G(-j\omega)H(-j\omega)$ are symmetrical with each other about the real axis. The semicircle with infinite radius maps into either the origin of the GH plane or a point on the real axis of the GH plane.

In the preceding discussion, $G(s)H(s)$ has been assumed to be the ratio of two polynomials in s. Thus, the transport lag e^{-Ts} has been excluded from the discussion. Note, however, that a similar discussion applies to systems with transport lag, although a proof of this is not given here. The stability of a system with transport lag can be determined from the open-loop frequency-response curves by examining the number of encirclements of the $-1 + j0$ point, just as in the case of a system whose open-loop transfer function is a ratio of two polynomials in s.

Nyquist Stability Criterion. The foregoing analysis, utilizing the encirclement of the $-1 + j0$ point by the $G(j\omega)H(j\omega)$ locus, is summarized in the following Nyquist stability criterion:

Nyquist stability criterion [for a special case when $G(s)H(s)$ has neither poles nor zeros on the $j\omega$ axis]: In the system shown in Figure 7–44, if the open-loop transfer function $G(s)H(s)$ has k poles in the right-half s plane and $\lim_{s \to \infty} G(s)H(s) = $ constant, then for stability, the $G(j\omega)H(j\omega)$ locus, as ω varies from $-\infty$ to ∞, must encircle the $-1 + j0$ point k times in the counterclockwise direction.

Remarks on the Nyquist Stability Criterion

1. This criterion can be expressed as

$$Z = N + P$$

where Z = number of zeros of $1 + G(s)H(s)$ in the right-half s plane
N = number of clockwise encirclements of the $-1 + j0$ point
P = number of poles of $G(s)H(s)$ in the right-half s plane

If P is not zero, for a stable control system, we must have $Z = 0$, or $N = -P$, which means that we must have P counterclockwise encirclements of the $-1 + j0$ point.

If $G(s)H(s)$ does not have any poles in the right-half s plane, then $Z = N$. Thus, for stability there must be no encirclement of the $-1 + j0$ point by the $G(j\omega)H(j\omega)$ locus. In this case it is not necessary to consider the locus for the entire $j\omega$ axis, only for the positive-frequency portion. The stability of such a system can be determined by seeing if the $-1 + j0$ point is enclosed by the Nyquist plot of $G(j\omega)H(j\omega)$. The region enclosed by the Nyquist plot is shown in Figure 7–49. For stability, the $-1 + j0$ point must lie outside the shaded region.

2. We must be careful when testing the stability of multiple-loop systems since they may include poles in the right-half s plane. (Note that although an inner loop may be unstable, the entire closed-loop system can be made stable by proper design.) Simple inspection of encirclements of the $-1 + j0$ point by the $G(j\omega)H(j\omega)$ locus is not sufficient to detect instability in multiple-loop systems. In such cases, however, whether any pole of $1 + G(s)H(s)$ is in the right-half s plane can be determined easily by applying the Routh stability criterion to the denominator of $G(s)H(s)$.

If transcendental functions, such as transport lag e^{-Ts}, are included in $G(s)H(s)$, they must be approximated by a series expansion before the Routh stability criterion can be applied.

3. If the locus of $G(j\omega)H(j\omega)$ passes through the $-1 + j0$ point, then zeros of the characteristic equation, or closed-loop poles, are located on the $j\omega$ axis. This is not desirable for practical control systems. For a well-designed closed-loop system, none of the roots of the characteristic equation should lie on the $j\omega$ axis.

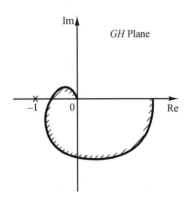

Figure 7–49
Region enclosed by a
Nyquist plot.

Special Case when $G(s)H(s)$ Involves Poles and/or Zeros on the $j\omega$ Axis. In the previous discussion, we assumed that the open-loop transfer function $G(s)H(s)$ has neither poles nor zeros at the origin. We now consider the case where $G(s)H(s)$ involves poles and/or zeros on the $j\omega$ axis.

Since the Nyquist path must not pass through poles or zeros of $G(s)H(s)$, if the function $G(s)H(s)$ has poles or zeros at the origin (or on the $j\omega$ axis at points other than the origin), the contour in the s plane must be modified. The usual way of modifying the contour near the origin is to use a semicircle with the infinitesimal radius ε, as shown in Figure 7–50. [Note that this semicircle may lie in the right-half s plane or in the left-half s plane. Here we take the semicircle in the right-half s plane.] A representative point s moves along the negative $j\omega$ axis from $-j\infty$ to $j0-$. From $s = j0-$ to $s = j0+$, the point moves along the semicircle of radius ε (where $\varepsilon \ll 1$) and then moves along the positive $j\omega$ axis from $j0+$ to $j\infty$. From $s = j\infty$, the contour follows a semicircle with infinite radius, and the representative point moves back to the starting point, $s = -j\infty$. The area that the modified closed contour avoids is very small and approaches zero as the radius ε approaches zero. Therefore, all the poles and zeros, if any, in the right-half s plane are enclosed by this contour.

Consider, for example, a closed-loop system whose open-loop transfer function is given by

$$G(s)H(s) = \frac{K}{s(Ts + 1)}$$

The points corresponding to $s = j0+$ and $s = j0-$ on the locus of $G(s)H(s)$ in the $G(s)H(s)$ plane are $-j\infty$ and $j\infty$, respectively. On the semicircular path with radius ε (where $\varepsilon \ll 1$), the complex variable s can be written

$$s = \varepsilon e^{j\theta}$$

where θ varies from $-90°$ to $+90°$. Then $G(s)H(s)$ becomes

$$G(\varepsilon e^{j\theta})H(\varepsilon e^{j\theta}) = \frac{K}{\varepsilon e^{j\theta}} = \frac{K}{\varepsilon}e^{-j\theta}$$

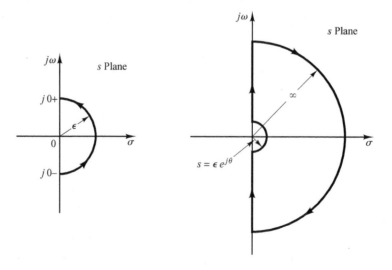

Figure 7–50
Contour near the origin of the s plane and closed contour in the s plane avoiding poles and zeros at the origin.

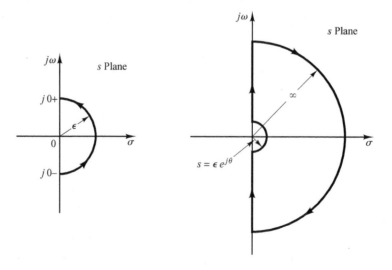

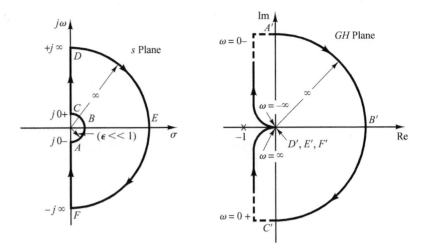

Figure 7–51
s-Plane contour and the $G(s)H(s)$ locus in the GH plane, where $G(s)H(s) = K/[s(Ts + 1)]$.

The value K/ε approaches infinity as ε approaches zero, and $-\theta$ varies from 90° to −90° as a representative point s moves along the semicircle in the s plane. Thus, the points $G(j0-)H(j0-) = j\infty$ and $G(j0+)H(j0+) = -j\infty$ are joined by a semicircle of infinite radius in the right-half GH plane. The infinitesimal semicircular detour around the origin in the s plane maps into the GH plane as a semicircle of infinite radius. Figure 7–51 shows the s-plane contour and the $G(s)H(s)$ locus in the GH plane. Points $A, B,$ and C on the s-plane contour map into the respective points $A', B',$ and C' on the $G(s)H(s)$ locus. As seen from Figure 7–51, points $D, E,$ and F on the semicircle of infinite radius in the s plane map into the origin of the GH plane. Since there is no pole in the right-half s plane and the $G(s)H(s)$ locus does not encircle the $-1 + j0$ point, there are no zeros of the function $1 + G(s)H(s)$ in the right-half s plane. Therefore, the system is stable.

For an open-loop transfer function $G(s)H(s)$ involving a $1/s^n$ factor (where $n = 2, 3,\dots$), the plot of $G(s)H(s)$ has n clockwise semicircles of infinite radius about the origin as a representative point s moves along the semicircle of radius ε (where $\varepsilon \ll 1$). For example, consider the following open-loop transfer function:

$$G(s)H(s) = \frac{K}{s^2(Ts + 1)}$$

Then

$$\lim_{s \to \varepsilon e^{j\theta}} G(s)H(s) = \frac{K}{\varepsilon^2 e^{2j\theta}} = \frac{K}{\varepsilon^2} e^{-2j\theta}$$

As θ varies from −90° to 90° in the s plane, the angle of $G(s)H(s)$ varies from 180° to −180°, as shown in Figure 7–52. Since there is no pole in the right-half s plane and the locus encircles the $-1 + j0$ point twice clockwise for any positive value of K, there are two zeros of $1 + G(s)H(s)$ in the right-half s plane. Therefore, this system is always unstable.

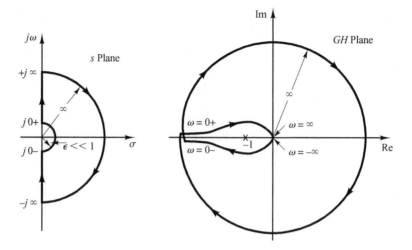

Figure 7–52
s-Plane contour and the $G(s)H(s)$ locus in the GH plane, where $G(s)H(s) = K/[s^2(Ts + 1)]$.

Note that a similar analysis can be made if $G(s)H(s)$ involves poles and/or zeros on the $j\omega$ axis. The Nyquist stability criterion can now be generalized as follows:

Nyquist stability criterion [for a general case when $G(s)H(s)$ has poles and/or zeros on the $j\omega$ axis]: In the system shown in Figure 7–44, if the open-loop transfer function $G(s)H(s)$ has k poles in the right-half s plane, then for stability the $G(s)H(s)$ locus, as a representative point s traces on the modified Nyquist path in the clockwise direction, must encircle the $-1 + j0$ point k times in the counterclockwise direction.

7–6 STABILITY ANALYSIS

In this section, we shall present several illustrative examples of the stability analysis of control systems using the Nyquist stability criterion.

If the Nyquist path in the s plane encircles Z zeros and P poles of $1 + G(s)H(s)$ and does not pass through any poles or zeros of $1 + G(s)H(s)$ as a representative point s moves in the clockwise direction along the Nyquist path, then the corresponding contour in the $G(s)H(s)$ plane encircles the $-1 + j0$ point $N = Z - P$ times in the clockwise direction. (Negative values of N imply counterclockwise encirclements.)

In examining the stability of linear control systems using the Nyquist stability criterion, we see that three possibilities can occur:

1. There is no encirclement of the $-1 + j0$ point. This implies that the system is stable if there are no poles of $G(s)H(s)$ in the right-half s plane; otherwise, the system is unstable.

2. There are one or more counterclockwise encirclements of the $-1 + j0$ point. In this case the system is stable if the number of counterclockwise encirclements is the same as the number of poles of $G(s)H(s)$ in the right-half s plane; otherwise, the system is unstable.

3. There are one or more clockwise encirclements of the $-1 + j0$ point. In this case the system is unstable.

In the following examples, we assume that the values of the gain K and the time constants (such as T, T_1, and T_2) are all positive.

EXAMPLE 7–14   Consider a closed-loop system whose open-loop transfer function is given by

$$G(s)H(s) = \frac{K}{(T_1 s + 1)(T_2 s + 1)}$$

Examine the stability of the system.

A plot of $G(j\omega)H(j\omega)$ is shown in Figure 7–53. Since $G(s)H(s)$ does not have any poles in the right-half s plane and the $-1 + j0$ point is not encircled by the $G(j\omega)H(j\omega)$ locus, this system is stable for any positive values of K, T_1, and T_2.

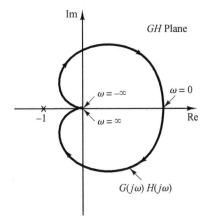

Figure 7–53
Polar plot of
$G(j\omega)H(j\omega)$
considered in
Example 7–14.

EXAMPLE 7–15   Consider the system with the following open-loop transfer function:

$$G(s)H(s) = \frac{K}{s(T_1 s + 1)(T_2 s + 1)}$$

Determine the stability of the system for two cases: (1) the gain K is small and (2) K is large.

The Nyquist plots of the open-loop transfer function with a small value of K and a large value of K are shown in Figure 7–54. The number of poles of $G(s)H(s)$ in the right-half s plane is zero.

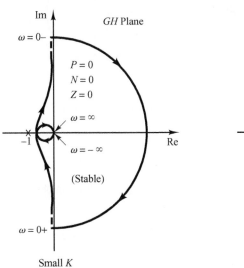
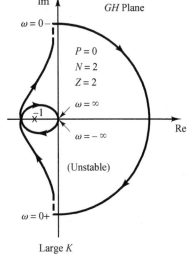

Figure 7–54
Polar plots of the
system considered in
Example 7–15.

Therefore, for this system to be stable, it is necessary that $N = Z = 0$ or that the $G(s)H(s)$ locus not encircle the $-1 + j0$ point.

For small values of K, there is no encirclement of the $-1 + j0$ point. Hence, the system is stable for small values of K. For large values of K, the locus of $G(s)H(s)$ encircles the $-1 + j0$ point twice in the clockwise direction, indicating two closed-loop poles in the right-half s plane, and the system is unstable. (For good accuracy, K should be large. From the stability viewpoint, however, a large value of K causes poor stability or even instability. To compromise between accuracy and stability, it is necessary to insert a compensation network into the system. Compensating techniques in the frequency domain are discussed in Sections 7–11 through 7–13.)

EXAMPLE 7–16 The stability of a closed-loop system with the following open-loop transfer function

$$G(s)H(s) = \frac{K(T_2 s + 1)}{s^2(T_1 s + 1)}$$

depends on the relative magnitudes of T_1 and T_2. Draw Nyquist plots and determine the stability of the system.

Plots of the locus $G(s)H(s)$ for three cases, $T_1 < T_2, T_1 = T_2$, and $T_1 > T_2$, are shown in Figure 7–55. For $T_1 < T_2$, the locus of $G(s)H(s)$ does not encircle the $-1 + j0$ point, and the closed-loop system is stable. For $T_1 = T_2$, the $G(s)H(s)$ locus passes through the $-1 + j0$ point, which indicates that the closed-loop poles are located on the $j\omega$ axis. For $T_1 > T_2$, the locus of $G(s)H(s)$ encircles the $-1 + j0$ point twice in the clockwise direction. Thus, the closed-loop system has two closed-loop poles in the right-half s plane, and the system is unstable.

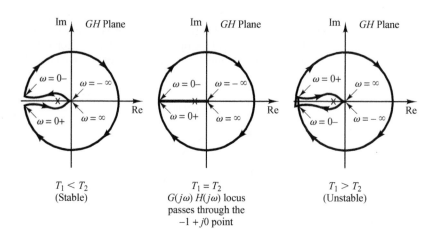

Figure 7–55
Polar plots of the system considered in Example 7–16.

$T_1 < T_2$
(Stable)

$T_1 = T_2$
$G(j\omega)\,H(j\omega)$ locus passes through the $-1 + j0$ point

$T_1 > T_2$
(Unstable)

EXAMPLE 7–17 Consider the closed-loop system having the following open-loop transfer function:

$$G(s)H(s) = \frac{K}{s(Ts - 1)}$$

Determine the stability of the system.

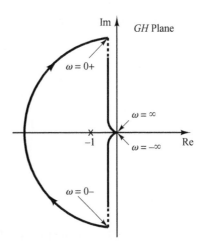

Figure 7–56
Polar plot of the system considered in Example 7–17.

The function $G(s)H(s)$ has one pole $(s = 1/T)$ in the right-half s plane. Therefore, $P = 1$. The Nyquist plot shown in Figure 7–56 indicates that the $G(s)H(s)$ plot encircles the $-1 + j0$ point once clockwise. Thus, $N = 1$. Since $Z = N + P$, we find that $Z = 2$. This means that the closed-loop system has two closed-loop poles in the right-half s plane and is unstable.

EXAMPLE 7–18   Investigate the stability of a closed-loop system with the following open-loop transfer function:

$$G(s)H(s) = \frac{K(s + 3)}{s(s - 1)} \quad (K > 1)$$

The open-loop transfer function has one pole $(s = 1)$ in the right-half s plane, or $P = 1$. The open-loop system is unstable. The Nyquist plot shown in Figure 7–57 indicates that the $-1 + j0$ point is encircled by the $G(s)H(s)$ locus once in the counterclockwise direction. Therefore, $N = -1$. Thus, Z is found from $Z = N + P$ to be zero, which indicates that there is no zero of $1 + G(s)H(s)$ in the right-half s plane, and the closed-loop system is stable. This is one of the examples for which an unstable open-loop system becomes stable when the loop is closed.

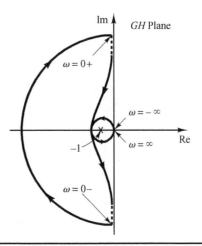

Figure 7–57
Polar plot of the system considered in Example 7–18.

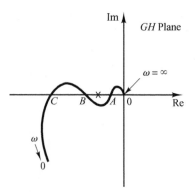

Figure 7–58
Polar plot of a
conditionally stable
system.

Conditionally Stable Systems.   Figure 7–58 shows an example of a $G(j\omega)H(j\omega)$ locus for which the closed-loop system can be made unstable by varying the open-loop gain. If the open-loop gain is increased sufficiently, the $G(j\omega)H(j\omega)$ locus encloses the $-1 + j0$ point twice, and the system becomes unstable. If the open-loop gain is decreased sufficiently, again the $G(j\omega)H(j\omega)$ locus encloses the $-1 + j0$ point twice. For stable operation of the system considered here, the critical point $-1 + j0$ must not be located in the regions between OA and BC shown in Figure 7–58. Such a system that is stable only for limited ranges of values of the open-loop gain for which the $-1 + j0$ point is completely outside the $G(j\omega)H(j\omega)$ locus is a conditionally stable system.

A conditionally stable system is stable for the value of the open-loop gain lying between critical values, but it is unstable if the open-loop gain is either increased or decreased sufficiently. Such a system becomes unstable when large input signals are applied, since a large signal may cause saturation, which in turn reduces the open-loop gain of the system. It is advisable to avoid such a situation.

Multiple-Loop System.   Consider the system shown in Figure 7–59. This is a multiple-loop system. The inner loop has the transfer function

$$G(s) = \frac{G_2(s)}{1 + G_2(s)H_2(s)}$$

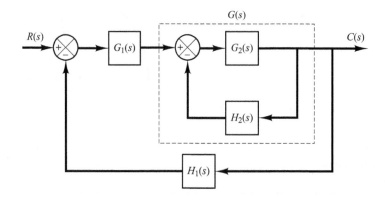

Figure 7–59
Multiple-loop
system.

If $G(s)$ is unstable, the effects of instability are to produce a pole or poles in the right-half s plane. Then the characteristic equation of the inner loop, $1 + G_2(s)H_2(s) = 0$, has a zero or zeros in the right-half s plane. If $G_2(s)$ and $H_2(s)$ have P_1 poles here, then the number Z_1 of right-half plane zeros of $1 + G_2(s)H_2(s)$ can be found from $Z_1 = N_1 + P_1$, where N_1 is the number of clockwise encirclements of the $-1 + j0$ point by the $G_2(s)H_2(s)$ locus. Since the open-loop transfer function of the entire system is given by $G_1(s)G(s)H_1(s)$, the stability of this closed-loop system can be found from the Nyquist plot of $G_1(s)G(s)H_1(s)$ and knowledge of the right-half plane poles of $G_1(s)G(s)H_1(s)$.

Notice that if a feedback loop is eliminated by means of block diagram reductions, there is a possibility that unstable poles are introduced; if the feedforward branch is eliminated by means of block diagram reductions, there is a possibility that right-half plane zeros are introduced. Therefore, we must note all right-half plane poles and zeros as they appear from subsidiary loop reductions. This knowledge is necessary in determining the stability of multiple-loop systems.

EXAMPLE 7–19 Consider the control system shown in Figure 7–60. The system involves two loops. Determine the range of gain K for stability of the system by the use of the Nyquist stability criterion. (The gain K is positive.)

To examine the stability of the control system, we need to sketch the Nyquist locus of $G(s)$, where

$$G(s) = G_1(s)G_2(s)$$

However, the poles of $G(s)$ are not known at this point. Therefore, we need to examine the minor loop if there are right-half s-plane poles. This can be done easily by use of the Routh stability criterion. Since

$$G_2(s) = \frac{1}{s^3 + s^2 + 1}$$

the Routh array becomes as follows:

$$
\begin{array}{ccc}
s^3 & 1 & 0 \\
s^2 & 1 & 1 \\
s^1 & -1 & 0 \\
s^0 & 1 &
\end{array}
$$

Notice that there are two sign changes in the first column. Hence, there are two poles of $G_2(s)$ in the right-half s plane.

Once we find the number of right-half s plane poles of $G_2(s)$, we proceed to sketch the Nyquist locus of $G(s)$, where

$$G(s) = G_1(s)G_2(s) = \frac{K(s + 0.5)}{s^3 + s^2 + 1}$$

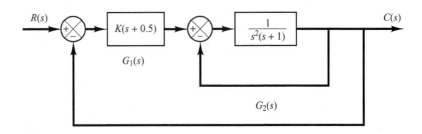

Figure 7–60
Control system.

Our problem is to determine the range of the gain K for stability. Hence, instead of plotting Nyquist loci of $G(j\omega)$ for various values of K, we plot the Nyquist locus of $G(j\omega)/K$. Figure 7–61 shows the Nyquist plot or polar plot of $G(j\omega)/K$.

Since $G(s)$ has two poles in the right-half s plane, we have $P = 2$. Noting that

$$Z = N + P$$

for stability, we require $Z = 0$ or $N = -2$. That is, the Nyquist locus of $G(j\omega)$ must encircle the $-1 + j0$ point twice counterclockwise. From Figure 7–61, we see that, if the critical point lies between 0 and -0.5, then the $G(j\omega)/K$ locus encircles the critical point twice counterclockwise. Therefore, we require

$$-0.5K < -1$$

The range of the gain K for stability is

$$2 < K$$

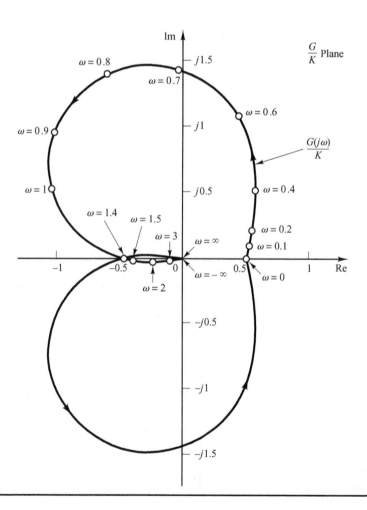

Figure 7–61
Polar plot of
$G(j\omega)/K$.

Nyquist Stability Criterion Applied to Inverse Polar Plots. In the previous analyses, the Nyquist stability criterion was applied to polar plots of the open-loop transfer function $G(s)H(s)$.

In analyzing multiple-loop systems, the inverse transfer function may sometimes be used in order to permit graphical analysis; this avoids much of the numerical calculation. (The Nyquist stability criterion can be applied equally well to inverse polar plots. The mathematical derivation of the Nyquist stability criterion for inverse polar plots is the same as that for direct polar plots.)

The inverse polar plot of $G(j\omega)H(j\omega)$ is a graph of $1/\big[G(j\omega)H(j\omega)\big]$ as a function of ω. For example, if $G(j\omega)H(j\omega)$ is

$$G(j\omega)H(j\omega) = \frac{j\omega T}{1 + j\omega T}$$

then

$$\frac{1}{G(j\omega)H(j\omega)} = \frac{1}{j\omega T} + 1$$

The inverse polar plot for $\omega \geq 0$ is the lower half of the vertical line starting at the point $(1, 0)$ on the real axis.

The Nyquist stability criterion applied to inverse plots may be stated as follows: For a closed-loop system to be stable, the encirclement, if any, of the $-1 + j0$ point by the $1/\big[G(s)H(s)\big]$ locus (as s moves along the Nyquist path) must be counterclockwise, and the number of such encirclements must be equal to the number of poles of $1/\big[G(s)H(s)\big]$ [that is, the zeros of $G(s)H(s)$] that lie in the right-half s plane. [The number of zeros of $G(s)H(s)$ in the right-half s plane may be determined by the use of the Routh stability criterion.] If the open-loop transfer function $G(s)H(s)$ has no zeros in the right-half s plane, then for a closed-loop system to be stable, the number of encirclements of the $-1 + j0$ point by the $1/\big[G(s)H(s)\big]$ locus must be zero.

Note that although the Nyquist stability criterion can be applied to inverse polar plots, if experimental frequency-response data are incorporated, counting the number of encirclements of the $1/\big[G(s)H(s)\big]$ locus may be difficult because the phase shift corresponding to the infinite semicircular path in the s plane is difficult to measure. For example, if the open-loop transfer function $G(s)H(s)$ involves transport lag such that

$$G(s)H(s) = \frac{Ke^{-j\omega L}}{s(Ts + 1)}$$

then the number of encirclements of the $-1 + j0$ point by the $1/\big[G(s)H(s)\big]$ locus becomes infinite, and the Nyquist stability criterion cannot be applied to the inverse polar plot of such an open-loop transfer function.

In general, if experimental frequency-response data cannot be put into analytical form, both the $G(j\omega)H(j\omega)$ and $1/\big[G(j\omega)H(j\omega)\big]$ loci must be plotted. In addition, the number of right-half plane zeros of $G(s)H(s)$ must be determined. It is more difficult to determine the right-half plane zeros of $G(s)H(s)$ (in other words, to determine whether a given component is minimum phase) than it is to determine the right-half plane poles of $G(s)H(s)$ (in other words, to determine whether the component is stable).

Depending on whether the data are graphical or analytical and whether nonminimum-phase components are included, an appropriate stability test must be used for multiple-loop systems. If the data are given in analytical form or if mathematical expressions for all the components are known, the application of the Nyquist stability criterion to inverse polar plots causes no difficulty, and multiple-loop systems may be analyzed and designed in the inverse GH plane. (See Problem **A–7–15**.)

7–7 RELATIVE STABILITY ANALYSIS

Relative Stability.   In designing a control system, we require that the system be stable. Furthermore, it is necessary that the system have adequate relative stability.

In this section, we shall show that the Nyquist plot indicates not only whether a system is stable, but also the degree of stability of a stable system. The Nyquist plot also gives information as to how stability may be improved, if this is necessary.

In the following discussion, we shall assume that the systems considered have unity feedback. Note that it is always possible to reduce a system with feedback elements to a unity-feedback system, as shown in Figure 7–62. Hence, the extension of relative stability analysis for the unity-feedback system to nonunity-feedback systems is possible.

We shall also assume that, unless otherwise stated, the systems are minimum-phase systems; that is, the open-loop transfer function has neither poles nor zeros in the right-half s plane.

Relative Stability Analysis by Conformal Mapping.   One of the important problems in analyzing a control system is to find all closed-loop poles or at least those closest to the $j\omega$ axis (or the dominant pair of closed-loop poles). If the open-loop frequency-response characteristics of a system are known, it may be possible to estimate the closed-loop poles closest to the $j\omega$ axis. It is noted that the Nyquist locus $G(j\omega)$ need not be an analytically known function of ω. The entire Nyquist locus may be experimentally obtained. The technique to be presented here is essentially graphical and is based on a conformal mapping of the s plane into the $G(s)$ plane.

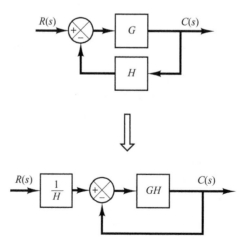

Figure 7–62
Modification of a system with feedback elements to a unity-feedback system.

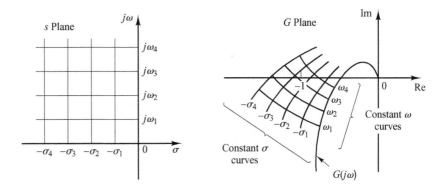

Figure 7–63
Conformal mapping
of s-plane grids into
the $G(s)$ plane.

Consider the conformal mapping of constant-σ lines (lines $s = \sigma + j\omega$, where σ is constant and ω varies) and constant-ω lines (lines $s = \sigma + j\omega$, where ω is constant and σ varies) in the s plane. The $\sigma = 0$ line (the $j\omega$ axis) in the s plane maps into the Nyquist plot in the $G(s)$ plane. The constant-σ lines in the s plane map into curves that are similar to the Nyquist plot and are in a sense parallel to the Nyquist plot, as shown in Figure 7–63. The constant-ω lines in the s plane map into curves, also shown in Figure 7–63.

Although the shapes of constant-σ and constant-ω loci in the $G(s)$ plane and the closeness of approach of the $G(j\omega)$ locus to the $-1 + j0$ point depend on a particular $G(s)$, the closeness of approach of the $G(j\omega)$ locus to the $-1 + j0$ point is an indication of the relative stability of a stable system. In general, we may expect that the closer the $G(j\omega)$ locus is to the $-1 + j0$ point, the larger the maximum overshoot is in the step transient response and the longer it takes to damp out.

Consider the two systems shown in Figures 7–64(a) and (b). (In Figure 7–64, the $\times$'s indicate closed-loop poles.) System (a) is obviously more stable than system (b) because the closed-loop poles of system (a) are located farther left than those of system (b). Figures 7–65(a) and (b) show the conformal mapping of s-plane grids into the $G(s)$ plane. The closer the closed-loop poles are located to the $j\omega$ axis, the closer the $G(j\omega)$ locus is to the $-1 + j0$ point.

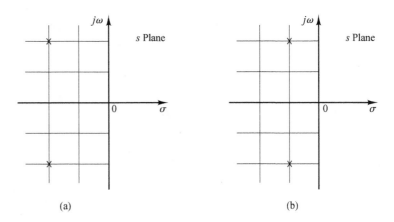

Figure 7–64
Two systems with
two closed-loop
poles each.

(a)   (b)

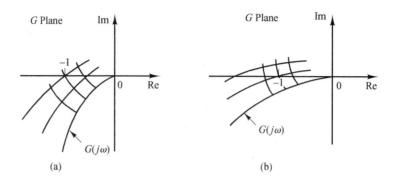

Figure 7–65
Conformal mappings
of s-plane grids for
the systems shown in
Figure 7–64 into the
$G(s)$ plane.

G Plane   Im

-1

0   Re

$G(j\omega)$

(a)

G Plane   Im

-1

0   Re

$G(j\omega)$

(b)

Phase and Gain Margins. Figure 7–66 shows the polar plots of $G(j\omega)$ for three different values of the open-loop gain K. For a large value of the gain K, the system is unstable. As the gain is decreased to a certain value, the $G(j\omega)$ locus passes through the $-1 + j0$ point. This means that with this gain value the system is on the verge of instability, and the system will exhibit sustained oscillations. For a small value of the gain K, the system is stable.

In general, the closer the $G(j\omega)$ locus comes to encircling the $-1 + j0$ point, the more oscillatory is the system response. The closeness of the $G(j\omega)$ locus to the $-1 + j0$ point can be used as a measure of the margin of stability. (This does not apply, however, to conditionally stable systems.) It is common practice to represent the closeness in terms of phase margin and gain margin.

Phase margin: The phase margin is that amount of additional phase lag at the gain crossover frequency required to bring the system to the verge of instability. The gain crossover frequency is the frequency at which $|G(j\omega)|$, the magnitude of the open-loop transfer function, is unity. The phase margin γ is 180° plus the phase angle ϕ of the open-loop transfer function at the gain crossover frequency, or

$$\gamma = 180° + \phi$$

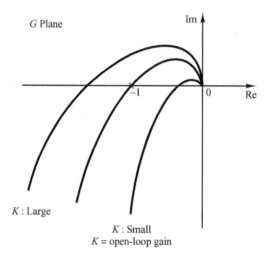

G Plane

Im

-1

0   Re

K : Large

K : Small
K = open-loop gain

Figure 7–66
Polar plots of
$$\frac{K(1 + j\omega T_a)(1 + j\omega T_b)\cdots}{(j\omega)(1 + j\omega T_1)(1 + j\omega T_2)\cdots}.$$

Figures 7–67(a), (b), and (c) illustrate the phase margin of both a stable system and an unstable system in Bode diagrams, polar plots, and log-magnitude-versus-phase plots. In the polar plot, a line may be drawn from the origin to the point at which the unit circle crosses the $G(j\omega)$ locus. If this line lies below (above) the negative real axis, then the

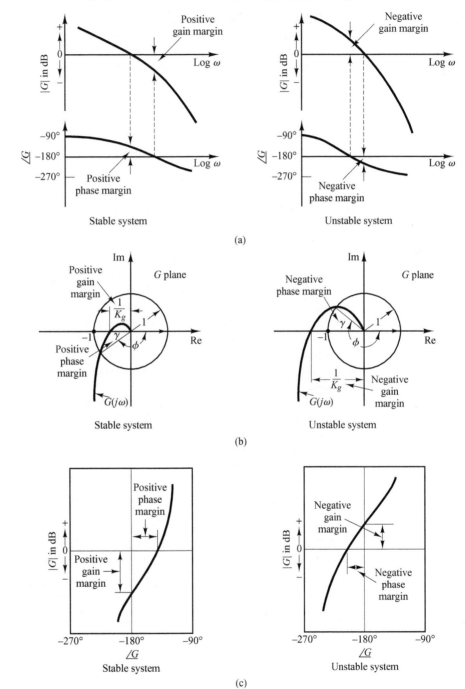

Figure 7–67
Phase and gain margins of stable and unstable systems.
(a) Bode diagrams;
(b) polar plots;
(c) log-magnitude-versus-phase plots.

angle γ is positive (negative). The angle from the negative real axis to this line is the phase margin. The phase margin is positive for $\gamma > 0$ and negative for $\gamma < 0$. For a minimum-phase system to be stable, the phase margin must be positive. In the logarithmic plots, the critical point corresponds to the intersection of the 0-dB and $-180°$ lines.

Gain margin: The gain margin is the reciprocal of the magnitude $|G(j\omega)|$ at the frequency at which the phase angle is $-180°$. Defining the phase crossover frequency ω_1 to be the frequency at which the phase angle of the open-loop transfer function equals $-180°$ gives the gain margin K_g:

$$K_g = \frac{1}{|G(j\omega_1)|}$$

In terms of decibels,

$$K_g \text{ dB} = 20 \log K_g = -20 \log |G(j\omega_1)|$$

The gain margin expressed in decibels is positive if K_g is greater than unity and negative if K_g is smaller than unity. Thus, a positive gain margin (in decibels) means that the system is stable, and a negative gain margin (in decibels) means that the system is unstable. The gain margin is shown in Figures 7–67(a), (b), and (c).

For a stable minimum-phase system, the gain margin indicates how much the gain can be increased before the system becomes unstable. For an unstable system, the gain margin is indicative of how much the gain must be decreased to make the system stable.

The gain margin of a first- or second-order system is infinite since the polar plots for such systems do not cross the negative real axis. Thus, theoretically, first- or second-order systems cannot be unstable. (Note, however, that so-called first- or second-order systems are only approximations in the sense that small time lags are neglected in deriving the system equations and are thus not truly first- or second-order systems. If these small lags are accounted for, the so-called first- or second-order systems may become unstable.)

It is noted that for a nonminimum-phase system with unstable open loop the stability condition will not be satisfied unless the $G(j\omega)$ plot encircles the $-1 + j0$ point. Hence, such a stable nonminimum-phase system will have negative phase and gain margins.

It is also important to point out that conditionally stable systems will have two or more phase crossover frequencies, and some higher-order systems with complicated numerator dynamics may also have two or more gain crossover frequencies, as shown in Figure 7–68. For stable systems having two or more gain crossover frequencies, the phase margin is measured at the highest gain crossover frequency.

A Few Comments on Phase and Gain Margins. The phase and gain margins of a control system are a measure of the closeness of the polar plot to the $-1 + j0$ point. Therefore, these margins may be used as design criteria.

It should be noted that either the gain margin alone or the phase margin alone does not give a sufficient indication of the relative stability. Both should be given in the determination of relative stability.

For a minimum-phase system, both the phase and gain margins must be positive for the system to be stable. Negative margins indicate instability.

Proper phase and gain margins ensure us against variations in the system components and are specified for definite positive values. The two values bound the behavior of the

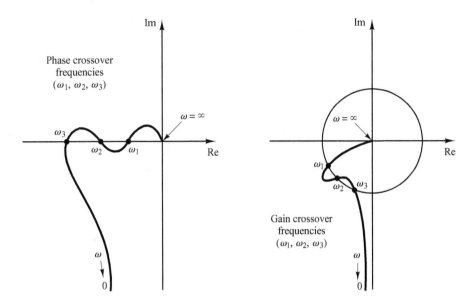

Figure 7–68
Polar plots showing more than two phase or gain crossover frequencies.

closed-loop system near the resonant frequency. For satisfactory performance, the phase margin should be between 30° and 60°, and the gain margin should be greater than 6 dB. With these values, a minimum-phase system has guaranteed stability, even if the open-loop gain and time constants of the components vary to a certain extent. Although the phase and gain margins give only rough estimates of the effective damping ratio of the closed-loop system, they do offer a convenient means for designing control systems or adjusting the gain constants of systems.

For minimum-phase systems, the magnitude and phase characteristics of the open-loop transfer function are definitely related. The requirement that the phase margin be between 30° and 60° means that in a Bode diagram the slope of the log-magnitude curve at the gain crossover frequency should be more gradual than −40 dB/decade. In most practical cases, a slope of −20 dB/decade is desirable at the gain crossover frequency for stability. If it is −40 dB/decade, the system could be either stable or unstable. (Even if the system is stable, however, the phase margin is small.) If the slope at the gain crossover frequency is −60 dB/decade or steeper, the system is most likely unstable.

For nonminimum-phase systems, the correct interpretation of stability margins requires careful study. The best way to determine the stability of nonminimum-phase systems is to use the Nyquist diagram approach rather than Bode diagram approach.

EXAMPLE 7–20   Obtain the phase and gain margins of the system shown in Figure 7–69 for the two cases where $K = 10$ and $K = 100$.

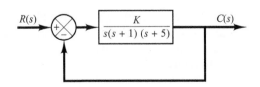

Figure 7–69
Control system.

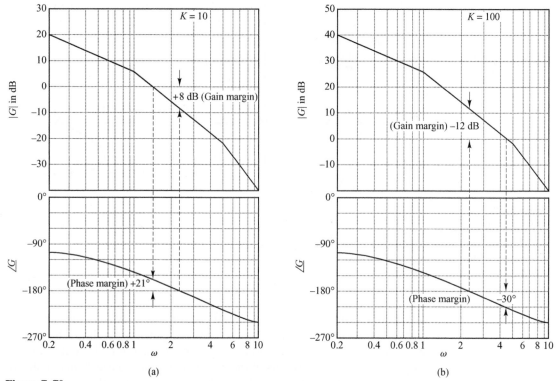

Figure 7–70
Bode diagrams of the system shown in Figure 7–69; (a) with $K = 10$ and (b) with $K = 100$.

The phase and gain margins can easily be obtained from the Bode diagram. A Bode diagram of the given open-loop transfer function with $K = 10$ is shown in Figure 7–70(a). The phase and gain margins for $K = 10$ are

$$\text{Phase margin} = 21°, \qquad \text{Gain margin} = 8 \text{ dB}$$

Therefore, the system gain may be increased by 8 dB before the instability occurs.

Increasing the gain from $K = 10$ to $K = 100$ shifts the 0-dB axis down by 20 dB, as shown in Figure 7–70(b). The phase and gain margins are

$$\text{Phase margin} = -30°, \qquad \text{Gain margin} = -12 \text{ dB}$$

Thus, the system is stable for $K = 10$, but unstable for $K = 100$.

Notice that one of the very convenient aspects of the Bode diagram approach is the ease with which the effects of gain changes can be evaluated. Note that to obtain satisfactory performance, we must increase the phase margin to $30° \sim 60°$. This can be done by decreasing the gain K. Decreasing K is not desirable, however, since a small value of K will yield a large error for the ramp input. This suggests that reshaping of the open-loop frequency-response curve by adding compensation may be necessary. Compensation techniques are discussed in detail in Sections 7–11 through 7–13.

Obtaining Gain Margin, Phase Margin, Phase-Crossover Frequency, and Gain-Crossover Frequency with MATLAB. The gain margin, phase margin, phase-crossover frequency, and gain-crossover frequency can be obtained easily with MATLAB. The command to be used is

$$[\text{Gm,pm,wcp,wcg}] = \text{margin(sys)}$$

where Gm is the gain margin, pm is the phase margin, wcp is the phase-crossover frequency, and wcg is the gain-crossover frequency. For details of how to use this command, see Example 7–21.

EXAMPLE 7–21   Draw a Bode diagram of the open-loop transfer function $G(s)$ of the closed-loop system shown in Figure 7–71. Determine the gain margin, phase margin, phase-crossover frequency, and gain-crossover frequency with MATLAB.

A MATLAB program to plot a Bode diagram and to obtain the gain margin, phase margin, phase-crossover frequency, and gain-crossover frequency is shown in MATLAB Program 7–11. The Bode diagram of $G(s)$ is shown in Figure 7–72.

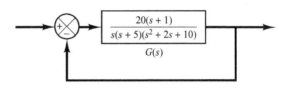

Figure 7–71
Closed-loop system.

MATLAB Program 7–11

```
num = [20  20];
den = conv([1  5  0],[1  2  10]);
sys = tf(num,den);
w = logspace(-1,2,100);
bode(sys,w)
[Gm,pm,wcp,wcg] = margin(sys);
GmdB = 20*log10(Gm);
[GmdB  pm  wcp  wcg]

ans =

       9.9293  103.6573  4.0131  0.4426
```

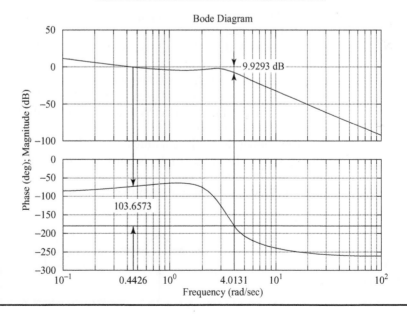

Figure 7–72
Bode diagram of $G(s)$ shown in Figure 7–71.

Resonant Peak Magnitude M_r and Resonant Frequency ω_r. Consider the standard second-order system shown in Figure 7–73. The closed-loop transfer function is

$$\frac{C(s)}{R(s)} = \frac{\omega_n^2}{s^2 + 2\zeta\omega_n s + \omega_n^2} \tag{7-16}$$

where ζ and ω_n are the damping ratio and the undamped natural frequency, respectively. The closed-loop frequency response is

$$\frac{C(j\omega)}{R(j\omega)} = \frac{1}{\left(1 - \dfrac{\omega^2}{\omega_n^2}\right) + j2\zeta\dfrac{\omega}{\omega_n}} = Me^{j\alpha}$$

where

$$M = \frac{1}{\sqrt{\left(1 - \dfrac{\omega^2}{\omega_n^2}\right)^2 + \left(2\zeta\dfrac{\omega}{\omega_n}\right)^2}}, \qquad \alpha = -\tan^{-1}\frac{2\zeta\dfrac{\omega}{\omega_n}}{1 - \dfrac{\omega^2}{\omega_n^2}}$$

As given by Equation (7–12), for $0 \le \zeta \le 0.707$, the maximum value of M occurs at the frequency ω_r, where

$$\omega_r = \omega_n\sqrt{1 - 2\zeta^2} \tag{7-17}$$

The frequency ω_r is the resonant frequency. At the resonant frequency, the value of M is maximum and is given by Equation (7–13), rewritten

$$M_r = \frac{1}{2\zeta\sqrt{1 - \zeta^2}} \tag{7-18}$$

where M_r is defined as the *resonant peak magnitude*. The resonant peak magnitude is related to the damping of the system.

The magnitude of the resonant peak gives an indication of the relative stability of the system. A large resonant peak magnitude indicates the presence of a pair of dominant closed-loop poles with small damping ratio, which will yield an undesirable transient response. A smaller resonant peak magnitude, on the other hand, indicates the absence of a pair of dominant closed-loop poles with small damping ratio, meaning that the system is well damped.

Remember that ω_r is real only if $\zeta < 0.707$. Thus, there is no closed-loop resonance if $\zeta > 0.707$. [The value of M_r is unity only if $\zeta > 0.707$. See Equation (7–14).] Since the values of M_r and ω_r can be easily measured in a physical system, they are quite useful for checking agreement between theoretical and experimental analyses.

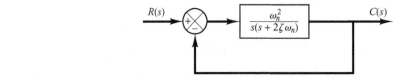

Figure 7–73
Standard second-order system.

470

It is noted, however, that in practical design problems the phase margin and gain margin are more frequently specified than the resonant peak magnitude to indicate the degree of damping in a system.

Correlation between Step Transient Response and Frequency Response in the Standard Second-Order System. The maximum overshoot in the unit-step response of the standard second-order system, as shown in Figure 7–73, can be exactly correlated with the resonant peak magnitude in the frequency response. Hence, essentially the same information about the system dynamics is contained in the frequency response as is in the transient response.

For a unit-step input, the output of the system shown in Figure 7–73 is given by Equation (5–12), or

$$c(t) = 1 - e^{-\zeta \omega_n t} \left(\cos \omega_d t + \frac{\zeta}{\sqrt{1 - \zeta^2}} \sin \omega_d t \right), \qquad \text{for } t \geq 0$$

where

$$\omega_d = \omega_n \sqrt{1 - \zeta^2} \tag{7–19}$$

On the other hand, the maximum overshoot M_p for the unit-step response is given by Equation (5–21), or

$$M_p = e^{-(\zeta / \sqrt{1 - \zeta^2})\pi} \tag{7–20}$$

This maximum overshoot occurs in the transient response that has the damped natural frequency $\omega_d = \omega_n \sqrt{1 - \zeta^2}$. The maximum overshoot becomes excessive for values of $\zeta < 0.4$.

Since the second-order system shown in Figure 7–73 has the open-loop transfer function

$$G(s) = \frac{\omega_n^2}{s(s + 2\zeta\omega_n)}$$

for sinusoidal operation, the magnitude of $G(j\omega)$ becomes unity when

$$\omega = \omega_n \sqrt{\sqrt{1 + 4\zeta^4} - 2\zeta^2}$$

which can be obtained by equating $|G(j\omega)|$ to unity and solving for ω. At this frequency, the phase angle of $G(j\omega)$ is

$$\underline{/G(j\omega)} = -\underline{/j\omega} - \underline{/j\omega + 2\zeta\omega_n} = -90° - \tan^{-1} \frac{\sqrt{\sqrt{1 + 4\zeta^4} - 2\zeta^2}}{2\zeta}$$

Thus, the phase margin γ is

$$\gamma = 180° + \underline{/G(j\omega)}$$

$$= 90° - \tan^{-1} \frac{\sqrt{\sqrt{1 + 4\zeta^4} - 2\zeta^2}}{2\zeta}$$

$$= \tan^{-1} \frac{2\zeta}{\sqrt{\sqrt{1 + 4\zeta^4} - 2\zeta^2}} \tag{7–21}$$

Equation (7–21) gives the relationship between the damping ratio ζ and the phase margin γ. (Notice that the phase margin γ is a function *only* of the damping ratio ζ.)

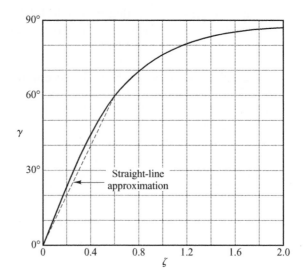

Figure 7–74
Curve γ (phase margin) versus ζ for the system shown in Figure 7–73.

In the following, we shall summarize the correlation between the step transient response and frequency response of the standard second-order system given by Equation (7–16):

1. The phase margin and the damping ratio are directly related. Figure 7–74 shows a plot of the phase margin γ as a function of the damping ratio ζ. It is noted that for the standard second-order system shown in Figure 7–73, the phase margin γ and the damping ratio ζ are related approximately by a straight line for $0 \le \zeta \le 0.6$, as follows:

$$\zeta = \frac{\gamma}{100^\circ}$$

Thus a phase margin of 60° corresponds to a damping ratio of 0.6. For higher-order systems having a dominant pair of closed-loop poles, this relationship may be used as a rule of thumb in estimating the relative stability in the transient response (that is, the damping ratio) from the frequency response.

2. Referring to Equations (7–17) and (7–19), we see that the values of ω_r and ω_d are almost the same for small values of ζ. Thus, for small values of ζ, the value of ω_r is indicative of the speed of the transient response of the system.

3. From Equations (7–18) and (7–20), we note that the smaller the value of ζ is, the larger the values of M_r and M_p are. The correlation between M_r and M_p as a function of ζ is shown in Figure 7–75. A close relationship between M_r and M_p can be seen for $\zeta > 0.4$. For very small values of ζ, M_r becomes very large $(M_r \gg 1)$, while the value of M_p does not exceed 1.

Correlation between Step Transient Response and Frequency Response in General Systems. The design of control systems is very often carried out on the basis of the frequency response. The main reason for this is the relative simplicity of this approach compared with others. Since in many applications it is the transient response of the system to aperiodic inputs rather than the steady-state response to sinusoidal inputs that is of primary concern, the question of correlation between transient response and frequency response arises.

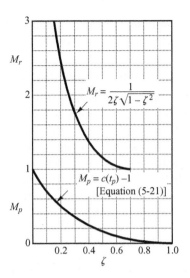

$$M_r = \frac{1}{2\zeta\sqrt{1-\zeta^2}}$$

$$M_p = c(t_p) - 1$$
[Equation (5-21)]

Figure 7–75
Curves M_r versus ζ
and M_p versus ζ for
the system shown in
Figure 7–73.

For the standard second-order system shown in Figure 7–73, mathematical relationships correlating the step transient response and frequency response can be obtained easily. The time response of the standard second-order system can be predicted exactly from a knowledge of the M_r and ω_r of its closed-loop frequency response.

For nonstandard second-order systems and higher-order systems, the correlation is more complex, and the transient response may not be predicted easily from the frequency response because additional zeros and/or poles may change the correlation between the step transient response and the frequency response existing for the standard second-order system. Mathematical techniques for obtaining the exact correlation are available, but they are very laborious and of little practical value.

The applicability of the transient-response–frequency-response correlation existing for the standard second-order system shown in Figure 7–73 to higher-order systems depends on the presence of a dominant pair of complex-conjugate closed-loop poles in the latter systems. Clearly, if the frequency response of a higher-order system is dominated by a pair of complex-conjugate closed-loop poles, the transient-response–frequency-response correlation existing for the standard second-order system can be extended to the higher-order system.

For linear, time-invariant, higher-order systems having a dominant pair of complex-conjugate closed-loop poles, the following relationships generally exist between the step transient response and frequency response:

1. The value of M_r is indicative of the relative stability. Satisfactory transient performance is usually obtained if the value of M_r is in the range $1.0 < M_r < 1.4$ $(0 \text{ dB} < M_r < 3 \text{ dB})$, which corresponds to an effective damping ratio of $0.4 < \zeta < 0.7$. For values of M_r greater than 1.5, the step transient response may exhibit several overshoots. (Note that, in general, a large value of M_r corresponds to a large overshoot in the step transient response. If the system is subjected to noise signals whose frequencies are near the resonant frequency ω_r, the noise will be amplified in the output and will present serious problems.)

2. The magnitude of the resonant frequency ω_r is indicative of the speed of the transient response. The larger the value of ω_r, the faster the time response is. In other words, the rise time varies inversely with ω_r. In terms of the open-loop frequency

response, the damped natural frequency in the transient response is somewhere between the gain crossover frequency and phase crossover frequency.

3. The resonant peak frequency ω_r and the damped natural frequency ω_d for the step transient response are very close to each other for lightly damped systems.

The three relationships just listed are useful for correlating the step transient response with the frequency response of higher-order systems, provided that they can be approximated by the standard second-order system or a pair of complex-conjugate closed-loop poles. If a higher-order system satisfies this condition, a set of time-domain specifications may be translated into frequency-domain specifications. This simplifies greatly the design work or compensation work of higher-order systems.

In addition to the phase margin, gain margin, resonant peak M_r, and resonant frequency ω_r, there are other frequency-domain quantities commonly used in performance specifications. They are the cutoff frequency, bandwidth, and the cutoff rate. These will be defined in what follows.

Cutoff Frequency and Bandwidth.   Referring to Figure 7–76, the frequency ω_b at which the magnitude of the closed-loop frequency response is 3 dB below its zero-frequency value is called the *cutoff frequency*. Thus

$$\left|\frac{C(j\omega)}{R(j\omega)}\right| < \left|\frac{C(j0)}{R(j0)}\right| - 3 \text{ dB}, \qquad \text{for } \omega > \omega_b$$

For systems in which $\left|C(j0)/R(j0)\right| = 0$ dB,

$$\left|\frac{C(j\omega)}{R(j\omega)}\right| < -3 \text{ dB}, \qquad \text{for } \omega > \omega_b$$

The closed-loop system filters out the signal components whose frequencies are greater than the cutoff frequency and transmits those signal components with frequencies lower than the cutoff frequency.

The frequency range $0 \le \omega \le \omega_b$ in which the magnitude of $C(j\omega)/R(j\omega)$ is greater than -3 dB is called the *bandwidth* of the system. The bandwidth indicates the frequency where the gain starts to fall off from its low-frequency value. Thus, the bandwidth indicates how well the system will track an input sinusoid. Note that for a given ω_n, the rise time increases with increasing damping ratio ζ. On the other hand, the bandwidth decreases with the increase in ζ. Therefore, the rise time and the bandwidth are inversely proportional to each other.

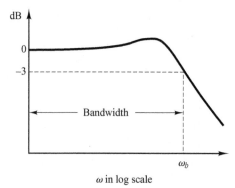

Figure 7–76
Plot of a closed-loop frequency response curve showing cutoff frequency ω_b and bandwidth.

The specification of the bandwidth may be determined by the following factors:

1. The ability to reproduce the input signal. A large bandwidth corresponds to a small rise time, or fast response. Roughly speaking, we can say that the bandwidth is proportional to the speed of response. (For example, to decrease the rise time in the step response by a factor of 2, the bandwidth must be increased by approximately a factor of 2.)

2. The necessary filtering characteristics for high-frequency noise.

For the system to follow arbitrary inputs accurately, it must have a large bandwidth. From the viewpoint of noise, however, the bandwidth should not be too large. Thus, there are conflicting requirements on the bandwidth, and a compromise is usually necessary for good design. Note that a system with large bandwidth requires high-performance components, so the cost of components usually increases with the bandwidth.

Cutoff Rate. The cutoff rate is the slope of the log-magnitude curve near the cutoff frequency. The cutoff rate indicates the ability of a system to distinguish the signal from noise.

It is noted that a closed-loop frequency response curve with a steep cutoff characteristic may have a large resonant peak magnitude, which implies that the system has a relatively small stability margin.

EXAMPLE 7–22   Consider the following two systems:

$$\text{System I:} \quad \frac{C(s)}{R(s)} = \frac{1}{s+1}, \quad \text{System II:} \quad \frac{C(s)}{R(s)} = \frac{1}{3s+1}$$

Compare the bandwidths of these two systems. Show that the system with the larger bandwidth has a faster speed of response and can follow the input much better than the one with the smaller bandwidth.

Figure 7–77(a) shows the closed-loop frequency-response curves for the two systems. (Asymptotic curves are shown by dashed lines.) We find that the bandwidth of system I is $0 \le \omega \le 1$ rad/sec and that of system II is $0 \le \omega \le 0.33$ rad/sec. Figures 7–77(b) and (c) show, respectively, the unit-step response and unit-ramp response curves for the two systems. Clearly, system I, whose bandwidth is three times wider than that of system II, has a faster speed of response and can follow the input much better.

Figure 7–77
Comparison of dynamic characteristics of the two systems considered in Example 7–22. (a) Closed-loop frequency-response curves; (b) unit-step response curves; (c) unit-ramp response curves.

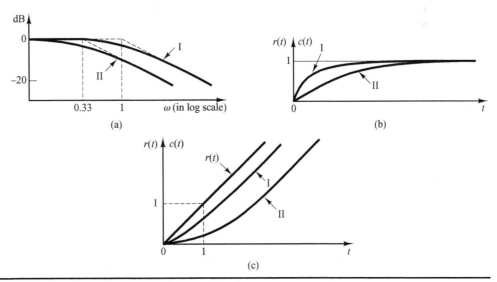

MATLAB Approach to Get Resonant Peak, Resonant Frequency, and Bandwidth. The resonant peak is the value of the maximum magnitude (in decibels) of the closed-loop frequency response. The resonant frequency is the frequency that yields the maximum magnitude. MATLAB commands to be used for obtaining the resonant peak and resonant frequency are as follows:

```
[mag,phase,w] = bode(num,den,w);    or    [mag,phase,w] = bode(sys,w);
[Mp,k] = max(mag);
resonant_peak = 20*log10(Mp);
resonant_frequency = w(k)
```

The bandwidth can be obtained by entering the following lines in the program:

```
n = 1;
while 20*log10(mag(n)) > = -3; n = n + 1;
end
bandwidth = w(n)
```

For a detailed MATLAB program, see Example 7–23.

EXAMPLE 7–23   Consider the system shown in Figure 7–78. Using MATLAB, obtain a Bode diagram for the closed-loop transfer function. Obtain also the resonant peak, resonant frequency, and bandwidth.

MATLAB Program 7–12 produces a Bode diagram for the closed-loop system as well as the resonant peak, resonant frequency, and bandwidth. The resulting Bode diagram is shown in

---

MATLAB Program 7–12

```
nump = [1];
denp = [0.5  1.5  1  0];
sysp = tf(nump,denp);
sys = feedback(sysp,1);
w = logspace(-1,1);
bode(sys,w)
[mag,phase,w] = bode(sys,w);
[Mp,k] = max(mag);
resonant_peak = 20*log10(Mp)
```

```
resonant_peak =

    5.2388
```

```
resonant_frequency = w(k)
```

```
resonant_frequency =

    0.7906
```

```
n = 1;
while 20*log(mag(n))> = -3; n = n + 1;
end
bandwidth = w(n)
```

```
bandwidth =

    1.2649
```

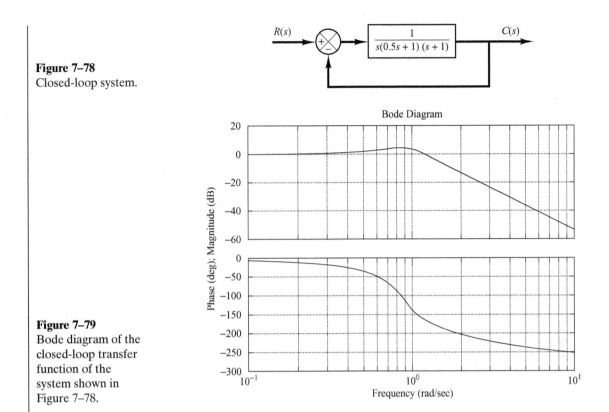

Figure 7–78
Closed-loop system.

Figure 7–79
Bode diagram of the
closed-loop transfer
function of the
system shown in
Figure 7–78.

Figure 7–79. The resonant peak is obtained as 5.2388 dB. The resonant frequency is 0.7906 rad/sec. The bandwidth is 1.2649 rad/sec. These values can be verified from Figure 7–79.

7–8 CLOSED-LOOP FREQUENCY RESPONSE OF UNITY-FEEDBACK SYSTEMS

Closed-Loop Frequency Response.   For a stable, unity-feedback closed-loop system, the closed-loop frequency response can be obtained easily from that of the open-loop frequency response. Consider the unity-feedback system shown in Figure 7–80(a). The closed-loop transfer function is

$$\frac{C(s)}{R(s)} = \frac{G(s)}{1 + G(s)}$$

In the Nyquist or polar plot shown in Figure 7–80(b), the vector $\overrightarrow{OA}$ represents $G(j\omega_1)$, where ω_1 is the frequency at point A. The length of the vector $\overrightarrow{OA}$ is $|G(j\omega_1)|$ and the angle of the vector $\overrightarrow{OA}$ is $\underline{/G(j\omega_1)}$. The vector $\overrightarrow{PA}$, the vector from the $-1 + j0$ point to the Nyquist locus, represents $1 + G(j\omega_1)$. Therefore, the ratio of $\overrightarrow{OA}$, to $\overrightarrow{PA}$ represents the closed-loop frequency response, or

$$\frac{\overrightarrow{OA}}{\overrightarrow{PA}} = \frac{G(j\omega_1)}{1 + G(j\omega_1)} = \frac{C(j\omega_1)}{R(j\omega_1)}$$

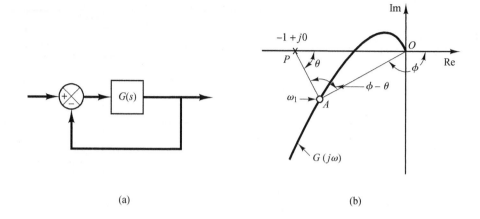

Figure 7–80
(a) Unity-feedback system;
(b) determination of closed-loop frequency response from open-loop frequency response.

(a)

(b)

The magnitude of the closed-loop transfer function at $\omega = \omega_1$ is the ratio of the magnitudes of $\overrightarrow{OA}$ to $\overrightarrow{PA}$. The phase angle of the closed-loop transfer function at $\omega = \omega_1$ is the angle formed by the vectors $\overrightarrow{OA}$ to $\overrightarrow{PA}$—that is $\phi - \theta$, shown in Figure 7–80(b). By measuring the magnitude and phase angle at different frequency points, the closed-loop frequency-response curve can be obtained.

Let us define the magnitude of the closed-loop frequency response as M and the phase angle as α, or

$$\frac{C(j\omega)}{R(j\omega)} = Me^{j\alpha}$$

In the following, we shall find the constant-magnitude loci and constant-phase-angle loci. Such loci are convenient in determining the closed-loop frequency response from the polar plot or Nyquist plot.

Constant-Magnitude Loci (M circles). To obtain the constant-magnitude loci, let us first note that $G(j\omega)$ is a complex quantity and can be written as follows:

$$G(j\omega) = X + jY$$

where X and Y are real quantities. Then M is given by

$$M = \frac{|X + jY|}{|1 + X + jY|}$$

and M^2 is

$$M^2 = \frac{X^2 + Y^2}{(1 + X)^2 + Y^2}$$

Hence

$$X^2(1 - M^2) - 2M^2X - M^2 + (1 - M^2)Y^2 = 0 \qquad (7\text{--}22)$$

If $M = 1$, then from Equation (7–22), we obtain $X = -\frac{1}{2}$. This is the equation of a straight line parallel to the Y axis and passing through the point $\left(-\frac{1}{2}, 0\right)$.

If $M \neq 1$, Equation (7–22) can be written

$$X^2 + \frac{2M^2}{M^2 - 1} X + \frac{M^2}{M^2 - 1} + Y^2 = 0$$

If the term $M^2/(M^2 - 1)^2$ is added to both sides of this last equation, we obtain

$$\left(X + \frac{M^2}{M^2 - 1}\right)^2 + Y^2 = \frac{M^2}{(M^2 - 1)^2} \tag{7–23}$$

Equation (7–23) is the equation of a circle with center at $X = -M^2/(M^2 - 1)$, $Y = 0$ and with radius $\left|M/(M^2 - 1)\right|$.

The constant M loci on the $G(s)$ plane are thus a family of circles. The center and radius of the circle for a given value of M can be easily calculated. For example, for $M = 1.3$, the center is at $(-2.45, 0)$ and the radius is 1.88. A family of constant M circles is shown in Figure 7–81. It is seen that as M becomes larger compared with 1, the M circles become smaller and converge to the $-1 + j0$ point. For $M > 1$, the centers of the M circles lie to the left of the $-1 + j0$ point. Similarly, as M becomes smaller compared with 1, the M circle becomes smaller and converges to the origin. For $0 < M < 1$, the centers of the M circles lie to the right of the origin. $M = 1$ corresponds to the locus of points equidistant from the origin and from the $-1 + j0$ point. As stated earlier, it is a straight line passing through the point $\left(-\frac{1}{2}, 0\right)$ and parallel to the imaginary axis. (The constant M circles corresponding to $M > 1$ lie to the left of the $M = 1$ line, and those corresponding to $0 < M < 1$ lie to the right of the $M = 1$ line.) The M circles are symmetrical with respect to the straight line corresponding to $M = 1$ and with respect to the real axis.

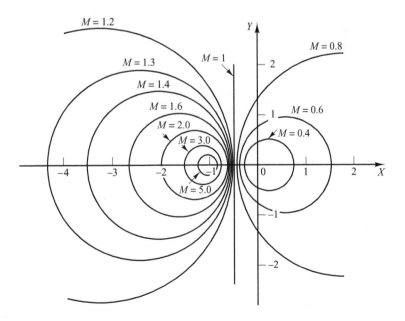

Figure 7–81
A family of constant
M circles.

Constant-Phase-Angle Loci (N Circles). We shall obtain the phase angle α in terms of X and Y. Since

$$\underline{/e^{j\alpha}} = \left/ \frac{X + jY}{1 + X + jY}\right.$$

the phase angle α is

$$\alpha = \tan^{-1}\left(\frac{Y}{X}\right) - \tan^{-1}\left(\frac{Y}{1 + X}\right)$$

If we define

$$\tan \alpha = N$$

then

$$N = \tan\left[\tan^{-1}\left(\frac{Y}{X}\right) - \tan^{-1}\left(\frac{Y}{1 + X}\right)\right]$$

Since

$$\tan(A - B) = \frac{\tan A - \tan B}{1 + \tan A \tan B}$$

we obtain

$$N = \frac{\dfrac{Y}{X} - \dfrac{Y}{1 + X}}{1 + \dfrac{Y}{X}\left(\dfrac{Y}{1 + X}\right)} = \frac{Y}{X^2 + X + Y^2}$$

or

$$X^2 + X + Y^2 - \frac{1}{N}Y = 0$$

The addition of $(\frac{1}{4}) + 1/(2N)^2$ to both sides of this last equation yields

$$\left(X + \frac{1}{2}\right)^2 + \left(Y - \frac{1}{2N}\right)^2 = \frac{1}{4} + \left(\frac{1}{2N}\right)^2 \tag{7-24}$$

This is an equation of a circle with center at $X = -\frac{1}{2}$, $Y = 1/(2N)$ and with radius $\sqrt{\frac{1}{4} + 1/(2N)^2}$. For example, if $\alpha = 30°$, then $N = \tan \alpha = 0.577$, and the center and the radius of the circle corresponding to $\alpha = 30°$ are found to be $(-0.5, 0.866)$ and unity, respectively. Since Equation (7–24) is satisfied when $X = Y = 0$ and $X = -1, Y = 0$ regardless of the value of N, each circle passes through the origin and the $-1 + j0$ point. The constant α loci can be drawn easily, once the value of N is given. A family of constant N circles is shown in Figure 7–82 with α as a parameter.

It should be noted that the constant N locus for a given value of α is actually not the entire circle, but only an arc. In other words, the $\alpha = 30°$ and $\alpha = -150°$ arcs are parts of the same circle. This is so because the tangent of an angle remains the same if $\pm 180°$ (or multiples thereof) is added to the angle.

The use of the M and N circles enables us to find the entire closed-loop frequency response from the open-loop frequency response $G(j\omega)$ without calculating the magnitude and phase of the closed-loop transfer function at each frequency. The intersections

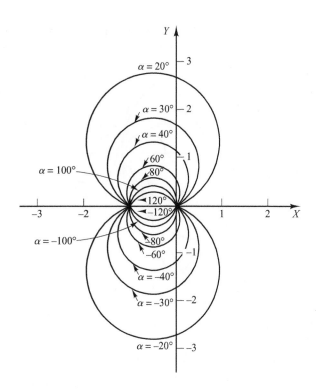

Figure 7–82
A family of constant
N circles.

of the $G(j\omega)$ locus and the M circles and N circles give the values of M and N at frequency points on the $G(j\omega)$ locus.

The N circles are multivalued in the sense that the circle for $\alpha = \alpha_1$ and that for $\alpha = \alpha_1 \pm 180°n$ ($n = 1, 2, \ldots$) are the same. In using the N circles for the determination of the phase angle of closed-loop systems, we must interpret the proper value of α. To avoid any error, start at zero frequency, which corresponds to $\alpha = 0°$, and proceed to higher frequencies. The phase-angle curve must be continuous.

Graphically, the intersections of the $G(j\omega)$ locus and M circles give the values of M at the frequencies denoted on the $G(j\omega)$ locus. Thus, the constant M circle with the smallest radius that is tangent to the $G(j\omega)$ locus gives the value of the resonant peak magnitude M_r. If it is desired to keep the resonant peak value less than a certain value, then the system should not enclose the critical point ($-1 + j0$ point) and, at the same time, there should be no intersections with the particular M circle and the $G(j\omega)$ locus.

Figure 7–83(a) shows the $G(j\omega)$ locus superimposed on a family of M circles. Figure 7–83(b) shows the $G(j\omega)$ locus superimposed on a family of N circles. From these plots, it is possible to obtain the closed-loop frequency response by inspection. It is seen that the $M = 1.1$ circle intersects the $G(j\omega)$ locus at frequency point $\omega = \omega_1$. This means that at this frequency the magnitude of the closed-loop transfer function is 1.1. In Figure 7–83(a), the $M = 2$ circle is just tangent to the $G(j\omega)$ locus. Thus, there is only one point on the $G(j\omega)$ locus for which $|C(j\omega)/R(j\omega)|$ is equal to 2. Figure 7–83(c) shows the closed-loop frequency-response curve for the system. The upper curve is the M-versus-frequency ω curve, and the lower curve is the phase angle α-versus-frequency ω curve.

The resonant peak value is the value of M corresponding to the M circle of smallest radius that is tangent to the $G(j\omega)$ locus. Thus, in the Nyquist diagram, the resonant

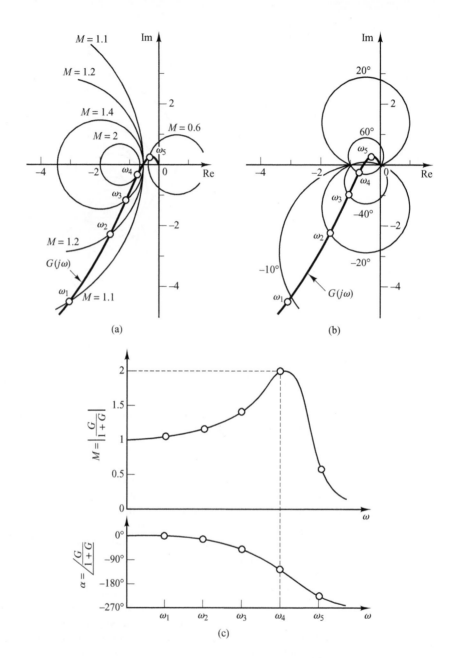

Figure 7–83
(a) $G(j\omega)$ locus
superimposed on a
family of M circles;
(b) $G(j\omega)$ locus
superimposed on a
family of N circles;
(c) closed-loop
frequency-response
curves.

peak value M_r and the resonant frequency ω_r can be found from the M-circle tangency
to the $G(j\omega)$ locus. (In the present example, $M_r = 2$ and $\omega_r = \omega_4$.)

Nichols Chart. In dealing with design problems, we find it convenient to construct
the M and N loci in the log-magnitude-versus-phase plane. The chart consisting of the
M and N loci in the log-magnitude-versus-phase diagram is called the Nichols chart.
The $G(j\omega)$ locus drawn on the Nichols chart gives both the gain characteristics and

Chapter 7 / Control Systems Analysis and Design by the Frequency-Response Method

phase characteristics of the closed-loop transfer function at the same time. The Nichols chart is shown in Figure 7–84, for phase angles between 0° and −240°.

Note that the critical point (−1 + j0 point) is mapped to the Nichols chart as the point (0 dB, −180°). The Nichols chart contains curves of constant closed-loop magnitude and phase angle. The designer can graphically determine the phase margin, gain margin, resonant peak magnitude, resonant frequency, and bandwidth of the closed-loop system from the plot of the open-loop locus, $G(j\omega)$.

The Nichols chart is symmetric about the −180° axis. The M and N loci repeat for every 360°, and there is symmetry at every 180° interval. The M loci are centered about the critical point (0 dB, −180°). The Nichols chart is useful for determining the frequency response of the closed loop from that of the open loop. If the open-loop frequency-response curve is superimposed on the Nichols chart, the intersections of the open-loop frequency-response curve $G(j\omega)$ and the M and N loci give the values of the magnitude M and phase angle α of the closed-loop frequency response at each frequency point. If the $G(j\omega)$ locus does not intersect the $M = M_r$ locus, but is tangent to it, then the resonant peak value of M of the closed-loop frequency response is given by M_r. The resonant frequency is given by the frequency at the point of tangency.

As an example, consider the unity-feedback system with the following open-loop transfer function:

$$G(j\omega) = \frac{K}{s(s + 1)(0.5s + 1)}, \qquad K = 1$$

To find the closed-loop frequency response by use of the Nichols chart, the $G(j\omega)$ locus is constructed in the log-magnitude-versus-phase plane by use of MATLAB or from

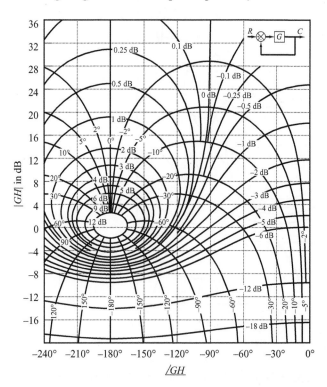

Figure 7–84
Nichols chart.

the Bode diagram. Figure 7–85(a) shows the $G(j\omega)$ locus together with the M and N loci. The closed-loop frequency-response curves may be constructed by reading the magnitudes and phase angles at various frequency points on the $G(j\omega)$ locus from the M and N loci, as shown in Figure 7–85(b). Since the largest magnitude contour touched by the $G(j\omega)$ locus is 5 dB, the resonant peak magnitude M_r is 5 dB. The corresponding resonant peak frequency is 0.8 rad/sec.

Notice that the phase crossover point is the point where the $G(j\omega)$ locus intersects the $-180°$ axis (for the present system, $\omega = 1.4$ rad/sec), and the gain crossover point is the point where the locus intersects the 0-dB axis (for the present system, $\omega = 0.76$ rad/sec). The phase margin is the horizontal distance (measured in degrees) between the gain crossover point and the critical point (0 dB, $-180°$). The gain margin is the distance (in decibels) between the phase crossover point and the critical point.

The bandwidth of the closed-loop system can easily be found from the $G(j\omega)$ locus in the Nichols diagram. The frequency at the intersection of the $G(j\omega)$ locus and the $M = -3$ dB locus gives the bandwidth.

If the open-loop gain K is varied, the shape of the $G(j\omega)$ locus in the log-magnitude-versus-phase diagram remains the same, but it is shifted up (for increasing K) or down (for decreasing K) along the vertical axis. Therefore, the $G(j\omega)$ locus intersects the M

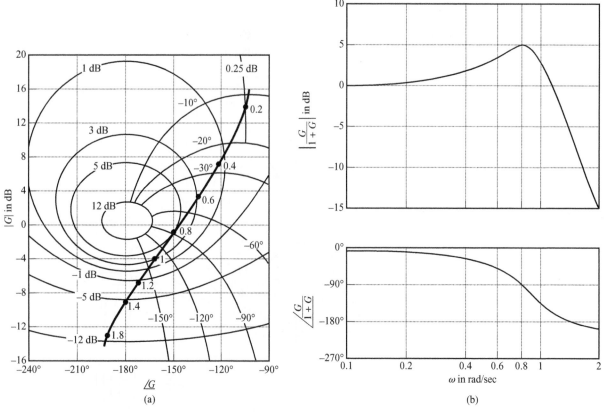

Figure 7–85
(a) Plot of $G(j\omega)$ superimposed on Nichols chart; (b) closed-loop frequency-response curves.

and N loci differently, resulting in a different closed-loop frequency-response curve. For a small value of the gain K, the $G(j\omega)$ locus will not be tangent to any of the M loci, which means that there is no resonance in the closed-loop frequency response.

EXAMPLE 7–24   Consider the unity-feedback control system whose open-loop transfer function is

$$G(j\omega) = \frac{K}{j\omega(1 + j\omega)}$$

Determine the value of the gain K so that $M_r = 1.4$.

The first step in the determination of the gain K is to sketch the polar plot of

$$\frac{G(j\omega)}{K} = \frac{1}{j\omega(1 + j\omega)}$$

Figure 7–86 shows the $M_r = 1.4$ locus and the $G(j\omega)/K$ locus. Changing the gain has no effect on the phase angle, but merely moves the curve vertically up for $K > 1$ and down for $K < 1$.

In Figure 7–86, the $G(j\omega)/K$ locus must be raised by 4 dB in order that it be tangent to the desired M_r locus and that the entire $G(j\omega)/K$ locus be outside the $M_r = 1.4$ locus. The amount of vertical shift of the $G(j\omega)/K$ locus determines the gain necessary to yield the desired value of M_r. Thus, by solving

$$20 \log K = 4$$

we obtain

$$K = 1.59$$

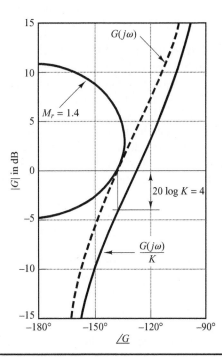

Figure 7–86
Determination of the gain K using the Nichols chart.

The first step in the analysis and design of a control system is to derive a mathematical model of the plant under consideration. Obtaining a model analytically may be quite difficult. We may have to obtain it by means of experimental analysis. The importance of the frequency-response methods is that the transfer function of the plant, or any other component of a system, may be determined by simple frequency-response measurements.

If the amplitude ratio and phase shift have been measured at a sufficient number of frequencies within the frequency range of interest, they may be plotted on the Bode diagram. Then the transfer function can be determined by asymptotic approximations. We build up asymptotic log-magnitude curves consisting of several segments. With some trial-and-error juggling of the corner frequencies, it is usually possible to find a very close fit to the curve. (Note that if the frequency is plotted in cycles per second rather than radians per second, the corner frequencies must be converted to radians per second before computing the time constants.)

Sinusoidal-Signal Generators. In performing a frequency-response test, suitable sinusoidal-signal generators must be available. The signal may have to be in mechanical, electrical, or pneumatic form. The frequency ranges needed for the test are approximately 0.001 to 10 Hz for large-time-constant systems and 0.1 to 1000 Hz for small-time-constant systems. The sinusoidal signal must be reasonably free from harmonics or distortion.

For very low frequency ranges (below 0.01 Hz), a mechanical signal generator (together with a suitable pneumatic or electrical transducer if necessary) may be used. For the frequency range from 0.01 to 1000 Hz, a suitable electrical-signal generator (together with a suitable transducer if necessary) may be used.

Determination of Minimum-Phase Transfer Functions from Bode Diagrams. As stated previously, whether a system is minimum phase can be determined from the frequency-response curves by examining the high-frequency characteristics.

To determine the transfer function, we first draw asymptotes to the experimentally obtained log-magnitude curve. The asymptotes must have slopes of multiples of ± 20 dB/decade. If the slope of the experimentally obtained log-magnitude curve changes from -20 to -40 dB/decade at $\omega = \omega_1$, it is clear that a factor $1/[1 + j(\omega/\omega_1)]$ exists in the transfer function. If the slope changes by -40 dB/decade at $\omega = \omega_2$, there must be a quadratic factor of the form

$$\frac{1}{1 + 2\zeta\left(j\dfrac{\omega}{\omega_2}\right) + \left(j\dfrac{\omega}{\omega_2}\right)^2}$$

in the transfer function. The undamped natural frequency of this quadratic factor is equal to the corner frequency ω_2. The damping ratio ζ can be determined from the experimentally obtained log-magnitude curve by measuring the amount of resonant peak near the corner frequency ω_2 and comparing this with the curves shown in Figure 7-9.

Once the factors of the transfer function $G(j\omega)$ have been determined, the gain can be determined from the low-frequency portion of the log-magnitude curve. Since such terms as $1 + j(\omega/\omega_1)$ and $1 + 2\zeta(j\omega/\omega_2) + (j\omega/\omega_2)^2$ become unity as ω approaches zero, at very low frequencies the sinusoidal transfer function $G(j\omega)$ can be written

$$\lim_{\omega \to 0} G(j\omega) = \frac{K}{(j\omega)^\lambda}$$

In many practical systems, λ equals 0, 1, or 2.

1. For $\lambda = 0$, or type 0 systems,

$$G(j\omega) = K, \quad \text{for } \omega \ll 1$$

or

$$20 \log|G(j\omega)| = 20 \log K, \quad \text{for } \omega \ll 1$$

The low-frequency asymptote is a horizontal line at $20 \log K$ dB. The value of K can thus be found from this horizontal asymptote.

2. For $\lambda = 1$, or type 1 systems,

$$G(j\omega) = \frac{K}{j\omega}, \quad \text{for } \omega \ll 1$$

or

$$20 \log|G(j\omega)| = 20 \log K - 20 \log \omega, \quad \text{for } \omega \ll 1$$

which indicates that the low-frequency asymptote has the slope -20 dB/decade. The frequency at which the low-frequency asymptote (or its extension) intersects the 0-dB line is numerically equal to K.

3. For $\lambda = 2$, or type 2 systems,

$$G(j\omega) = \frac{K}{(j\omega)^2}, \quad \text{for } \omega \ll 1$$

or

$$20 \log|G(j\omega)| = 20 \log K - 40 \log \omega, \quad \text{for } \omega \ll 1$$

The slope of the low-frequency asymptote is -40 dB/decade. The frequency at which this asymptote (or its extension) intersects the 0-dB line is numerically equal to $\sqrt{K}$.

Examples of log-magnitude curves for type 0, type 1, and type 2 systems are shown in Figure 7–87, together with the frequency to which the gain K is related.

The experimentally obtained phase-angle curve provides a means of checking the transfer function obtained from the log-magnitude curve. For a minimum-phase system, the experimental phase-angle curve should agree reasonably well with the theoretical phase-angle curve obtained from the transfer function just determined. These two phase-angle curves should agree exactly in both the very low and very high frequency ranges. If the experimentally obtained phase angle at very high frequencies (compared with the corner frequencies) is not equal to $-90°(q - p)$, where p and q are the degrees of the numerator and denominator polynomials of the transfer function, respectively, then the transfer function must be a nonminimum-phase transfer function.

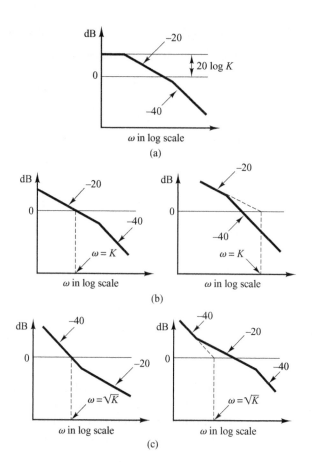

Figure 7–87
(a) Log-magnitude
curve of a type 0
system; (b) log-
magnitude curves of
type 1 systems;
(c) log-magnitude
curves of type 2
systems. (The slopes
shown are in
dB/decade.)

Nonminimum-Phase Transfer Functions.   If, at the high-frequency end, the computed phase lag is 180° less than the experimentally obtained phase lag, then one of the zeros of the transfer function should have been in the right-half s plane instead of the left-half s plane.

If the computed phase lag differed from the experimentally obtained phase lag by a constant rate of change of phase, then transport lag, or dead time, is present. If we assume the transfer function to be of the form

$$G(s)e^{-Ts}$$

where $G(s)$ is a ratio of two polynomials in s, then

$$\lim_{\omega \to \infty} \frac{d}{d\omega} \underline{/G(j\omega)e^{-j\omega T}} = \lim_{\omega \to \infty} \frac{d}{d\omega} \left[\underline{/G(j\omega)} + \underline{/e^{-j\omega T}} \right]$$

$$= \lim_{\omega \to \infty} \frac{d}{d\omega} \left[\underline{/G(j\omega)} - \omega T \right]$$

$$= 0 - T = -T$$

where we used the fact that $\lim_{\omega \to \infty} \underline{/G(j\omega)} =$ constant. Thus, from this last equation, we can evaluate the magnitude of the transport lag T.

A Few Remarks on the Experimental Determination of Transfer Functions

1. It is usually easier to make accurate amplitude measurements than accurate phase-shift measurements. Phase-shift measurements may involve errors that may be caused by instrumentation or by misinterpretation of the experimental records.

2. The frequency response of measuring equipment used to measure the system output must have a nearly flat magnitude-versus-frequency curve. In addition, the phase angle must be nearly proportional to the frequency.

3. Physical systems may have several kinds of nonlinearities. Therefore, it is necessary to consider carefully the amplitude of input sinusoidal signals. If the amplitude of the input signal is too large, the system will saturate, and the frequency-response test will yield inaccurate results. On the other hand, a small signal will cause errors due to dead zone. Hence, a careful choice of the amplitude of the input sinusoidal signal must be made. It is necessary to sample the waveform of the system output to make sure that the waveform is sinusoidal and that the system is operating in the linear region during the test period. (The waveform of the system output is not sinusoidal when the system is operating in its nonlinear region.)

4. If the system under consideration is operating continuously for days and weeks, then normal operation need not be stopped for frequency-response tests. The sinusoidal test signal may be superimposed on the normal inputs. Then, for linear systems, the output due to the test signal is superimposed on the normal output. For the determination of the transfer function while the system is in normal operation, stochastic signals (white noise signals) also are often used. By use of correlation functions, the transfer function of the system can be determined without interrupting normal operation.

EXAMPLE 7–25   Determine the transfer function of the system whose experimental frequency-response curves are as shown in Figure 7–88.

The first step in determining the transfer function is to approximate the log-magnitude curve by asymptotes with slopes ± 20 dB/decade and multiples thereof, as shown in Figure 7–88. We then estimate the corner frequencies. For the system shown in Figure 7–88, the following form of the transfer function is estimated:

$$G(j\omega) = \frac{K(1 + 0.5j\omega)}{j\omega(1 + j\omega)\left[1 + 2\zeta\left(j\dfrac{\omega}{8}\right) + \left(j\dfrac{\omega}{8}\right)^2\right]}$$

The value of the damping ratio ζ is estimated by examining the peak resonance near $\omega = 6$ rad/sec. Referring to Figure 7–9, ζ is determined to be 0.5. The gain K is numerically equal to the frequency at the intersection of the extension of the low-frequency asymptote that has 20 dB/decade slope and the 0-dB line. The value of K is thus found to be 10. Therefore, $G(j\omega)$ is tentatively determined as

$$G(j\omega) = \frac{10(1 + 0.5j\omega)}{j\omega(1 + j\omega)\left[1 + \left(j\dfrac{\omega}{8}\right) + \left(j\dfrac{\omega}{8}\right)^2\right]}$$

or

$$G(s) = \frac{320(s + 2)}{s(s + 1)(s^2 + 8s + 64)}$$

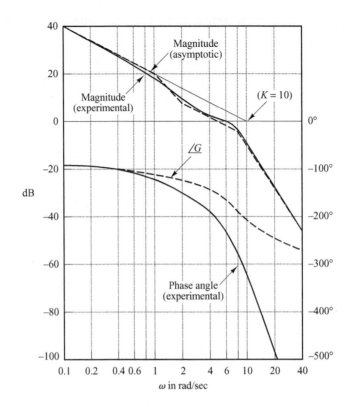

Figure 7–88
Bode diagram of a
system. (Solid curves
are experimentally
obtained curves.)

This transfer function is tentative because we have not examined the phase-angle curve yet.

Once the corner frequencies are noted on the log-magnitude curve, the corresponding phase-angle curve for each component factor of the transfer function can easily be drawn. The sum of these component phase-angle curves is that of the assumed transfer function. The phase-angle curve for $G(j\omega)$ is denoted by $\underline{/G}$ in Figure 7–88. From Figure 7–88, we clearly notice a discrepancy between the computed phase-angle curve and the experimentally obtained phase-angle curve. The difference between the two curves at very high frequencies appears to be a constant rate of change. Thus, the discrepancy in the phase-angle curves must be caused by transport lag.

Hence, we assume the complete transfer function to be $G(s)e^{-Ts}$. Since the discrepancy between the computed and experimental phase angles is -0.2ω rad for very high frequencies, we can determine the value of T as follows:

$$\lim_{\omega \to \infty} \frac{d}{d\omega} \underline{/G(j\omega)e^{-j\omega T}} = -T = -0.2$$

or

$$T = 0.2 \text{ sec.}$$

The presence of transport lag can thus be determined, and the complete transfer function determined from the experimental curves is

$$G(s)e^{-Ts} = \frac{320(s + 2)e^{-0.2s}}{s(s + 1)(s^2 + 8s + 64)}$$

7-10 CONTROL SYSTEMS DESIGN BY FREQUENCY-RESPONSE APPROACH

In Chapter 6 we presented root-locus analysis and design. The root-locus method was shown to be very useful to reshape the transient-response characteristics of closed-loop control systems. The root-locus approach gives us direct information on the transient response of the closed-loop system. The frequency-response approach, on the other hand, gives us this information only indirectly. However, as we shall see in the remaining three sections of this chapter, the frequency-response approach is very useful in designing control systems.

For any design problem, the designer will do well to use both approaches to the design and choose the compensator that most closely produces the desired closed-loop response.

In most control systems design, transient-response performance is usually very important. In the frequency-response approach, we specify the transient-response performance in an indirect manner. That is, the transient-response performance is specified in terms of the phase margin, gain margin, resonant peak magnitude (they give a rough estimate of the system damping); the gain crossover frequency, resonant frequency, bandwidth (they give a rough estimate of the speed of transient response); and static error constants (they give the steady-state accuracy). Although the correlation between the transient response and frequency response is indirect, the frequency-domain specifications can be easily met in the Bode diagram approach.

After the open loop has been designed, the closed-loop poles and zeros can be determined. Then, the transient-response characteristics must be checked to see whether the designed system satisfies the requirements in the time domain. If it does not, then the compensator must be modified and the analysis repeated until a satisfactory result is obtained.

Design in the frequency domain is simple and straightforward. The frequency-response plot indicates clearly the manner in which the system should be modified, although the exact quantitative prediction of the transient-response characteristics cannot be made. The frequency-response approach can be applied to systems or components whose dynamic characteristics are given in the form of frequency-response data. Note that because of difficulty in deriving the equations governing certain components, such as pneumatic and hydraulic components, the dynamic characteristics of such components are usually determined experimentally through frequency-response tests. The experimentally obtained frequency-response plots can be combined easily with other such plots when the Bode diagram approach is used. Note also that in dealing with high-frequency noises we find that the frequency-response approach is more convenient than other approaches.

There are basically two approaches in the frequency-domain design. One is the polar plot approach and the other is the Bode diagram approach. When a compensator is added, the polar plot does not retain the original shape, and, therefore, we need to draw a new polar plot, which will take time and is thus inconvenient. On the other hand, a Bode diagram of the compensator can be simply added to the original Bode diagram, and thus plotting the complete Bode diagram is a simple matter. Also, if the open-loop gain is varied, the magnitude curve is shifted up or down without changing the slope of the curve, and the phase curve remains the same. For design purposes, therefore, it is best to work with the Bode diagram.

A common approach to the design based on the Bode diagram is that we first adjust the open-loop gain so that the requirement on the steady-state accuracy is met. Then the magnitude and phase curves of the uncompensated open loop (with the open-loop gain just adjusted) are plotted. If the specifications on the phase margin and gain margin are not satisfied, then a suitable compensator that will reshape the open-loop transfer function is determined. Finally, if there are any other requirements to be met, we try to satisfy them, unless some of them are mutually contradictory.

Information Obtainable from Open-Loop Frequency Response. The low-frequency region (the region far below the gain crossover frequency) of the locus indicates the steady-state behavior of the closed-loop system. The medium-frequency region (the region near the gain crossover frequency) of the locus indicates relative stability. The high-frequency region (the region far above the gain crossover frequency) indicates the complexity of the system.

Requirements on Open-Loop Frequency Response. We might say that, in many practical cases, compensation is essentially a compromise between steady-state accuracy and relative stability.

To have a high value of the velocity error constant and yet satisfactory relative stability, we find it necessary to reshape the open-loop frequency-response curve.

The gain in the low-frequency region should be large enough, and near the gain crossover frequency, the slope of the log-magnitude curve in the Bode diagram should be −20 dB/decade. This slope should extend over a sufficiently wide frequency band to assure a proper phase margin. For the high-frequency region, the gain should be attenuated as rapidly as possible to minimize the effects of noise.

Examples of generally desirable and undesirable open-loop and closed-loop frequency-response curves are shown in Figure 7–89.

Referring to Figure 7–90, we see that the reshaping of the open-loop frequency-response curve may be done if the high-frequency portion of the locus follows the $G_1(j\omega)$ locus, while the low-frequency portion of the locus follows the $G_2(j\omega)$ locus. The reshaped locus $G_c(j\omega)G(j\omega)$ should have reasonable phase and gain margins or should be tangent to a proper M circle, as shown.

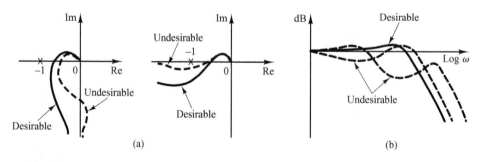

Figure 7–89
(a) Examples of desirable and undesirable open-loop frequency-response curves;
(b) examples of desirable and undesirable closed-loop frequency-response curves.

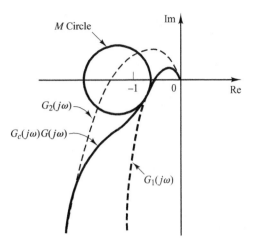

Figure 7–90
Reshaping of the open-loop frequency-response curve.

Basic Characteristics of Lead, Lag, and Lag–Lead Compensation. Lead compensation essentially yields an appreciable improvement in transient response and a small change in steady-state accuracy. It may accentuate high-frequency noise effects. Lag compensation, on the other hand, yields an appreciable improvement in steady-state accuracy at the expense of increasing the transient-response time. Lag compensation will suppress the effects of high-frequency noise signals. Lag–lead compensation combines the characteristics of both lead compensation and lag compensation. The use of a lead or lag compensator raises the order of the system by 1 (unless cancellation occurs between the zero of the compensator and a pole of the uncompensated open-loop transfer function). The use of a lag–lead compensator raises the order of the system by 2 [unless cancellation occurs between zero(s) of the lag–lead compensator and pole(s) of the uncompensated open-loop transfer function], which means that the system becomes more complex and it is more difficult to control the transient-response behavior. The particular situation determines the type of compensation to be used.

7–11 LEAD COMPENSATION

We shall first examine the frequency characteristics of the lead compensator. Then we present a design technique for the lead compensator by use of the Bode diagram.

Characteristics of Lead Compensators. Consider a lead compensator having the following transfer function:

$$K_c \alpha \frac{Ts + 1}{\alpha Ts + 1} = K_c \frac{s + \dfrac{1}{T}}{s + \dfrac{1}{\alpha T}} \qquad (0 < \alpha < 1)$$

where α is the attenuation factor of the lead compensator. It has a zero at $s = -1/T$ and a pole at $s = -1/(\alpha T)$. Since $0 < \alpha < 1$, we see that the zero is always located to the right of the pole in the complex plane. Note that for a small value of α the pole is located far to the left. The minimum value of α is limited by the physical construction of

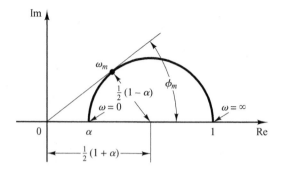

Figure 7–91
Polar plot of a lead
compensator
$\alpha(j\omega T + 1)/(j\omega\alpha T + 1)$,
where $0 < \alpha < 1$.

the lead compensator. The minimum value of α is usually taken to be about 0.05. (This means that the maximum phase lead that may be produced by a lead compensator is about 65°.) [See Equation (7–25).]

Figure 7–91 shows the polar plot of

$$K_c\alpha \frac{j\omega T + 1}{j\omega\alpha T + 1} \qquad (0 < \alpha < 1)$$

with $K_c = 1$. For a given value of α, the angle between the positive real axis and the tangent line drawn from the origin to the semicircle gives the maximum phase-lead angle, ϕ_m. We shall call the frequency at the tangent point ω_m. From Figure 7–91 the phase angle at $\omega = \omega_m$ is ϕ_m, where

$$\sin\phi_m = \frac{\dfrac{1-\alpha}{2}}{\dfrac{1+\alpha}{2}} = \frac{1-\alpha}{1+\alpha} \tag{7–25}$$

Equation (7–25) relates the maximum phase-lead angle and the value of α.

Figure 7–92 shows the Bode diagram of a lead compensator when $K_c = 1$ and $\alpha = 0.1$. The corner frequencies for the lead compensator are $\omega = 1/T$ and $\omega = 1/(\alpha T) = 10/T$. By examining Figure 7–92, we see that ω_m is the geometric mean of the two corner frequencies, or

$$\log\omega_m = \frac{1}{2}\left(\log\frac{1}{T} + \log\frac{1}{\alpha T}\right)$$

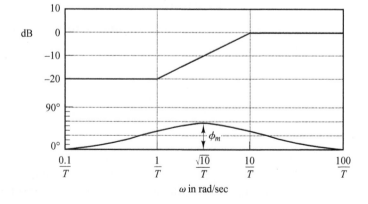

Figure 7–92
Bode diagram of a
lead compensator
$\alpha(j\omega T + 1)/(j\omega\alpha T + 1)$,
where $\alpha = 0.1$.

Hence,

$$\omega_m = \frac{1}{\sqrt{\alpha}T} \qquad (7\text{–}26)$$

As seen from Figure 7–92, the lead compensator is basically a high-pass filter. (The high frequencies are passed, but low frequencies are attenuated.)

Lead Compensation Techniques Based on the Frequency-Response Approach. The primary function of the lead compensator is to reshape the frequency-response curve to provide sufficient phase-lead angle to offset the excessive phase lag associated with the components of the fixed system.

Consider the system shown in Figure 7–93. Assume that the performance specifications are given in terms of phase margin, gain margin, static velocity error constants, and so on. The procedure for designing a lead compensator by the frequency-response approach may be stated as follows:

1. Assume the following lead compensator:

$$G_c(s) = K_c \alpha \frac{Ts + 1}{\alpha Ts + 1} = K_c \frac{s + \dfrac{1}{T}}{s + \dfrac{1}{\alpha T}} \qquad (0 < \alpha < 1)$$

Define

$$K_c \alpha = K$$

Then

$$G_c(s) = K \frac{Ts + 1}{\alpha Ts + 1}$$

The open-loop transfer function of the compensated system is

$$G_c(s)G(s) = K \frac{Ts + 1}{\alpha Ts + 1} G(s) = \frac{Ts + 1}{\alpha Ts + 1} KG(s) = \frac{Ts + 1}{\alpha Ts + 1} G_1(s)$$

where

$$G_1(s) = KG(s)$$

Determine gain K to satisfy the requirement on the given static error constant.

2. Using the gain K thus determined, draw a Bode diagram of $G_1(j\omega)$, the gain-adjusted but uncompensated system. Evaluate the phase margin.

3. Determine the necessary phase-lead angle to be added to the system. Add an additional 5° to 12° to the phase-lead angle required, because the addition of the

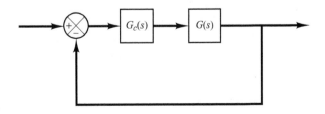

Figure 7–93
Control system.

lead compensator shifts the gain crossover frequency to the right and decreases the phase margin.

4. Determine the attenuation factor α by use of Equation (7–25). Determine the frequency where the magnitude of the uncompensated system $G_1(j\omega)$ is equal to $-20 \log (1/\sqrt{\alpha})$. Select this frequency as the new gain crossover frequency. This frequency corresponds to $\omega_m = 1/(\sqrt{\alpha}T)$, and the maximum phase shift ϕ_m occurs at this frequency.

5. Determine the corner frequencies of the lead compensator as follows:

$$\text{Zero of lead compensator:} \quad \omega = \frac{1}{T}$$

$$\text{Pole of lead compensator:} \quad \omega = \frac{1}{\alpha T}$$

6. Using the value of K determined in step 1 and that of α determined in step 4, calculate constant K_c from

$$K_c = \frac{K}{\alpha}$$

7. Check the gain margin to be sure it is satisfactory. If not, repeat the design process by modifying the pole–zero location of the compensator until a satisfactory result is obtained.

EXAMPLE 7–26   Consider the system shown in Figure 7–94. The open-loop transfer function is

$$G(s) = \frac{4}{s(s + 2)}$$

It is desired to design a compensator for the system so that the static velocity error constant K_v is 20 sec^{-1}, the phase margin is at least 50°, and the gain margin is at least 10 dB. We shall use a lead compensator of the form

$$G_c(s) = K_c\alpha \frac{Ts + 1}{\alpha Ts + 1} = K_c \frac{s + \dfrac{1}{T}}{s + \dfrac{1}{\alpha T}}$$

The compensated system will have the open-loop transfer function $G_c(s)G(s)$. Define

$$G_1(s) = KG(s) = \frac{4K}{s(s + 2)}$$

where $K = K_c\alpha$.

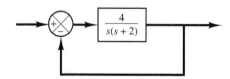

Figure 7–94
Control system.

The first step in the design is to adjust the gain K to meet the steady-state performance specification or to provide the required static velocity error constant. Since this constant is given as $20\ \text{sec}^{-1}$, we obtain

$$K_v = \lim_{s\to 0} s G_c(s) G(s) = \lim_{s\to 0} s \frac{Ts + 1}{\alpha Ts + 1} G_1(s) = \lim_{s\to 0} \frac{s 4K}{s(s + 2)} = 2K = 20$$

or

$$K = 10$$

With $K = 10$, the compensated system will satisfy the steady-state requirement.

We shall next plot the Bode diagram of

$$G_1(j\omega) = \frac{40}{j\omega(j\omega + 2)} = \frac{20}{j\omega(0.5j\omega + 1)}$$

Figure 7–95 shows the magnitude and phase-angle curves of $G_1(j\omega)$. From this plot, the phase and gain margins of the system are found to be 17° and $+\infty$ dB, respectively. (A phase margin of 17° implies that the system is quite oscillatory. Thus, satisfying the specification on the steady state yields a poor transient-response performance.) The specification calls for a phase margin of at least 50°. We thus find the additional phase lead necessary to satisfy the relative stability requirement is 33°. To achieve a phase margin of 50° without decreasing the value of K, the lead compensator must contribute the required phase angle.

Noting that the addition of a lead compensator modifies the magnitude curve in the Bode diagram, we realize that the gain crossover frequency will be shifted to the right. We must offset the increased phase lag of $G_1(j\omega)$ due to this increase in the gain crossover frequency. Considering the shift of the gain crossover frequency, we may assume that ϕ_m, the maximum phase lead required, is approximately 38°. (This means that 5° has been added to compensate for the shift in the gain crossover frequency.)

Since

$$\sin \phi_m = \frac{1 - \alpha}{1 + \alpha}$$

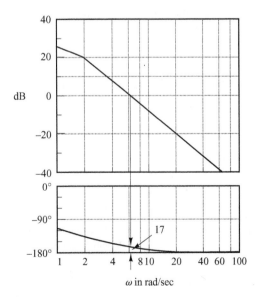

Figure 7–95
Bode diagram for
$G_1(j\omega) = 10G(j\omega)$
$= 40/\big[j\omega(j\omega + 2)\big]$

$\phi_m = 38°$ corresponds to $\alpha = 0.24$. Once the attenuation factor α has been determined on the basis of the required phase-lead angle, the next step is to determine the corner frequencies $\omega = 1/T$ and $\omega = 1/(\alpha T)$ of the lead compensator. To do so, we first note that the maximum phase-lead angle ϕ_m occurs at the geometric mean of the two corner frequencies, or $\omega = 1/(\sqrt{\alpha}T)$. [See Equation (7–26).] The amount of the modification in the magnitude curve at $\omega = 1/(\sqrt{\alpha}T)$ due to the inclusion of the term $(Ts + 1)/(\alpha Ts + 1)$ is

$$\left| \frac{1 + j\omega T}{1 + j\omega \alpha T} \right|_{\omega = 1/(\sqrt{\alpha}T)} = \left| \frac{1 + j\dfrac{1}{\sqrt{\alpha}}}{1 + j\alpha \dfrac{1}{\sqrt{\alpha}}} \right| = \frac{1}{\sqrt{\alpha}}$$

Note that

$$\frac{1}{\sqrt{\alpha}} = \frac{1}{\sqrt{0.24}} = \frac{1}{0.49} = 6.2 \text{ dB}$$

and $|G_1(j\omega)| = -6.2$ dB corresponds to $\omega = 9$ rad/sec. We shall select this frequency to be the new gain crossover frequency ω_c. Noting that this frequency corresponds to $1/(\sqrt{\alpha}T)$, or $\omega_c = 1/(\sqrt{\alpha}T)$, we obtain

$$\frac{1}{T} = \sqrt{\alpha}\omega_c = 4.41$$

and

$$\frac{1}{\alpha T} = \frac{\omega_c}{\sqrt{\alpha}} = 18.4$$

The lead compensator thus determined is

$$G_c(s) = K_c \frac{s + 4.41}{s + 18.4} = K_c \alpha \frac{0.227s + 1}{0.054s + 1}$$

where the value of K_c is determined as

$$K_c = \frac{K}{\alpha} = \frac{10}{0.24} = 41.7$$

Thus, the transfer function of the compensator becomes

$$G_c(s) = 41.7 \frac{s + 4.41}{s + 18.4} = 10 \frac{0.227s + 1}{0.054s + 1}$$

Note that

$$\frac{G_c(s)}{K} G_1(s) = \frac{G_c(s)}{10} 10G(s) = G_c(s)G(s)$$

The magnitude curve and phase-angle curve for $G_c(j\omega)/10$ are shown in Figure 7–96. The compensated system has the following open-loop transfer function:

$$G_c(s)G(s) = 41.7 \frac{s + 4.41}{s + 18.4} \frac{4}{s(s + 2)}$$

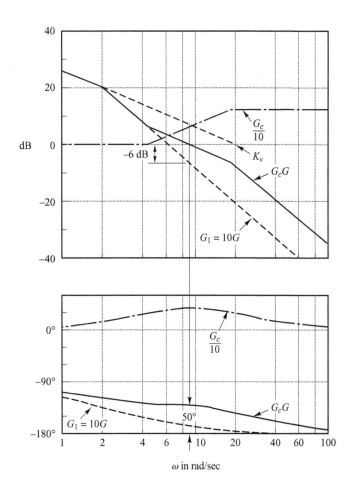

Figure 7–96
Bode diagram for the compensated system.

The solid curves in Figure 7–96 show the magnitude curve and phase-angle curve for the compensated system. Note that the bandwidth is approximately equal to the gain crossover frequency. The lead compensator causes the gain crossover frequency to increase from 6.3 to 9 rad/sec. The increase in this frequency means an increase in bandwidth. This implies an increase in the speed of response. The phase and gain margins are seen to be approximately 50° and +∞ dB, respectively. The compensated system shown in Figure 7–97 therefore meets both the steady-state and the relative-stability requirements.

Note that for type 1 systems, such as the system just considered, the value of the static velocity error constant K_v is merely the value of the frequency corresponding to the intersection of the extension of the initial −20-dB/decade slope line and the 0-dB line, as shown in Figure 7–96. Note also that we have changed the slope of the magnitude curve near the gain crossover frequency from −40 dB/decade to −20 dB/decade.

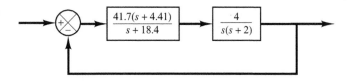

Figure 7–97
Compensated system.

Figure 7–98 shows the polar plots of the gain-adjusted but uncompensated open-loop transfer function $G_1(j\omega) = 10\,G(j\omega)$ and the compensated open-loop transfer function $G_c(j\omega)G(j\omega)$. From Figure 7–98, we see that the resonant frequency of the uncompensated system is about 6 rad/sec and that of the compensated system is about 7 rad/sec. (This also indicates that the bandwidth has been increased.)

From Figure 7–98, we find that the value of the resonant peak M_r for the uncompensated system with $K = 10$ is 3. The value of M_r for the compensated system is found to be 1.29. This clearly shows that the compensated system has improved relative stability.

Note that, if the phase angle of $G_1(j\omega)$ decreases rapidly near the gain crossover frequency, lead compensation becomes ineffective because the shift in the gain crossover frequency to the right makes it difficult to provide enough phase lead at the new gain crossover frequency. This means that, to provide the desired phase margin, we must use a very small value for α. The value of α, however, should not be too small (smaller than 0.05) nor should the maximum phase lead ϕ_m be too large (larger than 65°), because such values will require an additional gain of excessive value. [If more than 65° is needed, two (or more) lead networks may be used in series with an isolating amplifier.]

Finally, we shall examine the transient-response characteristics of the designed system. We shall obtain the unit-step response and unit-ramp response curves of the compensated and uncompensated systems with MATLAB. Note that the closed-loop transfer functions of the uncompensated and compensated systems are given, respectively, by

$$\frac{C(s)}{R(s)} = \frac{4}{s^2 + 2s + 4}$$

and

$$\frac{C(s)}{R(s)} = \frac{166.8s + 735.588}{s^3 + 20.4s^2 + 203.6s + 735.588}$$

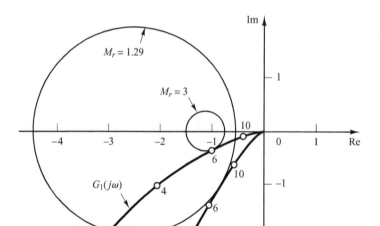

Figure 7–98
Polar plots of the gain-adjusted but uncompensated open-loop transfer function G_1 and compensated open-loop transfer function G_cG.

Chapter 7 / Control Systems Analysis and Design by the Frequency-Response Method

MATLAB programs for obtaining the unit-step response and unit-ramp response curves are given in MATLAB Program 7–13. Figure 7–99 shows the unit-step response curves of the system before and after compensation. Also, Figure 7–100 depicts the unit-ramp response curves before and after compensation. These response curves indicate that the designed system is satisfactory.

MATLAB Program 7–13

```
%*****Unit-step responses*****

num = [4];
den = [1  2  4];
numc = [166.8  735.588];
denc = [1  20.4  203.6  735.588];
t = 0:0.02:6;
[c1,x1,t] = step(num,den,t);
[c2,x2,t] = step(numc,denc,t);
plot (t,c1,'.',t,c2,'-')
grid
title('Unit-Step Responses of Compensated and Uncompensated Systems')
xlabel('t Sec')
ylabel('Outputs')
text(0.4,1.31,'Compensated system')
text(1.55,0.88,'Uncompensated system')

%*****Unit-ramp responses*****

num1 = [4];
den1 = [1  2  4  0];
num1c = [166.8  735.588];
den1c = [1  20.4  203.6  735.588  0];
t = 0:0.02:5;
[y1,z1,t] = step(num1,den1,t);
[y2,z2,t] = step(num1c,den1c,t);
plot(t,y1,'.',t,y2,'-',t,t,'--')
grid
title('Unit-Ramp Responses of Compensated and Uncompensated Systems')
xlabel('t Sec')
ylabel('Outputs')
text(0.89,3.7,'Compensated system')
text(2.25,1.1,'Uncompensated system')
```

It is noted that the closed-loop poles for the compensated system are located as follows:

$$s = -6.9541 \pm j8.0592$$

$$s = -6.4918$$

Because the dominant closed-loop poles are located far from the $j\omega$ axis, the response damps out quickly.

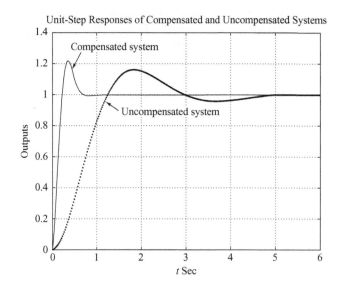

Figure 7–99
Unit-step response curves of the compensated and uncompensated systems.

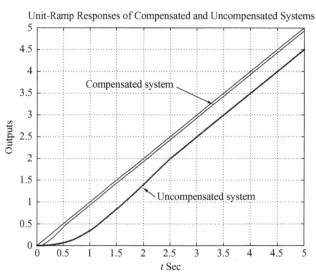

Figure 7–100
Unit-ramp response curves of the compensated and uncompensated systems.

7–12 LAG COMPENSATION

In this section we first discuss the Nyquist plot and Bode diagram of the lag compensator. Then we present lag compensation techniques based on the frequency-response approach.

Characteristics of Lag Compensators.   Consider a lag compensator having the following transfer function:

$$G_c(s) = K_c\beta\,\frac{Ts + 1}{\beta Ts + 1} = K_c\,\frac{s + \dfrac{1}{T}}{s + \dfrac{1}{\beta T}} \qquad (\beta > 1)$$

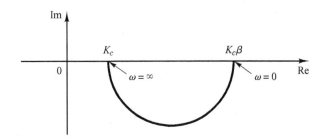

Figure 7–101
Polar plot of a lag
compensator
$K_c\beta(j\omega T + 1)/(j\omega\beta T + 1)$.

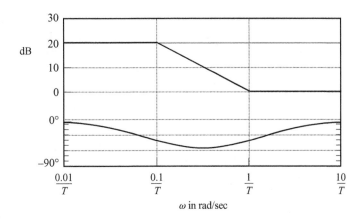

Figure 7–102
Bode diagram of a
lag compensator
$\beta(j\omega T + 1)/(j\omega\beta T + 1)$,
with $\beta = 10$.

In the complex plane, a lag compensator has a zero at $s = -1/T$ and a pole at $s = -1/(\beta T)$. The pole is located to the right of the zero.

Figure 7–101 shows a polar plot of the lag compensator. Figure 7–102 shows a Bode diagram of the compensator, where $K_c = 1$ and $\beta = 10$. The corner frequencies of the lag compensator are at $\omega = 1/T$ and $\omega = 1/(\beta T)$. As seen from Figure 7–102, where the values of K_c and β are set equal to 1 and 10, respectively, the magnitude of the lag compensator becomes 10 (or 20 dB) at low frequencies and unity (or 0 dB) at high frequencies. Thus, the lag compensator is essentially a low-pass filter.

Lag Compensation Techniques Based on the Frequency-Response Approach. The primary function of a lag compensator is to provide attenuation in the high-frequency range to give a system sufficient phase margin. The phase-lag characteristic is of no consequence in lag compensation.

The procedure for designing lag compensators for the system shown in Figure 7–93 by the frequency-response approach may be stated as follows:

1. Assume the following lag compensator:

$$G_c(s) = K_c\beta\frac{Ts + 1}{\beta Ts + 1} = K_c\frac{s + \dfrac{1}{T}}{s + \dfrac{1}{\beta T}} \qquad (\beta > 1)$$

Define

$$K_c \beta = K$$

Then

$$G_c(s) = K \frac{Ts + 1}{\beta Ts + 1}$$

The open-loop transfer function of the compensated system is

$$G_c(s)G(s) = K \frac{Ts + 1}{\beta Ts + 1} G(s) = \frac{Ts + 1}{\beta Ts + 1} KG(s) = \frac{Ts + 1}{\beta Ts + 1} G_1(s)$$

where

$$G_1(s) = KG(s)$$

Determine gain K to satisfy the requirement on the given static velocity error constant.

2. If the gain-adjusted but uncompensated system $G_1(j\omega) = KG(j\omega)$ does not satisfy the specifications on the phase and gain margins, then find the frequency point where the phase angle of the open-loop transfer function is equal to $-180°$ plus the required phase margin. The required phase margin is the specified phase margin plus $5°$ to $12°$. (The addition of $5°$ to $12°$ compensates for the phase lag of the lag compensator.) Choose this frequency as the new gain crossover frequency.

3. To prevent detrimental effects of phase lag due to the lag compensator, the pole and zero of the lag compensator must be located substantially lower than the new gain crossover frequency. Therefore, choose the corner frequency $\omega = 1/T$ (corresponding to the zero of the lag compensator) 1 octave to 1 decade below the new gain crossover frequency. (If the time constants of the lag compensator do not become too large, the corner frequency $\omega = 1/T$ may be chosen 1 decade below the new gain crossover frequency.)

Notice that we choose the compensator pole and zero sufficiently small. Thus the phase lag occurs at the low-frequency region so that it will not affect the phase margin.

4. Determine the attenuation necessary to bring the magnitude curve down to 0 dB at the new gain crossover frequency. Noting that this attenuation is $-20 \log \beta$, determine the value of β. Then the other corner frequency (corresponding to the pole of the lag compensator) is determined from $\omega = 1/(\beta T)$.

5. Using the value of K determined in step 1 and that of β determined in step 4, calculate constant K_c from

$$K_c = \frac{K}{\beta}$$

EXAMPLE 7–27 Consider the system shown in Figure 7–103. The open-loop transfer function is given by

$$G(s) = \frac{1}{s(s + 1)(0.5s + 1)}$$

It is desired to compensate the system so that the static velocity error constant K_v is 5 sec^{-1}, the phase margin is at least 40°, and the gain margin is at least 10 dB.

We shall use a lag compensator of the form

$$G_c(s) = K_c\beta\frac{Ts + 1}{\beta Ts + 1} = K_c\frac{s + \dfrac{1}{T}}{s + \dfrac{1}{\beta T}} \qquad (\beta > 1)$$

Define

$$K_c\beta = K$$

Define also

$$G_1(s) = KG(s) = \frac{K}{s(s + 1)(0.5s + 1)}$$

The first step in the design is to adjust the gain K to meet the required static velocity error constant. Thus,

$$K_v = \lim_{s \to 0} sG_c(s)G(s) = \lim_{s \to 0} s\frac{Ts + 1}{\beta Ts + 1}G_1(s) = \lim_{s \to 0} sG_1(s)$$

$$= \lim_{s \to 0} \frac{sK}{s(s + 1)(0.5s + 1)} = K = 5$$

or

$$K = 5$$

With $K = 5$, the compensated system satisfies the steady-state performance requirement.

We shall next plot the Bode diagram of

$$G_1(j\omega) = \frac{5}{j\omega(j\omega + 1)(0.5j\omega + 1)}$$

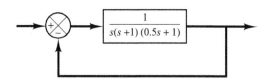

Figure 7–103
Control system.

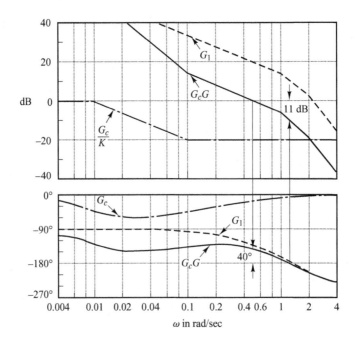

Figure 7–104
Bode diagrams for G_1 (gain-adjusted but uncompensated open-loop transfer function), G_c (compensator), and $G_c G$ (compensated open-loop transfer function).

The magnitude curve and phase-angle curve of $G_1(j\omega)$ are shown in Figure 7–104. From this plot, the phase margin is found to be $-20°$, which means that the gain-adjusted but uncompensated system is unstable.

Noting that the addition of a lag compensator modifies the phase curve of the Bode diagram, we must allow 5° to 12° to the specified phase margin to compensate for the modification of the phase curve. Since the frequency corresponding to a phase margin of 40° is 0.7 rad/sec, the new gain crossover frequency (of the compensated system) must be chosen near this value. To avoid overly large time constants for the lag compensator, we shall choose the corner frequency $\omega = 1/T$ (which corresponds to the zero of the lag compensator) to be 0.1 rad/sec. Since this corner frequency is not too far below the new gain crossover frequency, the modification in the phase curve may not be small. Hence, we add about 12° to the given phase margin as an allowance to account for the lag angle introduced by the lag compensator. The required phase margin is now 52°. The phase angle of the uncompensated open-loop transfer function is $-128°$ at about $\omega = 0.5$ rad/sec. So we choose the new gain crossover frequency to be 0.5 rad/sec. To bring the magnitude curve down to 0 dB at this new gain crossover frequency, the lag compensator must give the necessary attenuation, which in this case is -20 dB. Hence,

$$20\log\frac{1}{\beta} = -20$$

or

$$\beta = 10$$

The other corner frequency $\omega = 1(\beta T)$, which corresponds to the pole of the lag compensator, is then determined as

$$\frac{1}{\beta T} = 0.01 \text{ rad/sec}$$

Thus, the transfer function of the lag compensator is

$$G_c(s) = K_c(10) \frac{10s + 1}{100s + 1} = K_c \frac{s + \dfrac{1}{10}}{s + \dfrac{1}{100}}$$

Since the gain K was determined to be 5 and β was determined to be 10, we have

$$K_c = \frac{K}{\beta} = \frac{5}{10} = 0.5$$

The open-loop transfer function of the compensated system is

$$G_c(s)G(s) = \frac{5(10s + 1)}{s(100s + 1)(s + 1)(0.5s + 1)}$$

The magnitude and phase-angle curves of $G_c(j\omega)G(j\omega)$ are also shown in Figure 7–104.

The phase margin of the compensated system is about 40°, which is the required value. The gain margin is about 11 dB, which is quite acceptable. The static velocity error constant is 5 sec^{-1}, as required. The compensated system, therefore, satisfies the requirements on both the steady state and the relative stability.

Note that the new gain crossover frequency is decreased from approximately 1 to 0.5 rad/sec. This means that the bandwidth of the system is reduced.

To further show the effects of lag compensation, the log-magnitude-versus-phase plots of the gain-adjusted but uncompensated system $G_1(j\omega)$ and of the compensated system $G_c(j\omega)G(j\omega)$ are shown in Figure 7–105. The plot of $G_1(j\omega)$ clearly shows that the gain-adjusted but uncompensated system is unstable. The addition of the lag compensator stabilizes the system. The plot of $G_c(j\omega)G(j\omega)$ is tangent to the $M = 3$ dB locus. Thus, the resonant peak value is 3 dB, or 1.4, and this peak occurs at $\omega = 0.5$ rad/sec.

Compensators designed by different methods or by different designers (even using the same approach) may look sufficiently different. Any of the well-designed systems, however, will give similar transient and steady-state performance. The best among many alternatives may be chosen from the economic consideration that the time constants of the lag compensator should not be too large.

Figure 7–105
Log-magnitude-versus-phase plots of G_1 (gain-adjusted but uncompensated open-loop transfer function) and $G_c G$ (compensated open-loop transfer function).

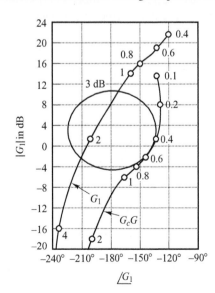

Finally, we shall examine the unit-step response and unit-ramp response of the compensated system and the original uncompensated system without gain adjustment. The closed-loop transfer functions of the compensated and uncompensated systems are

$$\frac{C(s)}{R(s)} = \frac{50s + 5}{50s^4 + 150.5s^3 + 101.5s^2 + 51s + 5}$$

and

$$\frac{C(s)}{R(s)} = \frac{1}{0.5s^3 + 1.5s^2 + s + 1}$$

respectively. MATLAB Program 7–14 will produce the unit-step and unit-ramp responses of the compensated and uncompensated systems. The resulting unit-step response curves and unit-ramp response curves are shown in Figures 7–106 and 7–107, respectively. From the response curves we find that the designed system satisfies the given specifications and is satisfactory.

MATLAB Program 7–14

```
%*****Unit-step response*****

num = [1];
den = [0.5  1.5  1  1];
numc = [50  5];
denc = [50  150.5  101.5  51  5];
t = 0:0.1:40;
[c1,x1,t] = step(num,den,t);
[c2,x2,t] = step(numc,denc,t);
plot(t,c1,'.',t,c2,'-')
grid
title('Unit-Step Responses of Compensated and Uncompensated Systems')
xlabel('t Sec')
ylabel('Outputs')
text(12.7,1.27,'Compensated system')
text(12.2,0.7,'Uncompensated system')

%*****Unit-ramp response*****

num1 = [1];
den1 = [0.5  1.5  1  1  0];
num1c = [50  5];
den1c = [50  150.5  101.5  51  5  0];
t = 0:0.1:20;
[y1,z1,t] = step(num1,den1,t);
[y2,z2,t] = step(num1c,den1c,t);
plot(t,y1,'.',t,y2,'-',t,t,'--');
grid
title('Unit-Ramp Responses of Compensated and Uncompensated Systems')
xlabel('t Sec')
ylabel('Outputs')
text(8.3,3,'Compensated system')
text(8.3,5,'Uncompensated system')
```

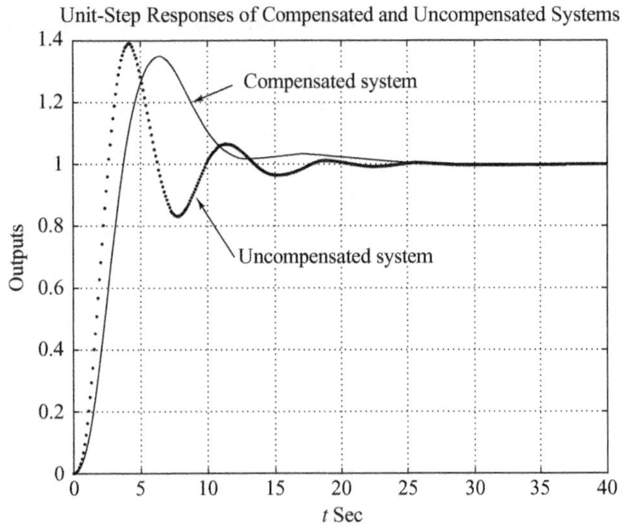

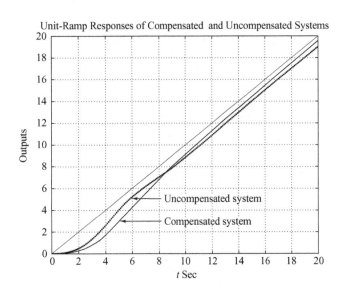

Figure 7–106
Unit-step response curves for the compensated and uncompensated systems (Example 7–27).

Unit-Step Responses of Compensated and Uncompensated Systems

Compensated system

Uncompensated system

Figure 7–107
Unit-ramp response curves for the compensated and uncompensated systems (Example 7–27).

Unit-Ramp Responses of Compensated and Uncompensated Systems

Uncompensated system

Compensated system

Note that the zero and poles of the designed closed-loop system are as follows:

Zero at $s = -0.1$

Poles at $s = -0.2859 \pm j0.5196$, $s = -0.1228$, $s = -2.3155$

The dominant closed-loop poles are very close to the $j\omega$ axis with the result that the response is slow. Also, a pair of the closed-loop pole at $s = -0.1228$ and the zero at $s = -0.1$ produces a slowly decreasing tail of small amplitude.

A Few Comments on Lag Compensation.

1. Lag compensators are essentially low-pass filters. Therefore, lag compensation permits a high gain at low frequencies (which improves the steady-state performance) and reduces gain in the higher critical range of frequencies so as to improve the phase margin. Note that in lag compensation we utilize the attenuation characteristic of the lag compensator at high frequencies rather than the phase-lag characteristic. (The phase-lag characteristic is of no use for compensation purposes.)

2. Suppose that the zero and pole of a lag compensator are located at $s = -z$ and $s = -p$, respectively. Then the exact locations of the zero and pole are not critical provided that they are close to the origin and the ratio z/p is equal to the required multiplication factor of the static velocity error constant.

It should be noted, however, that the zero and pole of the lag compensator should not be located unnecessarily close to the origin, because the lag compensator will create an additional closed-loop pole in the same region as the zero and pole of the lag compensator.

The closed-loop pole located near the origin gives a very slowly decaying transient response, although its magnitude will become very small because the zero of the lag compensator will almost cancel the effect of this pole. However, the transient response (decay) due to this pole is so slow that the settling time will be adversely affected.

It is also noted that in the system compensated by a lag compensator the transfer function between the plant disturbance and the system error may not involve a zero that is near this pole. Therefore, the transient response to the disturbance input may last very long.

3. The attenuation due to the lag compensator will shift the gain crossover frequency to a lower frequency point where the phase margin is acceptable. Thus, the lag compensator will reduce the bandwidth of the system and will result in slower transient response. [The phase angle curve of $G_c(j\omega)G(j\omega)$ is relatively unchanged near and above the new gain crossover frequency.]

4. Since the lag compensator tends to integrate the input signal, it acts approximately as a proportional-plus-integral controller. Because of this, a lag-compensated system tends to become less stable. To avoid this undesirable feature, the time constant T should be made sufficiently larger than the largest time constant of the system.

5. Conditional stability may occur when a system having saturation or limiting is compensated by use of a lag compensator. When the saturation or limiting takes place in the system, it reduces the effective loop gain. Then the system becomes less stable and unstable operation may even result, as shown in Figure 7–108. To avoid this, the system must be designed so that the effect of lag compensation becomes significant only when the amplitude of the input to the saturating element is small. (This can be done by means of minor feedback-loop compensation.)

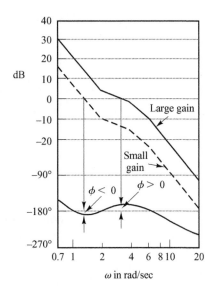

Figure 7–108
Bode diagram of a
conditionally stable
system.

7–13 LAG–LEAD COMPENSATION

We shall first examine the frequency-response characteristics of the lag–lead compensator. Then we present the lag–lead compensation technique based on the frequency-response approach.

Characteristic of Lag–Lead Compensator. Consider the lag–lead compensator given by

$$G_c(s) = K_c \left(\frac{s + \dfrac{1}{T_1}}{s + \dfrac{\gamma}{T_1}} \right) \left(\frac{s + \dfrac{1}{T_2}}{s + \dfrac{1}{\beta T_2}} \right) \tag{7–27}$$

where $\gamma > 1$ and $\beta > 1$. The term

$$\frac{s + \dfrac{1}{T_1}}{s + \dfrac{\gamma}{T_1}} = \frac{1}{\gamma} \left(\frac{T_1 s + 1}{\dfrac{T_1}{\gamma} s + 1} \right) \qquad (\gamma > 1)$$

produces the effect of the lead network, and the term

$$\frac{s + \dfrac{1}{T_2}}{s + \dfrac{1}{\beta T_2}} = \beta \left(\frac{T_2 s + 1}{\beta T_2 s + 1} \right) \qquad (\beta > 1)$$

produces the effect of the lag network.

Figure 7–109
Polar plot of a
lag–lead
compensator given
by Equation (7–27),
with $K_c = 1$ and
$\gamma = \beta$.

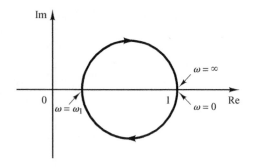

In designing a lag–lead compensator, we frequently chose $\gamma = \beta$. (This is not necessary. We can, of course, choose $\gamma \neq \beta$.) In what follows, we shall consider the case where $\gamma = \beta$. The polar plot of the lag–lead compensator with $K_c = 1$ and $\gamma = \beta$ becomes as shown in Figure 7–109. It can be seen that, for $0 < \omega < \omega_1$, the compensator acts as a lag compensator, while for $\omega_1 < \omega < \infty$ it acts as a lead compensator. The frequency ω_1 is the frequency at which the phase angle is zero. It is given by

$$\omega_1 = \frac{1}{\sqrt{T_1 T_2}}$$

(To derive this equation, see Problem **A–7–21**.)

Figure 7–110 shows the Bode diagram of a lag–lead compensator when $K_c = 1$, $\gamma = \beta = 10$, and $T_2 = 10T_1$. Notice that the magnitude curve has the value 0 dB at the low- and high-frequency regions.

Figure 7–110
Bode diagram of a
lag–lead
compensator given
by Equation (7–27)
with $K_c = 1$,
$\gamma = \beta = 10$, and
$T_2 = 10T_1$.

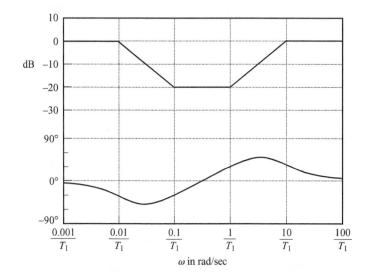

Chapter 7 / Control Systems Analysis and Design by the Frequency-Response Method

Lag–Lead Compensation Based on the Frequency-Response Approach. The design of a lag–lead compensator by the frequency-response approach is based on the combination of the design techniques discussed under lead compensation and lag compensation.

Let us assume that the lag–lead compensator is of the following form:

$$G_c(s) = K_c \frac{(T_1 s + 1)(T_2 s + 1)}{\left(\dfrac{T_1}{\beta} s + 1\right)(\beta T_2 s + 1)} = K_c \frac{\left(s + \dfrac{1}{T_1}\right)\left(s + \dfrac{1}{T_2}\right)}{\left(s + \dfrac{\beta}{T_1}\right)\left(s + \dfrac{1}{\beta T_2}\right)} \qquad (7\text{--}28)$$

where $\beta > 1$. The phase-lead portion of the lag–lead compensator (the portion involving T_1) alters the frequency-response curve by adding phase-lead angle and increasing the phase margin at the gain crossover frequency. The phase-lag portion (the portion involving T_2) provides attenuation near and above the gain crossover frequency and thereby allows an increase of gain at the low-frequency range to improve the steady-state performance.

We shall illustrate the details of the procedures for designing a lag–lead compensator by an example.

EXAMPLE 7–28   Consider the unity-feedback system whose open-loop transfer function is

$$G(s) = \frac{K}{s(s + 1)(s + 2)}$$

It is desired that the static velocity error constant be 10 sec^{-1}, the phase margin be 50°, and the gain margin be 10 dB or more.

Assume that we use the lag–lead compensator given by Equation (7–28). [Note that the phase-lead portion increases both the phase margin and the system bandwidth (which implies increasing the speed of response). The phase-lag portion maintains the low-frequency gain.]

The open-loop transfer function of the compensated system is $G_c(s)G(s)$. Since the gain K of the plant is adjustable, let us assume that $K_c = 1$. Then, $\lim_{s\to 0} G_c(s) = 1$.

From the requirement on the static velocity error constant, we obtain

$$K_v = \lim_{s\to 0} sG_c(s)G(s) = \lim_{s\to 0} sG_c(s)\frac{K}{s(s + 1)(s + 2)} = \frac{K}{2} = 10$$

Hence,

$$K = 20$$

We shall next draw the Bode diagram of the uncompensated system with $K = 20$, as shown in Figure 7–111. The phase margin of the gain-adjusted but uncompensated system is found to be $-32°$, which indicates that the gain-adjusted but uncompensated system is unstable.

The next step in the design of a lag–lead compensator is to choose a new gain crossover frequency. From the phase-angle curve for $G(j\omega)$, we notice that $\underline{/G(j\omega)} = -180°$ at $\omega = 1.5$ rad/sec. It is convenient to choose the new gain crossover frequency to be 1.5 rad/sec so that the phase-lead angle required at $\omega = 1.5$ rad/sec is about 50°, which is quite possible by use of a single lag–lead network.

Once we choose the gain crossover frequency to be 1.5 rad/sec, we can determine the corner frequency of the phase-lag portion of the lag–lead compensator. Let us choose the corner frequency $\omega = 1/T_2$ (which corresponds to the zero of the phase-lag portion of the compensator) to be 1 decade below the new gain crossover frequency, or at $\omega = 0.15$ rad/sec.

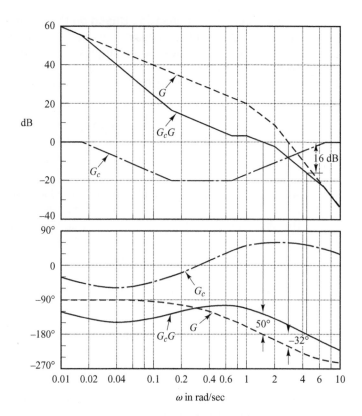

Figure 7–111
Bode diagrams for G (gain-adjusted but uncompensated open-loop transfer function), G_c (compensator), and G_cG (compensated open-loop transfer function).

Recall that for the lead compensator the maximum phase-lead angle ϕ_m is given by Equation (7–25), where α is $1/\beta$ in the present case. By substituting $\alpha = 1/\beta$ in Equation (7–25), we have

$$\sin\phi_m = \frac{1 - \dfrac{1}{\beta}}{1 + \dfrac{1}{\beta}} = \frac{\beta - 1}{\beta + 1}$$

Notice that $\beta = 10$ corresponds to $\phi_m = 54.9°$. Since we need a 50° phase margin, we may choose $\beta = 10$. (Note that we will be using several degrees less than the maximum angle, 54.9°.) Thus,

$$\beta = 10$$

Then the corner frequency $\omega = 1/\beta T_2$ (which corresponds to the pole of the phase-lag portion of the compensator) becomes $\omega = 0.015$ rad/sec. The transfer function of the phase-lag portion of the lag–lead compensator then becomes

$$\frac{s + 0.15}{s + 0.015} = 10\left(\frac{6.67s + 1}{66.7s + 1}\right)$$

The phase-lead portion can be determined as follows: Since the new gain crossover frequency is $\omega = 1.5$ rad/sec, from Figure 7–111, $G(j1.5)$ is found to be 13 dB. Hence, if the lag–lead compensator contributes -13 dB at $\omega = 1.5$ rad/sec, then the new gain crossover frequency is as desired. From this requirement, it is possible to draw a straight line of slope 20 dB/decade, passing through the point (1.5 rad/sec, -13 dB). The intersections of this line and the 0-dB line and -20-dB line determine the corner frequencies. Thus, the corner frequencies for the lead portion are $\omega = 0.7$ rad/sec and $\omega = 7$ rad/sec. Thus, the transfer function of the lead portion of the lag–lead compensator becomes

$$\frac{s + 0.7}{s + 7} = \frac{1}{10}\left(\frac{1.43s + 1}{0.143s + 1}\right)$$

Combining the transfer functions of the lag and lead portions of the compensator, we obtain the transfer function of the lag–lead compensator. Since we chose $K_c = 1$, we have

$$G_c(s) = \left(\frac{s + 0.7}{s + 7}\right)\left(\frac{s + 0.15}{s + 0.015}\right) = \left(\frac{1.43s + 1}{0.143s + 1}\right)\left(\frac{6.67s + 1}{66.7s + 1}\right)$$

The magnitude and phase-angle curves of the lag–lead compensator just designed are shown in Figure 7–111. The open-loop transfer function of the compensated system is

$$G_c(s)G(s) = \frac{(s + 0.7)(s + 0.15)20}{(s + 7)(s + 0.015)s(s + 1)(s + 2)}$$

$$= \frac{10(1.43s + 1)(6.67s + 1)}{s(0.143s + 1)(66.7s + 1)(s + 1)(0.5s + 1)} \quad (7\text{–}29)$$

The magnitude and phase-angle curves of the system of Equation (7–29) are also shown in Figure 7–111. The phase margin of the compensated system is 50°, the gain margin is 16 dB, and the static velocity error constant is 10 sec^{-1}. All the requirements are therefore met, and the design has been completed.

Figure 7–112 shows the polar plots of $G(j\omega)$ (gain-adjusted but uncompensated open-loop transfer function) and $G_c(j\omega)G(j\omega)$ (compensated open-loop transfer function). The $G_c(j\omega)G(j\omega)$ locus is tangent to the $M = 1.2$ circle at about $\omega = 2$ rad/sec. Clearly, this indicates that the compensated system has satisfactory relative stability. The bandwidth of the compensated system is slightly larger than 2 rad/sec.

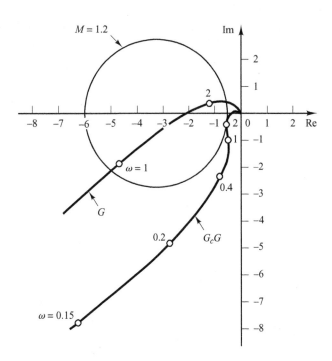

Figure 7–112
Polar plots of G (gain adjusted) and G_cG.

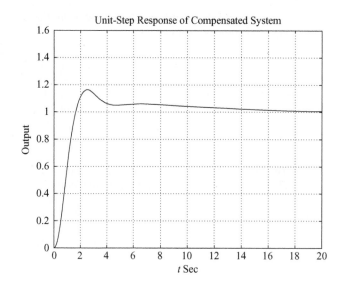

Figure 7–113
Unit-step response of
the compensated
system (Example
7–28).

In the following we shall examine the transient-response characteristics of the compensated system. (The gain-adjusted but uncompensated system is unstable.) The closed-loop transfer function of the compensated system is

$$\frac{C(s)}{R(s)} = \frac{95.381s^2 + 81s + 10}{4.7691s^5 + 47.7287s^4 + 110.3026s^3 + 163.724s^2 + 82s + 10}$$

The unit-step and unit-ramp response curves obtained with MATLAB are shown in Figures 7–113 and 7–114, respectively.

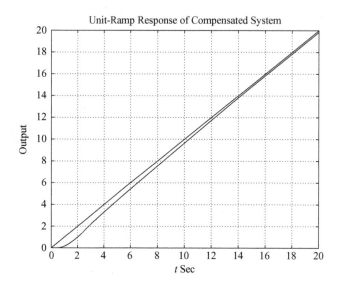

Figure 7–114
Unit-ramp response
of the compensated
system (Example
7–28).

Note that the designed closed-loop control system has the following closed-loop zeros and poles:

$$\text{Zeros at } s = -0.1499, \qquad s = -0.6993$$

$$\text{Poles at } s = -0.8973 \pm j1.4439$$

$$s = -0.1785, \qquad s = -0.5425, \qquad s = -7.4923$$

The pole at $s = -0.1785$ and zero at $s = -0.1499$ are located very close to each other. Such a pair of pole and zero produces a long tail of small amplitude in the step response, as seen in Figure 7–113. Also, the pole at $s = -0.5425$ and zero at $s = -0.6993$ are located fairly close to each other. This pair adds amplitude to the long tail.

Summary of Control Systems Design by Frequency-Response Approach. The last three sections presented detailed procedures for designing lead, lag, and lag–lead compensators by the use of simple examples. We have shown that the design of a compensator to satisfy the given specifications (in terms of the phase margin and gain margin) can be carried out in the Bode diagram in a simple and straightforward manner. It is noted that not every system can be compensated with a lead, lag, or lag–lead compensator. In some cases compensators with complex poles and zeros may be used. For systems that cannot be designed by use of the root-locus or frequency-response methods, the pole-placement method may be used. (See Chapter 10.) In a given design problem if both conventional design methods and the pole-placement method can be used, conventional methods (root-locus or frequency-response methods) usually result in a lower-order stable compensator. Note that a satisfactory design of a compensator for a complex system may require a creative application of all available design methods.

Comparison of Lead, Lag, and Lag–Lead Compensation

1. Lead compensation is commonly used for improving stability margins. Lag compensation is used to improve the steady-state performance. Lead compensation achieves the desired result through the merits of its phase-lead contribution, whereas lag compensation accomplishes the result through the merits of its attenuation property at high frequencies.

2. In some design problems both lead compensation and lag compensation may satisfy the specifications. Lead compensation yields a higher gain crossover frequency than is possible with lag compensation. The higher gain crossover frequency means a larger bandwidth. A large bandwidth means reduction in the settling time. The bandwidth of a system with lead compensation is always greater than that with lag compensation. Therefore, if a large bandwidth or fast response is desired, lead compensation should be employed. If, however, noise signals are present, then a large bandwidth may not be desirable, since it makes the system more susceptible to noise signals because of an increase in the high-frequency gain. Hence, lag compensation should be used for such a case.

3. Lead compensation requires an additional increase in gain to offset the attenuation inherent in the lead network. This means that lead compensation will require a larger gain than that required by lag compensation. A larger gain, in most cases, implies larger space, greater weight, and higher cost.

4. Lead compensation may generate large signals in the system. Such large signals are not desirable because they will cause saturation in the system.

5. Lag compensation reduces the system gain at higher frequencies without reducing the system gain at lower frequencies. Since the system bandwidth is reduced, the system has a slower speed to respond. Because of the reduced high-frequency gain, the total system gain can be increased, and thereby low-frequency gain can be increased and the steady-state accuracy can be improved. Also, any high-frequency noises involved in the system can be attenuated.

6. Lag compensation will introduce a pole-zero combination near the origin that will generate a long tail with small amplitude in the transient response.

7. If both fast responses and good static accuracy are desired, a lag–lead compensator may be employed. By use of the lag–lead compensator, the low-frequency gain can be increased (which means an improvement in steady-state accuracy), while at the same time the system bandwidth and stability margins can be increased.

8. Although a large number of practical compensation tasks can be accomplished with lead, lag, or lag–lead compensators, for complicated systems, simple compensation by use of these compensators may not yield satisfactory results. Then, different compensators having different pole–zero configurations must be employed.

Graphical Comparison.   Figure 7–115(a) shows a unit-step response curve and unit-ramp response curve of an uncompensated system. Typical unit-step response and unit-ramp response curves for the compensated system using a lead, lag, and lag–lead compensator, respectively, are shown in Figures 7–115(b), (c), and (d). The system with a lead compensator exhibits the fastest response, while that with a lag compensator exhibits the slowest response, but with marked improvements in the unit-ramp response. The system with a lag–lead compensator will give a compromise; reasonable improve-

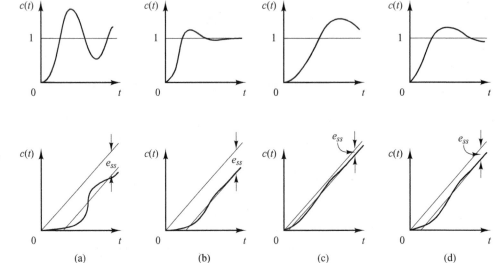

Figure 7–115
Unit-step response curves and unit-ramp response curves.
(a) Uncompensated system; (b) lead compensated system; (c) lag compensated system; (d) lag–lead compensated system.

ments in both the transient response and steady-state response can be expected. The response curves shown depict the nature of improvements that may be expected from using different types of compensators.

Feedback Compensation. A tachometer is one of the rate feedback devices. Another common rate feedback device is the rate gyro. Rate gyros are commonly used in aircraft autopilot systems.

Velocity feedback using a tachometer is very commonly used in positional servo systems. It is noted that, if the system is subjected to noise signals, velocity feedback may generate some difficulty if a particular velocity feedback scheme performs differentiation of the output signal. (The result is the accentuation of the noise effects.)

Cancellation of Undesirable Poles. Since the transfer function of elements in cascade is the product of their individual transfer functions, it is possible to cancel some undesirable poles or zeros by placing a compensating element in cascade, with its poles and zeros being adjusted to cancel the undesirable poles or zeros of the original system. For example, a large time constant T_1 may be canceled by use of the lead network $(T_1 s + 1)/(T_2 s + 1)$ as follows:

$$\left(\frac{1}{T_1 s + 1}\right)\left(\frac{T_1 s + 1}{T_2 s + 1}\right) = \frac{1}{T_2 s + 1}$$

If T_2 is much smaller than T_1, we can effectively eliminate the large time constant T_1. Figure 7–116 shows the effect of canceling a large time constant in step transient response.

If an undesirable pole in the original system lies in the right-half s plane, this compensation scheme should not be used since, although mathematically it is possible to cancel the undesirable pole with an added zero, exact cancellation is physically impossible because of inaccuracies involved in the location of the poles and zeros. A pole in the right-half s plane not exactly canceled by the compensator zero will eventually lead to unstable operation, because the response will involve an exponential term that increases with time.

It is noted that if a left-half plane pole is almost canceled but not exactly canceled, as is almost always the case, the uncanceled pole-zero combination will cause the response to have a small amplitude but long-lasting transient-response component. If the cancellation is not exact but is reasonably good, then this component will be small.

It should be noted that the ideal control system is not the one that has a transfer function of unity. Physically, such a control system cannot be built since it cannot

Figure 7–116
Step-response curves showing the effect of canceling a large time constant.

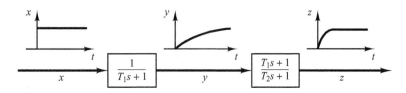

instantaneously transfer energy from the input to the output. In addition, since noise is almost always present in one form or another, a system with a unity transfer function is not desirable. A desired control system, in many practical cases, may have one set of dominant complex-conjugate closed-loop poles with a reasonable damping ratio and undamped natural frequency. The determination of the significant part of the closed-loop pole-zero configuration, such as the location of the dominant closed-loop poles, is based on the specifications that give the required system performance.

Cancellation of Undesirable Complex-Conjugate Poles. If the transfer function of a plant contains one or more pairs of complex-conjugate poles, then a lead, lag, or lag–lead compensator may not give satisfactory results. In such a case, a network that has two zeros and two poles may prove to be useful. If the zeros are chosen so as to cancel the undesirable complex-conjugate poles of the plant, then we can essentially replace the undesirable poles by acceptable poles. That is, if the undesirable complex-conjugate poles are in the left-half s plane and are in the form

$$\frac{1}{s^2 + 2\zeta_1\omega_1 s + \omega_1^2}$$

then the insertion of a compensating network having the transfer function

$$\frac{s^2 + 2\zeta_1\omega_1 s + \omega_1^2}{s^2 + 2\zeta_2\omega_2 s + \omega_2^2}$$

will result in an effective change of the undesirable complex-conjugate poles to acceptable poles. Note that even though the cancellation may not be exact, the compensated system will exhibit better response characteristics. (As stated earlier, this approach cannot be used if the undesirable complex-conjugate poles are in the right-half s plane.)

Familiar networks consisting only of RC components whose transfer functions possess two zeros and two poles are the bridged-T networks. Examples of bridged-T networks and their transfer functions are shown in Figure 7–117. (The derivations of the transfer functions of the bridged-T networks were given in Problem **A–3–5**.)

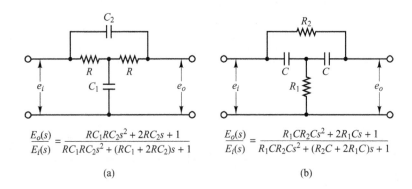

$$\frac{E_o(s)}{E_i(s)} = \frac{RC_1RC_2s^2 + 2RC_2s + 1}{RC_1RC_2s^2 + (RC_1 + 2RC_2)s + 1}$$

$$\frac{E_o(s)}{E_i(s)} = \frac{R_1CR_2Cs^2 + 2R_1Cs + 1}{R_1CR_2Cs^2 + (R_2C + 2R_1C)s + 1}$$

(a)

(b)

Figure 7–117
Bridged-T networks.

Concluding Comments. In the design examples presented in this chapter, we have been primarily concerned only with the transfer functions of compensators. In actual design problems, we must choose the hardware. Thus, we must satisfy additional design constraints such as cost, size, weight, and reliability.

The system designed may meet the specifications under normal operating conditions but may deviate considerably from the specifications when environmental changes are considerable. Since the changes in the environment affect the gain and time constants of the system, it is necessary to provide automatic or manual means to adjust the gain to compensate for such environmental changes, for nonlinear effects that were not taken into account in the design, and also to compensate for manufacturing tolerances from unit to unit in the production of system components. (The effects of manufacturing tolerances are suppressed in a closed-loop system; therefore, the effects may not be critical in closed-loop operation but critical in open-loop operation.) In addition to this, the designer must remember that any system is subject to small variations due mainly to the normal deterioration of the system.

EXAMPLE PROBLEMS AND SOLUTIONS

A–7–1. Consider a system whose closed-loop transfer function is

$$\frac{C(s)}{R(s)} = \frac{10(s + 1)}{(s + 2)(s + 5)}$$

Clearly, the closed-loop poles are located at $s = -2$ and $s = -5$, and the system is not oscillatory.

Show that the closed-loop frequency response of this system will exhibit a resonant peak, although the damping ratio of the closed-loop poles is greater than unity.

Solution. Figure 7–118 shows the Bode diagram for the system. The resonant peak value is approximately 3.5 dB. (Note that, in the absence of a zero, the second-order system with $\zeta > 0.7$ will not exhibit a resonant peak; however, the presence of a closed-loop zero will cause such a peak.)

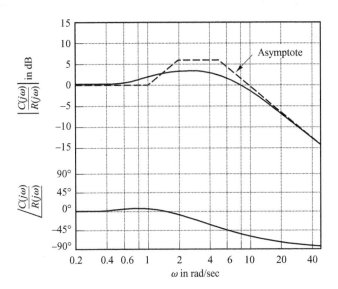

Figure 7–118
Bode diagram for
$10(1 + j\omega)/[(2 + j\omega)(5 + j\omega)]$.

A–7–2. Consider the system defined by

$$\begin{bmatrix} \dot{x}_1 \\ \dot{x}_2 \end{bmatrix} = \begin{bmatrix} 0 & 1 \\ -25 & -4 \end{bmatrix} \begin{bmatrix} x_1 \\ x_2 \end{bmatrix} + \begin{bmatrix} 1 & 1 \\ 0 & 1 \end{bmatrix} \begin{bmatrix} u_1 \\ u_2 \end{bmatrix}$$

$$\begin{bmatrix} y_1 \\ y_2 \end{bmatrix} = \begin{bmatrix} 1 & 0 \\ 0 & 1 \end{bmatrix} \begin{bmatrix} x_1 \\ x_2 \end{bmatrix}$$

Obtain the sinusoidal transfer functions $Y_1(j\omega)/U_1(j\omega)$, $Y_2(j\omega)/U_1(j\omega)$, $Y_1(j\omega)/U_2(j\omega)$, and $Y_2(j\omega)/U_2(j\omega)$. In deriving $Y_1(j\omega)/U_1(j\omega)$ and $Y_2(j\omega)/U_1(j\omega)$, we assume that $U_2(j\omega) = 0$. Similarly, in obtaining $Y_1(j\omega)/U_2(j\omega)$ and $Y_2(j\omega)/U_2(j\omega)$, we assume that $U_1(j\omega) = 0$.

Solution. The transfer matrix expression for the system defined by

$$\dot{\mathbf{x}} = \mathbf{Ax} + \mathbf{Bu}$$
$$\dot{\mathbf{y}} = \mathbf{Cx} + \mathbf{Du}$$

is given by

$$\mathbf{Y}(s) = \mathbf{G}(s)\mathbf{U}(s)$$

where $\mathbf{G}(s)$ is the transfer matrix and is given by

$$\mathbf{G}(s) = \mathbf{C}(s\mathbf{I} - \mathbf{A})^{-1}\mathbf{B} + \mathbf{D}$$

For the system considered here, the transfer matrix becomes

$$\mathbf{C}(s\mathbf{I} - \mathbf{A})^{-1}\mathbf{B} + \mathbf{D} = \begin{bmatrix} 1 & 0 \\ 0 & 1 \end{bmatrix} \begin{bmatrix} s & -1 \\ 25 & s+4 \end{bmatrix}^{-1} \begin{bmatrix} 1 & 1 \\ 0 & 1 \end{bmatrix}$$

$$= \frac{1}{s^2 + 4s + 25} \begin{bmatrix} s+4 & 1 \\ -25 & s \end{bmatrix} \begin{bmatrix} 1 & 1 \\ 0 & 1 \end{bmatrix}$$

$$= \begin{bmatrix} \dfrac{s+4}{s^2 + 4s + 25} & \dfrac{s+5}{s^2 + 4s + 25} \\ \dfrac{-25}{s^2 + 4s + 25} & \dfrac{s-25}{s^2 + 4s + 25} \end{bmatrix}$$

Hence

$$\begin{bmatrix} Y_1(s) \\ Y_2(s) \end{bmatrix} = \begin{bmatrix} \dfrac{s+4}{s^2 + 4s + 25} & \dfrac{s+5}{s^2 + 4s + 25} \\ \dfrac{-25}{s^2 + 4s + 25} & \dfrac{s-25}{s^2 + 4s + 25} \end{bmatrix} \begin{bmatrix} U_1(s) \\ U_2(s) \end{bmatrix}$$

Assuming that $U_2(j\omega) = 0$, we find $Y_1(j\omega)/U_1(j\omega)$ and $Y_2(j\omega)/U_1(j\omega)$ as follows:

$$\frac{Y_1(j\omega)}{U_1(j\omega)} = \frac{j\omega + 4}{(j\omega)^2 + 4j\omega + 25}$$

$$\frac{Y_2(j\omega)}{U_1(j\omega)} = \frac{-25}{(j\omega)^2 + 4j\omega + 25}$$

Similarly, assuming that $U_1(j\omega) = 0$, we find $Y_1(j\omega)/U_2(j\omega)$ and $Y_2(j\omega)/U_2(j\omega)$ as follows:

$$\frac{Y_1(j\omega)}{U_2(j\omega)} = \frac{j\omega + 5}{(j\omega)^2 + 4j\omega + 25}$$

$$\frac{Y_2(j\omega)}{U_2(j\omega)} = \frac{j\omega - 25}{(j\omega)^2 + 4j\omega + 25}$$

Notice that $Y_2(j\omega)/U_2(j\omega)$ is a nonminimum-phase transfer function.

A–7–3. Referring to Problem A–7–2, plot Bode diagrams for the system, using MATLAB.

Solution. MATLAB Program 7–15 produces Bode diagrams for the system. There are four sets of Bode diagrams: two for input 1 and two for input 2. These Bode diagrams are shown in Figure 7–119.

MATLAB Program 7–15
A = [0 1;-25 -4];
B = [1 1;0 1];
C = [1 0;0 1];
D = [0 0;0 0];
bode(A,B,C,D)

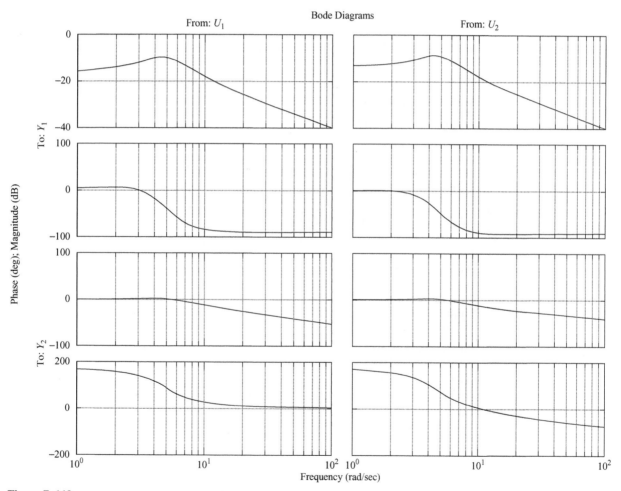

Figure 7–119
Bode diagrams.

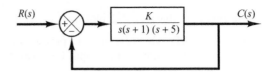

Figure 7–120
Closed-loop system.

A–7–4. Using MATLAB, plot Bode diagrams for the closed-loop system shown in Figure 7–120 for $K = 1$, $K = 10$, and $K = 20$. Plot three magnitude curves in one diagram and three phase-angle curves in another diagram.

Solution. The closed-loop transfer function of the system is given by

$$\frac{C(s)}{R(s)} = \frac{K}{s(s + 1)(s + 5) + K}$$

$$= \frac{K}{s^3 + 6s^2 + 5s + K}$$

Hence the numerator and denominator of $C(s)/R(s)$ are

$$\text{num} = [K]$$
$$\text{den} = [1 \quad 6 \quad 5 \quad K]$$

A possible MATLAB program is shown in MATLAB Program 7–16. The resulting Bode diagrams are shown in Figures 7–121(a) and (b).

MATLAB Program 7–16

```
w = logspace(-1,2,200);
for i = 1:3;
   if i = 1; K = 1;[mag,phase,w] = bode([K],[1  6  5  K],w);
      mag1dB = 20*log10(mag); phase1 = phase; end;
   if i = 2; K = 10;[mag,phase,w] = bode([K],[1  6  5  K],w);
      mag2dB = 20*log10(mag); phase2 = phase; end;
   if i = 3; K = 20;[mag,phase,w] = bode([K],[1  6  5  K],w);
      mag3dB = 20*log10(mag); phase3 = phase; end;
end
semilogx(w,mag1dB,'-',w,mag2dB,'-',w,mag3dB,'-')
grid
title('Bode Diagrams of K/[s(s + 1)(s + 5) + K], where K = 1, K = 10, and K = 20')
xlabel('Frequency (rad/sec)')
ylabel('Gain (dB)')
text(1.2,-31,'K = 1')
text(1.1,-8,'K = 10')
text(11,-31,'K = 20')
semilogx(w,phase1,'-',w,phase2,'-',w,phase3,'-')
grid
xlabel('Frequency (rad/sec)')
ylabel('Phase (deg)')
text(0.2,-90,'K = 1')
text(0.2,-20,'K =10')
text(1.6,-20,'K = 20')
```

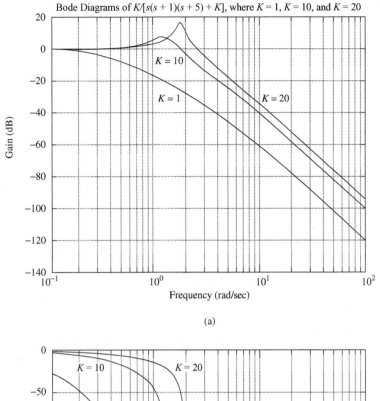

Bode Diagrams of $K/[s(s+1)(s+5) + K]$, where $K = 1$, $K = 10$, and $K = 20$

(a)

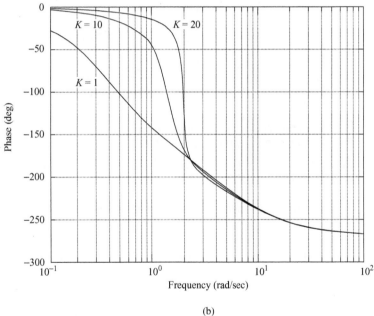

Figure 7–121
Bode diagrams:
(a) Magnitude-
versus-frequency
curves; (b) phase-
angle-versus-
frequency curves.

(b)

A–7–5. Prove that the polar plot of the sinusoidal transfer function

$$G(j\omega) = \frac{j\omega T}{1 + j\omega T}, \qquad \text{for } 0 \le \omega \le \infty$$

is a semicircle. Find the center and radius of the circle.

Example Problems and Solutions 525

Solution. The given sinusoidal transfer function $G(j\omega)$ can be written as follows:

$$G(j\omega) = X + jY$$

where

$$X = \frac{\omega^2 T^2}{1 + \omega^2 T^2}, \qquad Y = \frac{\omega T}{1 + \omega^2 T^2}$$

Then

$$\left(X - \frac{1}{2}\right)^2 + Y^2 = \frac{(\omega^2 T^2 - 1)^2}{4(1 + \omega^2 T^2)^2} + \frac{\omega^2 T^2}{(1 + \omega^2 T^2)^2} = \frac{1}{4}$$

Hence, we see that the plot of $G(j\omega)$ is a circle centered at $(0.5, 0)$ with radius equal to 0.5. The upper semicircle corresponds to $0 \leq \omega \leq \infty$, and the lower semicircle corresponds to $-\infty \leq \omega \leq 0$.

A–7–6. Prove the following mapping theorem: Let $F(s)$ be a ratio of polynomials in s. Let P be the number of poles and Z be the number of zeros of $F(s)$ that lie inside a closed contour in the s plane, with multiplicity accounted for. Let the closed contour be such that it does not pass through any poles or zeros of $F(s)$. The closed contour in the s plane then maps into the $F(s)$ plane as a closed curve. The number N of clockwise encirclements of the origin of the $F(s)$ plane, as a representative point s traces out the entire contour in the s plane in the clockwise direction, is equal to $Z - P$.

Solution. To prove this theorem, we use Cauchy's theorem and the residue theorem. Cauchy's theorem states that the integral of $F(s)$ around a closed contour in the s plane is zero if $F(s)$ is analytic[#] within and on the closed contour, or

$$\oint F(s)\, ds = 0$$

Suppose that $F(s)$ is given by

$$F(s) = \frac{(s + z_1)^{k_1}(s + z_2)^{k_2} \cdots}{(s + p_1)^{m_1}(s + p_2)^{m_2} \cdots} X(s)$$

where $X(s)$ is analytic in the closed contour in the s plane and all the poles and zeros are located in the contour. Then the ratio $F'(s)/F(s)$ can be written

$$\frac{F'(s)}{F(s)} = \left(\frac{k_1}{s + z_1} + \frac{k_2}{s + z_2} + \cdots\right) - \left(\frac{m_1}{s + p_1} + \frac{m_2}{s + p_2} + \cdots\right) + \frac{X'(s)}{X(s)} \qquad (7\text{–}30)$$

This may be seen from the following consideration: If $\hat{F}(s)$ is given by

$$\hat{F}(s) = (s + z_1)^k X(s)$$

then $\hat{F}(s)$ has a zero of kth order at $s = -z_1$. Differentiating $F(s)$ with respect to s yields

$$\hat{F}'(s) = k(s + z_1)^{k-1} X(s) + (s + z_1)^k X'(s)$$

Hence,

$$\frac{\hat{F}'(s)}{\hat{F}(s)} = \frac{k}{s + z_1} + \frac{X'(s)}{X(s)} \qquad (7\text{–}31)$$

We see that by taking the ratio $\hat{F}'(s)/\hat{F}(s)$, the kth-order zero of $\hat{F}(s)$ becomes a simple pole of $\hat{F}'(s)/\hat{F}(s)$.

[#]For the definition of an analytic function, see the footnote on page 447.

If the last term on the right-hand side of Equation (7–31) does not contain any poles or zeros in the closed contour in the s plane, $F'(s)/F(s)$ is analytic in this contour except at point $s = -z_1$. Then, referring to Equation (7–30) and using the residue theorem, which states that the integral of $F'(s)/F(s)$ taken in the clockwise direction around a closed contour in the s plane is equal to $-2\pi j$ times the residues at the simple poles of $F'(s)/F(s)$, or

$$\oint \frac{F'(s)}{F(s)} ds = -2\pi j \left(\sum \text{residues} \right)$$

we have

$$\oint \frac{F'(s)}{F(s)} ds = -2\pi j \left[(k_1 + k_2 + \cdots) - (m_1 + m_2 + \cdots) \right] = -2\pi j(Z - P)$$

where $Z = k_1 + k_2 + \cdots$ = total number of zeros of $F(s)$ enclosed in the closed contour in the s plane

$\quad\quad\quad P = m_1 + m_2 + \cdots$ = total number of poles of $F(s)$ enclosed in the closed contour in the s plane

[The k multiple zeros (or poles) are considered k zeros (or poles) located at the same point.] Since $F(s)$ is a complex quantity, $F(s)$ can be written

$$F(s) = |F|e^{j\theta}$$

and

$$\ln F(s) = \ln|F| + j\theta$$

Noting that $F'(s)/F(s)$ can be written

$$\frac{F'(s)}{F(s)} = \frac{d \ln F(s)}{ds}$$

we obtain

$$\frac{F'(s)}{F(s)} = \frac{d \ln|F|}{ds} + j\frac{d\theta}{ds}$$

If the closed contour in the s plane is mapped into the closed contour Γ in the $F(s)$ plane, then

$$\oint \frac{F'(s)}{F(s)} ds = \oint_\Gamma d \ln|F| + j\oint_\Gamma d\theta = j\int d\theta = 2\pi j(P - Z)$$

The integral $\oint_\Gamma d \ln|F|$ is zero since the magnitude $\ln|F|$ is the same at the initial point and the final point of the contour Γ. Thus we obtain

$$\frac{\theta_2 - \theta_1}{2\pi} = P - Z$$

The angular difference between the final and initial values of θ is equal to the total change in the phase angle of $F'(s)/F(s)$ as a representative point in the s plane moves along the closed contour. Noting that N is the number of clockwise encirclements of the origin of the $F(s)$ plane and $\theta_2 - \theta_1$ is zero or a multiple of 2π rad, we obtain

$$\frac{\theta_2 - \theta_1}{2\pi} = -N$$

Example Problems and Solutions

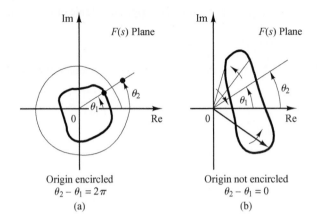

Figure 7–122
Determination of
encirclement of the
origin of $F(s)$ plane.

Origin encircled
$\theta_2 - \theta_1 = 2\pi$
(a)

Origin not encircled
$\theta_2 - \theta_1 = 0$
(b)

Thus, we have the relationship

$$N = Z - P$$

This proves the theorem.

Note that by this mapping theorem, the exact numbers of zeros and of poles cannot be found—only their difference. Note also that, from Figures 7–122(a) and (b), we see that if θ does not change through 2π rad, then the origin of the $F(s)$ plane cannot be encircled.

A–7–7. The Nyquist plot (polar plot) of the open-loop frequency response of a unity-feedback control system is shown in Figure 7–123(a). Assuming that the Nyquist path in the s plane encloses the entire right-half s plane, draw a complete Nyquist plot in the G plane. Then answer the following questions:

(a) If the open-loop transfer function has no poles in the right-half s plane, is the closed-loop system stable?

(b) If the open-loop transfer function has one pole and no zeros in right-half s plane, is the closed-loop system stable?

(c) If the open-loop transfer function has one zero and no poles in the right-half s plane, is the closed-loop system stable?

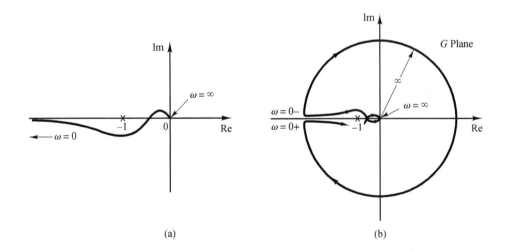

Figure 7–123
(a) Nyquist plot;
(b) complete Nyquist
plot in the G plane.

(a)

(b)

Chapter 7 / **Control Systems Analysis and Design by the Frequency-Response Method**

Solution. Figure 7–123(b) shows a complete Nyquist plot in the G plane. The answers to the three questions are as follows:

(a) The closed-loop system is stable, because the critical point $(-1 + j0)$ is not encircled by the Nyquist plot. That is, since $P = 0$ and $N = 0$, we have $Z = N + P = 0$.

(b) The open-loop transfer function has one pole in the right-half s plane. Hence, $P = 1$. (The open-loop system is unstable.) For the closed-loop system to be stable, the Nyquist plot must encircle the critical point $(-1 + j0)$ once counterclockwise. However, the Nyquist plot does not encircle the critical point. Hence, $N = 0$. Therefore, $Z = N + P = 1$. The closed-loop system is unstable.

(c) Since the open-loop transfer function has one zero, but no poles, in the right-half s plane, we have $Z = N + P = 0$. Thus, the closed-loop system is stable. (Note that the zeros of the open-loop transfer function do not affect the stability of the closed-loop system.)

A–7–8. Is a closed-loop system with the following open-loop transfer function and with $K = 2$ stable?

$$G(s)H(s) = \frac{K}{s(s + 1)(2s + 1)}$$

Find the critical value of the gain K for stability.

Solution. The open-loop transfer function is

$$G(j\omega)H(j\omega) = \frac{K}{j\omega(j\omega + 1)(2j\omega + 1)}$$

$$= \frac{K}{-3\omega^2 + j\omega(1 - 2\omega^2)}$$

This open-loop transfer function has no poles in the right-half s plane. Thus, for stability, the $-1 + j0$ point should not be encircled by the Nyquist plot. Let us find the point where the Nyquist plot crosses the negative real axis. Let the imaginary part of $G(j\omega)H(j\omega)$ be zero, or

$$1 - 2\omega^2 = 0$$

from which

$$\omega = \pm \frac{1}{\sqrt{2}}$$

Substituting $\omega = 1/\sqrt{2}$ into $G(j\omega)H(j\omega)$, we obtain

$$G\left(j\frac{1}{\sqrt{2}}\right)H\left(j\frac{1}{\sqrt{2}}\right) = -\frac{2K}{3}$$

The critical value of the gain K is obtained by equating $-2K/3$ to -1, or

$$-\frac{2}{3}K = -1$$

Hence,

$$K = \frac{3}{2}$$

The system is stable if $0 < K < \frac{3}{2}$. Hence, the system with $K = 2$ is unstable.

Example Problems and Solutions

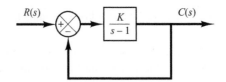

Figure 7–124
Closed-loop system.

A–7–9. Consider the closed-loop system shown in Figure 7–124. Determine the critical value of K for stability by the use of the Nyquist stability criterion.

Solution. The polar plot of

$$G(j\omega) = \frac{K}{j\omega - 1}$$

is a circle with center at $-K/2$ on the negative real axis and radius $K/2$, as shown in Figure 7–125(a). As ω is increased from $-\infty$ to ∞, the $G(j\omega)$ locus makes a counterclockwise rotation. In this system, $P = 1$ because there is one pole of $G(s)$ in the right-half s plane. For the closed-loop system to be stable, Z must be equal to zero. Therefore, $N = Z - P$ must be equal to -1, or there must be one counterclockwise encirclement of the $-1 + j0$ point for stability. (If there is no encirclement of the $-1 + j0$ point, the system is unstable.) Thus, for stability, K must be greater than unity, and $K = 1$ gives the stability limit. Figure 7–125(b) shows both stable and unstable cases of $G(j\omega)$ plots.

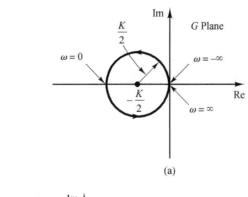

(a)

Figure 7–125
(a) Polar plot of
$K/(j\omega - 1)$;
(b) polar plots of
$K/(j\omega - 1)$ for
stable and unstable
cases.

P = 1
N = −1
Z = 0

(Stable)

K > 1

P = 1
N = 0
Z = 1

(Unstable)

K < 1

(b)

A–7–10. Consider a unity-feedback system whose open-loop transfer function is

$$G(s) = \frac{Ke^{-0.8s}}{s + 1}$$

Using the Nyquist plot, determine the critical value of K for stability.

Solution. For this system,

$$G(j\omega) = \frac{Ke^{-0.8j\omega}}{j\omega + 1}$$

$$= \frac{K(\cos 0.8\omega - j \sin 0.8\omega)(1 - j\omega)}{1 + \omega^2}$$

$$= \frac{K}{1 + \omega^2}\left[(\cos 0.8\omega - \omega \sin 0.8\omega) - j(\sin 0.8\omega + \omega \cos 0.8\omega)\right]$$

The imaginary part of $G(j\omega)$ is equal to zero if

$$\sin 0.8\omega + \omega \cos 0.8\omega = 0$$

Hence,

$$\omega = -\tan 0.8\omega$$

Solving this equation for the smallest positive value of ω, we obtain

$$\omega = 2.4482$$

Substituting $\omega = 2.4482$ into $G(j\omega)$, we obtain

$$G(j2.4482) = \frac{K}{1 + 2.4482^2}(\cos 1.9586 - 2.4482 \sin 1.9586) = -0.378K$$

The critical value of K for stability is obtained by letting $G(j2.4482)$ equal -1. Hence,

$$0.378K = 1$$

or

$$K = 2.65$$

Figure 7–126 shows the Nyquist or polar plots of $2.65e^{-0.8j\omega}/(1 + j\omega)$ and $2.65/(1 + j\omega)$. The first-order system without transport lag is stable for all values of K, but the one with a transport lag of 0.8 sec becomes unstable for $K > 2.65$.

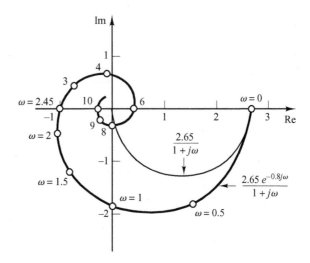

Figure 7–126
Polar plots of
$2.65e^{-0.8j\omega}/(1 + j\omega)$
and $2.65/(1 + j\omega)$.

A–7–11. Consider a unity-feedback system with the following open-loop transfer function:

$$G(s) = \frac{20(s^2 + s + 0.5)}{s(s + 1)(s + 10)}$$

Draw a Nyquist plot with MATLAB and examine the stability of the closed-loop system.

Solution. MATLAB Program 7–17 produces the Nyquist diagram shown in Figure 7–127. From this figure, we see that the Nyquist plot does not encircle the $-1 + j0$ point. Hence, $N = 0$ in the Nyquist stability criterion. Since no open-loop poles lie in the right-half s plane, $P = 0$. Therefore, $Z = N + P = 0$. The closed-loop system is stable.

MATLAB Program 7–17

```
num = [20  20  10];
den = [1  11  10  0];
nyquist(num,den)
v = [-2  3  -3  3]; axis(v)
grid
```

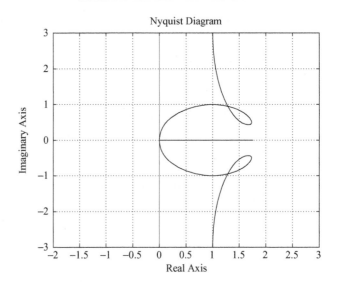

Figure 7–127
Nyquist plot of
$$G(s) = \frac{20(s^2 + s + 0.5)}{s(s + 1)(s + 10)}.$$

A–7–12. Consider the same system as discussed in Problem **A–7–11**. Draw the Nyquist plot for only the positive-frequency region.

Solution. Drawing a Nyquist plot for only the positive-frequency region can be done by the use of the following command:

$$[re,im,w] = nyquist(num,den,w)$$

The frequency region may be divided into several subregions by using different increments. For example, the frequency region of interest may be divided into three subregions as follows:

$$w1 = 0.1{:}0.1{:}10;$$
$$w2 = 10{:}2{:}100;$$
$$w3 = 100{:}10{:}500;$$
$$w = [w1 \ w2 \ w3]$$

MATLAB Program 7–18 uses this frequency region. Using this program, we obtain the Nyquist plot shown in Figure 7–128.

MATLAB Program 7–18

```
num = [20 20 10];
den = [1 11 10 0];
w1 = 0.1:0.1:10; w2 = 10:2:100; w3 = 100:10:500;
w = [w1 w2 w3];
[re,im,w] = nyquist(num,den,w);
plot(re,im)
v = [-3 3 -5 1]; axis(v);
grid
title('Nyquist Plot of G(s) = 20(s^2 + s + 0.5)/[s(s + 1)(s + 10)]')
xlabel('Real Axis')
ylabel('Imag Axis')
```

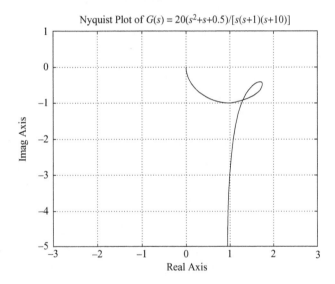

Figure 7–128
Nyquist plot for the positive-frequency region.

A–7–13. Referring to Problem **A–7–12**, plot the polar locus of $G(s)$ where

$$G(s) = \frac{20(s^2 + s + 0.5)}{s(s + 1)(s + 10)}$$

Locate on the polar locus frequency points where $\omega = 0.2, 0.3, 0.5, 1, 2, 6, 10$, and 20 rad/sec. Also, find the magnitudes and phase angles of $G(j\omega)$ at the specified frequency points.

Solution. In MATLAB Program 7–18 we used the frequency vector w, which consists of three frequency subvectors: w1, w2, and w3. Instead of such a w, we may simply use the frequency vector w = logspace(d_1, d_2, n). MATLAB Program 7–19 uses the following frequency vector:

$$w = \text{logspace}(-1, 2, 100)$$

This MATLAB program plots the polar locus and locates the specified frequency points on the polar locus, as shown in Figure 7–129.

MATLAB Program 7–19

```
num = [20  20  10];
den = [1  11  10  0];
ww = logspace(-1,2,100);
nyquist(num,den,ww)
v = [-2  3  -5  0]; axis(v);
grid
hold
Current plot held
w = [0.2  0.3  0.5  1  2  6  10  20];
[re,im,w] = nyquist(num,den,w);
plot(re,im,'o')
text(1.1,-4.8,'w = 0.2')
text(1.1,-3.1,'0.3')
text(1.25,-1.7,'0.5')
text(1.37,-0.4,'1')
text(1.8,-0.3,'2')
text(1.4,-1.1,'6')
text(0.77,-0.8,'10')
text(0.037,-0.8,'20')

% ----- To get the values of magnitude and phase (in degrees) of G(jw)
% at the specified w values, enter the command [mag,phase,w]
% = bode(num,den,w) ------

[mag,phase,w] = bode(num,den,w);

% ----- The following table shows the specified frequency values w and
% the corresponding values of magnitude and phase (in degrees) -----

[w  mag  phase]

ans =

    0.2000  4.9176  -78.9571
    0.3000  3.2426  -72.2244
    0.5000  1.9975  -55.9925
    1.0000  1.5733  -24.1455
    2.0000  1.7678  -14.4898
    6.0000  1.6918  -31.0946
   10.0000  1.4072  -45.0285
   20.0000  0.8933  -63.4385
```

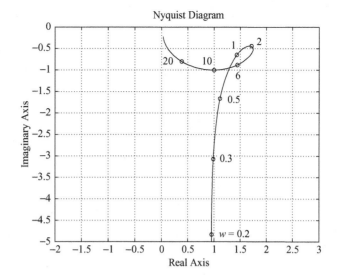

Figure 7–129
Polar plot of $G(j\omega)$
given in Problem
A–7–13.

A–7–14. Consider a unity-feedback, positive-feedback system with the following open-loop transfer function:

$$G(s) = \frac{s^2 + 4s + 6}{s^2 + 5s + 4}$$

Draw a Nyquist plot.

Solution. The Nyquist plot of the positive-feedback system can be obtained by defining num and den as

$$\text{num} = [-1 \ \ -4 \ \ -6]$$
$$\text{den} = [1 \ \ 5 \ \ 4]$$

and using the command nyquist(num,den). MATLAB Program 7–20 produces the Nyquist plot, as shown in Figure 7–130.

This system is unstable, because the $-1 + j0$ point is encircled once clockwise. Note that this is a special case where the Nyquist plot passes through $-1 + j0$ point and also encircles this point once clockwise. This means that the closed-loop system is degenerate; the system behaves as if it were an unstable first-order system. See the following closed-loop transfer function of the positive-feedback system:

$$\frac{C(s)}{R(s)} = \frac{s^2 + 4s + 6}{s^2 + 5s + 4 - \left(s^2 + 4s + 6\right)}$$

$$= \frac{s^2 + 4s + 6}{s - 2}$$

MATLAB Program 7–20

```
num = [-1  -4  -6];
den = [1  5  4];
nyquist(num,den);
grid
title('Nyquist Plot of G(s) = -(s^2 + 4s + 6)/(s^2 + 5s + 4)')
```

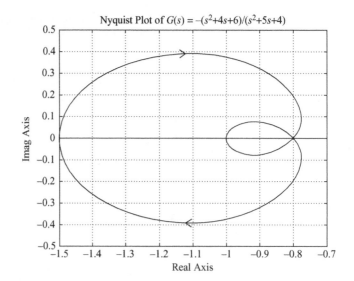

Figure 7–130
Nyquist plot for
positive-feedback
system.

Note that the Nyquist plot for the positive-feedback case is a mirror image about the imaginary axis of the Nyquist plot for the negative-feedback case. This may be seen from Figure 7–131, which was obtained by use of MATLAB Program 7–21. (Note that the positive-feedback case is unstable, but the negative-feedback case is stable.)

MATLAB Program 7–21

```
num1 = [1  4  6];
den1 = [1  5  4];
num2 = [-1  -4  -6];
den2 = [1  5  4];
nyquist(num1,den1);
hold on
nyquist(num2,den2);
v = [-2  2  -1  1];
axis(v);
grid
title('Nyquist Plots of G(s) and -G(s)')
text(1.0,0.5,'G(s)')
text(0.57,-0.48,'Use this Nyquist')
text(0.57,-0.61,'plot for negative')
text(0.57,-0.73,'feedback system')
text(-1.3,0.5,'-G(s)')
text(-1.7,-0.48,'Use this Nyquist')
text(-1.7,-0.61,'plot for positive')
text(-1.7,-0.73,'feedback system')
```

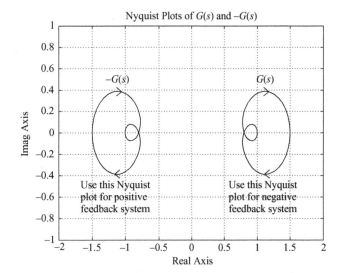

Figure 7–131
Nyquist plots for
positive-feedback
system and negative-
feedback system.

A–7–15. Consider the control system shown in Figure 7–60. (Refer to Example 7–19.) Using the inverse polar plot, determine the range of gain K for stability.

Solution. Since

$$G_2(s) = \frac{1}{s^3 + s^2 + 1}$$

we have

$$G(s) = G_1(s)G_2(s) = \frac{K(s + 0.5)}{s^3 + s^2 + 1}$$

Hence, the inverse of the feedforward transfer function is

$$\frac{1}{G(s)} = \frac{s^3 + s^2 + 1}{K(s + 0.5)}$$

Notice that $1/G(s)$ has a pole at $s = -0.5$. It does not have any pole in the right-half s plane. Therefore, the Nyquist stability equation

$$Z = N + P$$

reduces to $Z = N$ since $P = 0$. The reduced equation states that the number Z of the zeros of $1 + \left[1/G(s)\right]$ in the right-half s plane is equal to N, the number of clockwise encirclements of the $-1 + j0$ point. For stability, N must be equal to zero, or there should be no encirclement. Figure 7–132 shows the Nyquist plot or polar plot of $K/G(j\omega)$.

Notice that since

$$\frac{K}{G(j\omega)} = \left[\frac{(j\omega)^3 + (j\omega)^2 + 1}{j\omega + 0.5}\right]\left(\frac{0.5 - j\omega}{0.5 - j\omega}\right)$$

$$= \frac{0.5 - 0.5\omega^2 - \omega^4 + j\omega(-1 + 0.5\omega^2)}{0.25 + \omega^2}$$

Example Problems and Solutions

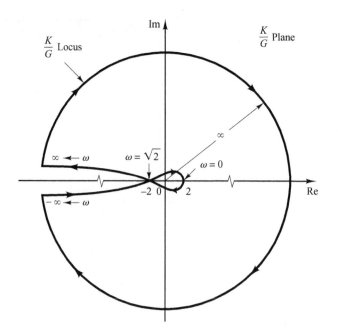

Figure 7–132
Polar plot of
$K/G(j\omega)$.

the $K/G(j\omega)$ locus crosses the negative real axis at $\omega = \sqrt{2}$, and the crossing point at the negative real axis is -2.

From Figure 7–132, we see that if the critical point lies in the region between -2 and $-\infty$, then the critical point is not encircled. Hence, for stability, we require

$$-1 < \frac{-2}{K}$$

Thus, the range of gain K for stability is

$$2 < K$$

which is the same result as we obtained in Example 7–19.

A–7–16. Figure 7–133 shows a block diagram of a space-vehicle control system. Determine the gain K such that the phase margin is $50°$. What is the gain margin in this case?

Solution. Since

$$G(j\omega) = \frac{K(j\omega + 2)}{(j\omega)^2}$$

we have

$$\underline{/G(j\omega)} = \underline{/j\omega + 2} - 2\underline{/j\omega} = \tan^{-1}\frac{\omega}{2} - 180°$$

The requirement that the phase margin be $50°$ means that $\underline{/G(j\omega_c)}$ must be equal to $-130°$, where ω_c is the gain crossover frequency, or

$$\underline{/G(j\omega_c)} = -130°$$

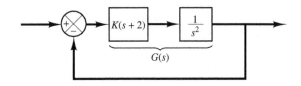

Figure 7–133
Space-vehicle control
system.

Hence, we set

$$\tan^{-1}\frac{\omega_c}{2} = 50°$$

from which we obtain

$$\omega_c = 2.3835 \text{ rad/sec}$$

Since the phase curve never crosses the $-180°$ line, the gain margin is $+\infty$ dB. Noting that the magnitude of $G(j\omega)$ must be equal to 0 dB at $\omega = 2.3835$, we have

$$\left|\frac{K(j\omega + 2)}{(j\omega)^2}\right|_{\omega=2.3835} = 1$$

from which we get

$$K = \frac{2.3835^2}{\sqrt{2^2 + 2.3835^2}} = 1.8259$$

This K value will give the phase margin of 50°.

A–7–17. For the standard second-order system

$$\frac{C(s)}{R(s)} = \frac{\omega_n^2}{s^2 + 2\zeta\omega_n s + \omega_n^2}$$

show that the bandwidth ω_b is given by

$$\omega_b = \omega_n\left(1 - 2\zeta^2 + \sqrt{4\zeta^4 - 4\zeta^2 + 2}\right)^{1/2}$$

Note that ω_b/ω_n is a function only of ζ. Plot a curve of ω_b/ω_n versus ζ.

Solution. The bandwidth ω_b is determined from $|C(j\omega_b)/R(j\omega_b)| = -3$ dB. Quite often, instead of -3 dB, we use -3.01 dB, which is equal to 0.707. Thus,

$$\left|\frac{C(j\omega_b)}{R(j\omega_b)}\right| = \left|\frac{\omega_n^2}{(j\omega_b)^2 + 2\zeta\omega_n(j\omega_b) + \omega_n^2}\right| = 0.707$$

Then

$$\frac{\omega_n^2}{\sqrt{(\omega_n^2 - \omega_b^2)^2 + (2\zeta\omega_n\omega_b)^2}} = 0.707$$

from which we get

$$\omega_n^4 = 0.5\left[(\omega_n^2 - \omega_b^2)^2 + 4\zeta^2\omega_n^2\omega_b^2\right]$$

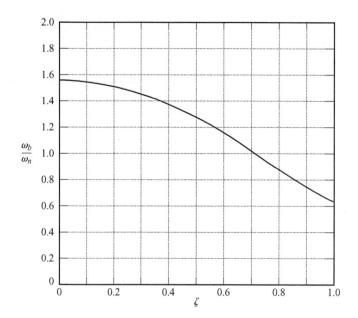

Figure 7–134
Curve of ω_b/ω_n versus ζ, where ω_b is the bandwidth.

By dividing both sides of this last equation by ω_n^4, we obtain

$$1 = 0.5\left\{\left[1 - \left(\frac{\omega_b}{\omega_n}\right)^2\right]^2 + 4\zeta^2\left(\frac{\omega_b}{\omega_n}\right)^2\right\}$$

Solving this last equation for $(\omega_b/\omega_n)^2$ yields

$$\left(\frac{\omega_b}{\omega_n}\right)^2 = -2\zeta^2 + 1 \pm \sqrt{4\zeta^4 - 4\zeta^2 + 2}$$

Since $(\omega_b/\omega_n)^2 > 0$, we take the plus sign in this last equation. Then

$$\omega_b^2 = \omega_n^2\left(1 - 2\zeta^2 + \sqrt{4\zeta^4 - 4\zeta^2 + 2}\right)$$

or

$$\omega_b = \omega_n\left(1 - 2\zeta^2 + \sqrt{4\zeta^4 - 4\zeta^2 + 2}\right)^{1/2}$$

Figure 7–134 shows a curve relating ω_b/ω_n versus ζ.

A–7–18. A Bode diagram of the open-loop transfer function $G(s)$ of a unity-feedback control system is shown in Figure 7–135. It is known that the open-loop transfer function is minimum phase. From the diagram, it can be seen that there is a pair of complex-conjugate poles at $\omega = 2$ rad/sec. Determine the damping ratio of the quadratic term involving these complex-conjugate poles. Also, determine the transfer function $G(s)$.

Solution. Referring to Figure 7–9 and examining the Bode diagram of Figure 7–135, we find the damping ratio ζ and undamped natural frequency ω_n of the quadratic term to be

$$\zeta = 0.1, \qquad \omega_n = 2 \text{ rad/sec}$$

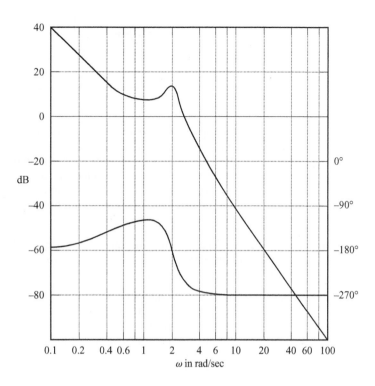

Figure 7–135
Bode diagram of the
open-loop transfer
function of a unity-
feedback control
system.

Noting that there is another corner frequency at $\omega = 0.5$ rad/sec and the slope of the magnitude curve in the low-frequency region is -40 dB/decade, $G(j\omega)$ can be tentatively determined as follows:

$$G(j\omega) = \frac{K\left(\dfrac{j\omega}{0.5} + 1\right)}{(j\omega)^2\left[\left(\dfrac{j\omega}{2}\right)^2 + 0.1(j\omega) + 1\right]}$$

Since, from Figure 7–135 we find $|G(j0.1)| = 40$ dB, the gain value K can be determined to be unity. Also, the calculated phase curve, $\underline{/G(j\omega)}$ versus ω, agrees with the given phase curve. Hence, the transfer function $G(s)$ can be determined to be

$$G(s) = \frac{4(2s + 1)}{s^2(s^2 + 0.4s + 4)}$$

A–7–19. A closed-loop control system may include an unstable element within the loop. When the Nyquist stability criterion is to be applied to such a system, the frequency-response curves for the unstable element must be obtained.

How can we obtain experimentally the frequency-response curves for such an unstable element? Suggest a possible approach to the experimental determination of the frequency response of an unstable linear element.

Solution. One possible approach is to measure the frequency-response characteristics of the unstable element by using it as a part of a stable system.

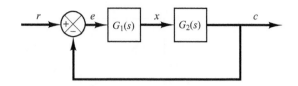

Figure 7–136
Control system.

Consider the system shown in Figure 7–136. Suppose that the element $G_1(s)$ is unstable. The complete system may be made stable by choosing a suitable linear element $G_2(s)$. We apply a sinusoidal signal at the input. At steady state, all signals in the loop will be sinusoidal. We measure the signals $e(t)$, the input to the unstable element, and $x(t)$, the output of the unstable element. By changing the frequency [and possibly the amplitude for the convenience of measuring $e(t)$ and $x(t)$] of the input sinusoid and repeating this process, it is possible to obtain the frequency response of the unstable linear element.

A–7–20. Show that the lead network and lag network inserted in cascade in an open loop act as proportional-plus-derivative control (in the region of small ω) and proportional-plus-integral control (in the region of large ω), respectively.

Solution. In the region of small ω, the polar plot of the lead network is approximately the same as that of the proportional-plus-derivative controller. This is shown in Figure 7–137(a).

Similarly, in the region of large ω, the polar plot of the lag network approximates the proportional-plus-integral controller, as shown in Figure 7–137(b).

A–7–21. Consider a lag–lead compensator $G_c(s)$ defined by

$$G_c(s) = K_c \frac{\left(s + \dfrac{1}{T_1}\right)\left(s + \dfrac{1}{T_2}\right)}{\left(s + \dfrac{\beta}{T_1}\right)\left(s + \dfrac{1}{\beta T_2}\right)}$$

Show that at frequency ω_1, where

$$\omega_1 = \frac{1}{\sqrt{T_1 T_2}}$$

the phase angle of $G_c(j\omega)$ becomes zero. (This compensator acts as a lag compensator for $0 < \omega < \omega_1$ and acts as a lead compensator for $\omega_1 < \omega < \infty$.) (Refer to Figure 7–109.)

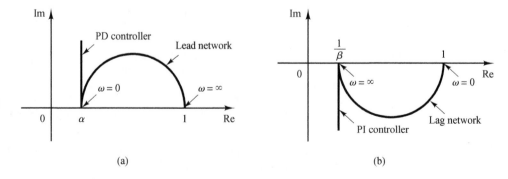

Figure 7–137
(a) Polar plots of a lead network and a proportional-plus-derivative controller;
(b) polar plots of a lag network and a proportional-plus-integral controller.

Solution. The angle of $G_c(j\omega)$ is given by

$$\underline{/G_c(j\omega)} = \underline{/j\omega + \frac{1}{T_1}} + \underline{/j\omega + \frac{1}{T_2}} - \underline{/j\omega + \frac{\beta}{T_1}} - \underline{/j\omega + \frac{1}{\beta T_2}}$$

$$= \tan^{-1}\omega T_1 + \tan^{-1}\omega T_2 - \tan^{-1}\omega T_1/\beta - \tan^{-1}\omega T_2\beta$$

At $\omega = \omega_1 = 1/\sqrt{T_1 T_2}$, we have

$$\underline{/G_c(j\omega_1)} = \tan^{-1}\sqrt{\frac{T_1}{T_2}} + \tan^{-1}\sqrt{\frac{T_2}{T_1}} - \tan^{-1}\frac{1}{\beta}\sqrt{\frac{T_1}{T_2}} - \tan^{-1}\beta\sqrt{\frac{T_2}{T_1}}$$

Since

$$\tan\left(\tan^{-1}\sqrt{\frac{T_1}{T_2}} + \tan^{-1}\sqrt{\frac{T_2}{T_1}}\right) = \frac{\sqrt{\dfrac{T_1}{T_2}} + \sqrt{\dfrac{T_2}{T_1}}}{1 - \sqrt{\dfrac{T_1}{T_2}}\sqrt{\dfrac{T_2}{T_1}}} = \infty$$

or

$$\tan^{-1}\sqrt{\frac{T_1}{T_2}} + \tan^{-1}\sqrt{\frac{T_2}{T_1}} = 90°$$

and also

$$\tan^{-1}\frac{1}{\beta}\sqrt{\frac{T_1}{T_2}} + \tan^{-1}\beta\sqrt{\frac{T_2}{T_1}} = 90°$$

we have

$$\underline{/G_c(j\omega_1)} = 0°$$

Thus, the angle of $G_c(j\omega_1)$ becomes $0°$ at $\omega = \omega_1 = 1/\sqrt{T_1 T_2}$.

A–7–22. Consider the control system shown in Figure 7–138. Determine the value of gain K such that the phase margin is $60°$. What is the gain margin with this value of gain K?

Solution. The open-loop transfer function is

$$G(s) = K\frac{s + 0.1}{s + 0.5}\frac{10}{s(s + 1)}$$

$$= \frac{K(10s + 1)}{s^3 + 1.5s^2 + 0.5s}$$

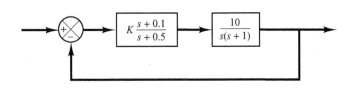

Figure 7–138
Control system.

Let us plot the Bode diagram of $G(s)$ when $K = 1$. MATLAB Program 7–22 may be used for this purpose. Figure 7–139 shows the Bode diagram produced by this program. From this diagram the required phase margin of 60° occurs at the frequency $\omega = 1.15$ rad/sec. The magnitude of $G(j\omega)$ at this frequency is found to be 14.5 dB. Then gain K must satisfy the following equation:

$$20 \log K = -14.5 \text{ dB}$$

or

$$K = 0.188$$

MATLAB Program 7–22

```
num = [10 1];
den = [1  1.5  0.5  0];
bode(num,den)
title('Bode Diagram of G(s) = (10s + 1)/[s(s + 0.5)(s + 1)]')
```

Thus, we have determined the value of gain K. Since the angle curve does not cross the $-180°$ line, the gain margin is $+\infty$ dB.

To verify the results, let us draw a Nyquist plot of G for the frequency range

$$w = 0.5:0.01:1.15$$

The end point of the locus ($\omega = 1.15$ rad/sec) will be on a unit circle in the Nyquist plane. To check the phase margin, it is convenient to draw the Nyquist plot on a polar diagram, using polar grids.

To draw the Nyquist plot on a polar diagram, first define a complex vector z by

$$z = \text{re} + i*\text{im} = \text{re}^{i\theta}$$

where r and θ (theta) are given by

$$r = \text{abs}(z)$$
$$\text{theta} = \text{angle}(z)$$

The abs means the square root of the sum of the real part squared and imaginary part squared; angle means $\tan^{-1}$ (imaginary part/real part).

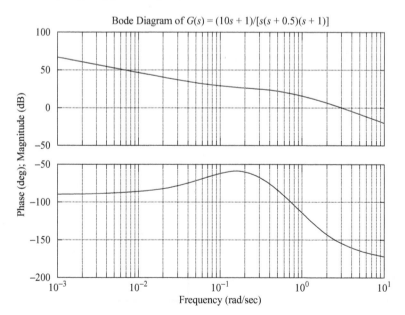

Figure 7–139
Bode diagram of
$$G(s) = \frac{10s + 1}{s(s + 0.5)(s + 1)}.$$

If we use the command

$$\text{polar(theta,r)}$$

MATLAB will produce a plot in the polar coordinates. Subsequent use of the grid command draws polar grid lines and grid circles.

MATLAB Program 7–23 produces the Nyquist plot of $G(j\omega)$, where ω is between 0.5 and 1.15 rad/sec. The resulting plot is shown in Figure 7–140. Notice that point $G(j1.15)$ lies on the unit

---

MATLAB Program 7–23

```
%*****Nyquist plot in rectangular coordinates*****

num = [1.88  0.188];
den = [1  1.5  0.5  0];
w = 0.5:0.01:1.15;
[re,im,w] = nyquist(num,den,w);

%*****Convert rectangular coordinates into polar coordinates
% by defining z, r, theta as follows*****

z = re + i*im;
r = abs(z);
theta = angle(z);

%*****To draw polar plot, enter command 'polar(theta,r)'*****

polar(theta,r)
text(-1,3,'Check of Phase Margin')
text(0.3,-1.7,'Nyquist plot')
text(-2.2,-0.75,'Phase margin')
text(-2.2,-1.1,'is 60 degrees')
text(1.45,-0.7,'Unit circle')
```

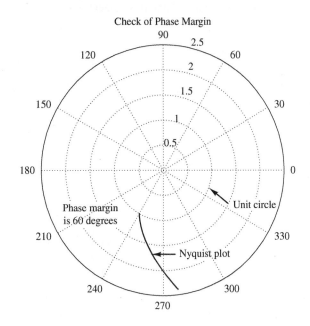

Figure 7–140
Nyquist plot of $G(j\omega)$ showing that the phase margin is 60°.

circle, and the phase angle of this point is $-120°$. Hence, the phase margin is $60°$. The fact that point $G(j1.15)$ is on the unit circle verifies that at $\omega = 1.15$ rad/sec the magnitude is equal to 1 or 0 dB. (Thus, $\omega = 1.15$ is the gain crossover frequency.) Thus, $K = 0.188$ gives the desired phase margin of $60°$.

Note that in writing 'text' in the polar diagram we enter the text command as follows:

$$text(x,y,'\ ')$$

For example, to write 'Nyquist plot' starting at point $(0.3, -1.7)$, enter the command

$$text(0.3, -1.7,'Nyquist plot')$$

The text is written horizontally on the screen.

A–7–23. If the open-loop transfer function $G(s)$ involves lightly damped complex-conjugant poles, then more than one M locus may be tangent to the $G(j\omega)$ locus.

Consider the unity-feedback system whose open-loop transfer function is

$$G(s) = \frac{9}{s(s + 0.5)(s^2 + 0.6s + 10)} \tag{7–32}$$

Draw the Bode diagram for this open-loop transfer function. Draw also the log-magnitude-versus-phase plot, and show that two M loci are tangent to the $G(j\omega)$ locus. Finally, plot the Bode diagram for the closed-loop transfer function.

Solution. Figure 7–141 shows the Bode diagram of $G(j\omega)$. Figure 7–142 shows the log-magnitude-versus-phase plot of $G(j\omega)$. It is seen that the $G(j\omega)$ locus is tangent to the $M = 8$-dB locus at $\omega = 0.97$ rad/sec, and it is tangent to the $M = -4$-dB locus at $\omega = 2.8$ rad/sec.

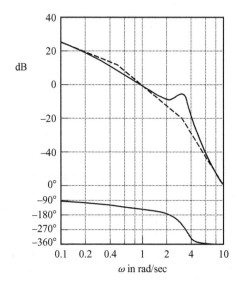

Figure 7–141
Bode diagram of $G(s)$ given by Equation (7–32).

546   Chapter 7 / Control Systems Analysis and Design by the Frequency-Response Method

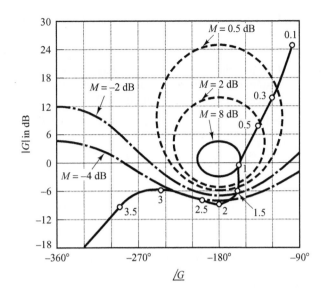

Figure 7–142
Log-magnitude-versus-phase plot of $G(s)$ given by Equation (7–32).

Figure 7–143 shows the Bode diagram of the closed-loop transfer function. The magnitude curve of the closed-loop frequency response shows two resonant peaks. Note that such a case occurs when the closed-loop transfer function involves the product of two lightly damped second-order terms and the two corresponding resonant frequencies are sufficiently separated from each other. As a matter of fact, the closed-loop transfer function of this system can be written

$$\frac{C(s)}{R(s)} = \frac{G(s)}{1 + G(s)}$$

$$= \frac{9}{(s^2 + 0.487s + 1)(s^2 + 0.613s + 9)}$$

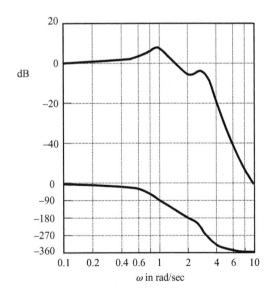

Figure 7–143
Bode diagram of $G(s)/[1 + G(s)]$, where $G(s)$ is given by Equation (7–32).

Clearly, the denominator of the closed-loop transfer function is a product of two lightly damped second-order terms (the damping ratios are 0.243 and 0.102), and the two resonant frequencies are sufficiently separated.

A–7–24. Consider the system shown in Figure 7–144(a). Design a compensator such that the closed-loop system will satisfy the requirements that the static velocity error constant $= 20 \text{ sec}^{-1}$, phase margin $= 50°$, and gain margin $\geq 10 \text{ dB}$.

Solution. To satisfy the requirements, we shall try a lead compensator $G_c(s)$ of the form

$$G_c(s) = K_c \alpha \frac{Ts + 1}{\alpha Ts + 1}$$

$$= K_c \frac{s + \dfrac{1}{T}}{s + \dfrac{1}{\alpha T}}$$

(If the lead compensator does not work, then we need to employ a compensator of different form.) The compensated system is shown in Figure 7–144(b).

Define

$$G_1(s) = KG(s) = \frac{10K}{s(s + 1)}$$

where $K = K_c \alpha$. The first step in the design is to adjust the gain K to meet the steady-state performance specification or to provide the required static velocity error constant. Since the static velocity error constant K_v is given as 20 sec^{-1}, we have

$$K_v = \lim_{s \to 0} sG_c(s)G(s)$$

$$= \lim_{s \to 0} s \frac{Ts + 1}{\alpha Ts + 1} G_1(s)$$

$$= \lim_{s \to 0} \frac{s10K}{s(s + 1)}$$

$$= 10K = 20$$

or

$$K = 2$$

With $K = 2$, the compensated system will satisfy the steady-state requirement.
We shall next plot the Bode diagram of

$$G_1(s) = \frac{20}{s(s + 1)}$$

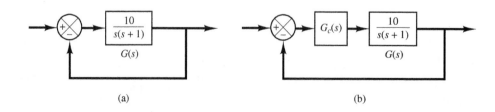

Figure 7–144
(a) Control system;
(b) compensated system.

(a)

(b)

Chapter 7 / Control Systems Analysis and Design by the Frequency-Response Method

MATLAB Program 7–24 produces the Bode diagram shown in Figure 7–145. From this plot, the phase margin is found to be 14°. The gain margin is +∞ dB.

MATLAB Program 7–24

```
num = [20];
den = [1  1  0];
w = logspace(-1,2,100);
bode(num,den,w)
title('Bode Diagram of G1(s) = 20/[s(s + 1)]')
```

Since the specification calls for a phase margin of 50°, the additional phase lead necessary to satisfy the phase-margin requirement is 36°. A lead compensator can contribute this amount.

Noting that the addition of a lead compensator modifies the magnitude curve in the Bode diagram, we realize that the gain crossover frequency will be shifted to the right. We must offset the increased phase lag of $G_1(j\omega)$ due to this increase in the gain crossover frequency. Taking the shift of the gain crossover frequency into consideration, we may assume that ϕ_m, the maximum phase lead required, is approximately 41°. (This means that approximately 5° has been added to compensate for the shift in the gain crossover frequency.) Since

$$\sin \phi_m = \frac{1 - \alpha}{1 + \alpha}$$

$\phi_m = 41°$ corresponds to $\alpha = 0.2077$. Note that $\alpha = 0.21$ corresponds to $\phi_m = 40.76°$. Whether we choose $\phi_m = 41°$ or $\phi_m = 40.76°$ does not make much difference in the final solution. Hence, let us choose $\alpha = 0.21$.

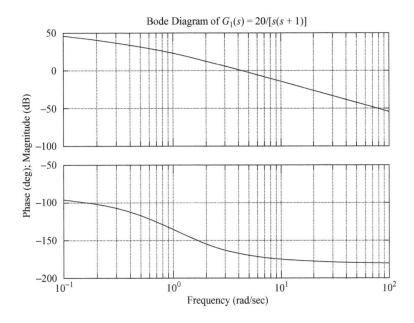

Figure 7–145
Bode diagram of $G_1(s)$.

Once the attenuation factor α has been determined on the basis of the required phase-lead angle, the next step is to determine the corner frequencies $\omega = 1/T$ and $\omega = 1/(\alpha T)$ of the lead compensator. Notice that the maximum phase-lead angle ϕ_m occurs at the geometric mean of the two corner frequencies, or $\omega = 1/(\sqrt{\alpha} T)$.

The amount of the modification in the magnitude curve at $\omega = 1/(\sqrt{\alpha} T)$ due to the inclusion of the term $(Ts + 1)/(\alpha Ts + 1)$ is

$$\left| \frac{1 + j\omega T}{1 + j\omega\alpha T} \right|_{\omega=\frac{1}{\sqrt{\alpha}T}} = \left| \frac{1 + j\dfrac{1}{\sqrt{\alpha}}}{1 + j\alpha \dfrac{1}{\sqrt{\alpha}}} \right| = \frac{1}{\sqrt{\alpha}}$$

Note that

$$\frac{1}{\sqrt{\alpha}} = \frac{1}{\sqrt{0.21}} = 6.7778 \text{ dB}$$

We need to find the frequency point where, when the lead compensator is added, the total magnitude becomes 0 dB.

From Figure 7–145 we see that the frequency point where the magnitude of $G_1(j\omega)$ is -6.7778 dB occurs between $\omega = 1$ and 10 rad/sec. Hence, we plot a new Bode diagram of $G_1(j\omega)$ in the frequency range between $\omega = 1$ and 10 to locate the exact point where $G_1(j\omega) = -6.7778$ dB. MATLAB Program 7–25 produces the Bode diagram in this frequency range, which is shown in Figure 7–146. From this diagram, we find the frequency point where $|G_1(j\omega)| = -6.7778$ dB occurs at $\omega = 6.5686$ rad/sec. Let us select this frequency to be the new gain crossover frequency, or $\omega_c = 6.5686$ rad/sec. Noting that this frequency corresponds to $1/(\sqrt{\alpha} T)$, or

$$\omega_c = \frac{1}{\sqrt{\alpha} T}$$

we obtain

$$\frac{1}{T} = \omega_c \sqrt{\alpha} = 6.5686 \sqrt{0.21} = 3.0101$$

and

$$\frac{1}{\alpha T} = \frac{\omega_c}{\sqrt{\alpha}} = \frac{6.5686}{\sqrt{0.21}} = 14.3339$$

MATLAB Program 7–25

```
num = [20];
den = [1  1  0];
w = logspace(0,1,100);
bode(num,den,w)
title('Bode Diagram of G1(s) = 20/[s(s + 1)]')
```

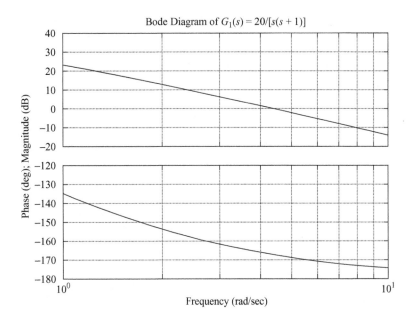

Figure 7–146
Bode diagram of $G_1(s)$.

The lead compensator thus determined is

$$G_c(s) = K_c \frac{s + 3.0101}{s + 14.3339} = K_c \alpha \frac{0.3322s + 1}{0.06976s + 1}$$

where K_c is determined as

$$K_c = \frac{K}{\alpha} = \frac{2}{0.21} = 9.5238$$

Thus, the transfer function of the compensator becomes

$$G_c(s) = 9.5238 \frac{s + 3.0101}{s + 14.3339} = 2 \frac{0.3322s + 1}{0.06976s + 1}$$

MATLAB Program 7–26 produces the Bode diagram of this lead compensator, which is shown in Figure 7–147.

MATLAB Program 7–26

numc = [9.5238 28.6676];
denc = [1 14.3339];
w = logspace(-1,3,100);
bode(numc,denc,w)
title('Bode Diagram of Gc(s) = 9.5238(s + 3.0101)/(s + 14.3339')

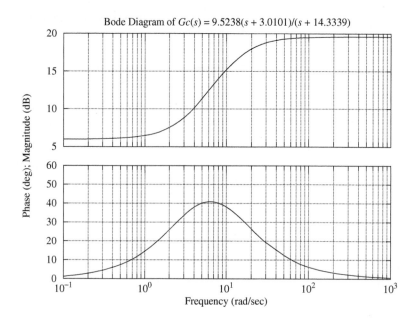

Figure 7–147
Bode diagram of $G_c(s)$.

The open-loop transfer function of the designed system is

$$G_c(s)G(s) = 9.5238 \frac{s + 3.0101}{s + 14.3339} \frac{10}{s(s + 1)}$$

$$= \frac{95.238s + 286.6759}{s^3 + 15.3339s^2 + 14.3339s}$$

MATLAB Program 7–27 will produce the Bode diagram of $G_c(s)G(s)$, which is shown in Figure 7–148.

MATLAB Program 7–27

```
num = [95.238  286.6759];
den = [1  15.3339  14.3339  0];
sys = tf(num,den);
w = logspace(−1,3,100);
bode(sys,w);
grid;
title('Bode Diagram of Gc(s)G(s)')
[Gm,pm,wcp,wcg] = margin(sys);
GmdB = 20*log10(Gm);
[Gmdb,pm,wcp,wcg]
ans =
     Inf   49.4164   Inf   6.5686
```

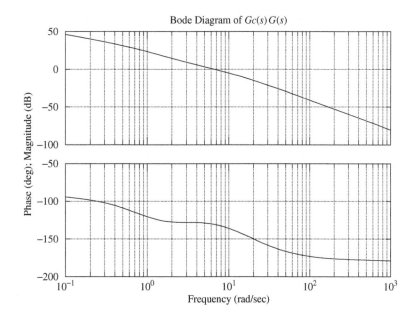

Figure 7–148
Bode diagram of $G_c(s)G(s)$.

From MATLAB Program 7–27 and Figure 7–148 it is clearly seen that the phase margin is approximately 50° and the gain margin is $+\infty$ dB. Since the static velocity error constant K_v is 20 sec^{-1}, all the specifications are met. Before we conclude this problem, we need to check the transient-response characteristics.

Unit-Step Response: We shall compare the unit-step response of the compensated system with that of the original uncompensated system.
    The closed-loop transfer function of the original uncompensated system is

$$\frac{C(s)}{R(s)} = \frac{10}{s^2 + s + 10}$$

The closed-loop transfer function of the compensated system is

$$\frac{C(s)}{R(s)} = \frac{95.238s + 286.6759}{s^3 + 15.3339s^2 + 110.5719s + 286.6759}$$

MATLAB Program 7–28 produces the unit-step responses of the uncompensated and compensated systems. The resulting response curves are shown in Figure 7–149. Clearly, the compensated system exhibits a satisfactory response. Note that the closed-loop zero and poles are located as follows:

$$\text{Zero at } s = -3.0101$$

$$\text{Poles at } s = -5.2880 \pm j5.6824, \quad s = -4.7579$$

Unit-Ramp Response: It is worthwhile to check the unit-ramp response of the compensated system. Since $K_v = 20$ sec^{-1}, the steady-state error following the unit-ramp input will be

$1/K_v = 0.05$. The static velocity error constant of the uncompensated system is 10 sec^{-1}. Hence, the original uncompensated system will have twice as large a steady-state error in following the unit-ramp input.

MATLAB Program 7–28

```
%*****Unit-step responses*****

num1 = [10];
den1 = [1  1  10];
num2 = [95.238  286.6759];
den2 = [1  15.3339  110.5719  286.6759];
t = 0:0.01:6;
[c1,x1,t] = step(num1,den1,t);
[c2,x2,t] = step(num2,den2,t);
plot(t,c1,'.',t,c2,'-')
grid;
title('Unit-Step Responses of Uncompensated System and Compensated System')
xlabel('t Sec');
ylabel('Outputs')
text(1.70,1.45,'Uncompensated System')
text(1.1,0.5,'Compensated System')
```

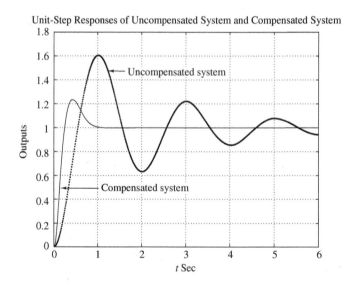

Figure 7–149
Unit-step responses
of the uncompensated
and compensated
systems.

MATLAB Program 7–29 produces the unit-ramp response curves. [Note that the unit-ramp response is obtained as the unit-step response of $C(s)/sR(s)$.] The resulting curves are shown in Figure 7–150. The compensated system has a steady-state error equal to one-half that of the original uncompensated system.

MATLAB Program 7–29

```
%*****Unit-ramp responses*****

num1 = [10];
den1 = [1 1 10 0];
num2 = [95.238 286.6759];
den2 = [1 15.3339 110.5719 286.6759 0];
t = 0:0.01:3;
[c1,x1,t] = step(num1,den1,t);
[c2,x2,t] = step(num2,den2,t);
plot(t,c1,'.',t,c2,'-',t,t,'--');
grid;
title('Unit-Ramp Responses of Uncompensated System and Compensated System');
xlabel('t Sec');
ylabel('Outputs')
text(1.2,0.65,'Uncompensated System')
text(0.1,1.3,'Compensated System')
```

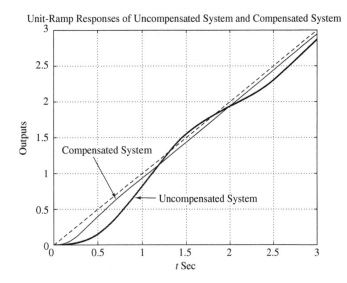

Figure 7–150
Unit-ramp responses
of the uncompensated
and compensated
systems.

A–7–25. Consider a unity-feedback system whose open-loop transfer function is

$$G(s) = \frac{K}{s(s + 1)(s + 4)}$$

Design a lag–lead compensator $G_c(s)$ such that the static velocity error constant is 10 sec^{-1}, the phase margin is 50°, and the gain margin is 10 dB or more.

Solution. We shall design a lag–lead compensator of the form

$$G_c(s) = K_c \frac{\left(s + \dfrac{1}{T_1}\right)\left(s + \dfrac{1}{T_2}\right)}{\left(s + \dfrac{\beta}{T_1}\right)\left(s + \dfrac{1}{\beta T_2}\right)}$$

Then the open-loop transfer function of the compensated system is $G_c(s)G(s)$. Since the gain K of the plant is adjustable, let us assume that $K_c = 1$. Then $\lim_{s \to 0} G_c(s) = 1$. From the requirement on the static velocity error constant, we obtain

$$K_v = \lim_{s \to 0} sG_c(s)G(s) = \lim_{s \to 0} sG_c(s) \frac{K}{s(s+1)(s+4)}$$

$$= \frac{K}{4} = 10$$

Hence,

$$K = 40$$

We shall first plot a Bode diagram of the uncompensated system with $K = 40$. MATLAB Program 7–30 may be used to plot this Bode diagram. The diagram obtained is shown in Figure 7–151.

MATLAB Program 7–30

```
num = [40];
den = [1  5  4  0];
w = logspace(-1,1,100);
bode(num,den,w)
title('Bode Diagram of G(s) = 40/[s(s + 1)(s + 4)]')
```

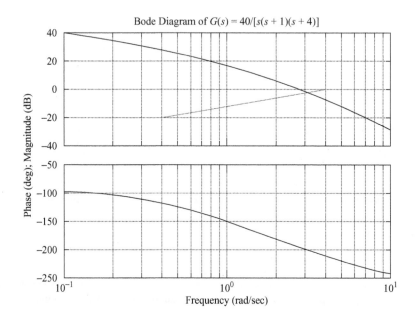

Figure 7–151
Bode diagram of
$G(s) = 40/[s(s + 1)(s + 4)]$.

From Figure 7–151, the phase margin of the gain-adjusted but uncompensated system is found to be −16°, which indicates that this system is unstable. The next step in the design of a lag–lead compensator is to choose a new gain crossover frequency. From the phase-angle curve for $G(j\omega)$, we notice that the phase crossover frequency is $\omega = 2$ rad/sec. We may choose the new gain crossover frequency to be 2 rad/sec so that the phase-lead angle required at $\omega = 2$ rad/sec is about 50°. A single lag–lead compensator can provide this amount of phase-lead angle quite easily.

Once we choose the gain crossover frequency to be 2 rad/sec, we can determine the corner frequencies of the phase-lag portion of the lag–lead compensator. Let us choose the corner frequency $\omega = 1/T_2$ (which corresponds to the zero of the phase-lag portion of the compensator) to be 1 decade below the new gain crossover frequency, or at $\omega = 0.2$ rad/sec. For another corner frequency $\omega = 1/(\beta T_2)$, we need the value of β. The value of β can be determined from the consideration of the lead portion of the compensator, as shown next.

For the lead compensator, the maximum phase-lead angle ϕ_m is given by

$$\sin\phi_m = \frac{\beta - 1}{\beta + 1}$$

Notice that $\beta = 10$ corresponds to $\phi_m = 54.9°$. Since we need a 50° phase margin, we may choose $\beta = 10$. (Note that we will be using several degrees less than the maximum angle, 54.9°.) Thus,

$$\beta = 10$$

Then the corner frequency $\omega = 1/(\beta T_2)$ (which corresponds to the pole of the phase-lag portion of the compensator) becomes

$$\omega = 0.02$$

The transfer function of the phase-lag portion of the lag–lead compensator becomes

$$\frac{s + 0.2}{s + 0.02} = 10\left(\frac{5s + 1}{50s + 1}\right)$$

The phase-lead portion can be determined as follows: Since the new gain crossover frequency is $\omega = 2$ rad/sec, from Figure 7–151, $|G(j2)|$ is found to be 6 dB. Hence, if the lag–lead compensator contributes −6 dB at $\omega = 2$ rad/sec, then the new gain crossover frequency is as desired. From this requirement, it is possible to draw a straight line of slope 20 dB/decade passing through the point (2 rad/sec, −6 dB). (Such a line has been manually drawn on Figure 7–151.) The intersections of this line and the 0-dB line and −20-dB line determine the corner frequencies. From this consideration, the corner frequencies for the lead portion can be determined as $\omega = 0.4$ rad/sec and $\omega = 4$ rad/sec. Thus, the transfer function of the lead portion of the lag–lead compensator becomes

$$\frac{s + 0.4}{s + 4} = \frac{1}{10}\left(\frac{2.5s + 1}{0.25s + 1}\right)$$

Combining the transfer functions of the lag and lead portions of the compensator, we can obtain the transfer function $G_c(s)$ of the lag–lead compensator. Since we chose $K_c = 1$, we have

$$G_c(s) = \frac{s + 0.4}{s + 4}\frac{s + 0.2}{s + 0.02} = \frac{(2.5s + 1)(5s + 1)}{(0.25s + 1)(50s + 1)}$$

Example Problems and Solutions

557

The Bode diagram of the lag–lead compensator $G_c(s)$ can be obtained by entering MATLAB Program 7–31 into the computer. The resulting plot is shown in Figure 7–152.

MATLAB Program 7–31

```
numc = [1  0.6  0.08];
denc = [1  4.02  0.08];
bode(numc,denc)
title('Bode Diagram of Lag–Lead Compensator')
```

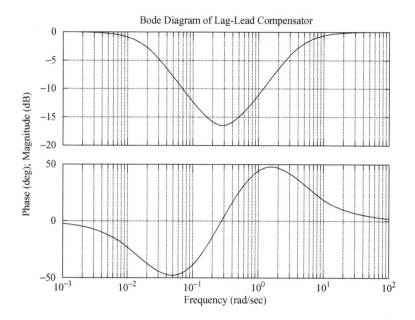

Figure 7–152
Bode diagram of the designed lag–lead compensator.

The open-loop transfer function of the compensated system is

$$G_c(s)G(s) = \frac{(s + 0.4)(s + 0.2)}{(s + 4)(s + 0.02)} \frac{40}{s(s + 1)(s + 4)}$$

$$= \frac{40s^2 + 24s + 3.2}{s^5 + 9.02s^4 + 24.18s^3 + 16.48s^2 + 0.32s}$$

Using MATLAB Program 7–32 the magnitude and phase-angle curves of the designed open-loop transfer function $G_c(s)G(s)$ can be obtained as shown in Figure 7–153. Note that the denominator polynomial den1 was obtained using the conv command, as follows:

```
a = [1  4.02  0.08];
b = [1  5  4  0];
conv(a,b)

ans =

        1.0000  9.0200  24.1800  16.4800  0.320000  0
```

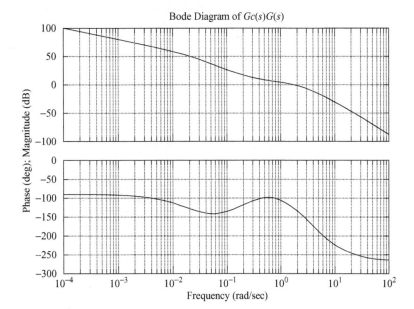

MATLAB Program 7–32

```
num1 = [40  24  3.2];
den1 = [1  9.02  24.18  16.48  0.32  0];
bode(num1,den1)
title('Bode Diagram of Gc(s)G(s)')
```

Figure 7–153
Bode diagram of the
open-loop transfer
function $G_c(s)G(s)$
of the compensated
system.

Since the phase margin of the compensated system is 50°, the gain margin is 12 dB, and the static velocity error constant is 10 sec^{-1}, all the requirements are met.

We shall next investigate the transient-response characteristics of the designed system.

Unit-Step Response: Noting that

$$G_c(s)G(s) = \frac{40(s + 0.4)(s + 0.2)}{(s + 4)(s + 0.02)s(s + 1)(s + 4)}$$

we have

$$\frac{C(s)}{R(s)} = \frac{G_c(s)G(s)}{1 + G_c(s)G(s)}$$

$$= \frac{40(s + 0.4)(s + 0.2)}{(s + 4)(s + 0.02)s(s + 1)(s + 4) + 40(s + 0.4)(s + 0.2)}$$

To determine the denominator polynomial with MATLAB, we may proceed as follows:
Define

$$a(s) = (s + 4)(s + 0.02) = s^2 + 4.02s + 0.08$$

$$b(s) = s(s + 1)(s + 4) = s^3 + 5s^2 + 4s$$

$$c(s) = 40(s + 0.4)(s + 0.2) = 40s^2 + 24s + 3.2$$

Example Problems and Solutions

559

Then we have

$$a = [1 \quad 4.02 \quad 0.08]$$
$$b = [1 \quad 5 \quad 4 \quad 0]$$
$$c = [40 \quad 24 \quad 3.2]$$

Using the following MATLAB program, we obtain the denominator polynomial.

```
a = [1  4.02  0.08];
b = [1  5  4  0];
c = [40  24  3.2];
p = [conv(a,b)] + [0  0  0  c]
p =
    1.0000  9.0200  24.1800  56.4800  24.3200  3.2000
```

MATLAB Program 7–33 is used to obtain the unit-step response of the compensated system. The resulting unit-step response curve is shown in Figure 7–154. (Note that the gain-adjusted but uncompensated system is unstable.)

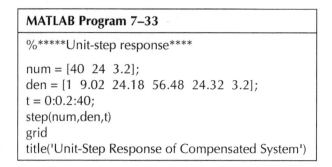

MATLAB Program 7–33

```
%*****Unit-step response****

num = [40  24  3.2];
den = [1  9.02  24.18  56.48  24.32  3.2];
t = 0:0.2:40;
step(num,den,t)
grid
title('Unit-Step Response of Compensated System')
```

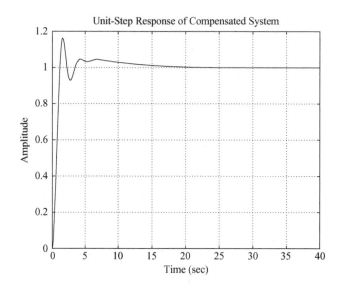

Figure 7–154
Unit-step response curve of the compensated system.

Unit-Ramp Response: The unit-ramp response of the compensated system may be obtained by entering MATLAB Program 7–34 into the computer. Here we converted the unit-ramp response of $G_cG/(1 + G_cG)$ into the unit-step response of $G_cG/[s(1 + G_cG)]$. The unit-ramp response curve obtained using this program is shown in Figure 7–155.

MATLAB Program 7–34

```
%*****Unit-ramp response*****

num = [40  24  3.2];
den = [1  9.02  24.18  56.48  24.32  3.2  0];
t = 0:0.05:20;
c = step(num,den,t);
plot(t,c,'-',t,t,'.')
grid
title('Unit-Ramp Response of Compensated System')
xlabel('Time (sec)')
ylabel('Unit-Ramp Input and Output c(t)')
```

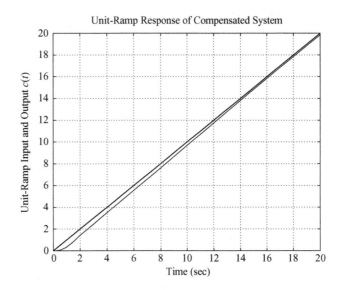

Figure 7–155
Unit-ramp response
of the compensated
system.

PROBLEMS

B–7–1. Consider the unity-feedback system with the open-loop transfer function:

$$G(s) = \frac{10}{s + 1}$$

Obtain the steady-state output of the system when it is subjected to each of the following inputs:

(a) $r(t) = \sin(t + 30°)$

(b) $r(t) = 2\cos(2t - 45°)$

(c) $r(t) = \sin(t + 30°) - 2\cos(2t - 45°)$

B–7–2. Consider the system whose closed-loop transfer function is

$$\frac{C(s)}{R(s)} = \frac{K(T_2 s + 1)}{T_1 s + 1}$$

Obtain the steady-state output of the system when it is subjected to the input $r(t) = R \sin \omega t$.

B–7–3. Using MATLAB, plot Bode diagrams of $G_1(s)$ and $G_2(s)$ given below.

$$G_1(s) = \frac{1 + s}{1 + 2s}$$

$$G_2(s) = \frac{1 - s}{1 + 2s}$$

$G_1(s)$ is a minimum-phase system and $G_2(s)$ is a nonminimum-phase system.

B–7–4. Plot the Bode diagram of

$$G(s) = \frac{10(s^2 + 0.4s + 1)}{s(s^2 + 0.8s + 9)}$$

B–7–5. Given

$$G(s) = \frac{\omega_n^2}{s^2 + 2\zeta\omega_n s + \omega_n^2}$$

show that

$$|G(j\omega_n)| = \frac{1}{2\zeta}$$

B–7–6. Consider a unity-feedback control system with the following open-loop transfer function:

$$G(s) = \frac{s + 0.5}{s^3 + s^2 + 1}$$

This is a nonminimum-phase system. Two of the three open-loop poles are located in the right-half s plane as follows:

Open-loop poles at $s = -1.4656$

$$s = 0.2328 + j0.7926$$

$$s = 0.2328 - j0.7926$$

Plot the Bode diagram of $G(s)$ with MATLAB. Explain why the phase-angle curve starts from $0°$ and approaches $+180°$.

B–7–7. Sketch the polar plots of the open-loop transfer function

$$G(s)H(s) = \frac{K(T_a s + 1)(T_b s + 1)}{s^2(Ts + 1)}$$

for the following two cases:

(a) $T_a > T > 0,$ $T_b > T > 0$

(b) $T > T_a > 0,$ $T > T_b > 0$

B–7–8. Draw a Nyquist locus for the unity-feedback control system with the open-loop transfer function

$$G(s) = \frac{K(1 - s)}{s + 1}$$

Using the Nyquist stability criterion, determine the stability of the closed-loop system.

B–7–9. A system with the open-loop transfer function

$$G(s)H(s) = \frac{K}{s^2(T_1 s + 1)}$$

is inherently unstable. This system can be stabilized by adding derivative control. Sketch the polar plots for the open-loop transfer function with and without derivative control.

B–7–10. Consider the closed-loop system with the following open-loop transfer function:

$$G(s)H(s) = \frac{10K(s + 0.5)}{s^2(s + 2)(s + 10)}$$

Plot both the direct and inverse polar plots of $G(s)H(s)$ with $K = 1$ and $K = 10$. Apply the Nyquist stability criterion to the plots, and determine the stability of the system with these values of K.

B–7–11. Consider the closed-loop system whose open-loop transfer function is

$$G(s)H(s) = \frac{Ke^{-2s}}{s}$$

Find the maximum value of K for which the system is stable.

B–7–12. Draw a Nyquist plot of the following $G(s)$:

$$G(s) = \frac{1}{s(s^2 + 0.8s + 1)}$$

B–7–13. Consider a unity-feedback control system with the following open-loop transfer function:

$$G(s) = \frac{1}{s^3 + 0.2s^2 + s + 1}$$

Draw a Nyquist plot of $G(s)$ and examine the stability of the system.

B–7–14. Consider a unity-feedback control system with the following open-loop transfer function:

$$G(s) = \frac{s^2 + 2s + 1}{s^3 + 0.2s^2 + s + 1}$$

Draw a Nyquist plot of $G(s)$ and examine the stability of the closed-loop system.

B–7–15. Consider the unity-feedback system with the following $G(s)$:

$$G(s) = \frac{1}{s(s - 1)}$$

Suppose that we choose the Nyquist path as shown in Figure 7–156. Draw the corresponding $G(j\omega)$ locus in the $G(s)$ plane. Using the Nyquist stability criterion, determine the stability of the system.

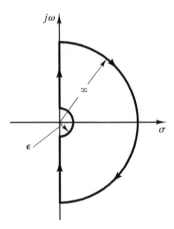

Figure 7–156
Nyquist path.

B–7–16. Consider the closed-loop system shown in Figure 7–157. $G(s)$ has no poles in the right-half s plane.

If the Nyquist plot of $G(s)$ is as shown in Figure 7–158(a), is this system stable?

If the Nyquist plot is as shown in Figure 7–158(b), is this system stable?

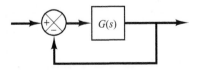

Figure 7–157
Closed-loop system.

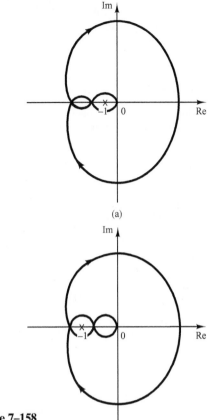

(a)

(b)

Figure 7–158
Nyquist plots.

B–7–17. A Nyquist plot of a unity-feedback system with the feedforward transfer function $G(s)$ is shown in Figure 7–159.

If $G(s)$ has one pole in the right-half s plane, is the system stable?

If $G(s)$ has no pole in the right-half s plane, but has one zero in the right-half s plane, is the system stable?

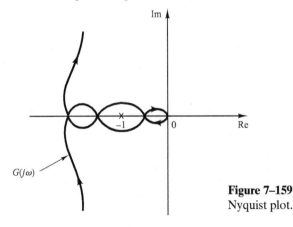

Figure 7–159
Nyquist plot.

B–7–18. Consider the unity-feedback control system with the following open-loop transfer function $G(s)$:

$$G(s) = \frac{K(s + 2)}{s(s + 1)(s + 10)}$$

Plot Nyquist diagrams of $G(s)$ for $K = 1, 10,$ and $100.$

B–7–19. Consider a negative-feedback system with the following open-loop transfer function:

$$G(s) = \frac{2}{s(s + 1)(s + 2)}$$

Plot the Nyquist diagram of $G(s)$. If the system were a positive-feedback one with the same open-loop transfer function $G(s)$, what would the Nyquist diagram look like?

B–7–20. Consider the control system shown in Figure 7–160. Plot Nyquist diagrams of $G(s)$, where

$$G(s) = \frac{10}{s\big[(s + 1)(s + 5) + 10k\big]}$$

$$= \frac{10}{s^3 + 6s^2 + (5 + 10k)s}$$

for $k = 0.3, 0.5,$ and $0.7.$

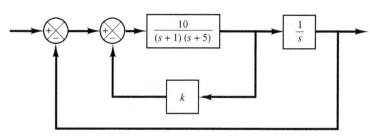

Figure 7–160
Control system.

B–7–21. Consider the system defined by

$$\begin{bmatrix} \dot{x}_1 \\ \dot{x}_2 \end{bmatrix} = \begin{bmatrix} -1 & -1 \\ 6.5 & 0 \end{bmatrix} \begin{bmatrix} x_1 \\ x_2 \end{bmatrix} + \begin{bmatrix} 1 & 1 \\ 1 & 0 \end{bmatrix} \begin{bmatrix} u_1 \\ u_2 \end{bmatrix}$$

$$\begin{bmatrix} y_1 \\ y_2 \end{bmatrix} = \begin{bmatrix} 1 & 0 \\ 0 & 1 \end{bmatrix} \begin{bmatrix} x_1 \\ x_2 \end{bmatrix} + \begin{bmatrix} 0 & 0 \\ 0 & 0 \end{bmatrix} \begin{bmatrix} u_1 \\ u_2 \end{bmatrix}$$

There are four individual Nyquist plots involved in this system. Draw two Nyquist plots for the input u_1 in one diagram and two Nyquist plots for the input u_2 in another diagram. Write a MATLAB program to obtain these two diagrams.

B–7–22. Referring to Problem **B–7–21**, it is desired to plot only $Y_1(j\omega)/U_1(j\omega)$ for $\omega > 0.$ Write a MATLAB program to produce such a plot.

If it is desired to plot $Y_1(j\omega)/U_1(j\omega)$ for $-\infty < \omega < \infty,$ what changes must be made in the MATLAB program?

B–7–23. Consider the unity-feedback control system whose open-loop transfer function is

$$G(s) = \frac{as + 1}{s^2}$$

Determine the value of a so that the phase margin is 45°.

B–7–24. Consider the system shown in Figure 7–161. Draw a Bode diagram of the open-loop transfer function $G(s)$. Determine the phase margin and gain margin.

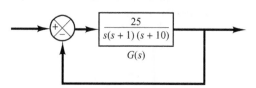

Figure 7–161
Control system.

B–7–25. Consider the system shown in Figure 7–162. Draw a Bode diagram of the open-loop transfer function $G(s)$. Determine the phase margin and gain margin with MATLAB.

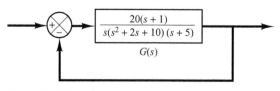

Figure 7–162
Control system.

B–7–26. Consider a unity-feedback control system with the open-loop transfer function

$$G(s) = \frac{K}{s(s^2 + s + 4)}$$

Determine the value of the gain K such that the phase margin is 50°. What is the gain margin with this gain K?

B–7–27. Consider the system shown in Figure 7–163. Draw a Bode diagram of the open-loop transfer function, and determine the value of the gain K such that the phase margin is 50°. What is the gain margin of this system with this gain K?

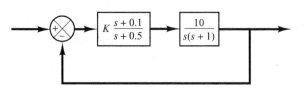

Figure 7–163
Control system.

B–7–28. Consider a unity-feedback control system whose open-loop transfer function is

$$G(s) = \frac{K}{s(s^2 + s + 0.5)}$$

Determine the value of the gain K such that the resonant peak magnitude in the frequency response is 2 dB, or $M_r = 2$ dB.

B–7–29. A Bode diagram of the open-loop transfer function $G(s)$ of a unity-feedback control system is shown in Figure 7–164. It is known that the open-loop transfer function is minimum phase. From the diagram, it can be seen that there is a pair of complex-conjugate poles at $\omega = 2$ rad/sec. Determine the damping ratio of the quadratic term involving these complex-conjugate poles. Also, determine the transfer function $G(s)$.

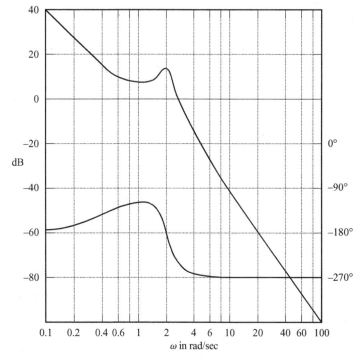

Figure 7–164
Bode diagram of the open-loop transfer function of a unity-feedback control system.

B–7–30. Draw Bode diagrams of the PI controller given by

$$G_c(s) = 5\left(1 + \frac{1}{2s}\right)$$

and the PD controller given by

$$G_c(s) = 5(1 + 0.5s)$$

B–7–31. Figure 7–165 shows a block diagram of a space-vehicle attitude-control system. Determine the proportional gain constant K_p and derivative time T_d such that the bandwidth of the closed-loop system is 0.4 to 0.5 rad/sec. (Note that the closed-loop bandwidth is close to the gain crossover frequency.) The system must have an adequate phase margin. Plot both the open-loop and closed-loop frequency response curves on Bode diagrams.

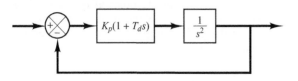

Figure 7–165
Block diagram of space-vehicle attitude-control system.

B–7–32. Referring to the closed-loop system shown in Figure 7–166, design a lead compensator $G_c(s)$ such that the phase margin is 45°, gain margin is not less than 8 dB, and the static velocity error constant K_v is 4.0 sec^{-1}. Plot unit-step and unit-ramp response curves of the compensated system with MATLAB.

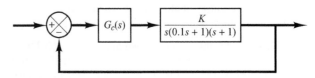

Figure 7–166
Closed-loop system.

B–7–33. Consider the system shown in Figure 7–167. It is desired to design a compensator such that the static velocity error constant is 4 sec^{-1}, phase margin is 50°, and gain margin is 8 dB or more. Plot the unit-step and unit-ramp response curves of the compensated system with MATLAB.

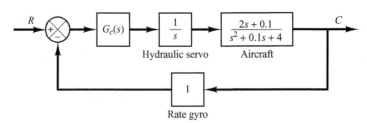

Figure 7–167
Control system.

B–7–34. Consider the system shown in Figure 7–168. Design a lag–lead compensator such that the static velocity error constant K_v is 20 sec^{-1}, phase margin is 60°, and gain margin is not less than 8 dB. Plot the unit-step and unit-ramp response curves of the compensated system with MATLAB.

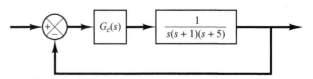

Figure 7–168
Control system.

PID Controllers and Modified PID Controllers

8–1 INTRODUCTION

In previous chapters, we occasionally discussed the basic PID controllers. For example, we presented electronic, hydraulic, and pneumatic PID controllers. We also designed control systems where PID controllers were involved.

It is interesting to note that more than half of the industrial controllers in use today are PID controllers or modified PID controllers.

Because most PID controllers are adjusted on-site, many different types of tuning rules have been proposed in the literature. Using these tuning rules, delicate and fine tuning of PID controllers can be made on-site. Also, automatic tuning methods have been developed and some of the PID controllers may possess on-line automatic tuning capabilities. Modified forms of PID control, such as I-PD control and multi-degrees-of-freedom PID control, are currently in use in industry. Many practical methods for bumpless switching (from manual operation to automatic operation) and gain scheduling are commercially available.

The usefulness of PID controls lies in their general applicability to most control systems. In particular, when the mathematical model of the plant is not known and therefore analytical design methods cannot be used, PID controls prove to be most useful. In the field of process control systems, it is well known that the basic and modified PID control schemes have proved their usefulness in providing satisfactory control, although in many given situations they may not provide optimal control.

In this chapter we first present the design of a PID controlled system using Ziegler and Nichols tuning rules. We next discuss a design of PID controller with the conventional frequency-response approach, followed by the computational optimization approach to design PID controllers. Then we introduce modified PID controls such as PI-D control and I-PD control. Then we introduce multi-degrees-of-freedom control systems, which can satisfy conflicting requirements that single-degree-of-freedom control systems cannot. (For the definition of multi-degrees-of-freedom control systems, see Section 8–6.)

In practical cases, there may be one requirement on the response to disturbance input and another requirement on the response to reference input. Often these two requirements conflict with each other and cannot be satisfied in the single-degree-of-freedom case. By increasing the degrees of freedom, we are able to satisfy both. In this chapter we present two-degrees-of-freedom control systems in detail.

The computational optimization approach presented in this chapter to design control systems (such as to search optimal sets of parameter values to satisfy given transient response specifications) can be used to design both single-degree-of-freedom control systems and multi-degrees-of-freedom control systems, provided a fairly precice mathematical model of the plant is known.

Outline of the Chapter. Section 8–1 has presented introductory material for the chapter. Section 8–2 deals with a design of a PID controller with Ziegler–Nichols Rules. Section 8–3 treats a design of a PID controller with the frequency-response approach. Section 8–4 presents a computational optimization approach to obtain optimal parameter values of PID controllers. Section 8–5 discusses multi-degrees-of-freedom control systems including modified PID control systems.

8–2 ZIEGLER–NICHOLS RULES FOR TUNING PID CONTROLLERS

PID Control of Plants. Figure 8–1 shows a PID control of a plant. If a mathematical model of the plant can be derived, then it is possible to apply various design techniques for determining parameters of the controller that will meet the transient and steady-state specifications of the closed-loop system. However, if the plant is so complicated that its mathematical model cannot be easily obtained, then an analytical or computational approach to the design of a PID controller is not possible. Then we must resort to experimental approaches to the tuning of PID controllers.

The process of selecting the controller parameters to meet given performance specifications is known as controller tuning. Ziegler and Nichols suggested rules for tuning PID controllers (meaning to set values K_p, T_i, and T_d) based on experimental step responses or based on the value of K_p that results in marginal stability when only proportional control action is used. Ziegler–Nichols rules, which are briefly presented in the following, are useful when mathematical models of plants are not known. (These rules can, of course, be applied to the design of systems with known mathematical

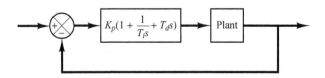

Figure 8–1
PID control
of a plant.

models.) Such rules suggest a set of values of K_p, T_i, and T_d that will give a stable operation of the system. However, the resulting system may exhibit a large maximum overshoot in the step response, which is unacceptable. In such a case we need series of fine tunings until an acceptable result is obtained. In fact, the Ziegler–Nichols tuning rules give an educated guess for the parameter values and provide a starting point for fine tuning, rather than giving the final settings for K_p, T_i, and T_d in a single shot.

Ziegler–Nichols Rules for Tuning PID Controllers. Ziegler and Nichols proposed rules for determining values of the proportional gain K_p, integral time T_i, and derivative time T_d based on the transient response characteristics of a given plant. Such determination of the parameters of PID controllers or tuning of PID controllers can be made by engineers on-site by experiments on the plant. (Numerous tuning rules for PID controllers have been proposed since the Ziegler–Nichols proposal. They are available in the literature and from the manufacturers of such controllers.)

There are two methods called Ziegler–Nichols tuning rules: the first method and the second method. We shall give a brief presentation of these two methods.

First Method. In the first method, we obtain experimentally the response of the plant to a unit-step input, as shown in Figure 8–2. If the plant involves neither integrator(s) nor dominant complex-conjugate poles, then such a unit-step response curve may look S-shaped, as shown in Figure 8–3. This method applies if the response to a step input exhibits an S-shaped curve. Such step-response curves may be generated experimentally or from a dynamic simulation of the plant.

The S-shaped curve may be characterized by two constants, delay time L and time constant T. The delay time and time constant are determined by drawing a tangent line at the inflection point of the S-shaped curve and determining the intersections of the tangent line with the time axis and line $c(t) = K$, as shown in Figure 8–3. The transfer

Figure 8–2
Unit-step response of a plant.

Figure 8–3
S-shaped response curve.

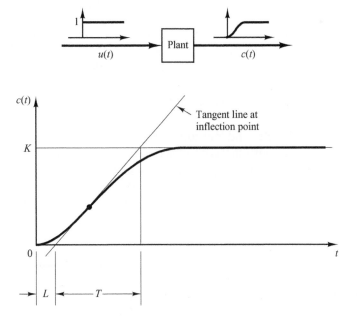

Type of Controller	K_p	T_i	T_d
P	$\dfrac{T}{L}$	∞	0
PI	$0.9\dfrac{T}{L}$	$\dfrac{L}{0.3}$	0
PID	$1.2\dfrac{T}{L}$	$2L$	$0.5L$

function $C(s)/U(s)$ may then be approximated by a first-order system with a transport lag as follows:

$$\frac{C(s)}{U(s)} = \frac{Ke^{-Ls}}{Ts + 1}$$

Ziegler and Nichols suggested to set the values of K_p, T_i, and T_d according to the formula shown in Table 8–1.

Notice that the PID controller tuned by the first method of Ziegler–Nichols rules gives

$$G_c(s) = K_p\left(1 + \frac{1}{T_i s} + T_d s\right)$$

$$= 1.2\frac{T}{L}\left(1 + \frac{1}{2Ls} + 0.5Ls\right)$$

$$= 0.6T\,\frac{\left(s + \dfrac{1}{L}\right)^2}{s}$$

Thus, the PID controller has a pole at the origin and double zeros at $s = -1/L$.

Second Method. In the second method, we first set $T_i = \infty$ and $T_d = 0$. Using the proportional control action only (see Figure 8–4), increase K_p from 0 to a critical value K_{cr} at which the output first exhibits sustained oscillations. (If the output does not exhibit sustained oscillations for whatever value K_p may take, then this method does not apply.) Thus, the critical gain K_{cr} and the corresponding period P_{cr} are experimentally

Figure 8–4
Closed-loop system
with a proportional
controller.

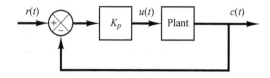

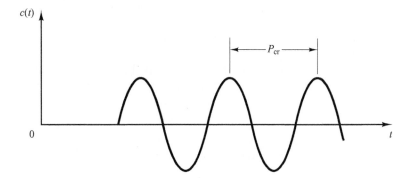

Figure 8–5
Sustained oscillation
with period P_{cr}.
(P_{cr} is measured in
sec.)

determined (see Figure 8–5). Ziegler and Nichols suggested that we set the values of the parameters K_p, T_i, and T_d according to the formula shown in Table 8–2.

Table 8–2   Ziegler–Nichols Tuning Rule Based on Critical Gain
K_{cr} and Critical Period P_{cr} (Second Method)

Type of Controller	K_p	T_i	T_d
P	$0.5K_{cr}$	∞	0
PI	$0.45K_{cr}$	$\dfrac{1}{1.2}P_{cr}$	0
PID	$0.6K_{cr}$	$0.5P_{cr}$	$0.125P_{cr}$

Notice that the PID controller tuned by the second method of Ziegler–Nichols rules gives

$$G_c(s) = K_p\left(1 + \frac{1}{T_i s} + T_d s\right)$$

$$= 0.6K_{cr}\left(1 + \frac{1}{0.5P_{cr}s} + 0.125P_{cr}s\right)$$

$$= 0.075K_{cr}P_{cr}\frac{\left(s + \dfrac{4}{P_{cr}}\right)^2}{s}$$

Thus, the PID controller has a pole at the origin and double zeros at $s = -4/P_{cr}$.

Note that if the system has a known mathematical model (such as the transfer function), then we can use the root-locus method to find the critical gain K_{cr} and the frequency of the sustained oscillations ω_{cr}, where $2\pi/\omega_{cr} = P_{cr}$. These values can be found from the crossing points of the root-locus branches with the $j\omega$ axis. (Obviously, if the root-locus branches do not cross the $j\omega$ axis, this method does not apply.)

Comments. Ziegler–Nichols tuning rules (and other tuning rules presented in the literature) have been widely used to tune PID controllers in process control systems where the plant dynamics are not precisely known. Over many years, such tuning rules proved to be very useful. Ziegler–Nichols tuning rules can, of course, be applied to plants whose dynamics are known. (If the plant dynamics are known, many analytical and graphical approaches to the design of PID controllers are available, in addition to Ziegler–Nichols tuning rules.)

EXAMPLE 8–1 Consider the control system shown in Figure 8–6 in which a PID controller is used to control the system. The PID controller has the transfer function

$$G_c(s) = K_p\left(1 + \frac{1}{T_i s} + T_d s\right)$$

Although many analytical methods are available for the design of a PID controller for the present system, let us apply a Ziegler–Nichols tuning rule for the determination of the values of parameters K_p, T_i, and T_d. Then obtain a unit-step response curve and check to see if the designed system exhibits approximately 25% maximum overshoot. If the maximum overshoot is excessive (40% or more), make a fine tuning and reduce the amount of the maximum overshoot to approximately 25% or less.

Since the plant has an integrator, we use the second method of Ziegler–Nichols tuning rules. By setting $T_i = \infty$ and $T_d = 0$, we obtain the closed-loop transfer function as follows:

$$\frac{C(s)}{R(s)} = \frac{K_p}{s(s+1)(s+5) + K_p}$$

The value of K_p that makes the system marginally stable so that sustained oscillation occurs can be obtained by use of Routh's stability criterion. Since the characteristic equation for the closed-loop system is

$$s^3 + 6s^2 + 5s + K_p = 0$$

the Routh array becomes as follows:

$$
\begin{array}{c|cc}
s^3 & 1 & 5 \\
s^2 & 6 & K_p \\
s^1 & \dfrac{30 - K_p}{6} & \\
s^0 & K_p &
\end{array}
$$

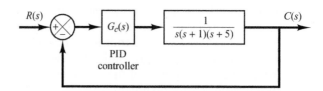

Figure 8–6
PID-controlled system.

Examining the coefficients of the first column of the Routh table, we find that sustained oscillation will occur if $K_p = 30$. Thus, the critical gain K_{cr} is

$$K_{cr} = 30$$

With gain K_p set equal to $K_{cr} (= 30)$, the characteristic equation becomes

$$s^3 + 6s^2 + 5s + 30 = 0$$

To find the frequency of the sustained oscillation, we substitute $s = j\omega$ into this characteristic equation as follows:

$$(j\omega)^3 + 6(j\omega)^2 + 5(j\omega) + 30 = 0$$

or

$$6(5 - \omega^2) + j\omega(5 - \omega^2) = 0$$

from which we find the frequency of the sustained oscillation to be $\omega^2 = 5$ or $\omega = \sqrt{5}$. Hence, the period of sustained oscillation is

$$P_{cr} = \frac{2\pi}{\omega} = \frac{2\pi}{\sqrt{5}} = 2.8099$$

Referring to Table 8–2, we determine K_p, T_i, and T_d as follows:

$$K_p = 0.6K_{cr} = 18$$

$$T_i = 0.5P_{cr} = 1.405$$

$$T_d = 0.125P_{cr} = 0.35124$$

The transfer function of the PID controller is thus

$$G_c(s) = K_p\left(1 + \frac{1}{T_i s} + T_d s\right)$$

$$= 18\left(1 + \frac{1}{1.405s} + 0.35124s\right)$$

$$= \frac{6.3223(s + 1.4235)^2}{s}$$

The PID controller has a pole at the origin and double zero at $s = -1.4235$. A block diagram of the control system with the designed PID controller is shown in Figure 8–7.

Figure 8–7
Block diagram of the system with PID controller designed by use of the Ziegler–Nichols tuning rule (second method).

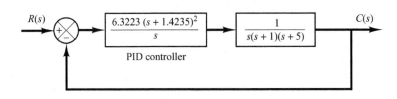

Next, let us examine the unit-step response of the system. The closed-loop transfer function $C(s)/R(s)$ is given by

$$\frac{C(s)}{R(s)} = \frac{6.3223s^2 + 18s + 12.811}{s^4 + 6s^3 + 11.3223s^2 + 18s + 12.811}$$

The unit-step response of this system can be obtained easily with MATLAB. See MATLAB Program 8–1. The resulting unit-step response curve is shown in Figure 8–8. The maximum overshoot in the unit-step response is approximately 62%. The amount of maximum overshoot is excessive. It can be reduced by fine tuning the controller parameters. Such fine tuning can be made on the computer. We find that by keeping $K_p = 18$ and by moving the double zero of the PID controller to $s = -0.65$—that is, using the PID controller

$$G_c(s) = 18\left(1 + \frac{1}{3.077s} + 0.7692s\right) = 13.846\,\frac{(s + 0.65)^2}{s} \tag{8-1}$$

the maximum overshoot in the unit-step response can be reduced to approximately 18% (see Figure 8–9). If the proportional gain K_p is increased to 39.42, without changing the location of the double zero ($s = -0.65$), that is, using the PID controller

$$G_c(s) = 39.42\left(1 + \frac{1}{3.077s} + 0.7692s\right) = 30.322\,\frac{(s + 0.65)^2}{s} \tag{8-2}$$

MATLAB Program 8–1

```
% ---------- Unit-step response ----------

num = [6.3223  18  12.811];
den = [1  6  11.3223  18  12.811];
step(num,den)
grid
title('Unit-Step Response')
```

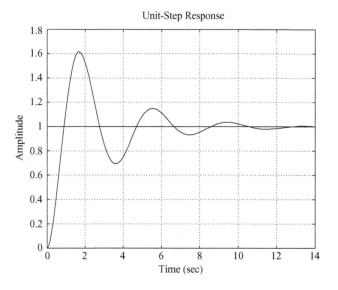

Figure 8–8
Unit-step response curve of PID-controlled system designed by use of the Ziegler–Nichols tuning rule (second method).

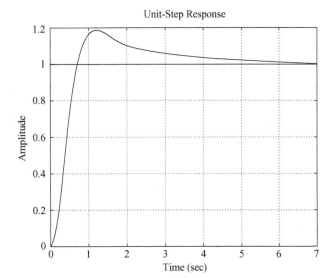

Figure 8–9
Unit-step response of
the system shown in
Figure 8–6 with PID
controller having
parameters $K_p = 18$,
$T_i = 3.077$, and
$T_d = 0.7692$.

then the speed of response is increased, but the maximum overshoot is also increased to approximately 28%, as shown in Figure 8–10. Since the maximum overshoot in this case is fairly close to 25% and the response is faster than the system with $G_c(s)$ given by Equation (8–1), we may consider $G_c(s)$ as given by Equation (8–2) as acceptable. Then the tuned values of K_p, T_i, and T_d become

$$K_p = 39.42, \qquad T_i = 3.077, \qquad T_d = 0.7692$$

It is interesting to observe that these values respectively are approximately twice the values suggested by the second method of the Ziegler–Nichols tuning rule. The important thing to note here is that the Ziegler–Nichols tuning rule has provided a starting point for fine tuning.

It is instructive to note that, for the case where the double zero is located at $s = -1.4235$, increasing the value of K_p increases the speed of response, but as far as the percentage maximum overshoot is concerned, varying gain K_p has very little effect. The reason for this may be seen from

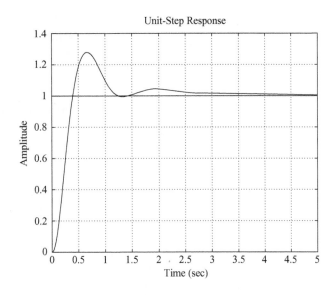

Figure 8–10
Unit-step response of
the system shown in
Figure 8–6 with PID
controller having
parameters
$K_p = 39.42$,
$T_i = 3.077$, and
$T_d = 0.7692$.

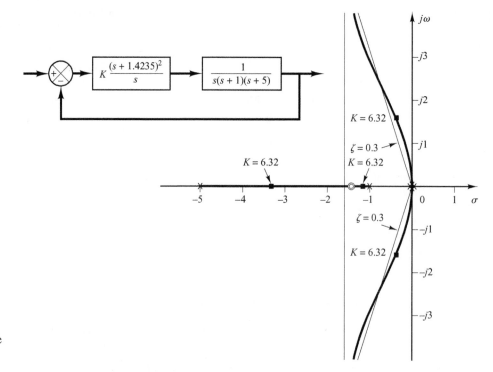

Figure 8–11
Root-locus diagram
of system when PID
controller has double
zero at $s = -1.4235$.

the root-locus analysis. Figure 8–11 shows the root-locus diagram for the system designed by use of the second method of Ziegler–Nichols tuning rules. Since the dominant branches of root loci are along the $\zeta = 0.3$ lines for a considerable range of K, varying the value of K (from 6 to 30) will not change the damping ratio of the dominant closed-loop poles very much. However, varying the location of the double zero has a significant effect on the maximum overshoot, because the damping ratio of the dominant closed-loop poles can be changed significantly. This can also be seen from the root-locus analysis. Figure 8–12 shows the root-locus diagram for the system where the PID controller has the double zero at $s = -0.65$. Notice the change of the root-locus configuration. This change in the configuration makes it possible to change the damping ratio of the dominant closed-loop poles.

In Figure 8–12, notice that, in the case where the system has gain $K = 30.322$, the closed-loop poles at $s = -2.35 \pm j4.82$ act as dominant poles. Two additional closed-loop poles are very near the double zero at $s = -0.65$, with the result that these closed-loop poles and the double zero almost cancel each other. The dominant pair of closed-loop poles indeed determines the nature of the response. On the other hand, when the system has $K = 13.846$, the closed-loop poles at $s = -2.35 \pm j2.62$ are not quite dominant because the two other closed-loop poles near the double zero at $s = -0.65$ have considerable effect on the response. The maximum overshoot in the step response in this case (18%) is much larger than the case where the system is of second order and having only dominant closed-loop poles. (In the latter case the maximum overshoot in the step response would be approximately 6%.)

It is possible to make a third, a fourth, and still further trials to obtain a better response. But this will take a lot of computations and time. If more trials are desired, it is desirable to use the computational approach presented in Section 10–3. Problem **A–8–12** solves this problem with the computational approach with MATLAB. It finds sets of parameter values that will yield the maximum overshoot of 10% or less and the settling time of 3 sec or less. A solution to the present problem obtained in Problem **A–8–12** is that for the PID controller defined by

$$G_c(s) = K\frac{(s + a)^2}{s}$$

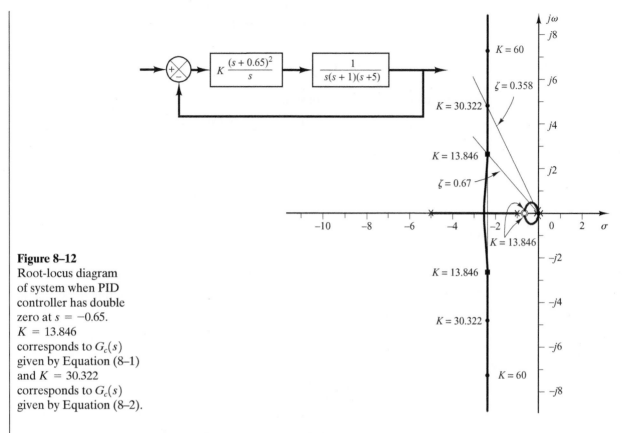

Figure 8–12
Root-locus diagram
of system when PID
controller has double
zero at $s = -0.65$.
$K = 13.846$
corresponds to $G_c(s)$
given by Equation (8–1)
and $K = 30.322$
corresponds to $G_c(s)$
given by Equation (8–2).

the values of K and a are

$$K = 29, \quad a = 0.25$$

with the maximum overshoot equal to 9.52% and settling time equal to 1.78 sec. Another possible solution obtained there is that

$$K = 27, \quad a = 0.2$$

with the 5.5% maximum overshoot and 2.89 sec of settling time. See Problem **A–8–12** for details.

8–3 DESIGN OF PID CONTROLLERS WITH FREQUENCY-RESPONSE APPROACH

In this section we present a design of a PID controller based on the frequency-response approach.

Consider the system shown in Figure 8–13. Using a frequency-response approach, design a PID controller such that the static velocity error constant is 4 sec^{-1}, phase margin is 50° or more, and gain margin is 10 dB or more. Obtain the unit-step and unit-ramp response curves of the PID controlled system with MATLAB.

Let us choose the PID controller to be

$$G_c(s) = \frac{K(as + 1)(bs + 1)}{s}$$

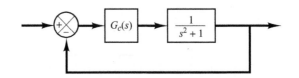

Figure 8–13
Control system.

Since the static velocity error constant K_v is specified as 4 $\sec^{-1}$, we have

$$K_v = \lim_{s \to 0} sG_c(s) \frac{1}{s^2 + 1} = \lim_{s \to 0} s \frac{K(as + 1)(bs + 1)}{s} \frac{1}{s^2 + 1}$$

$$= K = 4$$

Thus

$$G_c(s) = \frac{4(as + 1)(bs + 1)}{s}$$

Next, we plot a Bode diagram of

$$G(s) = \frac{4}{s(s^2 + 1)}$$

MATLAB Program 8–2 produces a Bode diagram of $G(s)$. The resulting Bode diagram is shown in Figure 8–14.

MATLAB Program 8–2

```
num = [4];
den = [1  0.00000000001  1  0];
w = logspace(-1,1,200);
bode(num,den,w)
title('Bode Diagram of 4/[s(s^2+1)]')
```

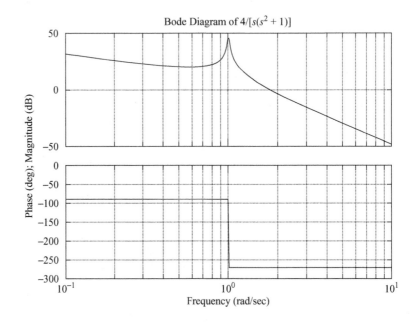

Figure 8–14
Bode diagram of
$4/[s(s^2 + 1)]$.

We need the phase margin of at least 50° and gain margin of 10 dB or more. From the Bode diagram of Figure 8–14, we notice that the gain crossover frequency is approximately $\omega = 1.8$ rad/sec. Let us assume the gain crossover frequency of the compensated system to be somewhere between $\omega = 1$ and $\omega = 10$ rad/sec. Noting that

$$G_c(s) = \frac{4(as + 1)(bs + 1)}{s}$$

we choose $a = 5$. Then, $(as + 1)$ will contribute up to 90° phase lead in the high-frequency region. MATLAB Program 8–3 produces the Bode diagram of

$$\frac{4(5s + 1)}{s(s^2 + 1)}$$

The resulting Bode diagram is shown in Figure 8–15.

MATLAB Program 8–3

```
num = [20  4];
den = [1 0.00000000001  1  0];
w = logspace(-2,1,101);
bode(num,den,w)
title('Bode Diagram of G(s) = 4(5s+1)/[s(s^2+1)]')
```

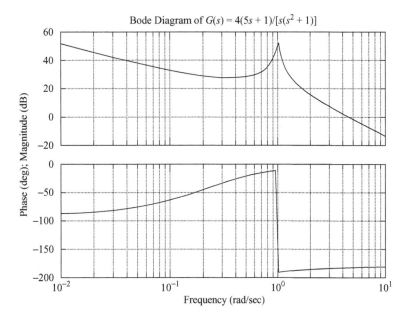

Figure 8–15
Bode diagram of $G(s) = 4(5s + 1)/\left[s(s^2 + 1)\right]$.

Based on the Bode diagram of Figure 8–15, we choose the value of b. The term $(bs + 1)$ needs to give the phase margin of at least $50°$. By simple MATLAB trials, we find $b = 0.25$ to give the phase margin of at least $50°$ and gain margin of $+\infty$ dB. Therefore, by choosing $b = 0.25$, we have

$$G_c(s) = \frac{4(5s + 1)(0.25s + 1)}{s}$$

and the open-loop transfer function of the designed system becomes

$$\text{Open-loop transfer function} = \frac{4(5s + 1)(0.25s + 1)}{s} \frac{1}{s^2 + 1}$$

$$= \frac{5s^2 + 21s + 4}{s^3 + s}$$

MATLAB Program 8–4 produces the Bode diagram of the open-loop transfer function. The resulting Bode diagram is shown in Figure 8–16. From it we see that the static velocity error constant is 4 sec^{-1}, the phase margin is $55°$, and the gain margin is $+\infty$ dB.

MATLAB Program 8–4

```
num = [5  21  4];
den = [1  0  1  0];
w = logspace(-2,2,100);
bode(num,den,w)
title('Bode Diagram of 4(5s+1)(0.25s+1)/[s(s^2+1)]')
```

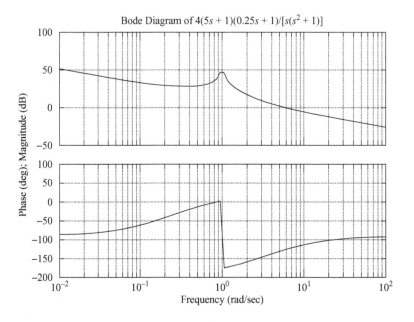

Figure 8–16
Bode diagram of
$4(5s + 1)(0.25s + 1)/$
$[s(s^2 + 1)]$.

Therefore, the designed system satisfies all the requirements. Thus, the designed system is acceptable. (Note that there exist infinitely many systems that satisfy all the requirements. The present system is just one of them.)

Next, we shall obtain the unit-step response and the unit-ramp response of the designed system. The closed-loop transfer function is

$$\frac{C(s)}{R(s)} = \frac{5s^2 + 21s + 4}{s^3 + 5s^2 + 22s + 4}$$

Note that the closed-loop zeros are located at

$$s = -4, \qquad s = -0.2$$

The closed-loop poles are located at

$$s = -2.4052 + j3.9119$$
$$s = -2.4052 - j3.9119$$
$$s = -0.1897$$

Notice that the complex-conjugate closed-loop poles have the damping ratio of 0.5237. MATLAB Program 8–5 produces the unit-step response and the unit-ramp response.

MATLAB Program 8–5

```
%***** Unit-step response *****

num = [5  21  4];
den = [1  5  22  4];
t = 0:0.01:14;
c = step(num,den,t);
plot(t,c)
grid
title('Unit-Step Response of Compensated System')
xlabel('t (sec)')
ylabel('Output c(t)')

%***** Unit-ramp response *****

num1 = [5  21  4];
den1 = [1  5  22  4  0];
t = 0:0.02:20;
c = step(num1,den1,t);
plot(t,c,'-',t,t,'--')
title('Unit-Ramp Response of Compensated System')
xlabel('t (sec)')
ylabel('Unit-Ramp Input and Output c(t)')
text(10.8,8,'Compensated System')
```

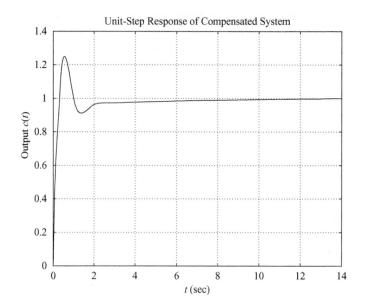

Figure 8–17
Unit-step response
curve.

The resulting unit-step response curve is shown in Figure 8–17 and the unit-ramp response curve in Figure 8–18. Notice that the closed-loop pole at $s = -0.1897$ and the zero at $s = -0.2$ produce a long tail of small amplitude in the unit-step response.

For an additional example of design of a PID controller based on the frequency-response approach, see Problem **A–8–7**.

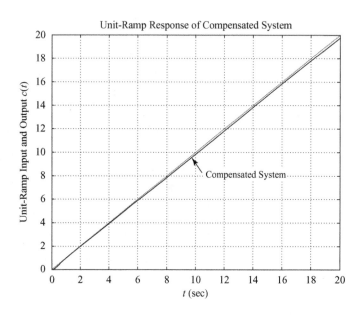

Figure 8–18
Unit-ramp input and
the output curve.

8-4 DESIGN OF PID CONTROLLERS WITH COMPUTATIONAL OPTIMIZATION APPROACH

In this section we shall explore how to obtain an optimal set (or optimal sets) of parameter values of PID controllers to satisfy the transient response specifications by use of MATLAB. We shall present two examples to illustrate the approach in this section.

EXAMPLE 8-2   Consider the PID-controlled system shown in Figure 8–19. The PID controller is given by

$$G_c(s) = K\frac{(s + a)^2}{s}$$

It is desired to find a combination of K and a such that the closed-loop system will have 10% (or less) maximum overshoot in the unit-step response. (We will not include any other condition in this problem. But other conditions can easily be included, such as that the settling time be less than a specified value. See, for example, Example 8–3.)

There may be more than one set of parameters that satisfy the specifications. In this example, we shall obtain all sets of parameters that satisfy the given specifications.

To solve this problem with MATLAB, we first specify the region to search for appropriate K and a. We then write a MATLAB program that, in the unit-step response, will find a combination of K and a which will satisfy the criterion that the maximum overshoot is 10% or less.

Note that the gain K should not be too large, so as to avoid the possibility that the system require an unnecessarily large power unit.

Assume that the region to search for K and a is

$$2 \le K \le 3 \quad \text{and} \quad 0.5 \le a \le 1.5$$

If a solution does not exist in this region, then we need to expand it. In some problems, however, there is no solution, no matter what the search region might be.

In the computational approach, we need to determine the step size for each of K and a. In the actual design process, we need to choose step sizes small enough. However, in this example, to avoid an overly large number of computations, we choose the step sizes to be reasonable—say, 0.2 for both K and a.

To solve this problem it is possible to write many different MATLAB programs. We present here one such program, MATLAB Program 8–6. In this program, notice that we use two "for" loops. We start the program with the outer loop to vary the "K" values. Then we vary the "a" values in the inner loop. We proceed by writing the MATLAB program such that the nested loops in the program begin with the lowest values of "K" and "a" and step toward the highest. Note that, depending on the system and the ranges of search for "K" and "a" and the step sizes chosen, it may take from several seconds to a few minutes for MATLAB to compute the desired sets of the values.

In this program the statement

$$\text{solution(k,:)} = [K(i) \ a(j) \ m]$$

will produce a table of K, a, m values. (In the present system there are 15 sets of K and a that will exhibit $m < 1.10$—that is, the maximum overshoot is less than 10%.)

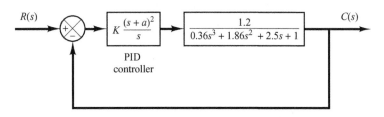

Figure 8–19
PID-controlled system.

To sort out the solution sets in the order of the magnitude of the maximum overshoot (starting from the smallest value of *m* and ending at the largest value of *m* in the table), we use the command

$$\text{sortsolution} = \text{sortrows(solution,3)}$$

MATLAB Program 8–6

```
%'K' and 'a' values to test

K = [2.0 2.2 2.4 2.6 2.8 3.0];
a = [0.5 0.7 0.9 1.1 1.3 1.5];

% Evaluate closed-loop unit-step response at each 'K' and 'a' combination
% that will yield the maximum overshoot less than 10%

t = 0:0.01:5;
g = tf([1.2],[0.36 1.86 2.5 1]);
k = 0;
for i = 1:6;
   for j = 1:6;
      gc = tf(K(i)*[1  2*a(j)  a(j)^2], [1  0]);  % controller
         G = gc*g/(1 + gc*g);  % closed-loop transfer function
         y = step(G,t);
         m = max(y);
         if m < 1.10
         k = k+1;
         solution(k,:) = [K(i)  a(j)  m];
         end
      end
   end
solution  % Print solution table

solution =

   2.0000  0.5000  0.9002
   2.0000  0.7000  0.9807
   2.0000  0.9000  1.0614
   2.2000  0.5000  0.9114
   2.2000  0.7000  0.9837
   2.2000  0.9000  1.0772
   2.4000  0.5000  0.9207
   2.4000  0.7000  0.9859
   2.4000  0.9000  1.0923
   2.6000  0.5000  0.9283
   2.6000  0.7000  0.9877
   2.8000  0.5000  0.9348
   2.8000  0.7000  1.0024
   3.0000  0.5000  0.9402
   3.0000  0.7000  1.0177
sortsolution = sortrows(solution,3)  % Print solution table sorted by
                                     % column 3
```

(continues on next page)

```
sortsolution =

    2.0000  0.5000  0.9002
    2.2000  0.5000  0.9114
    2.4000  0.5000  0.9207
    2.6000  0.5000  0.9283
    2.8000  0.5000  0.9348
    3.0000  0.5000  0.9402
    2.0000  0.7000  0.9807
    2.2000  0.7000  0.9837
    2.4000  0.7000  0.9859
    2.6000  0.7000  0.9877
    2.8000  0.7000  1.0024
    3.0000  0.7000  1.0177
    2.0000  0.9000  1.0614
    2.2000  0.9000  1.0772
    2.4000  0.9000  1.0923
```

% Plot the response with the largest overshoot that is less than 10%

K = sortsolution(k,1)

K =

    2.4000

a = sortsolution(k,2)

a =

    0.9000

```
gc = tf(K*[1  2*a  a^2], [1  0]);
G = gc*g/(1 + gc*g);
step(G,t)
grid  % See Figure 8–20
```

% If you wish to plot the response with the smallest overshoot that is
% greater than 0%, then enter the following values of 'K' and 'a'

K = sortsolution(11,1)

K =

    2.8000

a = sortsolution(11,2)

a =

    0.7000

```
gc = tf(K*[1  2*a  a^2], [1  0]);
G = gc*g/(1 + gc*g);
step(G,t)
grid  % See Figure 8–21
```

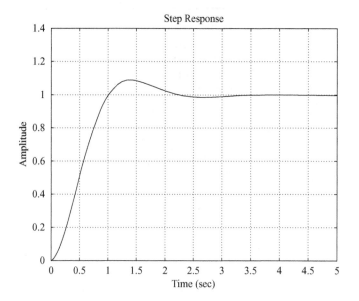

Figure 8–20
Unit-step response of
the system with
$K = 2.4$ and $a = 0.9$.
(The maximum
overshoot is 9.23%.)

To plot the unit-step response curve of the last set of the K and a values in the sorted table, we enter the commands

$$K = \text{sortsolution } (k,1)$$
$$a = \text{sortsolution } (k,2)$$

and use the step command. (The resulting unit-step response curve is shown in Figure 8–20.) To plot the unit-step response curve with the smallest overshoot that is greater than 0% found in the sorted table, enter the commands

$$K = \text{sortsolution } (11,1)$$
$$a = \text{sortsolution } (11,2)$$

and use the step command. (The resulting unit-step response curve is shown in Figure 8–21.)

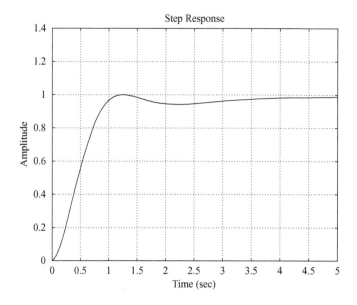

Figure 8–21
Unit-step response of
the system with
$K = 2.8$ and $a = 0.7$.
(The maximum
overshoot is 0.24%.)

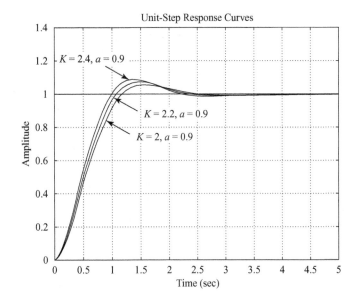

Unit-Step Response Curves

Figure 8–22
Unit-step response
curves of system with
$K = 2, a = 0.9$;
$K = 2.2, a = 0.9$;
and $K = 2.4$,
$a = 0.9$.

To plot the unit-step response curve of the system with any set shown in the sorted table, we specify the K and a values by entering an appropriate sortsolution command.

Note that for a specification that the maximum overshoot be between 10% and 5%, there would be three sets of solutions:

$$K = 2.0000, \quad a = 0.9000, \quad m = 1.0614$$

$$K = 2.2000, \quad a = 0.9000, \quad m = 1.0772$$

$$K = 2.4000, \quad a = 0.9000, \quad m = 1.0923$$

Unit-step response curves for these three cases are shown in Figure 8–22. Notice that the system with a larger gain K has a smaller rise time and larger maximum overshoot. Which one of these three systems is best depends on the system's objective.

EXAMPLE 8–3 Consider the system shown in Figure 8–23. We want to find all combinations of K and a values such that the closed-loop system has a maximum overshoot of less than 15%, but more than 10%, in the unit-step response. In addition, the settling time should be less than 3 sec. In this problem, assume that the search region is

$$3 \le K \le 5 \quad \text{and} \quad 0.1 \le a \le 3$$

Determine the best choice of the parameters K and a.

Figure 8–23
PID-controlled
system with a
simplified PID
controller.

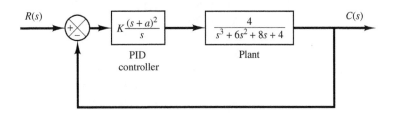

In this problem, we choose the step sizes to be reasonable, — say 0.2 for K and 0.1 for a. MATLAB Program 8–7 gives the solution to this problem. From the sortsolution table, it looks like the first row is a good choice. Figure 8–24 shows the unit step response curve for $K = 3.2$ and $a = 0.9$. Since this choice requires a smaller K value than most other choices, we may decide that the first row is the best choice.

MATLAB Program 8–7

```
t = 0:0.01:8;
k = 0;
for K = 3:0.2:5;
   for a = 0.1:0.1:3;
      num = [4*K  8*K*a  4*K*a^2];
      den = [1  6  8+4*K  4+8*K*a  4*K*a^2];
         y = step(num,den,t);
         s = 801;while y(s)>0.98 & y(s)<1.02; s = s – 1;end;
      ts = (s–1)*0.01; % ts = settling time;
      m = max(y);
      if m<1.15 & m>1.10; if ts<3.00;
         k = k+1;
         solution(k,:) = [K  a  m  ts];
      end
      end
   end
end
solution
```

```
solution =
      3.0000    1.0000    1.1469    2.7700
      3.2000    0.9000    1.1065    2.8300
      3.4000    0.9000    1.1181    2.7000
      3.6000    0.9000    1.1291    2.5800
      3.8000    0.9000    1.1396    2.4700
      4.0000    0.9000    1.1497    2.3800
      4.2000    0.8000    1.1107    2.8300
      4.4000    0.8000    1.1208    2.5900
      4.6000    0.8000    1.1304    2.4300
      4.8000    0.8000    1.1396    2.3100
      5.0000    0.8000    1.1485    2.2100
```

```
sortsolution = sortrows(solution,3)
```

```
sortsolution =
      3.2000    0.9000    1.1065    2.8300
      4.2000    0.8000    1.1107    2.8300
      3.4000    0.9000    1.1181    2.7000
      4.4000    0.8000    1.1208    2.5900
      3.6000    0.9000    1.1291    2.5800
      4.6000    0.8000    1.1304    2.4300
      4.8000    0.8000    1.1396    2.3100
      3.8000    0.9000    1.1396    2.4700
```

(continues on next page)

```
      3.0000   1.0000   1.1469   2.7700
      5.0000   0.8000   1.1485   2.2100
      4.0000   0.9000   1.1497   2.3800
```

% Plot the response curve with the smallest overshoot shown in sortsolution table.

```
  K = sortsolution(1,1), a = sortsolution(1,2)
```

K =

  3.2000

a =

  0.9000

```
  num = [4*K    8*K*a    4*K*a^2];
  den = [1   6   8+4*K   4+8*K*a   4*K*a^2];
  num
```

num =

  12.8000   23.0400   10.3680

  den

den =

  1.0000   6.0000   20.8000   27.0400   10.3680

```
  y = step(num,den,t);
  plot(t,y) % See Figure 8–24.
  grid
  title('Unit-Step Response')
  xlabel('t sec')
  ylabel('Output y(t)')
```

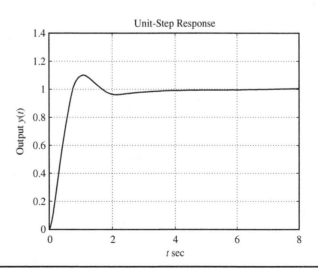

Figure 8–24
Unit-step response curve of the system with $K = 3.2$ and $a = 0.9$.

8–5 MODIFICATIONS OF PID CONTROL SCHEMES

Consider the basic PID control system shown in Figure 8–25(a), where the system is subjected to disturbances and noises. Figure 8–25(b) is a modified block diagram of the same system. In the basic PID control system such as the one shown in Figure 8–25(b), if the reference input is a step function, then, because of the presence of the derivative term in the control action, the manipulated variable $u(t)$ will involve an impulse function (delta function). In an actual PID controller, instead of the pure derivative term $T_d s$, we employ

$$\frac{T_d s}{1 + \gamma T_d s}$$

where the value of γ is somewhere around 0.1. Therefore, when the reference input is a step function, the manipulated variable $u(t)$ will not involve an impulse function, but will involve a sharp pulse function. Such a phenomenon is called *set-point kick*.

PI-D Control. To avoid the set-point kick phenomenon, we may wish to operate the derivative action only in the feedback path so that differentiation occurs only on the feedback signal and not on the reference signal. The control scheme arranged in this way is called the PI-D control. Figure 8–26 shows a PI-D-controlled system.

From Figure 8–26, it can be seen that the manipulated signal $U(s)$ is given by

$$U(s) = K_p\left(1 + \frac{1}{T_i s}\right)R(s) - K_p\left(1 + \frac{1}{T_i s} + T_d s\right)B(s)$$

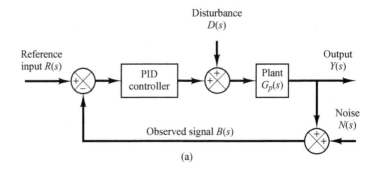

(a)

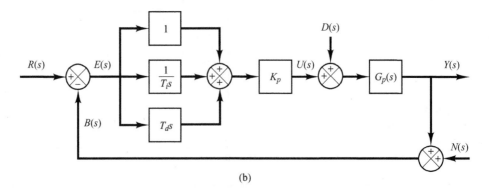

(b)

Figure 8–25
(a) PID-controlled system;
(b) equivalent block diagram.

Chapter 8 / **PID Controllers and Modified PID Controllers**

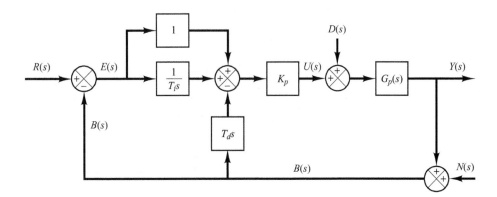

Figure 8–26
PI-D-controlled
system.

Notice that in the absence of the disturbances and noises, the closed-loop transfer function of the basic PID control system [shown in Figure 8–25(b)] and the PI-D control system (shown in Figure 8–26) are given, respectively, by

$$\frac{Y(s)}{R(s)} = \left(1 + \frac{1}{T_i s} + T_d s\right) \frac{K_p G_p(s)}{1 + \left(1 + \frac{1}{T_i s} + T_d s\right) K_p G_p(s)}$$

and

$$\frac{Y(s)}{R(s)} = \left(1 + \frac{1}{T_i s}\right) \frac{K_p G_p(s)}{1 + \left(1 + \frac{1}{T_i s} + T_d s\right) K_p G_p(s)}$$

It is important to point out that in the absence of the reference input and noises, the closed-loop transfer function between the disturbance $D(s)$ and the output $Y(s)$ in either case is the same and is given by

$$\frac{Y(s)}{D(s)} = \frac{G_p(s)}{1 + K_p G_p(s)\left(1 + \frac{1}{T_i s} + T_d s\right)}$$

I-PD Control. Consider the case where the reference input is a step function. Both PID control and PI-D control involve a step function in the manipulated signal. Such a step change in the manipulated signal may not be desirable in many occasions. Therefore, it may be advantageous to move the proportional action and derivative action to the feedback path so that these actions affect the feedback signal only. Figure 8–27 shows such a control scheme. It is called the I-PD control. The manipulated signal is given by

$$U(s) = K_p \frac{1}{T_i s} R(s) - K_p\left(1 + \frac{1}{T_i s} + T_d s\right) B(s)$$

Notice that the reference input $R(s)$ appears only in the integral control part. Thus, in I-PD control, it is imperative to have the integral control action for proper operation of the control system.

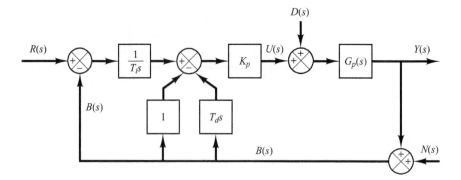

Figure 8–27
I-PD-controlled
system.

The closed-loop transfer function $Y(s)/R(s)$ in the absence of the disturbance input and noise input is given by

$$\frac{Y(s)}{R(s)} = \left(\frac{1}{T_i s}\right) \frac{K_p G_p(s)}{1 + K_p G_p(s)\left(1 + \dfrac{1}{T_i s} + T_d s\right)}$$

It is noted that in the absence of the reference input and noise signals, the closed-loop transfer function between the disturbance input and the output is given by

$$\frac{Y(s)}{D(s)} = \frac{G_p(s)}{1 + K_p G_p(s)\left(1 + \dfrac{1}{T_i s} + T_d s\right)}$$

This expression is the same as that for PID control or PI-D control.

Two-Degrees-of-Freedom PID Control. We have shown that PI-D control is obtained by moving the derivative control action to the feedback path, and I-PD control is obtained by moving the proportional control and derivative control actions to the feedback path. Instead of moving the entire derivative control action or proportional control action to the feedback path, it is possible to move only portions of these control actions to the feedback path, retaining the remaining portions in the feedforward path. In the literature, PI-PD control has been proposed. The characteristics of this control scheme lie between PID control and I-PD control. Similarly, PID-PD control can be considered. In these control schemes, we have a controller in the feedforward path and another controller in the feedback path. Such control schemes lead us to a more general two-degrees-of-freedom control scheme. We shall discuss details of such a two-degrees-of-freedom control scheme in subsequent sections of this chapter.

8–6 TWO-DEGREES-OF-FREEDOM CONTROL

Consider the system shown in Figure 8–28, where the system is subjected to the disturbance input $D(s)$ and noise input $N(s)$, in addition to the reference input $R(s)$. $G_c(s)$ is the transfer function of the controller and $G_p(s)$ is the transfer function of the plant. We assume that $G_p(s)$ is fixed and unalterable.

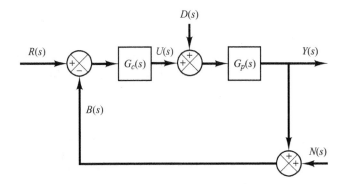

Figure 8–28
One-degree-of-
freedom control
system.

For this system, three closed-loop transfer functions $Y(s)/R(s) = G_{yr}$, $Y(s)/D(s) = G_{yd}$, and $Y(s)/N(s) = G_{yn}$ may be derived. They are

$$G_{yr} = \frac{Y(s)}{R(s)} = \frac{G_c G_p}{1 + G_c G_p}$$

$$G_{yd} = \frac{Y(s)}{D(s)} = \frac{G_p}{1 + G_c G_p}$$

$$G_{yn} = \frac{Y(s)}{N(s)} = -\frac{G_c G_p}{1 + G_c G_p}$$

[In deriving $Y(s)/R(s)$, we assumed $D(s) = 0$ and $N(s) = 0$. Similar comments apply to the derivations of $Y(s)/D(s)$ and $Y(s)/N(s)$.] The degrees of freedom of the control system refers to how many of these closed-loop transfer functions are independent. In the present case, we have

$$G_{yr} = \frac{G_p - G_{yd}}{G_p}$$

$$G_{yn} = \frac{G_{yd} - G_p}{G_p}$$

Among the three closed-loop transfer functions G_{yr}, G_{yn}, and G_{yd}, if one of them is given, the remaining two are fixed. This means that the system shown in Figure 8–28 is a one-degree-of-freedom control system.

Next consider the system shown in Figure 8–29, where $G_p(s)$ is the transfer function of the plant. For this system, closed-loop transfer functions G_{yr}, G_{yn}, and G_{yd} are given, respectively, by

$$G_{yr} = \frac{Y(s)}{R(s)} = \frac{G_{c1} G_p}{1 + (G_{c1} + G_{c2})G_p}$$

$$G_{yd} = \frac{Y(s)}{D(s)} = \frac{G_p}{1 + (G_{c1} + G_{c2})G_p}$$

$$G_{yn} = \frac{Y(s)}{N(s)} = -\frac{(G_{c1} + G_{c2})G_p}{1 + (G_{c1} + G_{c2})G_p}$$

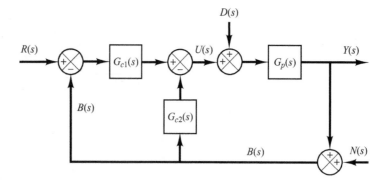

Figure 8–29
Two-degrees-of-freedom control system.

Hence, we have

$$G_{yr} = G_{c1}G_{yd}$$

$$G_{yn} = \frac{G_{yd} - G_p}{G_p}$$

In this case, if G_{yd} is given, then G_{yn} is fixed, but G_{yr} is not fixed, because G_{c1} is independent of G_{yd}. Thus, two closed-loop transfer functions among three closed-loop transfer functions G_{yr}, G_{yd}, and G_{yn} are independent. Hence, this system is a two-degrees-of-freedom control system.

Similarly, the system shown in Figure 8–30 is also a two-degrees-of-freedom control system, because for this system

$$G_{yr} = \frac{Y(s)}{R(s)} = \frac{G_{c1}G_p}{1 + G_{c1}G_p} + \frac{G_{c2}G_p}{1 + G_{c1}G_p}$$

$$G_{yd} = \frac{Y(s)}{D(s)} = \frac{G_p}{1 + G_{c1}G_p}$$

$$G_{yn} = \frac{Y(s)}{N(s)} = -\frac{G_{c1}G_p}{1 + G_{c1}G_p}$$

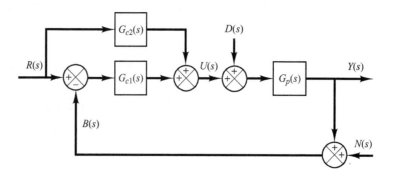

Figure 8–30
Two-degrees-of-freedom control system.

Hence,

$$G_{yr} = G_{c2}G_{yd} + \frac{G_p - G_{yd}}{G_p}$$

$$G_{yn} = \frac{G_{yd} - G_p}{G_p}$$

Clearly, if G_{yd} is given, then G_{yn} is fixed, but G_{yr} is not fixed, because G_{c2} is independent of G_{yd}.

It will be seen in Section 8–7 that, in such a two-degrees-of-freedom control system, both the closed-loop characteristics and the feedback characteristics can be adjusted independently to improve the system response performance.

8–7 ZERO-PLACEMENT APPROACH TO IMPROVE RESPONSE CHARACTERISTICS

We shall show here that by use of the zero-placement approach presented later in this section, we can achieve the following:

The responses to the ramp reference input and acceleration reference input exhibit no steady-state errors.

In high-performance control systems it is always desired that the system output follow the changing input with minimum error. For step, ramp, and acceleration inputs, it is desired that the system output exhibit no steady-state error.

In what follows, we shall demonstrate how to design control systems that will exhibit no steady-state errors in following ramp and acceleration inputs and at the same time force the response to the step disturbance input to approach zero quickly.

Consider the two-degrees-of-freedom control system shown in Figure 8–31. Assume that the plant transfer function $G_p(s)$ is a minimum-phase transfer function and is given by

$$G_p(s) = K\frac{A(s)}{B(s)}$$

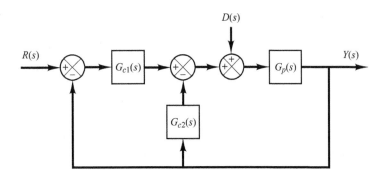

Figure 8–31
Two-degrees-of-freedom control system.

where

$$A(s) = (s + z_1)(s + z_2) \cdots (s + z_m)$$

$$B(s) = s^N(s + p_{N+1})(s + p_{N+2}) \cdots (s + p_n)$$

where N may be 0, 1, 2 and $n \geq m$. Assume also that G_{c1} is a PID controller followed by a filter $1/A(s)$, or

$$G_{c1}(s) = \frac{\alpha_1 s + \beta_1 + \gamma_1 s^2}{s} \frac{1}{A(s)}$$

and G_{c2} is a PID, PI, PD, I, D, or P controller followed by a filter $1/A(s)$. That is

$$G_{c2}(s) = \frac{\alpha_2 s + \beta_2 + \gamma_2 s^2}{s} \frac{1}{A(s)}$$

where some of $\alpha_2, \beta_2,$ and γ_2 may be zero. Then it is possible to write $G_{c1} + G_{c2}$ as

$$G_{c1} + G_{c2} = \frac{\alpha s + \beta + \gamma s^2}{s} \frac{1}{A(s)} \tag{8-3}$$

where $\alpha, \beta,$ and γ are constants. Then

$$\frac{Y(s)}{D(s)} = \frac{G_p}{1 + (G_{c1} + G_{c2})G_p} = \frac{K \dfrac{A(s)}{B(s)}}{1 + \dfrac{\alpha s + \beta + \gamma s^2}{s} \dfrac{K}{B(s)}}$$

$$= \frac{sKA(s)}{sB(s) + (\alpha s + \beta + \gamma s^2)K}$$

Because of the presence of s in the numerator, the response $y(t)$ to a step disturbance input approaches zero as t approaches infinity, as shown below. Since

$$Y(s) = \frac{sKA(s)}{sB(s) + (\alpha s + \beta + \gamma s^2)K} D(s)$$

if the disturbance input is a step function of magnitude d, or

$$D(s) = \frac{d}{s}$$

and assuming the system is stable, then

$$y(\infty) = \lim_{s \to 0} s \left[\frac{sKA(s)}{sB(s) + (\alpha s + \beta + \gamma s^2)K} \right] \frac{d}{s}$$

$$= \lim_{s \to 0} \frac{sKA(0)d}{sB(0) + \beta K}$$

$$= 0$$

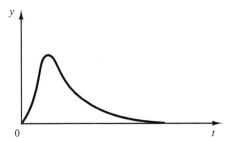

Figure 8–32
Typical response
curve to a step
disturbance input.

The response $y(t)$ to a step disturbance input will have the general form shown in Figure 8–32.

Note that $Y(s)/R(s)$ and $Y(s)/D(s)$ are given by

$$\frac{Y(s)}{R(s)} = \frac{G_{c1}G_p}{1 + (G_{c1} + G_{c2})G_p}, \qquad \frac{Y(s)}{D(s)} = \frac{G_p}{1 + (G_{c1} + G_{c2})G_p}$$

Notice that the denominators of $Y(s)/R(s)$ and $Y(s)/D(s)$ are the same. Before we choose the poles of $Y(s)/R(s)$, we need to place the zeros of $Y(s)/R(s)$.

Zero Placement.   Consider the system

$$\frac{Y(s)}{R(s)} = \frac{p(s)}{s^{n+1} + a_n s^n + a_{n-1}s^{n-1} + \cdots + a_2 s^2 + a_1 s + a_0}$$

If we choose $p(s)$ as

$$p(s) = a_2 s^2 + a_1 s + a_0 = a_2(s + s_1)(s + s_2)$$

that is, choose the zeros $s = -s_1$ and $s = -s_2$ such that, together with a_2, the numerator polynomial $p(s)$ is equal to the sum of the last three terms of the denominator polynomial—then the system will exhibit no steady-state errors in response to the step input, ramp input, and acceleration input.

Requirement Placed on System Response Characteristics.   Suppose that it is desired that the maximum overshoot in the response to the unit-step reference input be between arbitrarily selected upper and lower limits—for example,

$$2\% < \text{maximum overshoot} < 10\%$$

where we choose the lower limit to be slightly above zero to avoid having overdamped systems. The smaller the upper limit, the harder it is to determine the coefficient a's. In some cases, no combination of the a's may exist to satisfy the specification, so we must allow a higher upper limit for the maximum overshoot. We use MATLAB to search at least one set of the a's to satisfy the specification. As a practical computational matter, instead of searching for the a's, we try to obtain acceptable closed-loop poles by searching a reasonable region in the left-half s plane for each closed-loop pole. Once we determine all closed-loop poles, then all coefficients $a_n, a_{n-1}, \ldots, a_1, a_0$ will be determined.

Determination of G_{c2}.   Now that the coefficients of the transfer function $Y(s)/R(s)$ are all known and $Y(s)/R(s)$ is given by

$$\frac{Y(s)}{R(s)} = \frac{a_2 s^2 + a_1 s + a_0}{s^{n+1} + a_n s^n + a_{n-1} s^{n-1} + \cdots + a_2 s^2 + a_1 s + a_0} \tag{8-4}$$

we have

$$\frac{Y(s)}{R(s)} = G_{c1} \frac{Y(s)}{D(s)}$$

$$= \frac{G_{c1} s K A(s)}{s B(s) + (\alpha s + \beta + \gamma s^2) K}$$

$$= \frac{G_{c1} s K A(s)}{s^{n+1} + a_n s^n + a_{n-1} s^{n-1} + \cdots + a_2 s^2 + a_1 s + a_0}$$

Since G_{c1} is a PID controller and is given by

$$G_{c1} = \frac{\alpha_1 s + \beta_1 + \gamma_1 s^2}{s} \frac{1}{A(s)}$$

$Y(s)/R(s)$ can be written as

$$\frac{Y(s)}{R(s)} = \frac{K(\alpha_1 s + \beta_1 + \gamma_1 s^2)}{s^{n+1} + a_n s^n + a_{n-1} s^{n-1} + \cdots + a_2 s^2 + a_1 s + a_0}$$

Therefore, we choose

$$K\gamma_1 = a_2, \qquad K\alpha_1 = a_1, \qquad K\beta_1 = a_0$$

so that

$$G_{c1} = \frac{a_1 s + a_0 + a_2 s^2}{Ks} \frac{1}{A(s)} \tag{8-5}$$

The response of this system to the unit-step reference input can be made to exhibit the maximum overshoot between the chosen upper and lower limits, such as

$$2\% < \text{maximum overshoot} < 10\%$$

The response of the system to the ramp reference input or acceleration reference input can be made to exhibit no steady-state error. The characteristic of the system of Equation (8-4) is that it generally exhibits a short settling time. If we wish to further shorten the settling time, then we need to allow a larger maximum overshoot—for example,

$$2\% < \text{maximum overshoot} < 20\%$$

The controller G_{c2} can now be determined from Equations (8-3) and (8-5). Since

$$G_{c1} + G_{c2} = \frac{\alpha s + \beta + \gamma s^2}{s} \frac{1}{A(s)}$$

we have

$$
\begin{aligned}
G_{c2} &= \left[\frac{\alpha s + \beta + \gamma s^2}{s} - \frac{a_1 s + a_0 + a_2 s^2}{Ks} \right] \frac{1}{A(s)} \\
&= \frac{(K\alpha - a_1)s + (K\beta - a_0) + (K\gamma - a_2)s^2}{Ks} \frac{1}{A(s)}
\end{aligned}
\tag{8-6}
$$

The two controllers G_{c1} and G_{c2} can be determined from Equations (8–5) and (8–6).

EXAMPLE 8–4 Consider the two-degrees-of-freedom control system shown in Figure 8–33. The plant transfer function $G_p(s)$ is given by

$$
G_p(s) = \frac{10}{s(s + 1)}
$$

Design controllers $G_{c1}(s)$ and $G_{c2}(s)$ such that the maximum overshoot in the response to the unit-step reference input be less than 19%, but more than 2%, and the settling time be less than 1 sec. It is desired that the steady-state errors in following the ramp reference input and acceleration reference input be zero. The response to the unit-step disturbance input should have a small amplitude and settle to zero quickly.

To design suitable controllers $G_{c1}(s)$ and $G_{c2}(s)$, first note that

$$
\frac{Y(s)}{D(s)} = \frac{G_p}{1 + G_p(G_{c1} + G_{c2})}
$$

To simplify the notation, let us define

$$
G_c = G_{c1} + G_{c2}
$$

Then

$$
\begin{aligned}
\frac{Y(s)}{D(s)} = \frac{G_p}{1 + G_p G_c} &= \frac{\dfrac{10}{s(s + 1)}}{1 + \dfrac{10}{s(s + 1)} G_c} \\
&= \frac{10}{s(s + 1) + 10 G_c}
\end{aligned}
$$

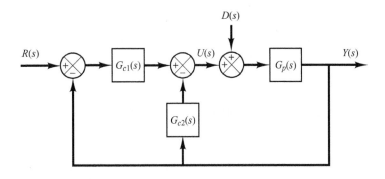

Figure 8–33
Two-degrees-of-freedom control system.

Second, note that

$$\frac{Y(s)}{R(s)} = \frac{G_p G_{c1}}{1 + G_p G_c} = \frac{10 G_{c1}}{s(s + 1) + 10 G_c}$$

Notice that the characteristic equation for $Y(s)/D(s)$ and the one for $Y(s)/R(s)$ are identical.

We may be tempted to choose a zero of $G_c(s)$ at $s = -1$ to cancel a pole at $s = -1$ of the plant $G_p(s)$. However, the canceled pole $s = -1$ becomes a closed-loop pole of the entire system, as seen below. If we define $G_c(s)$ as a PID controller such that

$$G_c(s) = \frac{K(s + 1)(s + \beta)}{s} \tag{8–7}$$

then

$$\frac{Y(s)}{D(s)} = \frac{10}{s(s + 1) + \dfrac{10K(s + 1)(s + \beta)}{s}}$$

$$= \frac{10s}{(s + 1)\left[s^2 + 10K(s + \beta)\right]}$$

The closed-loop pole at $s = -1$ is a slow-response pole, and if this closed-loop pole is included in the system, the settling time will not be less than 1 sec. Therefore, we should not choose $G_c(s)$ as given by Equation (8–7).

The design of controllers $G_{c1}(s)$ and $G_{c2}(s)$ consists of two steps.

Design Step 1: We design $G_c(s)$ to satisfy the requirements on the response to the step-disturbance input $D(s)$. In this design stage, we assume that the reference input is zero.

Suppose that we assume that $G_c(s)$ is a PID controller of the form

$$G_c(s) = \frac{K(s + \alpha)(s + \beta)}{s}$$

Then the closed-loop transfer function $Y(s)/D(s)$ becomes

$$\frac{Y(s)}{D(s)} = \frac{10}{s(s + 1) + 10 G_c}$$

$$= \frac{10}{s(s + 1) + \dfrac{10K(s + \alpha)(s + \beta)}{s}}$$

$$= \frac{10s}{s^2(s + 1) + 10K(s + \alpha)(s + \beta)}$$

Note that the presence of "s" in the numerator of $Y(s)/D(s)$ assures that the steady-state response to the step disturbance input is zero.

Let us assume that the desired dominant closed-loop poles are complex conjugates and are given by

$$s = -a \pm jb$$

and the remaining closed-loop pole is real and is located at

$$s = -c$$

Note that in this problem there are three requirements. The first requirement is that the response to the step disturbance input damp out quickly. The second requirement is that the maximum overshoot in the response to the unit-step reference input be between 19% and 2% and the settling time be less than 1 sec. The third requirement is that the steady-state errors in the responses to both the ramp and acceleration reference inputs be zero.

A set (or sets) of reasonable values of a, b, and c must be searched using a computational approach. To satisfy the first requirement, we choose the search region for a, b, and c to be

$$2 \le a \le 6, \qquad 2 \le b \le 6, \qquad 6 \le c \le 12$$

This region is shown in Figure 8–34. If the dominant closed-loop poles $s = -a \pm jb$ are located anywhere in the shaded region, the response to a step disturbance input will damp out quickly. (The first requirement will be met.)

Notice that the denominator of $Y(s)/D(s)$ can be written as

$$s^2(s + 1) + 10K(s + \alpha)(s + \beta)$$

$$= s^3 + (1 + 10K)s^2 + 10K(\alpha + \beta)s + 10K\alpha\beta$$

$$= (s + a + jb)(s + a - jb)(s + c)$$

$$= s^3 + (2a + c)s^2 + (a^2 + b^2 + 2ac)s + (a^2 + b^2)c$$

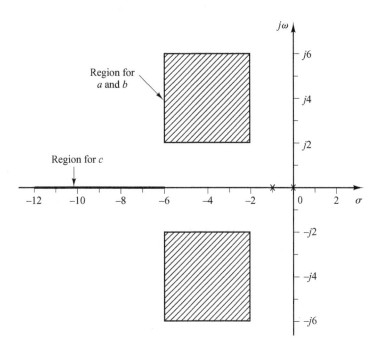

Figure 8–34
Search regions for
a, b, and c.

Since the denominators of $Y(s)/D(s)$ and $Y(s)/R(s)$ are the same, the denominator of $Y(s)/D(s)$ determines also the response characteristics for the reference input. To satisfy the third requirement, we refer to the zero-placement method and choose the closed-loop transfer function $Y(s)/R(s)$ to be of the following form:

$$\frac{Y(s)}{R(s)} = \frac{(2a + c)s^2 + (a^2 + b^2 + 2ac)s + (a^2 + b^2)c}{s^3 + (2a + c)s^2 + (a^2 + b^2 + 2ac)s + (a^2 + b^2)c}$$

in which case the third requirement is automatically satisfied.

Our problem then becomes a search of a set or sets of desired closed-loop poles in terms of a, b, and c in the specified region, such that the system will satisfy the requirement on the response to the unit-step reference input that the maximum overshoot be between 19% and 2% and the settling time be less than 1 sec. (If an acceptable set cannot be found in the search region, we need to widen the region.)

In the computational search, we need to assume a reasonable step size. In this problem, we assume it to be 0.2.

MATLAB Program 8–8 produces a table of sets of acceptable values of a, b, and c. Using this program, we find that the requirement on the response to the unit-step reference input is met by any of the 23 sets shown in the table in MATLAB Program 8–8. Note that the last row in the table corresponds to the last search point. This point does not satisfy the requirement and thus it should simply be ignored. (In the program written, the last search point produces the last row in the table whether or not it satisfies the requirement.)

```
MATLAB Program 8–8

t = 0:0.01:4;
k = 0;
for i = 1:21;
   a(i) = 6.2-i*0.2;
   for j = 1:21;
      b(j) = 6.2-j*0.2;
      for h = 1:31;
        c(h) = 12.2-h*0.2;
      num = [0  2*a(i)+c(h)  a(i)^2+b(j)^2+2*a(i)*c(h)  (a(i)^2+b(j)^2)*c(h)];
      den = [1  2*a(i)+c(h)  a(i)^2+b(j)^2+2*a(i)*c(h)  (a(i)^2+b(j)^2)*c(h)];
        y = step(num,den,t);
        m = max(y);
        s = 401; while y(s) > 0.98 & y(s) < 1.02;
        s = s-1; end;
        ts = (s-1)*0.01;
      if m < 1.19 & m > 1.02 & ts < 1.0;
      k = k+1;
      table(k,:) = [a(i)  b(j)  c(h)  m  ts];
         end
      end
   end
end
```

(continues on next page)

```
table(k,:) = [a(i)  b(j)  c(h)  m  ts]
table =

   4.2000  2.0000  12.0000  1.1896  0.8500
   4.0000  2.0000  12.0000  1.1881  0.8700
   4.0000  2.0000  11.8000  1.1890  0.8900
   4.0000  2.0000  11.6000  1.1899  0.9000
   3.8000  2.2000  12.0000  1.1883  0.9300
   3.8000  2.2000  11.8000  1.1894  0.9400
   3.8000  2.0000  12.0000  1.1861  0.8900
   3.8000  2.0000  11.8000  1.1872  0.9100
   3.8000  2.0000  11.6000  1.1882  0.9300
   3.8000  2.0000  11.4000  1.1892  0.9400
   3.6000  2.4000  12.0000  1.1893  0.9900
   3.6000  2.2000  12.0000  1.1867  0.9600
   3.6000  2.2000  11.8000  1.1876  0.9800
   3.6000  2.2000  11.6000  1.1886  0.9900
   3.6000  2.0000  12.0000  1.1842  0.9200
   3.6000  2.0000  11.8000  1.1852  0.9400
   3.6000  2.0000  11.6000  1.1861  0.9500
   3.6000  2.0000  11.4000  1.1872  0.9700
   3.6000  2.0000  11.2000  1.1883  0.9800
   3.4000  2.0000  12.0000  1.1820  0.9400
   3.4000  2.0000  11.8000  1.1831  0.9600
   3.4000  2.0000  11.6000  1.1842  0.9800
   3.2000  2.0000  12.0000  1.1797  0.9600
   2.0000  2.0000   6.0000  1.2163  1.8900
```

As noted above, 23 sets of variables a, b, and c satisfy the requirement. Unit-step response curves of the system with any of the 23 sets are about the same. The unit-step response curve with

$$a = 4.2, \qquad b = 2, \qquad c = 12$$

is shown in Figure 8–35(a). The maximum overshoot is 18.96% and the settling time is 0.85 sec. Using these values of a, b, and c, the desired closed-loop poles are located at

$$s = -4.2 \pm j2, \qquad s = -12$$

Using these closed-loop poles, the denominator of $Y(s)/D(s)$ becomes

$$s^2(s + 1) + 10K(s + \alpha)(s + \beta) = (s + 4.2 + j2)(s + 4.2 - j2)(s + 12)$$

or

$$s^3 + (1 + 10K)s^2 + 10K(\alpha + \beta)s + 10K\alpha\beta = s^3 + 20.4s^2 + 122.44s + 259.68$$

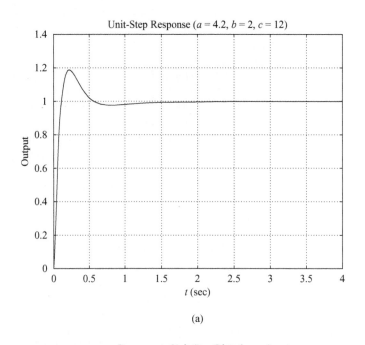

Unit-Step Response ($a = 4.2$, $b = 2$, $c = 12$)

(a)

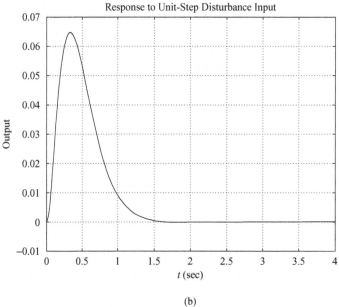

Response to Unit-Step Disturbance Input

Figure 8–35
(a) Response to unit-step reference input
($a = 4.2, b = 2, c = 12$);
(b) response to unit-step disturbance input
($a = 4.2, b = 2, c = 12$).

(b)

By equating the coefficients of equal powers of s on both sides of this last equation, we obtain

$$1 + 10K = 20.4$$

$$10K(\alpha + \beta) = 122.44$$

$$10K\alpha\beta = 259.68$$

Hence

$$K = 1.94, \qquad \alpha + \beta = \frac{122.44}{19.4}, \qquad \alpha\beta = \frac{259.68}{19.4}$$

Then $G_c(s)$ can be written as

$$G_c(s) = K \frac{(s + \alpha)(s + \beta)}{s}$$

$$= \frac{K[s^2 + (\alpha + \beta)s + \alpha\beta]}{s}$$

$$= \frac{1.94s^2 + 12.244s + 25.968}{s}$$

The closed-loop transfer function $Y(s)/D(s)$ becomes

$$\frac{Y(s)}{D(s)} = \frac{10}{s(s + 1) + 10G_c}$$

$$= \frac{10}{s(s + 1) + 10 \dfrac{1.94s^2 + 12.244s + 25.968}{s}}$$

$$= \frac{10s}{s^3 + 20.4s^2 + 122.44s + 259.68}$$

Using this expression, the response $y(t)$ to a unit-step disturbance input can be obtained as shown in Figure 8–35(b).

Figure 8–36(a) shows the response of the system to the unit-step reference input when a, b, and c are chosen as

$$a = 3.2, \qquad b = 2, \qquad c = 12$$

Figure 8–36(b) shows the response of this system when it is subjected to a unit-step disturbance input. Comparing Figures 8–35(a) and Figure 8–36(a), we find that they are about the same. However, comparing Figures 8–35(b) and 8–36(b), we find the former to be a little bit better than the latter. Comparing the responses of systems with each set in the table, we conclude the first set of values ($a = 4.2$, $b = 2$, $c = 12$) to be one of the best. Therefore, as the solution to this problem, we choose

$$a = 4.2, \qquad b = 2, \qquad c = 12$$

Design Step 2: Next, we determine G_{c1}. Since $Y(s)/R(s)$ can be given by

$$\frac{Y(s)}{R(s)} = \frac{G_p G_{c1}}{1 + G_p G_c}$$

$$= \frac{\dfrac{10}{s(s + 1)} G_{c1}}{1 + \dfrac{10}{s(s + 1)} \dfrac{1.94s^2 + 12.244s + 25.968}{s}}$$

$$= \frac{10sG_{c1}}{s^3 + 20.4s^2 + 122.44s + 259.68}$$

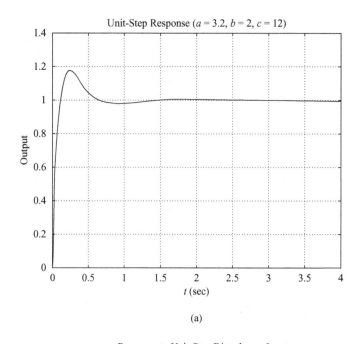

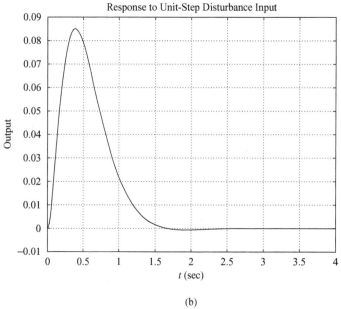

Figure 8–36
(a) Response to unit-step
reference input
($a = 3.2, b = 2, c = 12$);
(b) response to unit-step
disturbance input
($a = 3.2, b = 2, c = 12$).

our problem becomes that of designing $G_{c1}(s)$ to satisfy the requirements on the responses to the step, ramp, and acceleration inputs.

Since the numerator involves "s", $G_{c1}(s)$ must include an integrator to cancel this "s". [Although we want "s" in the numerator of the closed-loop transfer function $Y(s)/D(s)$ to obtain zero steady-state error to the step disturbance input, we do not want to have "s" in the numerator of the closed-loop transfer function $Y(s)/R(s)$.] To eliminate the offset in the response to the step reference input and eliminate the steady-state errors in following the ramp reference input and acceleration reference input, the numerator of $Y(s)/R(s)$ must be equal to the last three terms of the denominator, as mentioned earlier. That is,

$$10sG_{c1}(s) = 20.4s^2 + 122.44s + 259.68$$

or

$$G_{c1}(s) = 2.04s + 12.244 + \frac{25.968}{s}$$

Thus, $G_{c1}(s)$ is a PID controller. Since $G_c(s)$ is given as

$$G_c(s) = G_{c1}(s) + G_{c2}(s) = \frac{1.94s^2 + 12.244s + 25.968}{s}$$

we obtain

$$G_{c2}(s) = G_c(s) - G_{c1}(s)$$

$$= \left(1.94s + 12.244 + \frac{25.968}{s} \right) - \left(2.04s + 12.244 + \frac{25.968}{s} \right)$$

$$= -0.1s$$

Thus, $G_{c2}(s)$ is a derivative controller. A block diagram of the designed system is shown in Figure 8–37.

The closed-loop transfer function $Y(s)/R(s)$ now becomes

$$\frac{Y(s)}{R(s)} = \frac{20.4s^2 + 122.44s + 259.68}{s^3 + 20.4s^2 + 122.44s + 259.68}$$

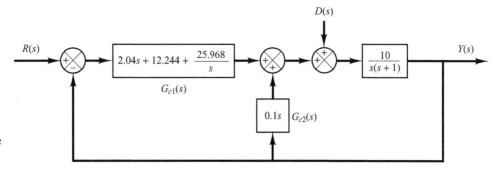

Figure 8–37
Block diagram of the designed system.

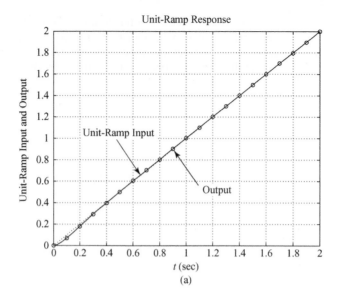

Unit-Ramp Response

Unit-Ramp Input

Output

t (sec)

(a)

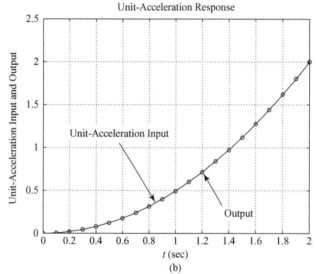

Unit-Acceleration Response

Unit-Acceleration Input

Output

t (sec)

(b)

Figure 8–38
(a) Response to unit-ramp reference input; (b) response to unit-acceleration reference input.

The response to the unit-ramp reference input and that to the unit-acceleration reference input are shown in Figures 8–38(a) and (b), respectively. The steady-state errors in following the ramp input and acceleration input are zero. Thus, all the requirements of the problem are satisfied. Hence, the designed controllers $G_{c1}(s)$ and $G_{c2}(s)$ are acceptable.

EXAMPLE 8–5 Consider the control system shown in Figure 8–39. This is a two-degrees-of-freedom system. In the design problem considered here, we assume that the noise input $N(s)$ is zero. Assume that the plant transfer function $G_p(s)$ is given by

$$G_p(s) = \frac{5}{(s+1)(s+5)}$$

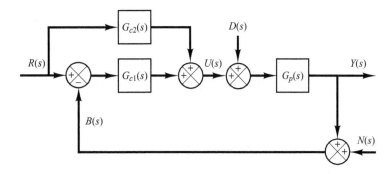

Figure 8–39
Two-degrees-of-freedom control system.

Assume also that the controller $G_{c1}(s)$ is of PID type. That is,

$$G_{c1}(s) = K_p\left(1 + \frac{1}{T_i s} + T_d s\right)$$

The controller $G_{c2}(s)$ is of P or PD type. [If $G_{c2}(s)$ involves integral control action, then this will introduce a ramp component in the input signal, which is not desirable. Therefore, $G_{c2}(s)$ should not include the integral control action.] Thus, we assume that

$$G_{c2}(s) = \hat{K}_p\left(1 + \hat{T}_d s\right)$$

where $\hat{T}_d$ may be zero.

Let us design controllers $G_{c1}(s)$ and $G_{c2}(s)$ such that the responses to the step-disturbance input and the step-reference input are of "desirable characteristics" in the sense that

1. The response to the step-disturbance input will have a small peak and eventually approach zero. (That is, there will be no steady-state error.)
2. The response to the step reference input will exhibit less than 25% overshoot with a settling time less than 2 sec. The steady-state errors to the ramp reference input and acceleration reference input should be zero.

The design of this two-degrees-of-freedom control system may be carried out by following the steps **1** and **2** below.

1. Determine $G_{c1}(s)$ so that the response to the step-disturbance input is of desirable characteristics.
2. Design $G_{c2}(s)$ so that the responses to the reference inputs are of desirable characteristics without changing the response to the step disturbance considered in step **1**.

Design of $G_{c1}(s)$: First, note that we assumed the noise input $N(s)$ to be zero. To obtain the response to the step-disturbance input, we assume that the reference input is zero. Then the block diagram which relates $Y(s)$ and $D(s)$ can be drawn as shown in Figure 8–40. The transfer function $Y(s)/D(s)$ is given by

$$\frac{Y(s)}{D(s)} = \frac{G_p}{1 + G_{c1}G_p}$$

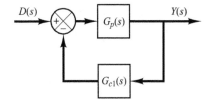

Figure 8–40
Control system.

where

$$G_{c1}(s) = K_p\left(1 + \frac{1}{T_i s} + T_d s\right)$$

This controller involves one pole at the origin and two zeros. If we assume that the two zeros are located at the same place (a double zero), then $G_{c1}(s)$ can be written as

$$G_{c1}(s) = K \frac{(s + a)^2}{s}$$

Then the characteristic equation for the system becomes

$$1 + G_{c1}(s)G_p(s) = 1 + \frac{K(s + a)^2}{s} \frac{5}{(s + 1)(s + 5)} = 0$$

or

$$s(s + 1)(s + 5) + 5K(s + a)^2 = 0$$

which can be rewritten as

$$s^3 + (6 + 5K)s^2 + (5 + 10Ka)s + 5Ka^2 = 0 \qquad (8\text{–}8)$$

If we place the double zero between $s = -3$ and $s = -6$, then the root-locus plot of $G_{c1}(s)G_p(s)$ may look like the one shown in Figure 8–41. The speed of response should be fast, but not faster than necessary, because faster response generally implies larger or more expensive components. Therefore, we may choose the dominant closed-loop poles at

$$s = -3 \pm j2$$

(Note that this choice is not unique. There are infinitely many possible closed-loop poles that we may choose from.)

Since the system is of third order, there are three closed-loop poles. The third one is located on the negative real axis to the left of point $s = -5$.

Let us substitute $s = -3 + j2$ into Equation (8–8).

$$(-3 + j2)^3 + (6 + 5K)(-3 + j2)^2 + (5 + 10Ka)(-3 + j2) + 5Ka^2 = 0$$

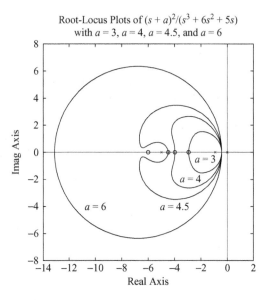

Root-Locus Plots of $(s + a)^2/(s^3 + 6s^2 + 5s)$
with $a = 3$, $a = 4$, $a = 4.5$, and $a = 6$

Figure 8–41
Root-locus plots of $5K(s + a)^2/[s(s + 1)(s + 5)]$ when $a = 3$, $a = 4$, $a = 4.5$, and $a = 6$.

which can be simplified to

$$24 + 25K - 30Ka + 5Ka^2 + j(-16 - 60K + 20Ka) = 0$$

By equating the real part and imaginary part to zero, respectively, we obtain

$$24 + 25K - 30Ka + 5Ka^2 = 0 \tag{8-9}$$

$$-16 - 60K + 20Ka = 0 \tag{8-10}$$

From Equation (8–10), we have

$$K = \frac{4}{5a - 15} \tag{8-11}$$

Substituting Equation (8–11) into Equation (8–9), we get

$$a^2 = 13$$

or $a = 3.6056$ or -3.6056. Notice that the values of K become

$$K = 1.3210 \qquad \text{for } a = 3.6056$$

$$K = -0.1211 \qquad \text{for } a = -3.6056$$

Since $G_{c1}(s)$ is in the feedforward path, the gain K should be positive. Hence, we choose

$$K = 1.3210, \qquad a = 3.6056$$

Then $G_{c1}(s)$ can be given by

$$G_{c1}(s) = K\frac{(s + a)^2}{s}$$

$$= 1.3210\frac{(s + 3.6056)^2}{s}$$

$$= \frac{1.3210s^2 + 9.5260s + 17.1735}{s}$$

To determine K_p, T_i, and T_d, we proceed as follows:

$$G_{c1}(s) = \frac{1.3210(s^2 + 7.2112s + 13)}{s}$$

$$= 9.5260\left(1 + \frac{1}{0.5547s} + 0.1387s\right) \tag{8-12}$$

Thus,

$$K_p = 9.5260, \qquad T_i = 0.5547, \qquad T_d = 0.1387$$

To check the response to a unit-step disturbance input, we obtain the closed-loop transfer function $Y(s)/D(s)$.

$$\frac{Y(s)}{D(s)} = \frac{G_p}{1 + G_{c1}G_p}$$

$$= \frac{5s}{s(s + 1)(s + 5) + 5K(s + a)^2}$$

$$= \frac{5s}{s^3 + 12.605s^2 + 52.63s + 85.8673}$$

The response to the unit-step disturbance input is shown in Figure 8–42. The response curve seems good and acceptable. Note that the closed-loop poles are located at $s = -3 \pm j2$ and $s = -6.6051$. The complex-conjugate closed-loop poles act as dominant closed-loop poles.

Design of $G_{c2}(s)$:   We now design $G_{c2}(s)$ to obtain the desired responses to the reference inputs. The closed-loop transfer function $Y(s)/R(s)$ can be given by

$$\frac{Y(s)}{R(s)} = \frac{(G_{c1} + G_{c2})G_p}{1 + G_{c1}G_p}$$

$$= \frac{\left[\dfrac{1.321s^2 + 9.526s + 17.1735}{s} + \hat{K}_p(1 + \hat{T}_d s)\right]\dfrac{5}{(s+1)(s+5)}}{1 + \dfrac{1.321s^2 + 9.526s + 17.1735}{s}\dfrac{5}{(s+1)(s+5)}}$$

$$= \frac{(6.6051 + 5\hat{K}_p\hat{T}_d)s^2 + (47.63 + 5\hat{K}_p)s + 85.8673}{s^3 + 12.6051s^2 + 52.63s + 85.8673}$$

Zero placement.   We place two zeros together with the dc gain constant such that the numerator is the same as the sum of the last three terms of the denominator. That is,

$$(6.6051 + 5\hat{K}_p\hat{T}_d)s^2 + (47.63 + 5\hat{K}_p)s + 85.8673 = 12.6051s^2 + 52.63s + 85.8673$$

By equating the coefficients of s^2 terms and s terms on both sides of this last equation,

$$6.6051 + 5\hat{K}_p\hat{T}_d = 12.6051$$

$$47.63 + 5\hat{K}_p = 52.63$$

from which we get

$$\hat{K}_p = 1, \qquad \hat{T}_d = 1.2$$

Therefore,

$$G_{c2}(s) = 1 + 1.2s \qquad\qquad (8\text{–}13)$$

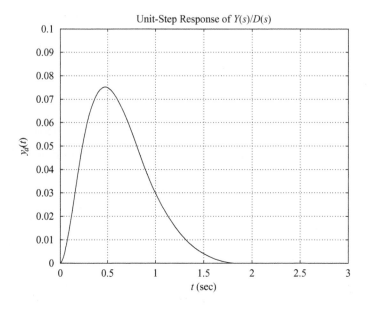

Figure 8–42
Response to unit-step disturbance input.

With this controller $G_{c2}(s)$, the closed-loop transfer function $Y(s)/R(s)$ becomes

$$\frac{Y(s)}{R(s)} = \frac{12.6051s^2 + 52.63s + 85.8673}{s^3 + 12.6051s^2 + 52.63s + 85.8673}$$

The response to the unit-step reference input becomes as shown in Figure 8–43(a).

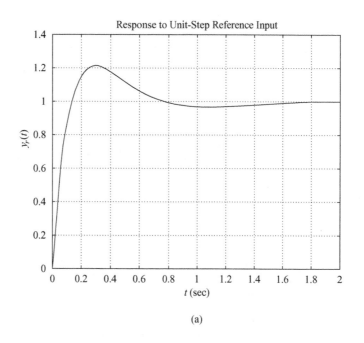

(a)

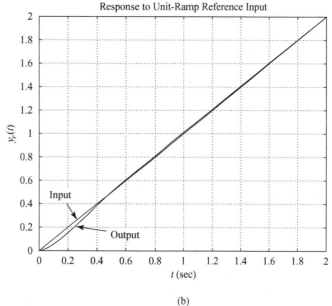

Figure 8–43
(a) Response to unit-step reference input; (b) response to unit-ramp reference input; (c) response to unit-acceleration reference input.

(b)

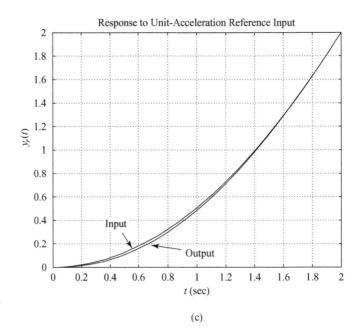

Figure 8–43
(continued)

(c)

The response exhibits the maximum overshoot of 21% and the settling time is approximately 1.6 sec. Figures 8–43(b) and (c) show the ramp response and acceleration response. The steady-state errors in both responses are zero. The response to the step disturbance was satisfactory. Thus, the designed controllers $G_{c1}(s)$ and $G_{c2}(s)$ given by Equations (8–12) and (8–13), respectively, are satisfactory.

If the response characteristics to the unit-step reference input are not satisfactory, we need to change the location of the dominant closed-loop poles and repeat the design process. The dominant closed-loop poles should lie in a certain region in the left-half s plane (such as $2 \le a \le 6$, $2 \le b \le 6$, $6 \le c \le 12$). If the computational search is desired, write a computer program (similar to MATLAB Program 8–8) and execute the search process. Then a desired set or sets of values of a, b, and c may be found such that the system response to the unit-step reference input satisfies all requirements on maximum overshoot and settling time.

EXAMPLE PROBLEMS AND SOLUTIONS

A–8–1. Describe briefly the dynamic characteristics of the PI controller, PD controller, and PID controller.

Solution. The PI controller is characterized by the transfer function

$$G_c(s) = K_p\left(1 + \frac{1}{T_i s}\right)$$

The PI controller is a lag compensator. It possesses a zero at $s = -1/T_i$ and a pole at $s = 0$. Thus, the characteristic of the PI controller is infinite gain at zero frequency. This improves the steady-state characteristics. However, inclusion of the PI control action in the system increases the type number of the compensated system by 1, and this causes the compensated system to be less stable or even makes the system unstable. Therefore, the values of K_p and T_i must be chosen carefully to ensure a proper transient response. By properly designing the PI controller, it is possible to make the transient response to a step input exhibit relatively small or no overshoot. The speed of response, however, becomes much slower. This is because the PI controller, being a low-pass filter, attenuates the high-frequency components of the signal.

The PD controller is a simplified version of the lead compensator. The PD controller has the transfer function $G_c(s)$, where

$$G_c(s) = K_p(1 + T_d s)$$

The value of K_p is usually determined to satisfy the steady-state requirement. The corner frequency $1/T_d$ is chosen such that the phase lead occurs in the neighborhood of the gain crossover frequency. Although the phase margin can be increased, the magnitude of the compensator continues to increase for the frequency region $1/T_d < \omega$. (Thus, the PD controller is a high-pass filter.) Such a continued increase of the magnitude is undesirable, since it amplifies high-frequency noises that may be present in the system. Lead compensation can provide a sufficient phase lead, while the increase of the magnitude for the high-frequency region is very much smaller than that for PD control. Therefore, lead compensation is preferred over PD control.

Because the transfer function of the PD controller involves one zero, but no pole, it is not possible to electrically realize it by passive RLC elements only. Realization of the PD controller using op amps, resistors, and capacitors is possible, but because the PD controller is a high-pass filter, as mentioned earlier, the differentiation process involved may cause serious noise problems in some cases. There is, however, no problem if the PD controller is realized by use of the hydraulic or pneumatic elements.

The PD control, as in the case of the lead compensator, improves the transient-response characteristics, improves system stability, and increases the system bandwidth, which implies fast rise time.

The PID controller is a combination of the PI and PD controllers. It is a lag–lead compensator. Note that the PI control action and PD control action occur in different frequency regions. The PI control action occurs at the low-frequency region and PD control action occurs at the high-frequency region. The PID control may be used when the system requires improvements in both transient and steady-state performances.

A–8–2. Show that the transfer function $U(s)/E(s)$ of the PID controller shown in Figure 8–44 is

$$\frac{U(s)}{E(s)} = K_0 \frac{T_1 + T_2}{T_1}\left[1 + \frac{1}{(T_1 + T_2)s} + \frac{T_1 T_2 s}{T_1 + T_2}\right]$$

Assume that the gain K is very large compared with unity, or $K \gg 1$.

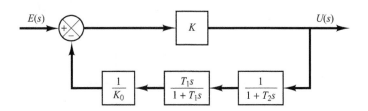

Figure 8–44
PID controller.

Solution

$$\frac{U(s)}{E(s)} = \frac{K}{1 + K\left(\dfrac{1}{K_0}\dfrac{T_1 s}{1 + T_1 s}\dfrac{1}{1 + T_2 s}\right)}$$

$$\doteq \frac{K}{K\left(\dfrac{1}{K_0}\dfrac{T_1 s}{1 + T_1 s}\dfrac{1}{1 + T_2 s}\right)}$$

$$= \frac{K_0(1 + T_1 s)(1 + T_2 s)}{T_1 s}$$

$$= K_0\left(1 + \frac{1}{T_1 s}\right)(1 + T_2 s)$$

$$= K_0\left(1 + \frac{1}{T_1 s} + T_2 s + \frac{T_2}{T_1}\right)$$

$$= K_0\frac{T_1 + T_2}{T_1}\left[1 + \frac{1}{(T_1 + T_2)s} + \frac{T_1 T_2 s}{T_1 + T_2}\right]$$

A–8–3. Consider the electronic circuit involving two operational amplifiers shown in Figure 8–45. This is a modified PID controller in that the transfer function involves an integrator and a first-order lag term. Obtain the transfer function of this PID controller.

Solution. Since

$$Z_1 = \frac{1}{\dfrac{1}{R_1} + C_1 s} + R_3 = \frac{R_1 + R_3 + R_1 R_3 C_1 s}{1 + R_1 C_1 s}$$

and

$$Z_2 = R_2 + \frac{1}{C_2 s}$$

we have

$$\frac{E(s)}{E_i(s)} = -\frac{Z_2}{Z_1} = -\frac{(R_2 C_2 s + 1)(R_1 C_1 s + 1)}{C_2 s(R_1 + R_3 + R_1 R_3 C_1 s)}$$

Also,

$$\frac{E_o(s)}{E(s)} = -\frac{R_5}{R_4}$$

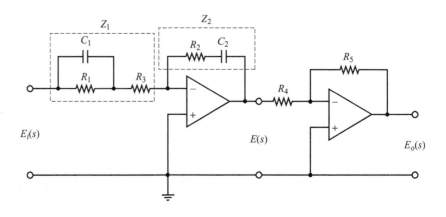

Figure 8–45
Modified PID
controller.

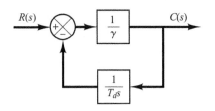

Figure 8–46
Approximate differentiator.

Consequently,

$$\frac{E_o(s)}{E_i(s)} = \frac{E_o(s)}{E(s)} \frac{E(s)}{E_i(s)} = \frac{R_5}{R_4(R_1 + R_3)C_2} \frac{(R_1C_1s + 1)(R_2C_2s + 1)}{s\left(\dfrac{R_1R_3}{R_1 + R_3}C_1s + 1\right)}$$

$$= \frac{R_5R_2}{R_4R_3} \frac{\left(s + \dfrac{1}{R_1C_1}\right)\left(s + \dfrac{1}{R_2C_2}\right)}{s\left(s + \dfrac{R_1 + R_3}{R_1R_3C_1}\right)}$$

Notice that R_1C_1 and R_2C_2 determine the locations of the zeros of the controller, while R_1, R_3, and C_1 affect the location of the pole on the negative real axis. R_5/R_4 adjusts the gain of the controller.

A–8–4. In practice, it is impossible to realize the true differentiator. Hence, we always have to approximate the true differentiator $T_d s$ by something like

$$\frac{T_d s}{1 + \gamma T_d s}$$

One way to realize such an approximate differentiator is to utilize an integrator in the feedback path. Show that the closed-loop transfer function of the system shown in Figure 8–46 is given by the preceding expression. (In the commercially available differentiator, the value of γ may be set as 0.1.)

Solution. The closed-loop transfer function of the system shown in Figure 8–46 is

$$\frac{C(s)}{R(s)} = \frac{\dfrac{1}{\gamma}}{1 + \dfrac{1}{\gamma T_d s}} = \frac{T_d s}{1 + \gamma T_d s}$$

Note that such a differentiator with first-order delay reduces the bandwidth of the closed-loop control system and reduces the detrimental effect of noise signals.

A–8–5. Consider the system shown in Figure 8–47. This is a PID control of a second-order plant $G(s)$. Assume that disturbances $D(s)$ enter the system as shown in the diagram. It is assumed that the reference input $R(s)$ is normally held constant, and the response characteristics to disturbances are a very important consideration in this system.

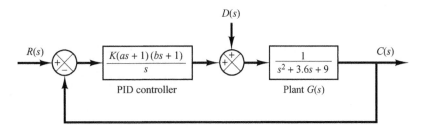

Figure 8–47
PID-controlled system.

Design a control system such that the response to any step disturbance will be damped out quickly (in 2 to 3 sec in terms of the 2% settling time). Choose the configuration of the closed-loop poles such that there is a pair of dominant closed-loop poles. Then obtain the response to the unit-step disturbance input. Also, obtain the response to the unit-step reference input.

Solution. The PID controller has the transfer function

$$G_c(s) = \frac{K(as + 1)(bs + 1)}{s}$$

For the disturbance input in the absence of the reference input, the closed-loop transfer function becomes

$$\frac{C_d(s)}{D(s)} = \frac{s}{s(s^2 + 3.6s + 9) + K(as + 1)(bs + 1)}$$

$$= \frac{s}{s^3 + (3.6 + Kab)s^2 + (9 + Ka + Kb)s + K} \tag{8-14}$$

The specification requires that the response to the unit-step disturbance be such that the settling time be 2 to 3 sec and the system have a reasonable damping. We may interpret the specification as $\zeta = 0.5$ and $\omega_n = 4$ rad/sec for the dominant closed-loop poles. We may choose the third pole at $s = -10$ so that the effect of this real pole on the response is small. Then the desired characteristic equation can be written as

$$(s + 10)(s^2 + 2 \times 0.5 \times 4s + 4^2) = (s + 10)(s^2 + 4s + 16) = s^3 + 14s^2 + 56s + 160$$

The characteristic equation for the system given by Equation (8–14) is

$$s^3 + (3.6 + Kab)s^2 + (9 + Ka + Kb)s + K = 0$$

Hence, we require

$$3.6 + Kab = 14$$

$$9 + Ka + Kb = 56$$

$$K = 160$$

which yields

$$ab = 0.065, \qquad a + b = 0.29375$$

The PID controller now becomes

$$G_c(s) = \frac{K[abs^2 + (a + b)s + 1]}{s}$$

$$= \frac{160(0.065s^2 + 0.29375s + 1)}{s}$$

$$= \frac{10.4(s^2 + 4.5192s + 15.385)}{s}$$

With this PID controller, the response to the disturbance is given by

$$C_d(s) = \frac{s}{s^3 + 14s^2 + 56s + 160} D(s)$$

$$= \frac{s}{(s + 10)(s^2 + 4s + 16)} D(s)$$

Clearly, for a unit-step disturbance input, the steady-state output is zero, since

$$\lim_{t \to \infty} c_d(t) = \lim_{s \to 0} s C_d(s) = \lim_{s \to 0} \frac{s^2}{(s + 10)(s^2 + 4s + 16)} \frac{1}{s} = 0$$

The response to a unit-step disturbance input can be obtained easily with MATLAB. MATLAB Program 8–9 produces a response curve as shown in Figure 8–48(a). From the response curve, we see that the settling time is approximately 2.7 sec. The response damps out quickly. Therefore, the system designed here is acceptable.

MATLAB Program 8–9

```
% ***** Response to unit-step disturbance input *****

numd = [1  0];
dend = [1  14  56  160];
t = 0:0.01:5;
[c1,x1,t] = step(numd,dend,t);
plot(t,c1)
grid
title('Response to Unit-Step Disturbance Input')
xlabel('t Sec')
ylabel('Output to Disturbance Input')

% ***** Response to unit-step reference input *****

numr = [10.4  47  160];
denr = [1  14  56  160];
[c2,x2,t] = step(numr,denr,t);
plot(t,c2)
grid
title('Response to Unit-Step Reference Input')
xlabel('t Sec')
ylabel('Output to Reference Input')
```

For the reference input $r(t)$, the closed-loop transfer function is

$$\frac{C_r(s)}{R(s)} = \frac{10.4(s^2 + 4.5192s + 15.385)}{s^3 + 14s^2 + 56s + 160}$$

$$= \frac{10.4s^2 + 47s + 160}{s^3 + 14s^2 + 56s + 160}$$

The response to a unit-step reference input can also be obtained by use of MATLAB Program 8–9. The resulting response curve is shown in Figure 8–48(b). The response curve shows that the maximum overshoot is 7.3% and the settling time is 1.2 sec. The system has quite acceptable response characteristics.

Example Problems and Solutions

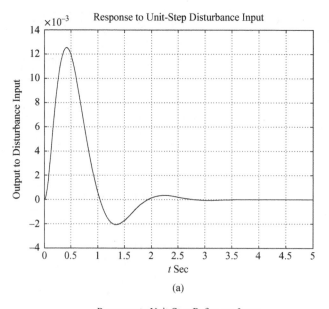

Response to Unit-Step Disturbance Input

(a)

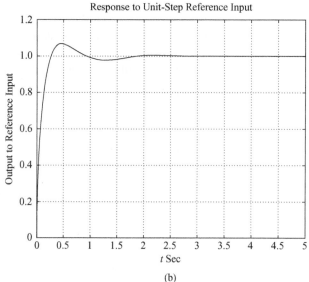

Response to Unit-Step Reference Input

(b)

Figure 8–48
(a) Response to
unit-step disturbance
input; (b) response to
unit-step reference
input.

A–8–6. Consider the system shown in Figure 8–49. It is desired to design a PID controller $G_c(s)$ such that the dominant closed-loop poles are located at $s = -1 \pm j\sqrt{3}$. For the PID controller, choose $a = 1$ and then determine the values of K and b. Sketch the root-locus diagram for the designed system.

Solution. Since

$$G_c(s)G(s) = K \frac{(s+1)(s+b)}{s} \frac{1}{s^2 + 1}$$

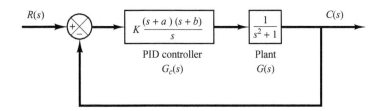

Figure 8–49
PID-controlled system.

the sum of the angles at $s = -1 + j\sqrt{3}$, one of the desired closed-loop poles, from the zero at $s = -1$ and poles at $s = 0$, $s = j$, and $s = -j$ is

$$90° - 143.794° - 120° - 110.104° = -283.898°$$

Hence the zero at $s = -b$ must contribute $103.898°$. This requires that the zero be located at

$$b = 0.5714$$

The gain constant K can be determined from the magnitude condition.

$$\left| K \frac{(s + 1)(s + 0.5714)}{s} \frac{1}{s^2 + 1} \right|_{s=-1+j\sqrt{3}} = 1$$

or

$$K = 2.3333$$

Then the compensator can be written as follows:

$$G_c(s) = 2.3333 \frac{(s + 1)(s + 0.5714)}{s}$$

The open-loop transfer function becomes

$$G_c(s)G(s) = \frac{2.3333(s + 1)(s + 0.5714)}{s} \frac{1}{s^2 + 1}$$

From this equation a root-locus plot for the compensated system can be drawn. Figure 8–50 is a root-locus plot.

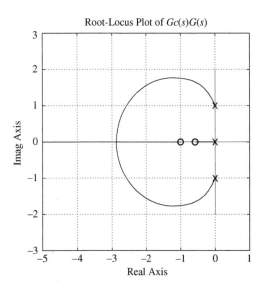

Figure 8–50
Root-locus plot of the compensated system.

Unit-Step Response of Compensated System

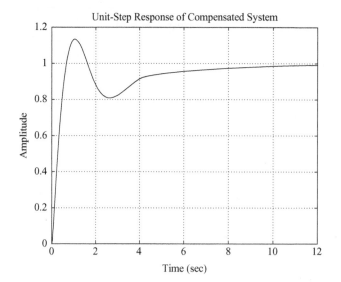

Figure 8–51
Unit-step response of
the compensated
system.

The closed-loop transfer function is given by

$$\frac{C(s)}{R(s)} = \frac{2.3333(s + 1)(s + 0.5714)}{s^3 + s + 2.3333(s + 1)(s + 0.5714)}$$

The closed-loop poles are located at $s = -1 \pm j\sqrt{3}$ and $s = -0.3333$. A unit-step response curve
is shown in Figure 8–51. The closed-loop pole at $s = -0.3333$ and a zero at $s = -0.5714$ produce
a long tail of small amplitude.

A–8–7. Consider the system shown in Figure 8–52. Design a compensator such that the static velocity
error constant is 4 sec^{-1}, phase margin is 50°, and gain margin is 10 dB or more. Plot unit-step and
unit-ramp response curves of the compensated system with MATLAB. Also, draw a Nyquist plot
of the compensated system with MATLAB. Using the Nyquist stability criterion, verify that the
designed system is stable.

Solution. Since the plant does not have an integrator, it is necessary to have an integrator in the
compensator. Let us choose the compensator to be

$$G_c(s) = \frac{K}{s}\hat{G}_c(s), \quad \lim_{s \to 0}\hat{G}_c(s) = 1$$

where $\hat{G}_c(s)$ is to be determined later. Since the static velocity error constant is specified as
4 sec^{-1}, we have

$$K_v = \lim_{s \to 0} sG_c(s)\frac{s + 0.1}{s^2 + 1} = \lim_{s \to 0} s\frac{K}{s}\hat{G}_c(s)\frac{s + 0.1}{s^2 + 1} = 0.1K = 4$$

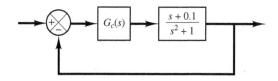

Figure 8–52
Control system.

Thus, $K = 40$. Hence

$$G_c(s) = \frac{40}{s}\hat{G}_c(s)$$

Next, we plot a Bode diagram of

$$G(s) = \frac{40(s + 0.1)}{s(s^2 + 1)}$$

MATLAB Program 8–10 produces a Bode diagram of $G(s)$ as shown in Figure 8–53.

MATLAB Program 8–10

```
% ***** Bode Diagram *****

num = [40  4];
den = [1  0.000000001  1  0];
bode(num,den)
title('Bode Diagram of G(s) = 40(s+0.1)/[s(s^2+1)]')
```

We need the phase margin of 50° and gain margin of 10 dB or more. Let us choose $\hat{G}_c(s)$ to be

$$\hat{G}_c(s) = as + 1 \quad (a > 0)$$

Then $G_c(s)$ will contribute up to 90° phase lead in the high-frequency region. By simple MATLAB trials, we find that $a = 0.1526$ gives the phase margin of 50° and gain margin of $+\infty$ dB.

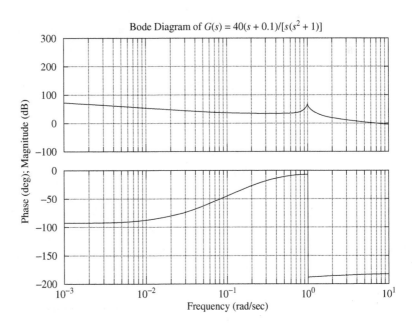

Figure 8–53
Bode diagram of
$G(s) =$
$40(s + 0.1)/[s(s^2 + 1)]$.

See MATLAB Program 8–11 and the resulting Bode diagram shown in Figure 8–54. From this Bode diagram we see that the static velocity error constant is 4 sec^{-1}, phase margin is 50° and gain margin is +∞ dB. Therefore, the designed system satisfies all the requirements.

MATLAB Program 8–11

```
% ***** Bode Diagram *****

num = conv([40 4],[0.1526 1]);
den = [1 0.000000001 1 0];
sys = tf(num,den);
w = logspace(-2,2,100);
bode(sys,w)
[Gm,pm,wcp,wcg] = margin(sys);
GmdB = 20*log10(Gm);
[GmdB,pm,wcp,wcg]

ans =

    Inf   50.0026   NaN   8.0114

title('Bode Diagram of G(s) = 40(s+0.1)(0.1526s+1)/[s(s^2+1)]')
```

The designed compensator has the following transfer function:

$$G_c(s) = \frac{40}{s} \hat{G}_c(s) = \frac{40(0.1526s + 1)}{s}$$

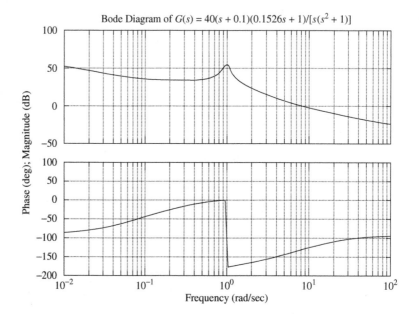

Figure 8–54
Bode diagram of
$G(s) = 40(s + 0.1)$
$(0.1526s + 1)/$
$[s(s^2 + 1)]$.

Chapter 8 / PID Controllers and Modified PID Controllers

The open-loop transfer function of the designed system is

$$\text{Open-loop transfer function} = \frac{40(0.1526s + 1)}{s} \frac{s + 0.1}{s^2 + 1}$$

$$= \frac{6.104s^2 + 40.6104s + 4}{s(s^2 + 1)}.$$

We shall next check the unit-step response and the unit-ramp response of the designed system. The closed-loop transfer function is

$$\frac{C(s)}{R(s)} = \frac{6.104s^2 + 40.6104s + 4}{s^3 + 6.104s^2 + 41.6104s + 4}$$

The closed-loop poles are located at

$$s = -3.0032 + j5.6573$$
$$s = -3.0032 - j5.6573$$
$$s = -0.0975$$

MATLAB Program 8–12 will produce the unit-step response curve of the designed system. The resulting unit-step response curve is shown in Figure 8–55. Notice that the closed-loop pole at $s = -0.0975$ and the plant zero at $s = -0.1$ produce a long tail of small amplitude.

MATLAB Program 8–12

```
% ***** Unit-Step Response *****

num = [6.104  40.6104  4];
den = [1  6.104  41.6104  4];
t = 0:0.01:10;
step(num,den,t)
grid
```

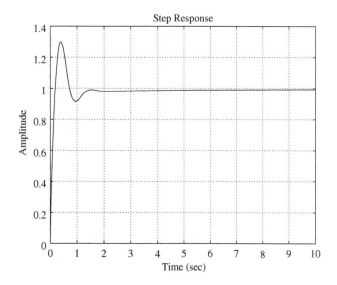

Figure 8–55
Unit-step response of
$C(s)/R(s) = (6.104s^2 +$
$40.6104s + 4)/(s^3 +$
$6.104s^2 + 41.6104s + 4)$.

Example Problems and Solutions

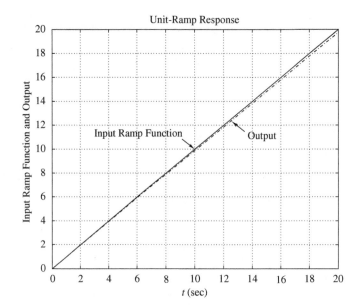

Figure 8–56
Unit-ramp response
of $C(s)/R(s) =$
$(6.104s^2 + 40.6104s +$
$4)/(s^3 + 6.104s^2 +$
$41.6104s + 4)$.

MATLAB Program 8–13 produces the unit-ramp response curve of the designed system. The resulting response curve is shown in Figure 8–56.

MATLAB Program 8–13

```
% ***** Unit-Ramp Response *****

num = [0  0  6.104  40.6104  4];
den = [1  6.104  41.6104  4  0];
t = 0:0.01:20;
c = step(num,den,t);
plot(t,c,'-.',t,t,'-')
title('Unit-Ramp Response')
xlabel('t(sec)')
ylabel('Input Ramp Function and Output')
text(3,11.5,'Input Ramp Function')
text(13.8,11.2,'Output')
```

Nyquist Plot. Earlier we found that the three closed-loop poles of the designed system are all in the left-half s plane. Hence the designed system is stable. The purpose of plotting Nyquist diagram here is not to test the stability of the system, but to enhance our understanding of Nyquist stability analysis. For a complicated system, Nyquist plot will look complicated enough that it is not easy to count the number of encirclements of the $-1 + j0$ point.

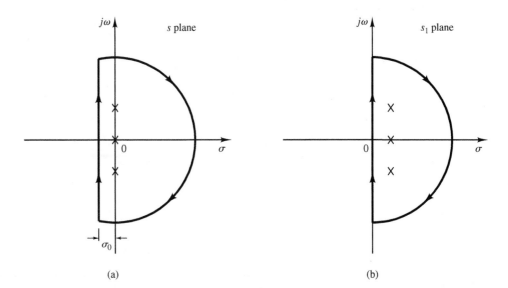

Figure 8–57
(a) Modified
Nyquist path in
the s plane;
(b) Nyquist path in
the s_1 plane.

(a)   (b)

Because the designed system involves three open-loop poles on the $j\omega$ axis, the Nyquist diagram will look quite complicated as we will see in what follows:

Define the open-loop transfer function of the designed system as $G(s)$. Then

$$G(s) = G_c(s)\frac{s + 0.1}{s^2 + 1} = \frac{6.104s^2 + 40.6104s + 4}{s(s^2 + 1)}$$

Let us choose a modified Nyquist path in the s plane as shown in Figure 8–57(a). The modified path encloses three open-loop poles ($s = 0$, $s = j1$, $s = -j1$). Now define $s_1 = s + \sigma_0$. Then, the Nyquist path in the s_1 plane becomes as shown in Figure 8–57(b). In the s_1 plane, the open-loop transfer function has three poles in the right-half s_1 plane.

Let us choose $\sigma_0 = 0.01$. Since $s = s_1 - \sigma_0$, we have

$$G(s) = G(s_1 - 0.01)$$

Open-loop transfer function in the s_1 plane

$$= \frac{6.104(s_1^2 - 0.02s_1 + 0.0001) + 40.6104(s_1 - 0.01) + 4}{(s_1 - 0.01)(s_1^2 - 0.02s_1 + 1.0001)}$$

$$= \frac{6.104s_1^2 + 40.48832s_1 + 3.5945064}{s_1^3 - 0.03s_1^2 + 1.0003s_1 - 0.010001}$$

A MATLAB program to obtain the Nyquist plot is shown in MATLAB Program 8–14. The resulting Nyquist plot is shown in Figure 8–58.

MATLAB Program 8–14

```
% ***** Nyquist Plot *****

num = [6.104  40.48832  3.5945064];
den = [1  -0.03  1.0003  -0.010001];
nyquist(num,den)
v = [-1500  1500  -2500  2500]; axis(v)
```

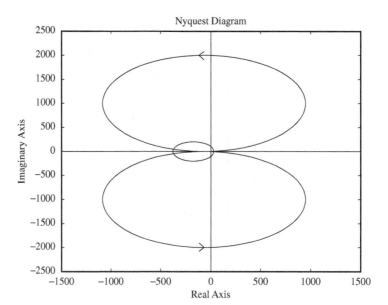

Nyquest Diagram

Figure 8–58
Nyquist plot.

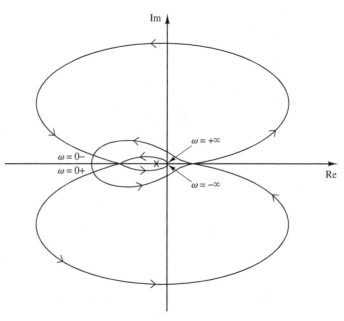

Figure 8–59
Redrawn Nyquist
plot.

Using the Nyquist plot obtained here, it is not easy to determine the encirclements of the $-1 + j0$ point by the Nyquist locus. Therefore, we need to redraw this Nyquist plot qualitatively to show the details near the $-1 + j0$ point. Such a redrawn Nyquist diagram is shown in Figure 8–59.

From this diagram we find that the $-1 + j0$ point is encircled counterclockwise three times. Hence, $N = -3$. Since the open-loop transfer function has three poles in the right-half s_1 plane, we have $P = 3$. Then, we have $Z = N + P = 0$. This means that there are no closed-loop poles in the right-half s_1 plane. The system is therefore stable.

A–8–8. Show that the I-PD-controlled system shown in Figure 8–60(a) is equivalent to the PID-controlled system with input filter shown in Figure 8–60(b).

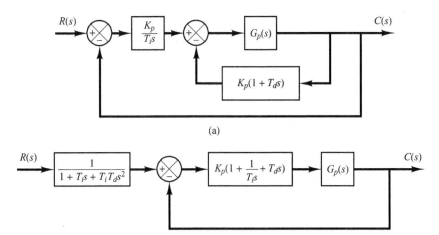

(a)

(b)

Figure 8–60
(a) I-PD-controlled
system;
(b) PID-controlled
system with input
filter.

Solution. The closed-loop transfer function $C(s)/R(s)$ of the I-PD-controlled system is

$$\frac{C(s)}{R(s)} = \frac{\dfrac{K_p}{T_i s} G_p(s)}{1 + K_p\left(1 + \dfrac{1}{T_i s} + T_d s\right)G_p(s)}$$

The closed-loop transfer function $C(s)/R(s)$ of the PID-controlled system with input filter shown in Figure 8–60(b) is

$$\frac{C(s)}{R(s)} = \frac{1}{1 + T_i s + T_i T_d s^2}\frac{K_p\left(1 + \dfrac{1}{T_i s} + T_d s\right)G_p(s)}{1 + K_p\left(1 + \dfrac{1}{T_i s} + T_d s\right)G_p(s)}$$

$$= \frac{\dfrac{K_p}{T_i s} G_p(s)}{1 + K_p\left(1 + \dfrac{1}{T_i s} + T_d s\right)G_p(s)}$$

The closed-loop transfer functions of both systems are the same. Thus, the two systems are equivalent.

A–8–9. The basic idea of the I-PD control is to avoid large control signals (which will cause a saturation phenomenon) within the system. By bringing the proportional and derivative control actions to the feedback path, it is possible to choose larger values for K_p and T_d than those possible by the PID control scheme.

Compare, qualitatively, the responses of the PID-controlled system and I-PD-controlled system to the disturbance input and to the reference input.

Solution. Consider first the response of the I-PD-controlled system to the disturbance input. Since, in the I-PD control of a plant, it is possible to select larger values for K_p and T_d than those of the PID-controlled case, the I-PD-controlled system will attenuate the effect of disturbance faster than the PID-controlled case.

Next, consider the response of the I-PD-controlled system to a reference input. Since the I-PD-controlled system is equivalent to the PID-controlled system with input filter (refer to Problem **A–8–8**), the PID-controlled system will have faster responses than the corresponding I-PD-controlled system, provided a saturation phenomenon does not occur in the PID-controlled system.

Example Problems and Solutions

629

A–8–10. In some cases it is desirable to provide an input filter as shown in Figure 8–61(a). Notice that the input filter $G_f(s)$ is outside the loop. Therefore, it does not affect the stability of the closed-loop portion of the system. An advantage of having the input filter is that the zeros of the closed-loop transfer function can be modified (canceled or replaced by other zeros) so that the closed-loop response is acceptable.

Show that the configuration in Figure 8–61(a) can be modified to that shown in Figure 8–61(b), where $G_d(s) = [G_f(s) - 1]G_c(s)$. The compensation structure shown in Figure 8–61(b) is sometimes called command compensation.

Solution. For the system of Figure 8–61(a), we have

$$\frac{C(s)}{R(s)} = G_f(s) \frac{G_c(s)G_p(s)}{1 + G_c(s)G_p(s)} \qquad (8\text{–}15)$$

For the system of Figure 8–61(b), we have

$$U(s) = G_d(s)R(s) + G_c(s)E(s)$$
$$E(s) = R(s) - C(s)$$
$$C(s) = G_p(s)U(s)$$

Thus

$$C(s) = G_p(s)\{G_d(s)R(s) + G_c(s)[R(s) - C(s)]\}$$

or

$$\frac{C(s)}{R(s)} = \frac{[G_d(s) + G_c(s)]G_p(s)}{1 + G_c(s)G_p(s)} \qquad (8\text{–}16)$$

By substituting $G_d(s) = [G_f(s) - 1]G_c(s)$ into Equation (8–16), we obtain

$$\frac{C(s)}{R(s)} = \frac{[G_f(s)G_c(s) - G_c(s) + G_c(s)]G_p(s)}{1 + G_c(s)G_p(s)}$$

$$= G_f(s) \frac{G_c(s)G_p(s)}{1 + G_c(s)G_p(s)}$$

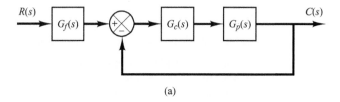

(a)

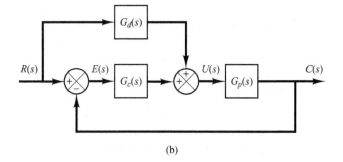

Figure 8–61
(a) Block diagram of control system with input filter;
(b) modified block diagram.

(b)

which is the same as Equation (8–15). Hence, we have shown that the systems shown in Figures 8–61(a) and (b) are equivalent.

It is noted that the system shown in Figure 8–61(b) has a feedforward controller $G_d(s)$. In such a case, $G_d(s)$ does not affect the stability of the closed-loop portion of the system.

A–8–11. A closed-loop system has the characteristic that the closed-loop transfer function is nearly equal to the inverse of the feedback transfer function whenever the open-loop gain is much greater than unity.

The open-loop characteristic may be modified by adding an internal feedback loop with a characteristic equal to the inverse of the desired open-loop characteristic. Suppose that a unity-feedback system has the open-loop transfer function

$$G(s) = \frac{K}{(T_1 s + 1)(T_2 s + 1)}$$

Determine the transfer function $H(s)$ of the element in the internal feedback loop so that the inner loop becomes ineffective at both low and high frequencies.

Solution. Figure 8–62(a) shows the original system. Figure 8–62(b) shows the addition of the internal feedback loop around $G(s)$. Since

$$\frac{C(s)}{E(s)} = \frac{G(s)}{1 + G(s)H(s)} = \frac{1}{H(s)} \frac{G(s)H(s)}{1 + G(s)H(s)}$$

if the gain around the inner loop is large compared with unity, then $G(s)H(s)/[1 + G(s)H(s)]$ is approximately equal to unity, and the transfer function $C(s)/E(s)$ is approximately equal to $1/H(s)$.

On the other hand, if the gain $|G(s)H(s)|$ is much less than unity, the inner loop becomes ineffective and $C(s)/E(s)$ becomes approximately equal to $G(s)$.

To make the inner loop ineffective at both the low- and high-frequency ranges, we require that

$$|G(j\omega)H(j\omega)| \ll 1, \quad \text{for } \omega \ll 1 \text{ and } \omega \gg 1$$

Since, in this problem,

$$G(j\omega) = \frac{K}{(1 + j\omega T_1)(1 + j\omega T_2)}$$

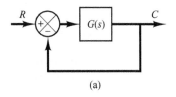

(a)

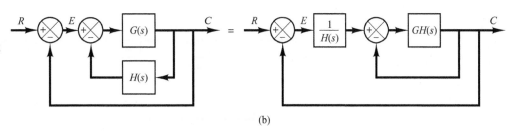

(b)

Figure 8–62
(a) Control system;
(b) addition of the internal feedback loop to modify the closed-loop characteristic.

Example Problems and Solutions

the requirement can be satisfied if $H(s)$ is chosen to be

$$H(s) = ks$$

because

$$\lim_{\omega \to 0} G(j\omega)H(j\omega) = \lim_{\omega \to 0} \frac{Kkj\omega}{(1 + j\omega T_1)(1 + j\omega T_2)} = 0$$

$$\lim_{\omega \to \infty} G(j\omega)H(j\omega) = \lim_{\omega \to \infty} \frac{Kkj\omega}{(1 + j\omega T_1)(1 + j\omega T_2)} = 0$$

Thus, with $H(s) = ks$ (velocity feedback), the inner loop becomes ineffective at both the low- and high-frequency regions. It becomes effective only in the intermediate-frequency region.

A–8–12. Consider the control system shown in Figure 8–63. This is the same system as that considered in Example 8–1. In that example we designed a PID controller $G_c(s)$, starting with the second method of the Ziegler–Nichols tuning rule. Here we design a PID controller using the computational approach with MATLAB. We shall determine the values of K and a of the PID controller

$$G_c(s) = K \frac{(s + a)^2}{s}$$

such that the unit-step response will exhibit the maximum overshoot between 10% and 2% ($1.02 \le$ maximum output ≤ 1.10) and the settling time will be less than 3 sec. The search region is

$$2 \le K \le 50, \qquad 0.05 \le a \le 2$$

Let us choose the step size for K to be 1 and that for a to be 0.05.

Write a MATLAB program to find the first set of variables K and a that will satisfy the given specifications. Also, write a MATLAB program to find all possible sets of variables K and a that will satisfy the given specifications. Plot the unit-step response curves of the designed system with the chosen sets of variables K and a.

Solution. The transfer function of the plant is

$$G_p(s) = \frac{1}{s^3 + 6s^2 + 5s}$$

The closed-loop transfer function $C(s)/R(s)$ is given by

$$\frac{C(s)}{R(s)} = \frac{Ks^2 + 2Kas + Ka^2}{s^4 + 6s^3 + (5 + K)s^2 + 2Kas + Ka^2}$$

A possible MATLAB program that will produce the first set of variables K and a that will satisfy the given specifications is given in MATLAB Program 8–15. In this program we

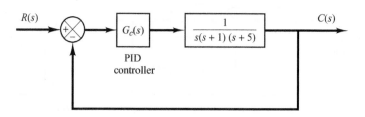

Figure 8–63
Control system.

use two 'for' loops. The specification for the settling time is interpreted by the following four lines:

$$s = 501; \text{while } y(s) > 0.98 \text{ and } y(s) < 1.02;$$

$$s = s - 1; \text{end};$$

$$ts = (s - 1) * 0.01$$

$$ts < 3.0$$

Note that for $t = 0:0.01:5$, we have 501 computing time points. $s = 501$ corresponds to the last computing time point.

The solution obtained by this program is

$$K = 32, \qquad a = 0.2$$

with the maximum overshoot equal to 9.69% and the settling time equal to 2.64 sec. The resulting unit-step response curve is shown in Figure 8–64.

MATLAB Program 8–15

```
t = 0:0.01:5;
for K = 50:-1:2;
   for a = 2:-0.05:0.05;
      num = [K  2*K*a  K*a^2];
      den = [1  6  5+K  2*K*a  K*a^2];
       y = step(num,den,t);
       m = max(y);
       s = 501; while y(s) > 0.98 & y(s) < 1.02;
       s = s-1; end;
       ts = (s-1)*0.01;
      if m < 1.10 & m > 1.02 & ts < 3.0
      break;
      end
      end
   if m < 1.10 & m > 1.02 & ts < 3.0
   break
   end
   end
plot(t,y)
grid
title('Unit-Step Response')
xlabel('t sec')
ylabel('Output')
solution = [K;a;m;ts]

solution =

32.0000
 0.2000
 1.0969
 2.6400
```

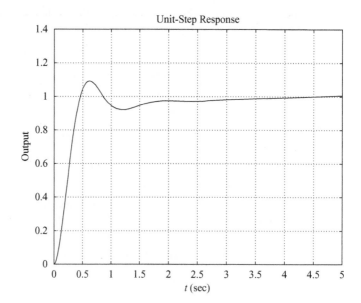

Figure 8–64
Unit-step response
curve.

Next, we shall consider the case where we want to find all sets of variables that will satisfy the given specifications. A possible MATLAB program for this purpose is given in MATLAB Program 8–16. Note that in the table shown in the program, the last row of the table (k, :) or the first row of the sorttable should be ignored. (These are the last K and a values for searching purposes.)

```
MATLAB Program 8–16

t = 0:0.01:5;
k = 0;
for i = 1:49;
   K(i) = 51-i*1;
      for j = 1:40;
      a(j) = 2.05-j*0.05;
      num = [K(i)  2*K(i)*a(j)  K(i)*a(j)*a(j)];
      den = [1  6  5+K(i)  2*K(i)*a(j)  K(i)*a(j)*a(j)];
         y = step(num,den,t);
         m = max(y);
         s = 501; while y(s) > 0.98 & y(s) < 1.02;
         s = s-1; end;
         ts = (s-1)*0.01;
         if m < 1.10 & m > 1.02 & ts < 3.0
         k = k+1;
         table(k,:) = [K(i)  a(j)  m  ts];
         end
      end
   end
table(k,:) = [K(i)  a(j)  m  ts]

table =
```

(continues on next page)

```
 32.0000  0.2000  1.0969  2.6400
 31.0000  0.2000  1.0890  2.6900
 30.0000  0.2000  1.0809  2.7300
 29.0000  0.2500  1.0952  1.7800
 29.0000  0.2000  1.0726  2.7800
 28.0000  0.2000  1.0639  2.8300
 27.0000  0.2000  1.0550  2.8900
  2.0000  0.0500  0.3781  5.0000

sorttable = sortrows(table,3)

sorttable =

  2.0000  0.0500  0.3781  5.0000
 27.0000  0.2000  1.0550  2.8900
 28.0000  0.2000  1.0639  2.8300
 29.0000  0.2000  1.0726  2.7800
 30.0000  0.2000  1.0809  2.7300
 31.0000  0.2000  1.0890  2.6900
 29.0000  0.2500  1.0952  1.7800
 32.0000  0.2000  1.0969  2.6400

K = sorttable(7,1)

K =

   29

a = sorttable(7,2)

a=

   0.2500

num = [K  2*K*a  K*a^2];
den = [1  6  5+K  2*K*a  K*a^2];
y = step(num,den,t);
plot(t,y)
grid
hold
Current plot held
K = sorttable(2,1)

K=

   27

a = sorttable(2,2)

a=

   0.2000
```

(continues on next page)

```
num = [K  2*K*a  K*a^2];
den = [1  6  5+K  2*K*a  K*a^2];
y = step(num,den,t);
plot(t,y)
title('Unit-Step Response Curves')
xlabel('t (sec)')
ylabel('Output')
text(1.22,1.22,'K = 29, a = 0.25')
text(1.22,0.72,'K = 27, a = 0.2')
```

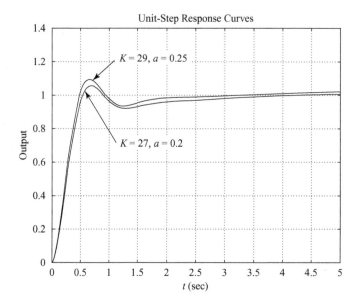

Figure 8–65
Unit-step response
curves.

From the sorttable, it seems that

$$K = 29, a = 0.25 \text{ (max overshoot = 9.52\%, settling time = 1.78 sec)}$$

and

$$K = 27, a = 0.2 \text{ (max overshoot = 5.5\%, settling time = 2.89 sec)}$$

are two of the best choices. The unit-step response curves for these two cases are shown in Figure 8–65. From these curves, we might conclude that the best choice depends on the system objective. If a small maximum overshoot is desired, $K = 27, a = 0.2$ will be the best choice. If the shorter settling time is more important than a small maximum overshoot, then $K = 29, a = 0.25$ will be the best choice.

A–8–13. Consider the two-degrees-of-freedom control system shown in Figure 8–66. The plant $G_p(s)$ is given by

$$G_p(s) = \frac{100}{s(s + 1)}$$

Assuming that the noise input $N(s)$ is zero, design controllers $G_{c1}(s)$ and $G_{c2}(s)$ such that the designed system satisfies the following:

1. The response to the step disturbance input has a small amplitude and settles to zero quickly (on the order of 1 sec to 2 sec).

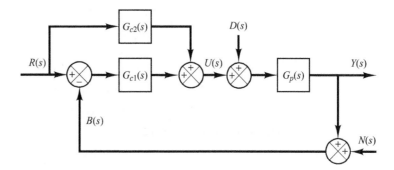

Figure 8–66
Two-degrees-of-freedom control system.

2. The response to the unit-step reference input has a maximum overshoot of 25% or less, and the settling time is 1 sec or less.

3. The steady-state errors in following ramp reference input and acceleration reference input are zero.

Solution. The closed-loop transfer functions for the disturbance input and reference input are given, respectively, by

$$\frac{Y(s)}{D(s)} = \frac{G_p(s)}{1 + G_{c1}(s)G_p(s)}$$

$$\frac{Y(s)}{R(s)} = \frac{[G_{c1}(s) + G_{c2}(s)]G_p(s)}{1 + G_{c1}(s)G_p(s)}$$

Let us assume that $G_{c1}(s)$ is a **PID** controller and has the following form:

$$G_{c1}(s) = \frac{K(s + a)^2}{s}$$

The characteristic equation for the system is

$$1 + G_{c1}(s)G_p(s) = 1 + \frac{K(s + a)^2}{s}\frac{100}{s(s + 1)}$$

Notice that the open-loop poles are located at $s = 0$ (a double pole) and $s = -1$. The zeros are located at $s = -a$ (a double zero).

In what follows, we shall use the root-locus approach to determine the values of a and K. Let us choose the dominant closed-loop poles at $s = -5 \pm j5$. Then, the angle deficiency at the desired closed-loop pole at $s = -5 + j5$ is

$$-135° - 135° - 128.66° + 180° = -218.66°$$

The double zero at $s = -a$ must contribute 218.66°. (Each zero must contribute 109.33°.) By a simple calculation, we find

$$a = -3.2460$$

The controller $G_{c1}(s)$ is then determined as

$$G_{c1}(s) = \frac{K(s + 3.2460)^2}{s}$$

The constant K must be determined by use of the magnitude condition. This condition is

$$\left| G_{c1}(s)G_p(s) \right|_{s=-5+j5} = 1$$

Since

$$G_{c1}(s)G_p(s) = \frac{K(s + 3.2460)^2}{s} \frac{100}{s(s + 1)}$$

we obtain

$$K = \left| \frac{s^2(s + 1)}{100(s + 3.2460)^2} \right|_{s=-5+j5}$$

$$= 0.11403$$

The controller $G_{c1}(s)$ thus becomes

$$G_{c1}(s) = \frac{0.11403(s + 3.2460)^2}{s}$$

$$= \frac{0.11403s^2 + 0.74028s + 1.20148}{s}$$

$$= 0.74028 + \frac{1.20148}{s} + 0.11403s \qquad (8\text{--}17)$$

Then, the closed-loop transfer function $Y(s)/D(s)$ is obtained as follows:

$$\frac{Y(s)}{D(s)} = \frac{G_p(s)}{1 + G_{c1}(s)G_p(s)}$$

$$= \frac{\dfrac{100}{s(s + 1)}}{1 + \dfrac{0.11403(s + 3.2460)^2}{s} \dfrac{100}{s(s + 1)}}$$

$$= \frac{100s}{s^3 + 12.403s^2 + 74.028s + 120.148}$$

The response curve when $D(s)$ is a unit-step disturbance is shown in Figure 8–67.

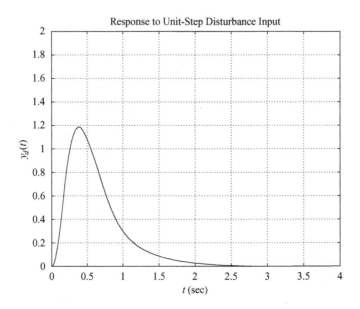

Figure 8–67
Response to unit-step disturbance input.

Next, we consider the responses to reference inputs. The closed-loop transfer function $Y(s)/R(s)$ is

$$\frac{Y(s)}{R(s)} = \frac{[G_{c1}(s) + G_{c2}(s)]G_p(s)}{1 + G_{c1}(s)G_p(s)}$$

Let us define

$$G_{c1}(s) + G_{c2}(s) = G_c(s)$$

Then

$$\frac{Y(s)}{R(s)} = \frac{G_c(s)G_p(s)}{1 + G_{c1}(s)G_p(s)}$$

$$= \frac{100sG_c(s)}{s^3 + 12.403s^2 + 74.028s + 120.148}$$

To satisfy the requirements on the responses to the ramp reference input and acceleration reference input, we use the zero-placement approach. That is, we choose the numerator of $Y(s)/R(s)$ to be the sum of the last three terms of the denominator, or

$$100sG_c(s) = 12.403s^2 + 74.028s + 120.148$$

from which we get

$$G_c(s) = \frac{0.12403s^2 + 0.74028s + 1.20148}{s}$$

$$= 0.74028 + \frac{1.20148}{s} + 0.12403s \qquad (8\text{--}18)$$

Hence, the closed-loop transfer function $Y(s)/R(s)$ becomes as

$$\frac{Y(s)}{R(s)} = \frac{12.403s^2 + 74.028s + 120.148}{s^3 + 12.403s^2 + 74.028s + 120.148}$$

The response curves to the unit-step reference input, unit-ramp reference input, and unit-acceleration reference input are shown in Figures 8–68(a), (b), and (c), respectively. The maximum

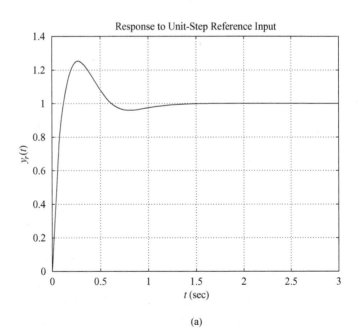

Figure 8–68
(a) Response to unit-step reference input; (b) response to unit-ramp reference input; (c) response to unit-acceleration reference input.

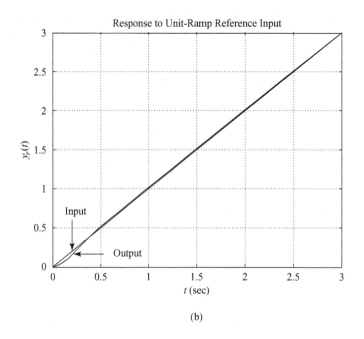

Response to Unit-Ramp Reference Input

(b)

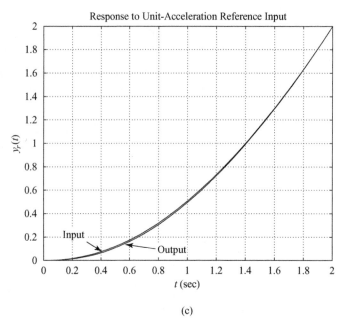

Response to Unit-Acceleration Reference Input

(c)

Figure 8–68
(continued)

overshoot in the unit-step response is approximately 25% and the settling time is approximately 1.2 sec. The steady-state errors in the ramp response and acceleration response are zero. Therefore, the designed controller $G_c(s)$ given by Equation (8–18) is satisfactory.

Finally, we determine $G_{c2}(s)$. Noting that

$$G_{c2}(s) = G_c(s) - G_{c1}(s)$$

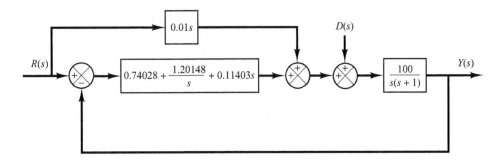

Figure 8–69
Block diagram of the
designed system.

and from Equation (8–17)

$$G_{c1}(s) = 0.74028 + \frac{1.20148}{s} + 0.11403s$$

we obtain

$$G_{c2}(s) = \left(0.74028 + \frac{1.20148}{s} + 0.12403s\right)$$

$$- \left(0.74028 + \frac{1.20148}{s} + 0.11403s\right)$$

$$= 0.01s \qquad\qquad (8\text{–}19)$$

Equations (8–17) and (8–19) give the transfer functions of the controllers $G_{c1}(s)$ and $G_{c2}(s)$, respectively. The block diagram of the designed system is shown in Figure 8–69.

Note that if the maximum overshoot were much higher than 25% and/or the settling time were much larger than 1.2 sec, then we might assume a search region (such as $3 \le a \le 6$, $3 \le b \le 6$, and $6 \le c \le 12$) and use the computational method presented in Example 8–4 to find a set or sets of variables that would give the desired response to the unit-step reference input.

PROBLEMS

B–8–1. Consider the electronic PID controller shown in Figure 8–70. Determine the values of R_1, R_2, R_3, R_4, C_1, and C_2 of the controller such that the transfer function $G_c(s) = E_o(s)/E_i(s)$ is

$$G_c(s) = 39.42\left(1 + \frac{1}{3.077s} + 0.7692s\right)$$

$$= 30.3215 \frac{(s + 0.65)^2}{s}$$

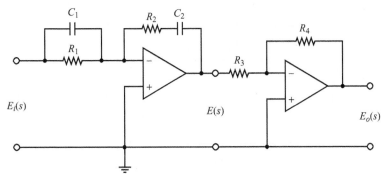

Figure 8–70
Electronic PID controller.

B–8–2. Consider the system shown in Figure 8–71. Assume that disturbances $D(s)$ enter the system as shown in the diagram. Determine parameters K, a, and b such that the response to the unit-step disturbance input and the response to the unit-step reference input satisfy the following specifications: The response to the step disturbance input should attenuate rapidly with no steady-state error, and the response to the step reference input exhibits a maximum overshoot of 20% or less and a settling time of 2 sec.

B–8–3. Show that the PID-controlled system shown in Figure 8–72(a) is equivalent to the I-PD-controlled system with feedforward control shown in Figure 8–72(b).

B–8–4. Consider the systems shown in Figures 8–73(a) and (b). The system shown in Figure 8–73(a) is the system designed in Example 8–1. The response to the unit-step reference input in the absence of the disturbance input is shown in Figure 8–10. The system shown in Figure 8–73(b) is the I-PD-controlled system using the same K_p, T_i, and T_d as the system shown in Figure 8–73(a).

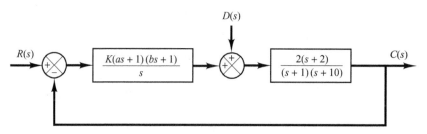

Figure 8–71
Control system.

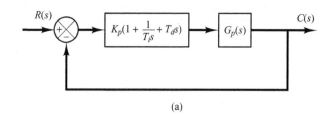

(a)

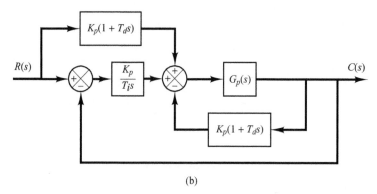

(b)

Figure 8–72
(a) PID-controlled system; (b) I-PD-controlled system with feedforward control.

Obtain the response of the I-PD-controlled system to the unit-step reference input with MATLAB. Compare the unit-step response curves of the two systems.

B-8-5. Referring to Problem B-8-4, obtain the response of the PID-controlled system shown in Figure 8-73(a) to the unit-step disturbance input.

Show that for the disturbance input, the responses of the PID-controlled system shown in Figure 8-73(a) and of the I-PD-controlled system shown in Figure 8-73(b) are

exactly the same. [When considering $D(s)$ to be the input, assume that the reference input $R(s)$ is zero, and vice versa.] Also, compare the closed-loop transfer function $C(s)/R(s)$ of both systems.

B-8-6. Consider the system shown in Figure 8-74. This system is subjected to three input signals: the reference input, disturbance input, and noise input. Show that the characteristic equation of this system is the same regardless of which input signal is chosen as input.

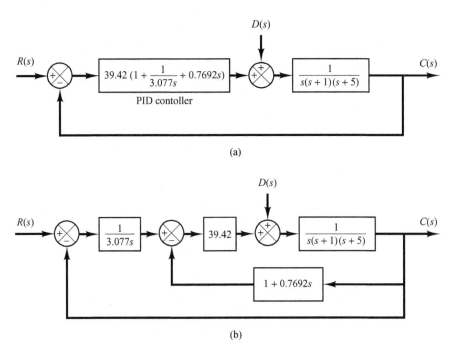

(a)

(b)

Figure 8-73
(a) PID-controlled system; (b) I-PD-controlled system.

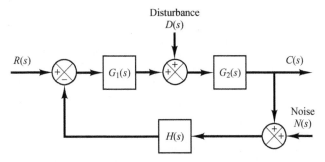

Figure 8-74
Control system.

B–8–7. Consider the system shown in Figure 8–75. Obtain the closed-loop transfer function $C(s)/R(s)$ for the reference input and the closed-loop transfer function $C(s)/D(s)$ for the disturbance input. When considering $R(s)$ as the input, assume that $D(s)$ is zero, and vice versa.

B–8–8. Consider the system shown in Figure 8–76(a), where K is an adjustable gain and $G(s)$ and $H(s)$ are fixed components. The closed-loop transfer function for the disturbance is

$$\frac{C(s)}{D(s)} = \frac{1}{1 + KG(s)H(s)}$$

To minimize the effect of disturbances, the adjustable gain K should be chosen as large as possible.

Is this true for the system in Figure 8–76(b), too?

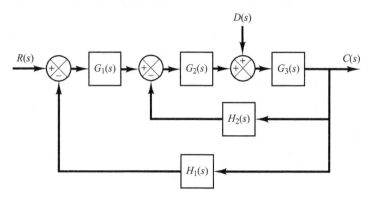

Figure 8–75
Control system.

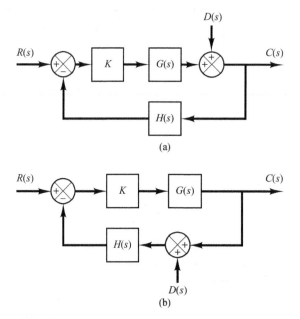

(a)

(b)

Figure 8–76
(a) Control system with disturbance entering in the feedforward path; (b) control system with disturbance entering in the feedback path.

B–8–9. Show that the control systems shown in Figures 8–77(a), (b), and (c) are two-degrees-of-freedom systems. In the diagrams, G_{c1} and G_{c2} are controllers and G_p is the plant.

B–8–10. Show that the control system shown in Figure 8–78 is a three-degrees-of freedom system. The transfer functions G_{c1}, G_{c2}, and G_{c3} are controllers. The plant consists of transfer functions G_1 and G_2.

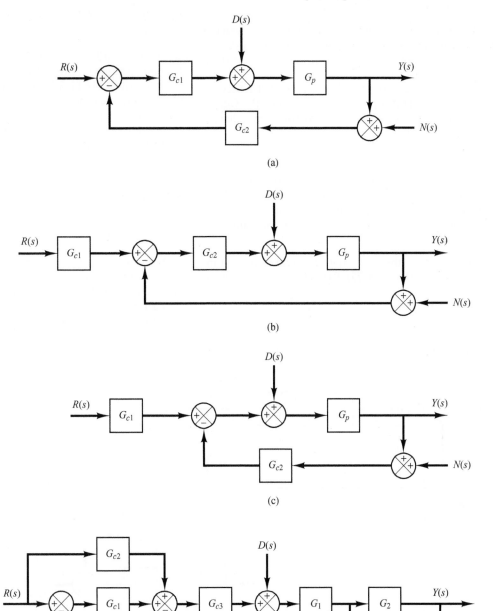

Figure 8–77
(a), (b), (c) Two degrees-of-freedom systems.

Figure 8–78
Three-degrees-of-freedom system.

B–8–11. Consider the control system shown in Figure 8–79. Assume that the PID controller is given by

$$G_c(s) = K \frac{(s + a)^2}{s}$$

It is desired that the unit-step response of the system exhibit the maximum overshoot of less than 10%, but more than 2% (to avoid an almost overdamped system), and the settling time be less than 2 sec.

Using the computational approach presented in Section 8–4, write a MATLAB program to determine the values of K and a that will satisfy the given specifications. Choose the search region to be

$$1 \le K \le 4, \qquad 0.4 \le a \le 4$$

Choose the step size for K and a to be 0.05. Write the program such that the nested loops start with the highest values of K and a and step toward the lowest.

Using the first-found solution, plot the unit-step response curve.

B–8–12. Consider the same control system as treated in Problem **B–8–11** (Figure 8–79). The PID controller is given by

$$G_c(s) = K \frac{(s + a)^2}{s}$$

It is desired to determine the values of K and a such that the unit-step response of the system exhibits the maximum overshoot of less than 8%, but more than 3%, and the settling time is less than 2 sec. Choose the search region to be

$$2 \le K \le 4, \qquad 0.5 \le a \le 3$$

Choose the step size for K and a to be 0.05.

First, write a MATLAB program such that the nested loops in the program start with the highest values of K and a and step toward the lowest and the computation stops when a successful set of K and a is found for the first time.

Next, write a MATLAB program that will find all possible sets of K and a that will satisfy the given specifications.

Among multiple sets of K and a that satisfy the given specifications, determine the best choice. Then, plot the unit-step response curve of the system with the best choice of K and a.

B–8–13. Consider the two-degrees-of-freedom control system shown in Figure 8–80. The plant $G_p(s)$ is given by

$$G_p(s) = \frac{3(s + 5)}{s(s + 1)(s^2 + 4s + 13)}$$

Design controllers $G_{c1}(s)$ and $G_{c2}(s)$ such that the response to the unit-step disturbance input should have small amplitude and settle to zero quickly (in approximately 2 sec). The response to the unit-step reference input should be such that the maximum overshoot is 25% (or less) and the settling time is 2 sec. Also, the steady-state errors in the response to the ramp and acceleration reference inputs should be zero.

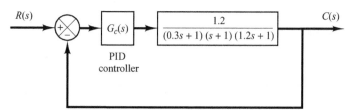

Figure 8–79
Control system.

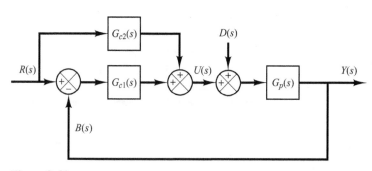

Figure 8–80
Two-degrees-of-freedom control system.

B–8–14. Consider the system shown in Figure 8–81. The plant $G_p(s)$ is given by

$$G_p(s) = \frac{2(s + 1)}{s(s + 3)(s + 5)}$$

Determine the controllers $G_{c1}(s)$ and $G_{c2}(s)$ such that, for the step disturbance input, the response shows a small amplitude and approaches zero quickly (in a matter of 1 to 2 sec). For the response to the unit-step reference input, it is desired that the maximum overshoot be 20% or less and the settling time 1 sec or less. For the ramp reference input and acceleration reference input, the steady-state errors should be zero.

B–8–15. Consider the two-degrees-of-freedom control system shown in Figure 8–82. Design controllers $G_{c1}(s)$ and $G_{c2}(s)$ such that the response to the step disturbance input shows a small amplitude and settles to zero quickly (in 1 to 2 sec) and the response to the step reference input exhibits 25% or less maximum overshoot and the settling time is less than 1 sec. The steady-state error in following the ramp reference input or acceleration reference input should be zero.

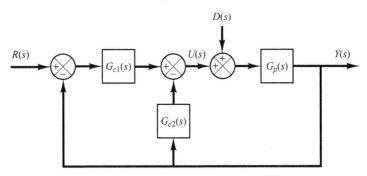

Figure 8–81
Two-degrees-of-freedom control system.

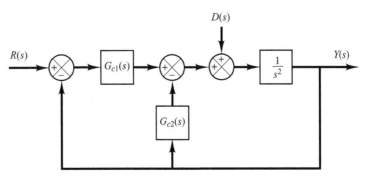

Figure 8–82
Two-degrees-of-freedom control system.

Control Systems Analysis in State Space

9-1 INTRODUCTION*

A modern complex system may have many inputs and many outputs, and these may be interrelated in a complicated manner. To analyze such a system, it is essential to reduce the complexity of the mathematical expressions, as well as to resort to computers for most of the tedious computations necessary in the analysis. The state-space approach to system analysis is best suited from this viewpoint.

While conventional control theory is based on the input–output relationship, or transfer function, modern control theory is based on the description of system equations in terms of n first-order differential equations, which may be combined into a first-order vector-matrix differential equation. The use of vector-matrix notation greatly simplifies the mathematical representation of systems of equations. The increase in the number of state variables, the number of inputs, or the number of outputs does not increase the complexity of the equations. In fact, the analysis of complicated multiple-input, multiple-output systems can be carried out by procedures that are only slightly more complicated than those required for the analysis of systems of first-order scalar differential equations.

This chapter and the next deal with the state-space analysis and design of control systems. Basic materials of state-space analysis, including the state-space representation of

---

*It is noted that in this book an asterisk used as a superscript of a matrix, such as $\mathbf{A}^*$, implies that it is a **conjugate transpose** of matrix $\mathbf{A}$. The conjugate transpose is the conjugate of the transpose of a matrix. For a real matrix (a matrix whose elements are all real), the conjugate transpose $\mathbf{A}^*$ is the same as the transpose $\mathbf{A}^T$.

systems, controllability, and observability are presented in this chapter. Useful design methods based on state-feedback control are given in Chapter 10.

Outline of the Chapter. Section 9–1 has presented an introduction to state-space analysis of control systems. Section 9–2 deals with the state-space representation of transfer-function systems. Here we present various canonical forms of state-space equations. Section 9–3 discusses the transformation of system models (such as from transfer-function to state-space models, and vice versa) with MATLAB. Section 9–4 presents the solution of time-invariant state equations. Section 9–5 gives some useful results in vector-matrix analysis that are necessary in studying the state-space analysis of control systems. Section 9–6 discusses the controllability of control systems and Section 9–7 treats the observability of control systems.

9–2 STATE-SPACE REPRESENTATIONS OF TRANSFER-FUNCTION SYSTEMS

Many techniques are available for obtaining state-space representations of transfer-function systems. In Chapter 2 we presented a few such methods. This section presents state-space representations in the controllable, observable, diagonal, or Jordan canonical forms. (Methods for obtaining such state-space representations from transfer functions are discussed in detail in Problems **A–9–1** through **A–9–4**.)

State-Space Representations in Canonical Forms. Consider a system defined by

$$\overset{(n)}{y} + a_1 \overset{(n-1)}{y} + \cdots + a_{n-1} \dot{y} + a_n y = b_0 \overset{(n)}{u} + b_1 \overset{(n-1)}{u} + \cdots + b_{n-1} \dot{u} + b_n u \qquad (9\text{–}1)$$

where u is the input and y is the output. This equation can also be written as

$$\frac{Y(s)}{U(s)} = \frac{b_0 s^n + b_1 s^{n-1} + \cdots + b_{n-1} s + b_n}{s^n + a_1 s^{n-1} + \cdots + a_{n-1} s + a_n} \qquad (9\text{–}2)$$

In what follows we shall present state-space representations of the system defined by Equation (9–1) or (9–2) in controllable canonical form, observable canonical form, and diagonal (or Jordan) canonical form.

Controllable Canonical Form. The following state-space representation is called a controllable canonical form:

$$
\begin{bmatrix} \dot{x}_1 \\ \dot{x}_2 \\ \cdot \\ \cdot \\ \cdot \\ \dot{x}_{n-1} \\ \dot{x}_n \end{bmatrix}
=
\begin{bmatrix} 0 & 1 & 0 & \cdots & 0 \\ 0 & 0 & 1 & \cdots & 0 \\ \cdot & \cdot & \cdot & & \cdot \\ \cdot & \cdot & \cdot & & \cdot \\ \cdot & \cdot & \cdot & & \cdot \\ 0 & 0 & 0 & \cdots & 1 \\ -a_n & -a_{n-1} & -a_{n-2} & \cdots & -a_1 \end{bmatrix}
\begin{bmatrix} x_1 \\ x_2 \\ \cdot \\ \cdot \\ \cdot \\ x_{n-1} \\ x_n \end{bmatrix}
+
\begin{bmatrix} 0 \\ 0 \\ \cdot \\ \cdot \\ \cdot \\ 0 \\ 1 \end{bmatrix} u \qquad (9\text{–}3)
$$

$$
y = \begin{bmatrix} b_n - a_n b_0 & \vdots & b_{n-1} - a_{n-1} b_0 & \vdots & \cdots & \vdots & b_1 - a_1 b_0 \end{bmatrix} \begin{bmatrix} x_1 \\ x_2 \\ \cdot \\ \cdot \\ \cdot \\ x_n \end{bmatrix} + b_0 u \qquad (9\text{–}4)
$$

The controllable canonical form is important in discussing the pole-placement approach to control systems design.

Observable Canonical Form. The following state-space representation is called an observable canonical form:

$$
\begin{bmatrix} \dot{x}_1 \\ \dot{x}_2 \\ \cdot \\ \cdot \\ \cdot \\ \dot{x}_n \end{bmatrix} = \begin{bmatrix} 0 & 0 & \cdots & 0 & -a_n \\ 1 & 0 & \cdots & 0 & -a_{n-1} \\ \cdot & \cdot & & \cdot & \cdot \\ \cdot & \cdot & & \cdot & \cdot \\ \cdot & \cdot & & \cdot & \cdot \\ 0 & 0 & \cdots & 1 & -a_1 \end{bmatrix} \begin{bmatrix} x_1 \\ x_2 \\ \cdot \\ \cdot \\ \cdot \\ x_n \end{bmatrix} + \begin{bmatrix} b_n - a_n b_0 \\ b_{n-1} - a_{n-1} b_0 \\ \cdot \\ \cdot \\ \cdot \\ b_1 - a_1 b_0 \end{bmatrix} u \qquad (9\text{–}5)
$$

$$
y = \begin{bmatrix} 0 & 0 & \cdots & 0 & 1 \end{bmatrix} \begin{bmatrix} x_1 \\ x_2 \\ \cdot \\ \cdot \\ \cdot \\ x_{n-1} \\ x_n \end{bmatrix} + b_0 u \qquad (9\text{–}6)
$$

Note that the $n \times n$ state matrix of the state equation given by Equation (9–5) is the transpose of that of the state equation defined by Equation (9–3).

Diagonal Canonical Form. Consider the transfer-function system defined by Equation (9–2). Here we consider the case where the denominator polynomial involves only distinct roots. For the distinct-roots case, Equation (9–2) can be written as

$$
\frac{Y(s)}{U(s)} = \frac{b_0 s^n + b_1 s^{n-1} + \cdots + b_{n-1} s + b_n}{(s + p_1)(s + p_2) \cdots (s + p_n)}
$$

$$
= b_0 + \frac{c_1}{s + p_1} + \frac{c_2}{s + p_2} + \cdots + \frac{c_n}{s + p_n} \qquad (9\text{–}7)
$$

The diagonal canonical form of the state-space representation of this system is given by

$$
\begin{bmatrix} \dot{x}_1 \\ \dot{x}_2 \\ \cdot \\ \cdot \\ \cdot \\ \dot{x}_n \end{bmatrix} = \begin{bmatrix} -p_1 & & & & 0 \\ & -p_2 & & & \\ & & \cdot & & \\ & & & \cdot & \\ & & & & \cdot \\ 0 & & & & -p_n \end{bmatrix} \begin{bmatrix} x_1 \\ x_2 \\ \cdot \\ \cdot \\ \cdot \\ x_n \end{bmatrix} + \begin{bmatrix} 1 \\ 1 \\ \cdot \\ \cdot \\ \cdot \\ 1 \end{bmatrix} u \tag{9-8}
$$

$$
y = \begin{bmatrix} c_1 & c_2 & \cdots & c_n \end{bmatrix} \begin{bmatrix} x_1 \\ x_2 \\ \cdot \\ \cdot \\ \cdot \\ x_n \end{bmatrix} + b_0 u \tag{9-9}
$$

Jordan Canonical Form. Next we shall consider the case where the denominator polynomial of Equation (9–2) involves multiple roots. For this case, the preceding diagonal canonical form must be modified into the Jordan canonical form. Suppose, for example, that the p_i's are different from one another, except that the first three p_i's are equal, or $p_1 = p_2 = p_3$. Then the factored form of $Y(s)/U(s)$ becomes

$$
\frac{Y(s)}{U(s)} = \frac{b_0 s^n + b_1 s^{n-1} + \cdots + b_{n-1} s + b_n}{(s + p_1)^3 (s + p_4)(s + p_5) \cdots (s + p_n)}
$$

The partial-fraction expansion of this last equation becomes

$$
\frac{Y(s)}{U(s)} = b_0 + \frac{c_1}{(s + p_1)^3} + \frac{c_2}{(s + p_1)^2} + \frac{c_3}{s + p_1} + \frac{c_4}{s + p_4} + \cdots + \frac{c_n}{s + p_n}
$$

A state-space representation of this system in the Jordan canonical form is given by

$$
\begin{bmatrix} \dot{x}_1 \\ \dot{x}_2 \\ \dot{x}_3 \\ \dot{x}_4 \\ \cdot \\ \cdot \\ \cdot \\ \dot{x}_n \end{bmatrix} = \begin{bmatrix} -p_1 & 1 & 0 & 0 & \cdots & 0 \\ 0 & -p_1 & 1 & \vdots & & \vdots \\ 0 & 0 & -p_1 & 0 & \cdots & 0 \\ 0 & \cdots & 0 & -p_4 & & 0 \\ \cdot & & \cdot & & \cdot & \\ \cdot & & \cdot & & & \cdot \\ \cdot & & \cdot & & & \cdot \\ 0 & \cdots & 0 & 0 & & -p_n \end{bmatrix} \begin{bmatrix} x_1 \\ x_2 \\ x_3 \\ x_4 \\ \cdot \\ \cdot \\ \cdot \\ x_n \end{bmatrix} + \begin{bmatrix} 0 \\ 0 \\ 1 \\ 1 \\ \cdot \\ \cdot \\ \cdot \\ 1 \end{bmatrix} u \tag{9-10}
$$

$$
y = \begin{bmatrix} c_1 & c_2 & \cdots & c_n \end{bmatrix} \begin{bmatrix} x_1 \\ x_2 \\ \cdot \\ \cdot \\ \cdot \\ x_n \end{bmatrix} + b_0 u \tag{9-11}
$$

EXAMPLE 9–1   Consider the system given by

$$\frac{Y(s)}{U(s)} = \frac{s+3}{s^2+3s+2}$$

Obtain state-space representations in the controllable canonical form, observable canonical form, and diagonal canonical form.

Controllable Canonical Form:

$$\begin{bmatrix} \dot{x}_1(t) \\ \dot{x}_2(t) \end{bmatrix} = \begin{bmatrix} 0 & 1 \\ -2 & -3 \end{bmatrix} \begin{bmatrix} x_1(t) \\ x_2(t) \end{bmatrix} + \begin{bmatrix} 0 \\ 1 \end{bmatrix} u(t)$$

$$y(t) = \begin{bmatrix} 3 & 1 \end{bmatrix} \begin{bmatrix} x_1(t) \\ x_2(t) \end{bmatrix}$$

Observable Canonical Form:

$$\begin{bmatrix} \dot{x}_1(t) \\ \dot{x}_2(t) \end{bmatrix} = \begin{bmatrix} 0 & -2 \\ 1 & -3 \end{bmatrix} \begin{bmatrix} x_1(t) \\ x_2(t) \end{bmatrix} + \begin{bmatrix} 3 \\ 1 \end{bmatrix} u(t)$$

$$y(t) = \begin{bmatrix} 0 & 1 \end{bmatrix} \begin{bmatrix} x_1(t) \\ x_2(t) \end{bmatrix}$$

Diagonal Canonical Form:

$$\begin{bmatrix} \dot{x}_1(t) \\ \dot{x}_2(t) \end{bmatrix} = \begin{bmatrix} -1 & 0 \\ 0 & -2 \end{bmatrix} \begin{bmatrix} x_1(t) \\ x_2(t) \end{bmatrix} + \begin{bmatrix} 1 \\ 1 \end{bmatrix} u(t)$$

$$y(t) = \begin{bmatrix} 2 & -1 \end{bmatrix} \begin{bmatrix} x_1(t) \\ x_2(t) \end{bmatrix}$$

Eigenvalues of an $n \times n$ Matrix A.   The eigenvalues of an $n \times n$ matrix $\mathbf{A}$ are the roots of the characteristic equation

$$|\lambda \mathbf{I} - \mathbf{A}| = 0$$

The eigenvalues are also called the characteristic roots.

Consider, for example, the following matrix $\mathbf{A}$:

$$\mathbf{A} = \begin{bmatrix} 0 & 1 & 0 \\ 0 & 0 & 1 \\ -6 & -11 & -6 \end{bmatrix}$$

The characteristic equation is

$$|\lambda \mathbf{I} - \mathbf{A}| = \begin{vmatrix} \lambda & -1 & 0 \\ 0 & \lambda & -1 \\ 6 & 11 & \lambda + 6 \end{vmatrix}$$

$$= \lambda^3 + 6\lambda^2 + 11\lambda + 6$$

$$= (\lambda + 1)(\lambda + 2)(\lambda + 3) = 0$$

The eigenvalues of $\mathbf{A}$ are the roots of the characteristic equation, or -1, -2, and -3.

Diagonalization of $n \times n$ Matrix.   Note that if an $n \times n$ matrix $\mathbf{A}$ with distinct eigenvalues is given by

$$\mathbf{A} = \begin{bmatrix} 0 & 1 & 0 & \cdots & 0 \\ 0 & 0 & 1 & \cdots & 0 \\ \cdot & \cdot & \cdot & & \cdot \\ \cdot & \cdot & \cdot & & \cdot \\ \cdot & \cdot & \cdot & & \cdot \\ 0 & 0 & 0 & \cdots & 1 \\ -a_n & -a_{n-1} & -a_{n-2} & \cdots & -a_1 \end{bmatrix} \tag{9-12}$$

the transformation $\mathbf{x} = \mathbf{Pz}$, where

$$\mathbf{P} = \begin{bmatrix} 1 & 1 & \cdots & 1 \\ \lambda_1 & \lambda_2 & \cdots & \lambda_n \\ \lambda_1^2 & \lambda_2^2 & \cdots & \lambda_n^2 \\ \cdot & \cdot & & \cdot \\ \cdot & \cdot & & \cdot \\ \cdot & \cdot & & \cdot \\ \lambda_1^{n-1} & \lambda_2^{n-1} & \cdots & \lambda_n^{n-1} \end{bmatrix}$$

$\lambda_1, \lambda_2, \ldots, \lambda_n = n$ distinct eigenvalues of $\mathbf{A}$

will transform $\mathbf{P}^{-1}\mathbf{AP}$ into the diagonal matrix, or

$$\mathbf{P}^{-1}\mathbf{AP} = \begin{bmatrix} \lambda_1 & & & & 0 \\ & \lambda_2 & & & \\ & & \cdot & & \\ & & & \cdot & \\ & & & & \cdot \\ 0 & & & & \lambda_n \end{bmatrix}$$

If the matrix $\mathbf{A}$ defined by Equation (9-12) involves multiple eigenvalues, then diagonalization is impossible. For example, if the 3×3 matrix $\mathbf{A}$, where

$$\mathbf{A} = \begin{bmatrix} 0 & 1 & 0 \\ 0 & 0 & 1 \\ -a_3 & -a_2 & -a_1 \end{bmatrix}$$

has the eigenvalues $\lambda_1, \lambda_1, \lambda_3$, then the transformation $\mathbf{x} = \mathbf{Sz}$, where

$$\mathbf{S} = \begin{bmatrix} 1 & 0 & 1 \\ \lambda_1 & 1 & \lambda_3 \\ \lambda_1^2 & 2\lambda_1 & \lambda_3^2 \end{bmatrix}$$

will yield

$$\mathbf{S}^{-1}\mathbf{AS} = \begin{bmatrix} \lambda_1 & 1 & 0 \\ 0 & \lambda_1 & 0 \\ 0 & 0 & \lambda_3 \end{bmatrix}$$

This is in the Jordan canonical form.

EXAMPLE 9–2 Consider the following state-space representation of a system.

$$\begin{bmatrix} \dot{x}_1 \\ \dot{x}_2 \\ \dot{x}_3 \end{bmatrix} = \begin{bmatrix} 0 & 1 & 0 \\ 0 & 0 & 1 \\ -6 & -11 & -6 \end{bmatrix} \begin{bmatrix} x_1 \\ x_2 \\ x_3 \end{bmatrix} + \begin{bmatrix} 0 \\ 0 \\ 6 \end{bmatrix} u \qquad (9\text{--}13)$$

$$y = \begin{bmatrix} 1 & 0 & 0 \end{bmatrix} \begin{bmatrix} x_1 \\ x_2 \\ x_3 \end{bmatrix} \qquad (9\text{--}14)$$

Equations (9–13) and (9–14) can be put in a standard form as

$$\dot{x} = Ax + Bu \qquad (9\text{--}15)$$

$$y = Cx \qquad (9\text{--}16)$$

where

$$A = \begin{bmatrix} 0 & 1 & 0 \\ 0 & 0 & 1 \\ -6 & -11 & -6 \end{bmatrix}, \qquad B = \begin{bmatrix} 0 \\ 0 \\ 6 \end{bmatrix}, \qquad C = \begin{bmatrix} 1 & 0 & 0 \end{bmatrix}$$

The eigenvalues of matrix A are

$$\lambda_1 = -1, \qquad \lambda_2 = -2, \qquad \lambda_3 = -3$$

Thus, three eigenvalues are distinct. If we define a set of new state variables z_1, z_2, and z_3 by the transformation

$$\begin{bmatrix} x_1 \\ x_2 \\ x_3 \end{bmatrix} = \begin{bmatrix} 1 & 1 & 1 \\ -1 & -2 & -3 \\ 1 & 4 & 9 \end{bmatrix} \begin{bmatrix} z_1 \\ z_2 \\ z_3 \end{bmatrix}$$

or

$$x = Pz \qquad (9\text{--}17)$$

where

$$P = \begin{bmatrix} 1 & 1 & 1 \\ \lambda_1 & \lambda_2 & \lambda_3 \\ \lambda_1^2 & \lambda_2^2 & \lambda_3^2 \end{bmatrix} = \begin{bmatrix} 1 & 1 & 1 \\ -1 & -2 & -3 \\ 1 & 4 & 9 \end{bmatrix} \qquad (9\text{--}18)$$

then, by substituting Equation (9–17) into Equation (9–15), we obtain

$$P\dot{z} = APz + Bu$$

By premultiplying both sides of this last equation by P^{-1}, we get

$$\dot{z} = P^{-1}APz + P^{-1}Bu \qquad (9\text{--}19)$$

or

$$\begin{bmatrix} \dot{z}_1 \\ \dot{z}_2 \\ \dot{z}_3 \end{bmatrix} = \begin{bmatrix} 3 & 2.5 & 0.5 \\ -3 & -4 & -1 \\ 1 & 1.5 & 0.5 \end{bmatrix} \begin{bmatrix} 0 & 1 & 0 \\ 0 & 0 & 1 \\ -6 & -11 & -6 \end{bmatrix} \begin{bmatrix} 1 & 1 & 1 \\ -1 & -2 & -3 \\ 1 & 4 & 9 \end{bmatrix} \begin{bmatrix} z_1 \\ z_2 \\ z_3 \end{bmatrix}$$

$$+ \begin{bmatrix} 3 & 2.5 & 0.5 \\ -3 & -4 & -1 \\ 1 & 1.5 & 0.5 \end{bmatrix} \begin{bmatrix} 0 \\ 0 \\ 6 \end{bmatrix} u$$

Simplifying gives

$$
\begin{bmatrix} \dot{z}_1 \\ \dot{z}_2 \\ \dot{z}_3 \end{bmatrix} = \begin{bmatrix} -1 & 0 & 0 \\ 0 & -2 & 0 \\ 0 & 0 & -3 \end{bmatrix} \begin{bmatrix} z_1 \\ z_2 \\ z_3 \end{bmatrix} + \begin{bmatrix} 3 \\ -6 \\ 3 \end{bmatrix} u \tag{9-20}
$$

Equation (9–20) is also a state equation that describes the same system as defined by Equation (9–13).

The output equation, Equation (9–16), is modified to

$$ y = \mathbf{CPz} $$

or

$$
y = \begin{bmatrix} 1 & 0 & 0 \end{bmatrix} \begin{bmatrix} 1 & 1 & 1 \\ -1 & -2 & -3 \\ 1 & 4 & 9 \end{bmatrix} \begin{bmatrix} z_1 \\ z_2 \\ z_3 \end{bmatrix}
$$

$$
= \begin{bmatrix} 1 & 1 & 1 \end{bmatrix} \begin{bmatrix} z_1 \\ z_2 \\ z_3 \end{bmatrix} \tag{9-21}
$$

Notice that the transformation matrix $\mathbf{P}$, defined by Equation (9–18), modifies the coefficient matrix of $\mathbf{z}$ into the diagonal matrix. As is clearly seen from Equation (9–20), the three scalar state equations are uncoupled. Notice also that the diagonal elements of the matrix $\mathbf{P}^{-1}\mathbf{AP}$ in Equation (9–19) are identical with the three eigenvalues of $\mathbf{A}$. It is very important to note that the eigenvalues of $\mathbf{A}$ and those of $\mathbf{P}^{-1}\mathbf{AP}$ are identical. We shall prove this for a general case in what follows.

Invariance of Eigenvalues. To prove the invariance of the eigenvalues under a linear transformation, we must show that the characteristic polynomials $|\lambda\mathbf{I} - \mathbf{A}|$ and $|\lambda\mathbf{I} - \mathbf{P}^{-1}\mathbf{AP}|$ are identical.

Since the determinant of a product is the product of the determinants, we obtain

$$
|\lambda\mathbf{I} - \mathbf{P}^{-1}\mathbf{AP}| = |\lambda\mathbf{P}^{-1}\mathbf{P} - \mathbf{P}^{-1}\mathbf{AP}|
$$

$$
= |\mathbf{P}^{-1}(\lambda\mathbf{I} - \mathbf{A})\mathbf{P}|
$$

$$
= |\mathbf{P}^{-1}||\lambda\mathbf{I} - \mathbf{A}||\mathbf{P}|
$$

$$
= |\mathbf{P}^{-1}||\mathbf{P}||\lambda\mathbf{I} - \mathbf{A}|
$$

Noting that the product of the determinants $|\mathbf{P}^{-1}|$ and $|\mathbf{P}|$ is the determinant of the product $|\mathbf{P}^{-1}\mathbf{P}|$, we obtain

$$
|\lambda\mathbf{I} - \mathbf{P}^{-1}\mathbf{AP}| = |\mathbf{P}^{-1}\mathbf{P}||\lambda\mathbf{I} - \mathbf{A}|
$$

$$
= |\lambda\mathbf{I} - \mathbf{A}|
$$

Thus, we have proved that the eigenvalues of $\mathbf{A}$ are invariant under a linear transformation.

Nonuniqueness of a Set of State Variables. It has been stated that a set of state variables is not unique for a given system. Suppose that $x_1, x_2, \ldots, x_n$ are a set of state variables.

Then we may take as another set of state variables any set of functions

$$\hat{x}_1 = X_1(x_1, x_2, \dots, x_n)$$
$$\hat{x}_2 = X_2(x_1, x_2, \dots, x_n)$$

.

.

.

$$\hat{x}_n = X_n(x_1, x_2, \dots, x_n)$$

provided that, for every set of values $\hat{x}_1, \hat{x}_2, \dots, \hat{x}_n$, there corresponds a unique set of values $x_1, x_2, \dots, x_n$, and vice versa. Thus, if $\mathbf{x}$ is a state vector, then $\hat{\mathbf{x}}$, where

$$\hat{\mathbf{x}} = \mathbf{Px}$$

is also a state vector, provided the matrix $\mathbf{P}$ is nonsingular. Different state vectors convey the same information about the system behavior.

9–3 TRANSFORMATION OF SYSTEM MODELS WITH MATLAB

In this section we shall consider the transformation of the system model from transfer function to state space, and vice versa. We shall begin our discussion with the transformation from transfer function to state space.

Let us write the closed-loop transfer function as

$$\frac{Y(s)}{U(s)} = \frac{\text{numerator polynomial in } s}{\text{denominator polynomial in } s} = \frac{\text{num}}{\text{den}}$$

Once we have this transfer-function expression, the MATLAB command

$$[A, B, C, D] = \text{tf2ss(num,den)}$$

will give a state-space representation. It is important to note that the state-space representation for any system is not unique. There are many (indeed, infinitely many) state-space representations for the same system. The MATLAB command gives one possible such state-space representation.

State-Space Formulation of Transfer-Function Systems. Consider the transfer-function system

$$\frac{Y(s)}{U(s)} = \frac{10s + 10}{s^3 + 6s^2 + 5s + 10} \tag{9–22}$$

There are many (again, infinitely many) possible state-space representations for this system. One possible state-space representation is

$$\begin{bmatrix} \dot{x}_1 \\ \dot{x}_2 \\ \dot{x}_3 \end{bmatrix} = \begin{bmatrix} 0 & 1 & 0 \\ 0 & 0 & 1 \\ -10 & -5 & -6 \end{bmatrix} \begin{bmatrix} x_1 \\ x_2 \\ x_3 \end{bmatrix} + \begin{bmatrix} 0 \\ 10 \\ -50 \end{bmatrix} u$$

$$y = \begin{bmatrix} 1 & 0 & 0 \end{bmatrix} \begin{bmatrix} x_1 \\ x_2 \\ x_3 \end{bmatrix} + \begin{bmatrix} 0 \end{bmatrix} u$$

Another possible state-space representation (among infinitely many alternatives) is

$$
\begin{bmatrix} \dot{x}_1 \\ \dot{x}_2 \\ \dot{x}_3 \end{bmatrix} = \begin{bmatrix} -6 & -5 & -10 \\ 1 & 0 & 0 \\ 0 & 1 & 0 \end{bmatrix} \begin{bmatrix} x_1 \\ x_2 \\ x_3 \end{bmatrix} + \begin{bmatrix} 1 \\ 0 \\ 0 \end{bmatrix} u \tag{9–23}
$$

$$
y = \begin{bmatrix} 0 & 10 & 10 \end{bmatrix} \begin{bmatrix} x_1 \\ x_2 \\ x_3 \end{bmatrix} + \begin{bmatrix} 0 \end{bmatrix} u \tag{9–24}
$$

MATLAB transforms the transfer function given by Equation (9–22) into the state-space representation given by Equations (9–23) and (9–24). For the example system considered here, MATLAB Program 9–1 will produce matrices **A**, **B**, **C**, and *D*.

MATLAB Program 9–1

```
num = [10  10];
den = [1  6  5  10];
[A,B,C,D] = tf2ss(num,den)

A =

      -6   -5   -10
       1    0     0
       0    1     0

B =

       1
       0
       0

C =

       0   10   10

D =

       0
```

Transformation from State Space to Transfer Function.   To obtain the transfer function from state-space equations, use the following command:

$$[num,den] = ss2tf(A,B,C,D,iu)$$

iu must be specified for systems with more than one input. For example, if the system has three inputs ($u1$, $u2$, $u3$), then iu must be either 1, 2, or 3, where 1 implies $u1$, 2 implies $u2$, and 3 implies $u3$.

If the system has only one input, then either

$$[num,den] = ss2tf(A,B,C,D)$$

or

$$[num,den] = ss2tf(A,B,C,D,1)$$

may be used. (See Example 9–3 and MATLAB Program 9–2.)

For the case where the system has multiple inputs and multiple outputs, see Example 9–4.

EXAMPLE 9–3 Obtain the transfer function of the system defined by the following state-space equations:

$$
\begin{bmatrix} \dot{x}_1 \\ \dot{x}_2 \\ \dot{x}_3 \end{bmatrix} = \begin{bmatrix} 0 & 1 & 0 \\ 0 & 0 & 1 \\ -5.008 & -25.1026 & -5.03247 \end{bmatrix} \begin{bmatrix} x_1 \\ x_2 \\ x_3 \end{bmatrix} + \begin{bmatrix} 0 \\ 25.04 \\ -121.005 \end{bmatrix} u
$$

$$
y = \begin{bmatrix} 1 & 0 & 0 \end{bmatrix} \begin{bmatrix} x_1 \\ x_2 \\ x_3 \end{bmatrix}
$$

MATLAB Program 9–2 will produce the transfer function for the given system. The transfer function obtained is given by

$$
\frac{Y(s)}{U(s)} = \frac{25.04s + 5.008}{s^3 + 5.0325s^2 + 25.1026s + 5.008}
$$

MATLAB Program 9–2

```
A = [0 1 0;0 0 1;-5.008 -25.1026 -5.03247];
B = [0;25.04; -121.005];
C = [1 0 0];
D = [0];
[num,den] = ss2tf(A,B,C,D)

num =

       0   -0.0000   25.0400   5.0080

den =

   1.0000   5.0325   25.1026   5.0080
```
% ***** The same result can be obtained by entering the following command *****

```
[num,den] = ss2tf(A,B,C,D,1)

num =

       0   -0.0000   25.0400   5.0080

den =

   1.0000   5.0325   25.1026   5.0080
```

EXAMPLE 9–4 Consider a system with multiple inputs and multiple outputs. When the system has more than one output, the command

$$[NUM,den] = ss2tf(A,B,C,D,iu)$$

produces transfer functions for all outputs to each input. (The numerator coefficients are returned to matrix NUM with as many rows as there are outputs.)

Consider the system defined by

$$\begin{bmatrix} \dot{x}_1 \\ \dot{x}_2 \end{bmatrix} = \begin{bmatrix} 0 & 1 \\ -25 & -4 \end{bmatrix} \begin{bmatrix} x_1 \\ x_2 \end{bmatrix} + \begin{bmatrix} 1 & 1 \\ 0 & 1 \end{bmatrix} \begin{bmatrix} u_1 \\ u_2 \end{bmatrix}$$

$$\begin{bmatrix} y_1 \\ y_2 \end{bmatrix} = \begin{bmatrix} 1 & 0 \\ 0 & 1 \end{bmatrix} \begin{bmatrix} x_1 \\ x_2 \end{bmatrix} + \begin{bmatrix} 0 & 0 \\ 0 & 0 \end{bmatrix} \begin{bmatrix} u_1 \\ u_2 \end{bmatrix}$$

This system involves two inputs and two outputs. Four transfer functions are involved: $Y_1(s)/U_1(s)$, $Y_2(s)/U_1(s)$, $Y_1(s)/U_2(s)$, and $Y_2(s)/U_2(s)$. (When considering input u_1, we assume that input u_2 is zero and vice versa.) See the output of MATLAB Program 9–3.

MATLAB Program 9–3

```
A = [0 1;-25 -4];
B = [1 1;0 1];
C = [1 0;0 1];
D = [0 0;0 0];
[NUM,den] = ss2tf(A,B,C,D,1)

NUM =

    0   1    4
    0   0  -25

den =

    1  4  25

[NUM,den] = ss2tf(A,B,C,D,2)

NUM =

    0   1.0000     5.0000
    0   1.0000   -25.0000

den =

    1    4    25
```

This is the MATLAB representation of the following four transfer functions:

$$\frac{Y_1(s)}{U_1(s)} = \frac{s + 4}{s^2 + 4s + 25}, \qquad \frac{Y_2(s)}{U_1(s)} = \frac{-25}{s^2 + 4s + 25}$$

$$\frac{Y_1(s)}{U_2(s)} = \frac{s + 5}{s^2 + 4s + 25}, \qquad \frac{Y_2(s)}{U_2(s)} = \frac{s - 25}{s^2 + 4s + 25}$$

In this section, we shall obtain the general solution of the linear time-invariant state equation. We shall first consider the homogeneous case and then the nonhomogeneous case.

Solution of Homogeneous State Equations. Before we solve vector-matrix differential equations, let us review the solution of the scalar differential equation

$$\dot{x} = ax \tag{9–25}$$

In solving this equation, we may assume a solution $x(t)$ of the form

$$x(t) = b_0 + b_1 t + b_2 t^2 + \cdots + b_k t^k + \cdots \tag{9–26}$$

By substituting this assumed solution into Equation (9–25), we obtain

$$b_1 + 2b_2 t + 3b_3 t^2 + \cdots + kb_k t^{k-1} + \cdots$$
$$= a\left(b_0 + b_1 t + b_2 t^2 + \cdots + b_k t^k + \cdots\right) \tag{9–27}$$

If the assumed solution is to be the true solution, Equation (9–27) must hold for any t. Hence, equating the coefficients of the equal powers of t, we obtain

$$b_1 = ab_0$$

$$b_2 = \frac{1}{2} ab_1 = \frac{1}{2} a^2 b_0$$

$$b_3 = \frac{1}{3} ab_2 = \frac{1}{3 \times 2} a^3 b_0$$

$$\cdot$$
$$\cdot$$
$$\cdot$$

$$b_k = \frac{1}{k!} a^k b_0$$

The value of b_0 is determined by substituting $t = 0$ into Equation (9–26), or

$$x(0) = b_0$$

Hence, the solution $x(t)$ can be written as

$$x(t) = \left(1 + at + \frac{1}{2!} a^2 t^2 + \cdots + \frac{1}{k!} a^k t^k + \cdots\right) x(0)$$
$$= e^{at} x(0)$$

We shall now solve the vector-matrix differential equation

$$\dot{x} = Ax \tag{9–28}$$

where $x = n$-vector
$\qquad A = n \times n$ constant matrix

By analogy with the scalar case, we assume that the solution is in the form of a vector power series in t, or

$$x(t) = b_0 + b_1 t + b_2 t^2 + \cdots + b_k t^k + \cdots \tag{9–29}$$

By substituting this assumed solution into Equation (9–28), we obtain

$$\mathbf{b}_1 + 2\mathbf{b}_2 t + 3\mathbf{b}_3 t^2 + \cdots + k\mathbf{b}_k t^{k-1} + \cdots$$

$$= \mathbf{A}\big(\mathbf{b}_0 + \mathbf{b}_1 t + \mathbf{b}_2 t^2 + \cdots + \mathbf{b}_k t^k + \cdots \big) \qquad (9\text{–}30)$$

If the assumed solution is to be the true solution, Equation (9–30) must hold for all t. Thus, by equating the coefficients of like powers of t on both sides of Equation (9–30), we obtain

$$\mathbf{b}_1 = \mathbf{A}\mathbf{b}_0$$

$$\mathbf{b}_2 = \frac{1}{2}\mathbf{A}\mathbf{b}_1 = \frac{1}{2}\mathbf{A}^2\mathbf{b}_0$$

$$\mathbf{b}_3 = \frac{1}{3}\mathbf{A}\mathbf{b}_2 = \frac{1}{3 \times 2}\mathbf{A}^3\mathbf{b}_0$$

$$\vdots$$

$$\mathbf{b}_k = \frac{1}{k!}\mathbf{A}^k\mathbf{b}_0$$

By substituting $t = 0$ into Equation (9–29), we obtain

$$\mathbf{x}(0) = \mathbf{b}_0$$

Thus, the solution $\mathbf{x}(t)$ can be written as

$$\mathbf{x}(t) = \left(\mathbf{I} + \mathbf{A}t + \frac{1}{2!}\mathbf{A}^2 t^2 + \cdots + \frac{1}{k!}\mathbf{A}^k t^k + \cdots \right)\mathbf{x}(0)$$

The expression in the parentheses on the right-hand side of this last equation is an $n \times n$ matrix. Because of its similarity to the infinite power series for a scalar exponential, we call it the matrix exponential and write

$$\mathbf{I} + \mathbf{A}t + \frac{1}{2!}\mathbf{A}^2 t^2 + \cdots + \frac{1}{k!}\mathbf{A}^k t^k + \cdots = e^{\mathbf{A}t}$$

In terms of the matrix exponential, the solution of Equation (9–28) can be written as

$$\mathbf{x}(t) = e^{\mathbf{A}t}\mathbf{x}(0) \qquad (9\text{–}31)$$

Since the matrix exponential is very important in the state-space analysis of linear systems, we shall next examine its properties.

Matrix Exponential. It can be proved that the matrix exponential of an $n \times n$ matrix $\mathbf{A}$,

$$e^{\mathbf{A}t} = \sum_{k=0}^{\infty} \frac{\mathbf{A}^k t^k}{k!}$$

converges absolutely for all finite t. (Hence, computer calculations for evaluating the elements of $e^{\mathbf{A}t}$ by using the series expansion can be easily carried out.)

Because of the convergence of the infinite series $\sum_{k=0}^{\infty} \mathbf{A}^k t^k / k!$, the series can be differentiated term by term to give

$$\frac{d}{dt} e^{\mathbf{A}t} = \mathbf{A} + \mathbf{A}^2 t + \frac{\mathbf{A}^3 t^2}{2!} + \cdots + \frac{\mathbf{A}^k t^{k-1}}{(k-1)!} + \cdots$$

$$= \mathbf{A}\left[\mathbf{I} + \mathbf{A}t + \frac{\mathbf{A}^2 t^2}{2!} + \cdots + \frac{\mathbf{A}^{k-1} t^{k-1}}{(k-1)!} + \cdots\right] = \mathbf{A}e^{\mathbf{A}t}$$

$$= \left[\mathbf{I} + \mathbf{A}t + \frac{\mathbf{A}^2 t^2}{2!} + \cdots + \frac{\mathbf{A}^{k-1} t^{k-1}}{(k-1)!} + \cdots\right]\mathbf{A} = e^{\mathbf{A}t}\mathbf{A}$$

The matrix exponential has the property that

$$e^{\mathbf{A}(t+s)} = e^{\mathbf{A}t} e^{\mathbf{A}s}$$

This can be proved as follows:

$$e^{\mathbf{A}t} e^{\mathbf{A}s} = \left(\sum_{k=0}^{\infty} \frac{\mathbf{A}^k t^k}{k!}\right)\left(\sum_{k=0}^{\infty} \frac{\mathbf{A}^k s^k}{k!}\right)$$

$$= \sum_{k=0}^{\infty} \mathbf{A}^k \left(\sum_{i=0}^{\infty} \frac{t^i s^{k-i}}{i!\,(k-i)!}\right)$$

$$= \sum_{k=0}^{\infty} \mathbf{A}^k \frac{(t+s)^k}{k!}$$

$$= e^{\mathbf{A}(t+s)}$$

In particular, if $s = -t$, then

$$e^{\mathbf{A}t} e^{-\mathbf{A}t} = e^{-\mathbf{A}t} e^{\mathbf{A}t} = e^{\mathbf{A}(t-t)} = \mathbf{I}$$

Thus, the inverse of $e^{\mathbf{A}t}$ is $e^{-\mathbf{A}t}$. Since the inverse of $e^{\mathbf{A}t}$ always exists, $e^{\mathbf{A}t}$ is nonsingular.

It is very important to remember that

$$e^{(\mathbf{A}+\mathbf{B})t} = e^{\mathbf{A}t} e^{\mathbf{B}t}, \qquad \text{if } \mathbf{AB} = \mathbf{BA}$$

$$e^{(\mathbf{A}+\mathbf{B})t} \neq e^{\mathbf{A}t} e^{\mathbf{B}t}, \qquad \text{if } \mathbf{AB} \neq \mathbf{BA}$$

To prove this, note that

$$e^{(\mathbf{A}+\mathbf{B})t} = \mathbf{I} + (\mathbf{A}+\mathbf{B})t + \frac{(\mathbf{A}+\mathbf{B})^2}{2!} t^2 + \frac{(\mathbf{A}+\mathbf{B})^3}{3!} t^3 + \cdots$$

$$e^{\mathbf{A}t} e^{\mathbf{B}t} = \left(\mathbf{I} + \mathbf{A}t + \frac{\mathbf{A}^2 t^2}{2!} + \frac{\mathbf{A}^3 t^3}{3!} + \cdots\right)\left(\mathbf{I} + \mathbf{B}t + \frac{\mathbf{B}^2 t^2}{2!} + \frac{\mathbf{B}^3 t^3}{3!} + \cdots\right)$$

$$= \mathbf{I} + (\mathbf{A}+\mathbf{B})t + \frac{\mathbf{A}^2 t^2}{2!} + \mathbf{AB}t^2 + \frac{\mathbf{B}^2 t^2}{2!} + \frac{\mathbf{A}^3 t^3}{3!}$$

$$+ \frac{\mathbf{A}^2 \mathbf{B}t^3}{2!} + \frac{\mathbf{AB}^2 t^3}{2!} + \frac{\mathbf{B}^3 t^3}{3!} + \cdots$$

Hence,

$$e^{(\mathbf{A}+\mathbf{B})t} - e^{\mathbf{A}t}e^{\mathbf{B}t} = \frac{\mathbf{BA} - \mathbf{AB}}{2!}t^2$$

$$+ \frac{\mathbf{BA}^2 + \mathbf{ABA} + \mathbf{B}^2\mathbf{A} + \mathbf{BAB} - 2\mathbf{A}^2\mathbf{B} - 2\mathbf{AB}^2}{3!}t^3 + \cdots$$

The difference between $e^{(\mathbf{A}+\mathbf{B})t}$ and $e^{\mathbf{A}t}e^{\mathbf{B}t}$ vanishes if $\mathbf{A}$ and $\mathbf{B}$ commute.

Laplace Transform Approach to the Solution of Homogeneous State Equations. Let us first consider the scalar case:

$$\dot{x} = ax \tag{9–32}$$

Taking the Laplace transform of Equation (9–32), we obtain

$$sX(s) - x(0) = aX(s) \tag{9–33}$$

where $X(s) = \mathcal{L}[x]$. Solving Equation (9–33) for $X(s)$ gives

$$X(s) = \frac{x(0)}{s - a} = (s - a)^{-1}x(0)$$

The inverse Laplace transform of this last equation gives the solution

$$x(t) = e^{at}x(0)$$

The foregoing approach to the solution of the homogeneous scalar differential equation can be extended to the homogeneous state equation:

$$\dot{\mathbf{x}}(t) = \mathbf{A}\mathbf{x}(t) \tag{9–34}$$

Taking the Laplace transform of both sides of Equation (9–34), we obtain

$$s\mathbf{X}(s) - \mathbf{x}(0) = \mathbf{A}\mathbf{X}(s)$$

where $\mathbf{X}(s) = \mathcal{L}[\mathbf{x}]$. Hence,

$$(s\mathbf{I} - \mathbf{A})\mathbf{X}(s) = \mathbf{x}(0)$$

Premultiplying both sides of this last equation by $(s\mathbf{I} - \mathbf{A})^{-1}$, we obtain

$$\mathbf{X}(s) = (s\mathbf{I} - \mathbf{A})^{-1}\mathbf{x}(0)$$

The inverse Laplace transform of $\mathbf{X}(s)$ gives the solution $\mathbf{x}(t)$. Thus,

$$\mathbf{x}(t) = \mathcal{L}^{-1}\big[(s\mathbf{I} - \mathbf{A})^{-1}\big]\mathbf{x}(0) \tag{9–35}$$

Note that

$$(s\mathbf{I} - \mathbf{A})^{-1} = \frac{\mathbf{I}}{s} + \frac{\mathbf{A}}{s^2} + \frac{\mathbf{A}^2}{s^3} + \cdots$$

Hence, the inverse Laplace transform of $(s\mathbf{I} - \mathbf{A})^{-1}$ gives

$$\mathcal{L}^{-1}\big[(s\mathbf{I} - \mathbf{A})^{-1}\big] = \mathbf{I} + \mathbf{A}t + \frac{\mathbf{A}^2 t^2}{2!} + \frac{\mathbf{A}^3 t^3}{3!} + \cdots = e^{\mathbf{A}t} \tag{9–36}$$

(The inverse Laplace transform of a matrix is the matrix consisting of the inverse Laplace transforms of all elements.) From Equations (9–35) and (9–36), the solution of Equation (9–34) is obtained as

$$\mathbf{x}(t) = e^{\mathbf{A}t}\mathbf{x}(0)$$

The importance of Equation (9–36) lies in the fact that it provides a convenient means for finding the closed solution for the matrix exponential.

State-Transition Matrix. We can write the solution of the homogeneous state equation

$$\dot{\mathbf{x}} = \mathbf{A}\mathbf{x} \tag{9–37}$$

as

$$\mathbf{x}(t) = \mathbf{\Phi}(t)\mathbf{x}(0) \tag{9–38}$$

where $\mathbf{\Phi}(t)$ is an $n \times n$ matrix and is the unique solution of

$$\dot{\mathbf{\Phi}}(t) = \mathbf{A}\mathbf{\Phi}(t), \qquad \mathbf{\Phi}(0) = \mathbf{I}$$

To verify this, note that

$$\mathbf{x}(0) = \mathbf{\Phi}(0)\mathbf{x}(0) = \mathbf{x}(0)$$

and

$$\dot{\mathbf{x}}(t) = \dot{\mathbf{\Phi}}(t)\mathbf{x}(0) = \mathbf{A}\mathbf{\Phi}(t)\mathbf{x}(0) = \mathbf{A}\mathbf{x}(t)$$

We thus confirm that Equation (9–38) is the solution of Equation (9–37).
    From Equations (9–31), (9–35), and (9–38), we obtain

$$\mathbf{\Phi}(t) = e^{\mathbf{A}t} = \mathscr{L}^{-1}\big[(s\mathbf{I} - \mathbf{A})^{-1}\big]$$

Note that

$$\mathbf{\Phi}^{-1}(t) = e^{-\mathbf{A}t} = \mathbf{\Phi}(-t)$$

From Equation (9–38), we see that the solution of Equation (9–37) is simply a transformation of the initial condition. Hence, the unique matrix $\mathbf{\Phi}(t)$ is called the state-transition matrix. The state-transition matrix contains all the information about the free motions of the system defined by Equation (9–37).
    If the eigenvalues $\lambda_1, \lambda_2, \ldots, \lambda_n$ of the matrix $\mathbf{A}$ are distinct, than $\mathbf{\Phi}(t)$ will contain the n exponentials

$$e^{\lambda_1 t}, e^{\lambda_2 t}, \ldots, e^{\lambda_n t}$$

In particular, if the matrix $\mathbf{A}$ is diagonal, then

$$\mathbf{\Phi}(t) = e^{\mathbf{A}t} = \begin{bmatrix} e^{\lambda_1 t} & & & & 0 \\ & e^{\lambda_2 t} & & & \\ & & \cdot & & \\ & & & \cdot & \\ 0 & & & & e^{\lambda_n t} \end{bmatrix} \qquad (\mathbf{A}: \text{diagonal})$$

If there is a multiplicity in the eigenvalues—for example, if the eigenvalues of $\mathbf{A}$ are

$$\lambda_1, \lambda_1, \lambda_1, \lambda_4, \lambda_5, \ldots, \lambda_n,$$

then $\mathbf{\Phi}(t)$ will contain, in addition to the exponentials $e^{\lambda_1 t}, e^{\lambda_4 t}, e^{\lambda_5 t}, \ldots, e^{\lambda_n t}$, terms like $te^{\lambda_1 t}$ and $t^2 e^{\lambda_1 t}$.

Properties of State-Transition Matrices. We shall now summarize the important properties of the state-transition matrix $\mathbf{\Phi}(t)$. For the time-invariant system

$$\dot{\mathbf{x}} = \mathbf{A}\mathbf{x}$$

for which

$$\mathbf{\Phi}(t) = e^{\mathbf{A}t}$$

we have the following:

1. $\mathbf{\Phi}(0) = e^{\mathbf{A}0} = \mathbf{I}$
2. $\mathbf{\Phi}(t) = e^{\mathbf{A}t} = \left(e^{-\mathbf{A}t}\right)^{-1} = \left[\mathbf{\Phi}(-t)\right]^{-1}$ or $\mathbf{\Phi}^{-1}(t) = \mathbf{\Phi}(-t)$
3. $\mathbf{\Phi}(t_1 + t_2) = e^{\mathbf{A}(t_1 + t_2)} = e^{\mathbf{A}t_1}e^{\mathbf{A}t_2} = \mathbf{\Phi}(t_1)\mathbf{\Phi}(t_2) = \mathbf{\Phi}(t_2)\mathbf{\Phi}(t_1)$
4. $\left[\mathbf{\Phi}(t)\right]^n = \mathbf{\Phi}(nt)$
5. $\mathbf{\Phi}(t_2 - t_1)\mathbf{\Phi}(t_1 - t_0) = \mathbf{\Phi}(t_2 - t_0) = \mathbf{\Phi}(t_1 - t_0)\mathbf{\Phi}(t_2 - t_1)$

EXAMPLE 9–5 Obtain the state-transition matrix $\mathbf{\Phi}(t)$ of the following system:

$$\begin{bmatrix} \dot{x}_1 \\ \dot{x}_2 \end{bmatrix} = \begin{bmatrix} 0 & 1 \\ -2 & -3 \end{bmatrix} \begin{bmatrix} x_1 \\ x_2 \end{bmatrix}$$

Obtain also the inverse of the state-transition matrix, $\mathbf{\Phi}^{-1}(t)$.
For this system,

$$\mathbf{A} = \begin{bmatrix} 0 & 1 \\ -2 & -3 \end{bmatrix}$$

The state-transition matrix $\mathbf{\Phi}(t)$ is given by

$$\mathbf{\Phi}(t) = e^{\mathbf{A}t} = \mathcal{L}^{-1}\left[(s\mathbf{I} - \mathbf{A})^{-1}\right]$$

Since

$$s\mathbf{I} - \mathbf{A} = \begin{bmatrix} s & 0 \\ 0 & s \end{bmatrix} - \begin{bmatrix} 0 & 1 \\ -2 & -3 \end{bmatrix} = \begin{bmatrix} s & -1 \\ 2 & s+3 \end{bmatrix}$$

the inverse of $(s\mathbf{I} - \mathbf{A})$ is given by

$$(s\mathbf{I} - \mathbf{A})^{-1} = \frac{1}{(s+1)(s+2)}\begin{bmatrix} s+3 & 1 \\ -2 & s \end{bmatrix}$$

$$= \begin{bmatrix} \dfrac{s+3}{(s+1)(s+2)} & \dfrac{1}{(s+1)(s+2)} \\ \dfrac{-2}{(s+1)(s+2)} & \dfrac{s}{(s+1)(s+2)} \end{bmatrix}$$

Hence,

$$\Phi(t) = e^{\mathbf{A}t} = \mathcal{L}^{-1}\left[(s\mathbf{I} - \mathbf{A})^{-1}\right]$$

$$= \begin{bmatrix} 2e^{-t} - e^{-2t} & e^{-t} - e^{-2t} \\ -2e^{-t} + 2e^{-2t} & -e^{-t} + 2e^{-2t} \end{bmatrix}$$

Noting that $\Phi^{-1}(t) = \Phi(-t)$, we obtain the inverse of the state-transition matrix as follows:

$$\Phi^{-1}(t) = e^{-\mathbf{A}t} = \begin{bmatrix} 2e^{t} - e^{2t} & e^{t} - e^{2t} \\ -2e^{t} + 2e^{2t} & -e^{t} + 2e^{2t} \end{bmatrix}$$

Solution of Nonhomogeneous State Equations. We shall begin by considering the scalar case

$$\dot{x} = ax + bu \qquad\qquad (9\text{--}39)$$

Let us rewrite Equation (9–39) as

$$\dot{x} - ax = bu$$

Multiplying both sides of this equation by e^{-at}, we obtain

$$e^{-at}\left[\dot{x}(t) - ax(t)\right] = \frac{d}{dt}\left[e^{-at}x(t)\right] = e^{-at}bu(t)$$

Integrating this equation between 0 and t gives

$$e^{-at}x(t) - x(0) = \int_0^t e^{-a\tau}bu(\tau)\,d\tau$$

or

$$x(t) = e^{at}x(0) + e^{at}\int_0^t e^{-a\tau}bu(\tau)\,d\tau$$

The first term on the right-hand side is the response to the initial condition and the second term is the response to the input $u(t)$.

Let us now consider the nonhomogeneous state equation described by

$$\dot{\mathbf{x}} = \mathbf{A}\mathbf{x} + \mathbf{B}\mathbf{u} \qquad\qquad (9\text{--}40)$$

where $\mathbf{x} = n$-vector
   $\mathbf{u} = r$-vector
   $\mathbf{A} = n \times n$ constant matrix
   $\mathbf{B} = n \times r$ constant matrix

By writing Equation (9–40) as

$$\dot{\mathbf{x}}(t) - \mathbf{A}\mathbf{x}(t) = \mathbf{B}\mathbf{u}(t)$$

and premultiplying both sides of this equation by $e^{-\mathbf{A}t}$, we obtain

$$e^{-\mathbf{A}t}\left[\dot{\mathbf{x}}(t) - \mathbf{A}\mathbf{x}(t)\right] = \frac{d}{dt}\left[e^{-\mathbf{A}t}\mathbf{x}(t)\right] = e^{-\mathbf{A}t}\mathbf{B}\mathbf{u}(t)$$

Integrating the preceding equation between 0 and t gives

$$e^{-\mathbf{A}t}\mathbf{x}(t) - \mathbf{x}(0) = \int_0^t e^{-\mathbf{A}\tau}\mathbf{B}\mathbf{u}(\tau)\,d\tau$$

or

$$\mathbf{x}(t) = e^{\mathbf{A}t}\mathbf{x}(0) + \int_0^t e^{\mathbf{A}(t-\tau)}\mathbf{B}\mathbf{u}(\tau)\,d\tau \tag{9-41}$$

Equation (9–41) can also be written as

$$\mathbf{x}(t) = \mathbf{\Phi}(t)\mathbf{x}(0) + \int_0^t \mathbf{\Phi}(t-\tau)\mathbf{B}\mathbf{u}(\tau)\,d\tau \tag{9-42}$$

where $\mathbf{\Phi}(t) = e^{\mathbf{A}t}$. Equation (9–41) or (9–42) is the solution of Equation (9–40). The solution $\mathbf{x}(t)$ is clearly the sum of a term consisting of the transition of the initial state and a term arising from the input vector.

Laplace Transform Approach to the Solution of Nonhomogeneous State Equations. The solution of the nonhomogeneous state equation

$$\dot{\mathbf{x}} = \mathbf{A}\mathbf{x} + \mathbf{B}\mathbf{u}$$

can also be obtained by the Laplace transform approach. The Laplace transform of this last equation yields

$$s\mathbf{X}(s) - \mathbf{x}(0) = \mathbf{A}\mathbf{X}(s) + \mathbf{B}\mathbf{U}(s)$$

or

$$(s\mathbf{I} - \mathbf{A})\mathbf{X}(s) = \mathbf{x}(0) + \mathbf{B}\mathbf{U}(s)$$

Premultiplying both sides of this last equation by $(s\mathbf{I} - \mathbf{A})^{-1}$, we obtain

$$\mathbf{X}(s) = (s\mathbf{I} - \mathbf{A})^{-1}\mathbf{x}(0) + (s\mathbf{I} - \mathbf{A})^{-1}\mathbf{B}\mathbf{U}(s)$$

Using the relationship given by Equation (9–36) gives

$$\mathbf{X}(s) = \mathcal{L}\left[e^{\mathbf{A}t}\right]\mathbf{x}(0) + \mathcal{L}\left[e^{\mathbf{A}t}\right]\mathbf{B}\mathbf{U}(s)$$

The inverse Laplace transform of this last equation can be obtained by use of the convolution integral as follows:

$$\mathbf{x}(t) = e^{\mathbf{A}t}\mathbf{x}(0) + \int_0^t e^{\mathbf{A}(t-\tau)}\mathbf{B}\mathbf{u}(\tau)\,d\tau$$

Solution in Terms of $\mathbf{x}(t_0)$. Thus far we have assumed the initial time to be zero. If, however, the initial time is given by t_0 instead of 0, then the solution to Equation (9–40) must be modified to

$$\mathbf{x}(t) = e^{\mathbf{A}(t-t_0)}\mathbf{x}(t_0) + \int_{t_0}^t e^{\mathbf{A}(t-\tau)}\mathbf{B}\mathbf{u}(\tau)\,d\tau \tag{9-43}$$

EXAMPLE 9–6 Obtain the time response of the following system:

$$\begin{bmatrix} \dot{x}_1 \\ \dot{x}_2 \end{bmatrix} = \begin{bmatrix} 0 & 1 \\ -2 & -3 \end{bmatrix} \begin{bmatrix} x_1 \\ x_2 \end{bmatrix} + \begin{bmatrix} 0 \\ 1 \end{bmatrix} u$$

where $u(t)$ is the unit-step function occurring at $t = 0$, or

$$u(t) = 1(t)$$

For this system,

$$\mathbf{A} = \begin{bmatrix} 0 & 1 \\ -2 & -3 \end{bmatrix}, \qquad \mathbf{B} = \begin{bmatrix} 0 \\ 1 \end{bmatrix}$$

The state-transition matrix $\mathbf{\Phi}(t) = e^{\mathbf{A}t}$ was obtained in Example 9–5 as

$$\mathbf{\Phi}(t) = e^{\mathbf{A}t} = \begin{bmatrix} 2e^{-t} - e^{-2t} & e^{-t} - e^{-2t} \\ -2e^{-t} + 2e^{-2t} & -e^{-t} + 2e^{-2t} \end{bmatrix}$$

The response to the unit-step input is then obtained as

$$\mathbf{x}(t) = e^{\mathbf{A}t}\mathbf{x}(0) + \int_0^t \begin{bmatrix} 2e^{-(t-\tau)} - e^{-2(t-\tau)} & e^{-(t-\tau)} - e^{-2(t-\tau)} \\ -2e^{-(t-\tau)} + 2e^{-2(t-\tau)} & -e^{-(t-\tau)} + 2e^{-2(t-\tau)} \end{bmatrix} \begin{bmatrix} 0 \\ 1 \end{bmatrix}[1]\, d\tau$$

or

$$\begin{bmatrix} x_1(t) \\ x_2(t) \end{bmatrix} = \begin{bmatrix} 2e^{-t} - e^{-2t} & e^{-t} - e^{-2t} \\ -2e^{-t} + 2e^{-2t} & -e^{-t} + 2e^{-2t} \end{bmatrix} \begin{bmatrix} x_1(0) \\ x_2(0) \end{bmatrix} + \begin{bmatrix} \frac{1}{2} - e^{-t} + \frac{1}{2}e^{-2t} \\ e^{-t} - e^{-2t} \end{bmatrix}$$

If the initial state is zero, or $\mathbf{x}(0) = \mathbf{0}$, then $\mathbf{x}(t)$ can be simplified to

$$\begin{bmatrix} x_1(t) \\ x_2(t) \end{bmatrix} = \begin{bmatrix} \dfrac{1}{2} - e^{-t} + \dfrac{1}{2}e^{-2t} \\ e^{-t} - e^{-2t} \end{bmatrix}$$

9–5 SOME USEFUL RESULTS IN VECTOR-MATRIX ANALYSIS

In this section we present some useful results in vector-matrix analysis that we use in Section 9–6. Specifically, we present the Cayley–Hamilton theorem, the minimal polynomial, Sylvester's interpolation method for calculating $e^{\mathbf{A}t}$, and the linear independence of vectors.

Cayley–Hamilton Theorem. The Cayley–Hamilton theorem is very useful in proving theorems involving matrix equations or solving problems involving matrix equations.

Consider an $n \times n$ matrix $\mathbf{A}$ and its characteristic equation:

$$|\lambda\mathbf{I} - \mathbf{A}| = \lambda^n + a_1\lambda^{n-1} + \cdots + a_{n-1}\lambda + a_n = 0$$

The Cayley–Hamilton theorem states that the matrix $\mathbf{A}$ satisfies its own characteristic equation, or that

$$\mathbf{A}^n + a_1\mathbf{A}^{n-1} + \cdots + a_{n-1}\mathbf{A} + a_n\mathbf{I} = \mathbf{0} \tag{9–44}$$

To prove this theorem, note that $\text{adj}(\lambda\mathbf{I} - \mathbf{A})$ is a polynomial in λ of degree $n - 1$. That is,

$$\text{adj}(\lambda\mathbf{I} - \mathbf{A}) = \mathbf{B}_1\lambda^{n-1} + \mathbf{B}_2\lambda^{n-2} + \cdots + \mathbf{B}_{n-1}\lambda + \mathbf{B}_n$$

where $\mathbf{B}_1 = \mathbf{I}$. Since

$$(\lambda \mathbf{I} - \mathbf{A}) \, \mathrm{adj}(\lambda \mathbf{I} - \mathbf{A}) = [\mathrm{adj}(\lambda \mathbf{I} - \mathbf{A})](\lambda \mathbf{I} - \mathbf{A}) = |\lambda \mathbf{I} - \mathbf{A}|\mathbf{I}$$

we obtain

$$
\begin{aligned}
|\lambda \mathbf{I} - \mathbf{A}|\mathbf{I} &= \mathbf{I}\lambda^n + a_1 \mathbf{I}\lambda^{n-1} + \cdots + a_{n-1}\mathbf{I}\lambda + a_n \mathbf{I} \\
&= (\lambda \mathbf{I} - \mathbf{A})(\mathbf{B}_1 \lambda^{n-1} + \mathbf{B}_2 \lambda^{n-2} + \cdots + \mathbf{B}_{n-1}\lambda + \mathbf{B}_n) \\
&= (\mathbf{B}_1 \lambda^{n-1} + \mathbf{B}_2 \lambda^{n-2} + \cdots + \mathbf{B}_{n-1}\lambda + \mathbf{B}_n)(\lambda \mathbf{I} - \mathbf{A})
\end{aligned}
$$

From this equation, we see that $\mathbf{A}$ and $\mathbf{B}_i$ $(i = 1, 2, \ldots, n)$ commute. Hence, the product of $(\lambda \mathbf{I} - \mathbf{A})$ and $\mathrm{adj}(\lambda \mathbf{I} - \mathbf{A})$ becomes zero if either of these is zero. If $\mathbf{A}$ is substituted for λ in this last equation, then clearly $\lambda \mathbf{I} - \mathbf{A}$ becomes zero. Hence, we obtain

$$\mathbf{A}^n + a_1 \mathbf{A}^{n-1} + \cdots + a_{n-1}\mathbf{A} + a_n \mathbf{I} = \mathbf{0}$$

This proves the Cayley–Hamilton theorem, or Equation (9–44).

Minimal Polynomial. Referring to the Cayley–Hamilton theorem, every $n \times n$ matrix $\mathbf{A}$ satisfies its own characteristic equation. The characteristic equation is not, however, necessarily the scalar equation of least degree that $\mathbf{A}$ satisfies. The least-degree polynomial having $\mathbf{A}$ as a root is called the *minimal polynomial*. That is, the minimal polynomial of an $n \times n$ matrix $\mathbf{A}$ is defined as the polynomial $\phi(\lambda)$ of least degree,

$$\phi(\lambda) = \lambda^m + a_1 \lambda^{m-1} + \cdots + a_{m-1}\lambda + a_m, \qquad m \leq n$$

such that $\phi(\mathbf{A}) = \mathbf{0}$, or

$$\phi(\mathbf{A}) = \mathbf{A}^m + a_1 \mathbf{A}^{m-1} + \cdots + a_{m-1}\mathbf{A} + a_m \mathbf{I} = \mathbf{0}$$

The minimal polynomial plays an important role in the computation of polynomials in an $n \times n$ matrix.

Let us suppose that $d(\lambda)$, a polynomial in λ, is the greatest common divisor of all the elements of $\mathrm{adj}(\lambda \mathbf{I} - \mathbf{A})$. We can show that if the coefficient of the highest-degree term in λ of $d(\lambda)$ is chosen as 1, then the minimal polynomial $\phi(\lambda)$ is given by

$$\phi(\lambda) = \frac{|\lambda \mathbf{I} - \mathbf{A}|}{d(\lambda)} \tag{9–45}$$

[See Problem A–9–8 for the derivation of Equation (9–45).]

It is noted that the minimal polynomial $\phi(\lambda)$ of an $n \times n$ matrix $\mathbf{A}$ can be determined by the following procedure:

1. Form $\mathrm{adj}(\lambda \mathbf{I} - \mathbf{A})$ and write the elements of $\mathrm{adj}(\lambda \mathbf{I} - \mathbf{A})$ as factored polynomials in λ.
2. Determine $d(\lambda)$ as the greatest common divisor of all the elements of $\mathrm{adj}(\lambda \mathbf{I} - \mathbf{A})$. Choose the coefficient of the highest-degree term in λ of $d(\lambda)$ to be 1. If there is no common divisor, $d(\lambda) = 1$.
3. The minimal polynomial $\phi(\lambda)$ is then given as $|\lambda \mathbf{I} - \mathbf{A}|$ divided by $d(\lambda)$.

Matrix Exponential $e^{\mathbf{A}t}$. In solving control engineering problems, it often becomes necessary to compute $e^{\mathbf{A}t}$. If matrix $\mathbf{A}$ is given with all elements in numerical values, MATLAB provides a simple way to compute $e^{\mathbf{A}T}$, where T is a constant.

Aside from computational methods, several analytical methods are available for the computation of $e^{\mathbf{A}t}$. We shall present three methods here.

Computation of $e^{\mathbf{A}t}$: Method 1. If matrix $\mathbf{A}$ can be transformed into a diagonal form, then $e^{\mathbf{A}t}$ can be given by

$$e^{\mathbf{A}t} = \mathbf{P}e^{\mathbf{D}t}\mathbf{P}^{-1} = \mathbf{P}\begin{bmatrix} e^{\lambda_1 t} & & & & 0 \\ & e^{\lambda_2 t} & & & \\ & & \cdot & & \\ & & & \cdot & \\ 0 & & & & e^{\lambda_n t} \end{bmatrix}\mathbf{P}^{-1} \qquad (9\text{--}46)$$

where $\mathbf{P}$ is a diagonalizing matrix for $\mathbf{A}$. [For the derivation of Equation (9–46), see Problem **A–9–11**.]

If matrix $\mathbf{A}$ can be transformed into a Jordan canonical form, then $e^{\mathbf{A}t}$ can be given by

$$e^{\mathbf{A}t} = \mathbf{S}e^{\mathbf{J}t}\mathbf{S}^{-1}$$

where $\mathbf{S}$ is a transformation matrix that transforms matrix $\mathbf{A}$ into a Jordan canonical form $\mathbf{J}$.

As an example, consider the following matrix $\mathbf{A}$:

$$\mathbf{A} = \begin{bmatrix} 0 & 1 & 0 \\ 0 & 0 & 1 \\ 1 & -3 & 3 \end{bmatrix}$$

The characteristic equation is

$$|\lambda\mathbf{I} - \mathbf{A}| = \lambda^3 - 3\lambda^2 + 3\lambda - 1 = (\lambda - 1)^3 = 0$$

Thus, matrix $\mathbf{A}$ has a multiple eigenvalue of order 3 at $\lambda = 1$. It can be shown that matrix $\mathbf{A}$ has a multiple eigenvector of order 3. The transformation matrix that will transform matrix $\mathbf{A}$ into a Jordan canonical form can be given by

$$\mathbf{S} = \begin{bmatrix} 1 & 0 & 0 \\ 1 & 1 & 0 \\ 1 & 2 & 1 \end{bmatrix}$$

The inverse of matrix $\mathbf{S}$ is

$$\mathbf{S}^{-1} = \begin{bmatrix} 1 & 0 & 0 \\ -1 & 1 & 0 \\ 1 & -2 & 1 \end{bmatrix}$$

Then it can be seen that

$$\mathbf{S}^{-1}\mathbf{A}\mathbf{S} = \begin{bmatrix} 1 & 0 & 0 \\ -1 & 1 & 0 \\ 1 & -2 & 1 \end{bmatrix}\begin{bmatrix} 0 & 1 & 0 \\ 0 & 0 & 1 \\ 1 & -3 & 3 \end{bmatrix}\begin{bmatrix} 1 & 0 & 0 \\ 1 & 1 & 0 \\ 1 & 2 & 1 \end{bmatrix}$$

$$= \begin{bmatrix} 1 & 1 & 0 \\ 0 & 1 & 1 \\ 0 & 0 & 1 \end{bmatrix} = \mathbf{J}$$

Noting that

$$e^{\mathbf{J}t} = \begin{bmatrix} e^t & te^t & \frac{1}{2}t^2 e^t \\ 0 & e^t & te^t \\ 0 & 0 & e^t \end{bmatrix}$$

we find

$$e^{\mathbf{A}t} = \mathbf{S}e^{\mathbf{J}t}\mathbf{S}^{-1}$$

$$= \begin{bmatrix} 1 & 0 & 0 \\ 1 & 1 & 0 \\ 1 & 2 & 1 \end{bmatrix} \begin{bmatrix} e^t & te^t & \frac{1}{2}t^2 e^t \\ 0 & e^t & te^t \\ 0 & 0 & e^t \end{bmatrix} \begin{bmatrix} 1 & 0 & 0 \\ -1 & 1 & 0 \\ 1 & -2 & 1 \end{bmatrix}$$

$$= \begin{bmatrix} e^t - te^t + \frac{1}{2}t^2 e^t & te^t - t^2 e^t & \frac{1}{2}t^2 e^t \\ \frac{1}{2}t^2 e^t & e^t - te^t - t^2 e^t & te^t + \frac{1}{2}t^2 e^t \\ te^t + \frac{1}{2}t^2 e^t & -3te^t - t^2 e^t & e^t + 2te^t + \frac{1}{2}t^2 e^t \end{bmatrix}$$

Computation of $e^{\mathbf{A}t}$: Method 2. The second method of computing $e^{\mathbf{A}t}$ uses the Laplace transform approach. Referring to Equation (9–36), $e^{\mathbf{A}t}$ can be given as follows:

$$e^{\mathbf{A}t} = \mathcal{L}^{-1}\left[(s\mathbf{I} - \mathbf{A})^{-1}\right]$$

Thus, to obtain $e^{\mathbf{A}t}$, first invert the matrix $(s\mathbf{I} - \mathbf{A})$. This results in a matrix whose elements are rational functions of s. Then take the inverse Laplace transform of each element of the matrix.

EXAMPLE 9–7   Consider the following matrix $\mathbf{A}$:

$$\mathbf{A} = \begin{bmatrix} 0 & 1 \\ 0 & -2 \end{bmatrix}$$

Compute $e^{\mathbf{A}t}$ by use of the two analytical methods presented previously.

Method 1.   The eigenvalues of $\mathbf{A}$ are 0 and -2 ($\lambda_1 = 0$, $\lambda_2 = -2$). A necessary transformation matrix $\mathbf{P}$ may be obtained as

$$\mathbf{P} = \begin{bmatrix} 1 & 1 \\ 0 & -2 \end{bmatrix}$$

Then, from Equation (9–46), $e^{\mathbf{A}t}$ is obtained as follows:

$$e^{\mathbf{A}t} = \begin{bmatrix} 1 & 1 \\ 0 & -2 \end{bmatrix} \begin{bmatrix} e^0 & 0 \\ 0 & e^{-2t} \end{bmatrix} \begin{bmatrix} 1 & \frac{1}{2} \\ 0 & -\frac{1}{2} \end{bmatrix} = \begin{bmatrix} 1 & \frac{1}{2}(1 - e^{-2t}) \\ 0 & e^{-2t} \end{bmatrix}$$

Method 2.   Since

$$s\mathbf{I} - \mathbf{A} = \begin{bmatrix} s & 0 \\ 0 & s \end{bmatrix} - \begin{bmatrix} 0 & 1 \\ 0 & -2 \end{bmatrix} = \begin{bmatrix} s & -1 \\ 0 & s+2 \end{bmatrix}$$

we obtain

$$(s\mathbf{I} - \mathbf{A})^{-1} = \begin{bmatrix} \dfrac{1}{s} & \dfrac{1}{s(s+2)} \\ 0 & \dfrac{1}{s+2} \end{bmatrix}$$

Hence,

$$e^{\mathbf{A}t} = \mathscr{L}^{-1}\left[(s\mathbf{I} - \mathbf{A})^{-1}\right] = \begin{bmatrix} 1 & \frac{1}{2}\left(1 - e^{-2t}\right) \\ 0 & e^{-2t} \end{bmatrix}$$

Computation of $e^{\mathbf{A}t}$: Method 3. The third method is based on Sylvester's interpolation method. (For Sylvester's interpolation formula, see Problem A–9–12.) We shall first consider the case where the roots of the minimal polynomial $\phi(\lambda)$ of $\mathbf{A}$ are distinct. Then we shall deal with the case of multiple roots.

Case 1: Minimal Polynomial of $\mathbf{A}$ Involves Only Distinct Roots. We shall assume that the degree of the minimal polynomial of $\mathbf{A}$ is m. By using Sylvester's interpolation formula, it can be shown that $e^{\mathbf{A}t}$ can be obtained by solving the following determinant equation:

$$\begin{vmatrix} 1 & \lambda_1 & \lambda_1^2 & \cdots & \lambda_1^{m-1} & e^{\lambda_1 t} \\ 1 & \lambda_2 & \lambda_2^2 & \cdots & \lambda_2^{m-1} & e^{\lambda_2 t} \\ \cdot & \cdot & \cdot & & \cdot & \cdot \\ \cdot & \cdot & \cdot & & \cdot & \cdot \\ \cdot & \cdot & \cdot & & \cdot & \cdot \\ 1 & \lambda_m & \lambda_m^2 & \cdots & \lambda_m^{m-1} & e^{\lambda_m t} \\ \mathbf{I} & \mathbf{A} & \mathbf{A}^2 & \cdots & \mathbf{A}^{m-1} & e^{\mathbf{A}t} \end{vmatrix} = \mathbf{0} \qquad (9\text{–}47)$$

By solving Equation (9–47) for $e^{\mathbf{A}t}$, $e^{\mathbf{A}t}$ can be obtained in terms of the $\mathbf{A}^k$ ($k = 0, 1, 2, \ldots, m - 1$) and the $e^{\lambda_i t}$ ($i = 1, 2, 3, \ldots, m$). [Equation (9–47) may be expanded, for example, about the last column.]

Notice that solving Equation (9–47) for $e^{\mathbf{A}t}$ is the same as writing

$$e^{\mathbf{A}t} = \alpha_0(t)\mathbf{I} + \alpha_1(t)\mathbf{A} + \alpha_2(t)\mathbf{A}^2 + \cdots + \alpha_{m-1}(t)\mathbf{A}^{m-1} \qquad (9\text{–}48)$$

and determining the $\alpha_k(t)$ ($k = 0, 1, 2, \ldots, m - 1$) by solving the following set of m equations for the $\alpha_k(t)$:

$$\alpha_0(t) + \alpha_1(t)\lambda_1 + \alpha_2(t)\lambda_1^2 + \cdots + \alpha_{m-1}(t)\lambda_1^{m-1} = e^{\lambda_1 t}$$
$$\alpha_0(t) + \alpha_1(t)\lambda_2 + \alpha_2(t)\lambda_2^2 + \cdots + \alpha_{m-1}(t)\lambda_2^{m-1} = e^{\lambda_2 t}$$
$$\cdot$$
$$\cdot$$
$$\cdot$$
$$\alpha_0(t) + \alpha_1(t)\lambda_m + \alpha_2(t)\lambda_m^2 + \cdots + \alpha_{m-1}(t)\lambda_m^{m-1} = e^{\lambda_m t}$$

If $\mathbf{A}$ is an $n \times n$ matrix and has distinct eigenvalues, then the number of $\alpha_k(t)$'s to be determined is $m = n$. If $\mathbf{A}$ involves multiple eigenvalues, but its minimal polynomial has only simple roots, however, then the number m of $\alpha_k(t)$'s to be determined is less than n.

Case 2: Minimal Polynomial of $\mathbf{A}$ Involves Multiple Roots. As an example, consider the case where the minimal polynomial of $\mathbf{A}$ involves three equal roots $(\lambda_1 = \lambda_2 = \lambda_3)$ and has other roots $(\lambda_4, \lambda_5, \ldots, \lambda_m)$ that are all distinct. By applying Sylvester's interpolation formula, it can be shown that $e^{\mathbf{A}t}$ can be obtained from the following determinant equation:

$$
\begin{vmatrix}
0 & 0 & 1 & 3\lambda_1 & \cdots & \dfrac{(m-1)(m-2)}{2}\lambda_1^{m-3} & \dfrac{t^2}{2}e^{\lambda_1 t} \\[2ex]
0 & 1 & 2\lambda_1 & 3\lambda_1^2 & \cdots & (m-1)\lambda_1^{m-2} & te^{\lambda_1 t} \\[1ex]
1 & \lambda_1 & \lambda_1^2 & \lambda_1^3 & \cdots & \lambda_1^{m-1} & e^{\lambda_1 t} \\[1ex]
1 & \lambda_4 & \lambda_4^2 & \lambda_4^3 & \cdots & \lambda_4^{m-1} & e^{\lambda_4 t} \\[1ex]
\cdot & \cdot & \cdot & \cdot & \cdots & \cdot & \cdot \\
\cdot & \cdot & \cdot & \cdot & \cdots & \cdot & \cdot \\
\cdot & \cdot & \cdot & \cdot & \cdots & \cdot & \cdot \\
1 & \lambda_m & \lambda_m^2 & \lambda_m^3 & \cdots & \lambda_m^{m-1} & e^{\lambda_m t} \\[1ex]
\mathbf{I} & \mathbf{A} & \mathbf{A}^2 & \mathbf{A}^3 & \cdots & \mathbf{A}^{m-1} & e^{\mathbf{A}t}
\end{vmatrix} = \mathbf{0}
\qquad (9\text{--}49)
$$

Equation (9–49) can be solved for $e^{\mathbf{A}t}$ by expanding it about the last column.

It is noted that, just as in case 1, solving Equation (9–49) for $e^{\mathbf{A}t}$ is the same as writing

$$
e^{\mathbf{A}t} = \alpha_0(t)\mathbf{I} + \alpha_1(t)\mathbf{A} + \alpha_2(t)\mathbf{A}^2 + \cdots + \alpha_{m-1}(t)\mathbf{A}^{m-1}
\qquad (9\text{--}50)
$$

and determining the $\alpha_k(t)$'s $(k = 0, 1, 2, \ldots, m-1)$ from

$$
\alpha_2(t) + 3\alpha_3(t)\lambda_1 + \cdots + \frac{(m-1)(m-2)}{2}\alpha_{m-1}(t)\lambda_1^{m-3} = \frac{t^2}{2}e^{\lambda_1 t}
$$

$$
\alpha_1(t) + 2\alpha_2(t)\lambda_1 + 3\alpha_3(t)\lambda_1^2 + \cdots + (m-1)\alpha_{m-1}(t)\lambda_1^{m-2} = te^{\lambda_1 t}
$$

$$
\alpha_0(t) + \alpha_1(t)\lambda_1 + \alpha_2(t)\lambda_1^2 + \cdots + \alpha_{m-1}(t)\lambda_1^{m-1} = e^{\lambda_1 t}
$$

$$
\alpha_0(t) + \alpha_1(t)\lambda_4 + \alpha_2(t)\lambda_4^2 + \cdots + \alpha_{m-1}(t)\lambda_4^{m-1} = e^{\lambda_4 t}
$$

$$
\cdot
$$
$$
\cdot
$$
$$
\cdot
$$

$$
\alpha_0(t) + \alpha_1(t)\lambda_m + \alpha_2(t)\lambda_m^2 + \cdots + \alpha_{m-1}(t)\lambda_m^{m-1} = e^{\lambda_m t}
$$

The extension to other cases where, for example, there are two or more sets of multiple roots will be apparent. Note that if the minimal polynomial of $\mathbf{A}$ is not found, it is possible to substitute the characteristic polynomial for the minimal polynomial. The number of computations may, of course, be increased.

EXAMPLE 9–8 Consider the matrix

$$
\mathbf{A} = \begin{bmatrix} 0 & 1 \\ 0 & -2 \end{bmatrix}
$$

Compute $e^{\mathbf{A}t}$ using Sylvester's interpolation formula.

From Equation (9–47), we get

$$
\begin{vmatrix}
1 & \lambda_1 & e^{\lambda_1 t} \\
1 & \lambda_2 & e^{\lambda_2 t} \\
\mathbf{I} & \mathbf{A} & e^{\mathbf{A}t}
\end{vmatrix} = \mathbf{0}
$$

Substituting 0 for λ_1 and -2 for λ_2 in this last equation, we obtain

$$\begin{vmatrix} 1 & 0 & 1 \\ 1 & -2 & e^{-2t} \\ \mathbf{I} & \mathbf{A} & e^{\mathbf{A}t} \end{vmatrix} = \mathbf{0}$$

Expanding the determinant, we obtain

$$-2e^{\mathbf{A}t} + \mathbf{A} + 2\mathbf{I} - \mathbf{A}e^{-2t} = \mathbf{0}$$

or

$$e^{\mathbf{A}t} = \tfrac{1}{2}\left(\mathbf{A} + 2\mathbf{I} - \mathbf{A}e^{-2t}\right)$$

$$= \frac{1}{2}\left\{\begin{bmatrix} 0 & 1 \\ 0 & -2 \end{bmatrix} + \begin{bmatrix} 2 & 0 \\ 0 & 2 \end{bmatrix} - \begin{bmatrix} 0 & 1 \\ 0 & -2 \end{bmatrix}e^{-2t}\right\}$$

$$= \begin{bmatrix} 1 & \tfrac{1}{2}\left(1 - e^{-2t}\right) \\ 0 & e^{-2t} \end{bmatrix}$$

An alternative approach is to use Equation (9–48). We first determine $\alpha_0(t)$ and $\alpha_1(t)$ from

$$\alpha_0(t) + \alpha_1(t)\lambda_1 = e^{\lambda_1 t}$$

$$\alpha_0(t) + \alpha_1(t)\lambda_2 = e^{\lambda_2 t}$$

Since $\lambda_1 = 0$ and $\lambda_2 = -2$, the last two equations become

$$\alpha_0(t) = 1$$

$$\alpha_0(t) - 2\alpha_1(t) = e^{-2t}$$

Solving for $\alpha_0(t)$ and $\alpha_1(t)$ gives

$$\alpha_0(t) = 1, \qquad \alpha_1(t) = \frac{1}{2}\left(1 - e^{-2t}\right)$$

Then $e^{\mathbf{A}t}$ can be written as

$$e^{\mathbf{A}t} = \alpha_0(t)\mathbf{I} + \alpha_1(t)\mathbf{A} = \mathbf{I} + \frac{1}{2}\left(1 - e^{-2t}\right)\mathbf{A} = \begin{bmatrix} 1 & \tfrac{1}{2}\left(1 - e^{-2t}\right) \\ 0 & e^{-2t} \end{bmatrix}$$

Linear Independence of Vectors. The vectors $\mathbf{x}_1, \mathbf{x}_2, \dots, \mathbf{x}_n$ are said to be linearly independent if

$$c_1\mathbf{x}_1 + c_2\mathbf{x}_2 + \cdots + c_n\mathbf{x}_n = \mathbf{0}$$

where $c_1, c_2, \dots, c_n$ are constants, implies that

$$c_1 = c_2 = \cdots = c_n = 0$$

Conversely, the vectors $\mathbf{x}_1, \mathbf{x}_2, \dots, \mathbf{x}_n$ are said to be linearly dependent if and only if $\mathbf{x}_i$ can be expressed as a linear combination of $\mathbf{x}_j$ $(j = 1, 2, \dots, n; j \neq i)$, or

$$\mathbf{x}_i = \sum_{\substack{j=1 \\ j \neq i}}^{n} c_j \mathbf{x}_j$$

for some set of constants c_j. This means that if $\mathbf{x}_i$ can be expressed as a linear combination of the other vectors in the set, it is linearly dependent on them or it is not an independent member of the set.

EXAMPLE 9–9   The vectors

$$\mathbf{x}_1 = \begin{bmatrix} 1 \\ 2 \\ 3 \end{bmatrix}, \qquad \mathbf{x}_2 = \begin{bmatrix} 1 \\ 0 \\ 1 \end{bmatrix}, \qquad \mathbf{x}_3 = \begin{bmatrix} 2 \\ 2 \\ 4 \end{bmatrix}$$

are linearly dependent since

$$\mathbf{x}_1 + \mathbf{x}_2 - \mathbf{x}_3 = \mathbf{0}$$

The vectors

$$\mathbf{y}_1 = \begin{bmatrix} 1 \\ 2 \\ 3 \end{bmatrix}, \qquad \mathbf{y}_2 = \begin{bmatrix} 1 \\ 0 \\ 1 \end{bmatrix}, \qquad \mathbf{y}_3 = \begin{bmatrix} 2 \\ 2 \\ 2 \end{bmatrix}$$

are linearly independent since

$$c_1 \mathbf{y}_1 + c_2 \mathbf{y}_2 + c_3 \mathbf{y}_3 = \mathbf{0}$$

implies that

$$c_1 = c_2 = c_3 = 0$$

Note that if an $n \times n$ matrix is nonsingular (that is, the matrix is of rank n or the determinant is nonzero) then n column (or row) vectors are linearly independent. If the $n \times n$ matrix is singular (that is, the rank of the matrix is less than n or the determinant is zero), then n column (or row) vectors are linearly dependent. To demonstrate this, notice that

$$\begin{bmatrix} \mathbf{x}_1 & \vdots & \mathbf{x}_2 & \vdots & \mathbf{x}_3 \end{bmatrix} = \begin{bmatrix} 1 & 1 & 2 \\ 2 & 0 & 2 \\ 3 & 1 & 4 \end{bmatrix} = \text{singular}$$

$$\begin{bmatrix} \mathbf{y}_1 & \vdots & \mathbf{y}_2 & \vdots & \mathbf{y}_3 \end{bmatrix} = \begin{bmatrix} 1 & 1 & 2 \\ 2 & 0 & 2 \\ 3 & 1 & 2 \end{bmatrix} = \text{nonsingular}$$

9–6 CONTROLLABILITY

Controllability and Observability.   A system is said to be controllable at time t_0 if it is possible by means of an unconstrained control vector to transfer the system from any initial state $\mathbf{x}(t_0)$ to any other state in a finite interval of time.

A system is said to be observable at time t_0 if, with the system in state $\mathbf{x}(t_0)$, it is possible to determine this state from the observation of the output over a finite time interval.

The concepts of controllability and observability were introduced by Kalman. They play an important role in the design of control systems in state space. In fact, the conditions of controllability and observability may govern the existence of a complete solution to the control system design problem. The solution to this problem may not exist if the system considered is not controllable. Although most physical systems are controllable and observable, corresponding mathematical models may not possess the property of controllability and observability. Then it is necessary to know the conditions under which a system is controllable and observable. This section deals with controllability and the next section discusses observability.

In what follows, we shall first derive the condition for complete state controllability. Then we derive alternative forms of the condition for complete state controllability followed by discussions of complete output controllability. Finally, we present the concept of stabilizability.

Complete State Controllability of Continuous-Time Systems. Consider the continuous-time system.

$$\dot{\mathbf{x}} = \mathbf{A}\mathbf{x} + \mathbf{B}u \tag{9–51}$$

where
$\mathbf{x} =$ state vector (n-vector)
$u =$ control signal (scalar)
$\mathbf{A} = n \times n$ matrix
$\mathbf{B} = n \times 1$ matrix

The system described by Equation (9–51) is said to be state controllable at $t = t_0$ if it is possible to construct an unconstrained control signal that will transfer an initial state to any final state in a finite time interval $t_0 \leq t \leq t_1$. If every state is controllable, then the system is said to be completely state controllable.

We shall now derive the condition for complete state controllability. Without loss of generality, we can assume that the final state is the origin of the state space and that the initial time is zero, or $t_0 = 0$.

The solution of Equation (9–51) is

$$\mathbf{x}(t) = e^{\mathbf{A}t}\mathbf{x}(0) + \int_0^t e^{\mathbf{A}(t-\tau)}\mathbf{B}u(\tau)\,d\tau$$

Applying the definition of complete state controllability just given, we have

$$\mathbf{x}(t_1) = \mathbf{0} = e^{\mathbf{A}t_1}\mathbf{x}(0) + \int_0^{t_1} e^{\mathbf{A}(t_1-\tau)}\mathbf{B}u(\tau)\,d\tau$$

or

$$\mathbf{x}(0) = -\int_0^{t_1} e^{-\mathbf{A}\tau}\mathbf{B}u(\tau)\,d\tau \tag{9–52}$$

Referring to Equation (9–48) or (9–50), $e^{-\mathbf{A}\tau}$ can be written

$$e^{-\mathbf{A}\tau} = \sum_{k=0}^{n-1} \alpha_k(\tau)\mathbf{A}^k \tag{9–53}$$

Substituting Equation (9–53) into Equation (9–52) gives

$$\mathbf{x}(0) = -\sum_{k=0}^{n-1} \mathbf{A}^k\mathbf{B}\int_0^{t_1} \alpha_k(\tau)u(\tau)\,d\tau \tag{9–54}$$

Let us put

$$\int_0^{t_1} \alpha_k(\tau)u(\tau)\, d\tau = \beta_k$$

Then Equation (9–54) becomes

$$\mathbf{x}(0) = -\sum_{k=0}^{n-1} \mathbf{A}^k \mathbf{B}\beta_k$$

$$= -\begin{bmatrix} \mathbf{B} & \vdots & \mathbf{AB} & \vdots & \cdots & \vdots & \mathbf{A}^{n-1}\mathbf{B} \end{bmatrix} \begin{bmatrix} \beta_0 \\ \hline \beta_1 \\ \hline \vdots \\ \vdots \\ \hline \beta_{n-1} \end{bmatrix} \qquad (9\text{–}55)$$

If the system is completely state controllable, then, given any initial state $\mathbf{x}(0)$, Equation (9–55) must be satisfied. This requires that the rank of the $n \times n$ matrix

$$\begin{bmatrix} \mathbf{B} & \vdots & \mathbf{AB} & \vdots & \cdots & \vdots & \mathbf{A}^{n-1}\mathbf{B} \end{bmatrix}$$

be n.

From this analysis, we can state the condition for complete state controllability as follows: The system given by Equation (9–51) is completely state controllable if and only if the vectors $\mathbf{B}, \mathbf{AB}, \ldots, \mathbf{A}^{n-1}\mathbf{B}$ are linearly independent, or the $n \times n$ matrix

$$\begin{bmatrix} \mathbf{B} & \vdots & \mathbf{AB} & \vdots & \cdots & \vdots & \mathbf{A}^{n-1}\mathbf{B} \end{bmatrix}$$

is of rank n.

The result just obtained can be extended to the case where the control vector $\mathbf{u}$ is r-dimensional. If the system is described by

$$\dot{\mathbf{x}} = \mathbf{Ax} + \mathbf{Bu}$$

where $\mathbf{u}$ is an r-vector, then it can be proved that the condition for complete state controllability is that the $n \times nr$ matrix

$$\begin{bmatrix} \mathbf{B} & \vdots & \mathbf{AB} & \vdots & \cdots & \vdots & \mathbf{A}^{n-1}\mathbf{B} \end{bmatrix}$$

be of rank n, or contain n linearly independent column vectors. The matrix

$$\begin{bmatrix} \mathbf{B} & \vdots & \mathbf{AB} & \vdots & \cdots & \vdots & \mathbf{A}^{n-1}\mathbf{B} \end{bmatrix}$$

is commonly called the *controllability matrix*.

EXAMPLE 9–10 Consider the system given by

$$\begin{bmatrix} \dot{x}_1 \\ \dot{x}_2 \end{bmatrix} = \begin{bmatrix} 1 & 1 \\ 0 & -1 \end{bmatrix} \begin{bmatrix} x_1 \\ x_2 \end{bmatrix} + \begin{bmatrix} 1 \\ 0 \end{bmatrix} u$$

Since

$$\begin{bmatrix} \mathbf{B} & \vdots & \mathbf{AB} \end{bmatrix} = \begin{bmatrix} 1 & 1 \\ 0 & 0 \end{bmatrix} = \text{singular}$$

the system is not completely state controllable.

EXAMPLE 9–11 Consider the system given by

$$\begin{bmatrix} \dot{x}_1 \\ \dot{x}_2 \end{bmatrix} = \begin{bmatrix} 1 & 1 \\ 2 & -1 \end{bmatrix} \begin{bmatrix} x_1 \\ x_2 \end{bmatrix} + \begin{bmatrix} 0 \\ 1 \end{bmatrix} [u]$$

For this case,

$$\begin{bmatrix} \mathbf{B} & \vdots & \mathbf{AB} \end{bmatrix} = \begin{bmatrix} 0 & 1 \\ 1 & -1 \end{bmatrix} = \text{nonsingular}$$

The system is therefore completely state controllable.

Alternative Form of the Condition for Complete State Controllability. Consider the system defined by

$$\dot{\mathbf{x}} = \mathbf{Ax} + \mathbf{Bu} \tag{9–56}$$

where $\mathbf{x}$ = state vector (n-vector)

$\mathbf{u}$ = control vector (r-vector)

$\mathbf{A}$ = $n \times n$ matrix

$\mathbf{B}$ = $n \times r$ matrix

If the eigenvectors of $\mathbf{A}$ are distinct, then it is possible to find a transformation matrix $\mathbf{P}$ such that

$$\mathbf{P}^{-1}\mathbf{AP} = \mathbf{D} = \begin{bmatrix} \lambda_1 & & & & 0 \\ & \lambda_2 & & & \\ & & \cdot & & \\ & & & \cdot & \\ & & & & \cdot \\ 0 & & & & \lambda_n \end{bmatrix}$$

Note that if the eigenvalues of $\mathbf{A}$ are distinct, then the eigenvectors of $\mathbf{A}$ are distinct; however, the converse is not true. For example, an $n \times n$ real symmetric matrix having multiple eigenvalues has n distinct eigenvectors. Note also that each column of the $\mathbf{P}$ matrix is an eigenvector of $\mathbf{A}$ associated with λ_i ($i = 1, 2, \ldots, n$).

Let us define

$$\mathbf{x} = \mathbf{Pz} \tag{9–57}$$

Substituting Equation (9–57) into Equation (9–56), we obtain

$$\dot{\mathbf{z}} = \mathbf{P}^{-1}\mathbf{APz} + \mathbf{P}^{-1}\mathbf{Bu} \tag{9–58}$$

By defining

$$\mathbf{P}^{-1}\mathbf{B} = \mathbf{F} = (f_{ij})$$

we can rewrite Equation (9–58) as

$$\dot{z}_1 = \lambda_1 z_1 + f_{11}u_1 + f_{12}u_2 + \cdots + f_{1r}u_r$$
$$\dot{z}_2 = \lambda_2 z_2 + f_{21}u_1 + f_{22}u_2 + \cdots + f_{2r}u_r$$

$$\cdot$$
$$\cdot$$
$$\cdot$$

$$\dot{z}_n = \lambda_n z_n + f_{n1}u_1 + f_{n2}u_2 + \cdots + f_{nr}u_r$$

If the elements of any one row of the $n \times r$ matrix $\mathbf{F}$ are all zero, then the corresponding state variable cannot be controlled by any of the u_i. Hence, the condition of complete state controllability is that if the eigenvectors of $\mathbf{A}$ are distinct, then the system is completely state controllable if and only if no row of $\mathbf{P}^{-1}\mathbf{B}$ has all zero elements. It is important to note that, to apply this condition for complete state controllability, we must put the matrix $\mathbf{P}^{-1}\mathbf{AP}$ in Equation (9–58) in diagonal form.

If the $\mathbf{A}$ matrix in Equation (9–56) does not possess distinct eigenvectors, then diagonalization is impossible. In such a case, we may transform $\mathbf{A}$ into a Jordan canonical form. If, for example, $\mathbf{A}$ has eigenvalues $\lambda_1, \lambda_1, \lambda_1, \lambda_4, \lambda_4, \lambda_6, \ldots, \lambda_n$ and has $n - 3$ distinct eigenvectors, then the Jordan canonical form of $\mathbf{A}$ is

$$\mathbf{J} = \begin{bmatrix} \lambda_1 & 1 & 0 & & & & & 0 \\ 0 & \lambda_1 & 1 & & & & & \\ 0 & 0 & \lambda_1 & & & & & \\ & & & \lambda_4 & 1 & & & \\ & & & 0 & \lambda_4 & & & \\ & & & & & \lambda_6 & & \\ & & & & & & \cdot & \\ & & & & & & & \cdot \\ 0 & & & & & & & \lambda_n \end{bmatrix}$$

The square submatrices on the main diagonal are called *Jordan blocks*.

Suppose that we can find a transformation matrix $\mathbf{S}$ such that

$$\mathbf{S}^{-1}\mathbf{AS} = \mathbf{J}$$

If we define a new state vector $\mathbf{z}$ by

$$\mathbf{x} = \mathbf{Sz} \tag{9–59}$$

then substitution of Equation (9–59) into Equation (9–56) yields

$$\dot{\mathbf{z}} = \mathbf{S}^{-1}\mathbf{ASz} + \mathbf{S}^{-1}\mathbf{Bu}$$

$$= \mathbf{Jz} + \mathbf{S}^{-1}\mathbf{Bu} \tag{9–60}$$

The condition for complete state controllability of the system of Equation (9–56) may then be stated as follows: The system is completely state controllable if and only if (1) no two Jordan blocks in $\mathbf{J}$ of Equation (9–60) are associated with the same eigenvalues, (2) the elements of any row of $\mathbf{S}^{-1}\mathbf{B}$ that correspond to the last row of each Jordan block are not all zero, and (3) the elements of each row of $\mathbf{S}^{-1}\mathbf{B}$ that correspond to distinct eigenvalues are not all zero.

EXAMPLE 9–12 The following systems are completely state controllable:

$$\begin{bmatrix} \dot{x}_1 \\ \dot{x}_2 \end{bmatrix} = \begin{bmatrix} -1 & 0 \\ 0 & -2 \end{bmatrix}\begin{bmatrix} x_1 \\ x_2 \end{bmatrix} + \begin{bmatrix} 2 \\ 5 \end{bmatrix}u$$

$$\begin{bmatrix} \dot{x}_1 \\ \dot{x}_2 \\ \dot{x}_3 \end{bmatrix} = \begin{bmatrix} -1 & 1 & 0 \\ 0 & -1 & 0 \\ 0 & 0 & -2 \end{bmatrix}\begin{bmatrix} x_1 \\ x_2 \\ x_3 \end{bmatrix} + \begin{bmatrix} 0 \\ 4 \\ 3 \end{bmatrix}u$$

$$\begin{bmatrix} \dot{x}_1 \\ \dot{x}_2 \\ \dot{x}_3 \\ \dot{x}_4 \\ \dot{x}_5 \end{bmatrix} = \begin{bmatrix} -2 & 1 & 0 & & 0 \\ 0 & -2 & 1 & & \\ 0 & 0 & -2 & & \\ \hline & & & -5 & 1 \\ 0 & & & 0 & -5 \end{bmatrix}\begin{bmatrix} x_1 \\ x_2 \\ x_3 \\ x_4 \\ x_5 \end{bmatrix} + \begin{bmatrix} 0 & 1 \\ 0 & 0 \\ 3 & 0 \\ 0 & 0 \\ 2 & 1 \end{bmatrix}\begin{bmatrix} u_1 \\ u_2 \end{bmatrix}$$

The following systems are not completely state controllable:

$$\begin{bmatrix} \dot{x}_1 \\ \dot{x}_2 \end{bmatrix} = \begin{bmatrix} -1 & 0 \\ 0 & -2 \end{bmatrix}\begin{bmatrix} x_1 \\ x_2 \end{bmatrix} + \begin{bmatrix} 2 \\ 0 \end{bmatrix}u$$

$$\begin{bmatrix} \dot{x}_1 \\ \dot{x}_2 \\ \dot{x}_3 \end{bmatrix} = \begin{bmatrix} -1 & 1 & 0 \\ 0 & -1 & 0 \\ 0 & 0 & -2 \end{bmatrix}\begin{bmatrix} x_1 \\ x_2 \\ x_3 \end{bmatrix} + \begin{bmatrix} 4 & 2 \\ 0 & 0 \\ 3 & 0 \end{bmatrix}\begin{bmatrix} u_1 \\ u_2 \end{bmatrix}$$

$$\begin{bmatrix} \dot{x}_1 \\ \dot{x}_2 \\ \dot{x}_3 \\ \dot{x}_4 \\ \dot{x}_5 \end{bmatrix} = \begin{bmatrix} -2 & 1 & 0 & & 0 \\ 0 & -2 & 1 & & \\ 0 & 0 & -2 & & \\ \hline & & & -5 & 1 \\ 0 & & & 0 & -5 \end{bmatrix}\begin{bmatrix} x_1 \\ x_2 \\ x_3 \\ x_4 \\ x_5 \end{bmatrix} + \begin{bmatrix} 4 \\ 2 \\ 1 \\ 3 \\ 0 \end{bmatrix}u$$

Condition for Complete State Controllability in the s Plane.   The condition for complete state controllability can be stated in terms of transfer functions or transfer matrices.

It can be proved that a necessary and sufficient condition for complete state controllability is that no cancellation occur in the transfer function or transfer matrix. If cancellation occurs, the system cannot be controlled in the direction of the canceled mode.

EXAMPLE 9–13 Consider the following transfer function:

$$\frac{X(s)}{U(s)} = \frac{s + 2.5}{(s + 2.5)(s - 1)}$$

Clearly, cancellation of the factor $(s + 2.5)$ occurs in the numerator and denominator of this transfer function. (Thus one degree of freedom is lost.) Because of this cancellation, this system is not completely state controllable.

The same conclusion can be obtained by writing this transfer function in the form of a state equation. A state-space representation is

$$\begin{bmatrix} \dot{x}_1 \\ \dot{x}_2 \end{bmatrix} = \begin{bmatrix} 0 & 1 \\ 2.5 & -1.5 \end{bmatrix} \begin{bmatrix} x_1 \\ x_2 \end{bmatrix} + \begin{bmatrix} 1 \\ 1 \end{bmatrix} u$$

Since

$$\begin{bmatrix} \mathbf{B} & \vdots & \mathbf{AB} \end{bmatrix} = \begin{bmatrix} 1 & 1 \\ 1 & 1 \end{bmatrix}$$

the rank of the matrix $\begin{bmatrix} \mathbf{B} & \vdots & \mathbf{AB} \end{bmatrix}$ is 1. Therefore, we arrive at the same conclusion: The system is not completely state controllable.

Output Controllability. In the practical design of a control system, we may want to control the output rather than the state of the system. Complete state controllability is neither necessary nor sufficient for controlling the output of the system. For this reason, it is desirable to define separately complete output controllability.

Consider the system described by

$$\dot{\mathbf{x}} = \mathbf{Ax} + \mathbf{Bu} \tag{9–61}$$

$$\mathbf{y} = \mathbf{Cx} + \mathbf{Du} \tag{9–62}$$

where $\mathbf{x}$ = state vector (n-vector)

$\mathbf{u}$ = control vector (r-vector)

$\mathbf{y}$ = output vector (m-vector)

$\mathbf{A}$ = $n \times n$ matrix

$\mathbf{B}$ = $n \times r$ matrix

$\mathbf{C}$ = $m \times n$ matrix

$\mathbf{D}$ = $m \times r$ matrix

The system described by Equations (9–61) and (9–62) is said to be completely output controllable if it is possible to construct an unconstrained control vector $\mathbf{u}(t)$ that will transfer any given initial output $\mathbf{y}(t_0)$ to any final output $\mathbf{y}(t_1)$ in a finite time interval $t_0 \leq t \leq t_1$.

It can be proved that the condition for complete output controllability is as follows: The system described by Equations (9–61) and (9–62) is completely output controllable if and only if the $m \times (n + 1)r$ matrix

$$\begin{bmatrix} \mathbf{CB} & \vdots & \mathbf{CAB} & \vdots & \mathbf{CA^2B} & \vdots & \cdots & \vdots & \mathbf{CA^{n-1}B} & \vdots & \mathbf{D} \end{bmatrix}$$

is of rank m. (For a proof, see Problem **A–9–15**.) Note that the presence of the **Du** term in Equation (9–62) always helps to establish output controllability.

Uncontrollable System. An uncontrollable system has a subsystem that is physically disconnected from the input.

Stabilizability.   For a partially controllable system, if the uncontrollable modes are stable and the unstable modes are controllable, the system is said to be stabilizable. For example, the system defined by

$$\begin{bmatrix} \dot{x}_1 \\ \dot{x}_2 \end{bmatrix} = \begin{bmatrix} 1 & 0 \\ 0 & -1 \end{bmatrix} \begin{bmatrix} x_1 \\ x_2 \end{bmatrix} + \begin{bmatrix} 1 \\ 0 \end{bmatrix} u$$

is not state controllable. The stable mode that corresponds to the eigenvalue of -1 is not controllable. The unstable mode that corresponds to the eigenvalue of 1 is controllable. Such a system can be made stable by the use of a suitable feedback. Thus this system is stabilizable.

9–7 OBSERVABILITY

In this section we discuss the observability of linear systems. Consider the unforced system described by the following equations:

$$\dot{x} = Ax \tag{9–63}$$

$$y = Cx \tag{9–64}$$

where x = state vector (n-vector)
   y = output vector (m-vector)
   A = $n \times n$ matrix
   C = $m \times n$ matrix

The system is said to be completely observable if every state $x(t_0)$ can be determined from the observation of $y(t)$ over a finite time interval, $t_0 \leq t \leq t_1$. The system is, therefore, completely observable if every transition of the state eventually affects every element of the output vector. The concept of observability is useful in solving the problem of reconstructing unmeasurable state variables from measurable variables in the minimum possible length of time. In this section we treat only linear, time-invariant systems. Therefore, without loss of generality, we can assume that $t_0 = 0$.

The concept of observability is very important because, in practice, the difficulty encountered with state feedback control is that some of the state variables are not accessible for direct measurement, with the result that it becomes necessary to estimate the unmeasurable state variables in order to construct the control signals. It will be shown in Section 10–5 that such estimates of state variables are possible if and only if the system is completely observable.

In discussing observability conditions, we consider the unforced system as given by Equations (9–63) and (9–64). The reason for this is as follows: If the system is described by

$$\dot{x} = Ax + Bu$$

$$y = Cx + Du$$

then

$$x(t) = e^{At}x(0) + \int_0^t e^{A(t-\tau)}Bu(\tau)\,d\tau$$

and $\mathbf{y}(t)$ is

$$\mathbf{y}(t) = \mathbf{C}e^{\mathbf{A}t}\mathbf{x}(0) + \mathbf{C}\int_0^t e^{\mathbf{A}(t-\tau)}\mathbf{B}\mathbf{u}(\tau)\,d\tau + \mathbf{D}\mathbf{u}$$

Since the matrices $\mathbf{A}, \mathbf{B}, \mathbf{C}$, and $\mathbf{D}$ are known and $\mathbf{u}(t)$ is also known, the last two terms on the right-hand side of this last equation are known quantities. Therefore, they may be subtracted from the observed value of $\mathbf{y}(t)$. Hence, for investigating a necessary and sufficient condition for complete observability, it suffices to consider the system described by Equations (9–63) and (9–64).

Complete Observability of Continuous-Time Systems. Consider the system described by Equations (9–63) and (9–64). The output vector $\mathbf{y}(t)$ is

$$\mathbf{y}(t) = \mathbf{C}e^{\mathbf{A}t}\mathbf{x}(0)$$

Referring to Equation (9–48) or (9–50), we have

$$e^{\mathbf{A}t} = \sum_{k=0}^{n-1} \alpha_k(t)\mathbf{A}^k$$

where n is the degree of the characteristic polynomial. [Note that Equations (9–48) and (9–50) with m replaced by n can be derived using the characteristic polynomial.]
Hence, we obtain

$$\mathbf{y}(t) = \sum_{k=0}^{n-1} \alpha_k(t)\mathbf{C}\mathbf{A}^k\mathbf{x}(0)$$

or

$$\mathbf{y}(t) = \alpha_0(t)\mathbf{C}\mathbf{x}(0) + \alpha_1(t)\mathbf{C}\mathbf{A}\mathbf{x}(0) + \cdots + \alpha_{n-1}(t)\mathbf{C}\mathbf{A}^{n-1}\mathbf{x}(0) \qquad (9\text{–}65)$$

If the system is completely observable, then, given the output $\mathbf{y}(t)$ over a time interval $0 \le t \le t_1$, $\mathbf{x}(0)$ is uniquely determined from Equation (9–65). It can be shown that this requires the rank of the $nm \times n$ matrix

$$\begin{bmatrix} \mathbf{C} \\ \mathbf{CA} \\ \cdot \\ \cdot \\ \mathbf{CA}^{n-1} \end{bmatrix}$$

to be n. (See Problem **A–9–18** for the derivation of this condition.)
From this analysis, we can state the condition for complete observability as follows: The system described by Equations (9–63) and (9–64) is completely observable if and only if the $n \times nm$ matrix

$$\begin{bmatrix} \mathbf{C}^* & \vdots & \mathbf{A}^*\mathbf{C}^* & \vdots & \cdots & \vdots & (\mathbf{A}^*)^{n-1}\mathbf{C}^* \end{bmatrix}$$

is of rank n or has n linearly independent column vectors. This matrix is called the *observability matrix*.

EXAMPLE 9–14   Consider the system described by

$$\begin{bmatrix} \dot{x}_1 \\ \dot{x}_2 \end{bmatrix} = \begin{bmatrix} 1 & 1 \\ -2 & -1 \end{bmatrix} \begin{bmatrix} x_1 \\ x_2 \end{bmatrix} + \begin{bmatrix} 0 \\ 1 \end{bmatrix} u$$

$$y = \begin{bmatrix} 1 & 0 \end{bmatrix} \begin{bmatrix} x_1 \\ x_2 \end{bmatrix}$$

Is this system controllable and observable?

Since the rank of the matrix

$$\begin{bmatrix} \mathbf{B} & \vdots & \mathbf{AB} \end{bmatrix} = \begin{bmatrix} 0 & 1 \\ 1 & -1 \end{bmatrix}$$

is 2, the system is completely state controllable.

For output controllability, let us find the rank of the matrix $\begin{bmatrix} \mathbf{CB} & \vdots & \mathbf{CAB} \end{bmatrix}$. Since

$$\begin{bmatrix} \mathbf{CB} & \vdots & \mathbf{CAB} \end{bmatrix} = \begin{bmatrix} 0 & 1 \end{bmatrix}$$

the rank of this matrix is 1. Hence, the system is completely output controllable.

To test the observability condition, examine the rank of $\begin{bmatrix} \mathbf{C}^* & \vdots & \mathbf{A}^*\mathbf{C}^* \end{bmatrix}$. Since

$$\begin{bmatrix} \mathbf{C}^* & \vdots & \mathbf{A}^*\mathbf{C}^* \end{bmatrix} = \begin{bmatrix} 1 & 1 \\ 0 & 1 \end{bmatrix}$$

the rank of $\begin{bmatrix} \mathbf{C}^* & \vdots & \mathbf{A}^*\mathbf{C}^* \end{bmatrix}$ is 2. Hence, the system is completely observable.

Conditions for Complete Observability in the s Plane.   The conditions for complete observability can also be stated in terms of transfer functions or transfer matrices. The necessary and sufficient conditions for complete observability is that no cancellation occur in the transfer function or transfer matrix. If cancellation occurs, the canceled mode cannot be observed in the output.

EXAMPLE 9–15   Show that the following system is not completely observable:

$$\dot{\mathbf{x}} = \mathbf{Ax} + \mathbf{Bu}$$
$$y = \mathbf{Cx}$$

where

$$\mathbf{x} = \begin{bmatrix} x_1 \\ x_2 \\ x_3 \end{bmatrix}, \qquad \mathbf{A} = \begin{bmatrix} 0 & 1 & 0 \\ 0 & 0 & 1 \\ -6 & -11 & -6 \end{bmatrix}, \qquad \mathbf{B} = \begin{bmatrix} 0 \\ 0 \\ 1 \end{bmatrix}, \qquad \mathbf{C} = \begin{bmatrix} 4 & 5 & 1 \end{bmatrix}$$

Note that the control function u does not affect the complete observability of the system. To examine complete observability, we may simply set $u = 0$. For this system, we have

$$\begin{bmatrix} \mathbf{C}^* & \vdots & \mathbf{A}^*\mathbf{C}^* & \vdots & (\mathbf{A}^*)^2\mathbf{C}^* \end{bmatrix} = \begin{bmatrix} 4 & -6 & 6 \\ 5 & -7 & 5 \\ 1 & -1 & -1 \end{bmatrix}$$

Note that

$$\begin{vmatrix} 4 & -6 & 6 \\ 5 & -7 & 5 \\ 1 & -1 & -1 \end{vmatrix} = 0$$

Hence, the rank of the matrix $\begin{bmatrix} \mathbf{C}^* & \vdots & \mathbf{A}^*\mathbf{C}^* & \vdots & (\mathbf{A}^*)^2\mathbf{C}^* \end{bmatrix}$ is less than 3. Therefore, the system is not completely observable.

In fact, in this system, cancellation occurs in the transfer function of the system. The transfer function between $X_1(s)$ and $U(s)$ is

$$\frac{X_1(s)}{U(s)} = \frac{1}{(s + 1)(s + 2)(s + 3)}$$

and the transfer function between $Y(s)$ and $X_1(s)$ is

$$\frac{Y(s)}{X_1(s)} = (s + 1)(s + 4)$$

Therefore, the transfer function between the output $Y(s)$ and the input $U(s)$ is

$$\frac{Y(s)}{U(s)} = \frac{(s + 1)(s + 4)}{(s + 1)(s + 2)(s + 3)}$$

Clearly, the two factors $(s + 1)$ cancel each other. This means that there are nonzero initial states $\mathbf{x}(0)$, which cannot be determined from the measurement of $y(t)$.

Comments. The transfer function has no cancellation if and only if the system is completely state controllable and completely observable. This means that the canceled transfer function does not carry along all the information characterizing the dynamic system.

Alternative Form of the Condition for Complete Observability. Consider the system described by Equations (9–63) and (9–64), rewritten

$$\dot{\mathbf{x}} = \mathbf{A}\mathbf{x} \qquad\qquad (9\text{–}66)$$

$$\mathbf{y} = \mathbf{C}\mathbf{x} \qquad\qquad (9\text{–}67)$$

Suppose that the transformation matrix $\mathbf{P}$ transforms $\mathbf{A}$ into a diagonal matrix, or

$$\mathbf{P}^{-1}\mathbf{A}\mathbf{P} = \mathbf{D}$$

where $\mathbf{D}$ is a diagonal matrix. Let us define

$$\mathbf{x} = \mathbf{P}\mathbf{z}$$

Then Equations (9–66) and (9–67) can be written

$$\dot{\mathbf{z}} = \mathbf{P}^{-1}\mathbf{A}\mathbf{P}\mathbf{z} = \mathbf{D}\mathbf{z}$$

$$\mathbf{y} = \mathbf{C}\mathbf{P}\mathbf{z}$$

Hence,

$$\mathbf{y}(t) = \mathbf{C}\mathbf{P}e^{\mathbf{D}t}\mathbf{z}(0)$$

or

$$\mathbf{y}(t) = \mathbf{CP} \begin{bmatrix} e^{\lambda_1 t} & & & & 0 \\ & e^{\lambda_2 t} & & & \\ & & \cdot & & \\ & & & \cdot & \\ & & & & \cdot \\ 0 & & & & e^{\lambda_n t} \end{bmatrix} \mathbf{z}(0) = \mathbf{CP} \begin{bmatrix} e^{\lambda_1 t} z_1(0) \\ e^{\lambda_2 t} z_2(0) \\ \cdot \\ \cdot \\ \cdot \\ e^{\lambda_n t} z_n(0) \end{bmatrix}$$

The system is completely observable if none of the columns of the $m \times n$ matrix $\mathbf{CP}$ consists of all zero elements. This is because, if the ith column of $\mathbf{CP}$ consists of all zero elements, then the state variable $z_i(0)$ will not appear in the output equation and therefore cannot be determined from observation of $\mathbf{y}(t)$. Thus, $\mathbf{x}(0)$, which is related to $\mathbf{z}(0)$ by the nonsingular matrix $\mathbf{P}$, cannot be determined. (Remember that this test applies only if the matrix $\mathbf{P}^{-1}\mathbf{AP}$ is in diagonal form.)

If the matrix $\mathbf{A}$ cannot be transformed into a diagonal matrix, then by use of a suitable transformation matrix $\mathbf{S}$, we can transform $\mathbf{A}$ into a Jordan canonical form, or

$$\mathbf{S}^{-1}\mathbf{AS} = \mathbf{J}$$

where $\mathbf{J}$ is in the Jordan canonical form.

Let us define

$$\mathbf{x} = \mathbf{Sz}$$

Then Equations (9–66) and (9–67) can be written

$$\dot{\mathbf{z}} = \mathbf{S}^{-1}\mathbf{ASz} = \mathbf{Jz}$$
$$\mathbf{y} = \mathbf{CSz}$$

Hence,

$$\mathbf{y}(t) = \mathbf{CS}e^{\mathbf{J}t}\mathbf{z}(0)$$

The system is completely observable if (1) no two Jordan blocks in $\mathbf{J}$ are associated with the same eigenvalues, (2) no columns of $\mathbf{CS}$ that correspond to the first row of each Jordan block consist of zero elements, and (3) no columns of $\mathbf{CS}$ that correspond to distinct eigenvalues consist of zero elements.

To clarify condition (2), in Example 9–16 we have encircled by dashed lines the columns of $\mathbf{CS}$ that correspond to the first row of each Jordan block.

EXAMPLE 9–16

The following systems are completely observable.

$$\begin{bmatrix} \dot{x}_1 \\ \dot{x}_2 \end{bmatrix} = \begin{bmatrix} -1 & 0 \\ 0 & -2 \end{bmatrix} \begin{bmatrix} x_1 \\ x_2 \end{bmatrix}, \qquad y = \begin{bmatrix} 1 & 3 \end{bmatrix} \begin{bmatrix} x_1 \\ x_2 \end{bmatrix}$$

$$\begin{bmatrix} \dot{x}_1 \\ \dot{x}_2 \\ \dot{x}_3 \end{bmatrix} = \begin{bmatrix} 2 & 1 & 0 \\ 0 & 2 & 1 \\ 0 & 0 & 2 \end{bmatrix} \begin{bmatrix} x_1 \\ x_2 \\ x_3 \end{bmatrix}, \qquad \begin{bmatrix} y_1 \\ y_2 \end{bmatrix} = \begin{bmatrix} 3 & 0 & 0 \\ 4 & 0 & 0 \end{bmatrix} \begin{bmatrix} x_1 \\ x_2 \\ x_3 \end{bmatrix}$$

$$\begin{bmatrix} \dot{x}_1 \\ \dot{x}_2 \\ \dot{x}_3 \\ \dot{x}_4 \\ \dot{x}_5 \end{bmatrix} = \begin{bmatrix} 2 & 1 & 0 & & 0 \\ 0 & 2 & 1 & & \\ 0 & 0 & 2 & & \\ \hline & & & -3 & 1 \\ 0 & & & 0 & -3 \end{bmatrix} \begin{bmatrix} x_1 \\ x_2 \\ x_3 \\ x_4 \\ x_5 \end{bmatrix}, \qquad \begin{bmatrix} y_1 \\ y_2 \end{bmatrix} = \begin{bmatrix} 1 & 1 & 1 & 0 & 0 \\ 0 & 1 & 1 & 1 & 0 \end{bmatrix} \begin{bmatrix} x_1 \\ x_2 \\ x_3 \\ x_4 \\ x_5 \end{bmatrix}$$

The following systems are not completely observable.

$$\begin{bmatrix} \dot{x}_1 \\ \dot{x}_2 \end{bmatrix} = \begin{bmatrix} -1 & 0 \\ 0 & -2 \end{bmatrix}\begin{bmatrix} x_1 \\ x_2 \end{bmatrix}, \qquad y = \begin{bmatrix} 0 & 1 \end{bmatrix}\begin{bmatrix} x_1 \\ x_2 \end{bmatrix}$$

$$\begin{bmatrix} \dot{x}_1 \\ \dot{x}_2 \\ \dot{x}_3 \end{bmatrix} = \begin{bmatrix} 2 & 1 & 0 \\ 0 & 2 & 1 \\ 0 & 0 & 2 \end{bmatrix}\begin{bmatrix} x_1 \\ x_2 \\ x_3 \end{bmatrix}, \qquad \begin{bmatrix} y_1 \\ y_2 \end{bmatrix} = \begin{bmatrix} 0 & 1 & 3 \\ 0 & 2 & 4 \end{bmatrix}\begin{bmatrix} x_1 \\ x_2 \\ x_3 \end{bmatrix}$$

$$\begin{bmatrix} \dot{x}_1 \\ \dot{x}_2 \\ \dot{x}_3 \\ \dot{x}_4 \\ \dot{x}_5 \end{bmatrix} = \begin{bmatrix} 2 & 1 & 0 & & 0 \\ 0 & 2 & 1 & & \\ 0 & 0 & 2 & & \\ \hline & & & -3 & 1 \\ 0 & & & 0 & -3 \end{bmatrix}\begin{bmatrix} x_1 \\ x_2 \\ x_3 \\ x_4 \\ x_5 \end{bmatrix}, \qquad \begin{bmatrix} y_1 \\ y_2 \end{bmatrix} = \begin{bmatrix} 1 & 1 & 1 & 0 & 0 \\ 0 & 1 & 1 & 0 & 0 \end{bmatrix}\begin{bmatrix} x_1 \\ x_2 \\ x_3 \\ x_4 \\ x_5 \end{bmatrix}$$

Principle of Duality.   We shall now discuss the relationship between controllability and observability. We shall introduce the principle of duality, due to Kalman, to clarify apparent analogies between controllability and observability.

Consider the system S_1 described by

$$\dot{x} = Ax + Bu$$

$$y = Cx$$

where x = state vector (n-vector)

u = control vector (r-vector)

y = output vector (m-vector)

$A = n \times n$ matrix

$B = n \times r$ matrix

$C = m \times n$ matrix

and the dual system S_2 defined by

$$\dot{z} = A^*z + C^*v$$

$$n = B^*z$$

where z = state vector (n-vector)

v = control vector (m-vector)

n = output vector (r-vector)

A^* = conjugate transpose of A

B^* = conjugate transpose of B

C^* = conjugate transpose of C

The principle of duality states that the system S_1 is completely state controllable (observable) if and only if system S_2 is completely observable (state controllable).

To verify this principle, let us write down the necessary and sufficient conditions for complete state controllability and complete observability for systems S_1 and S_2.

For system S_1:

1. A necessary and sufficient condition for complete state controllability is that the rank of the $n \times nr$ matrix

$$\begin{bmatrix} \mathbf{B} & \vdots & \mathbf{AB} & \vdots & \cdots & \vdots & \mathbf{A}^{n-1}\mathbf{B} \end{bmatrix}$$

   be n.

2. A necessary and sufficient condition for complete observability is that the rank of the $n \times nm$ matrix

$$\begin{bmatrix} \mathbf{C}^* & \vdots & \mathbf{A}^*\mathbf{C}^* & \vdots & \cdots & \vdots & (\mathbf{A}^*)^{n-1}\mathbf{C}^* \end{bmatrix}$$

   be n.

For system S_2:

1. A necessary and sufficient condition for complete state controllability is that the rank of the $n \times nm$ matrix

$$\begin{bmatrix} \mathbf{C}^* & \vdots & \mathbf{A}^*\mathbf{C}^* & \vdots & \cdots & \vdots & (\mathbf{A}^*)^{n-1}\mathbf{C}^* \end{bmatrix}$$

   be n.

2. A necessary and sufficient condition for complete observability is that the rank of the $n \times nr$ matrix

$$\begin{bmatrix} \mathbf{B} & \vdots & \mathbf{AB} & \vdots & \cdots & \vdots & \mathbf{A}^{n-1}\mathbf{B} \end{bmatrix}$$

   be n.

By comparing these conditions, the truth of this principle is apparent. By use of this principle, the observability of a given system can be checked by testing the state controllability of its dual.

Detectability. For a partially observable system, if the unobservable modes are stable and the observable modes are unstable, the system is said to be detectable. Note that the concept of detectability is dual to the concept of stabilizability.

EXAMPLE PROBLEMS AND SOLUTIONS

A–9–1. Consider the transfer function system defined by Equation (9–2), rewritten

$$\frac{Y(s)}{U(s)} = \frac{b_0 s^n + b_1 s^{n-1} + \cdots + b_{n-1} s + b_n}{s^n + a_1 s^{n-1} + \cdots + a_{n-1} s + a_n} \tag{9–68}$$

Derive the following controllable canonical form of the state-space representation for this transfer-function system:

$$\begin{bmatrix} \dot{x}_1 \\ \dot{x}_2 \\ \vdots \\ \vdots \\ \dot{x}_{n-1} \\ \dot{x}_n \end{bmatrix} = \begin{bmatrix} 0 & 1 & 0 & \cdots & 0 \\ 0 & 0 & 1 & \cdots & 0 \\ \vdots & \vdots & \vdots & & \vdots \\ \vdots & \vdots & \vdots & & \vdots \\ 0 & 0 & 0 & \cdots & 1 \\ -a_n & -a_{n-1} & -a_{n-2} & \cdots & -a_1 \end{bmatrix} \begin{bmatrix} x_1 \\ x_2 \\ \vdots \\ \vdots \\ x_{n-1} \\ x_n \end{bmatrix} + \begin{bmatrix} 0 \\ 0 \\ \vdots \\ \vdots \\ 0 \\ 1 \end{bmatrix} u \tag{9–69}$$

$$y = \begin{bmatrix} b_n - a_n b_0 & \vdots & b_{n-1} - a_{n-1} b_0 & \vdots & \cdots & \vdots & b_1 - a_1 b_0 \end{bmatrix} \begin{bmatrix} x_1 \\ x_2 \\ \cdot \\ \cdot \\ \cdot \\ x_n \end{bmatrix} + b_0 u \tag{9–70}$$

Solution. Equation (9–68) can be written as

$$\frac{Y(s)}{U(s)} = b_0 + \frac{(b_1 - a_1 b_0)s^{n-1} + \cdots + (b_{n-1} - a_{n-1} b_0)s + (b_n - a_n b_0)}{s^n + a_1 s^{n-1} + \cdots + a_{n-1} s + a_n}$$

which can be modified to

$$Y(s) = b_0 U(s) + \hat{Y}(s) \tag{9–71}$$

where

$$\hat{Y}(s) = \frac{(b_1 - a_1 b_0)s^{n-1} + \cdots + (b_{n-1} - a_{n-1} b_0)s + (b_n - a_n b_0)}{s^n + a_1 s^{n-1} + \cdots + a_{n-1} s + a_n} U(s)$$

Let us rewrite this last equation in the following form:

$$\frac{\hat{Y}(s)}{(b_1 - a_1 b_0)s^{n-1} + \cdots + (b_{n-1} - a_{n-1} b_0)s + (b_n - a_n b_0)}$$

$$= \frac{U(s)}{s^n + a_1 s^{n-1} + \cdots + a_{n-1} s + a_n} = Q(s)$$

From this last equation, the following two equations may be obtained:

$$s^n Q(s) = -a_1 s^{n-1} Q(s) - \cdots - a_{n-1} s Q(s) - a_n Q(s) + U(s) \tag{9–72}$$

$$\hat{Y}(s) = (b_1 - a_1 b_0)s^{n-1} Q(s) + \cdots + (b_{n-1} - a_{n-1} b_0)s Q(s)$$
$$+ (b_n - a_n b_0)Q(s) \tag{9–73}$$

Now define state variables as follows:

$$X_1(s) = Q(s)$$
$$X_2(s) = sQ(s)$$
$$\cdot$$
$$\cdot$$
$$\cdot$$
$$X_{n-1}(s) = s^{n-2}Q(s)$$
$$X_n(s) = s^{n-1}Q(s)$$

Then, clearly,

$$sX_1(s) = X_2(s)$$
$$sX_2(s) = X_3(s)$$
$$\cdot$$
$$\cdot$$
$$\cdot$$
$$sX_{n-1}(s) = X_n(s)$$

which may be rewritten as

$$\dot{x}_1 = x_2$$
$$\dot{x}_2 = x_3$$
$$\cdot$$
$$\cdot$$ \hspace{3cm} (9–74)
$$\cdot$$
$$\dot{x}_{n-1} = x_n$$

Noting that $s^n Q(s) = sX_n(s)$, we can rewrite Equation (9–72) as

$$sX_n(s) = -a_1 X_n(s) - \cdots - a_{n-1} X_2(s) - a_n X_1(s) + U(s)$$

or

$$\dot{x}_n = -a_n x_1 - a_{n-1} x_2 - \cdots - a_1 x_n + u \hspace{2cm} (9–75)$$

Also, from Equations (9–71) and (9–73), we obtain

$$Y(s) = b_0 U(s) + (b_1 - a_1 b_0)s^{n-1}Q(s) + \cdots + (b_{n-1} - a_{n-1}b_0)sQ(s)$$
$$+ (b_n - a_n b_0)Q(s)$$
$$= b_0 U(s) + (b_1 - a_1 b_0)X_n(s) + \cdots + (b_{n-1} - a_{n-1}b_0)X_2(s)$$
$$+ (b_n - a_n b_0)X_1(s)$$

The inverse Laplace transform of this output equation becomes

$$y = (b_n - a_n b_0)x_1 + (b_{n-1} - a_{n-1}b_0)x_2 + \cdots + (b_1 - a_1 b_0)x_n + b_0 u \hspace{1cm} (9–76)$$

Combining Equations (9–74) and (9–75) into one vector–matrix differential equation, we obtain Equation (9–69). Equation (9–76) can be rewritten as given by Equation (9–70). Equations (9–69) and (9–70) are said to be in the controllable canonical form. Figure 9–1 shows the block diagram representation of the system defined by Equations (9–69) and (9–70).

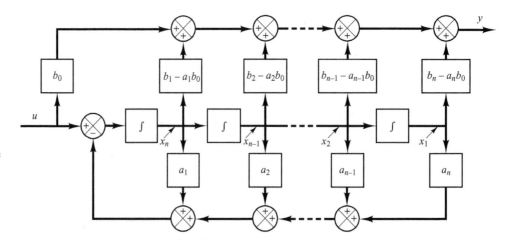

Figure 9–1
Block diagram
representation of the
system defined by
Equations (9–69)
and (9–70)
(controllable
canonical form).

Chapter 9 / **Control Systems Analysis in State Space**

A-9-2. Consider the following transfer-function system:

$$\frac{Y(s)}{U(s)} = \frac{b_0 s^n + b_1 s^{n-1} + \cdots + b_{n-1}s + b_n}{s^n + a_1 s^{n-1} + \cdots + a_{n-1}s + a_n} \tag{9-77}$$

Derive the following observable canonical form of the state-space representation for this transfer-function system:

$$\begin{bmatrix} \dot{x}_1 \\ \dot{x}_2 \\ \cdot \\ \cdot \\ \cdot \\ \dot{x}_n \end{bmatrix} = \begin{bmatrix} 0 & 0 & \cdots & 0 & -a_n \\ 1 & 0 & \cdots & 0 & -a_{n-1} \\ \cdot & \cdot & & \cdot & \cdot \\ \cdot & \cdot & & \cdot & \cdot \\ \cdot & \cdot & & \cdot & \cdot \\ 0 & 0 & \cdots & 1 & -a_1 \end{bmatrix} \begin{bmatrix} x_1 \\ x_2 \\ \cdot \\ \cdot \\ \cdot \\ x_n \end{bmatrix} + \begin{bmatrix} b_n - a_n b_0 \\ b_{n-1} - a_{n-1}b_0 \\ \cdot \\ \cdot \\ \cdot \\ b_1 - a_1 b_0 \end{bmatrix} u \tag{9-78}$$

$$y = \begin{bmatrix} 0 & 0 & \cdots & 0 & 1 \end{bmatrix} \begin{bmatrix} x_1 \\ x_2 \\ \cdot \\ \cdot \\ \cdot \\ x_{n-1} \\ x_n \end{bmatrix} + b_0 u \tag{9-79}$$

Solution. Equation (9–77) can be modified into the following form:

$$s^n \left[Y(s) - b_0 U(s) \right] + s^{n-1} \left[a_1 Y(s) - b_1 U(s) \right] + \cdots$$
$$+ s \left[a_{n-1} Y(s) - b_{n-1}U(s) \right] + a_n Y(s) - b_n U(s) = 0$$

By dividing the entire equation by s^n and rearranging, we obtain

$$Y(s) = b_0 U(s) + \frac{1}{s} \left[b_1 U(s) - a_1 Y(s) \right] + \cdots$$
$$+ \frac{1}{s^{n-1}} \left[b_{n-1}U(s) - a_{n-1}Y(s) \right] + \frac{1}{s^n} \left[b_n U(s) - a_n Y(s) \right] \tag{9-80}$$

Now define state variables as follows:

$$X_n(s) = \frac{1}{s} \left[b_1 U(s) - a_1 Y(s) + X_{n-1}(s) \right]$$

$$X_{n-1}(s) = \frac{1}{s} \left[b_2 U(s) - a_2 Y(s) + X_{n-2}(s) \right]$$

$$\cdot$$
$$\cdot \tag{9-81}$$
$$\cdot$$

$$X_2(s) = \frac{1}{s} \left[b_{n-1}U(s) - a_{n-1}Y(s) + X_1(s) \right]$$

$$X_1(s) = \frac{1}{s} \left[b_n U(s) - a_n Y(s) \right]$$

Example Problems and Solutions

691

Then Equation (9–80) can be written as

$$Y(s) = b_0 U(s) + X_n(s) \qquad\qquad (9\text{–}82)$$

By substituting Equation (9–82) into Equation (9–81) and multiplying both sides of the equations by s, we obtain

$$sX_n(s) = X_{n-1}(s) - a_1 X_n(s) + (b_1 - a_1 b_0)U(s)$$

$$sX_{n-1}(s) = X_{n-2}(s) - a_2 X_n(s) + (b_2 - a_2 b_0)U(s)$$

$$\cdot$$
$$\cdot$$
$$\cdot$$

$$sX_2(s) = X_1(s) - a_{n-1} X_n(s) + (b_{n-1} - a_{n-1} b_0)U(s)$$

$$sX_1(s) = -a_n X_n(s) + (b_n - a_n b_0)U(s)$$

Taking the inverse Laplace transforms of the preceding n equations and writing them in the reverse order, we get

$$\dot{x}_1 = -a_n x_n + (b_n - a_n b_0)u$$

$$\dot{x}_2 = x_1 - a_{n-1} x_n + (b_{n-1} - a_{n-1} b_0)u$$

$$\cdot$$
$$\cdot$$
$$\cdot$$

$$\dot{x}_{n-1} = x_{n-2} - a_2 x_n + (b_2 - a_2 b_0)u$$

$$\dot{x}_n = x_{n-1} - a_1 x_n + (b_1 - a_1 b_0)u$$

Also, the inverse Laplace transform of Equation (9–82) gives

$$y = x_n + b_0 u$$

Rewriting the state and output equations in the standard vector-matrix forms gives Equations (9–78) and (9–79). Figure 9–2 shows a block diagram representation of the system defined by Equations (9–78) and (9–79).

Figure 9–2
Block diagram representation of the system defined by Equations (9–78) and (9–79) (observable canonical form).

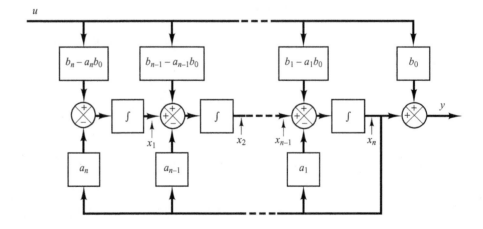

Chapter 9 / Control Systems Analysis in State Space

A-9-3. Consider the transfer-function system defined by

$$\frac{Y(s)}{U(s)} = \frac{b_0 s^n + b_1 s^{n-1} + \cdots + b_{n-1}s + b_n}{(s + p_1)(s + p_2)\cdots(s + p_n)}$$

$$= b_0 + \frac{c_1}{s + p_1} + \frac{c_2}{s + p_2} + \cdots + \frac{c_n}{s + p_n} \qquad (9\text{–}83)$$

where $p_i \neq p_j$. Derive the state-space representation of this system in the following diagonal canonical form:

$$\begin{bmatrix} \dot{x}_1 \\ \dot{x}_2 \\ \cdot \\ \cdot \\ \cdot \\ \dot{x}_n \end{bmatrix} = \begin{bmatrix} -p_1 & & & & 0 \\ & -p_2 & & & \\ & & \cdot & & \\ & & & \cdot & \\ 0 & & & & -p_n \end{bmatrix} \begin{bmatrix} x_1 \\ x_2 \\ \cdot \\ \cdot \\ \cdot \\ x_n \end{bmatrix} + \begin{bmatrix} 1 \\ 1 \\ \cdot \\ \cdot \\ \cdot \\ 1 \end{bmatrix} u \qquad (9\text{–}84)$$

$$y = \begin{bmatrix} c_1 & c_2 & \cdots & c_n \end{bmatrix} \begin{bmatrix} x_1 \\ x_2 \\ \cdot \\ \cdot \\ \cdot \\ x_n \end{bmatrix} + b_0 u \qquad (9\text{–}85)$$

Solution. Equation (9–83) may be written as

$$Y(s) = b_0 U(s) + \frac{c_1}{s + p_1} U(s) + \frac{c_2}{s + p_2} U(s) + \cdots + \frac{c_n}{s + p_n} U(s) \qquad (9\text{–}86)$$

Define the state variables as follows:

$$X_1(s) = \frac{1}{s + p_1} U(s)$$

$$X_2(s) = \frac{1}{s + p_2} U(s)$$

$$\cdot$$

$$\cdot$$

$$\cdot$$

$$X_n(s) = \frac{1}{s + p_n} U(s)$$

which may be rewritten as

$$sX_1(s) = -p_1 X_1(s) + U(s)$$
$$sX_2(s) = -p_2 X_2(s) + U(s)$$
$$\cdot$$
$$\cdot$$
$$\cdot$$
$$sX_n(s) = -p_n X_n(s) + U(s)$$

The inverse Laplace transforms of these equations give

$$\dot{x}_1 = -p_1 x_1 + u$$
$$\dot{x}_2 = -p_2 x_2 + u$$
$$\cdot$$
$$\cdot \qquad (9\text{--}87)$$
$$\cdot$$
$$\dot{x}_n = -p_n x_n + u$$

These n equations make up a state equation.

In terms of the state variables $X_1(s), X_2(s), \ldots, X_n(s)$, Equation (9–86) can be written as

$$Y(s) = b_0 U(s) + c_1 X_1(s) + c_2 X_2(s) + \cdots + c_n X_n(s)$$

The inverse Laplace transform of this last equation is

$$y = c_1 x_1 + c_2 x_2 + \cdots + c_n x_n + b_0 u \qquad (9\text{--}88)$$

which is the output equation.

Equation (9–87) can be put in the vector-matrix equation as given by Equation (9–84). Equation (9–88) can be put in the form of Equation (9–85).

Figure 9–3 shows a block diagram representation of the system defined by Equations (9–84) and (9–85).

It is noted that if we choose the state variables as

$$\hat{X}_1(s) = \frac{c_1}{s + p_1} U(s)$$

$$\hat{X}_2(s) = \frac{c_2}{s + p_2} U(s)$$

$$\cdot$$
$$\cdot$$
$$\cdot$$

$$\hat{X}_n(s) = \frac{c_n}{s + p_n} U(s)$$

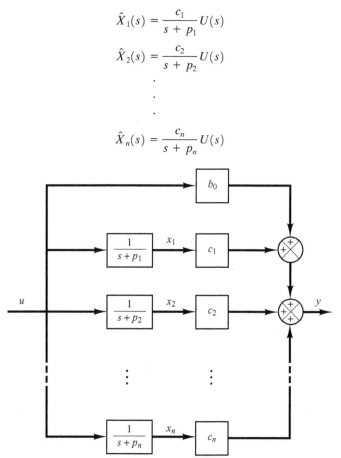

Figure 9–3
Block diagram
representation of the
system defined by
Equations (9–84)
and (9–85) (diagonal
canonical form).

then we get a slightly different state-space representation. This choice of state variables gives

$$s\hat{X}_1(s) = -p_1\hat{X}_1(s) + c_1U(s)$$
$$s\hat{X}_2(s) = -p_2\hat{X}_2(s) + c_2U(s)$$
$$\cdot$$
$$\cdot$$
$$\cdot$$
$$s\hat{X}_n(s) = -p_n\hat{X}_n(s) + c_nU(s)$$

from which we obtain

$$\dot{\hat{x}}_1 = -p_1\hat{x}_1 + c_1u$$
$$\dot{\hat{x}}_2 = -p_2\hat{x}_2 + c_2u$$
$$\cdot$$
$$\cdot \qquad\qquad\qquad (9\text{--}89)$$
$$\cdot$$
$$\dot{\hat{x}}_n = -p_n\hat{x}_n + c_nu$$

Referring to Equation (9–86), the output equation becomes

$$Y(s) = b_0U(s) + \hat{X}_1(s) + \hat{X}_2(s) + \cdots + \hat{X}_n(s)$$

from which we get

$$y = \hat{x}_1 + \hat{x}_2 + \cdots + \hat{x}_n + b_0u \qquad\qquad (9\text{--}90)$$

Equations (9–89) and (9–90) give the following state-space representation for the system:

$$\begin{bmatrix} \dot{\hat{x}}_1 \\ \dot{\hat{x}}_2 \\ \cdot \\ \cdot \\ \cdot \\ \dot{\hat{x}}_n \end{bmatrix} = \begin{bmatrix} -p_1 & & & 0 \\ & -p_2 & & \\ & & \cdot & \\ & & & \cdot \\ 0 & & & -p_n \end{bmatrix} \begin{bmatrix} \hat{x}_1 \\ \hat{x}_2 \\ \cdot \\ \cdot \\ \cdot \\ \hat{x}_n \end{bmatrix} + \begin{bmatrix} c_1 \\ c_2 \\ \cdot \\ \cdot \\ c_n \end{bmatrix} u$$

$$y = \begin{bmatrix} 1 & 1 & \cdots & 1 \end{bmatrix} \begin{bmatrix} \hat{x}_1 \\ \hat{x}_2 \\ \cdot \\ \cdot \\ \cdot \\ \hat{x}_n \end{bmatrix} + b_0u$$

A–9–4. Consider the system defined by

$$\frac{Y(s)}{U(s)} = \frac{b_0s^n + b_1s^{n-1} + \cdots + b_{n-1}s + b_n}{(s + p_1)^3(s + p_4)(s + p_5)\cdots(s + p_n)} \qquad (9\text{--}91)$$

where the system involves a triple pole at $s = -p_1$. (We assume that, except for the first three p_i's being equal, the p_i's are different from one another.) Obtain the Jordan canonical form of the state-space representation for this system.

Solution. The partial-fraction expansion of Equation (9–91) becomes

$$\frac{Y(s)}{U(s)} = b_0 + \frac{c_1}{(s + p_1)^3} + \frac{c_2}{(s + p_1)^2} + \frac{c_3}{s + p_1} + \frac{c_4}{s + p_4} + \cdots + \frac{c_n}{s + p_n}$$

which may be written as

$$Y(s) = b_0 U(s) + \frac{c_1}{(s + p_1)^3} U(s) + \frac{c_2}{(s + p_1)^2} U(s)$$

$$+ \frac{c_3}{s + p_1} U(s) + \frac{c_4}{s + p_4} U(s) + \cdots + \frac{c_n}{s + p_n} U(s) \qquad (9\text{–}92)$$

Define

$$X_1(s) = \frac{1}{(s + p_1)^3} U(s)$$

$$X_2(s) = \frac{1}{(s + p_1)^2} U(s)$$

$$X_3(s) = \frac{1}{s + p_1} U(s)$$

$$X_4(s) = \frac{1}{s + p_4} U(s)$$

$$\cdot$$
$$\cdot$$
$$\cdot$$

$$X_n(s) = \frac{1}{s + p_n} U(s)$$

Notice that the following relationships exist among $X_1(s), X_2(s)$, and $X_3(s)$:

$$\frac{X_1(s)}{X_2(s)} = \frac{1}{s + p_1}$$

$$\frac{X_2(s)}{X_3(s)} = \frac{1}{s + p_1}$$

Then, from the preceding definition of the state variables and the preceding relationships, we obtain

$$sX_1(s) = -p_1 X_1(s) + X_2(s)$$

$$sX_2(s) = -p_1 X_2(s) + X_3(s)$$

$$sX_3(s) = -p_1 X_3(s) + U(s)$$

$$sX_4(s) = -p_4 X_4(s) + U(s)$$

$$\cdot$$
$$\cdot$$
$$\cdot$$

$$sX_n(s) = -p_n X_n(s) + U(s)$$

The inverse Laplace transforms of the preceding n equations give

$$\dot{x}_1 = -p_1 x_1 + x_2$$
$$\dot{x}_2 = -p_1 x_2 + x_3$$
$$\dot{x}_3 = -p_1 x_3 + u$$
$$\dot{x}_4 = -p_4 x_4 + u$$

$$\cdot$$
$$\cdot$$
$$\cdot$$

$$\dot{x}_n = -p_n x_n + u$$

The output equation, Equation (9–92), can be rewritten as

$$Y(s) = b_0 U(s) + c_1 X_1(s) + c_2 X_2(s) + c_3 X_3(s) + c_4 X_4(s) + \cdots + c_n X_n(s)$$

The inverse Laplace transform of this output equation is

$$y = c_1 x_1 + c_2 x_2 + c_3 x_3 + c_4 x_4 + \cdots + c_n x_n + b_0 u$$

Thus, the state-space representation of the system for the case when the denominator polynomial involves a triple root $-p_1$ can be given as follows:

$$
\begin{bmatrix} \dot{x}_1 \\ \dot{x}_2 \\ \dot{x}_3 \\ \dot{x}_4 \\ \cdot \\ \cdot \\ \cdot \\ \dot{x}_n \end{bmatrix}
=
\left[
\begin{array}{ccc:c ccc}
-p_1 & 1 & 0 & 0 & \cdots & & 0 \\
0 & -p_1 & 1 & \cdot & & & \cdot \\
0 & 0 & -p_1 & 0 & \cdots & & 0 \\ \hdashline
0 & \cdots & 0 & -p_4 & & & \\
\cdot & & & & \cdot & & \\
\cdot & & & & & \cdot & \\
\cdot & & & & & & \cdot \\
0 & \cdots & 0 & 0 & & & -p_n
\end{array}
\right]
\begin{bmatrix} x_1 \\ x_2 \\ x_3 \\ x_4 \\ \cdot \\ \cdot \\ \cdot \\ x_n \end{bmatrix}
+
\begin{bmatrix} 0 \\ 0 \\ 1 \\ 1 \\ \cdot \\ \cdot \\ \cdot \\ 1 \end{bmatrix} u
\tag{9–93}
$$

$$
y = \begin{bmatrix} c_1 & c_2 & \cdots & c_n \end{bmatrix}
\begin{bmatrix} x_1 \\ x_2 \\ \cdot \\ \cdot \\ \cdot \\ x_n \end{bmatrix} + b_0 u
\tag{9–94}
$$

The state-space representation in the form given by Equations (9–93) and (9–94) is said to be in the Jordan canonical form. Figure 9–4 shows a block diagram representation of the system given by Equations (9–93) and (9–94).

A–9–5. Consider the transfer-function system

$$\frac{Y(s)}{U(s)} = \frac{25.04s + 5.008}{s^3 + 5.03247s^2 + 25.1026s + 5.008}$$

Obtain a state-space representation of this system with MATLAB.

Example Problems and Solutions

697

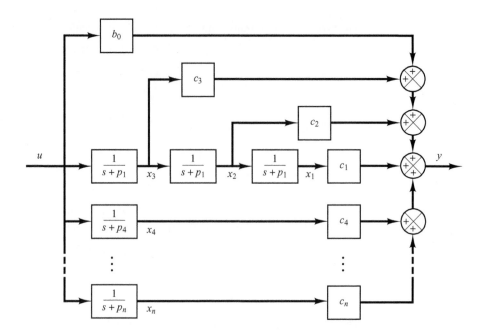

Figure 9–4
Block diagram
representation of the
system defined by
Equations (9–93)
and (9–94) (Jordan
canonical form).

Solution. MATLAB command

$$[A,B,C,D] = \text{tf2ss(num,den)}$$

will produce a state-space representation for the system. See MATLAB Program 9–4.

```
MATLAB Program 9–4

num = [25.04  5.008];
den = [1  5.03247  25.1026  5.008];
[A,B,C,D] = tf2ss(num,den)

A =

   -5.0325   -25.1026    -5.0080
    1.0000          0          0
         0     1.0000          0

B =

     1
     0
     0

C =

         0    25.0400     5.0080
D  =

     0
```

This is the MATLAB representation of the following state-space equations:

$$\begin{bmatrix} \dot{x}_1 \\ \dot{x}_2 \\ \dot{x}_3 \end{bmatrix} = \begin{bmatrix} -5.0325 & -25.1026 & -5.008 \\ 1 & 0 & 0 \\ 0 & 1 & 0 \end{bmatrix} \begin{bmatrix} x_1 \\ x_2 \\ x_3 \end{bmatrix} + \begin{bmatrix} 1 \\ 0 \\ 0 \end{bmatrix} u$$

$$y = \begin{bmatrix} 0 & 25.04 & 5.008 \end{bmatrix} \begin{bmatrix} x_1 \\ x_2 \\ x_3 \end{bmatrix} + [0]u$$

A–9–6. Consider the system defined by

$$\dot{\mathbf{x}} = \mathbf{A}\mathbf{x} + \mathbf{B}\mathbf{u}$$

where $\mathbf{x}$ = state vector (n-vector)

$\mathbf{u}$ = control vector (r-vector)

$\mathbf{A} = n \times n$ constant matrix

$\mathbf{B} = n \times r$ constant matrix

Obtain the response of the system to each of the following inputs:

(**a**) The r components of $\mathbf{u}$ are impulse functions of various magnitudes.

(**b**) The r components of $\mathbf{u}$ are step functions of various magnitudes.

(**c**) The r components of $\mathbf{u}$ are ramp functions of various magnitudes.

Solution.

(**a**) *Impulse response:* Referring to Equation (9–43), the solution to the given state equation is

$$\mathbf{x}(t) = e^{\mathbf{A}(t-t_0)}\mathbf{x}(t_0) + \int_{t_0}^{t} e^{\mathbf{A}(t-\tau)}\mathbf{B}\mathbf{u}(\tau)\,d\tau$$

Substituting $t_0 = 0-$ into this solution, we obtain

$$\mathbf{x}(t) = e^{\mathbf{A}t}\mathbf{x}(0-) + \int_{0-}^{t} e^{\mathbf{A}(t-\tau)}\mathbf{B}\mathbf{u}(\tau)\,d\tau$$

Let us write the impulse input $\mathbf{u}(t)$ as

$$\mathbf{u}(t) = \delta(t)\mathbf{w}$$

where $\mathbf{w}$ is a vector whose components are the magnitudes of r impulse functions applied at $t = 0$. The solution of the state equation when the impulse input $\delta(t)\mathbf{w}$ is given at $t = 0$ is

$$\mathbf{x}(t) = e^{\mathbf{A}t}\mathbf{x}(0-) + \int_{0-}^{t} e^{\mathbf{A}(t-\tau)}\mathbf{B}\delta(\tau)\mathbf{w}\,d\tau$$

$$= e^{\mathbf{A}t}\mathbf{x}(0-) + e^{\mathbf{A}t}\mathbf{B}\mathbf{w} \tag{9–95}$$

(**b**) *Step response:* Let us write the step input $\mathbf{u}(t)$ as

$$\mathbf{u}(t) = \mathbf{k}$$

where $\mathbf{k}$ is a vector whose components are the magnitudes of r step functions applied at $t = 0$. The solution to the step input at $t = 0$ is given by

$$\mathbf{x}(t) = e^{\mathbf{A}t}\mathbf{x}(0) + \int_0^t e^{\mathbf{A}(t-\tau)}\mathbf{Bk}\,d\tau$$

$$= e^{\mathbf{A}t}\mathbf{x}(0) + e^{\mathbf{A}t}\left[\int_0^t \left(\mathbf{I} - \mathbf{A}\tau + \frac{\mathbf{A}^2\tau^2}{2!} - \cdots\right)d\tau\right]\mathbf{Bk}$$

$$= e^{\mathbf{A}t}\mathbf{x}(0) + e^{\mathbf{A}t}\left(\mathbf{I}t - \frac{\mathbf{A}t^2}{2!} + \frac{\mathbf{A}^2t^3}{3!} - \cdots\right)\mathbf{Bk}$$

If $\mathbf{A}$ is nonsingular, then this last equation can be simplified to give

$$\mathbf{x}(t) = e^{\mathbf{A}t}\mathbf{x}(0) + e^{\mathbf{A}t}\left[-(\mathbf{A}^{-1})(e^{-\mathbf{A}t} - \mathbf{I})\right]\mathbf{Bk}$$

$$= e^{\mathbf{A}t}\mathbf{x}(0) + \mathbf{A}^{-1}(e^{\mathbf{A}t} - \mathbf{I})\mathbf{Bk} \tag{9–96}$$

(c) *Ramp response:* Let us write the ramp input $\mathbf{u}(t)$ as

$$\mathbf{u}(t) = t\mathbf{v}$$

where $\mathbf{v}$ is a vector whose components are magnitudes of ramp functions applied at $t = 0$. The solution to the ramp input $t\mathbf{v}$ given at $t = 0$ is

$$\mathbf{x}(t) = e^{\mathbf{A}t}\mathbf{x}(0) + \int_0^t e^{\mathbf{A}(t-\tau)}\mathbf{B}\tau\mathbf{v}\,d\tau$$

$$= e^{\mathbf{A}t}\mathbf{x}(0) + e^{\mathbf{A}t}\int_0^t e^{-\mathbf{A}\tau}\tau\,d\tau\,\mathbf{Bv}$$

$$= e^{\mathbf{A}t}\mathbf{x}(0) + e^{\mathbf{A}t}\left(\frac{\mathbf{I}}{2}t^2 - \frac{2\mathbf{A}}{3!}t^3 + \frac{3\mathbf{A}^2}{4!}t^4 - \frac{4\mathbf{A}^3}{5!}t^5 + \cdots\right)\mathbf{Bv}$$

If $\mathbf{A}$ is nonsingular, then this last equation can be simplified to give

$$\mathbf{x}(t) = e^{\mathbf{A}t}\mathbf{x}(0) + (\mathbf{A}^{-2})(e^{\mathbf{A}t} - \mathbf{I} - \mathbf{A}t)\mathbf{Bv}$$

$$= e^{\mathbf{A}t}\mathbf{x}(0) + \left[\mathbf{A}^{-2}(e^{\mathbf{A}t} - \mathbf{I}) - \mathbf{A}^{-1}t\right]\mathbf{Bv} \tag{9–97}$$

A–9–7. Obtain the response $y(t)$ of the following system:

$$\begin{bmatrix} \dot{x}_1 \\ \dot{x}_1 \end{bmatrix} = \begin{bmatrix} -1 & -0.5 \\ 1 & 0 \end{bmatrix}\begin{bmatrix} x_1 \\ x_2 \end{bmatrix} + \begin{bmatrix} 0.5 \\ 0 \end{bmatrix}u, \qquad \begin{bmatrix} x_1(0) \\ x_2(0) \end{bmatrix} = \begin{bmatrix} 0 \\ 0 \end{bmatrix}$$

$$y = \begin{bmatrix} 1 & 0 \end{bmatrix}\begin{bmatrix} x_1 \\ x_2 \end{bmatrix}$$

where $u(t)$ is the unit-step input occurring at $t = 0$, or

$$u(t) = 1(t)$$

Solution. For this system

$$\mathbf{A} = \begin{bmatrix} -1 & -0.5 \\ 1 & 0 \end{bmatrix}, \qquad \mathbf{B} = \begin{bmatrix} 0.5 \\ 0 \end{bmatrix}$$

The state transition matrix $\mathbf{\Phi}(t) = e^{\mathbf{A}t}$ can be obtained as follows:

$$\mathbf{\Phi}(t) = e^{\mathbf{A}t} = \mathcal{L}^{-1}\left[(s\mathbf{I} - \mathbf{A})^{-1}\right]$$

Since

$$(s\mathbf{I} - \mathbf{A})^{-1} = \begin{bmatrix} s+1 & 0.5 \\ -1 & s \end{bmatrix}^{-1} = \frac{1}{s^2 + s + 0.5} \begin{bmatrix} s & -0.5 \\ 1 & s+1 \end{bmatrix}$$

$$= \begin{bmatrix} \dfrac{s+0.5-0.5}{(s+0.5)^2 + 0.5^2} & \dfrac{-0.5}{(s+0.5)^2 + 0.5^2} \\ \dfrac{1}{(s+0.5)^2 + 0.5^2} & \dfrac{s+0.5+0.5}{(s+0.5)^2 + 0.5^2} \end{bmatrix}$$

we have

$$\boldsymbol{\Phi}(t) = e^{\mathbf{A}t} = \mathscr{L}^{-1}\left[(s\mathbf{I} - \mathbf{A})^{-1}\right]$$

$$= \begin{bmatrix} e^{-0.5t}(\cos 0.5t - \sin 0.5t) & -e^{-0.5t}\sin 0.5t \\ 2e^{-0.5t}\sin 0.5t & e^{-0.5t}(\cos 0.5t + \sin 0.5t) \end{bmatrix}$$

Since $\mathbf{x}(0) = \mathbf{0}$ and $k = 1$, referring to Equation (9–96), we have

$$\mathbf{x}(t) = e^{\mathbf{A}t}\mathbf{x}(0) + \mathbf{A}^{-1}(e^{\mathbf{A}t} - \mathbf{I})\mathbf{B}k$$

$$= \mathbf{A}^{-1}(e^{\mathbf{A}t} - \mathbf{I})\mathbf{B}$$

$$= \begin{bmatrix} 0 & 1 \\ -2 & -2 \end{bmatrix} \begin{bmatrix} 0.5e^{-0.5t}(\cos 0.5t - \sin 0.5t) - 0.5 \\ e^{-0.5t}\sin 0.5t \end{bmatrix}$$

$$= \begin{bmatrix} e^{-0.5t}\sin 0.5t \\ -e^{-0.5t}(\cos 0.5t + \sin 0.5t) + 1 \end{bmatrix}$$

Hence, the output $y(t)$ can be given by

$$y(t) = \begin{bmatrix} 1 & 0 \end{bmatrix} \begin{bmatrix} x_1 \\ x_2 \end{bmatrix} = x_1 = e^{-0.5t}\sin 0.5t$$

A–9–8. The Cayley–Hamilton theorem states that every $n \times n$ matrix $\mathbf{A}$ satisfies its own characteristic equation. The characteristic equation is not, however, necessarily the scalar equation of least degree that $\mathbf{A}$ satisfies. The least-degree polynomial having $\mathbf{A}$ as a root is called the *minimal polynomial*. That is, the minimal polynomial of an $n \times n$ matrix $\mathbf{A}$ is defined as the polynomial $\phi(\lambda)$ of least degree,

$$\phi(\lambda) = \lambda^m + a_1\lambda^{m-1} + \cdots + a_{m-1}\lambda + a_m, \qquad m \leq n$$

such that $\phi(\mathbf{A}) = \mathbf{0}$, or

$$\phi(\mathbf{A}) = \mathbf{A}^m + a_1\mathbf{A}^{m-1} + \cdots + a_{m-1}\mathbf{A} + a_m\mathbf{I} = \mathbf{0}$$

The minimal polynomial plays an important role in the computation of polynomials in an $n \times n$ matrix.

Let us suppose that $d(\lambda)$, a polynomial in λ, is the greatest common divisor of all the elements of $\text{adj}(\lambda\mathbf{I} - \mathbf{A})$. Show that, if the coefficient of the highest-degree term in λ of $d(\lambda)$ is chosen as 1, then the minimal polynomial $\phi(\lambda)$ is given by

$$\phi(\lambda) = \left| \frac{\lambda\mathbf{I} - \mathbf{A}}{d(\lambda)} \right|$$

Solution. By assumption, the greatest common divisor of the matrix $\text{adj}(\lambda\mathbf{I} - \mathbf{A})$ is $d(\lambda)$. Therefore,

$$\text{adj}(\lambda\mathbf{I} - \mathbf{A}) = d(\lambda)\mathbf{B}(\lambda)$$

Example Problems and Solutions

where the greatest common divisor of the n^2 elements (which are functions of λ) of $\mathbf{B}(\lambda)$ is unity. Since

$$(\lambda\mathbf{I} - \mathbf{A})\,\text{adj}(\lambda\mathbf{I} - \mathbf{A}) = |\lambda\mathbf{I} - \mathbf{A}|\mathbf{I}$$

we obtain

$$d(\lambda)(\lambda\mathbf{I} - \mathbf{A})\mathbf{B}(\lambda) = |\lambda\mathbf{I} - \mathbf{A}|\mathbf{I} \tag{9–98}$$

from which we find that $|\lambda\mathbf{I} - \mathbf{A}|$ is divisible by $d(\lambda)$. Let us put

$$|\lambda\mathbf{I} - \mathbf{A}| = d(\lambda)\psi(\lambda) \tag{9–99}$$

Because the coefficient of the highest-degree term in λ of $d(\lambda)$ has been chosen to be 1, the coefficient of the highest-degree term in λ of $\psi(\lambda)$ is also 1. From Equations (9–98) and (9–99), we have

$$(\lambda\mathbf{I} - \mathbf{A})\mathbf{B}(\lambda) = \psi(\lambda)\mathbf{I}$$

Hence,

$$\psi(\mathbf{A}) = \mathbf{0}$$

Note that $\psi(\lambda)$ can be written as

$$\psi(\lambda) = g(\lambda)\phi(\lambda) + \alpha(\lambda)$$

where $\alpha(\lambda)$ is of lower degree than $\phi(\lambda)$. Since $\psi(\mathbf{A}) = \mathbf{0}$ and $\phi(\mathbf{A}) = \mathbf{0}$, we must have $\alpha(\mathbf{A}) = \mathbf{0}$. Also, since $\phi(\lambda)$ is the minimal polynomial, $\alpha(\lambda)$ must be identically zero, or

$$\psi(\lambda) = g(\lambda)\phi(\lambda)$$

Note that because $\phi(\mathbf{A}) = \mathbf{0}$, we can write

$$\phi(\lambda)\mathbf{I} = (\lambda\mathbf{I} - \mathbf{A})\mathbf{C}(\lambda)$$

Hence,

$$\psi(\lambda)\mathbf{I} = g(\lambda)\phi(\lambda)\mathbf{I} = g(\lambda)(\lambda\mathbf{I} - \mathbf{A})\mathbf{C}(\lambda)$$

Noting that $(\lambda\mathbf{I} - \mathbf{A})\mathbf{B}(\lambda) = \psi(\lambda)\mathbf{I}$, we obtain

$$\mathbf{B}(\lambda) = g(\lambda)\mathbf{C}(\lambda)$$

Since the greatest common divisor of the n^2 elements of $\mathbf{B}(\lambda)$ is unity, we have

$$g(\lambda) = 1$$

Therefore,

$$\psi(\lambda) = \phi(\lambda)$$

Then, from this last equation and Equation (9–99), we obtain

$$\phi(\lambda) = \frac{|\lambda\mathbf{I} - \mathbf{A}|}{d(\lambda)}$$

A–9–9. If an $n \times n$ matrix $\mathbf{A}$ has n distinct eigenvalues, then the minimal polynomial of $\mathbf{A}$ is identical to the characteristic polynomial. Also, if the multiple eigenvalues of $\mathbf{A}$ are linked in a Jordan chain, the minimal polynomial and the characteristic polynomial are identical. If, however, the multiple eigenvalues of $\mathbf{A}$ are not linked in a Jordan chain, the minimal polynomial is of lower degree than the characteristic polynomial.

Using the following matrices $\mathbf{A}$ and $\mathbf{B}$ as examples, verify the foregoing statements about the minimal polynomial when multiple eigenvalues are involved:

$$\mathbf{A} = \begin{bmatrix} 2 & 1 & 4 \\ 0 & 2 & 0 \\ 0 & 3 & 1 \end{bmatrix}, \quad \mathbf{B} = \begin{bmatrix} 2 & 0 & 0 \\ 0 & 2 & 0 \\ 0 & 3 & 1 \end{bmatrix}$$

Solution. First, consider the matrix $\mathbf{A}$. The characteristic polynomial is given by

$$|\lambda\mathbf{I} - \mathbf{A}| = \begin{vmatrix} \lambda - 2 & -1 & -4 \\ 0 & \lambda - 2 & 0 \\ 0 & -3 & \lambda - 1 \end{vmatrix} = (\lambda - 2)^2(\lambda - 1)$$

Thus, the eigenvalues of $\mathbf{A}$ are 2, 2, and 1. It can be shown that the Jordan canonical form of $\mathbf{A}$ is

$$\begin{bmatrix} 2 & 1 & 0 \\ 0 & 2 & 0 \\ 0 & 0 & 1 \end{bmatrix}$$

and the multiple eigenvalues are linked in the Jordan chain as shown.

To determine the minimal polynomial, let us first obtain $\text{adj}(\lambda\mathbf{I} - \mathbf{A})$. It is given by

$$\text{adj}(\lambda\mathbf{I} - \mathbf{A}) = \begin{bmatrix} (\lambda - 2)(\lambda - 1) & (\lambda + 11) & 4(\lambda - 2) \\ 0 & (\lambda - 2)(\lambda - 1) & 0 \\ 0 & 3(\lambda - 2) & (\lambda - 2)^2 \end{bmatrix}$$

Notice that there is no common divisor of all the elements of $\text{adj}(\lambda\mathbf{I} - \mathbf{A})$. Hence, $d(\lambda) = 1$. Thus, the minimal polynomial $\phi(\lambda)$ is identical to the characteristic polynomial, or

$$\phi(\lambda) = |\lambda\mathbf{I} - \mathbf{A}| = (\lambda - 2)^2(\lambda - 1)$$
$$= \lambda^3 - 5\lambda^2 + 8\lambda - 4$$

A simple calculation proves that

$$\mathbf{A}^3 - 5\mathbf{A}^2 + 8\mathbf{A} - 4\mathbf{I}$$

$$= \begin{bmatrix} 8 & 72 & 28 \\ 0 & 8 & 0 \\ 0 & 21 & 1 \end{bmatrix} - 5\begin{bmatrix} 4 & 16 & 12 \\ 0 & 4 & 0 \\ 0 & 9 & 1 \end{bmatrix} + 8\begin{bmatrix} 2 & 1 & 4 \\ 0 & 2 & 0 \\ 0 & 3 & 1 \end{bmatrix} - 4\begin{bmatrix} 1 & 0 & 0 \\ 0 & 1 & 0 \\ 0 & 0 & 1 \end{bmatrix}$$

$$= \begin{bmatrix} 0 & 0 & 0 \\ 0 & 0 & 0 \\ 0 & 0 & 0 \end{bmatrix} = \mathbf{0}$$

but

$$\mathbf{A}^2 - 3\mathbf{A} + 2\mathbf{I}$$

$$= \begin{bmatrix} 4 & 16 & 12 \\ 0 & 4 & 0 \\ 0 & 9 & 1 \end{bmatrix} - 3\begin{bmatrix} 2 & 1 & 4 \\ 0 & 2 & 0 \\ 0 & 3 & 1 \end{bmatrix} + 2\begin{bmatrix} 1 & 0 & 0 \\ 0 & 1 & 0 \\ 0 & 0 & 1 \end{bmatrix}$$

$$= \begin{bmatrix} 0 & 13 & 0 \\ 0 & 0 & 0 \\ 0 & 0 & 0 \end{bmatrix} \neq \mathbf{0}$$

Thus, we have shown that the minimal polynomial and the characteristic polynomial of this matrix $\mathbf{A}$ are the same.

Next, consider the matrix $\mathbf{B}$. The characteristic polynomial is given by

$$|\lambda\mathbf{I} - \mathbf{B}| = \begin{vmatrix} \lambda - 2 & 0 & 0 \\ 0 & \lambda - 2 & 0 \\ 0 & -3 & \lambda - 1 \end{vmatrix} = (\lambda - 2)^2(\lambda - 1)$$

A simple computation reveals that matrix $\mathbf{B}$ has three eigenvectors, and the Jordan canonical form of $\mathbf{B}$ is given by

$$\begin{bmatrix} 2 & 0 & 0 \\ 0 & 2 & 0 \\ 0 & 0 & 1 \end{bmatrix}$$

Thus, the multiple eigenvalues are not linked. To obtain the minimal polynomial, we first compute $\mathrm{adj}(\lambda\mathbf{I} - \mathbf{B})$:

$$\mathrm{adj}(\lambda\mathbf{I} - \mathbf{B}) = \begin{bmatrix} (\lambda - 2)(\lambda - 1) & 0 & 0 \\ 0 & (\lambda - 2)(\lambda - 1) & 0 \\ 0 & 3(\lambda - 2) & (\lambda - 2)^2 \end{bmatrix}$$

from which it is evident that

$$d(\lambda) = \lambda - 2$$

Hence,

$$\phi(\lambda) = \frac{|\lambda\mathbf{I} - \mathbf{B}|}{d(\lambda)} = \frac{(\lambda - 2)^2(\lambda - 1)}{\lambda - 2} = \lambda^2 - 3\lambda + 2$$

As a check, let us compute $\phi(\mathbf{B})$:

$$\phi(\mathbf{B}) = \mathbf{B}^2 - 3\mathbf{B} + 2\mathbf{I} = \begin{bmatrix} 4 & 0 & 0 \\ 0 & 4 & 0 \\ 0 & 9 & 1 \end{bmatrix} - 3\begin{bmatrix} 2 & 0 & 0 \\ 0 & 2 & 0 \\ 0 & 3 & 1 \end{bmatrix} + 2\begin{bmatrix} 1 & 0 & 0 \\ 0 & 1 & 0 \\ 0 & 0 & 1 \end{bmatrix} = \begin{bmatrix} 0 & 0 & 0 \\ 0 & 0 & 0 \\ 0 & 0 & 0 \end{bmatrix} = \mathbf{0}$$

For the given matrix $\mathbf{B}$, the degree of the minimal polynomial is lower by 1 than that of the characteristic polynomial. As shown here, if the multiple eigenvalues of an $n \times n$ matrix are not linked in a Jordan chain, the minimal polynomial is of lower degree than the characteristic polynomial.

A-9-10. Show that by use of the minimal polynomial, the inverse of a nonsingular matrix $\mathbf{A}$ can be expressed as a polynomial in $\mathbf{A}$ with scalar coefficients as follows:

$$\mathbf{A}^{-1} = -\frac{1}{a_m}\left(\mathbf{A}^{m-1} + a_1\mathbf{A}^{m-2} + \cdots + a_{m-2}\mathbf{A} + a_{m-1}\mathbf{I}\right) \tag{9-100}$$

where $a_1, a_2, \ldots, a_m$ are coefficients of the minimal polynomial

$$\phi(\lambda) = \lambda^m + a_1\lambda^{m-1} + \cdots + a_{m-1}\lambda + a_m$$

Then obtain the inverse of the following matrix $\mathbf{A}$:

$$\mathbf{A} = \begin{bmatrix} 1 & 2 & 0 \\ 3 & -1 & -2 \\ 1 & 0 & -3 \end{bmatrix}$$

Solution. For a nonsingular matrix $\mathbf{A}$, its minimal polynomial $\phi(\mathbf{A})$ can be written as

$$\phi(\mathbf{A}) = \mathbf{A}^m + a_1\mathbf{A}^{m-1} + \cdots + a_{m-1}\mathbf{A} + a_m\mathbf{I} = 0$$

where $a_m \neq 0$. Hence,

$$\mathbf{I} = -\frac{1}{a_m}\left(\mathbf{A}^m + a_1\mathbf{A}^{m-1} + \cdots + a_{m-2}\mathbf{A}^2 + a_{m-1}\mathbf{A}\right)$$

Premultiplying by $\mathbf{A}^{-1}$, we obtain

$$\mathbf{A}^{-1} = -\frac{1}{a_m}\left(\mathbf{A}^{m-1} + a_1\mathbf{A}^{m-2} + \cdots + a_{m-2}\mathbf{A} + a_{m-1}\mathbf{I}\right)$$

which is Equation (9-100).

For the given matrix $\mathbf{A}$, $\mathrm{adj}(\lambda\mathbf{I} - \mathbf{A})$ can be given as

$$\mathrm{adj}(\lambda\mathbf{I} - \mathbf{A}) = \begin{bmatrix} \lambda^2 + 4\lambda + 3 & 2\lambda + 6 & -4 \\ 3\lambda + 7 & \lambda^2 + 2\lambda - 3 & -2\lambda + 2 \\ \lambda + 1 & 2 & \lambda^2 - 7 \end{bmatrix}$$

Clearly, there is no common divisor $d(\lambda)$ of all elements of $\mathrm{adj}(\lambda\mathbf{I} - \mathbf{A})$. Hence, $d(\lambda) = 1$. Consequently, the minimal polynomial $\phi(\lambda)$ is given by

$$\phi(\lambda) = \frac{|\lambda\mathbf{I} - \mathbf{A}|}{d(\lambda)} = |\lambda\mathbf{I} - \mathbf{A}|$$

Thus, the minimal polynomial $\phi(\lambda)$ is the same as the characteristic polynomial.
Since the characteristic polynomial is

$$|\lambda\mathbf{I} - \mathbf{A}| = \lambda^3 + 3\lambda^2 - 7\lambda - 17$$

we obtain

$$\phi(\lambda) = \lambda^3 + 3\lambda^2 - 7\lambda - 17$$

Example Problems and Solutions

By identifying the coefficients a_i of the minimal polynomial (which is the same as the characteristic polynomial in this case), we have

$$a_1 = 3, \qquad a_2 = -7, \qquad a_3 = -17$$

The inverse of $\mathbf{A}$ can then be obtained from Equation (9–100) as follows:

$$\mathbf{A}^{-1} = -\frac{1}{a_3}\left(\mathbf{A}^2 + a_1\mathbf{A} + a_2\mathbf{I}\right) = \frac{1}{17}\left(\mathbf{A}^2 + 3\mathbf{A} - 7\mathbf{I}\right)$$

$$= \frac{1}{17}\left\{ \begin{bmatrix} 7 & 0 & -4 \\ -2 & 7 & 8 \\ -2 & 2 & 9 \end{bmatrix} + 3\begin{bmatrix} 1 & 2 & 0 \\ 3 & -1 & -2 \\ 1 & 0 & -3 \end{bmatrix} - 7\begin{bmatrix} 1 & 0 & 0 \\ 0 & 1 & 0 \\ 0 & 0 & 1 \end{bmatrix} \right\}$$

$$= \frac{1}{17}\begin{bmatrix} 3 & 6 & -4 \\ 7 & -3 & 2 \\ 1 & 2 & -7 \end{bmatrix}$$

$$= \begin{bmatrix} \frac{3}{17} & \frac{6}{17} & -\frac{4}{17} \\ \frac{7}{17} & -\frac{3}{17} & \frac{2}{17} \\ \frac{1}{17} & \frac{2}{17} & -\frac{7}{17} \end{bmatrix}$$

A–9–11. Show that if matrix $\mathbf{A}$ can be diagonalized, then

$$e^{\mathbf{A}t} = \mathbf{P}e^{\mathbf{D}t}\mathbf{P}^{-1}$$

where $\mathbf{P}$ is a diagonalizing transformation matrix that transforms $\mathbf{A}$ into a diagonal matrix, or $\mathbf{P}^{-1}\mathbf{A}\mathbf{P} = \mathbf{D}$, where $\mathbf{D}$ is a diagonal matrix.

Show also that if matrix $\mathbf{A}$ can be transformed into a Jordan canonical form, then

$$e^{\mathbf{A}t} = \mathbf{S}e^{\mathbf{J}t}\mathbf{S}^{-1}$$

where $\mathbf{S}$ is a transformation matrix that transforms $\mathbf{A}$ into a Jordan canonical form $\mathbf{J}$, or $\mathbf{S}^{-1}\mathbf{A}\mathbf{S} = \mathbf{J}$.

Solution. Consider the state equation

$$\dot{\mathbf{x}} = \mathbf{A}\mathbf{x}$$

If a square matrix can be diagonalized, then a diagonalizing matrix (transformation matrix) exists and it can be obtained by a standard method. Let $\mathbf{P}$ be a diagonalizing matrix for $\mathbf{A}$. Let us define

$$\mathbf{x} = \mathbf{P}\hat{\mathbf{x}}$$

Then

$$\dot{\hat{\mathbf{x}}} = \mathbf{P}^{-1}\mathbf{A}\mathbf{P}\hat{\mathbf{x}} = \mathbf{D}\hat{\mathbf{x}}$$

where $\mathbf{D}$ is a diagonal matrix. The solution of this last equation is

$$\hat{\mathbf{x}}(t) = e^{\mathbf{D}t}\hat{\mathbf{x}}(0)$$

Hence,

$$\mathbf{x}(t) = \mathbf{P}\hat{\mathbf{x}}(t) = \mathbf{P}e^{\mathbf{D}t}\mathbf{P}^{-1}\mathbf{x}(0)$$

Noting that $\mathbf{x}(t)$ can also be given by the equation

$$\mathbf{x}(t) = e^{\mathbf{A}t}\mathbf{x}(0)$$

we obtain $e^{\mathbf{A}t} = \mathbf{P}e^{\mathbf{D}t}\mathbf{P}^{-1}$, or

$$e^{\mathbf{A}t} = \mathbf{P}e^{\mathbf{D}t}\mathbf{P}^{-1} = \mathbf{P}\begin{bmatrix} e^{\lambda_1 t} & & & & 0 \\ & e^{\lambda_2 t} & & & \\ & & \cdot & & \\ & & & \cdot & \\ & & & & \cdot \\ 0 & & & & e^{\lambda_n t} \end{bmatrix}\mathbf{P}^{-1} \tag{9-101}$$

Next, we shall consider the case where matrix $\mathbf{A}$ may be transformed into a Jordan canonical form. Consider again the state equation

$$\dot{\mathbf{x}} = \mathbf{A}\mathbf{x}$$

First obtain a transformation matrix $\mathbf{S}$ that will transform matrix $\mathbf{A}$ into a Jordan canonical form so that

$$\mathbf{S}^{-1}\mathbf{A}\mathbf{S} = \mathbf{J}$$

where $\mathbf{J}$ is a matrix in a Jordan canonical form. Now define

$$\mathbf{x} = \mathbf{S}\hat{\mathbf{x}}$$

Then

$$\dot{\hat{\mathbf{x}}} = \mathbf{S}^{-1}\mathbf{A}\mathbf{S}\hat{\mathbf{x}} = \mathbf{J}\hat{\mathbf{x}}$$

The solution of this last equation is

$$\hat{\mathbf{x}}(t) = e^{\mathbf{J}t}\hat{\mathbf{x}}(0)$$

Hence,

$$\mathbf{x}(t) = \mathbf{S}\hat{\mathbf{x}}(t) = \mathbf{S}e^{\mathbf{J}t}\mathbf{S}^{-1}\mathbf{x}(0)$$

Since the solution $\mathbf{x}(t)$ can also be given by the equation

$$\mathbf{x}(t) = e^{\mathbf{A}t}\mathbf{x}(0)$$

we obtain

$$e^{\mathbf{A}t} = \mathbf{S}e^{\mathbf{J}t}\mathbf{S}^{-1}$$

Note that $e^{\mathbf{J}t}$ is a triangular matrix [which means that the elements below (or above, as the case may be) the principal diagonal line are zeros] whose elements are $e^{\lambda t}, te^{\lambda t}, \frac{1}{2}t^2 e^{\lambda t}$, and so forth. For example, if matrix $\mathbf{J}$ has the following Jordan canonical form:

$$\mathbf{J} = \begin{bmatrix} \lambda_1 & 1 & 0 \\ 0 & \lambda_1 & 1 \\ 0 & 0 & \lambda_1 \end{bmatrix}$$

then

$$e^{\mathbf{J}t} = \begin{bmatrix} e^{\lambda_1 t} & te^{\lambda_1 t} & \frac{1}{2}t^2 e^{\lambda_1 t} \\ 0 & e^{\lambda_1 t} & te^{\lambda_1 t} \\ 0 & 0 & e^{\lambda_1 t} \end{bmatrix}$$

Similarly, if

$$\mathbf{J} = \begin{bmatrix} \lambda_1 & 1 & 0 & & & & 0 \\ 0 & \lambda_1 & 1 & & & & \\ 0 & 0 & \lambda_1 & & & & \\ & & & \lambda_4 & 1 & & \\ & & & 0 & \lambda_4 & & \\ & & & & & \lambda_6 & \\ 0 & & & & & & \lambda_7 \end{bmatrix}$$

then

$$e^{\mathbf{J}t} = \begin{bmatrix} e^{\lambda_1 t} & te^{\lambda_1 t} & \frac{1}{2}t^2 e^{\lambda_1 t} & & & & 0 \\ 0 & e^{\lambda_1 t} & te^{\lambda_1 t} & & & & \\ 0 & 0 & e^{\lambda_1 t} & & & & \\ & & & e^{\lambda_4 t} & te^{\lambda_4 t} & & \\ & & & 0 & e^{\lambda_4 t} & & \\ & & & & & e^{\lambda_6 t} & 0 \\ 0 & & & & & 0 & e^{\lambda_7 t} \end{bmatrix}$$

A–9–12. Consider the following polynomial in λ of degree $m - 1$, where we assume $\lambda_1, \lambda_2, \ldots, \lambda_m$ to be distinct:

$$p_k(\lambda) = \frac{(\lambda - \lambda_1) \cdots (\lambda - \lambda_{k-1})(\lambda - \lambda_{k+1}) \cdots (\lambda - \lambda_m)}{(\lambda_k - \lambda_1) \cdots (\lambda_k - \lambda_{k-1})(\lambda_k - \lambda_{k+1}) \cdots (\lambda_k - \lambda_m)}$$

where $k = 1, 2, \ldots, m$. Notice that

$$p_k(\lambda_i) = \begin{cases} 1, & \text{if } i = k \\ 0, & \text{if } i \neq k \end{cases}$$

Then the polynomial $f(\lambda)$ of degree $m - 1$,

$$f(\lambda) = \sum_{k=1}^{m} f(\lambda_k) p_k(\lambda)$$

$$= \sum_{k=1}^{m} f(\lambda_k) \frac{(\lambda - \lambda_1) \cdots (\lambda - \lambda_{k-1})(\lambda - \lambda_{k+1}) \cdots (\lambda - \lambda_m)}{(\lambda_k - \lambda_1) \cdots (\lambda_k - \lambda_{k-1})(\lambda_k - \lambda_{k+1}) \cdots (\lambda_k - \lambda_m)}$$

takes on the values $f(\lambda_k)$ at the points λ_k. This last equation is commonly called *Lagrange's interpolation formula*. The polynomial $f(\lambda)$ of degree $m - 1$ is determined from m independent data $f(\lambda_1)$, $f(\lambda_2)$, $\ldots$, $f(\lambda_m)$. That is, the polynomial $f(\lambda)$ passes through m points $f(\lambda_1), f(\lambda_2), \ldots, f(\lambda_m)$. Since $f(\lambda)$ is a polynomial of degree $m - 1$, it is uniquely determined. Any other representations of the polynomial of degree $m - 1$ can be reduced to the Lagrange polynomial $f(\lambda)$.

Chapter 9 / Control Systems Analysis in State Space

Assuming that the eigenvalues of an $n \times n$ matrix $\mathbf{A}$ are distinct, substitute $\mathbf{A}$ for λ in the polynomial $p_k(\lambda)$. Then we get

$$p_k(\mathbf{A}) = \frac{(\mathbf{A} - \lambda_1 \mathbf{I}) \cdots (\mathbf{A} - \lambda_{k-1}\mathbf{I})(\mathbf{A} - \lambda_{k+1}\mathbf{I}) \cdots (\mathbf{A} - \lambda_m \mathbf{I})}{(\lambda_k - \lambda_1) \cdots (\lambda_k - \lambda_{k-1})(\lambda_k - \lambda_{k+1}) \cdots (\lambda_k - \lambda_m)}$$

Notice that $p_k(\mathbf{A})$ is a polynomial in $\mathbf{A}$ of degree $m - 1$. Notice also that

$$p_k(\lambda_i \mathbf{I}) = \begin{cases} \mathbf{I}, & \text{if } i = k \\ \mathbf{0}, & \text{if } i \neq k \end{cases}$$

Now define

$$f(\mathbf{A}) = \sum_{k=1}^{m} f(\lambda_k) p_k(\mathbf{A})$$

$$= \sum_{k=1}^{m} f(\lambda_k) \frac{(\mathbf{A} - \lambda_1 \mathbf{I}) \cdots (\mathbf{A} - \lambda_{k-1}\mathbf{I})(\mathbf{A} - \lambda_{k+1}\mathbf{I}) \cdots (\mathbf{A} - \lambda_m \mathbf{I})}{(\lambda_k - \lambda_1) \cdots (\lambda_k - \lambda_{k-1})(\lambda_k - \lambda_{k+1}) \cdots (\lambda_k - \lambda_m)} \qquad (9\text{--}102)$$

Equation (9–102) is known as Sylvester's interpolation formula. Equation (9–102) is equivalent to the following equation:

$$\begin{vmatrix} 1 & 1 & \cdots & 1 & \mathbf{I} \\ \lambda_1 & \lambda_2 & \cdots & \lambda_m & \mathbf{A} \\ \lambda_1^2 & \lambda_2^2 & \cdots & \lambda_m^2 & \mathbf{A}^2 \\ \cdot & \cdot & & \cdot & \cdot \\ \cdot & \cdot & & \cdot & \cdot \\ \cdot & \cdot & & \cdot & \cdot \\ \lambda_1^{m-1} & \lambda_2^{m-1} & \cdots & \lambda_m^{m-1} & \mathbf{A}^{m-1} \\ f(\lambda_1) & f(\lambda_2) & \cdots & f(\lambda_m) & f(\mathbf{A}) \end{vmatrix} = \mathbf{0} \qquad (9\text{--}103)$$

Equations (9–102) and (9–103) are frequently used for evaluating functions $f(\mathbf{A})$ of matrix $\mathbf{A}$—for example, $(\lambda \mathbf{I} - \mathbf{A})^{-1}$, $e^{\mathbf{A}t}$, and so forth. Note that Equation (9–103) can also be written as

$$\begin{vmatrix} 1 & \lambda_1 & \lambda_1^2 & \cdots & \lambda_1^{m-1} & f(\lambda_1) \\ 1 & \lambda_2 & \lambda_2^2 & \cdots & \lambda_2^{m-1} & f(\lambda_2) \\ \cdot & \cdot & \cdot & & \cdot & \cdot \\ \cdot & \cdot & \cdot & & \cdot & \cdot \\ \cdot & \cdot & \cdot & \cdots & \cdot & \cdot \\ 1 & \lambda_m & \lambda_m^2 & \cdots & \lambda_m^{m-1} & f(\lambda_m) \\ \mathbf{I} & \mathbf{A} & \mathbf{A}^2 & \cdots & \mathbf{A}^{m-1} & f(\mathbf{A}) \end{vmatrix} = \mathbf{0} \qquad (9\text{--}104)$$

Show that Equations (9–102) and (9–103) are equivalent. To simplify the arguments, assume that $m = 4$.

Solution. Equation (9–103), where $m = 4$, can be expanded as follows:

$$\Delta = \begin{vmatrix} 1 & 1 & 1 & 1 & \mathbf{I} \\ \lambda_1 & \lambda_2 & \lambda_3 & \lambda_4 & \mathbf{A} \\ \lambda_1^2 & \lambda_2^2 & \lambda_3^2 & \lambda_4^2 & \mathbf{A}^2 \\ \lambda_1^3 & \lambda_2^3 & \lambda_3^3 & \lambda_4^3 & \mathbf{A}^3 \\ f(\lambda_1) & f(\lambda_2) & f(\lambda_3) & f(\lambda_4) & f(\mathbf{A}) \end{vmatrix}$$

$$= f(\mathbf{A}) \begin{vmatrix} 1 & 1 & 1 & 1 \\ \lambda_1 & \lambda_2 & \lambda_3 & \lambda_4 \\ \lambda_1^2 & \lambda_2^2 & \lambda_3^2 & \lambda_4^2 \\ \lambda_1^3 & \lambda_2^3 & \lambda_3^3 & \lambda_4^3 \end{vmatrix} - f(\lambda_4) \begin{vmatrix} 1 & 1 & 1 & \mathbf{I} \\ \lambda_1 & \lambda_2 & \lambda_3 & \mathbf{A} \\ \lambda_1^2 & \lambda_2^2 & \lambda_3^2 & \mathbf{A}^2 \\ \lambda_1^3 & \lambda_2^3 & \lambda_3^3 & \mathbf{A}^3 \end{vmatrix}$$

$$+ f(\lambda_3) \begin{vmatrix} 1 & 1 & 1 & \mathbf{I} \\ \lambda_1 & \lambda_2 & \lambda_4 & \mathbf{A} \\ \lambda_1^2 & \lambda_2^2 & \lambda_4^2 & \mathbf{A}^2 \\ \lambda_1^3 & \lambda_2^3 & \lambda_4^3 & \mathbf{A}^3 \end{vmatrix} - f(\lambda_2) \begin{vmatrix} 1 & 1 & 1 & \mathbf{I} \\ \lambda_1 & \lambda_3 & \lambda_4 & \mathbf{A} \\ \lambda_1^2 & \lambda_3^2 & \lambda_4^2 & \mathbf{A}^2 \\ \lambda_1^3 & \lambda_3^3 & \lambda_4^3 & \mathbf{A}^3 \end{vmatrix}$$

$$+ f(\lambda_1) \begin{vmatrix} 1 & 1 & 1 & \mathbf{I} \\ \lambda_2 & \lambda_3 & \lambda_4 & \mathbf{A} \\ \lambda_2^2 & \lambda_3^2 & \lambda_4^2 & \mathbf{A}^2 \\ \lambda_2^3 & \lambda_3^3 & \lambda_4^3 & \mathbf{A}^3 \end{vmatrix}$$

Since

$$\begin{vmatrix} 1 & 1 & 1 & 1 \\ \lambda_1 & \lambda_2 & \lambda_3 & \lambda_4 \\ \lambda_1^2 & \lambda_2^2 & \lambda_3^2 & \lambda_4^2 \\ \lambda_1^3 & \lambda_2^3 & \lambda_3^3 & \lambda_4^3 \end{vmatrix} = (\lambda_4 - \lambda_3)(\lambda_4 - \lambda_2)(\lambda_4 - \lambda_1)(\lambda_3 - \lambda_2)(\lambda_3 - \lambda_1)(\lambda_2 - \lambda_1)$$

and

$$\begin{vmatrix} 1 & 1 & 1 & \mathbf{I} \\ \lambda_i & \lambda_j & \lambda_k & \mathbf{A} \\ \lambda_i^2 & \lambda_j^2 & \lambda_k^2 & \mathbf{A}^2 \\ \lambda_i^3 & \lambda_j^3 & \lambda_k^3 & \mathbf{A}^3 \end{vmatrix} = (\mathbf{A} - \lambda_k \mathbf{I})(\mathbf{A} - \lambda_j \mathbf{I})(\mathbf{A} - \lambda_i \mathbf{I})(\lambda_k - \lambda_j)(\lambda_k - \lambda_i)(\lambda_j - \lambda_i)$$

we obtain

$$\begin{aligned} \Delta = \ & f(\mathbf{A})[(\lambda_4 - \lambda_3)(\lambda_4 - \lambda_2)(\lambda_4 - \lambda_1)(\lambda_3 - \lambda_2)(\lambda_3 - \lambda_1)(\lambda_2 - \lambda_1)] \\ & - f(\lambda_4)[(\mathbf{A} - \lambda_3 \mathbf{I})(\mathbf{A} - \lambda_2 \mathbf{I})(\mathbf{A} - \lambda_1 \mathbf{I})(\lambda_3 - \lambda_2)(\lambda_3 - \lambda_1)(\lambda_2 - \lambda_1)] \\ & + f(\lambda_3)[(\mathbf{A} - \lambda_4 \mathbf{I})(\mathbf{A} - \lambda_2 \mathbf{I})(\mathbf{A} - \lambda_1 \mathbf{I})(\lambda_4 - \lambda_2)(\lambda_4 - \lambda_1)(\lambda_2 - \lambda_1)] \\ & - f(\lambda_2)[(\mathbf{A} - \lambda_4 \mathbf{I})(\mathbf{A} - \lambda_3 \mathbf{I})(\mathbf{A} - \lambda_1 \mathbf{I})(\lambda_4 - \lambda_3)(\lambda_4 - \lambda_1)(\lambda_3 - \lambda_1)] \\ & + f(\lambda_1)[(\mathbf{A} - \lambda_4 \mathbf{I})(\mathbf{A} - \lambda_3 \mathbf{I})(\mathbf{A} - \lambda_2 \mathbf{I})(\lambda_4 - \lambda_3)(\lambda_4 - \lambda_2)(\lambda_3 - \lambda_2)] \\ = \ & \mathbf{0} \end{aligned}$$

Solving this last equation for $f(\mathbf{A})$, we obtain

$$f(\mathbf{A}) = f(\lambda_1)\frac{(\mathbf{A} - \lambda_2\mathbf{I})(\mathbf{A} - \lambda_3\mathbf{I})(\mathbf{A} - \lambda_4\mathbf{I})}{(\lambda_1 - \lambda_2)(\lambda_1 - \lambda_3)(\lambda_1 - \lambda_4)} + f(\lambda_2)\frac{(\mathbf{A} - \lambda_1\mathbf{I})(\mathbf{A} - \lambda_3\mathbf{I})(\mathbf{A} - \lambda_4\mathbf{I})}{(\lambda_2 - \lambda_1)(\lambda_2 - \lambda_3)(\lambda_2 - \lambda_4)}$$

$$+ f(\lambda_3)\frac{(\mathbf{A} - \lambda_1\mathbf{I})(\mathbf{A} - \lambda_2\mathbf{I})(\mathbf{A} - \lambda_4\mathbf{I})}{(\lambda_3 - \lambda_1)(\lambda_3 - \lambda_2)(\lambda_3 - \lambda_4)} + f(\lambda_4)\frac{(\mathbf{A} - \lambda_1\mathbf{I})(\mathbf{A} - \lambda_2\mathbf{I})(\mathbf{A} - \lambda_3\mathbf{I})}{(\lambda_4 - \lambda_1)(\lambda_4 - \lambda_2)(\lambda_4 - \lambda_3)}$$

$$= \sum_{k=1}^{m} f(\lambda_k)\frac{(\mathbf{A} - \lambda_1\mathbf{I})\cdots(\mathbf{A} - \lambda_{k-1}\mathbf{I})(\mathbf{A} - \lambda_{k+1}\mathbf{I})\cdots(\mathbf{A} - \lambda_m\mathbf{I})}{(\lambda_k - \lambda_1)\cdots(\lambda_k - \lambda_{k-1})(\lambda_k - \lambda_{k+1})\cdots(\lambda_k - \lambda_m)}$$

where $m = 4$. Thus, we have shown the equivalence of Equations (9–102) and (9–103). Although we assumed $m = 4$, the entire argument can be extended to an arbitrary positive integer m. (For the case when the matrix $\mathbf{A}$ involves multiple eigenvalues, refer to Problem **A–9–13**.)

A–9–13. Consider Sylvester's interpolation formula in the form given by Equation (9–104):

$$\begin{vmatrix} 1 & \lambda_1 & \lambda_1^2 & \cdots & \lambda_1^{m-1} & f(\lambda_1) \\ 1 & \lambda_2 & \lambda_2^2 & \cdots & \lambda_2^{m-1} & f(\lambda_2) \\ . & . & . & & . & . \\ . & . & . & & . & . \\ . & . & . & \cdots & . & . \\ 1 & \lambda_m & \lambda_m^2 & \cdots & \lambda_m^{m-1} & f(\lambda_m) \\ \mathbf{I} & \mathbf{A} & \mathbf{A}^2 & \cdots & \mathbf{A}^{m-1} & f(\mathbf{A}) \end{vmatrix} = \mathbf{0}$$

This formula for the determination of $f(\mathbf{A})$ applies to the case where the minimal polynomial of $\mathbf{A}$ involves only distinct roots.

Suppose that the minimal polynomial of $\mathbf{A}$ involves multiple roots. Then the rows in the determinant that correspond to the multiple roots become identical, and therefore modification of the determinant in Equation (9–104) becomes necessary.

Modify the form of Sylvester's interpolation formula given by Equation (9–104) when the minimal polynomial of $\mathbf{A}$ involves multiple roots. In deriving a modified determinant equation, assume that there are three equal roots $(\lambda_1 = \lambda_2 = \lambda_3)$ in the minimal polynomial of $\mathbf{A}$ and that there are other roots $(\lambda_4, \lambda_5, \ldots, \lambda_m)$ that are distinct.

Solution. Since the minimal polynomial of $\mathbf{A}$ involves three equal roots, the minimal polynomial $\phi(\lambda)$ can be written as

$$\phi(\lambda) = \lambda^m + a_1\lambda^{m-1} + \cdots + a_{m-1}\lambda + a_m$$

$$= (\lambda - \lambda_1)^3(\lambda - \lambda_4)(\lambda - \lambda_5)\cdots(\lambda - \lambda_m)$$

An arbitrary function $f(\mathbf{A})$ of an $n \times n$ matrix $\mathbf{A}$ can be written as

$$f(\mathbf{A}) = g(\mathbf{A})\phi(\mathbf{A}) + \alpha(\mathbf{A})$$

where the minimal polynomial $\phi(\mathbf{A})$ is of degree m and $\alpha(\mathbf{A})$ is a polynomial in $\mathbf{A}$ of degree $m - 1$ or less. Hence we have

$$f(\lambda) = g(\lambda)\phi(\lambda) + \alpha(\lambda)$$

where $\alpha(\lambda)$ is a polynomial in λ of degree $m - 1$ or less, which can thus be written as

$$\alpha(\lambda) = \alpha_0 + \alpha_1\lambda + \alpha_2\lambda^2 + \cdots + \alpha_{m-1}\lambda^{m-1} \tag{9–105}$$

In the present case we have

$$f(\lambda) = g(\lambda)\phi(\lambda) + \alpha(\lambda)$$

$$= g(\lambda)[(\lambda - \lambda_1)^3(\lambda - \lambda_4)\cdots(\lambda - \lambda_m)] + \alpha(\lambda) \qquad (9\text{-}106)$$

By substituting $\lambda_1, \lambda_4, \ldots, \lambda_m$ for λ in Equation (9–106), we obtain the following $m - 2$ equations:

$$f(\lambda_1) = \alpha(\lambda_1)$$

$$f(\lambda_4) = \alpha(\lambda_4)$$

$$\cdot$$
$$\cdot \qquad\qquad (9\text{-}107)$$
$$\cdot$$

$$f(\lambda_m) = \alpha(\lambda_m)$$

By differentiating Equation (9–106) with respect to λ, we obtain

$$\frac{d}{d\lambda}f(\lambda) = (\lambda - \lambda_1)^2 h(\lambda) + \frac{d}{d\lambda}\alpha(\lambda) \qquad (9\text{-}108)$$

where

$$(\lambda - \lambda_1)^2 h(\lambda) = \frac{d}{d\lambda}\left[g(\lambda)(\lambda - \lambda_1)^3(\lambda - \lambda_4)\cdots(\lambda - \lambda_m)\right]$$

Substitution of λ_1 for λ in Equation (9–108) gives

$$\left.\frac{d}{d\lambda}f(\lambda)\right|_{\lambda=\lambda_1} = f'(\lambda_1) = \left.\frac{d}{d\lambda}\alpha(\lambda)\right|_{\lambda=\lambda_1}$$

Referring to Equation (9–105), this last equation becomes

$$f'(\lambda_1) = \alpha_1 + 2\alpha_2\lambda_1 + \cdots + (m - 1)\alpha_{m-1}\lambda_1^{m-2} \qquad (9\text{-}109)$$

Similarly, differentiating Equation (9–106) twice with respect to λ and substituting λ_1 for λ, we obtain

$$\left.\frac{d^2}{d^2\lambda}f(\lambda)\right|_{\lambda=\lambda_1} = f''(\lambda_1) = \left.\frac{d^2}{d\lambda^2}\alpha(\lambda)\right|_{\lambda=\lambda_1}$$

This last equation can be written as

$$f''(\lambda_1) = 2\alpha_2 + 6\alpha_3\lambda_1 + \cdots + (m - 1)(m - 2)\alpha_{m-1}\lambda_1^{m-3} \qquad (9\text{-}110)$$

Rewriting Equations (9–110), (9–109), and (9–107), we get

$$\alpha_2 + 3\alpha_3\lambda_1 + \cdots + \frac{(m-1)(m-2)}{2}\alpha_{m-1}\lambda_1^{m-3} = \frac{f''(\lambda_1)}{2}$$

$$\alpha_1 + 2\alpha_2\lambda_1 + \cdots + (m-1)\alpha_{m-1}\lambda_1^{m-2} = f'(\lambda_1)$$

$$\alpha_0 + \alpha_1\lambda_1 + \alpha_2\lambda_1^2 + \cdots + \alpha_{m-1}\lambda_1^{m-1} = f(\lambda_1)$$

$$\alpha_0 + \alpha_1\lambda_4 + \alpha_2\lambda_4^2 + \cdots + \alpha_{m-1}\lambda_4^{m-1} = f(\lambda_4) \qquad (9\text{-}111)$$

$$\cdot$$
$$\cdot$$
$$\cdot$$

$$\alpha_0 + \alpha_1\lambda_m + \alpha_2\lambda_m^2 + \cdots + \alpha_{m-1}\lambda_m^{m-1} = f(\lambda_m)$$

These m simultaneous equations determine the α_k values (where $k = 0, 1, 2, \ldots, m - 1$). Noting that $\phi(\mathbf{A}) = \mathbf{0}$ because it is a minimal polynomial, we have $f(\mathbf{A})$ as follows:

$$f(\mathbf{A}) = g(\mathbf{A})\phi(\mathbf{A}) + \alpha(\mathbf{A}) = \alpha(\mathbf{A})$$

Hence, referring to Equation (9–105), we have

$$f(\mathbf{A}) = \alpha(\mathbf{A}) = \alpha_0 \mathbf{I} + \alpha_1 \mathbf{A} + \alpha_2 \mathbf{A}^2 + \cdots + \alpha_{m-1}\mathbf{A}^{m-1} \tag{9–112}$$

where the α_k values are given in terms of $f(\lambda_1), f'(\lambda_1), f''(\lambda_1), f(\lambda_4), f(\lambda_5), \ldots, f(\lambda_m)$. In terms of the determinant equation, $f(\mathbf{A})$ can be obtained by solving the following equation:

$$\begin{vmatrix} 0 & 0 & 1 & 3\lambda_1 & \cdots & \dfrac{(m-1)(m-2)}{2}\lambda_1^{m-3} & \dfrac{f''(\lambda_1)}{2} \\ 0 & 1 & 2\lambda_1 & 3\lambda_1^2 & \cdots & (m-1)\lambda_1^{m-2} & f'(\lambda_1) \\ 1 & \lambda_1 & \lambda_1^2 & \lambda_1^3 & \cdots & \lambda_1^{m-1} & f(\lambda_1) \\ 1 & \lambda_4 & \lambda_4^2 & \lambda_4^3 & \cdots & \lambda_4^{m-1} & f(\lambda_4) \\ \cdot & \cdot & \cdot & \cdot & & \cdot & \cdot \\ \cdot & \cdot & \cdot & \cdot & & \cdot & \cdot \\ \cdot & \cdot & \cdot & \cdot & & \cdot & \cdot \\ 1 & \lambda_m & \lambda_m^2 & \lambda_m^3 & \cdots & \lambda_m^{m-1} & f(\lambda_m) \\ \mathbf{I} & \mathbf{A} & \mathbf{A}^2 & \mathbf{A}^3 & \cdots & \mathbf{A}^{m-1} & f(\mathbf{A}) \end{vmatrix} = \mathbf{0} \tag{9–113}$$

Equation (9–113) shows the desired modification in the form of the determinant. This equation gives the form of Sylvester's interpolation formula when the minimal polynomial of $\mathbf{A}$ involves three equal roots. (The necessary modification of the form of the determinant for other cases will be apparent.)

A–9–14. Using Sylvester's interpolation formula, compute $e^{\mathbf{A}t}$, where

$$\mathbf{A} = \begin{bmatrix} 2 & 1 & 4 \\ 0 & 2 & 0 \\ 0 & 3 & 1 \end{bmatrix}$$

Solution. Referring to Problem **A–9–9**, the characteristic polynomial and the minimal polynomial are the same for this **A**. The minimal polynomial (characteristic polynomial) is given by

$$\phi(\lambda) = (\lambda - 2)^2(\lambda - 1)$$

Note that $\lambda_1 = \lambda_2 = 2$ and $\lambda_3 = 1$. Referring to Equation (9–112) and noting that $f(\mathbf{A})$ in this problem is $e^{\mathbf{A}t}$, we have

$$e^{\mathbf{A}t} = \alpha_0(t)\mathbf{I} + \alpha_1(t)\mathbf{A} + \alpha_2(t)\mathbf{A}^2$$

where $\alpha_0(t), \alpha_1(t)$, and $\alpha_2(t)$ are determined from the equations

$$\alpha_1(t) + 2\alpha_2(t)\lambda_1 = te^{\lambda_1 t}$$

$$\alpha_0(t) + \alpha_1(t)\lambda_1 + \alpha_2(t)\lambda_1^2 = e^{\lambda_1 t}$$

$$\alpha_0(t) + \alpha_1(t)\lambda_3 + \alpha_2(t)\lambda_3^2 = e^{\lambda_3 t}$$

Substituting $\lambda_1 = 2$, and $\lambda_3 = 1$ into these three equations gives

$$\alpha_1(t) + 4\alpha_2(t) = te^{2t}$$

$$\alpha_0(t) + 2\alpha_1(t) + 4\alpha_2(t) = e^{2t}$$

$$\alpha_0(t) + \alpha_1(t) + \alpha_2(t) = e^t$$

Solving for $\alpha_0(t)$, $\alpha_1(t)$, and $\alpha_2(t)$, we obtain

$$\alpha_0(t) = 4e^t - 3e^{2t} + 2te^{2t}$$

$$\alpha_1(t) = -4e^t + 4e^{2t} - 3te^{2t}$$

$$\alpha_2(t) = e^t - e^{2t} + te^{2t}$$

Hence,

$$e^{\mathbf{A}t} = \left(4e^t - 3e^{2t} + 2te^{2t}\right)\begin{bmatrix} 1 & 0 & 0 \\ 0 & 1 & 0 \\ 0 & 0 & 1 \end{bmatrix} + \left(-4e^t + 4e^{2t} - 3te^{2t}\right)\begin{bmatrix} 2 & 1 & 4 \\ 0 & 2 & 0 \\ 0 & 3 & 1 \end{bmatrix}$$

$$+ \left(e^t - e^{2t} + te^{2t}\right)\begin{bmatrix} 4 & 16 & 12 \\ 0 & 4 & 0 \\ 0 & 9 & 1 \end{bmatrix}$$

$$= \begin{bmatrix} e^{2t} & 12e^t - 12e^{2t} + 13te^{2t} & -4e^t + 4e^{2t} \\ 0 & e^{2t} & 0 \\ 0 & -3e^t + 3e^{2t} & e^t \end{bmatrix}$$

A–9–15. Show that the system described by

$$\dot{\mathbf{x}} = \mathbf{A}\mathbf{x} + \mathbf{B}\mathbf{u} \tag{9–114}$$

$$\mathbf{y} = \mathbf{C}\mathbf{x} \tag{9–115}$$

where $\mathbf{x}$ = state vector (n-vector)

   $\mathbf{u}$ = control vector (r-vector)

   $\mathbf{y}$ = output vector (m-vector) $(m \leq n)$

   $\mathbf{A}$ = $n \times n$ matrix

   $\mathbf{B}$ = $n \times r$ matrix

   $\mathbf{C}$ = $m \times n$ matrix

is completely output controllable if and only if the composite $m \times nr$ matrix $\mathbf{P}$, where

$$\mathbf{P} = \begin{bmatrix} \mathbf{CB} & \vdots & \mathbf{CAB} & \vdots & \mathbf{CA^2B} & \vdots & \cdots & \vdots & \mathbf{CA}^{n-1}\mathbf{B} \end{bmatrix}$$

is of rank m. (Notice that complete state controllability is neither necessary nor sufficient for complete output controllability.)

Solution. Suppose that the system is output controllable and the output $\mathbf{y}(t)$ starting from any $\mathbf{y}(0)$, the initial output, can be transferred to the origin of the output space in a finite time interval $0 \leq t \leq T$. That is,

$$\mathbf{y}(T) = \mathbf{C}\mathbf{x}(T) = \mathbf{0} \tag{9–116}$$

Since the solution of Equation (9–114) is

$$\mathbf{x}(t) = e^{\mathbf{A}t}\left[\mathbf{x}(0) + \int_0^t e^{-\mathbf{A}\tau}\mathbf{B}\mathbf{u}(\tau)\,d\tau\right]$$

at $t = T$, we have

$$\mathbf{x}(T) = e^{\mathbf{A}T}\left[\mathbf{x}(0) + \int_0^T e^{-\mathbf{A}\tau}\mathbf{B}\mathbf{u}(\tau)\,d\tau\right] \qquad (9\text{–}117)$$

Substituting Equation (9–117) into Equation (9–116), we obtain

$$\mathbf{y}(T) = \mathbf{C}\mathbf{x}(T)$$

$$= \mathbf{C}e^{\mathbf{A}T}\left[\mathbf{x}(0) + \int_0^T e^{-\mathbf{A}\tau}\mathbf{B}\mathbf{u}(\tau)\,d\tau\right] = \mathbf{0} \qquad (9\text{–}118)$$

On the other hand, $\mathbf{y}(0) = \mathbf{C}\mathbf{x}(0)$. Notice that the complete output controllability means that the vector $\mathbf{C}\mathbf{x}(0)$ spans the m-dimensional output space. Since $e^{\mathbf{A}T}$ is nonsingular, if $\mathbf{C}\mathbf{x}(0)$ spans the m-dimensional output space, so does $\mathbf{C}e^{\mathbf{A}T}\mathbf{x}(0)$, and vice versa. From Equation (9–118) we obtain

$$\mathbf{C}e^{\mathbf{A}T}\mathbf{x}(0) = -\mathbf{C}e^{\mathbf{A}T}\int_0^T e^{-\mathbf{A}\tau}\mathbf{B}\mathbf{u}(\tau)\,d\tau$$

$$= -\mathbf{C}\int_0^T e^{\mathbf{A}\tau}\mathbf{B}\mathbf{u}(T-\tau)\,d\tau$$

Note that $\int_0^T e^{\mathbf{A}\tau}\mathbf{B}\mathbf{u}(T-\tau)\,d\tau$ can be expressed as the sum of $\mathbf{A}^i\mathbf{B}_j$; that is,

$$\int_0^T e^{\mathbf{A}\tau}\mathbf{B}\mathbf{u}(T-\tau)\,d\tau = \sum_{i=0}^{p-1}\sum_{j=1}^{r}\gamma_{ij}\mathbf{A}^i\mathbf{B}_j$$

where

$$\gamma_{ij} = \int_0^T \alpha_i(\tau)u_j(T-\tau)\,d\tau = \text{scalar}$$

and $\alpha_i(\tau)$ satisfies

$$e^{\mathbf{A}\tau} = \sum_{i=0}^{p-1}\alpha_i(\tau)\mathbf{A}^i \qquad (p\text{: degree of the minimal polynomial of }\mathbf{A})$$

and $\mathbf{B}_j$ is the jth column of $\mathbf{B}$. Therefore, we can write $\mathbf{C}e^{\mathbf{A}T}\mathbf{x}(0)$ as

$$\mathbf{C}e^{\mathbf{A}T}\mathbf{x}(0) = -\sum_{i=0}^{p-1}\sum_{j=1}^{r}\gamma_{ij}\mathbf{C}\mathbf{A}^i\mathbf{B}_j$$

From this last equation, we see that $\mathbf{C}e^{\mathbf{A}T}\mathbf{x}(0)$ is a linear combination of $\mathbf{C}\mathbf{A}^i\mathbf{B}_j$ ($i = 0, 1, 2, \ldots, p-1; j = 1, 2, \ldots, r$). Note that if the rank of $\mathbf{Q}$, where

$$\mathbf{Q} = \begin{bmatrix} \mathbf{C}\mathbf{B} & \vdots & \mathbf{C}\mathbf{A}\mathbf{B} & \vdots & \mathbf{C}\mathbf{A}^2\mathbf{B} & \vdots & \cdots & \vdots & \mathbf{C}\mathbf{A}^{p-1}\mathbf{B}\end{bmatrix} \qquad (p \le n)$$

is m, then so is the rank of $\mathbf{P}$, and vice versa. [This is obvious if $p = n$. If $p < n$, then the $\mathbf{C}\mathbf{A}^h\mathbf{B}_j$ (where $p \le h \le n-1$) are linearly dependent on $\mathbf{C}\mathbf{B}_j, \mathbf{C}\mathbf{A}\mathbf{B}_j, \ldots, \mathbf{C}\mathbf{A}^{p-1}\mathbf{B}_j$. Hence, the rank of

P is equal to that of **Q**.] If the rank of **P** is m, then $Ce^{A^T}\mathbf{x}(0)$ spans the m-dimensional output space. This means that if the rank of **P** is m, then $C\mathbf{x}(0)$ also spans the m-dimensional output space and the system is completely output controllable.

Conversely, suppose that the system is completely output controllable, but the rank of **P** is k, where $k < m$. Then the set of all initial outputs that can be transferred to the origin is of k-dimensional space. Hence, the dimension of this set is less than m. This contradicts the assumption that the system is completely output controllable. This completes the proof.

Note that it can be immediately proved that, in the system of Equations (9–114) and (9–115), complete state controllability on $0 \le t \le T$ implies complete output controllability on $0 \le t \le T$ if and only if m rows of **C** are linearly independent.

A–9–16.   Discuss the state controllability of the following system:

$$\begin{bmatrix} \dot{x}_1 \\ \dot{x}_2 \end{bmatrix} = \begin{bmatrix} -3 & 1 \\ -2 & 1.5 \end{bmatrix} \begin{bmatrix} x_1 \\ x_2 \end{bmatrix} + \begin{bmatrix} 1 \\ 4 \end{bmatrix} u \tag{9–119}$$

Solution. For this system,

$$\mathbf{A} = \begin{bmatrix} -3 & 1 \\ -2 & 1.5 \end{bmatrix}, \qquad \mathbf{B} = \begin{bmatrix} 1 \\ 4 \end{bmatrix}$$

Since

$$\mathbf{AB} = \begin{bmatrix} -3 & 1 \\ -2 & 1.5 \end{bmatrix} \begin{bmatrix} 1 \\ 4 \end{bmatrix} = \begin{bmatrix} 1 \\ 4 \end{bmatrix}$$

we see that vectors **B** and **AB** are not linearly independent and the rank of the matrix $[\mathbf{B} \vdots \mathbf{AB}]$ is 1. Therefore, the system is not completely state controllable. In fact, elimination of x_2 from Equation (9–119), or the following two simultaneous equations,

$$\dot{x}_1 = -3x_1 + x_2 + u$$

$$\dot{x}_2 = -2x_1 + 1.5x_2 + 4u$$

yields

$$\ddot{x}_1 + 1.5\dot{x}_1 - 2.5x_1 = \dot{u} + 2.5u$$

or, in the form of a transfer function,

$$\frac{X_1(s)}{U(s)} = \frac{s + 2.5}{(s + 2.5)(s - 1)}$$

Notice that cancellation of the factor $(s + 2.5)$ occurs in the numerator and denominator of the transfer function. Because of this cancellation, this system is not completely state controllable. This is an unstable system. Remember that stability and controllability are quite different things. There are many systems that are unstable, but are completely state controllable.

A–9–17.   A state-space representation of a system in the controllable canonical form is given by

$$\begin{bmatrix} \dot{x}_1 \\ \dot{x}_2 \end{bmatrix} = \begin{bmatrix} 0 & 1 \\ -0.4 & -1.3 \end{bmatrix} \begin{bmatrix} x_1 \\ x_2 \end{bmatrix} + \begin{bmatrix} 0 \\ 1 \end{bmatrix} u \tag{9–120}$$

$$y = \begin{bmatrix} 0.8 & 1 \end{bmatrix} \begin{bmatrix} x_1 \\ x_2 \end{bmatrix} \tag{9–121}$$

The same system may be represented by the following state-space equation, which is in the observable canonical form:

$$\begin{bmatrix} \dot{x}_1 \\ \dot{x}_2 \end{bmatrix} = \begin{bmatrix} 0 & -0.4 \\ 1 & -1.3 \end{bmatrix} \begin{bmatrix} x_1 \\ x_2 \end{bmatrix} + \begin{bmatrix} 0.8 \\ 1 \end{bmatrix} u \qquad (9\text{--}122)$$

$$y = \begin{bmatrix} 0 & 1 \end{bmatrix} \begin{bmatrix} x_1 \\ x_2 \end{bmatrix} \qquad (9\text{--}123)$$

Show that the state-space representation given by Equations (9–120) and (9–121) gives a system that is state controllable, but not observable. Show, on the other hand, that the state-space representation defined by Equations (9–122) and (9–123) gives a system that is not completely state controllable, but is observable. Explain what causes the apparent difference in the controllability and observability of the same system.

Solution. Consider the system defined by Equations (9–120) and (9–121). The rank of the controllability matrix

$$\begin{bmatrix} \mathbf{B} & \vdots & \mathbf{AB} \end{bmatrix} = \begin{bmatrix} 0 & 1 \\ 1 & -1.3 \end{bmatrix}$$

is 2. Hence, the system is completely state controllable. The rank of the observability matrix

$$\begin{bmatrix} \mathbf{C}^* & \vdots & \mathbf{A}^*\mathbf{C}^* \end{bmatrix} = \begin{bmatrix} 0.8 & -0.4 \\ 1 & -0.5 \end{bmatrix}$$

is 1. Hence the system is not observable.

Next consider the system defined by Equations (9–122) and (9–123). The rank of the controllability matrix

$$\begin{bmatrix} \mathbf{B} & \vdots & \mathbf{AB} \end{bmatrix} = \begin{bmatrix} 0.8 & -0.4 \\ 1 & -0.5 \end{bmatrix}$$

is 1. Hence, the system is not completely state controllable. The rank of the observability matrix

$$\begin{bmatrix} \mathbf{C}^* & \vdots & \mathbf{A}^*\mathbf{C}^* \end{bmatrix} = \begin{bmatrix} 0 & 1 \\ 1 & -1.3 \end{bmatrix}$$

is 2. Hence, the system is observable.

The apparent difference in the controllability and observability of the same system is caused by the fact that the original system has a pole-zero cancellation in the transfer function. Referring to Equation (2–29), for $D = 0$ we have

$$G(s) = \mathbf{C}(s\mathbf{I} - \mathbf{A})^{-1}\mathbf{B}$$

If we use Equations (9–120) and (9–121), then

$$G(s) = \begin{bmatrix} 0.8 & 1 \end{bmatrix} \begin{bmatrix} s & -1 \\ 0.4 & s + 1.3 \end{bmatrix}^{-1} \begin{bmatrix} 0 \\ 1 \end{bmatrix}$$

$$= \frac{1}{s^2 + 1.3s + 0.4} \begin{bmatrix} 0.8 & 1 \end{bmatrix} \begin{bmatrix} s + 1.3 & 1 \\ -0.4 & s \end{bmatrix} \begin{bmatrix} 0 \\ 1 \end{bmatrix}$$

$$= \frac{s + 0.8}{(s + 0.8)(s + 0.5)}$$

[Note that the same transfer function can be obtained by using Equations (9–122) and (9–123).] Clearly, cancellation occurs in this transfer function.

Example Problems and Solutions

If a pole-zero cancellation occurs in the transfer function, then the controllability and observability vary, depending on how the state variables are chosen. Remember that, to be completely state controllable and observable, the transfer function must not have any pole-zero cancellations.

A–9–18. Prove that the system defined by

$$\dot{\mathbf{x}} = \mathbf{A}\mathbf{x}$$
$$\mathbf{y} = \mathbf{C}\mathbf{x}$$

where $\mathbf{x}$ = state vector (n-vector)

$\mathbf{y}$ = output vector (m-vector) $(m \leq n)$

$\mathbf{A} = n \times n$ matrix

$\mathbf{C} = m \times n$ matrix

is completely observable if and only if the composite $mn \times n$ matrix $\mathbf{P}$, where

$$\mathbf{P} = \begin{bmatrix} \mathbf{C} \\ \mathbf{CA} \\ \cdot \\ \cdot \\ \cdot \\ \mathbf{CA}^{n-1} \end{bmatrix}$$

is of rank n.

Solution. We shall first obtain the necessary condition. Suppose that

$$\text{rank } \mathbf{P} < n$$

Then there exists $\mathbf{x}(0)$ such that

$$\mathbf{P}\mathbf{x}(0) = \mathbf{0}$$

or

$$\mathbf{P}\mathbf{x}(0) = \begin{bmatrix} \mathbf{C} \\ \mathbf{CA} \\ \cdot \\ \cdot \\ \cdot \\ \mathbf{CA}^{n-1} \end{bmatrix} \mathbf{x}(0) = \begin{bmatrix} \mathbf{C}\mathbf{x}(0) \\ \mathbf{CA}\mathbf{x}(0) \\ \cdot \\ \cdot \\ \cdot \\ \mathbf{CA}^{n-1}\mathbf{x}(0) \end{bmatrix} = \mathbf{0}$$

Hence, we obtain, for a certain $\mathbf{x}(0)$,

$$\mathbf{CA}^i\mathbf{x}(0) = \mathbf{0}, \qquad \text{for } i = 0, 1, 2, \dots, n-1$$

Notice that from Equation (9–48) or (9–50), we have

$$e^{\mathbf{A}t} = \alpha_0(t)\mathbf{I} + \alpha_1(t)\mathbf{A} + \alpha_2(t)\mathbf{A}^2 + \cdots + \alpha_{m-1}(t)\mathbf{A}^{m-1}$$

where $m(m \leq n)$ is the degree of the minimal polynomial for $\mathbf{A}$. Hence, for a certain $\mathbf{x}(0)$, we have

$$\mathbf{C}e^{\mathbf{A}t}\mathbf{x}(0) = \mathbf{C}\big[\alpha_0(t)\mathbf{I} + \alpha_1(t)\mathbf{A} + \alpha_2(t)\mathbf{A}^2 + \cdots + \alpha_{m-1}(t)\mathbf{A}^{m-1}\big]\mathbf{x}(0) = \mathbf{0}$$

Consequently, for a certain $\mathbf{x}(0)$,

$$\mathbf{y}(t) = \mathbf{C}\mathbf{x}(t) = \mathbf{C}e^{\mathbf{A}t}\mathbf{x}(0) = \mathbf{0}$$

which implies that, for a certain $\mathbf{x}(0)$, $\mathbf{x}(0)$ cannot be determined from $\mathbf{y}(t)$. Therefore, the rank of matrix $\mathbf{P}$ must be equal to n.

Next we shall obtain the sufficient condition. Suppose that rank $\mathbf{P} = n$. Since

$$\mathbf{y}(t) = \mathbf{C}e^{\mathbf{A}t}\mathbf{x}(0)$$

by premultiplying both sides of this last equation by $e^{\mathbf{A}^*t}\mathbf{C}^*$, we get

$$e^{\mathbf{A}^*t}\mathbf{C}^*\mathbf{y}(t) = e^{\mathbf{A}^*t}\mathbf{C}^*\mathbf{C}e^{\mathbf{A}t}\mathbf{x}(0)$$

If we integrate this last equation from 0 to t, we obtain

$$\int_0^t e^{\mathbf{A}^*t}\mathbf{C}^*\mathbf{y}(t)\,dt = \int_0^t e^{\mathbf{A}^*t}\mathbf{C}^*\mathbf{C}e^{\mathbf{A}t}\mathbf{x}(0)\,dt \tag{9-124}$$

Notice that the left-hand side of this equation is a known quantity. Define

$$\mathbf{Q}(t) = \int_0^t e^{\mathbf{A}^*t}\mathbf{C}^*\mathbf{y}(t)\,dt = \text{known quantity} \tag{9-125}$$

Then, from Equations (9–124) and (9–125), we have

$$\mathbf{Q}(t) = \mathbf{W}(t)\mathbf{x}(0) \tag{9-126}$$

where

$$\mathbf{W}(t) = \int_0^t e^{\mathbf{A}^*\tau}\mathbf{C}^*\mathbf{C}e^{\mathbf{A}\tau}\,d\tau$$

It can be established that $\mathbf{W}(t)$ is a nonsingular matrix as follows: If $|\mathbf{W}(t)|$ were equal to 0, then

$$\mathbf{x}^*\mathbf{W}(t_1)\mathbf{x} = \int_0^{t_1}\|\mathbf{C}e^{\mathbf{A}t}\mathbf{x}\|^2\,dt = 0$$

which means that

$$\mathbf{C}e^{\mathbf{A}t}\mathbf{x} = \mathbf{0}, \qquad \text{for } 0 \le t \le t_1$$

which implies that rank $\mathbf{P} < n$. Therefore, $|\mathbf{W}(t)| \ne 0$, or $\mathbf{W}(t)$ is nonsingular. Then, from Equation (9–126), we obtain

$$\mathbf{x}(0) = [\mathbf{W}(t)]^{-1}\mathbf{Q}(t) \tag{9-127}$$

and $\mathbf{x}(0)$ can be determined from Equation (9–127).

Hence, we have proved that $\mathbf{x}(0)$ can be determined from $\mathbf{y}(t)$ if and only if rank $\mathbf{P} = n$. Note that $\mathbf{x}(0)$ and $\mathbf{y}(t)$ are related by

$$\mathbf{y}(t) = \mathbf{C}e^{\mathbf{A}t}\mathbf{x}(0) = \alpha_0(t)\mathbf{C}\mathbf{x}(0) + \alpha_1(t)\mathbf{C}\mathbf{A}\mathbf{x}(0) + \cdots + \alpha_{n-1}(t)\mathbf{C}\mathbf{A}^{n-1}\mathbf{x}(0)$$

Example Problems and Solutions

PROBLEMS

B–9–1. Consider the following transfer-function system:

$$\frac{Y(s)}{U(s)} = \frac{s + 6}{s^2 + 5s + 6}$$

Obtain the state-space representation of this system in (a) controllable canonical form and (b) observable canonical form.

B–9–2. Consider the following system:

$$\dddot{y} + 6\ddot{y} + 11\dot{y} + 6y = 6u$$

Obtain a state-space representation of this system in a diagonal canonical form.

B–9–3. Consider the system defined by

$$\dot{x} = Ax + Bu$$

$$y = Cx$$

where

$$A = \begin{bmatrix} 1 & 2 \\ -4 & -3 \end{bmatrix}, \quad B = \begin{bmatrix} 1 \\ 2 \end{bmatrix}, \quad C = \begin{bmatrix} 1 & 1 \end{bmatrix}$$

Transform the system equations into the controllable canonical form.

B–9–4. Consider the system defined by

$$\dot{x} = Ax + Bu$$

$$y = Cx$$

where

$$A = \begin{bmatrix} -1 & 0 & 1 \\ 1 & -2 & 0 \\ 0 & 0 & -3 \end{bmatrix}, \quad B = \begin{bmatrix} 0 \\ 0 \\ 1 \end{bmatrix}, \quad C = \begin{bmatrix} 1 & 1 & 0 \end{bmatrix}$$

Obtain the transfer function $Y(s)/U(s)$.

B–9–5. Consider the following matrix A:

$$A = \begin{bmatrix} 0 & 1 & 0 & 0 \\ 0 & 0 & 1 & 0 \\ 0 & 0 & 0 & 1 \\ 1 & 0 & 0 & 0 \end{bmatrix}$$

Obtain the eigenvalues $\lambda_1, \lambda_2, \lambda_3,$ and λ_4 of the matrix A. Then obtain a transformation matrix P such that

$$P^{-1}AP = \text{diag}(\lambda_1, \lambda_2, \lambda_3, \lambda_4)$$

B–9–6. Consider the following matrix A:

$$A = \begin{bmatrix} 0 & 1 \\ -2 & -3 \end{bmatrix}$$

Compute e^{At} by three methods.

B–9–7. Given the system equation

$$\begin{bmatrix} \dot{x}_1 \\ \dot{x}_2 \\ \dot{x}_3 \end{bmatrix} = \begin{bmatrix} 2 & 1 & 0 \\ 0 & 2 & 1 \\ 0 & 0 & 2 \end{bmatrix} \begin{bmatrix} x_1 \\ x_2 \\ x_3 \end{bmatrix}$$

find the solution in terms of the initial conditions $x_1(0)$, $x_2(0),$ and $x_3(0)$.

B–9–8. Find $x_1(t)$ and $x_2(t)$ of the system described by

$$\begin{bmatrix} \dot{x}_1 \\ \dot{x}_2 \end{bmatrix} = \begin{bmatrix} 0 & 1 \\ -3 & -2 \end{bmatrix} \begin{bmatrix} x_1 \\ x_2 \end{bmatrix}$$

where the initial conditions are

$$\begin{bmatrix} x_1(0) \\ x_2(0) \end{bmatrix} = \begin{bmatrix} 1 \\ -1 \end{bmatrix}$$

B–9–9. Consider the following state equation and output equation:

$$\begin{bmatrix} \dot{x}_1 \\ \dot{x}_2 \\ \dot{x}_3 \end{bmatrix} = \begin{bmatrix} -6 & 1 & 0 \\ -11 & 0 & 1 \\ -6 & 0 & 0 \end{bmatrix} \begin{bmatrix} x_1 \\ x_2 \\ x_3 \end{bmatrix} + \begin{bmatrix} 2 \\ 6 \\ 2 \end{bmatrix} u$$

$$y = \begin{bmatrix} 1 & 0 & 0 \end{bmatrix} \begin{bmatrix} x_1 \\ x_2 \\ x_3 \end{bmatrix}$$

Show that the state equation can be transformed into the following form by use of a proper transformation matrix:

$$\begin{bmatrix} \dot{z}_1 \\ \dot{z}_2 \\ \dot{z}_3 \end{bmatrix} = \begin{bmatrix} 0 & 0 & -6 \\ 1 & 0 & -11 \\ 0 & 1 & -6 \end{bmatrix} \begin{bmatrix} z_1 \\ z_2 \\ z_3 \end{bmatrix} + \begin{bmatrix} 1 \\ 0 \\ 0 \end{bmatrix} u$$

Then obtain the output y in terms of $z_1, z_2,$ and z_3.

B–9–10. Obtain a state-space representation of the following system with MATLAB:

$$\frac{Y(s)}{U(s)} = \frac{10.4s^2 + 47s + 160}{s^3 + 14s^2 + 56s + 160}$$

B–9–11. Obtain a transfer-function representation of the following system with MATLAB:

$$\begin{bmatrix} \dot{x}_1 \\ \dot{x}_2 \\ \dot{x}_3 \end{bmatrix} = \begin{bmatrix} 0 & 1 & 0 \\ -1 & -1 & 0 \\ 1 & 0 & 0 \end{bmatrix} \begin{bmatrix} x_1 \\ x_2 \\ x_3 \end{bmatrix} + \begin{bmatrix} 0 \\ 1 \\ 0 \end{bmatrix} u$$

$$y = \begin{bmatrix} 0 & 0 & 1 \end{bmatrix} \begin{bmatrix} x_1 \\ x_2 \\ x_3 \end{bmatrix}$$

B–9–12. Obtain a transfer-function representation of the following system with MATLAB:

$$\begin{bmatrix} \dot{x}_1 \\ \dot{x}_2 \\ \dot{x}_3 \end{bmatrix} = \begin{bmatrix} 2 & 1 & 0 \\ 0 & 2 & 0 \\ 0 & 1 & 3 \end{bmatrix} \begin{bmatrix} x_1 \\ x_2 \\ x_3 \end{bmatrix} + \begin{bmatrix} 0 & 1 \\ 1 & 0 \\ 0 & 1 \end{bmatrix} \begin{bmatrix} u_1 \\ u_2 \end{bmatrix}$$

$$y = \begin{bmatrix} 1 & 0 & 0 \end{bmatrix} \begin{bmatrix} x_1 \\ x_2 \\ x_3 \end{bmatrix}$$

B–9–13. Consider the system defined by

$$\begin{bmatrix} \dot{x}_1 \\ \dot{x}_2 \\ \dot{x}_3 \end{bmatrix} = \begin{bmatrix} -1 & -2 & -2 \\ 0 & -1 & 1 \\ 1 & 0 & -1 \end{bmatrix} \begin{bmatrix} x_1 \\ x_2 \\ x_3 \end{bmatrix} + \begin{bmatrix} 2 \\ 0 \\ 1 \end{bmatrix} u$$

$$y = \begin{bmatrix} 1 & 1 & 0 \end{bmatrix} \begin{bmatrix} x_1 \\ x_2 \\ x_3 \end{bmatrix}$$

Is the system completely state controllable and completely observable?

B–9–14. Consider the system given by

$$\begin{bmatrix} \dot{x}_1 \\ \dot{x}_2 \\ \dot{x}_3 \end{bmatrix} = \begin{bmatrix} 2 & 0 & 0 \\ 0 & 2 & 0 \\ 0 & 3 & 1 \end{bmatrix} \begin{bmatrix} x_1 \\ x_2 \\ x_3 \end{bmatrix} + \begin{bmatrix} 0 & 1 \\ 1 & 0 \\ 0 & 1 \end{bmatrix} \begin{bmatrix} u_1 \\ u_2 \end{bmatrix}$$

$$\begin{bmatrix} y_1 \\ y_2 \end{bmatrix} = \begin{bmatrix} 1 & 0 & 0 \\ 0 & 1 & 0 \end{bmatrix} \begin{bmatrix} x_1 \\ x_2 \\ x_3 \end{bmatrix}$$

Is the system completely state controllable and completely observable? Is the system completely output controllable?

B–9–15. Is the following system completely state controllable and completely observable?

$$\begin{bmatrix} \dot{x}_1 \\ \dot{x}_2 \\ \dot{x}_3 \end{bmatrix} = \begin{bmatrix} 0 & 1 & 0 \\ 0 & 0 & 1 \\ -6 & -11 & -6 \end{bmatrix} \begin{bmatrix} x_1 \\ x_2 \\ x_3 \end{bmatrix} + \begin{bmatrix} 0 \\ 0 \\ 1 \end{bmatrix} u$$

$$y = \begin{bmatrix} 20 & 9 & 1 \end{bmatrix} \begin{bmatrix} x_1 \\ x_2 \\ x_3 \end{bmatrix}$$

B–9–16. Consider the system defined by

$$\begin{bmatrix} \dot{x}_1 \\ \dot{x}_2 \\ \dot{x}_3 \end{bmatrix} = \begin{bmatrix} 0 & 1 & 0 \\ 0 & 0 & 1 \\ -6 & -11 & -6 \end{bmatrix} \begin{bmatrix} x_1 \\ x_2 \\ x_3 \end{bmatrix} + \begin{bmatrix} 0 \\ 0 \\ 1 \end{bmatrix} u$$

$$y = \begin{bmatrix} c_1 & c_2 & c_3 \end{bmatrix} \begin{bmatrix} x_1 \\ x_2 \\ x_3 \end{bmatrix}$$

Except for an obvious choice of $c_1 = c_2 = c_3 = 0$, find an example of a set of c_1, c_2, c_3 that will make the system unobservable.

B–9–17. Consider the system

$$\begin{bmatrix} \dot{x}_1 \\ \dot{x}_2 \\ \dot{x}_3 \end{bmatrix} = \begin{bmatrix} 2 & 0 & 0 \\ 0 & 2 & 0 \\ 0 & 3 & 1 \end{bmatrix} \begin{bmatrix} x_1 \\ x_2 \\ x_3 \end{bmatrix}$$

The output is given by

$$y = \begin{bmatrix} 1 & 1 & 1 \end{bmatrix} \begin{bmatrix} x_1 \\ x_2 \\ x_3 \end{bmatrix}$$

(a) Show that the system is not completely observable.

(b) Show that the system is completely observable if the output is given by

$$\begin{bmatrix} y_1 \\ y_2 \end{bmatrix} = \begin{bmatrix} 1 & 1 & 1 \\ 1 & 2 & 3 \end{bmatrix} \begin{bmatrix} x_1 \\ x_2 \\ x_3 \end{bmatrix}$$

Control Systems Design in State Space

10–1 INTRODUCTION

This chapter discusses state-space design methods based on the pole-placement method, observers, the quadratic optimal regulator systems, and introductory aspects of robust control systems. The pole-placement method is somewhat similar to the root-locus method in that we place closed-loop poles at desired locations. The basic difference is that in the root-locus design we place only the dominant closed-loop poles at the desired locations, while in the pole-placement design we place all closed-loop poles at desired locations.

We begin by presenting the basic materials on pole placement in regulator systems. We then discuss the design of state observers, followed by the design of regulator systems and control systems using the pole-placement-with-state-observer approach. Then, we discuss the quadratic optimal regulator systems. Finally, we present an introduction to robust control systems.

Outline of the Chapter.   Section 10–1 has presented introductory material. Section 10–2 discusses the pole-placement approach to the design of control systems. We begin with the derivation of the necessary and sufficient conditions for arbitrary pole placement. Then we derive equations for the state feedback gain matrix **K** for pole placement. Section 10–3 presents the solution of the pole-placement problem with MATLAB. Section 10–4 discusses the design of servo systems using the pole-placement approach. Section 10–5 presents state observers. We discuss both full-order and minimum-order state observers. Also, transfer functions of observer controllers are derived. Section 10–6 presents the design of regulator systems with observers. Section 10–7 treats the design of control

722

systems with observers. Section 10–8 discusses quadratic optimal regulator systems. Note that the state feedback gain matrix **K** can be obtained by both the pole-placement method and the quadratic optimal control method. Finally, Section 10–9 presents robust control systems. The discussions here are limited to introductory subjects only.

10–2 POLE PLACEMENT

In this section we shall present a design method commonly called the *pole-placement* or *pole-assignment technique*. We assume that all state variables are measurable and are available for feedback. It will be shown that if the system considered is completely state controllable, then poles of the closed-loop system may be placed at any desired locations by means of state feedback through an appropriate state feedback gain matrix.

The present design technique begins with a determination of the desired closed-loop poles based on the transient-response and/or frequency-response requirements, such as speed, damping ratio, or bandwidth, as well as steady-state requirements.

Let us assume that we decide that the desired closed-loop poles are to be at $s = \mu_1$, $s = \mu_2, \ldots, s = \mu_n$. By choosing an appropriate gain matrix for state feedback, it is possible to force the system to have closed-loop poles at the desired locations, provided that the original system is completely state controllable.

In this chapter we limit our discussions to single-input, single-output systems. That is, we assume the control signal $u(t)$ and output signal $y(t)$ to be scalars. In the derivation in this section we assume that the reference input $r(t)$ is zero. [In Section 10–7 we discuss the case where the reference input $r(t)$ is nonzero.]

In what follows we shall prove that a necessary and sufficient condition that the closed-loop poles can be placed at any arbitrary locations in the s plane is that the system be completely state controllable. Then we shall discuss methods for determining the required state feedback gain matrix.

It is noted that when the control signal is a vector quantity, the mathematical aspects of the pole-placement scheme become complicated. We shall not discuss such a case in this book. (When the control signal is a vector quantity, the state feedback gain matrix is not unique. It is possible to choose freely more than n parameters; that is, in addition to being able to place n closed-loop poles properly, we have the freedom to satisfy some or all of the other requirements, if any, of the closed-loop system.)

Design by Pole Placement. In the conventional approach to the design of a single-input, single-output control system, we design a controller (compensator) such that the dominant closed-loop poles have a desired damping ratio ζ and a desired undamped natural frequency ω_n. In this approach, the order of the system may be raised by 1 or 2 unless pole–zero cancellation takes place. Note that in this approach we assume the effects on the responses of nondominant closed-loop poles to be negligible.

Different from specifying only dominant closed-loop poles (the conventional design approach), the present pole-placement approach specifies all closed-loop poles. (There is a cost associated with placing all closed-loop poles, however, because placing all closed-loop poles requires successful measurements of all state variables or else requires the inclusion of a state observer in the system.) There is also a requirement on the part of the system for the closed-loop poles to be placed at arbitrarily chosen locations. The requirement is that the system be completely state controllable. We shall prove this fact in this section.

Consider a control system

$$\dot{\mathbf{x}} = \mathbf{A}\mathbf{x} + \mathbf{B}u \qquad (10\text{--}1)$$

$$y = \mathbf{C}\mathbf{x} + Du$$

where $\mathbf{x}$ = state vector (n-vector)
  y = output signal (scalar)
  u = control signal (scalar)
  $\mathbf{A} = n \times n$ constant matrix
  $\mathbf{B} = n \times 1$ constant matrix
  $\mathbf{C} = 1 \times n$ constant matrix
  D = constant (scalar)

We shall choose the control signal to be

$$u = -\mathbf{K}\mathbf{x} \qquad (10\text{--}2)$$

This means that the control signal u is determined by an instantaneous state. Such a scheme is called state feedback. The $1 \times n$ matrix $\mathbf{K}$ is called the state feedback gain matrix. We assume that all state variables are available for feedback. In the following analysis we assume that u is unconstrained. A block diagram for this system is shown in Figure 10–1.

This closed-loop system has no input. Its objective is to maintain the zero output. Because of the disturbances that may be present, the output will deviate from zero. The nonzero output will be returned to the zero reference input because of the state feedback scheme of the system. Such a system where the reference input is always zero is called a regulator system. (Note that if the reference input to the system is always a nonzero constant, the system is also called a regulator system.)

Substituting Equation (10–2) into Equation (10–1) gives

$$\dot{\mathbf{x}}(t) = (\mathbf{A} - \mathbf{B}\mathbf{K})\mathbf{x}(t)$$

The solution of this equation is given by

$$\mathbf{x}(t) = e^{(\mathbf{A}-\mathbf{B}\mathbf{K})t}\mathbf{x}(0) \qquad (10\text{--}3)$$

where $\mathbf{x}(0)$ is the initial state caused by external disturbances. The stability and transient-response characteristics are determined by the eigenvalues of matrix $\mathbf{A} - \mathbf{B}\mathbf{K}$. If matrix

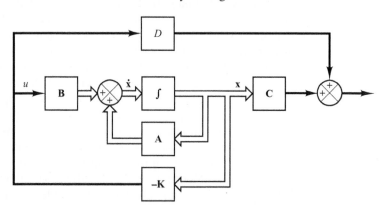

Figure 10–1
Closed-loop control system with $u = -\mathbf{K}\mathbf{x}$.

K is chosen properly, the matrix $\mathbf{A} - \mathbf{BK}$ can be made an asymptotically stable matrix, and for all $\mathbf{x}(0) \neq \mathbf{0}$, it is possible to make $\mathbf{x}(t)$ approach $\mathbf{0}$ as t approaches infinity. The eigenvalues of matrix $\mathbf{A} - \mathbf{BK}$ are called the regulator poles. If these regulator poles are placed in the left-half s plane, then $\mathbf{x}(t)$ approaches $\mathbf{0}$ as t approaches infinity. The problem of placing the regulator poles (closed-loop poles) at the desired location is called a pole-placement problem.

In what follows, we shall prove that arbitrary pole placement for a given system is possible if and only if the system is completely state controllable.

Necessary and Sufficient Condition for Arbitrary Pole Placement   We shall now prove that a necessary and sufficient condition for arbitrary pole placement is that the system be completely state controllable. We shall first derive the necessary condition. We begin by proving that if the system is not completely state controllable, then there are eigenvalues of matrix $\mathbf{A} - \mathbf{BK}$ that cannot be controlled by state feedback.

Suppose that the system of Equation (10–1) is not completely state controllable. Then the rank of the controllability matrix is less than n, or

$$\text{rank}\begin{bmatrix} \mathbf{B} & \vdots & \mathbf{AB} & \vdots & \cdots & \vdots & \mathbf{A}^{n-1}\mathbf{B} \end{bmatrix} = q < n$$

This means that there are q linearly independent column vectors in the controllability matrix. Let us define such q linearly independent column vectors as $\mathbf{f}_1, \mathbf{f}_2, \ldots, \mathbf{f}_q$. Also, let us choose $n - q$ additional n-vectors $\mathbf{v}_{q+1}, \mathbf{v}_{q+2}, \ldots, \mathbf{v}_n$ such that

$$\mathbf{P} = \begin{bmatrix} \mathbf{f}_1 & \vdots & \mathbf{f}_2 & \vdots & \cdots & \vdots & \mathbf{f}_q & \vdots & \mathbf{v}_{q+1} & \vdots & \mathbf{v}_{q+2} & \vdots & \cdots & \vdots & \mathbf{v}_n \end{bmatrix}$$

is of rank n. Then it can be shown that

$$\hat{\mathbf{A}} = \mathbf{P}^{-1}\mathbf{A}\mathbf{P} = \begin{bmatrix} \mathbf{A}_{11} & \mathbf{A}_{12} \\ \hline \mathbf{0} & \mathbf{A}_{22} \end{bmatrix}, \qquad \hat{\mathbf{B}} = \mathbf{P}^{-1}\mathbf{B} = \begin{bmatrix} \mathbf{B}_{11} \\ \hline \mathbf{0} \end{bmatrix}$$

(See Problem **A–10–1** for the derivation of these equations.) Now define

$$\hat{\mathbf{K}} = \mathbf{KP} = \begin{bmatrix} \mathbf{k}_1 & \vdots & \mathbf{k}_2 \end{bmatrix}$$

Then we have

$$\begin{aligned}
|s\mathbf{I} - \mathbf{A} + \mathbf{BK}| &= |\mathbf{P}^{-1}(s\mathbf{I} - \mathbf{A} + \mathbf{BK})\mathbf{P}| \\
&= |s\mathbf{I} - \mathbf{P}^{-1}\mathbf{A}\mathbf{P} + \mathbf{P}^{-1}\mathbf{BKP}| \\
&= |s\mathbf{I} - \hat{\mathbf{A}} + \hat{\mathbf{B}}\hat{\mathbf{K}}| \\
&= \left| s\mathbf{I} - \begin{bmatrix} \mathbf{A}_{11} & \mathbf{A}_{12} \\ \hline \mathbf{0} & \mathbf{A}_{22} \end{bmatrix} + \begin{bmatrix} \mathbf{B}_{11} \\ \hline \mathbf{0} \end{bmatrix}\begin{bmatrix} \mathbf{k}_1 & \vdots & \mathbf{k}_2 \end{bmatrix} \right| \\
&= \begin{vmatrix} s\mathbf{I}_q - \mathbf{A}_{11} + \mathbf{B}_{11}\mathbf{k}_1 & -\mathbf{A}_{12} + \mathbf{B}_{11}\mathbf{k}_2 \\ \mathbf{0} & s\mathbf{I}_{n-q} - \mathbf{A}_{22} \end{vmatrix} \\
&= |s\mathbf{I}_q - \mathbf{A}_{11} + \mathbf{B}_{11}\mathbf{k}_1| \cdot |s\mathbf{I}_{n-q} - \mathbf{A}_{22}| = 0
\end{aligned}$$

where $\mathbf{I}_q$ is a q-dimensional identity matrix and $\mathbf{I}_{n-q}$ is an $(n - q)$-dimensional identity matrix.

Notice that the eigenvalues of $\mathbf{A}_{22}$ do not depend on $\mathbf{K}$. Thus, if the system is not completely state controllable, then there are eigenvalues of matrix $\mathbf{A}$ that cannot be arbitrarily placed. Therefore, to place the eigenvalues of matrix $\mathbf{A} - \mathbf{BK}$ arbitrarily, the system must be completely state controllable (necessary condition).

Next we shall prove a sufficient condition: that is, if the system is completely state controllable, then all eigenvalues of matrix $\mathbf{A}$ can be arbitrarily placed.

In proving a sufficient condition, it is convenient to transform the state equation given by Equation (10–1) into the controllable canonical form.

Define a transformation matrix $\mathbf{T}$ by

$$\mathbf{T} = \mathbf{MW} \tag{10–4}$$

where $\mathbf{M}$ is the controllability matrix

$$\mathbf{M} = \begin{bmatrix} \mathbf{B} & \vdots & \mathbf{AB} & \vdots & \cdots & \vdots & \mathbf{A}^{n-1}\mathbf{B} \end{bmatrix} \tag{10–5}$$

and

$$\mathbf{W} = \begin{bmatrix} a_{n-1} & a_{n-2} & \cdots & a_1 & 1 \\ a_{n-2} & a_{n-3} & \cdots & 1 & 0 \\ \cdot & \cdot & & \cdot & \cdot \\ \cdot & \cdot & & \cdot & \cdot \\ \cdot & \cdot & & \cdot & \cdot \\ a_1 & 1 & \cdots & 0 & 0 \\ 1 & 0 & \cdots & 0 & 0 \end{bmatrix} \tag{10–6}$$

where the a_i's are coefficients of the characteristic polynomial

$$|s\mathbf{I} - \mathbf{A}| = s^n + a_1 s^{n-1} + \cdots + a_{n-1}s + a_n$$

Define a new state vector $\hat{\mathbf{x}}$ by

$$\mathbf{x} = \mathbf{T}\hat{\mathbf{x}}$$

If the rank of the controllability matrix $\mathbf{M}$ is n (meaning that the system is completely state controllable), then the inverse of matrix $\mathbf{T}$ exists, and Equation (10–1) can be modified to

$$\dot{\hat{\mathbf{x}}} = \mathbf{T}^{-1}\mathbf{AT}\hat{x} + \mathbf{T}^{-1}\mathbf{B}u \tag{10–7}$$

where

$$\mathbf{T}^{-1}\mathbf{AT} = \begin{bmatrix} 0 & 1 & 0 & \cdots & 0 \\ 0 & 0 & 1 & \cdots & 0 \\ \cdot & \cdot & \cdot & & \cdot \\ \cdot & \cdot & \cdot & & \cdot \\ \cdot & \cdot & \cdot & & \cdot \\ 0 & 0 & 0 & \cdots & 1 \\ -a_n & -a_{n-1} & -a_{n-2} & \cdots & -a_1 \end{bmatrix} \tag{10–8}$$

$$\mathbf{T^{-1}B} = \begin{bmatrix} 0 \\ 0 \\ \cdot \\ \cdot \\ \cdot \\ 0 \\ 1 \end{bmatrix} \qquad (10\text{–}9)$$

[See Problems **A–10–2** and **A–10–3** for the derivation of Equations (10–8) and (10–9).] Equation (10–7) is in the controllable canonical form. Thus, given a state equation, Equation (10–1), it can be transformed into the controllable canonical form if the system is completely state controllable and if we transform the state vector $\mathbf{x}$ into state vector $\hat{\mathbf{x}}$ by use of the transformation matrix $\mathbf{T}$ given by Equation (10–4).

Let us choose a set of the desired eigenvalues as $\mu_1, \mu_2, \ldots, \mu_n$. Then the desired characteristic equation becomes

$$(s - \mu_1)(s - \mu_2)\cdots(s - \mu_n) = s^n + \alpha_1 s^{n-1} + \cdots + \alpha_{n-1}s + \alpha_n = 0 \quad (10\text{–}10)$$

Let us write

$$\mathbf{KT} = \begin{bmatrix} \delta_n & \delta_{n-1} & \cdots & \delta_1 \end{bmatrix} \qquad (10\text{–}11)$$

When $u = -\mathbf{KT}\hat{\mathbf{x}}$ is used to control the system given by Equation (10–7), the system equation becomes

$$\dot{\hat{\mathbf{x}}} = \mathbf{T^{-1}AT}\hat{\mathbf{x}} - \mathbf{T^{-1}BKT}\hat{\mathbf{x}}$$

The characteristic equation is

$$\left| s\mathbf{I} - \mathbf{T^{-1}AT} + \mathbf{T^{-1}BKT} \right| = 0$$

This characteristic equation is the same as the characteristic equation for the system, defined by Equation (10–1), when $u = -\mathbf{Kx}$ is used as the control signal. This can be seen as follows: Since

$$\dot{\mathbf{x}} = \mathbf{Ax} + \mathbf{Bu} = (\mathbf{A} - \mathbf{BK})\mathbf{x}$$

the characteristic equation for this system is

$$\left| s\mathbf{I} - \mathbf{A} + \mathbf{BK} \right| = \left| \mathbf{T^{-1}}(s\mathbf{I} - \mathbf{A} + \mathbf{BK})\mathbf{T} \right| = \left| s\mathbf{I} - \mathbf{T^{-1}AT} + \mathbf{T^{-1}BKT} \right| = 0$$

Now let us simplify the characteristic equation of the system in the controllable canonical form. Referring to Equations (10–8), (10–9), and (10–11), we have

$$\left|s\mathbf{I} - \mathbf{T}^{-1}\mathbf{AT} + \mathbf{T}^{-1}\mathbf{BKT}\right|$$

$$= \left| s\mathbf{I} - \begin{bmatrix} 0 & 1 & \cdots & 0 \\ \cdot & \cdot & & \cdot \\ \cdot & \cdot & & \cdot \\ \cdot & \cdot & & \cdot \\ 0 & 0 & \cdots & 1 \\ -a_n & -a_{n-1} & \cdots & -a_1 \end{bmatrix} + \begin{bmatrix} 0 \\ \cdot \\ \cdot \\ \cdot \\ 0 \\ 1 \end{bmatrix} \begin{bmatrix} \delta_n & \delta_{n-1} & \cdots & \delta_1 \end{bmatrix} \right|$$

$$= \left| \begin{matrix} s & -1 & \cdots & 0 \\ 0 & s & \cdots & 0 \\ \cdot & \cdot & & \cdot \\ \cdot & \cdot & & \cdot \\ \cdot & \cdot & & \cdot \\ a_n + \delta_n & a_{n-1} + \delta_{n-1} & \cdots & s + a_1 + \delta_1 \end{matrix} \right|$$

$$= s^n + \left(a_1 + \delta_1\right)s^{n-1} + \cdots + \left(a_{n-1} + \delta_{n-1}\right)s + \left(a_n + \delta_n\right) = 0 \qquad (10\text{–}12)$$

This is the characteristic equation for the system with state feedback. Therefore, it must be equal to Equation (10–10), the desired characteristic equation. By equating the coefficients of like powers of s, we get

$$a_1 + \delta_1 = \alpha_1$$
$$a_2 + \delta_2 = \alpha_2$$
$$\cdot$$
$$\cdot$$
$$\cdot$$
$$a_n + \delta_n = \alpha_n$$

Solving the preceding equations for the δ_i's and substituting them into Equation (10–11), we obtain

$$\mathbf{K} = \begin{bmatrix} \delta_n & \delta_{n-1} & \cdots & \delta_1 \end{bmatrix}\mathbf{T}^{-1}$$
$$= \begin{bmatrix} \alpha_n - a_n & | & \alpha_{n-1} - a_{n-1} & | & \cdots & | & \alpha_2 - a_2 & | & \alpha_1 - a_1 \end{bmatrix}\mathbf{T}^{-1} \qquad (10\text{–}13)$$

Thus, if the system is completely state controllable, all eigenvalues can be arbitrarily placed by choosing matrix $\mathbf{K}$ according to Equation (10–13) (sufficient condition).

We have thus proved that a necessary and sufficient condition for arbitrary pole placement is that the system be completely state controllable.

It is noted that if the system is not completely state controllable, but is stabilizable, then it is possible to make the entire system stable by placing the closed-loop poles at desired locations for q controllable modes. The remaining $n - q$ uncontrollable modes are stable. So the entire system can be made stable.

Determination of Matrix K Using Transformation Matrix T. Suppose that the system is defined by

$$\dot{\mathbf{x}} = \mathbf{A}\mathbf{x} + \mathbf{B}u$$

and the control signal is given by

$$u = -\mathbf{K}\mathbf{x}$$

The feedback gain matrix $\mathbf{K}$ that forces the eigenvalues of $\mathbf{A} - \mathbf{BK}$ to be $\mu_1, \mu_2, \ldots, \mu_n$ (desired values) can be determined by the following steps (if μ_i is a complex eigenvalue, then its conjugate must also be an eigenvalue of $\mathbf{A} - \mathbf{BK}$):

Step 1: Check the controllability condition for the system. If the system is completely state controllable, then use the following steps:

Step 2: From the characteristic polynomial for matrix $\mathbf{A}$, that is,

$$|s\mathbf{I} - \mathbf{A}| = s^n + a_1 s^{n-1} + \cdots + a_{n-1}s + a_n$$

determine the values of $a_1, a_2, \ldots, a_n$.

Step 3: Determine the transformation matrix $\mathbf{T}$ that transforms the system state equation into the controllable canonical form. (If the given system equation is already in the controllable canonical form, then $\mathbf{T} = \mathbf{I}$.) It is not necessary to write the state equation in the controllable canonical form. All we need here is to find the matrix $\mathbf{T}$. The transformation matrix $\mathbf{T}$ is given by Equation (10–4), or

$$\mathbf{T} = \mathbf{MW}$$

where $\mathbf{M}$ is given by Equation (10–5) and $\mathbf{W}$ is given by Equation (10–6).

Step 4: Using the desired eigenvalues (desired closed-loop poles), write the desired characteristic polynomial:

$$(s - \mu_1)(s - \mu_2) \cdots (s - \mu_n) = s^n + \alpha_1 s^{n-1} + \cdots + \alpha_{n-1}s + \alpha_n$$

and determine the values of $\alpha_1, \alpha_2, \ldots, \alpha_n$.

Step 5: The required state feedback gain matrix $\mathbf{K}$ can be determined from Equation (10–13), rewritten thus:

$$\mathbf{K} = \begin{bmatrix} \alpha_n - a_n & \vdots & \alpha_{n-1} - a_{n-1} & \vdots & \cdots & \vdots & \alpha_2 - a_2 & \vdots & \alpha_1 - a_1 \end{bmatrix}\mathbf{T}^{-1}$$

Determination of Matrix K Using Direct Substitution Method. If the system is of low order ($n \leq 3$), direct substitution of matrix $\mathbf{K}$ into the desired characteristic polynomial may be simpler. For example, if $n = 3$, then write the state feedback gain matrix $\mathbf{K}$ as

$$\mathbf{K} = \begin{bmatrix} k_1 & k_2 & k_3 \end{bmatrix}$$

Substitute this $\mathbf{K}$ matrix into the desired characteristic polynomial $|s\mathbf{I} - \mathbf{A} + \mathbf{BK}|$ and equate it to $(s - \mu_1)(s - \mu_2)(s - \mu_3)$, or

$$|s\mathbf{I} - \mathbf{A} + \mathbf{BK}| = (s - \mu_1)(s - \mu_2)(s - \mu_3)$$

Since both sides of this characteristic equation are polynomials in s, by equating the coefficients of the like powers of s on both sides, it is possible to determine the values of k_1, k_2, and k_3. This approach is convenient if $n = 2$ or 3. (For $n = 4, 5, 6, \ldots$, this approach may become very tedious.)

Note that if the system is not completely controllable, matrix $\mathbf{K}$ cannot be determined. (No solution exists.)

Determination of Matrix K Using Ackermann's Formula.   There is a well-known formula, known as Ackermann's formula, for the determination of the state feedback gain matrix $\mathbf{K}$. We shall present this formula in what follows.

Consider the system

$$\dot{\mathbf{x}} = \mathbf{A}\mathbf{x} + \mathbf{B}u$$

where we use the state feedback control $u = -\mathbf{K}\mathbf{x}$. We assume that the system is completely state controllable. We also assume that the desired closed-loop poles are at $s = \mu_1, s = \mu_2, \ldots, s = \mu_n$.

Use of the state feedback control

$$u = -\mathbf{K}\mathbf{x}$$

modifies the system equation to

$$\dot{\mathbf{x}} = (\mathbf{A} - \mathbf{B}\mathbf{K})\mathbf{x} \tag{10-14}$$

Let us define

$$\tilde{\mathbf{A}} = \mathbf{A} - \mathbf{B}\mathbf{K}$$

The desired characteristic equation is

$$|s\mathbf{I} - \mathbf{A} + \mathbf{B}\mathbf{K}| = |s\mathbf{I} - \tilde{\mathbf{A}}| = (s - \mu_1)(s - \mu_2)\cdots(s - \mu_n)$$

$$= s^n + \alpha_1 s^{n-1} + \cdots + \alpha_{n-1}s + \alpha_n = 0$$

Since the Cayley–Hamilton theorem states that $\tilde{\mathbf{A}}$ satisfies its own characteristic equation, we have

$$\phi(\tilde{\mathbf{A}}) = \tilde{\mathbf{A}}^n + \alpha_1 \tilde{\mathbf{A}}^{n-1} + \cdots + \alpha_{n-1}\tilde{\mathbf{A}} + \alpha_n\mathbf{I} = \mathbf{0} \tag{10-15}$$

We shall utilize Equation (10–15) to derive Ackermann's formula. To simplify the derivation, we consider the case where $n = 3$. (For any other positive integer n, the following derivation can be easily extended.)

Consider the following identities:

$$\mathbf{I} = \mathbf{I}$$

$$\tilde{\mathbf{A}} = \mathbf{A} - \mathbf{B}\mathbf{K}$$

$$\tilde{\mathbf{A}}^2 = (\mathbf{A} - \mathbf{B}\mathbf{K})^2 = \mathbf{A}^2 - \mathbf{A}\mathbf{B}\mathbf{K} - \mathbf{B}\mathbf{K}\tilde{\mathbf{A}}$$

$$\tilde{\mathbf{A}}^3 = (\mathbf{A} - \mathbf{B}\mathbf{K})^3 = \mathbf{A}^3 - \mathbf{A}^2\mathbf{B}\mathbf{K} - \mathbf{A}\mathbf{B}\mathbf{K}\tilde{\mathbf{A}} - \mathbf{B}\mathbf{K}\tilde{\mathbf{A}}^2$$

Multiplying the preceding equations in order by $\alpha_3, \alpha_2, \alpha_1$, and α_0 (where $\alpha_0 = 1$), respectively, and adding the results, we obtain

$$\alpha_3 \mathbf{I} + \alpha_2 \widetilde{\mathbf{A}} + \alpha_1 \widetilde{\mathbf{A}}^2 + \widetilde{\mathbf{A}}^3$$
$$= \alpha_3 \mathbf{I} + \alpha_2 (\mathbf{A} - \mathbf{BK}) + \alpha_1 (\mathbf{A}^2 - \mathbf{ABK} - \mathbf{BK}\widetilde{\mathbf{A}}) + \mathbf{A}^3 - \mathbf{A}^2\mathbf{BK}$$
$$\quad - \mathbf{ABK}\widetilde{\mathbf{A}} - \mathbf{BK}\widetilde{\mathbf{A}}^2$$
$$= \alpha_3 \mathbf{I} + \alpha_2 \mathbf{A} + \alpha_1 \mathbf{A}^2 + \mathbf{A}^3 - \alpha_2 \mathbf{BK} - \alpha_1 \mathbf{ABK} - \alpha_1 \mathbf{BK}\widetilde{\mathbf{A}} - \mathbf{A}^2\mathbf{BK}$$
$$\quad - \mathbf{ABK}\widetilde{\mathbf{A}} - \mathbf{BK}\widetilde{\mathbf{A}}^2 \tag{10-16}$$

Referring to Equation (10–15), we have

$$\alpha_3 \mathbf{I} + \alpha_2 \widetilde{\mathbf{A}} + \alpha_1 \widetilde{\mathbf{A}}^2 + \widetilde{\mathbf{A}}^3 = \phi(\widetilde{\mathbf{A}}) = \mathbf{0}$$

Also, we have

$$\alpha_3 \mathbf{I} + \alpha_2 \mathbf{A} + \alpha_1 \mathbf{A}^2 + \mathbf{A}^3 = \phi(\mathbf{A}) \neq \mathbf{0}$$

Substituting the last two equations into Equation (10–16), we have

$$\phi(\widetilde{\mathbf{A}}) = \phi(\mathbf{A}) - \alpha_2 \mathbf{BK} - \alpha_1 \mathbf{BK}\widetilde{\mathbf{A}} - \mathbf{BK}\widetilde{\mathbf{A}}^2 - \alpha_1 \mathbf{ABK} - \mathbf{ABK}\widetilde{\mathbf{A}} - \mathbf{A}^2\mathbf{BK}$$

Since $\phi(\widetilde{\mathbf{A}}) = \mathbf{0}$, we obtain

$$\phi(\mathbf{A}) = \mathbf{B}(\alpha_2 \mathbf{K} + \alpha_1 \mathbf{K}\widetilde{\mathbf{A}} + \mathbf{K}\widetilde{\mathbf{A}}^2) + \mathbf{AB}(\alpha_1 \mathbf{K} + \mathbf{K}\widetilde{\mathbf{A}}) + \mathbf{A}^2\mathbf{BK}$$
$$= \begin{bmatrix} \mathbf{B} & \vdots & \mathbf{AB} & \vdots & \mathbf{A}^2\mathbf{B} \end{bmatrix} \begin{bmatrix} \alpha_2 \mathbf{K} + \alpha_1 \mathbf{K}\widetilde{\mathbf{A}} + \mathbf{K}\widetilde{\mathbf{A}}^2 \\ \alpha_1 \mathbf{K} + \mathbf{K}\widetilde{\mathbf{A}} \\ \mathbf{K} \end{bmatrix} \tag{10-17}$$

Since the system is completely state controllable, the inverse of the controllability matrix

$$\begin{bmatrix} \mathbf{B} & \vdots & \mathbf{AB} & \vdots & \mathbf{A}^2\mathbf{B} \end{bmatrix}$$

exists. Premultiplying both sides of Equation (10–17) by the inverse of the controllability matrix, we obtain

$$\begin{bmatrix} \mathbf{B} & \vdots & \mathbf{AB} & \vdots & \mathbf{A}^2\mathbf{B} \end{bmatrix}^{-1}\phi(\mathbf{A}) = \begin{bmatrix} \alpha_2 \mathbf{K} + \alpha_1 \mathbf{K}\widetilde{\mathbf{A}} + \mathbf{K}\widetilde{\mathbf{A}}^2 \\ \alpha_1 \mathbf{K} + \mathbf{K}\widetilde{\mathbf{A}} \\ \mathbf{K} \end{bmatrix}$$

Premultiplying both sides of this last equation by $\begin{bmatrix} 0 & 0 & 1 \end{bmatrix}$, we obtain

$$\begin{bmatrix} 0 & 0 & 1 \end{bmatrix} \begin{bmatrix} \mathbf{B} & \vdots & \mathbf{AB} & \vdots & \mathbf{A}^2\mathbf{B} \end{bmatrix}^{-1}\phi(\mathbf{A}) = \begin{bmatrix} 0 & 0 & 1 \end{bmatrix} \begin{bmatrix} \alpha_2 \mathbf{K} + \alpha_1 \mathbf{K}\widetilde{\mathbf{A}} + \mathbf{K}\widetilde{\mathbf{A}}^2 \\ \alpha_1 \mathbf{K} + \mathbf{K}\widetilde{\mathbf{A}} \\ \mathbf{K} \end{bmatrix} = \mathbf{K}$$

which can be rewritten as

$$\mathbf{K} = \begin{bmatrix} 0 & 0 & 1 \end{bmatrix} \begin{bmatrix} \mathbf{B} & \vdots & \mathbf{AB} & \vdots & \mathbf{A}^2\mathbf{B} \end{bmatrix}^{-1}\phi(\mathbf{A})$$

This last equation gives the required state feedback gain matrix $\mathbf{K}$.

For an arbitrary positive integer n, we have

$$\mathbf{K} = \begin{bmatrix} 0 & 0 & \cdots & 0 & 1 \end{bmatrix} \begin{bmatrix} \mathbf{B} & \vdots & \mathbf{AB} & \vdots & \cdots & \vdots & \mathbf{A}^{n-1}\mathbf{B} \end{bmatrix}^{-1}\phi(\mathbf{A}) \tag{10-18}$$

Equation (10–18) is known as Ackermann's formula for the determination of the state feedback gain matrix **K**.

Regulator Systems and Control Systems. Systems that include controllers can be divided into two categories: regulator systems (where the reference input is constant, including zero) and control systems (where the reference input is time varying). In what follows we shall consider regulator systems. Control systems will be treated in Section 10–7.

Choosing the Locations of Desired Closed-Loop Poles. The first step in the pole-placement design approach is to choose the locations of the desired closed-loop poles. The most frequently used approach is to choose such poles based on experience in the root-locus design, placing a dominant pair of closed-loop poles and choosing other poles so that they are far to the left of the dominant closed-loop poles.

Note that if we place the dominant closed-loop poles far from the $j\omega$ axis, so that the system response becomes very fast, the signals in the system become very large, with the result that the system may become nonlinear. This should be avoided.

Another approach is based on the quadratic optimal control approach. This approach will determine the desired closed-loop poles such that it balances between the acceptable response and the amount of control energy required. (See Section 10–8.) Note that requiring a high-speed response implies requiring large amounts of control energy. Also, in general, increasing the speed of response requires a larger, heavier actuator, which will cost more.

EXAMPLE 10–1   Consider the regulator system shown in Figure 10–2. The plant is given by

$$\dot{\mathbf{x}} = \mathbf{A}\mathbf{x} + \mathbf{B}u$$

where

$$\mathbf{A} = \begin{bmatrix} 0 & 1 & 0 \\ 0 & 0 & 1 \\ -1 & -5 & -6 \end{bmatrix}, \quad \mathbf{B} = \begin{bmatrix} 0 \\ 0 \\ 1 \end{bmatrix}$$

The system uses the state feedback control $\mathbf{u} = -\mathbf{Kx}$. Let us choose the desired closed-loop poles at

$$s = -2 + j4, \qquad s = -2 - j4, \qquad s = -10$$

(We make such a choice because we know from experience that such a set of closed-loop poles will result in a reasonable or acceptable transient response.) Determine the state feedback gain matrix **K**.

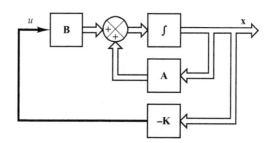

Figure 10–2
Regulator system.

First, we need to check the controllability matrix of the system. Since the controllability matrix **M** is given by

$$\mathbf{M} = \begin{bmatrix} \mathbf{B} & \vdots & \mathbf{AB} & \vdots & \mathbf{A}^2\mathbf{B} \end{bmatrix} = \begin{bmatrix} 0 & 0 & 1 \\ 0 & 1 & -6 \\ 1 & -6 & 31 \end{bmatrix}$$

we find that $|\mathbf{M}| = -1$, and therefore, rank $\mathbf{M} = 3$. Thus, the system is completely state controllable and arbitrary pole placement is possible.

Next, we shall solve this problem. We shall demonstrate each of the three methods presented in this chapter.

Method 1: The first method is to use Equation (10–13). The characteristic equation for the system is

$$|s\mathbf{I} - \mathbf{A}| = \begin{vmatrix} s & -1 & 0 \\ 0 & s & -1 \\ 1 & 5 & s+6 \end{vmatrix}$$

$$= s^3 + 6s^2 + 5s + 1$$

$$= s^3 + a_1s^2 + a_2s + a_3 = 0$$

Hence,

$$a_1 = 6, \qquad a_2 = 5, \qquad a_3 = 1$$

The desired characteristic equation is

$$(s + 2 - j4)(s + 2 + j4)(s + 10) = s^3 + 14s^2 + 60s + 200$$

$$= s^3 + \alpha_1s^2 + \alpha_2s + \alpha_3 = 0$$

Hence,

$$\alpha_1 = 14, \qquad \alpha_2 = 60, \qquad \alpha_3 = 200$$

Referring to Equation (10–13), we have

$$\mathbf{K} = \begin{bmatrix} \alpha_3 - a_3 & \vdots & \alpha_2 - a_2 & \vdots & \alpha_1 - a_1 \end{bmatrix} \mathbf{T}^{-1}$$

where $\mathbf{T} = \mathbf{I}$ for this problem because the given state equation is in the controllable canonical form. Then we have

$$\mathbf{K} = \begin{bmatrix} 200 - 1 & \vdots & 60 - 5 & \vdots & 14 - 6 \end{bmatrix}$$

$$= \begin{bmatrix} 199 & 55 & 8 \end{bmatrix}$$

Method 2: By defining the desired state feedback gain matrix **K** as

$$\mathbf{K} = \begin{bmatrix} k_1 & k_2 & k_3 \end{bmatrix}$$

and equating $|s\mathbf{I} - \mathbf{A} + \mathbf{BK}|$ with the desired characteristic equation, we obtain

$$|s\mathbf{I} - \mathbf{A} + \mathbf{BK}| = \begin{vmatrix} \begin{bmatrix} s & 0 & 0 \\ 0 & s & 0 \\ 0 & 0 & s \end{bmatrix} - \begin{bmatrix} 0 & 1 & 0 \\ 0 & 0 & 1 \\ -1 & -5 & -6 \end{bmatrix} + \begin{bmatrix} 0 \\ 0 \\ 1 \end{bmatrix} \begin{bmatrix} k_1 & k_2 & k_3 \end{bmatrix} \end{vmatrix}$$

$$= \begin{vmatrix} s & -1 & 0 \\ 0 & s & -1 \\ 1 + k_1 & 5 + k_2 & s + 6 + k_3 \end{vmatrix}$$

$$= s^3 + (6 + k_3)s^2 + (5 + k_2)s + 1 + k_1$$

$$= s^3 + 14s^2 + 60s + 200$$

Thus,

$$6 + k_3 = 14, \qquad 5 + k_2 = 60, \qquad 1 + k_1 = 200$$

from which we obtain

$$k_1 = 199, \qquad k_2 = 55, \qquad k_3 = 8$$

or

$$\mathbf{K} = \begin{bmatrix} 199 & 55 & 8 \end{bmatrix}$$

Method 3: The third method is to use Ackermann's formula. Referring to Equation (10–18), we have

$$\mathbf{K} = \begin{bmatrix} 0 & 0 & 1 \end{bmatrix} \begin{bmatrix} \mathbf{B} & \vdots & \mathbf{AB} & \vdots & \mathbf{A}^2\mathbf{B} \end{bmatrix}^{-1} \phi(\mathbf{A})$$

Since

$$\phi(\mathbf{A}) = \mathbf{A}^3 + 14\,\mathbf{A}^2 + 60\mathbf{A} + 200\mathbf{I}$$

$$= \begin{bmatrix} 0 & 1 & 0 \\ 0 & 0 & 1 \\ -1 & -5 & -6 \end{bmatrix}^3 + 14 \begin{bmatrix} 0 & 1 & 0 \\ 0 & 0 & 1 \\ -1 & -5 & -6 \end{bmatrix}^2$$

$$+ 60 \begin{bmatrix} 0 & 1 & 0 \\ 0 & 0 & 1 \\ -1 & -5 & -6 \end{bmatrix} + 200 \begin{bmatrix} 1 & 0 & 0 \\ 0 & 1 & 0 \\ 0 & 0 & 1 \end{bmatrix}$$

$$= \begin{bmatrix} 199 & 55 & 8 \\ -8 & 159 & 7 \\ -7 & -43 & 117 \end{bmatrix}$$

and

$$\begin{bmatrix} \mathbf{B} & \vdots & \mathbf{AB} & \vdots & \mathbf{A}^2\mathbf{B} \end{bmatrix} = \begin{bmatrix} 0 & 0 & 1 \\ 0 & 1 & -6 \\ 1 & -6 & 31 \end{bmatrix}$$

we obtain

$$\mathbf{K} = \begin{bmatrix} 0 & 0 & 1 \end{bmatrix} \begin{bmatrix} 0 & 0 & 1 \\ 0 & 1 & -6 \\ 1 & -6 & 31 \end{bmatrix}^{-1} \begin{bmatrix} 199 & 55 & 8 \\ -8 & 159 & 7 \\ -7 & -43 & 117 \end{bmatrix}$$

$$= \begin{bmatrix} 0 & 0 & 1 \end{bmatrix} \begin{bmatrix} 5 & 6 & 1 \\ 6 & 1 & 0 \\ 1 & 0 & 0 \end{bmatrix} \begin{bmatrix} 199 & 55 & 8 \\ -8 & 159 & 7 \\ -7 & -43 & 117 \end{bmatrix}$$

$$= \begin{bmatrix} 199 & 55 & 8 \end{bmatrix}$$

As a matter of course, the feedback gain matrix $\mathbf{K}$ obtained by the three methods are the same. With this state feedback, the closed-loop poles are placed at $s = -2 \pm j4$ and $s = -10$, as desired.

It is noted that if the order n of the system were 4 or higher, methods 1 and 3 are recommended, since all matrix computations can be carried out by a computer. If method 2 is used, hand computations become necessary because a computer may not handle the characteristic equation with unknown parameters $k_1, k_2, \ldots, k_n$.

Comments. It is important to note that matrix **K** is not unique for a given system, but depends on the desired closed-loop pole locations (which determine the speed and damping of the response) selected. Note that the selection of the desired closed-loop poles or the desired characteristic equation is a compromise between the rapidity of the response of the error vector and the sensitivity to disturbances and measurement noises. That is, if we increase the speed of error response, then the adverse effects of disturbances and measurement noises generally increase. If the system is of second order, then the system dynamics (response characteristics) can be precisely correlated to the location of the desired closed-loop poles and the zero(s) of the plant. For higher-order systems, the location of the closed-loop poles and the system dynamics (response characteristics) are not easily correlated. Hence, in determining the state feedback gain matrix **K** for a given system, it is desirable to examine by computer simulations the response characteristics of the system for several different matrices **K** (based on several different desired characteristic equations) and to choose the one that gives the best overall system performance.

10–3 SOLVING POLE-PLACEMENT PROBLEMS WITH MATLAB

Pole-placement problems can be solved easily with MATLAB. MATLAB has two commands—acker and place—for the computation of feedback-gain matrix **K**. The command acker is based on Ackermann's formula. This command applies to single-input systems only. The desired closed-loop poles can include multiple poles (poles located at the same place).

If the system involves multiple inputs, for a specified set of closed-loop poles the state-feedback gain matrix **K** is not unique and we have an additional freedom (or freedoms) to choose **K**. There are many approaches to constructively utilize this additional freedom (or freedoms) to determine **K**. One common use is to maximize the stability margin. The pole placement based on this approach is called the robust pole placement. The MATLAB command for the robust pole placement is place.

Although the command place can be used for both single-input and multiple-input systems, this command requires that the multiplicity of poles in the desired closed-loop poles be no greater than the rank of **B**. That is, if matrix **B** is an $n \times 1$ matrix, the command place requires that there be no multiple poles in the set of desired closed-loop poles.

For single-input systems, the commands acker and place yield the same **K**. (But for multiple-input systems, one must use the command place instead of acker.)

It is noted that when the single-input system is barely controllable, some computational problem may occur if the command acker is used. In such a case the use of the place command is preferred, provided that no multiple poles are involved in the desired set of closed-loop poles.

To use the command acker or place, we first enter the following matrices in the program:

$$\textbf{A} \text{ matrix,} \qquad \textbf{B} \text{ matrix,} \qquad \textbf{J} \text{ matrix}$$

where **J** matrix is the matrix consisting of the desired closed-loop poles such that

$$\mathbf{J} = \begin{bmatrix} \mu_1 & \mu_2 & \cdots & \mu_n \end{bmatrix}$$

Then we enter

$$K = \text{acker}(A,B,J)$$

or

$$K = \text{place}(A,B,J)$$

It is noted that the command eig (A-B*K) may be used to verify that K thus obtained gives the desired eigenvalues.

EXAMPLE 10–2   Consider the same system as treated in Example 10–1. The system equation is

$$\dot{\mathbf{x}} = \mathbf{A}\mathbf{x} + \mathbf{B}u$$

where

$$\mathbf{A} = \begin{bmatrix} 0 & 1 & 0 \\ 0 & 0 & 1 \\ -1 & -5 & -6 \end{bmatrix}, \quad \mathbf{B} = \begin{bmatrix} 0 \\ 0 \\ 1 \end{bmatrix}$$

By using state feedback control $u = -\mathbf{K}\mathbf{x}$, it is desired to have the closed-loop poles at $s = \mu_i$ ($i = 1, 2, 3$), where

$$\mu_1 = -2 + j4, \qquad \mu_2 = -2 - j4, \qquad \mu_3 = -10$$

Determine the state feedback-gain matrix **K** with MATLAB.

MATLAB programs that generate matrix **K** are shown in MATLAB Programs 10–1 and 10–2. MATLAB Program 10–1 uses command acker and MATLAB Program 10–2 uses command place.

MATLAB Program 10–1

```
A = [0 1 0;0 0 1;-1 -5 -6];
B = [0;0;1];
J = [-2+j*4 -2-j*4 -10];
K = acker(A,B,J)

K =

  199   55   8
```

MATLAB Program 10–2

```
A = [0 1 0;0 0 1;-1 -5 -6];
B = [0;0;1];
J = [-2+j*4 -2-j*4 -10];
K = place(A,B,J)
  place: ndigits = 15

K =

  199.0000   55.0000   8.0000
```

EXAMPLE 10–3   Consider the same system as discussed in Example 10–1. It is desired that this regulator system have closed-loop poles at

$$s = -2 + j4, \qquad s = -2 - j4, \qquad s = -10$$

The necessary state feedback gain matrix **K** was obtained in Example 10–1 as follows:

$$\mathbf{K} = [199 \quad 55 \quad 8]$$

Using MATLAB, obtain the response of the system to the following initial condition:

$$\mathbf{x}(0) = \begin{bmatrix} 1 \\ 0 \\ 0 \end{bmatrix}$$

Response to Initial Condition: To obtain the response to the given initial condition $\mathbf{x}(0)$, we substitute $u = -\mathbf{Kx}$ into the plant equation to get

$$\dot{\mathbf{x}} = (\mathbf{A} - \mathbf{BK})\mathbf{x}, \qquad \mathbf{x}(0) = \begin{bmatrix} 1 \\ 0 \\ 0 \end{bmatrix}$$

To plot the response curves (x_1 versus t, x_2 versus t, and x_3 versus t), we may use the command initial. We first define the state-space equations for the system as follows:

$$\dot{\mathbf{x}} = (\mathbf{A} - \mathbf{BK})\mathbf{x} + \mathbf{Iu}$$

$$\mathbf{y} = \mathbf{Ix} + \mathbf{Iu}$$

where we included **u** (a three-dimensional input vector). This **u** vector is considered **0** in the computation of the response to the initial condition. Then we define

sys = ss(A - BK, eye(3), eye(3), eye(3))

and use the initial command as follows:

x = initial(sys, [1;0;0],t)

where t is the time duration we want to use, such as

t = 0:0.01:4;

Then obtain x1, x2, and x3 as follows:

x1 = [1 0 0]*x';

x2 = [0 1 0]*x';

x3 = [0 0 1]*x';

and use the plot command. This program is shown in MATLAB Program 10–3. The resulting response curves are shown in Figure 10–3.

MATLAB Program 10–3

```
% Response to initial condition:

A = [0 1 0;0 0 1;-1 -5 -6];
B = [0;0;1];
K = [199 55 8];
sys = ss(A-B*K, eye(3), eye(3), eye(3));
t = 0:0.01:4;
x = initial(sys,[1;0;0],t);
x1 = [1 0 0]*x';
x2 = [0 1 0]*x';
x3 = [0 0 1]*x';

subplot(3,1,1); plot(t,x1), grid
title('Response to Initial Condition')
ylabel('state variable x1')

subplot(3,1,2); plot(t,x2),grid
ylabel('state variable x2')

subplot(3,1,3); plot(t,x3),grid
xlabel('t (sec)')
ylabel('state variable x3')
```

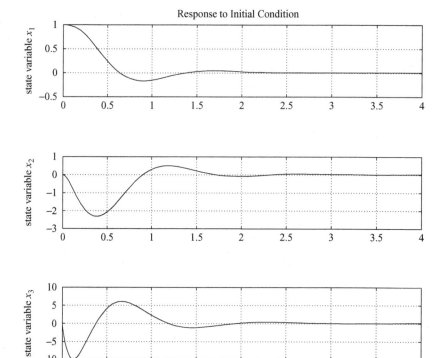

Figure 10–3
Response to initial
condition.

In this section we shall discuss the pole-placement approach to the design of type 1 servo systems. Here we shall limit our systems each to have a scalar control signal u and a scalar output y.

In what follows we shall first discuss a problem of designing a type 1 servo system when the plant involves an integrator. Then we shall discuss the design of a type 1 servo system when the plant has no integrator.

Design of Type 1 Servo System when the Plant Has An Integrator. Assume that the plant is defined by

$$\dot{\mathbf{x}} = \mathbf{A}\mathbf{x} + \mathbf{B}u \tag{10–19}$$

$$y = \mathbf{C}\mathbf{x} \tag{10–20}$$

where $\mathbf{x}$ = state vector for the plant (n-vector)
  u = control signal (scalar)
  y = output signal (scalar)
  $\mathbf{A}$ = $n \times n$ constant matrix
  $\mathbf{B}$ = $n \times 1$ constant matrix
  $\mathbf{C}$ = $1 \times n$ constant matrix

As stated earlier, we assume that both the control signal u and the output signal y are scalars. By a proper choice of a set of state variables, it is possible to choose the output to be equal to one of the state variables. (See the method presented in Chapter 2 for obtaining a state-space representation of the transfer function system in which the output y becomes equal to x_1.)

Figure 10–4 shows a general configuration of the type 1 servo system when the plant has an integrator. Here we assumed that $y = x_1$. In the present analysis we assume that

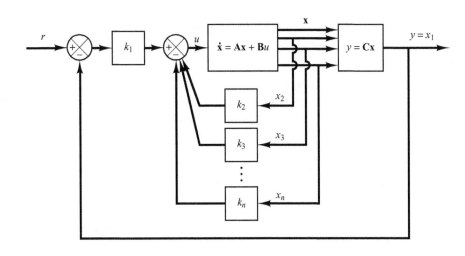

Figure 10–4
Type 1 servo system when the plant has an integrator.

the reference input r is a step function. In this system we use the following state-feedback control scheme:

$$u = -\begin{bmatrix} 0 & k_2 & k_3 & \cdots & k_n \end{bmatrix} \begin{bmatrix} x_1 \\ x_2 \\ \cdot \\ \cdot \\ \cdot \\ x_n \end{bmatrix} + k_1(r - x_1)$$

$$= -\mathbf{K}\mathbf{x} + k_1 r \qquad (10\text{--}21)$$

where

$$\mathbf{K} = \begin{bmatrix} k_1 & k_2 & \cdots & k_n \end{bmatrix}$$

Assume that the reference input (step function) is applied at $t = 0$. Then, for $t > 0$, the system dynamics can be described by Equations (10–19) and (10–21), or

$$\dot{\mathbf{x}} = \mathbf{A}\mathbf{x} + \mathbf{B}u = (\mathbf{A} - \mathbf{B}\mathbf{K})\mathbf{x} + \mathbf{B}k_1 r \qquad (10\text{--}22)$$

We shall design the type 1 servo system such that the closed-loop poles are located at desired positions. The designed system will be an asymptotically stable system, $y(\infty)$ will approach the constant value r, and $u(\infty)$ will approach zero. (r is a step input.)

Notice that at steady state we have

$$\dot{\mathbf{x}}(\infty) = (\mathbf{A} - \mathbf{B}\mathbf{K})\mathbf{x}(\infty) + \mathbf{B}k_1 r(\infty) \qquad (10\text{--}23)$$

Noting that $r(t)$ is a step input, we have $r(\infty) = r(t) = r$(constant) for $t > 0$. By subtracting Equation (10–23) from Equation (10–22), we obtain

$$\dot{\mathbf{x}}(t) - \dot{\mathbf{x}}(\infty) = (\mathbf{A} - \mathbf{B}\mathbf{K})\big[\mathbf{x}(t) - \mathbf{x}(\infty)\big] \qquad (10\text{--}24)$$

Define

$$\mathbf{x}(t) - \mathbf{x}(\infty) = \mathbf{e}(t)$$

Then Equation (10–24) becomes

$$\dot{\mathbf{e}} = (\mathbf{A} - \mathbf{B}\mathbf{K})\mathbf{e} \qquad (10\text{--}25)$$

Equation (10–25) describes the error dynamics.

The design of the type 1 servo system here is converted to the design of an asymptotically stable regulator system such that $\mathbf{e}(t)$ approaches zero, given any initial condition $\mathbf{e}(0)$. If the system defined by Equation (10–19) is completely state controllable, then, by specifying the desired eigenvalues $\mu_1, \mu_2, \ldots, \mu_n$ for the matrix $\mathbf{A} - \mathbf{B}\mathbf{K}$, matrix $\mathbf{K}$ can be determined by the pole-placement technique presented in Section 10–2.

The steady-state values of $\mathbf{x}(t)$ and $u(t)$ can be found as follows: At steady state $(t = \infty)$, we have, from Equation (10–22),

$$\dot{\mathbf{x}}(\infty) = \mathbf{0} = (\mathbf{A} - \mathbf{B}\mathbf{K})\mathbf{x}(\infty) + \mathbf{B}k_1 r$$

Since the desired eigenvalues of $\mathbf{A} - \mathbf{B}\mathbf{K}$ are all in the left-half s plane, the inverse of matrix $\mathbf{A} - \mathbf{B}\mathbf{K}$ exists. Consequently, $\mathbf{x}(\infty)$ can be determined as

$$\mathbf{x}(\infty) = -(\mathbf{A} - \mathbf{B}\mathbf{K})^{-1}\mathbf{B}k_1 r$$

Also, $u(\infty)$ can be obtained as

$$u(\infty) = -\mathbf{K}\mathbf{x}(\infty) + k_1 r = 0$$

(See Example 10–4 to verify this last equation.)

EXAMPLE 10–4   Design a type 1 servo system when the plant transfer function has an integrator. Assume that the plant transfer function is given by

$$\frac{Y(s)}{U(s)} = \frac{1}{s(s + 1)(s + 2)}$$

The desired closed-loop poles are $s = -2 \pm j2\sqrt{3}$ and $s = -10$. Assume that the system configuration is the same as that shown in Figure 10–4 and the reference input r is a step function. Obtain the unit-step response of the designed system.

Define state variables x_1, x_2, and x_3 as follows:

$$x_1 = y$$
$$x_2 = \dot{x}_1$$
$$x_3 = \dot{x}_2$$

Then the state-space representation of the system becomes

$$\dot{\mathbf{x}} = \mathbf{A}\mathbf{x} + \mathbf{B}u \tag{10–26}$$

$$y = \mathbf{C}\mathbf{x} \tag{10–27}$$

where

$$\mathbf{A} = \begin{bmatrix} 0 & 1 & 0 \\ 0 & 0 & 1 \\ 0 & -2 & -3 \end{bmatrix}, \quad \mathbf{B} = \begin{bmatrix} 0 \\ 0 \\ 1 \end{bmatrix}, \quad \mathbf{C} = \begin{bmatrix} 1 & 0 & 0 \end{bmatrix}$$

Referring to Figure 10–4 and noting that $n = 3$, the control signal u is given by

$$u = -(k_2 x_2 + k_3 x_3) + k_1(r - x_1) = -\mathbf{K}\mathbf{x} + k_1 r \tag{10–28}$$

where

$$\mathbf{K} = \begin{bmatrix} k_1 & k_2 & k_3 \end{bmatrix}$$

The state-feedback gain matrix $\mathbf{K}$ can be obtained easily with MATLAB. See MATLAB Program 10–4.

MATLAB Program 10–4

```
A = [0 1 0;0 0 1;0 -2 -3];
B = [0;0;1];
J = [-2+j*2*sqrt(3) -2-j*2*sqrt(3) -10];
K = acker(A,B,J)

K =

   160.0000   54.0000   11.0000
```

The state feedback gain matrix **K** is thus

$$K = [160 \quad 54 \quad 11]$$

Unit-Step Response of the Designed System: The unit-step response of the designed system can be obtained as follows:
   Since

$$\mathbf{A} - \mathbf{BK} = \begin{bmatrix} 0 & 1 & 0 \\ 0 & 0 & 1 \\ 0 & -2 & -3 \end{bmatrix} - \begin{bmatrix} 0 \\ 0 \\ 1 \end{bmatrix} [160 \quad 54 \quad 11] = \begin{bmatrix} 0 & 1 & 0 \\ 0 & 0 & 1 \\ -160 & -56 & -14 \end{bmatrix}$$

from Equation (10–22) the state equation for the designed system is

$$\begin{bmatrix} \dot{x}_1 \\ \dot{x}_2 \\ \dot{x}_3 \end{bmatrix} = \begin{bmatrix} 0 & 1 & 0 \\ 0 & 0 & 1 \\ -160 & -56 & -14 \end{bmatrix} \begin{bmatrix} x_1 \\ x_2 \\ x_3 \end{bmatrix} + \begin{bmatrix} 0 \\ 0 \\ 160 \end{bmatrix} r \tag{10–29}$$

and the output equation is

$$y = \begin{bmatrix} 1 & 0 & 0 \end{bmatrix} \begin{bmatrix} x_1 \\ x_2 \\ x_3 \end{bmatrix} \tag{10–30}$$

Solving Equations (10–29) and (10–30) for $y(t)$ when r is a unit-step function gives the unit-step response curve $y(t)$ versus t. MATLAB Program 10–5 yields the unit-step response. The resulting unit-step response curve is shown in Figure 10–5.

MATLAB Program 10–5

```
% ---------- Unit-step response ----------

% ***** Enter the state matrix, control matrix, output matrix,
% and direct transmission matrix of the designed system *****

AA = [0 1 0;0 0 1;-160 -56 -14];
BB = [0;0;160];
CC = [1 0 0];
DD = [0];

% ***** Enter step command and plot command *****

t = 0:0.01:5;
y = step(AA,BB,CC,DD,1,t);
plot(t,y)
grid
title('Unit-Step Response')
xlabel('t Sec')
ylabel('Output y')
```

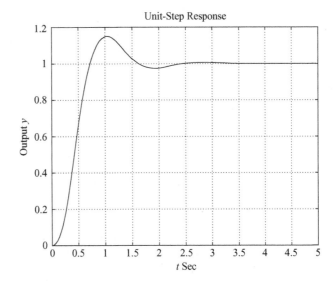

Figure 10–5
Unit-step response curve $y(t)$ versus t for the system designed in Example 10–4.

Note that since

$$u(\infty) = -\mathbf{K}\mathbf{x}(\infty) + k_1 r(\infty) = -\mathbf{K}\mathbf{x}(\infty) + k_1 r$$

we have

$$u(\infty) = -\begin{bmatrix} 160 & 54 & 11 \end{bmatrix} \begin{bmatrix} x_1(\infty) \\ x_2(\infty) \\ x_3(\infty) \end{bmatrix} + 160r$$

$$= -\begin{bmatrix} 160 & 54 & 11 \end{bmatrix} \begin{bmatrix} r \\ 0 \\ 0 \end{bmatrix} + 160r = 0$$

At steady state the control signal u becomes zero.

Design of Type 1 Servo System when the Plant Has No Integrator.   If the plant has no integrator (type 0 plant), the basic principle of the design of a type 1 servo system is to insert an integrator in the feedforward path between the error comparator and the plant, as shown in Figure 10–6. (The block diagram of Figure 10–6 is a basic form of the type 1 servo system where the plant has no integrator.) From the diagram, we obtain

$$\dot{\mathbf{x}} = \mathbf{A}\mathbf{x} + \mathbf{B}u \tag{10–31}$$

$$y = \mathbf{C}\mathbf{x} \tag{10–32}$$

$$u = -\mathbf{K}\mathbf{x} + k_I \xi \tag{10–33}$$

$$\dot{\xi} = r - y = r - \mathbf{C}\mathbf{x} \tag{10–34}$$

where $\mathbf{x}$ = state vector of the plant (n-vector)

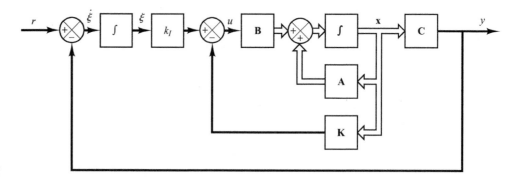

Figure 10–6
Type 1 servo system.

$$u = \text{control signal (scalar)}$$
$$y = \text{output signal (scalar)}$$
$$\xi = \text{output of the integrator (state variable of the system, scalar)}$$
$$r = \text{reference input signal (step function, scalar)}$$
$$\mathbf{A} = n \times n \text{ constant matrix}$$
$$\mathbf{B} = n \times 1 \text{ constant matrix}$$
$$\mathbf{C} = 1 \times n \text{ constant matrix}$$

We assume that the plant given by Equation (10–31) is completely state controllable. The transfer function of the plant can be given by

$$G_p(s) = \mathbf{C}(s\mathbf{I} - \mathbf{A})^{-1}\mathbf{B}$$

To avoid the possibility of the inserted integrator being canceled by the zero at the origin of the plant, we assume that $G_p(s)$ has no zero at the origin.

Assume that the reference input (step function) is applied at $t = 0$. Then, for $t > 0$, the system dynamics can be described by an equation that is a combination of Equations (10–31) and (10–34):

$$\begin{bmatrix} \dot{\mathbf{x}}(t) \\ \dot{\xi}(t) \end{bmatrix} = \begin{bmatrix} \mathbf{A} & \mathbf{0} \\ -\mathbf{C} & 0 \end{bmatrix} \begin{bmatrix} \mathbf{x}(t) \\ \xi(t) \end{bmatrix} + \begin{bmatrix} \mathbf{B} \\ 0 \end{bmatrix} u(t) + \begin{bmatrix} 0 \\ 1 \end{bmatrix} r(t) \qquad (10\text{–}35)$$

We shall design an asymptotically stable system such that $\mathbf{x}(\infty)$, $\xi(\infty)$, and $u(\infty)$ approach constant values, respectively. Then, at steady state, $\dot{\xi}(t) = 0$, and we get $y(\infty) = r$.

Notice that at steady state we have

$$\begin{bmatrix} \dot{\mathbf{x}}(\infty) \\ \dot{\xi}(\infty) \end{bmatrix} = \begin{bmatrix} \mathbf{A} & \mathbf{0} \\ -\mathbf{C} & 0 \end{bmatrix} \begin{bmatrix} \mathbf{x}(\infty) \\ \xi(\infty) \end{bmatrix} + \begin{bmatrix} \mathbf{B} \\ 0 \end{bmatrix} u(\infty) + \begin{bmatrix} 0 \\ 1 \end{bmatrix} r(\infty) \qquad (10\text{–}36)$$

Noting that $r(t)$ is a step input, we have $r(\infty) = r(t) = r$ (constant) for $t > 0$. By subtracting Equation (10–36) from Equation (10–35), we obtain

$$\begin{bmatrix} \dot{\mathbf{x}}(t) - \dot{\mathbf{x}}(\infty) \\ \dot{\xi}(t) - \dot{\xi}(\infty) \end{bmatrix} = \begin{bmatrix} \mathbf{A} & \mathbf{0} \\ -\mathbf{C} & 0 \end{bmatrix} \begin{bmatrix} \mathbf{x}(t) - \mathbf{x}(\infty) \\ \xi(t) - \xi(\infty) \end{bmatrix} + \begin{bmatrix} \mathbf{B} \\ 0 \end{bmatrix} [u(t) - u(\infty)] \qquad (10\text{–}37)$$

Define

$$\mathbf{x}(t) - \mathbf{x}(\infty) = \mathbf{x}_e(t)$$
$$\xi(t) - \xi(\infty) = \xi_e(t)$$
$$u(t) - u(\infty) = u_e(t)$$

Then Equation (10–37) can be written as

$$\begin{bmatrix} \dot{\mathbf{x}}_e(t) \\ \dot{\xi}_e(t) \end{bmatrix} = \begin{bmatrix} \mathbf{A} & \mathbf{0} \\ -\mathbf{C} & 0 \end{bmatrix} \begin{bmatrix} \mathbf{x}_e(t) \\ \xi_e(t) \end{bmatrix} + \begin{bmatrix} \mathbf{B} \\ 0 \end{bmatrix} u_e(t) \qquad (10\text{–}38)$$

where

$$u_e(t) = -\mathbf{K}\mathbf{x}_e(t) + k_I \xi_e(t) \qquad (10\text{–}39)$$

Define a new $(n + 1)$th-order error vector $\mathbf{e}(t)$ by

$$\mathbf{e}(t) = \begin{bmatrix} \mathbf{x}_e(t) \\ \xi_e(t) \end{bmatrix} = (n + 1)\text{-vector}$$

Then Equation (10–38) becomes

$$\dot{\mathbf{e}} = \hat{\mathbf{A}}\mathbf{e} + \hat{\mathbf{B}}u_e \qquad (10\text{–}40)$$

where

$$\hat{\mathbf{A}} = \begin{bmatrix} \mathbf{A} & \mathbf{0} \\ -\mathbf{C} & 0 \end{bmatrix}, \qquad \hat{\mathbf{B}} = \begin{bmatrix} \mathbf{B} \\ 0 \end{bmatrix}$$

and Equation (10–39) becomes

$$u_e = -\hat{\mathbf{K}}\mathbf{e} \qquad (10\text{–}41)$$

where

$$\hat{\mathbf{K}} = \begin{bmatrix} \mathbf{K} & \vdots & -k_I \end{bmatrix}$$

The state error equation can be obtained by substituting Equation (10–41) into Equation (10–40):

$$\dot{\mathbf{e}} = (\hat{\mathbf{A}} - \hat{\mathbf{B}}\hat{\mathbf{K}})\mathbf{e} \qquad (10\text{–}42)$$

If the desired eigenvalues of matrix $\hat{\mathbf{A}} - \hat{\mathbf{B}}\hat{\mathbf{K}}$ (that is, the desired closed-loop poles) are specified as $\mu_1, \mu_2, \ldots, \mu_{n+1}$, then the state-feedback gain matrix $\mathbf{K}$ and the integral gain constant k_I can be determined by the pole-placement technique presented in Section 10–2, provided that the system defined by Equation (10–40) is completely state controllable. Note that if the matrix

$$\begin{bmatrix} \mathbf{A} & \mathbf{B} \\ -\mathbf{C} & 0 \end{bmatrix}$$

has rank $n + 1$, then the system defined by Equation (10–40) is completely state controllable. (See Problem **A–10–12**.)

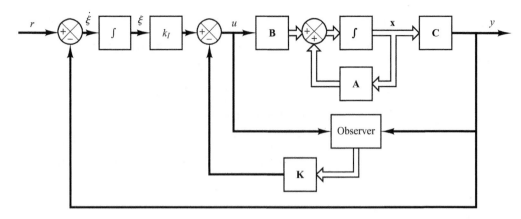

Figure 10–7
Type 1 servo system with state observer.

As is usually the case, not all state variables can be directly measurable. If this is the case, we need to use a state observer. Figure 10–7 shows a block diagram of a type 1 servo system with a state observer. [In the figure, each block with an integral symbol represents an integrator $(1/s)$.] Detailed discussions of state observers are given in Section 10–5.

EXAMPLE 10–5 Consider the inverted-pendulum control system shown in Figure 10–8. In this example, we are concerned only with the motion of the pendulum and motion of the cart in the plane of the page.

It is desired to keep the inverted pendulum upright as much as possible and yet control the position of the cart—for instance, move the cart in a step fashion. To control the position of the cart, we need to build a type 1 servo system. The inverted-pendulum system mounted on a cart does not have an integrator. Therefore, we feed the position signal y (which indicates the position of the cart) back to the input and insert an integrator in the feedforward path, as shown

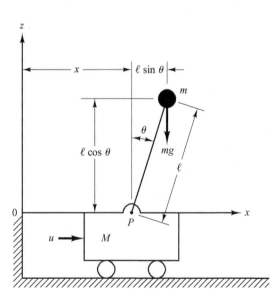

Figure 10–8
Inverted-pendulum control system.

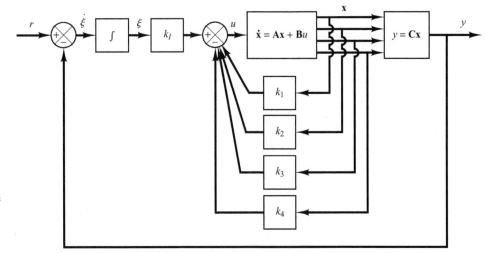

Figure 10–9
Inverted-pendulum
control system. (Type
1 servo system when
the plant has no
integrator.)

in Figure 10–9. We assume that the pendulum angle θ and the angular velocity $\dot\theta$ are small, so that $\sin\theta \doteq \theta$, $\cos\theta \doteq 1$, and $\theta\dot\theta^2 \doteq 0$. We also assume that the numerical values for M, m, and l are given as

$$M = 2 \text{ kg}, \qquad m = 0.1 \text{ kg}, \qquad l = 0.5 \text{ m}$$

Earlier in Example 3–6 we derived the equations for the inverted-pendulum system shown in Figure 3–6, which is the same as that in Figure 10–8. Referring to Figure 3–6, we started with the force-balance and torque-balance equations and ended up with Equations (3–20) and (3–21) to model the inverted-pendulum system. Referring to Equations (3–20) and (3–21), the equations for the inverted-pendulum control system shown in Figure 10–8 are

$$Ml\ddot\theta = (M + m)g\theta - u \tag{10–43}$$

$$M\ddot x = u - mg\theta \tag{10–44}$$

When the given numerical values are substituted, Equations (10–43) and (10–44) become

$$\ddot\theta = 20.601\theta - u \tag{10–45}$$

$$\ddot x = 0.5u - 0.4905\theta \tag{10–46}$$

Let us define the state variables x_1, x_2, x_3, and x_4 as

$$x_1 = \theta$$

$$x_2 = \dot\theta$$

$$x_3 = x$$

$$x_4 = \dot x$$

Then, referring to Equations (10–45) and (10–46) and Figure 10–9 and considering the cart position x as the output of the system, we obtain the equations for the system as follows:

$$\dot{\mathbf{x}} = \mathbf{Ax} + \mathbf{B}u \tag{10–47}$$

$$y = \mathbf{Cx} \tag{10–48}$$

$$u = -\mathbf{Kx} + k_I\xi \tag{10–49}$$

$$\dot\xi = r - y = r - \mathbf{Cx} \tag{10–50}$$

where

$$\mathbf{A} = \begin{bmatrix} 0 & 1 & 0 & 0 \\ 20.601 & 0 & 0 & 0 \\ 0 & 0 & 0 & 1 \\ -0.4905 & 0 & 0 & 0 \end{bmatrix}, \qquad \mathbf{B} = \begin{bmatrix} 0 \\ -1 \\ 0 \\ 0.5 \end{bmatrix}, \qquad \mathbf{C} = \begin{bmatrix} 0 & 0 & 1 & 0 \end{bmatrix}$$

For the type 1 servo system, we have the state error equation as given by Equation (10–40):

$$\dot{\mathbf{e}} = \hat{\mathbf{A}}\mathbf{e} + \hat{\mathbf{B}}u_e \tag{10–51}$$

where

$$\hat{\mathbf{A}} = \begin{bmatrix} \mathbf{A} & \mathbf{0} \\ -\mathbf{C} & 0 \end{bmatrix} = \begin{bmatrix} 0 & 1 & 0 & 0 & 0 \\ 20.601 & 0 & 0 & 0 & 0 \\ 0 & 0 & 0 & 1 & 0 \\ -0.4905 & 0 & 0 & 0 & 0 \\ 0 & 0 & -1 & 0 & 0 \end{bmatrix}, \qquad \hat{\mathbf{B}} = \begin{bmatrix} \mathbf{B} \\ 0 \end{bmatrix} = \begin{bmatrix} 0 \\ -1 \\ 0 \\ 0.5 \\ 0 \end{bmatrix}$$

and the control signal is given by Equation (10–41):

$$u_e = -\hat{\mathbf{K}}\mathbf{e}$$

where

$$\hat{\mathbf{K}} = \begin{bmatrix} \mathbf{K} & \vdots & -k_I \end{bmatrix} = \begin{bmatrix} k_1 & k_2 & k_3 & k_4 & \vdots & -k_I \end{bmatrix}$$

To obtain a reasonable speed and damping in the response of the designed system (for example, the settling time of approximately $4 \sim 5$ sec and the maximum overshoot of $15\% \sim 16\%$ in the step response of the cart), let us choose the desired closed-loop poles at $s = \mu_i$ ($i = 1, 2, 3, 4, 5$), where

$$\mu_1 = -1 + j\sqrt{3}, \qquad \mu_2 = -1 - j\sqrt{3}, \qquad \mu_3 = -5, \qquad \mu_4 = -5, \qquad \mu_5 = -5$$

We shall determine the necessary state-feedback gain matrix by the use of MATLAB.
Before we proceed further, we must examine the rank of matrix $\mathbf{P}$, where

$$\mathbf{P} = \begin{bmatrix} \mathbf{A} & \mathbf{B} \\ -\mathbf{C} & 0 \end{bmatrix}$$

Matrix $\mathbf{P}$ is given by

$$\mathbf{P} = \begin{bmatrix} \mathbf{A} & \mathbf{B} \\ -\mathbf{C} & 0 \end{bmatrix} = \begin{bmatrix} 0 & 1 & 0 & 0 & 0 \\ 20.601 & 0 & 0 & 0 & -1 \\ 0 & 0 & 0 & 1 & 0 \\ -0.4905 & 0 & 0 & 0 & 0.5 \\ 0 & 0 & -1 & 0 & 0 \end{bmatrix} \tag{10–52}$$

The rank of this matrix can be found to be 5. Therefore, the system defined by Equation (10–51) is completely state controllable, and arbitrary pole placement is possible. MATLAB Program 10–6 produces the state feedback gain matrix $\hat{\mathbf{K}}$.

```
MATLAB Program 10–6

A = [0 1 0 0; 20.601 0 0 0; 0 0 0 1; -0.4905 0 0 0];
B = [0;-1;0;0.5];
C = [0 0 1 0];
Ahat = [A  zeros(4,1); -C 0];
Bhat = [B;0];
J = [-1+j*sqrt(3) -1-j*sqrt(3) -5 -5 -5];
Khat = acker(Ahat,Bhat,J)

Khat =

   -157.6336 -35.3733 -56.0652 -36.7466 50.9684
```

Thus, we get

$$\mathbf{K} = \begin{bmatrix} k_1 & k_2 & k_3 & k_4 \end{bmatrix} = \begin{bmatrix} -157.6336 & -35.3733 & -56.0652 & -36.7466 \end{bmatrix}$$

and

$$k_I = -50.9684$$

Unit Step-Response Characteristics of the Designed System. Once we determine the feedback gain matrix $\mathbf{K}$ and the integral gain constant k_I, the step response in the cart position can be obtained by solving the following equation, which is obtained by substituting Equation (10–49) into Equation (10–35):

$$\begin{bmatrix} \dot{\mathbf{x}} \\ \dot{\xi} \end{bmatrix} = \begin{bmatrix} \mathbf{A} - \mathbf{BK} & \mathbf{B}k_I \\ -\mathbf{C} & 0 \end{bmatrix} \begin{bmatrix} \mathbf{x} \\ \xi \end{bmatrix} + \begin{bmatrix} \mathbf{0} \\ 1 \end{bmatrix} r \tag{10–53}$$

The output $y(t)$ of the system is $x_3(t)$, or

$$y = \begin{bmatrix} 0 & 0 & 1 & 0 & 0 \end{bmatrix} \begin{bmatrix} \mathbf{x} \\ \xi \end{bmatrix} + [0]r \tag{10–54}$$

Define the state matrix, control matrix, output matrix, and direct transmission matrix of the system given by Equations (10–53) and (10–54) as AA, BB, CC, and DD, respectively. MATLAB Program 10–7 may be used to obtain the step-response curves of the designed system. Notice that, to obtain the unit-step response, we entered the command

$$[y,x,t] = step(AA,BB,CC,DD,1,t)$$

Figure 10–10 shows curves x_1 versus t, x_2 versus t, x_3 (= output y) versus t, x_4 versus t, and x_5 (= ξ) versus t. Notice that $y(t) \left[= x_3(t)\right]$ has approximately 15% overshoot and the settling time is approximately 4.5 sec. $\xi(t) \left[= x_5(t)\right]$ approaches 1.1. This result can be derived as follows: Since

$$\dot{\mathbf{x}}(\infty) = \mathbf{0} = \mathbf{A}\mathbf{x}(\infty) + \mathbf{B}u(\infty)$$

or

$$\begin{bmatrix} 0 \\ 0 \\ 0 \\ 0 \end{bmatrix} = \begin{bmatrix} 0 & 1 & 0 & 0 \\ 20.601 & 0 & 0 & 0 \\ 0 & 0 & 0 & 1 \\ -0.4905 & 0 & 0 & 0 \end{bmatrix} \begin{bmatrix} 0 \\ 0 \\ r \\ 0 \end{bmatrix} + \begin{bmatrix} 0 \\ -1 \\ 0 \\ 0.5 \end{bmatrix} u(\infty)$$

MATLAB Program 10–7

```
%**** The following program is to obtain step response
% of the inverted-pendulum system just designed *****

A = [0  1  0  0;20.601  0  0  0;0  0  0  1;-0.4905  0  0  0];
B = [0;-1;0;0.5];
C = [0  0  1  0]
D = [0];
K = [-157.6336  -35.3733  -56.0652  -36.7466];
KI = -50.9684;
AA = [A - B*K  B*KI;-C  0];
BB = [0;0;0;0;1];
CC = [C  0];
DD = [0];

%***** To obtain response curves x1 versus t, x2 versus t,
% x3 versus t, x4 versus t, and x5 versus t, separately, enter
% the following command *****

t = 0:0.02:6;
[y,x,t] = step(AA,BB,CC,DD,1,t);

x1 = [1  0  0  0  0]*x';
x2 = [0  1  0  0  0]*x';
x3 = [0  0  1  0  0]*x';
x4 = [0  0  0  1  0]*x';
x5 = [0  0  0  0  1]*x';

subplot(3,2,1); plot(t,x1); grid
title('x1 versus t')
xlabel('t Sec'); ylabel('x1')

subplot(3,2,2); plot(t,x2); grid
title('x2 versus t')
xlabel('t Sec'); ylabel('x2')

subplot(3,2,3); plot(t,x3); grid
title('x3 versus t')
xlabel('t Sec'); ylabel('x3')

subplot(3,2,4); plot(t,x4); grid
title('x4 versus t')
xlabel('t Sec'); ylabel('x4')

subplot(3,2,5); plot(t,x5); grid
title('x5 versus t')
xlabel('t Sec'); ylabel('x5')
```

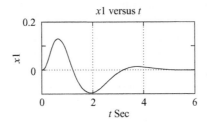

x1 versus t

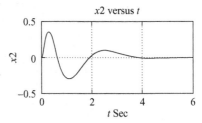

x2 versus t

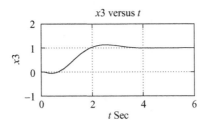

x3 versus t

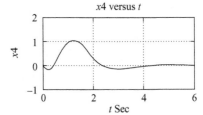

x4 versus t

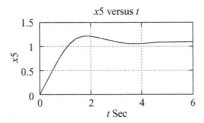

x5 versus t

Figure 10–10
Curves x_1 versus t, x_2 versus t, x_3 (= output y) versus t, x_4 versus t, and x_5 (= ξ) versus t.

we get

$$u(\infty) = 0$$

Since $u(\infty) = 0$, we have, from Equation (10–33),

$$u(\infty) = 0 = -\mathbf{Kx}(\infty) + k_I\xi(\infty)$$

and so

$$\xi(\infty) = \frac{1}{k_I}\left[\mathbf{Kx}(\infty)\right] = \frac{1}{k_I} k_3 x_3(\infty) = \frac{-56.0652}{-50.9684} r = 1.1r$$

Hence, for $r = 1$, we have

$$\xi(\infty) = 1.1$$

It is noted that, as in any design problem, if the speed and damping are not quite satisfactory, then we must modify the desired characteristic equation and determine a new matrix $\hat{\mathbf{K}}$. Computer simulations must be repeated until a satisfactory result is obtained.

10–5 STATE OBSERVERS

In the pole-placement approach to the design of control systems, we assumed that all state variables are available for feedback. In practice, however, not all state variables are available for feedback. Then we need to estimate unavailable state variables.

Estimation of unmeasurable state variables is commonly called *observation*. A device (or a computer program) that estimates or observes the state variables is called a *state observer*, or simply an *observer*. If the state observer observes all state variables of the system, regardless of whether some state variables are available for direct measurement, it is called a *full-order state observer*. There are times when this will not be necessary, when we will need observation of only the unmeasurable state variables, but not of those that are directly measurable as well. For example, since the output variables are observable and they are linearly related to the state variables, we need not observe all state variables, but observe only $n - m$ state variables, where n is the dimension of the state vector and m is the dimension of the output vector.

An observer that estimates fewer than n state variables, where n is the dimension of the state vector, is called a *reduced-order state observer* or, simply, a *reduced-order observer*. If the order of the reduced-order state observer is the minimum possible, the observer is called a *minimum-order state observer* or *minimum-order observer*. In this section, we shall discuss both the full-order state observer and the minimum-order state observer.

State Observer. A state observer estimates the state variables based on the measurements of the output and control variables. Here the concept of observability discussed in Section 9–7 plays an important role. As we shall see later, state observers can be designed if and only if the observability condition is satisfied.

In the following discussions of state observers, we shall use the notation $\tilde{\mathbf{x}}$ to designate the observed state vector. In many practical cases, the observed state vector $\tilde{\mathbf{x}}$ is used in the state feedback to generate the desired control vector.

Consider the plant defined by

$$\dot{\mathbf{x}} = \mathbf{A}\mathbf{x} + \mathbf{B}u \tag{10–55}$$

$$y = \mathbf{C}\mathbf{x} \tag{10–56}$$

The observer is a subsystem to reconstruct the state vector of the plant. The mathematical model of the observer is basically the same as that of the plant, except that we include an additional term that includes the estimation error to compensate for inaccuracies in matrices $\mathbf{A}$ and $\mathbf{B}$ and the lack of the initial error. The estimation error or observation error is the difference between the measured output and the estimated output. The initial error is the difference between the initial state and the initial estimated state. Thus, we define the mathematical model of the observer to be

$$\dot{\tilde{\mathbf{x}}} = \mathbf{A}\tilde{\mathbf{x}} + \mathbf{B}u + \mathbf{K}_e(y - \mathbf{C}\tilde{\mathbf{x}})$$

$$= (\mathbf{A} - \mathbf{K}_e\mathbf{C})\tilde{\mathbf{x}} + \mathbf{B}u + \mathbf{K}_e y \tag{10–57}$$

where $\tilde{\mathbf{x}}$ is the estimated state and $\mathbf{C}\tilde{\mathbf{x}}$ is the estimated output. The inputs to the observer are the output y and the control input u. Matrix $\mathbf{K}_e$, which is called the observer gain matrix, is a weighting matrix to the correction term involving the difference between the measured output y and the estimated output $\mathbf{C}\tilde{\mathbf{x}}$. This term continuously corrects the model output and improves the performance of the observer. Figure 10–11 shows the block diagram of the system and the full-order state observer.

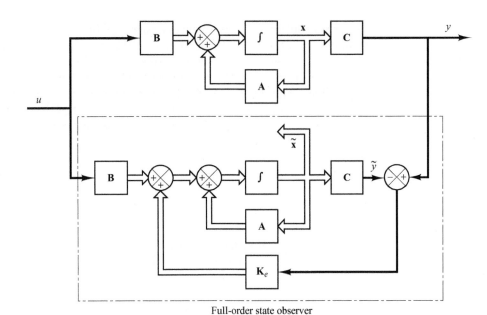

Figure 10–11
Block diagram of
system and full-order
state observer, when
input u and output y
are scalars.

Full-order state observer

Full-Order State Observer. The order of the state observer that will be discussed here is the same as that of the plant. Assume that the plant is defined by Equations (10–55) and (10–56) and the observer model is defined by Equation (10–57).

To obtain the observer error equation, let us subtract Equation (10–57) from Equation (10–55):

$$\dot{\mathbf{x}} - \dot{\tilde{\mathbf{x}}} = \mathbf{A}\mathbf{x} - \mathbf{A}\tilde{\mathbf{x}} - \mathbf{K}_e(\mathbf{C}\mathbf{x} - \mathbf{C}\tilde{\mathbf{x}})$$
$$= (\mathbf{A} - \mathbf{K}_e\mathbf{C})(\mathbf{x} - \tilde{\mathbf{x}}) \tag{10–58}$$

Define the difference between $\mathbf{x}$ and $\tilde{\mathbf{x}}$ as the error vector $\mathbf{e}$, or

$$\mathbf{e} = \mathbf{x} - \tilde{\mathbf{x}}$$

Then Equation (10–58) becomes

$$\dot{\mathbf{e}} = (\mathbf{A} - \mathbf{K}_e\mathbf{C})\mathbf{e} \tag{10–59}$$

From Equation (10–59), we see that the dynamic behavior of the error vector is determined by the eigenvalues of matrix $\mathbf{A} - \mathbf{K}_e\mathbf{C}$. If matrix $\mathbf{A} - \mathbf{K}_e\mathbf{C}$ is a stable matrix, the error vector will converge to zero for any initial error vector $\mathbf{e}(0)$. That is, $\tilde{\mathbf{x}}(t)$ will converge to $\mathbf{x}(t)$ regardless of the values of $\mathbf{x}(0)$ and $\tilde{\mathbf{x}}(0)$. If the eigenvalues of matrix $\mathbf{A} - \mathbf{K}_e\mathbf{C}$ are chosen in such a way that the dynamic behavior of the error vector is asymptotically stable and is adequately fast, then any error vector will tend to zero (the origin) with an adequate speed.

If the plant is completely observable, then it can be proved that it is possible to choose matrix $\mathbf{K}_e$ such that $\mathbf{A} - \mathbf{K}_e\mathbf{C}$ has arbitrarily desired eigenvalues. That is, the observer gain matrix $\mathbf{K}_e$ can be determined to yield the desired matrix $\mathbf{A} - \mathbf{K}_e\mathbf{C}$. We shall discuss this matter in what follows.

Dual Problem. The problem of designing a full-order observer becomes that of determining the observer gain matrix $\mathbf{K}_e$ such that the error dynamics defined by Equation (10–59) are asymptotically stable with sufficient speed of response. (The asymptotic stability and the speed of response of the error dynamics are determined by the eigenvalues of matrix $\mathbf{A} - \mathbf{K}_e\mathbf{C}$.) Hence, the design of the full-order observer becomes that of determining an appropriate $\mathbf{K}_e$ such that $\mathbf{A} - \mathbf{K}_e\mathbf{C}$ has desired eigenvalues. Thus, the problem here becomes the same as the pole-placement problem we discussed in Section 10–2. In fact, the two problems are mathematically the same. This property is called duality.

Consider the system defined by

$$\dot{\mathbf{x}} = \mathbf{A}\mathbf{x} + \mathbf{B}u$$

$$y = \mathbf{C}\mathbf{x}$$

In designing the full-order state observer, we may solve the dual problem, that is, solve the pole-placement problem for the dual system

$$\dot{\mathbf{z}} = \mathbf{A}^*\mathbf{z} + \mathbf{C}^*v$$

$$n = \mathbf{B}^*\mathbf{z}$$

assuming the control signal v to be

$$v = -\mathbf{K}\mathbf{z}$$

If the dual system is completely state controllable, then the state feedback gain matrix $\mathbf{K}$ can be determined such that matrix $\mathbf{A}^* - \mathbf{C}^*\mathbf{K}$ will yield a set of the desired eigenvalues.

If $\mu_1, \mu_2, \ldots, \mu_n$ are the desired eigenvalues of the state observer matrix, then by taking the same μ_i's as the desired eigenvalues of the state-feedback gain matrix of the dual system, we obtain

$$\left|s\mathbf{I} - (\mathbf{A}^* - \mathbf{C}^*\mathbf{K})\right| = (s - \mu_1)(s - \mu_2)\cdots(s - \mu_n)$$

Noting that the eigenvalues of $\mathbf{A}^* - \mathbf{C}^*\mathbf{K}$ and those of $\mathbf{A} - \mathbf{K}^*\mathbf{C}$ are the same, we have

$$\left|s\mathbf{I} - (\mathbf{A}^* - \mathbf{C}^*\mathbf{K})\right| = \left|s\mathbf{I} - (\mathbf{A} - \mathbf{K}^*\mathbf{C})\right|$$

Comparing the characteristic polynomial $\left|s\mathbf{I} - (\mathbf{A} - \mathbf{K}^*\mathbf{C})\right|$ and the characteristic polynomial $\left|s\mathbf{I} - (\mathbf{A} - \mathbf{K}_e\mathbf{C})\right|$ for the observer system [refer to Equation (10–57)], we find that $\mathbf{K}_e$ and $\mathbf{K}^*$ are related by

$$\mathbf{K}_e = \mathbf{K}^*$$

Thus, using the matrix $\mathbf{K}$ determined by the pole-placement approach in the dual system, the observer gain matrix $\mathbf{K}_e$ for the original system can be determined by using the relationship $\mathbf{K}_e = \mathbf{K}^*$. (See Problem **A–10–10** for the details.)

Necessary and Sufficient Condition for State Observation. As discussed, a necessary and sufficient condition for the determination of the observer gain matrix $\mathbf{K}_e$ for the desired eigenvalues of $\mathbf{A} - \mathbf{K}_e\mathbf{C}$ is that the dual of the original system

$$\dot{\mathbf{z}} = \mathbf{A}^*\mathbf{z} + \mathbf{C}^*v$$

be completely state controllable. The complete state controllability condition for this dual system is that the rank of

$$\begin{bmatrix} \mathbf{C}^* & \vdots & \mathbf{A}^*\mathbf{C}^* & \vdots & \cdots & \vdots & (\mathbf{A}^*)^{n-1}\mathbf{C}^* \end{bmatrix}$$

be n. This is the condition for complete observability of the original system defined by Equations (10–55) and (10–56). This means that a necessary and sufficient condition for the observation of the state of the system defined by Equations (10–55) and (10–56) is that the system be completely observable.

Once we select the desired eigenvalues (or desired characteristic equation), the full-order state observer can be designed, provided the plant is completely observable. The desired eigenvalues of the characteristic equation should be chosen so that the state observer responds at least two to five times faster than the closed-loop system considered. As stated earlier, the equation for the full-order state observer is

$$\dot{\tilde{\mathbf{x}}} = (\mathbf{A} - \mathbf{K}_e\mathbf{C})\tilde{\mathbf{x}} + \mathbf{B}u + \mathbf{K}_e y \tag{10–60}$$

It is noted that thus far we have assumed the matrices $\mathbf{A}$, $\mathbf{B}$, and $\mathbf{C}$ in the observer to be exactly the same as those of the physical plant. If there are discrepancies in $\mathbf{A}$, $\mathbf{B}$, and $\mathbf{C}$ in the observer and in the physical plant, the dynamics of the observer error are no longer governed by Equation (10–59). This means that the error may not approach zero as expected. Therefore, we need to choose $\mathbf{K}_e$ so that the observer is stable and the error remains acceptably small in the presence of small modeling errors.

Transformation Approach to Obtain State Observer Gain Matrix $\mathbf{K}_e$. By following the same approach as we used in deriving the equation for the state feedback gain matrix $\mathbf{K}$, we can obtain the following equation:

$$\mathbf{K}_e = \mathbf{Q}\begin{bmatrix} \alpha_n - a_n \\ \alpha_{n-1} - a_{n-1} \\ \cdot \\ \cdot \\ \cdot \\ \alpha_1 - a_1 \end{bmatrix} = (\mathbf{WN}^*)^{-1}\begin{bmatrix} \alpha_n - a_n \\ \alpha_{n-1} - a_{n-1} \\ \cdot \\ \cdot \\ \cdot \\ \alpha_1 - a_1 \end{bmatrix} \tag{10–61}$$

where $\mathbf{K}_e$ is an $n \times 1$ matrix,

$$\mathbf{Q} = (\mathbf{WN}^*)^{-1}$$

and

$$\mathbf{N} = \begin{bmatrix} \mathbf{C}^* & \vdots & \mathbf{A}^*\mathbf{C}^* & \vdots & \cdots & \vdots & (\mathbf{A}^*)^{n-1}\mathbf{C}^* \end{bmatrix}$$

$$\mathbf{W} = \begin{bmatrix} a_{n-1} & a_{n-2} & \cdots & a_1 & 1 \\ a_{n-2} & a_{n-3} & \cdots & 1 & 0 \\ \cdot & \cdot & & \cdot & \cdot \\ \cdot & \cdot & & \cdot & \cdot \\ \cdot & \cdot & & \cdot & \cdot \\ a_1 & 1 & \cdots & 0 & 0 \\ 1 & 0 & \cdots & 0 & 0 \end{bmatrix}$$

[Refer to Problem **A–10–10** for the derivation of Equation (10–61).]

Direct-Substitution Approach to Obtain State Observer Gain Matrix $\mathbf{K}_e$.
Similar to the case of pole placement, if the system is of low order, then direct substitution of matrix $\mathbf{K}_e$ into the desired characteristic polynomial may be simpler. For example, if $\mathbf{x}$ is a 3-vector, then write the observer gain matrix $\mathbf{K}_e$ as

$$\mathbf{K}_e = \begin{bmatrix} k_{e1} \\ k_{e2} \\ k_{e3} \end{bmatrix}$$

Substitute this $\mathbf{K}_e$ matrix into the desired characteristic polynomial:

$$\left| s\mathbf{I} - (\mathbf{A} - \mathbf{K}_e\mathbf{C}) \right| = (s - \mu_1)(s - \mu_2)(s - \mu_3)$$

By equating the coefficients of the like powers of s on both sides of this last equation, we can determine the values of k_{e1}, k_{e2}, and k_{e3}. This approach is convenient if $n = 1$, 2, or 3, where n is the dimension of the state vector $\mathbf{x}$. (Although this approach can be used when $n = 4, 5, 6, \ldots$, the computations involved may become very tedious.)

Another approach to the determination of the state observer gain matrix $\mathbf{K}_e$ is to use Ackermann's formula. This approach is presented in the following.

Ackermann's Formula.   Consider the system defined by

$$\dot{\mathbf{x}} = \mathbf{A}\mathbf{x} + \mathbf{B}u \tag{10-62}$$

$$y = \mathbf{C}\mathbf{x} \tag{10-63}$$

In Section 10–2 we derived Ackermann's formula for pole placement for the system defined by Equation (10–62). The result was given by Equation (10–18), rewritten thus:

$$\mathbf{K} = \begin{bmatrix} 0 & 0 & \cdots & 0 & 1 \end{bmatrix}\begin{bmatrix} \mathbf{B} & \vdots & \mathbf{AB} & \vdots & \cdots & \vdots & \mathbf{A}^{n-1}\mathbf{B} \end{bmatrix}^{-1}\phi(\mathbf{A})$$

For the dual of the system defined by Equations (10–62) and (10–63),

$$\dot{\mathbf{z}} = \mathbf{A}^*\mathbf{z} + \mathbf{C}^*v$$

$$n = \mathbf{B}^*\mathbf{z}$$

the preceding Ackermann's formula for pole placement is modified to

$$\mathbf{K} = \begin{bmatrix} 0 & 0 & \cdots & 0 & 1 \end{bmatrix}\begin{bmatrix} \mathbf{C}^* & \vdots & \mathbf{A}^*\mathbf{C}^* & \vdots & \cdots & \vdots & (\mathbf{A}^*)^{n-1}\mathbf{C}^* \end{bmatrix}^{-1}\phi(\mathbf{A}^*) \tag{10-64}$$

As stated earlier, the state observer gain matrix $\mathbf{K}_e$ is given by $\mathbf{K}^*$, where $\mathbf{K}$ is given by Equation (10–64). Thus,

$$\mathbf{K}_e = \mathbf{K}^* = \phi(\mathbf{A}^*)^*\begin{bmatrix} \mathbf{C} \\ \mathbf{CA} \\ \cdot \\ \cdot \\ \cdot \\ \mathbf{CA}^{n-2} \\ \mathbf{CA}^{n-1} \end{bmatrix}^{-1}\begin{bmatrix} 0 \\ 0 \\ \cdot \\ \cdot \\ \cdot \\ 0 \\ 1 \end{bmatrix} = \phi(\mathbf{A})\begin{bmatrix} \mathbf{C} \\ \mathbf{CA} \\ \cdot \\ \cdot \\ \cdot \\ \mathbf{CA}^{n-2} \\ \mathbf{CA}^{n-1} \end{bmatrix}^{-1}\begin{bmatrix} 0 \\ 0 \\ \cdot \\ \cdot \\ \cdot \\ 0 \\ 1 \end{bmatrix} \tag{10-65}$$

where $\phi(s)$ is the desired characteristic polynomial for the state observer, or

$$\phi(s) = (s - \mu_1)(s - \mu_2) \cdots (s - \mu_n)$$

where $\mu_1, \mu_2, \ldots, \mu_n$ are the desired eigenvalues. Equation (10–65) is called Ackermann's formula for the determination of the observer gain matrix $\mathbf{K}_e$.

Comments on Selecting the Best $\mathbf{K}_e$. Referring to Figure 10–11, notice that the feedback signal through the observer gain matrix $\mathbf{K}_e$ serves as a correction signal to the plant model to account for the unknowns in the plant. If significant unknowns are involved, the feedback signal through the matrix $\mathbf{K}_e$ should be relatively large. However, if the output signal is contaminated significantly by disturbances and measurement noises, then the output y is not reliable and the feedback signal through the matrix $\mathbf{K}_e$ should be relatively small. In determining the matrix $\mathbf{K}_e$, we should carefully examine the effects of disturbances and noises involved in the output y.

Remember that the observer gain matrix $\mathbf{K}_e$ depends on the desired characteristic equation

$$(s - \mu_1)(s - \mu_2) \cdots (s - \mu_n) = 0$$

The choice of a set of $\mu_1, \mu_2, \ldots, \mu_n$ is, in many instances, not unique. As a general rule, however, the observer poles must be two to five times faster than the controller poles to make sure the observation error (estimation error) converges to zero quickly. This means that the observer estimation error decays two to five times faster than does the state vector $\mathbf{x}$. Such faster decay of the observer error compared with the desired dynamics makes the controller poles dominate the system response.

It is important to note that if sensor noise is considerable, we may choose the observer poles to be slower than two times the controller poles, so that the bandwidth of the system will become lower and smooth the noise. In this case the system response will be strongly influenced by the observer poles. If the observer poles are located to the right of the controller poles in the left-half s plane, the system response will be dominated by the observer poles rather than by the control poles.

In the design of the state observer, it is desirable to determine several observer gain matrices $\mathbf{K}_e$ based on several different desired characteristic equations. For each of the several different matrices $\mathbf{K}_e$, simulation tests must be run to evaluate the resulting system performance. Then we select the best $\mathbf{K}_e$ from the viewpoint of overall system performance. In many practical cases, the selection of the best matrix $\mathbf{K}_e$ boils down to a compromise between speedy response and sensitivity to disturbances and noises.

EXAMPLE 10–6 Consider the system

$$\dot{\mathbf{x}} = \mathbf{A}\mathbf{x} + \mathbf{B}u$$

$$y = \mathbf{C}\mathbf{x}$$

where

$$\mathbf{A} = \begin{bmatrix} 0 & 20.6 \\ 1 & 0 \end{bmatrix}, \qquad \mathbf{B} = \begin{bmatrix} 0 \\ 1 \end{bmatrix}, \qquad \mathbf{C} = \begin{bmatrix} 0 & 1 \end{bmatrix}$$

We use the observed state feedback such that

$$u = -\mathbf{K}\tilde{\mathbf{x}}$$

Design a full-order state observer, assuming that the system configuration is identical to that shown in Figure 10–11. Assume that the desired eigenvalues of the observer matrix are

$$\mu_1 = -10, \qquad \mu_2 = -10$$

The design of the state observer reduces to the determination of an appropriate observer gain matrix $\mathbf{K}_e$.

Let us examine the observability matrix. The rank of

$$[\mathbf{C}^* \vdots \mathbf{A}^*\mathbf{C}^*] = \begin{bmatrix} 0 & 1 \\ 1 & 0 \end{bmatrix}$$

is 2. Hence, the system is completely observable and the determination of the desired observer gain matrix is possible. We shall solve this problem by three methods.

Method 1: We shall determine the observer gain matrix by use of Equation (10–61). The given system is already in the observable canonical form. Hence, the transformation matrix $\mathbf{Q} = (\mathbf{WN}^*)^{-1}$ is $\mathbf{I}$. Since the characteristic equation of the given system is

$$|s\mathbf{I} - \mathbf{A}| = \begin{vmatrix} s & -20.6 \\ -1 & s \end{vmatrix} = s^2 - 20.6 = s^2 + a_1 s + a_2 = 0$$

we have

$$a_1 = 0, \qquad a_2 = -20.6$$

The desired characteristic equation is

$$(s + 10)^2 = s^2 + 20s + 100 = s^2 + \alpha_1 s + \alpha_2 = 0$$

Hence,

$$\alpha_1 = 20, \qquad \alpha_2 = 100$$

Then the observer gain matrix $\mathbf{K}_e$ can be obtained from Equation (10–61) as follows:

$$\mathbf{K}_e = (\mathbf{WN}^*)^{-1} \begin{bmatrix} \alpha_2 - a_2 \\ \alpha_1 - a_1 \end{bmatrix} = \begin{bmatrix} 1 & 0 \\ 0 & 1 \end{bmatrix} \begin{bmatrix} 100 + 20.6 \\ 20 - 0 \end{bmatrix} = \begin{bmatrix} 120.6 \\ 20 \end{bmatrix}$$

Method 2: Referring to Equation (10–59):

$$\dot{\mathbf{e}} = (\mathbf{A} - \mathbf{K}_e \mathbf{C})\mathbf{e}$$

the characteristic equation for the observer becomes

$$|s\mathbf{I} - \mathbf{A} + \mathbf{K}_e \mathbf{C}| = 0$$

Define

$$\mathbf{K}_e = \begin{bmatrix} k_{e1} \\ k_{e2} \end{bmatrix}$$

Then the characteristic equation becomes

$$\left| \begin{bmatrix} s & 0 \\ 0 & s \end{bmatrix} - \begin{bmatrix} 0 & 20.6 \\ 1 & 0 \end{bmatrix} + \begin{bmatrix} k_{e1} \\ k_{e2} \end{bmatrix} [0 \quad 1] \right| = \begin{vmatrix} s & -20.6 + k_{e1} \\ -1 & s + k_{e2} \end{vmatrix}$$

$$= s^2 + k_{e2}s - 20.6 + k_{e1} = 0 \tag{10–66}$$

Since the desired characteristic equation is

$$s^2 + 20s + 100 = 0$$

by comparing Equation (10–66) with this last equation, we obtain

$$k_{e1} = 120.6, \qquad k_{e2} = 20$$

or

$$\mathbf{K}_e = \begin{bmatrix} 120.6 \\ 20 \end{bmatrix}$$

Method 3:   We shall use Ackermann's formula given by Equation (10–65):

$$\mathbf{K}_e = \phi(\mathbf{A}) \begin{bmatrix} \mathbf{C} \\ \mathbf{CA} \end{bmatrix}^{-1} \begin{bmatrix} 0 \\ 1 \end{bmatrix}$$

where

$$\phi(s) = (s - \mu_1)(s - \mu_2) = s^2 + 20s + 100$$

Thus,

$$\phi(\mathbf{A}) = \mathbf{A}^2 + 20\mathbf{A} + 100\mathbf{I}$$

and

$$\mathbf{K}_e = (\mathbf{A}^2 + 20\mathbf{A} + 100\mathbf{I}) \begin{bmatrix} 0 & 1 \\ 1 & 0 \end{bmatrix}^{-1} \begin{bmatrix} 0 \\ 1 \end{bmatrix}$$

$$= \begin{bmatrix} 120.6 & 412 \\ 20 & 120.6 \end{bmatrix} \begin{bmatrix} 0 & 1 \\ 1 & 0 \end{bmatrix} \begin{bmatrix} 0 \\ 1 \end{bmatrix} = \begin{bmatrix} 120.6 \\ 20 \end{bmatrix}$$

As a matter of course, we get the same $\mathbf{K}_e$ regardless of the method employed.

The equation for the full-order state observer is given by Equation (10–57),

$$\dot{\tilde{\mathbf{x}}} = (\mathbf{A} - \mathbf{K}_e \mathbf{C}) \tilde{\mathbf{x}} + \mathbf{B}u + \mathbf{K}_e y$$

or

$$\begin{bmatrix} \dot{\tilde{x}}_1 \\ \dot{\tilde{x}}_2 \end{bmatrix} = \begin{bmatrix} 0 & -100 \\ 1 & -20 \end{bmatrix} \begin{bmatrix} \tilde{x}_1 \\ \tilde{x}_2 \end{bmatrix} + \begin{bmatrix} 0 \\ 1 \end{bmatrix} u + \begin{bmatrix} 120.6 \\ 20 \end{bmatrix} y$$

Finally, it is noted that, similar to the case of pole placement, if the system order n is 4 or higher, methods 1 and 3 are preferred, because all matrix computations can be carried out by a computer, while method 2 always requires hand computation of the characteristic equation involving unknown parameters $k_{e1}, k_{e2}, \ldots, k_{en}$.

Effects of the Addition of the Observer on a Closed-Loop System.   In the pole-placement design process, we assumed that the actual state $\mathbf{x}(t)$ was available for feedback. In practice, however, the actual state $\mathbf{x}(t)$ may not be measurable, so we will need to design an observer and use the observed state $\tilde{\mathbf{x}}(t)$ for feedback as shown in Figure 10–12. The design process, therefore, becomes a two-stage process, the first stage being the determination of the feedback gain matrix $\mathbf{K}$ to yield the desired characteristic equation and the second stage being the determination of the observer gain matrix $\mathbf{K}_e$ to yield the desired observer characteristic equation.

Let us now investigate the effects of the use of the observed state $\tilde{\mathbf{x}}(t)$, rather than the actual state $\mathbf{x}(t)$, on the characteristic equation of a closed-loop control system.

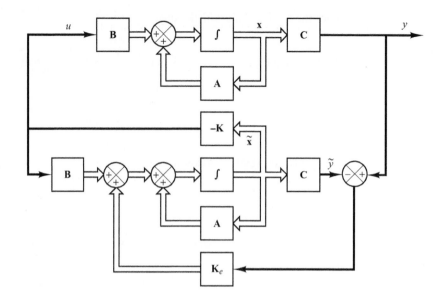

Figure 10–12
Observed-state
feedback control
system.

Consider the completely state controllable and completely observable system defined by the equations

$$\dot{\mathbf{x}} = \mathbf{A}\mathbf{x} + \mathbf{B}u$$

$$y = \mathbf{C}\mathbf{x}$$

For the state-feedback control based on the observed state $\tilde{\mathbf{x}}$,

$$u = -\mathbf{K}\tilde{\mathbf{x}}$$

With this control, the state equation becomes

$$\dot{\mathbf{x}} = \mathbf{A}\mathbf{x} - \mathbf{B}\mathbf{K}\tilde{\mathbf{x}} = (\mathbf{A} - \mathbf{B}\mathbf{K})\mathbf{x} + \mathbf{B}\mathbf{K}(\mathbf{x} - \tilde{\mathbf{x}}) \qquad (10\text{–}67)$$

The difference between the actual state $\mathbf{x}(t)$ and the observed state $\tilde{\mathbf{x}}(t)$ has been defined as the error $\mathbf{e}(t)$:

$$\mathbf{e}(t) = \mathbf{x}(t) - \tilde{\mathbf{x}}(t)$$

Substitution of the error vector $\mathbf{e}(t)$ into Equation (10–67) gives

$$\dot{\mathbf{x}} = (\mathbf{A} - \mathbf{B}\mathbf{K})\mathbf{x} + \mathbf{B}\mathbf{K}\mathbf{e} \qquad (10\text{–}68)$$

Note that the observer error equation was given by Equation (10–59), repeated here:

$$\dot{\mathbf{e}} = (\mathbf{A} - \mathbf{K}_e\mathbf{C})\mathbf{e} \qquad (10\text{–}69)$$

Combining Equations (10–68) and (10–69), we obtain

$$\begin{bmatrix} \dot{\mathbf{x}} \\ \dot{\mathbf{e}} \end{bmatrix} = \begin{bmatrix} \mathbf{A} - \mathbf{B}\mathbf{K} & \mathbf{B}\mathbf{K} \\ \mathbf{0} & \mathbf{A} - \mathbf{K}_e\mathbf{C} \end{bmatrix} \begin{bmatrix} \mathbf{x} \\ \mathbf{e} \end{bmatrix} \qquad (10\text{–}70)$$

Chapter 10 / Control Systems Design in State Space

Equation (10–70) describes the dynamics of the observed-state feedback control system. The characteristic equation for the system is

$$\begin{vmatrix} s\mathbf{I} - \mathbf{A} + \mathbf{BK} & -\mathbf{BK} \\ \mathbf{0} & s\mathbf{I} - \mathbf{A} + \mathbf{K}_e\mathbf{C} \end{vmatrix} = 0$$

or

$$|s\mathbf{I} - \mathbf{A} + \mathbf{BK}||s\mathbf{I} - \mathbf{A} + \mathbf{K}_e\mathbf{C}| = 0$$

Notice that the closed-loop poles of the observed-state feedback control system consist of the poles due to the pole-placement design alone and the poles due to the observer design alone. This means that the pole-placement design and the observer design are independent of each other. They can be designed separately and combined to form the observed-state feedback control system. Note that, if the order of the plant is n, then the observer is also of nth order (if the full-order state observer is used), and the resulting characteristic equation for the entire closed-loop system becomes of order $2n$.

Transfer Function of the Observer-Based Controller. Consider the plant defined by

$$\dot{\mathbf{x}} = \mathbf{A}\mathbf{x} + \mathbf{B}u$$

$$y = \mathbf{C}\mathbf{x}$$

Assume that the plant is completely observable. Assume that we use observed-state feedback control $u = -\mathbf{K}\tilde{\mathbf{x}}$. Then, the equations for the observer are given by

$$\dot{\tilde{\mathbf{x}}} = (\mathbf{A} - \mathbf{K}_e\mathbf{C} - \mathbf{BK})\tilde{\mathbf{x}} + \mathbf{K}_e y \tag{10–71}$$

$$u = -\mathbf{K}\tilde{\mathbf{x}} \tag{10–72}$$

where Equation (10–71) is obtained by substituting $u = -\mathbf{K}\tilde{\mathbf{x}}$ into Equation (10–57).

By taking the Laplace transform of Equation (10–71), assuming a zero initial condition, and solving for $\tilde{\mathbf{X}}(s)$, we obtain

$$\tilde{\mathbf{X}}(s) = (s\mathbf{I} - \mathbf{A} + \mathbf{K}_e\mathbf{C} + \mathbf{BK})^{-1}\mathbf{K}_e Y(s)$$

By substituting this $\tilde{\mathbf{X}}(s)$ into the Laplace transform of Equation (10–72), we obtain

$$U(s) = -\mathbf{K}(s\mathbf{I} - \mathbf{A} + \mathbf{K}_e\mathbf{C} + \mathbf{BK})^{-1}\mathbf{K}_e Y(s) \tag{10–73}$$

Then the transfer function $U(s)/Y(s)$ can be obtained as

$$\frac{U(s)}{Y(s)} = -\mathbf{K}(s\mathbf{I} - \mathbf{A} + \mathbf{K}_e\mathbf{C} + \mathbf{BK})^{-1}\mathbf{K}_e$$

Figure 10–13 shows the block diagram representation for the system. Notice that the transfer function

$$\mathbf{K}(s\mathbf{I} - \mathbf{A} + \mathbf{K}_e\mathbf{C} + \mathbf{BK})^{-1}\mathbf{K}_e$$

acts as a controller for the system. Hence, we call the transfer function

$$\frac{U(s)}{-Y(s)} = \frac{num}{den} = \mathbf{K}(s\mathbf{I} - \mathbf{A} + \mathbf{K}_e\mathbf{C} + \mathbf{BK})^{-1}\mathbf{K}_e \tag{10–74}$$

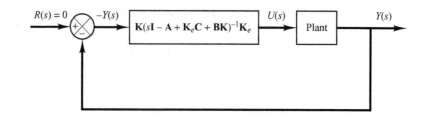

the observer-based controller transfer function or, simply, the observer-controller transfer function.

Note that the observer-controller matrix

$$\mathbf{A} - \mathbf{K}_e\mathbf{C} - \mathbf{B}\mathbf{K}$$

may or may not be stable, although $\mathbf{A} - \mathbf{B}\mathbf{K}$ and $\mathbf{A} - \mathbf{K}_e\mathbf{C}$ are chosen to be stable. In fact, in some cases the matrix $\mathbf{A} - \mathbf{K}_e\mathbf{C} - \mathbf{B}\mathbf{K}$ may be poorly stable or even unstable.

EXAMPLE 10–7 Consider the design of a regulator system for the following plant:

$$\dot{\mathbf{x}} = \mathbf{A}\mathbf{x} + \mathbf{B}u \qquad (10\text{–}75)$$

$$y = \mathbf{C}\mathbf{x} \qquad (10\text{–}76)$$

where

$$\mathbf{A} = \begin{bmatrix} 0 & 1 \\ 20.6 & 0 \end{bmatrix}, \qquad \mathbf{B} = \begin{bmatrix} 0 \\ 1 \end{bmatrix}, \qquad \mathbf{C} = [1 \quad 0]$$

Suppose that we use the pole-placement approach to the design of the system and that the desired closed-loop poles for this system are at $s = \mu_i$ $(i = 1, 2)$, where $\mu_1 = -1.8 + j2.4$ and $\mu_2 = -1.8 - j2.4$. The state-feedback gain matrix $\mathbf{K}$ for this case can be obtained as follows:

$$\mathbf{K} = [29.6 \quad 3.6]$$

Using this state-feedback gain matrix $\mathbf{K}$, the control signal u is given by

$$u = -\mathbf{K}\mathbf{x} = -[29.6 \quad 3.6]\begin{bmatrix} x_1 \\ x_2 \end{bmatrix}$$

Suppose that we use the observed-state feedback control instead of the actual-state feedback control, or

$$u = -\mathbf{K}\tilde{\mathbf{x}} = -[29.6 \quad 3.6]\begin{bmatrix} \tilde{x}_1 \\ \tilde{x}_2 \end{bmatrix}$$

where we choose the observer poles to be at

$$s = -8, \qquad s = -8$$

Obtain the observer gain matrix $\mathbf{K}_e$ and draw a block diagram for the observed-state feedback control system. Then obtain the transfer function $U(s)/[-Y(s)]$ for the observer controller, and draw another block diagram with the observer controller as a series controller in the feedforward path. Finally, obtain the response of the system to the following initial condition:

$$\mathbf{x}(0) = \begin{bmatrix} 1 \\ 0 \end{bmatrix}, \qquad \mathbf{e}(0) = \mathbf{x}(0) - \tilde{\mathbf{x}}(0) = \begin{bmatrix} 0.5 \\ 0 \end{bmatrix}$$

For the system defined by Equation (10–75), the characteristic polynomial is

$$|s\mathbf{I} - \mathbf{A}| = \begin{vmatrix} s & -1 \\ -20.6 & s \end{vmatrix} = s^2 - 20.6 = s^2 + a_1 s + a_2$$

Thus,

$$a_1 = 0, \qquad a_2 = -20.6$$

The desired characteristic polynomial for the observer is

$$(s - \mu_1)(s - \mu_2) = (s + 8)(s + 8) = s^2 + 16s + 64$$
$$= s^2 + \alpha_1 s + \alpha_2$$

Hence,

$$\alpha_1 = 16, \qquad \alpha_2 = 64$$

For the determination of the observer gain matrix, we use Equation (10–61), or

$$\mathbf{K}_e = (\mathbf{W}\mathbf{N}^*)^{-1} \begin{bmatrix} \alpha_2 - a_2 \\ \alpha_1 - a_1 \end{bmatrix}$$

where

$$\mathbf{N} = [\mathbf{C}^* \ \vdots \ \mathbf{A}^*\mathbf{C}^*] = \begin{bmatrix} 1 & 0 \\ 0 & 1 \end{bmatrix}$$

$$\mathbf{W} = \begin{bmatrix} a_1 & 1 \\ 1 & 0 \end{bmatrix} = \begin{bmatrix} 0 & 1 \\ 1 & 0 \end{bmatrix}$$

Hence,

$$\mathbf{K}_e = \left\{ \begin{bmatrix} 0 & 1 \\ 1 & 0 \end{bmatrix} \begin{bmatrix} 1 & 0 \\ 0 & 1 \end{bmatrix} \right\}^{-1} \begin{bmatrix} 64 + 20.6 \\ 16 - 0 \end{bmatrix}$$
$$= \begin{bmatrix} 0 & 1 \\ 1 & 0 \end{bmatrix} \begin{bmatrix} 84.6 \\ 16 \end{bmatrix} = \begin{bmatrix} 16 \\ 84.6 \end{bmatrix} \tag{10–77}$$

Equation (10–77) gives the observer gain matrix $\mathbf{K}_e$. The observer equation is given by Equation (10–60):

$$\dot{\tilde{\mathbf{x}}} = (\mathbf{A} - \mathbf{K}_e\mathbf{C})\tilde{\mathbf{x}} + \mathbf{B}u + \mathbf{K}_e y \tag{10–78}$$

Since

$$u = -\mathbf{K}\tilde{\mathbf{x}}$$

Equation (10–78) becomes

$$\dot{\tilde{\mathbf{x}}} = (\mathbf{A} - \mathbf{K}_e\mathbf{C} - \mathbf{B}\mathbf{K})\tilde{\mathbf{x}} + \mathbf{K}_e y$$

or

$$\begin{bmatrix} \dot{\tilde{x}}_1 \\ \dot{\tilde{x}}_2 \end{bmatrix} = \left\{ \begin{bmatrix} 0 & 1 \\ 20.6 & 0 \end{bmatrix} - \begin{bmatrix} 16 \\ 84.6 \end{bmatrix} \begin{bmatrix} 1 & 0 \end{bmatrix} - \begin{bmatrix} 0 \\ 1 \end{bmatrix} \begin{bmatrix} 29.6 & 3.6 \end{bmatrix} \right\} \begin{bmatrix} \tilde{x}_1 \\ \tilde{x}_2 \end{bmatrix} + \begin{bmatrix} 16 \\ 84.6 \end{bmatrix} y$$
$$= \begin{bmatrix} -16 & 1 \\ -93.6 & -3.6 \end{bmatrix} \begin{bmatrix} \tilde{x}_1 \\ \tilde{x}_2 \end{bmatrix} + \begin{bmatrix} 16 \\ 84.6 \end{bmatrix} y$$

The block diagram of the system with observed-state feedback is shown in Figure 10–14(a).

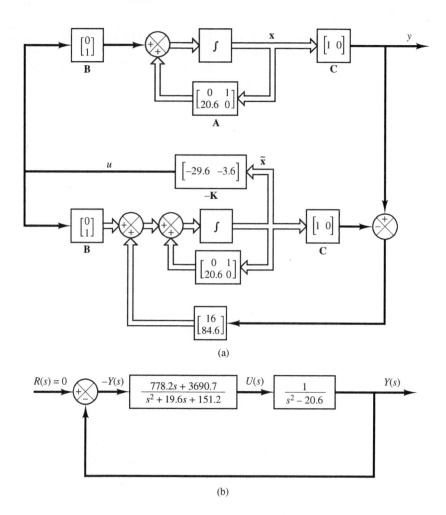

Figure 10–14
(a) Block diagram of system with observed-state feedback; (b) block diagram of transfer-function system.

(a)

(b)

Referring to Equation (10–74), the transfer function of the observer-controller is

$$\frac{U(s)}{-Y(s)} = \mathbf{K}(s\mathbf{I} - \mathbf{A} + \mathbf{K}_e\mathbf{C} + \mathbf{B}\mathbf{K})^{-1}\mathbf{K}_e$$

$$= [29.6 \quad 3.6]\begin{bmatrix} s + 16 & -1 \\ 93.6 & s + 3.6 \end{bmatrix}^{-1}\begin{bmatrix} 16 \\ 84.6 \end{bmatrix}$$

$$= \frac{778.2s + 3690.7}{s^2 + 19.6s + 151.2}$$

As a matter of course, the same transfer function can be obtained with MATLAB. For example, MATLAB Program 10–8 produces the transfer function of the observer controller. Figure 10–14(b) shows a block diagram of the system.

```
MATLAB Program 10–8

% Obtaining transfer function of observer controller --- full-order observer

A = [0 1;20.6 0];
B = [0;1];
C = [1 0];
K = [29.6 3.6];
Ke = [16;84.6];
AA = A-Ke*C-B*K;
BB = Ke;
CC = K;
DD = 0;
[num,den] = ss2tf(AA,BB,CC,DD)

num =

   1.0e+003*

     0  0.7782  3.6907

den =

   1.0000  19.6000  151.2000
```

The dynamics of the observed-state feedback control system just designed can be described by the following equations: For the plant,

$$\begin{bmatrix} \dot{x}_1 \\ \dot{x}_2 \end{bmatrix} = \begin{bmatrix} 0 & 1 \\ 20.6 & 0 \end{bmatrix} \begin{bmatrix} x_1 \\ x_2 \end{bmatrix} + \begin{bmatrix} 0 \\ 1 \end{bmatrix} u$$

$$y = \begin{bmatrix} 1 & 0 \end{bmatrix} \begin{bmatrix} x_1 \\ x_2 \end{bmatrix}$$

For the observer,

$$\begin{bmatrix} \dot{\tilde{x}}_1 \\ \dot{\tilde{x}}_2 \end{bmatrix} = \begin{bmatrix} -16 & 1 \\ -93.6 & -3.6 \end{bmatrix} \begin{bmatrix} \tilde{x}_1 \\ \tilde{x}_2 \end{bmatrix} + \begin{bmatrix} 16 \\ 84.6 \end{bmatrix} y$$

$$u = -\begin{bmatrix} 29.6 & 3.6 \end{bmatrix} \begin{bmatrix} \tilde{x}_1 \\ \tilde{x}_2 \end{bmatrix}$$

The system, as a whole, is of fourth order. The characteristic equation for the system is

$$|s\mathbf{I} - \mathbf{A} + \mathbf{BK}||s\mathbf{I} - \mathbf{A} + \mathbf{K}_e\mathbf{C}| = (s^2 + 3.6s + 9)(s^2 + 16s + 64)$$

$$= s^4 + 19.6s^3 + 130.6s^2 + 374.4s + 576 = 0$$

The characteristic equation can also be obtained from the block diagram for the system shown in Figure 10–14(b). Since the closed-loop transfer function is

$$\frac{Y(s)}{R(s)} = \frac{778.2s + 3690.7}{(s^2 + 19.6s + 151.2)(s^2 - 20.6) + 778.2s + 3690.7}$$

the characteristic equation is

$$(s^2 + 19.6s + 151.2)(s^2 - 20.6) + 778.2s + 3690.7$$
$$= s^4 + 19.6s^3 + 130.6s^2 + 374.4s + 576 = 0$$

As a matter of course, the characteristic equation is the same for the system in state-space representation and in transfer-function representation.

Finally, we shall obtain the response of the system to the following initial condition:

$$\mathbf{x}(0) = \begin{bmatrix} 1 \\ 0 \end{bmatrix}, \qquad \mathbf{e}(0) = \begin{bmatrix} 0.5 \\ 0 \end{bmatrix}$$

Referring to Equation (10–70), the response to the initial condition can be determined from

$$\begin{bmatrix} \dot{\mathbf{x}} \\ \dot{\mathbf{e}} \end{bmatrix} = \begin{bmatrix} \mathbf{A} - \mathbf{BK} & \mathbf{BK} \\ \mathbf{0} & \mathbf{A} - \mathbf{K}_e\mathbf{C} \end{bmatrix} \begin{bmatrix} \mathbf{x} \\ \mathbf{e} \end{bmatrix}, \qquad \begin{bmatrix} \mathbf{x}(0) \\ \mathbf{e}(0) \end{bmatrix} = \begin{bmatrix} 1 \\ 0 \\ 0.5 \\ 0 \end{bmatrix}$$

A MATLAB Program to obtain the response is shown in MATLAB Program 10–9. The resulting response curves are shown in Figure 10–15.

MATLAB Program 10–9

```
A = [0  1; 20.6  0];
B = [0;1];
C = [1  0];
K = [29.6  3.6];
Ke = [16; 84.6];
sys = ss([A-B*K  B*K; zeros(2,2)  A-Ke*C],eye(4),eye(4),eye(4));
t = 0:0.01:4;
z = initial(sys,[1;0;0.5;0],t);
x1 = [1  0  0  0]*z';
x2 = [0  1  0  0]*z';
e1 = [0  0  1  0]*z';
e2 = [0  0  0  1]*z';

subplot(2,2,1); plot(t,x1 ),grid
title('Response to Initial Condition')
ylabel('state variable x1')

subplot(2,2,2); plot(t,x2),grid
title('Response to Initial Condition')
ylabel('state variable x2')

subplot(2,2,3); plot(t,e1),grid
xlabel('t (sec)'), ylabel('error state variable e1')

subplot(2,2,4); plot(t,e2),grid
xlabel('t (sec)'), ylabel('error state variable e2')
```

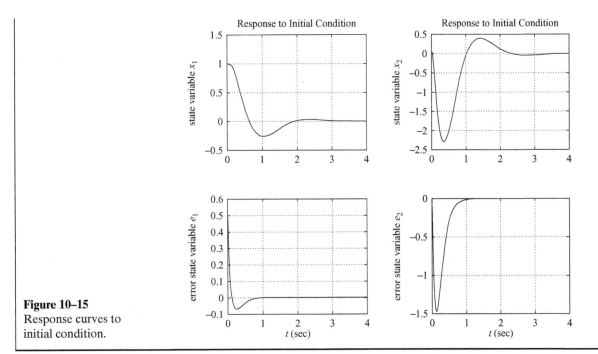

Figure 10–15
Response curves to
initial condition.

Minimum-Order Observer. The observers discussed thus far are designed to reconstruct all the state variables. In practice, some of the state variables may be accurately measured. Such accurately measurable state variables need not be estimated.

Suppose that the state vector **x** is an n-vector and the output vector **y** is an m-vector that can be measured. Since m output variables are linear combinations of the state variables, m state variables need not be estimated. We need to estimate only $n - m$ state variables. Then the reduced-order observer becomes an $(n - m)$th-order observer. Such an $(n - m)$th-order observer is the minimum-order observer. Figure 10–16 shows the block diagram of a system with a minimum-order observer.

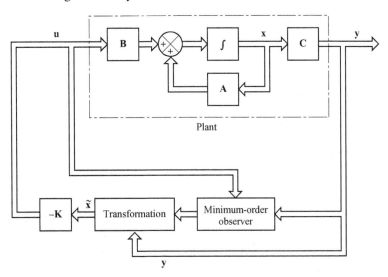

Figure 10–16
Observed-state
feedback control
system with a
minimum-order
observer.

It is important to note, however, that if the measurement of output variables involves significant noises and is relatively inaccurate, then the use of the full-order observer may result in a better system performance.

To present the basic idea of the minimum-order observer, without undue mathematical complications, we shall present the case where the output is a scalar (that is, $m = 1$) and derive the state equation for the minimum-order observer. Consider the system

$$\dot{\mathbf{x}} = \mathbf{A}\mathbf{x} + \mathbf{B}u \tag{10-79}$$

$$y = \mathbf{C}\mathbf{x} \tag{10-80}$$

where the state vector $\mathbf{x}$ can be partitioned into two parts x_a (a scalar) and $\mathbf{x}_b$ [an $(n-1)$-vector]. Here the state variable x_a is equal to the output y and thus can be directly measured, and $\mathbf{x}_b$ is the unmeasurable portion of the state vector. Then the partitioned state and output equations become

$$\left[\frac{\dot{x}_a}{\dot{\mathbf{x}}_b}\right] = \left[\begin{array}{c|c} A_{aa} & \mathbf{A}_{ab} \\ \hline \mathbf{A}_{ba} & \mathbf{A}_{bb} \end{array}\right]\left[\frac{x_a}{\mathbf{x}_b}\right] + \left[\frac{B_a}{\mathbf{B}_b}\right]u \tag{10-81}$$

$$y = \begin{bmatrix} 1 & \vdots & \mathbf{0} \end{bmatrix}\left[\frac{x_a}{\mathbf{x}_b}\right] \tag{10-82}$$

where $A_{aa} = $ scalar

$\mathbf{A}_{ab} = 1 \times (n-1)$ matrix

$\mathbf{A}_{ba} = (n-1) \times 1$ matrix

$\mathbf{A}_{bb} = (n-1) \times (n-1)$ matrix

$B_a = $ scalar

$\mathbf{B}_b = (n-1) \times 1$ matrix

From Equation (10-81), the equation for the measured portion of the state becomes

$$\dot{x}_a = A_{aa}x_a + \mathbf{A}_{ab}\mathbf{x}_b + B_a u$$

or

$$\dot{x}_a - A_{aa}x_a - B_a u = \mathbf{A}_{ab}\mathbf{x}_b \tag{10-83}$$

The terms on the left-hand side of Equation (10-83) can be measured. Equation (10-83) acts as the output equation. In designing the minimum-order observer, we consider the left-hand side of Equation (10-83) to be known quantities. Thus, Equation (10-83) relates the measurable quantities and unmeasurable quantities of the state.

From Equation (10-81), the equation for the unmeasured portion of the state becomes

$$\dot{\mathbf{x}}_b = \mathbf{A}_{ba}x_a + \mathbf{A}_{bb}\mathbf{x}_b + \mathbf{B}_b u \tag{10-84}$$

Noting that terms $\mathbf{A}_{ba}x_a$ and $\mathbf{B}_b u$ are known quantities, Equation (10-84) describes the dynamics of the unmeasured portion of the state.

In what follows we shall present a method for designing a minimum-order observer. The design procedure can be simplified if we utilize the design technique developed for the full-order state observer.

Let us compare the state equation for the full-order observer with that for the minimum-order observer. The state equation for the full-order observer is

$$\dot{x} = Ax + Bu$$

and the "state equation" for the minimum-order observer is

$$\dot{x}_b = A_{bb}x_b + A_{ba}x_a + B_b u$$

The output equation for the full-order observer is

$$y = Cx$$

and the "output equation" for the minimum-order observer is

$$\dot{x}_a - A_{aa}x_a - B_a u = A_{ab}x_b$$

The design of the minimum-order observer can be carried out as follows: First, note that the observer equation for the full-order observer was given by Equation (10–57), which we repeat here:

$$\dot{\tilde{x}} = (A - K_e C)\tilde{x} + Bu + K_e y \qquad (10\text{–}85)$$

Then, making the substitutions of Table 10–1 into Equation (10–85), we obtain

$$\dot{\tilde{x}}_b = (A_{bb} - K_e A_{ab})\tilde{x}_b + A_{ba}x_a + B_b u + K_e(\dot{x}_a - A_{aa}x_a - B_a u) \quad (10\text{–}86)$$

where the state observer gain matrix K_e is an $(n - 1) \times 1$ matrix. In Equation (10–86), notice that in order to estimate $\tilde{x}_b$, we need the derivative of x_a. This presents a difficulty, because differentiation amplifies noise. If $x_a (= y)$ is noisy, the use of $\dot{x}_a$ is unacceptable.

Table 10–1 List of Necessary Substitutions for Writing the Observer Equation for the Minimum-Order State Observer

Full-Order State Observer	Minimum-Order State Observer
$\tilde{x}$	$\tilde{x}_b$
A	A_{bb}
Bu	$A_{ba}x_a + B_b u$
y	$\dot{x}_a - A_{aa}x_a - B_a u$
C	A_{ab}
K_e $(n \times 1$ matrix)	K_e $[(n - 1) \times 1$ matrix]

To avoid this difficulty, we eliminate $\dot{x}_a$ in the following way. First rewrite Equation (10–86) as

$$\dot{\tilde{\mathbf{x}}}_b - \mathbf{K}_e \dot{x}_a = (\mathbf{A}_{bb} - \mathbf{K}_e \mathbf{A}_{ab})\tilde{\mathbf{x}}_b + (\mathbf{A}_{ba} - \mathbf{K}_e \mathbf{A}_{aa})y + (\mathbf{B}_b - \mathbf{K}_e \mathbf{B}_a)u$$

$$= (\mathbf{A}_{bb} - \mathbf{K}_e \mathbf{A}_{ab})(\tilde{\mathbf{x}}_b - \mathbf{K}_e y)$$

$$+ \left[(\mathbf{A}_{bb} - \mathbf{K}_e \mathbf{A}_{ab})\mathbf{K}_e + \mathbf{A}_{ba} - \mathbf{K}_e \mathbf{A}_{aa} \right] y$$

$$+ (\mathbf{B}_b - \mathbf{K}_e \mathbf{B}_a)u \qquad (10\text{–}87)$$

Define

$$\mathbf{x}_b - \mathbf{K}_e y = \mathbf{x}_b - \mathbf{K}_e x_a = \eta$$

and

$$\tilde{\mathbf{x}}_b - \mathbf{K}_e y = \tilde{\mathbf{x}}_b - \mathbf{K}_e x_a = \tilde{\eta} \qquad (10\text{–}88)$$

Then Equation (10–87) becomes

$$\dot{\tilde{\eta}} = (\mathbf{A}_{bb} - \mathbf{K}_e \mathbf{A}_{ab})\tilde{\eta} + \left[(\mathbf{A}_{bb} - \mathbf{K}_e \mathbf{A}_{ab})\mathbf{K}_e \right.$$

$$\left. + \mathbf{A}_{ba} - \mathbf{K}_e \mathbf{A}_{aa} \right] y + (\mathbf{B}_b - \mathbf{K}_e \mathbf{B}_a)u \qquad (10\text{–}89)$$

Define

$$\hat{\mathbf{A}} = \mathbf{A}_{bb} - \mathbf{K}_e \mathbf{A}_{ab}$$

$$\hat{\mathbf{B}} = \hat{\mathbf{A}}\mathbf{K}_e + \mathbf{A}_{ba} - \mathbf{K}_e \mathbf{A}_{aa}$$

$$\hat{\mathbf{F}} = \mathbf{B}_b - \mathbf{K}_e \mathbf{B}_a$$

Then Equation (10–89) becomes

$$\dot{\tilde{\eta}} = \hat{\mathbf{A}}\tilde{\eta} + \hat{\mathbf{B}}y + \hat{\mathbf{F}}u \qquad (10\text{–}90)$$

Equation (10–90) and Equation (10–88) together define the minimum-order observer. Since

$$y = \begin{bmatrix} 1 & \vdots & \mathbf{0} \end{bmatrix} \begin{bmatrix} x_a \\ \cdots \\ \mathbf{x}_b \end{bmatrix}$$

$$\tilde{\mathbf{x}} = \begin{bmatrix} x_a \\ \cdots \\ \tilde{\mathbf{x}}_b \end{bmatrix} = \begin{bmatrix} y \\ \cdots \\ \tilde{\mathbf{x}}_b \end{bmatrix} = \begin{bmatrix} \mathbf{0} \\ \cdots \\ \mathbf{I}_{n-1} \end{bmatrix} [\tilde{\mathbf{x}}_b - \mathbf{K}_e y] + \begin{bmatrix} 1 \\ \cdots \\ \mathbf{K}_e \end{bmatrix} y$$

where $\mathbf{0}$ is a row vector consisting of $(n - 1)$ zeros, if we define

$$\hat{\mathbf{C}} = \begin{bmatrix} \mathbf{0} \\ \cdots \\ \mathbf{I}_{n-1} \end{bmatrix}, \qquad \hat{\mathbf{D}} = \begin{bmatrix} 1 \\ \cdots \\ \mathbf{K}_e \end{bmatrix}$$

then we can write $\tilde{\mathbf{x}}$ in terms of $\tilde{\eta}$ and y as follows:

$$\tilde{\mathbf{x}} = \hat{\mathbf{C}}\tilde{\eta} + \hat{\mathbf{D}}y \qquad (10\text{–}91)$$

This equation gives the transformation from $\tilde{\eta}$ to $\tilde{\mathbf{x}}$.

Figure 10–17 shows the block diagram of the observed-state feedback control system with the minimum-order observer, based on Equations (10–79), (10–80), (10–90), (10–91) and $u = -\mathbf{K}\tilde{\mathbf{x}}$.

Next we shall derive the observer error equation. Using Equation (10–83), Equation (10–86) can be modified to

$$\dot{\tilde{\mathbf{x}}}_b = (\mathbf{A}_{bb} - \mathbf{K}_e \mathbf{A}_{ab})\tilde{\mathbf{x}}_b + \mathbf{A}_{ba} x_a + \mathbf{B}_b u + \mathbf{K}_e \mathbf{A}_{ab} \mathbf{x}_b \qquad (10\text{–}92)$$

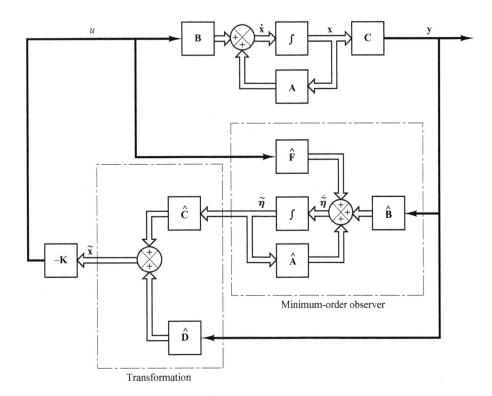

Figure 10–17
System with observed-state feedback, where the observer is the minimum-order observer.

By subtracting Equation (10–92) from Equation (10–84), we obtain

$$\dot{\mathbf{x}}_b - \dot{\tilde{\mathbf{x}}}_b = (\mathbf{A}_{bb} - \mathbf{K}_e \mathbf{A}_{ab})(\mathbf{x}_b - \tilde{\mathbf{x}}_b) \tag{10–93}$$

Define

$$\mathbf{e} = \mathbf{x}_b - \tilde{\mathbf{x}}_b = \boldsymbol{\eta} - \tilde{\boldsymbol{\eta}}$$

Then Equation (10–93) becomes

$$\dot{\mathbf{e}} = (\mathbf{A}_{bb} - \mathbf{K}_e \mathbf{A}_{ab})\mathbf{e} \tag{10–94}$$

This is the error equation for the minimum-order observer. Note that $\mathbf{e}$ is an $(n-1)$-vector.

The error dynamics can be chosen as desired by following the technique developed for the full-order observer, provided that the rank of matrix

$$\begin{bmatrix} \mathbf{A}_{ab} \\ \mathbf{A}_{ab}\mathbf{A}_{bb} \\ \cdot \\ \cdot \\ \cdot \\ \mathbf{A}_{ab}\mathbf{A}_{bb}^{n-2} \end{bmatrix}$$

is $n-1$. (This is the complete observability condition applicable to the minimum-order observer.)

The characteristic equation for the minimum-order observer is obtained from Equation (10–94) as follows:

$$\left| s\mathbf{I} - \mathbf{A}_{bb} + \mathbf{K}_e \mathbf{A}_{ab} \right| = (s - \mu_1)(s - \mu_2)\cdots(s - \mu_{n-1})$$

$$= s^{n-1} + \hat{\alpha}_1 s^{n-2} + \cdots + \hat{\alpha}_{n-2}s + \hat{\alpha}_{n-1} = 0 \quad (10\text{–}95)$$

where $\mu_1, \mu_2, \ldots, \mu_{n-1}$ are desired eigenvalues for the minimum-order observer. The observer gain matrix $\mathbf{K}_e$ can be determined by first choosing the desired eigenvalues for the minimum-order observer [that is, by placing the roots of the characteristic equation, Equation (10–95), at the desired locations] and then using the procedure developed for the full-order observer with appropriate modifications. For example, if the formula for determining matrix $\mathbf{K}_e$ given by Equation (10–61) is to be used, it should be modified to

$$\mathbf{K}_e = \hat{\mathbf{Q}} \begin{bmatrix} \hat{\alpha}_{n-1} - \hat{a}_{n-1} \\ \hat{\alpha}_{n-2} - \hat{a}_{n-2} \\ \cdot \\ \cdot \\ \cdot \\ \hat{\alpha}_1 - \hat{a}_1 \end{bmatrix} = (\hat{\mathbf{W}}\hat{\mathbf{N}}*)^{-1} \begin{bmatrix} \hat{\alpha}_{n-1} - \hat{a}_{n-1} \\ \hat{\alpha}_{n-2} - \hat{a}_{n-2} \\ \cdot \\ \cdot \\ \cdot \\ \hat{\alpha}_1 - \hat{a}_1 \end{bmatrix} \quad (10\text{–}96)$$

where $\mathbf{K}_e$ is an $(n - 1) \times 1$ matrix and

$$\hat{\mathbf{N}} = \left[\mathbf{A}_{ab}* \mid \mathbf{A}_{bb}*\mathbf{A}_{ab}* \mid \cdots \mid (\mathbf{A}_{bb}*)^{n-2}\mathbf{A}_{ab}* \right] = (n - 1) \times (n - 1) \text{ matrix}$$

$$\hat{\mathbf{W}} = \begin{bmatrix} \hat{a}_{n-2} & \hat{a}_{n-3} & \cdots & \hat{a}_1 & 1 \\ \hat{a}_{n-3} & \hat{a}_{n-4} & \cdots & 1 & 0 \\ \cdot & \cdot & & \cdot & \cdot \\ \cdot & \cdot & & \cdot & \cdot \\ \cdot & \cdot & & \cdot & \cdot \\ \hat{a}_1 & 1 & \cdots & 0 & 0 \\ 1 & 0 & \cdots & 0 & 0 \end{bmatrix} = (n - 1) \times (n - 1) \text{ matrix}$$

Note that $\hat{a}_1, \hat{a}_2, \ldots, \hat{a}_{n-2}$ are coefficients in the characteristic equation for the state equation

$$\left| s\mathbf{I} - \mathbf{A}_{bb} \right| = s^{n-1} + \hat{a}_1 s^{n-2} + \cdots + \hat{a}_{n-2}s + \hat{a}_{n-1} = 0$$

Also, if Ackermann's formula given by Equation (10–65) is to be used, then it should be modified to

$$\mathbf{K}_e = \phi(\mathbf{A}_{bb}) \begin{bmatrix} \mathbf{A}_{ab} \\ \mathbf{A}_{ab}\mathbf{A}_{bb} \\ \cdot \\ \cdot \\ \cdot \\ \mathbf{A}_{ab}\mathbf{A}_{bb}^{n-3} \\ \mathbf{A}_{ab}\mathbf{A}_{bb}^{n-2} \end{bmatrix}^{-1} \begin{bmatrix} 0 \\ 0 \\ \cdot \\ \cdot \\ \cdot \\ 0 \\ 1 \end{bmatrix} \quad (10\text{–}97)$$

where

$$\phi(\mathbf{A}_{bb}) = \mathbf{A}_{bb}^{n-1} + \hat{\alpha}_1 \mathbf{A}_{bb}^{n-2} + \cdots + \hat{\alpha}_{n-2} \mathbf{A}_{bb} + \hat{\alpha}_{n-1} \mathbf{I}$$

Observed-State Feedback Control System with Minimum-Order Observer.
For the case of the observed-state feedback control system with full-order state observer, we have shown that the closed-loop poles of the observed-state feedback control system consist of the poles due to the pole-placement design alone, plus the poles due to the observer design alone. Hence, the pole-placement design and the full-order observer design are independent of each other.

For the observed-state feedback control system with minimum-order observer, the same conclusion applies. The system characteristic equation can be derived as

$$|s\mathbf{I} - \mathbf{A} + \mathbf{BK}||s\mathbf{I} - \mathbf{A}_{bb} + \mathbf{K}_e \mathbf{A}_{ab}| = 0 \qquad (10\text{--}98)$$

(See Problem **A–10–11** for the details.) The closed-loop poles of the observed-state feedback control system with a minimum-order observer comprise the closed-loop poles due to pole placement [the eigenvalues of matrix $(\mathbf{A} - \mathbf{BK})$] and the closed-loop poles due to the minimum-order observer [the eigenvalues of matrix $(\mathbf{A}_{bb} - \mathbf{K}_e \mathbf{A}_{ab})$]. Therefore, the pole-placement design and the design of the minimum-order observer are independent of each other.

Determining Observer Gain Matrix $\mathbf{K}_e$ with MATLAB.   Because of the duality of pole-placement and observer design, the same algorithm can be applied to both the pole-placement problem and the observer-design problem. Thus, the commands acker and place can be used to determine the observer gain matrix $\mathbf{K}_e$.

The closed-loop poles of the observer are the eigenvalues of matrix $\mathbf{A} - \mathbf{K}_e \mathbf{C}$. The closed-loop poles of the pole-placement are the eigenvalues of matrix $\mathbf{A} - \mathbf{BK}$.

Referring to the duality problem between the pole-placement problem and observer-design problem, we can determine $\mathbf{K}_e$ by considering the pole-placement problem for the dual system. That is, we determine $\mathbf{K}_e$ by placing the eigenvalues of $\mathbf{A}^* - \mathbf{C}^*\mathbf{K}_e$ at the desired place. Since $\mathbf{K}_e = \mathbf{K}^*$, for the full-order observer we use the command

$$K_e = \text{acker}(A',C',L)'$$

where L is the vector of the desired eigenvalues for the observer. Similarly, for the full-order observer, we may use

$$K_e = \text{place}(A',C',L)'$$

provided L does not include multiple poles. [In the above commands, prime (') indicates the transpose.] For the minimum-order (or reduced-order) observers, use the following commands:

$$K_e = \text{acker}(Abb',Aab',L)'$$

or

$$K_e = \text{place}(Abb',Aab',L)'$$

EXAMPLE 10–8   Consider the system

$$\dot{\mathbf{x}} = \mathbf{Ax} + \mathbf{B}u$$

$$y = \mathbf{Cx}$$

where

$$\mathbf{A} = \begin{bmatrix} 0 & 1 & 0 \\ 0 & 0 & 1 \\ -6 & -11 & -6 \end{bmatrix}, \quad \mathbf{B} = \begin{bmatrix} 0 \\ 0 \\ 1 \end{bmatrix}, \quad \mathbf{C} = \begin{bmatrix} 1 & 0 & 0 \end{bmatrix}$$

Let us assume that we want to place the closed-loop poles at

$$s_1 = -2 + j2\sqrt{3}, \quad s_2 = -2 - j2\sqrt{3}, \quad s_3 = -6$$

Then the necessary state-feedback gain matrix **K** can be obtained as follows:

$$\mathbf{K} = \begin{bmatrix} 90 & 29 & 4 \end{bmatrix}$$

(See MATLAB Program 10–10 for a MATLAB computation of this matrix **K**.)

Next, let us assume that the output y can be measured accurately so that state variable x_1 (which is equal to y) need not be estimated. Let us design a minimum-order observer. (The minimum-order observer is of second order.) Assume that we choose the desired observer poles to be at

$$s = -10, \quad s = -10$$

Referring to Equation (10–95), the characteristic equation for the minimum-order observer is

$$|s\mathbf{I} - \mathbf{A}_{bb} + \mathbf{K}_e \mathbf{A}_{ab}| = (s - \mu_1)(s - \mu_2)$$

$$= (s + 10)(s + 10)$$

$$= s^2 + 20s + 100 = 0$$

In what follows, we shall use Ackermann's formula given by Equation (10–97).

$$\mathbf{K}_e = \phi(\mathbf{A}_{bb}) \begin{bmatrix} \mathbf{A}_{ab} \\ \hline \mathbf{A}_{ab} \mathbf{A}_{bb} \end{bmatrix}^{-1} \begin{bmatrix} 0 \\ 1 \end{bmatrix} \qquad (10\text{–}99)$$

where

$$\phi(\mathbf{A}_{bb}) = \mathbf{A}_{bb}^2 + \hat{\alpha}_1 \mathbf{A}_{bb} + \hat{\alpha}_2 \mathbf{I} = \mathbf{A}_{bb}^2 + 20\mathbf{A}_{bb} + 100\mathbf{I}$$

Since

$$\tilde{\mathbf{x}} = \begin{bmatrix} x_a \\ \hline \tilde{\mathbf{x}}_b \end{bmatrix} = \begin{bmatrix} x_1 \\ \hline \tilde{x}_2 \\ \tilde{x}_3 \end{bmatrix}, \quad \mathbf{A} = \begin{bmatrix} 0 & 1 & 0 \\ \hline 0 & 0 & 1 \\ -6 & -11 & -6 \end{bmatrix}, \quad \mathbf{B} = \begin{bmatrix} 0 \\ \hline 0 \\ 1 \end{bmatrix}$$

we have

$$A_{aa} = 0, \quad \mathbf{A}_{ab} = \begin{bmatrix} 1 & 0 \end{bmatrix}, \quad \mathbf{A}_{ba} = \begin{bmatrix} 0 \\ -6 \end{bmatrix}$$

$$\mathbf{A}_{bb} = \begin{bmatrix} 0 & 1 \\ -11 & -6 \end{bmatrix}, \quad B_a = 0, \quad \mathbf{B}_b = \begin{bmatrix} 0 \\ 1 \end{bmatrix}$$

Equation (10–99) now becomes

$$\mathbf{K}_e = \left\{ \begin{bmatrix} 0 & 1 \\ -11 & -6 \end{bmatrix}^2 + 20 \begin{bmatrix} 0 & 1 \\ -11 & -6 \end{bmatrix} + 100 \begin{bmatrix} 1 & 0 \\ 0 & 1 \end{bmatrix} \right\} \begin{bmatrix} 1 & 0 \\ 0 & 1 \end{bmatrix}^{-1} \begin{bmatrix} 0 \\ 1 \end{bmatrix}$$

$$= \begin{bmatrix} 89 & 14 \\ -154 & 5 \end{bmatrix} \begin{bmatrix} 0 \\ 1 \end{bmatrix} = \begin{bmatrix} 14 \\ 5 \end{bmatrix}$$

(A MATLAB computation of this $\mathbf{K}_e$ is given in MATLAB Program 10–10.)

MATLAB Program 10–10

```
A = [0  1  0;0  0  1;-6 -11 -6];
B = [0;0;1];
J = [-2+j*2*sqrt(3)  -2-j*2*sqrt(3)  -6];
K = acker(A,B,J)

K =

     90.0000    29.0000    4.0000

Abb = [0  1;-11  -6];
Aab = [1  0];
L = [-10  -10];
Ke = acker(Abb',Aab',L)'

Ke =

     14
      5
```

Referring to Equations (10–88) and (10–89), the equation for the minimum-order observer can be given by

$$\dot{\tilde{\boldsymbol{\eta}}} = \left(\mathbf{A}_{bb} - \mathbf{K}_e \mathbf{A}_{ab}\right)\tilde{\boldsymbol{\eta}} + \left[\left(\mathbf{A}_{bb} - \mathbf{K}_e \mathbf{A}_{ab}\right)\mathbf{K}_e + \mathbf{A}_{ba} - \mathbf{K}_e A_{aa}\right]y + \left(\mathbf{B}_b - \mathbf{K}_e B_a\right)u \qquad (10\text{–}100)$$

where

$$\tilde{\boldsymbol{\eta}} = \tilde{\mathbf{x}}_b - \mathbf{K}_e y = \tilde{\mathbf{x}}_b - \mathbf{K}_e x_1$$

Noting that

$$\mathbf{A}_{bb} - \mathbf{K}_e \mathbf{A}_{ab} = \begin{bmatrix} 0 & 1 \\ -11 & -6 \end{bmatrix} - \begin{bmatrix} 14 \\ 5 \end{bmatrix} \begin{bmatrix} 1 & 0 \end{bmatrix} = \begin{bmatrix} -14 & 1 \\ -16 & -6 \end{bmatrix}$$

the equation for the minimum-order observer, Equation (10–100), becomes

$$\begin{bmatrix} \dot{\tilde{\eta}}_2 \\ \dot{\tilde{\eta}}_3 \end{bmatrix} = \begin{bmatrix} -14 & 1 \\ -16 & -6 \end{bmatrix} \begin{bmatrix} \tilde{\eta}_2 \\ \tilde{\eta}_3 \end{bmatrix} + \left\{ \begin{bmatrix} -14 & 1 \\ -16 & -6 \end{bmatrix} \begin{bmatrix} 14 \\ 5 \end{bmatrix} \right.$$

$$\left. + \begin{bmatrix} 0 \\ -6 \end{bmatrix} - \begin{bmatrix} 14 \\ 5 \end{bmatrix} 0 \right\} y + \left\{ \begin{bmatrix} 0 \\ 1 \end{bmatrix} - \begin{bmatrix} 14 \\ 5 \end{bmatrix} 0 \right\} u$$

or

$$\begin{bmatrix} \dot{\tilde{\eta}}_2 \\ \dot{\tilde{\eta}}_3 \end{bmatrix} = \begin{bmatrix} -14 & 1 \\ -16 & -6 \end{bmatrix} \begin{bmatrix} \tilde{\eta}_2 \\ \tilde{\eta}_3 \end{bmatrix} + \begin{bmatrix} -191 \\ -260 \end{bmatrix} y + \begin{bmatrix} 0 \\ 1 \end{bmatrix} u$$

where

$$\begin{bmatrix} \tilde{\eta}_2 \\ \tilde{\eta}_3 \end{bmatrix} = \begin{bmatrix} \tilde{x}_2 \\ \tilde{x}_3 \end{bmatrix} - \mathbf{K}_e y$$

or

$$\begin{bmatrix} \tilde{x}_2 \\ \tilde{x}_3 \end{bmatrix} = \begin{bmatrix} \tilde{\eta}_2 \\ \tilde{\eta}_3 \end{bmatrix} + \mathbf{K}_e x_1$$

If the observed-state feedback is used, then the control signal u becomes

$$u = -\mathbf{K}\tilde{\mathbf{x}} = -\mathbf{K} \begin{bmatrix} x_1 \\ \tilde{x}_2 \\ \tilde{x}_3 \end{bmatrix}$$

where $\mathbf{K}$ is the state feedback gain matrix. Figure 10–18 is a block diagram showing the configuration of the system with observed-state feedback, where the observer is the minimum-order observer.

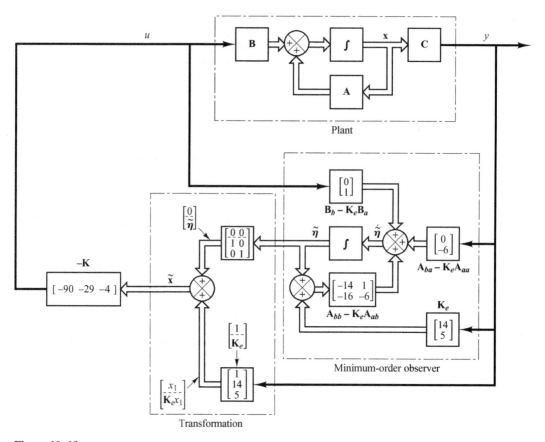

Figure 10–18
System with observed state feedback, where the observer is the minimum-order observer designed in Example 10–8.

Transfer Function of Minimum-Order Observer-Based Controller.

In the minimum-order observer equation given by Equation (10–89):

$$\dot{\tilde{\boldsymbol{\eta}}} = (\mathbf{A}_{bb} - \mathbf{K}_e\mathbf{A}_{ab})\tilde{\boldsymbol{\eta}} + \left[(\mathbf{A}_{bb} - \mathbf{K}_e\mathbf{A}_{ab})\mathbf{K}_e + \mathbf{A}_{ba} - \mathbf{K}_e\mathbf{A}_{aa}\right]y + (\mathbf{B}_b - \mathbf{K}_e B_a)u$$

define, similar to the case of the derivation of Equation (10–90),

$$\hat{\mathbf{A}} = \mathbf{A}_{bb} - \mathbf{K}_e\mathbf{A}_{ab}$$
$$\hat{\mathbf{B}} = \hat{\mathbf{A}}\mathbf{K}_e + \mathbf{A}_{ba} - \mathbf{K}_e A_{aa}$$
$$\hat{\mathbf{F}} = \mathbf{B}_b - \mathbf{K}_e B_a$$

Then, the following three equations define the minimum-order oberver:

$$\dot{\tilde{\boldsymbol{\eta}}} = \hat{\mathbf{A}}\tilde{\boldsymbol{\eta}} + \hat{\mathbf{B}}y + \hat{\mathbf{F}}u \qquad (10\text{–}101)$$
$$\tilde{\boldsymbol{\eta}} = \tilde{\mathbf{x}}_b - \mathbf{K}_e y \qquad (10\text{–}102)$$
$$u = -\mathbf{K}\tilde{\mathbf{x}} \qquad (10\text{–}103)$$

Since Equation (10–103) can be rewritten as

$$u = -\mathbf{K}\tilde{\mathbf{x}} = -\begin{bmatrix} K_a & \mathbf{K}_b \end{bmatrix}\begin{bmatrix} y \\ \tilde{\mathbf{x}}_b \end{bmatrix} = -K_a y - \mathbf{K}_b \tilde{\mathbf{x}}_b$$

$$= -\mathbf{K}_b\tilde{\boldsymbol{\eta}} - (K_a + \mathbf{K}_b\mathbf{K}_e)y \qquad (10\text{–}104)$$

by substituting Equation (10–104) into Equation (10–101), we obtain

$$\dot{\tilde{\boldsymbol{\eta}}} = \hat{\mathbf{A}}\tilde{\boldsymbol{\eta}} + \hat{\mathbf{B}}y + \hat{\mathbf{F}}\left[-\mathbf{K}_b\tilde{\boldsymbol{\eta}} - (K_a + \mathbf{K}_b\mathbf{K}_e)y\right]$$
$$= (\hat{\mathbf{A}} - \hat{\mathbf{F}}\mathbf{K}_b)\tilde{\boldsymbol{\eta}} + \left[\hat{\mathbf{B}} - \hat{\mathbf{F}}(K_a + \mathbf{K}_b\mathbf{K}_e)\right]y \qquad (10\text{–}105)$$

Define

$$\tilde{\mathbf{A}} = \hat{\mathbf{A}} - \hat{\mathbf{F}}\mathbf{K}_b$$
$$\tilde{\mathbf{B}} = \hat{\mathbf{B}} - \hat{\mathbf{F}}(K_a + \mathbf{K}_b\mathbf{K}_e)$$
$$\tilde{\mathbf{C}} = -\mathbf{K}_b$$
$$\tilde{D} = -(K_a + \mathbf{K}_b\mathbf{K}_e)$$

Then Equations (10–105) and (10–104) can be written as

$$\dot{\tilde{\boldsymbol{\eta}}} = \tilde{\mathbf{A}}\tilde{\boldsymbol{\eta}} + \tilde{\mathbf{B}}y \qquad (10\text{–}106)$$
$$u = \tilde{\mathbf{C}}\tilde{\boldsymbol{\eta}} + \tilde{D}y \qquad (10\text{–}107)$$

Equations (10–106) and (10–107) define the minimum-order observer-based controller. By considering u as the output and $-y$ as the input, $U(s)$ can be written as

$$U(s) = \left[\tilde{\mathbf{C}}(s\mathbf{I} - \tilde{\mathbf{A}})^{-1}\tilde{\mathbf{B}} + \tilde{D}\right]Y(s)$$
$$= -\left[\tilde{\mathbf{C}}(s\mathbf{I} - \tilde{\mathbf{A}})^{-1}\tilde{\mathbf{B}} + \tilde{D}\right][-Y(s)]$$

Since the input to the observer controller is $-Y(s)$, rather than $Y(s)$, the transfer function of the observer controller is

$$\frac{U(s)}{-Y(s)} = \frac{\text{num}}{\text{den}} = -\left[\tilde{\mathbf{C}}(s\mathbf{I} - \tilde{\mathbf{A}})^{-1}\tilde{\mathbf{B}} + \tilde{D}\right] \qquad (10\text{–}108)$$

This transfer function can be easily obtained by using the following MATLAB statement:

$$[\text{num,den}] = \text{ss2tf(Atilde, Btilde, -Ctilde, -Dtilde)} \qquad (10\text{–}109)$$

In this section we shall consider a problem of designing regulator systems by using the pole-placement-with-observer approach.

Consider the regulator system shown in Figure 10–19. (The reference input is zero.) The plant transfer function is

$$G(s) = \frac{10(s + 2)}{s(s + 4)(s + 6)}$$

Using the pole-placement approach, design a controller such that when the system is subjected to the following initial condition:

$$\mathbf{x}(0) = \begin{bmatrix} 1 \\ 0 \\ 0 \end{bmatrix}, \qquad \mathbf{e}(0) = \begin{bmatrix} 1 \\ 0 \end{bmatrix}$$

where $\mathbf{x}$ is the state vector for the plant and $\mathbf{e}$ is the observer error vector, the maximum undershoot of $y(t)$ is 25 to 35% and the settling time is about 4 sec. Assume that we use the minimum-order observer. (We assume that only the output y is measurable.)

We shall use the following design procedure:

1. Derive a state-space model of the plant.
2. Choose the desired closed-loop poles for pole placement. Choose the desired observer poles.
3. Determine the state feedback gain matrix $\mathbf{K}$ and the observer gain matrix $\mathbf{K}_e$.
4. Using the gain matrices $\mathbf{K}$ and $\mathbf{K}_e$ obtained in step 3, derive the transfer function of the observer controller. If it is a stable controller, check the response to the given initial condition. If the response is not acceptable, adjust the closed-loop pole location and/or observer pole location until an acceptable response is obtained.

Design step 1: We shall derive the state-space representation of the plant. Since the plant transfer function is

$$\frac{Y(s)}{U(s)} = \frac{10(s + 2)}{s(s + 4)(s + 6)}$$

the corresponding differential equation is

$$\dddot{y} + 10\ddot{y} + 24\dot{y} = 10\dot{u} + 20u$$

Referring to Section 2–5, let us define the state variables x_1, x_2, and x_3 as follows:

$$x_1 = y - \beta_0 u$$
$$x_2 = \dot{x}_1 - \beta_1 u$$
$$x_3 = \dot{x}_2 - \beta_2 u$$

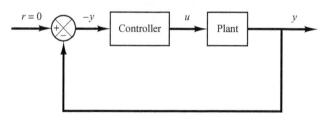

Figure 10–19
Regulator system.

Also, $\dot{x}_3$ is defined by

$$\dot{x}_3 = -a_3 x_1 - a_2 x_2 - a_1 x_3 + \beta_3 u$$
$$= -24 x_2 - 10 x_3 + \beta_3 u$$

where $\beta_0 = 0$, $\beta_1 = 0$, $\beta_2 = 10$, and $\beta_3 = -80$.

[See Equation (2–35) for the calculation of β's.] Then the state-space equation and output equation can be obtained as

$$
\begin{bmatrix} \dot{x}_1 \\ \dot{x}_2 \\ \dot{x}_3 \end{bmatrix} = \begin{bmatrix} 0 & 1 & 0 \\ 0 & 0 & 1 \\ 0 & -24 & -10 \end{bmatrix} \begin{bmatrix} x_1 \\ x_2 \\ x_3 \end{bmatrix} + \begin{bmatrix} 0 \\ 10 \\ -80 \end{bmatrix} u
$$

$$
y = \begin{bmatrix} 1 & 0 & 0 \end{bmatrix} \begin{bmatrix} x_1 \\ x_2 \\ x_3 \end{bmatrix} + [0] u
$$

Design step 2:   As the first trial, let us choose the desired closed-loop poles at

$$s = -1 + j2, \qquad s = -1 - j2, \qquad s = -5$$

and choose the desired observer poles at

$$s = -10, \qquad s = -10$$

Design step 3:   We shall use MATLAB to compute the state feedback gain matrix **K** and the observer gain matrix **K**$_e$. MATLAB Program 10–11 produces matrices **K** and **K**$_e$.

MATLAB Program 10–11

```
% Obtaining the state feedback gain matrix K

A = [0 1 0;0 0 1;0 -24 -10];
B = [0;10;-80];
C = [1 0 0];
J = [-1+j*2 -1-j*2 -5];
K = acker(A,B,J)

K =

   1.2500   1.2500   0.19375

% Obtaining the observer gain matrix Ke

Aaa = 0; Aab = [1 0]; Aba = [0;0]; Abb = [0 1;-24 -10];Ba = 0; Bb = [10;-80];
L = [-10 -10];
Ke = acker(Abb',Aab',L)'

Ke =

   10
  -24
```

In the program, matrices **J** and **L** represent the desired closed-loop poles for pole placement and the desired poles for the observer, respectively. The matrices **K** and $\mathbf{K}_e$ are obtained as

$$\mathbf{K} = \begin{bmatrix} 1.25 & 1.25 & 0.19375 \end{bmatrix}$$

$$\mathbf{K}_e = \begin{bmatrix} 10 \\ -24 \end{bmatrix}$$

Design step 4: We shall determine the transfer function of the observer controller. Referring to Equation (10–108), the transfer function of the observer controller can be given by

$$G_c(s) = \frac{U(s)}{-Y(s)} = \frac{\text{num}}{\text{den}} = -\left[\tilde{\mathbf{C}}(s\mathbf{I} - \tilde{\mathbf{A}})^{-1}\tilde{\mathbf{B}} + \tilde{D} \right]$$

We shall use MATLAB to calculate the transfer function of the observer controller. MATLAB Program 10–12 produces this transfer function. The result is

$$G_c(s) = \frac{9.1s^2 + 73.5s + 125}{s^2 + 17s - 30}$$

$$= \frac{9.1(s + 5.6425)(s + 2.4344)}{(s + 18.6119)(s - 1.6119)}$$

Define the system with this observer controller as System 1. Figure 10–20 shows the block diagram of System 1.

MATLAB Program 10–12

```
% Determination of transfer function of observer controller
A = [0  1  0;0  0  1;0 -24 -10];
B = [0;10;-80];
Aaa = 0; Aab = [1  0]; Aba = [0;0]; Abb = [0  1;-24  -10];
Ba = 0; Bb = [10;-80];
Ka = 1.25; Kb = [1.25    0.19375];
Ke = [10;-24];
Ahat = Abb - Ke*Aab;
Bhat = Ahat*Ke + Aba - Ke*Aaa;
Fhat = Bb - Ke*Ba;
Atilde = Ahat - Fhat*Kb;
Btilde = Bhat - Fhat*(Ka + Kb*Ke);
Ctilde = -Kb;
Dtilde = -(Ka + Kb*Ke);
[num,den] = ss2tf(Atilde, Btilde, -Ctilde, -Dtilde)

num =
    9.1000  73.5000  125.0000

den =

    1.0000  17.0000  -30.0000
```

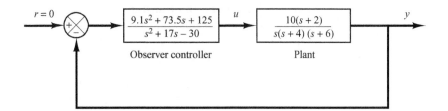

Figure 10–20
Block diagram of
System 1.

The observer controller has a pole in the right-half s plane ($s = 1.6119$). The existence of an open-loop right-half s plane pole in the observer controller means that the system is open-loop unstable, although the closed-loop system is stable. The latter can be seen from the characteristic equation for the system:

$$|s\mathbf{I} - \mathbf{A} + \mathbf{BK}| \cdot |s\mathbf{I} - \mathbf{A}_{bb} + \mathbf{K}_e\mathbf{A}_{ab}|$$
$$= s^5 + 27s^4 + 255s^3 + 1025s^2 + 2000s + 2500$$
$$= (s + 1 + j2)(s + 1 - j2)(s + 5)(s + 10)(s + 10) = 0$$

(See MATLAB Program 10–13 for the calculation of the characteristic equation.)

A disadvantage of using an unstable controller is that the system becomes unstable if the dc gain of the system becomes small. Such a control system is neither desirable nor acceptable. Hence, to get a satisfactory system, we need to modify the closed-loop pole location and/or observer pole location.

MATLAB Program 10–13

```
% Obtaining the characteristic equation

[num1,den1] = ss2tf(A-B*K,eye(3),eye(3),eye(3),1);
[num2,den2] = ss2tf(Abb-Ke*Aab,eye(2),eye(2),eye(2),1);
charact_eq = conv(den1,den2)

charact_eq =

  1.0e+003*

  0.0010   0.0270   0.2550   1.0250   2.0000   2.5000
```

Second trial: Let us keep the desired closed-loop poles for pole placement as before, but modify the observer pole locations as follows:

$$s = -4.5, \qquad s = -4.5$$

Thus,

$$\mathbf{L} = \begin{bmatrix} -4.5 & -4.5 \end{bmatrix}$$

Using MATLAB, we find the new $\mathbf{K}_e$ to be

$$\mathbf{K}_e = \begin{bmatrix} -1 \\ 6.25 \end{bmatrix}$$

Next, we shall obtain the transfer function of the observer controller. MATLAB Program 10–14 produces this transfer function as follows:

$$G_c(s) = \frac{1.2109s^2 + 11.2125s + 25.3125}{s^2 + 6s + 2.1406}$$

$$= \frac{1.2109(s + 5.3582)(s + 3.9012)}{(s + 5.619)(s + 0.381)}$$

MATLAB Program 10–14

```
% Determination of transfer function of observer controller.

A = [0  1  0;0  0  1;0 -24 -10];
B = [0;10;-80];
Aaa = 0; Aab = [1  0]; Aba = [0;0]; Abb = [0  1;-24  -10];
Ba = 0; Bb = [10;-80];
Ka = 1.25; Kb = [1.25  0.19375];
Ke = [-1;6.25];
Ahat = Abb - Ke*Aab;
Bhat = Ahat*Ke + Aba - Ke*Aaa;
Fhat = Bb - Ke*Ba;
Atilde = Ahat - Fhat*Kb;
Btilde = Bhat - Fhat*(Ka + Kb*Ke);
Ctilde = -Kb;
Dtilde = -(Ka + Kb*Ke);
[num,den] = ss2tf(Atilde,Btilde,-Ctilde,-Dtilde)

num =

    1.2109  11.2125  25.3125

den =

    1.0000  6.0000  2.1406
```

Notice that this is a stable controller. Define the system with this observer controller as System 2. We shall proceed to obtain the response of System 2 to the given initial condition:

$$\mathbf{x}(0) = \begin{bmatrix} 1 \\ 0 \\ 0 \end{bmatrix}, \qquad \mathbf{e}(0) = \begin{bmatrix} 1 \\ 0 \end{bmatrix}$$

By substituting $u = -\mathbf{K}\tilde{\mathbf{x}}$ into the state-space equation for the plant, we obtain

$$\dot{\mathbf{x}} = \mathbf{Ax} - \mathbf{BK}\tilde{\mathbf{x}} = \mathbf{Ax} - \mathbf{BK}\begin{bmatrix} x_a \\ \tilde{\mathbf{x}}_b \end{bmatrix} = \mathbf{Ax} - \mathbf{BK}\begin{bmatrix} x_a \\ \mathbf{x}_b - \mathbf{e} \end{bmatrix}$$

$$= \mathbf{Ax} - \mathbf{BK}\left\{ \mathbf{x} - \begin{bmatrix} 0 \\ \mathbf{e} \end{bmatrix} \right\} = \mathbf{Ax} - \mathbf{BKx} + \mathbf{B}\begin{bmatrix} K_a & \mathbf{K}_b \end{bmatrix}\begin{bmatrix} 0 \\ \mathbf{e} \end{bmatrix} \qquad (10\text{–}110)$$

The error equation for the minimum-order observer is

$$\dot{\mathbf{e}} = (\mathbf{A}_{bb} - \mathbf{K}_e \mathbf{A}_{ab})\mathbf{e} \qquad (10\text{–}111)$$

By combining Equations (10–110) and (10–111), we get

$$\begin{bmatrix} \dot{\mathbf{x}} \\ \dot{\mathbf{e}} \end{bmatrix} = \begin{bmatrix} \mathbf{A} - \mathbf{BK} & \mathbf{BK}_b \\ \mathbf{0} & \mathbf{A}_{bb} - \mathbf{K}_e \mathbf{A}_{ab} \end{bmatrix} \begin{bmatrix} \mathbf{x} \\ \mathbf{e} \end{bmatrix}$$

with the initial condition

$$\begin{bmatrix} \mathbf{x}(0) \\ \mathbf{e}(0) \end{bmatrix} = \begin{bmatrix} 1 \\ 0 \\ 0 \\ 1 \\ 0 \end{bmatrix}$$

MATLAB Program 10–15 produces the response to the given initial condition. The response curves are shown in Figure 10–21. They seem to be acceptable.

MATLAB Program 10–15

```
% Response to initial condition.

A = [0  1  0;0  0  1;0 -24 -10];
B = [0;10;-80];
K = [1.25  1.25  0.19375];
Kb = [1.25  0.19375];
Ke = [-1;6.25];
Aab = [1  0]; Abb = [0  1;-24  -10];
AA = [A-B*K  B*Kb; zeros(2,3)  Abb-Ke*Aab];
sys = ss(AA,eye(5),eye(5),eye(5));
t = 0:0.01:8;
x = initial(sys,[1;0;0;1;0],t);
x1 = [1  0  0  0  0]*x';
x2 = [0  1  0  0  0]*x';
x3 = [0  0  1  0  0]*x';
e1 = [0  0  0  1  0]*x';
e2 = [0  0  0  0  1]*x';

subplot(3,2,1); plot(t,x1); grid
xlabel ('t (sec)'); ylabel('x1')

subplot(3,2,2); plot(t,x2); grid
xlabel ('t (sec)'); ylabel('x2')

subplot(3,2,3); plot(t,x3); grid
xlabel ('t (sec)'); ylabel('x3')

subplot(3,2,4); plot(t,e1); grid
xlabel('t (sec)'); ylabel('e1')

subplot(3,2,5); plot(t,e2); grid
xlabel('t (sec)'); ylabel('e2')
```

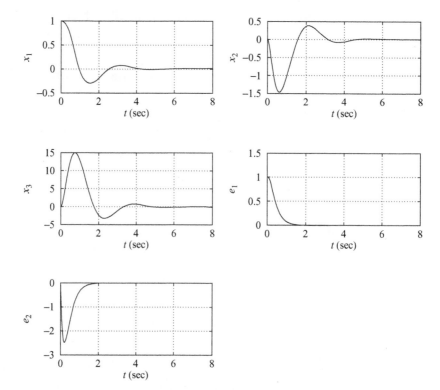

Figure 10–21
Response to the given initial condition; $x_1(0) = 1$, $x_2(0) = 0$, $x_3(0) = 0$, $e_1(0) = 1$, $e_2(0) = 0$.

Next, we shall check the frequency-response characteristics. The Bode diagram of the open-loop system just designed is shown in Figure 10–22. The phase margin is about 40° and the gain margin is $+\infty$ dB. The Bode diagram of the closed-loop system is shown in Figure 10–23. The bandwidth of the system is approximately 3.8 rad/sec.

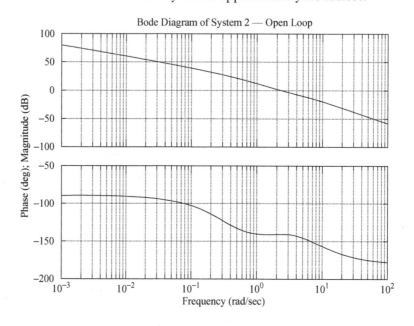

Figure 10–22
Bode diagram for the open-loop transfer function of System 2.

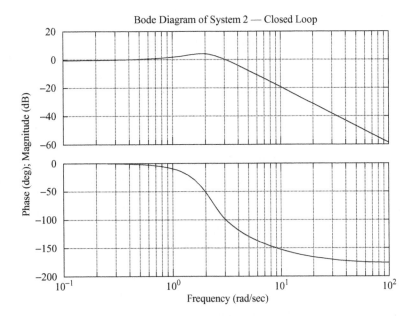

Figure 10–23
Bode diagram for the closed-loop transfer function of System 2.

Finally, we shall compare the root-locus plots of the first system with $L = \begin{bmatrix} -10 & -10 \end{bmatrix}$ and the second system with $L = \begin{bmatrix} -4.5 & -4.5 \end{bmatrix}$. The plot for the first system given in Figure 10–24(a) shows that the system is unstable for small dc gain and becomes stable for large dc gain. The plot for the second system given in Figure 10–24(b), on the other hand, shows that the system is stable for any positive dc gain.

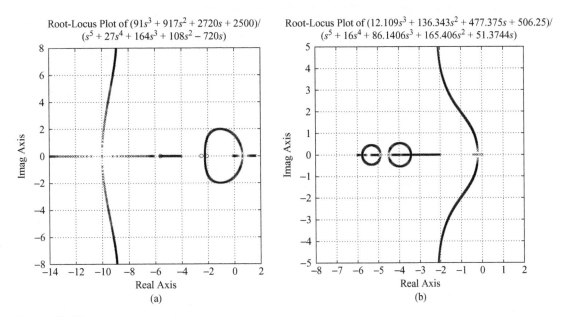

Figure 10–24
(a) Root-locus plot of the system with observer poles at $s = -10$ and $s = -10$; (b) root-locus plot of the system with observer poles at $s = -4.5$ and $s = -4.5$.

Comments

1. In designing regulator systems, note that if the dominant controller poles are placed far to the left of the $j\omega$ axis, the elements of the state feedback gain matrix $\mathbf{K}$ will become large. Large gain values will make the actuator output become large, so that saturation may take place. Then the designed system will not behave as designed.

2. Also, by placing the observer poles far to the left of the $j\omega$ axis, the observer controller becomes unstable, although the closed-loop system is stable. An unstable observer controller is not acceptable.

3. If the observer controller becomes unstable, move the observer poles to the right in the left-half s plane until the observer controller becomes stable. Also, the desired closed-loop pole locations may need to be modified.

4. Note that if the observer poles are placed far to the left of the $j\omega$ axis, the bandwidth of the observer will increase and will cause noise problems. If there is a serious noise problem, the observer poles should not be placed too far to the left of the $j\omega$ axis. The general requirement is that the bandwidth should be sufficiently low so that the sensor noise will not become a problem.

5. The bandwidth of the system with the minimum-order observer is higher than that of the system with the full-order observer, provided that the multiple observer poles are placed at the same place for both observers. If the sensor noise is a serious problem, use of a full-order observer is recomnended.

10–7 DESIGN OF CONTROL SYSTEMS WITH OBSERVERS

In Section 10–6 we discussed the design of regulator systems with observers. (The systems did not have reference or command inputs.) In this section we consider the design of control systems with observers when the systems have reference inputs or command inputs. The output of the control system must follow the input that is time varying. In following the command input, the system must exhibit satisfactory performance (a reasonable rise time, overshoot, settling time, and so on).

In this section we consider control systems that are designed by use of the pole-placement-with-observer approach. Specifically, we consider control systems using observer controllers. In Section 10–6 we discussed regulator systems, whose block diagram is shown in Figure 10–25. This system has no reference input, or $r = 0$. When the system has a reference input, several different block diagram configurations are conceivable, each having an observer controller. Two of these configurations are shown in Figures 10–26 (a) and (b); we shall consider them in this section.

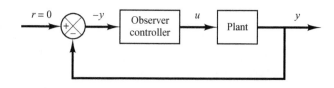

Figure 10–25
Regulator system.

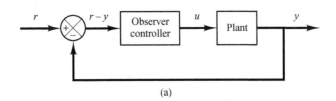

(a)

Figure 10–26
(a) Control system
with observer
controller in the
feedforward path;
(b) Control system
with observer
controller in the
feedback path.

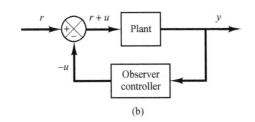

(b)

Configuration 1: Consider the system shown in Figure 10–27. In this system the reference input is simply added at the summing point. We would like to design the observer
controller such that in the unit-step response the maximum overshoot is less than 30%
and the settling time is about 5 sec.

In what follows we first design a regulator system. Then, using the observer controller
designed, we simply add the reference input r at the summing point.

Before we design the observer controller, we need to obtain a state-space representation of the plant. Since

$$\frac{Y(s)}{U(s)} = \frac{1}{s(s^2 + 1)}$$

we obtain

$$\ddot{y} + \dot{y} = u$$

By choosing the state variables as

$$x_1 = y$$
$$x_2 = \dot{y}$$
$$x_3 = \ddot{y}$$

we get

$$\dot{\mathbf{x}} = \mathbf{Ax} + \mathbf{B}u$$

$$y = \mathbf{Cx}$$

Figure 10–27
Control system with
observer controller
in the feedforward
path.

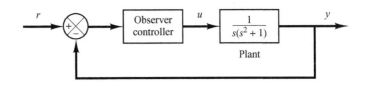

where

$$A = \begin{bmatrix} 0 & 1 & 0 \\ 0 & 0 & 1 \\ 0 & -1 & 0 \end{bmatrix}, \qquad B = \begin{bmatrix} 0 \\ 0 \\ 1 \end{bmatrix}, \qquad C = \begin{bmatrix} 1 & 0 & 0 \end{bmatrix}$$

Next, we choose the desired closed-loop poles for pole placement at

$$s = -1 + j, \qquad s = -1 - j, \qquad s = -8$$

and the desired observer poles at

$$s = -4, \qquad s = -4$$

The state feedback gain matrix **K** and the observer gain matrix K_e can be obtained as follows:

$$K = \begin{bmatrix} 16 & 17 & 10 \end{bmatrix}$$

$$K_e = \begin{bmatrix} 8 \\ 15 \end{bmatrix}$$

See MATLAB Program 10–16.

MATLAB Program 10–16

```
A = [0 1 0;0 0 1;0 -1 0];
B = [0;0;1];
J = [-1+j -1-j -8];
K = acker(A,B,J)

K =

  16 17 10

Aab = [1 0];
Abb = [0 1;-1 0];
L = [-4 -4];
Ke = acker(Abb',Aab',L)'

Ke =
  8
  15
```

The transfer function of the observer controller is obtained by use of MATLAB Program 10–17. The result is

$$G_c(s) = \frac{302s^2 + 303s + 256}{s^2 + 18s + 113}$$

$$= \frac{302(s + 0.5017 + j0.772)(s + 0.5017 - j0.772)}{(s + 9 + j5.6569)(s + 9 - j5.6569)}$$

```
MATLAB Program 10–17

% Determination of transfer function of observer controller
A = [0  1  0;0  0  1;0  -1  0];
B = [0;0;1];
Aaa = 0; Aab = [1  0]; Aba = [0;0]; Abb = [0  1;-1  0];
Ba = 0; Bb = [0;1];
Ka = 16; Kb=[17  10];
Ke = [8;15];
Ahat = Abb - Ke*Aab;
Bhat = Ahat*Ke + Aba - Ke*Aaa;
Fhat = Bb - Ke*Ba;
Atilde = Ahat - Fhat*Kb;
Btilde = Bhat - Fhat*(Ka + Kb*Ke);
Ctilde = -Kb;
Dtilde = -(Ka + Kb*Ke);
[num,den] = ss2tf(Atilde,Btilde,-Ctilde,-Dtilde)

num =

   302.0000  303.0000  256.0000
den =

   1  18  113
```

Figure 10–28 shows the block diagram of the regulator system just designed. Figure 10–29 shows the block diagram of a possible configuration of the control system based on the regulator system shown in Figure 10–28. The unit-step response curve for this control system is shown in Figure 10–30. The maximum overshoot is about 28% and the settling time is about 4.5 sec. Thus, the designed system satisfies the design requirements.

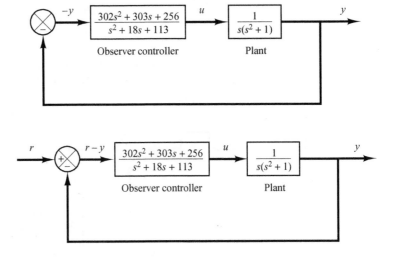

Figure 10–28
Regulator system with observer controller.

Figure 10–29
Control system with observer controller in the feedforward path.

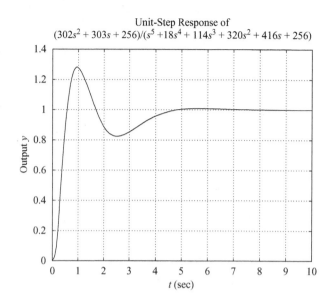

Unit-Step Response of
$$(302s^2 + 303s + 256)/(s^5 + 18s^4 + 114s^3 + 320s^2 + 416s + 256)$$

Figure 10–30
Unit-step response of
the control system
shown in Figure
10–29.

Configuration 2: A different configuration of the control system is shown in Figure
10–31. The observer controller is placed in the feedback path. The input r is introduced
into the closed-loop system through the box with gain N. From this block diagram, the
closed-loop transfer function is obtained as

$$\frac{Y(s)}{R(s)} = \frac{N(s^2 + 18s + 113)}{s(s^2 + 1)(s^2 + 18s + 113) + 302s^2 + 303s + 256}$$

We determine the value of constant N such that for a unit-step input r, the output y is
unity as t approaches infinity. Thus we choose

$$N = \frac{256}{113} = 2.2655$$

The unit-step response of the system is shown in Figure 10–32. Notice that the maximum overshoot is very small, approximately 4%. The settling time is about 5 sec.

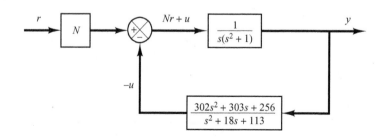

Figure 10–31
Control system with
observer controller
in the feedback path.

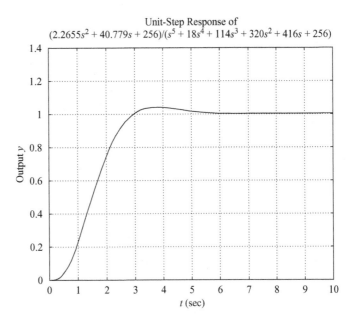

Figure 10–32
The unit-step response of the system shown in Figure 10–31. (The closed-loop poles for pole placement are at $s = -1 \pm j$, $s = -8$. The observer poles are at $s = -4$, $s = -4$.)

Unit-Step Response of
$(2.2655s^2 + 40.779s + 256)/(s^5 + 18s^4 + 114s^3 + 320s^2 + 416s + 256)$

Comments. We considered two possible configurations for the closed-loop control systems using observer controllers. As stated earlier, other configurations are possible.

The first configuration, which places the observer controller in the feedforward path, generally gives a fairly large overshoot. The second configuration, which places the observer controller in the feedback path, gives a smaller overshoot. This response curve is quite similar to that of the system designed by the pole-placement approach without using the observer controller. See the unit-step response curve of the system, shown in Figure 10–33, designed by the pole-placement approach without observer. Here the desired closed-loop poles used are

$$s = -1 + j, \qquad s = -1 - j, \qquad s = -8$$

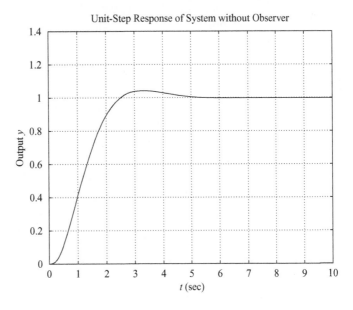

Unit-Step Response of System without Observer

Figure 10–33
The unit-step response of the control system designed by the pole placement approach without observer. (The closed-loop poles are at $s = -1 \pm j, s = -8$.)

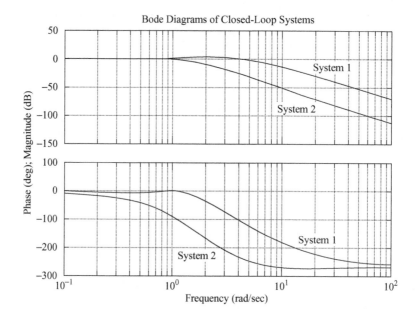

Figure 10–34
Bode diagrams of closed-loop system 1 (shown in Figure 10–29) and closed-loop system 2 (shown in Figure 10–31).

Note that, in these two systems, the rise time and settling time are determined primarily by the desired closed-loop poles for pole placement. (See Figures 10–32 and 10–33.)

The Bode diagrams of closed-loop system 1 (shown in Figure 10–29) and closed-loop system 2 (shown in Figure 10–31) are shown in Figure 10–34. From this figure, we find that the bandwidth of system 1 is 5 rad/sec and that of system 2 is 1.3 rad/sec.

Summary of State-Space Design Method

1. The state-space design method based on the pole-placement-combined-with-observer approach is very powerful. It is a time-domain method. The desired closed-loop poles can be arbitrarily placed, provided the plant is completely state controllable.

2. If not all state variables can be measured, an observer must be incorporated to estimate the unmeasurable state variables.

3. In designing a system using the pole-placement approach, several different sets of desired closed-loop poles need be considered, the response characteristics compared, and the best one chosen.

4. The bandwidth of the observer controller is generally large, because we choose observer poles far to the left in the s plane. A large bandwidth passes high-frequency noises and causes the noise problem.

5. Adding an observer to the system generally reduces the stability margin. In some cases, an observer controller may have zero(s) in the right-half s plane, which means that the controller may be stable but of nonminimum phase. In other cases, the controller may have pole(s) in the right-half s plane—that is, the controller is unstable. Then the designed system may become conditionally stable.

6. When the system is designed by the pole-placement-with-observer approach, it is advisable to check the stability margins (phase margin and gain margin), using a

frequency-response method. If the system designed has poor stability margins, it is possible that the designed system may become unstable if the mathematical model involves uncertainties.

7. Note that for nth-order systems, classical design methods (root-locus and frequency-response methods) yield low-order compensators (first or second order). Since the observer-based controllers are nth-order $\left[\text{or } (n - m)\text{th-order if the minimum-order observer is used}\right]$ for an nth-order system, the designed system will become $2n$th order $\left[\text{or } (2n - m)\text{th order}\right]$. Since lower-order compensators are cheaper than higher-order ones, the designer should first apply classical methods and, if no suitable compensators can be determined, then try the pole-placement-with-observer design approach presented in this chapter.

10–8 QUADRATIC OPTIMAL REGULATOR SYSTEMS

An advantage of the quadratic optimal control method over the pole-placement method is that the former provides a systematic way of computing the state feedback control gain matrix.

Quadratic Optimal Regulator Problems. We shall now consider the optimal regulator problem that, given the system equation

$$\dot{\mathbf{x}} = \mathbf{Ax} + \mathbf{Bu} \tag{10–112}$$

determines the matrix $\mathbf{K}$ of the optimal control vector

$$\mathbf{u}(t) = -\mathbf{Kx}(t) \tag{10–113}$$

so as to minimize the performance index

$$J = \int_0^\infty (\mathbf{x}^*\mathbf{Qx} + \mathbf{u}^*\mathbf{Ru})\, dt \tag{10–114}$$

where $\mathbf{Q}$ is a positive-definite (or positive-semidefinite) Hermitian or real symmetric matrix and $\mathbf{R}$ is a positive-definite Hermitian or real symmetric matrix. Note that the second term on the right-hand side of Equation (10–114) accounts for the expenditure of the energy of the control signals. The matrices $\mathbf{Q}$ and $\mathbf{R}$ determine the relative importance of the error and the expenditure of this energy. In this problem, we assume that the control vector $\mathbf{u}(t)$ is unconstrained.

As will be seen later, the linear control law given by Equation (10–113) is the optimal control law. Therefore, if the unknown elements of the matrix $\mathbf{K}$ are determined so as to minimize the performance index, then $\mathbf{u}(t) = -\mathbf{Kx}(t)$ is optimal for any initial state $\mathbf{x}(0)$. The block diagram showing the optimal configuration is shown in Figure 10–35.

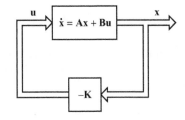

Figure 10–35
Optimal regulator system.

Now let us solve the optimization problem. Substituting Equation (10–113) into Equation (10–112), we obtain

$$\dot{\mathbf{x}} = \mathbf{A}\mathbf{x} - \mathbf{B}\mathbf{K}\mathbf{x} = (\mathbf{A} - \mathbf{B}\mathbf{K})\mathbf{x}$$

In the following derivations, we assume that the matrix $\mathbf{A} - \mathbf{B}\mathbf{K}$ is stable, or that the eigenvalues of $\mathbf{A} - \mathbf{B}\mathbf{K}$ have negative real parts.

Substituting Equation (10–113) into Equation (10–114) yields

$$J = \int_0^\infty (\mathbf{x}^*\mathbf{Q}\mathbf{x} + \mathbf{x}^*\mathbf{K}^*\mathbf{R}\mathbf{K}\mathbf{x})\, dt$$

$$= \int_0^\infty \mathbf{x}^*(\mathbf{Q} + \mathbf{K}^*\mathbf{R}\mathbf{K})\mathbf{x}\, dt$$

Let us set

$$\mathbf{x}^*(\mathbf{Q} + \mathbf{K}^*\mathbf{R}\mathbf{K})\mathbf{x} = -\frac{d}{dt}(\mathbf{x}^*\mathbf{P}\mathbf{x})$$

where $\mathbf{P}$ is a positive-definite Hermitian or real symmetric matrix. Then we obtain

$$\mathbf{x}^*(\mathbf{Q} + \mathbf{K}^*\mathbf{R}\mathbf{K})\mathbf{x} = -\dot{\mathbf{x}}^*\mathbf{P}\mathbf{x} - \mathbf{x}^*\mathbf{P}\dot{\mathbf{x}} = -\mathbf{x}^*\big[(\mathbf{A} - \mathbf{B}\mathbf{K})^*\mathbf{P} + \mathbf{P}(\mathbf{A} - \mathbf{B}\mathbf{K})\big]\mathbf{x}$$

Comparing both sides of this last equation and noting that this equation must hold true for any $\mathbf{x}$, we require that

$$(\mathbf{A} - \mathbf{B}\mathbf{K})^*\mathbf{P} + \mathbf{P}(\mathbf{A} - \mathbf{B}\mathbf{K}) = -(\mathbf{Q} + \mathbf{K}^*\mathbf{R}\mathbf{K}) \qquad (10\text{–}115)$$

It can be proved that if $\mathbf{A} - \mathbf{B}\mathbf{K}$ is a stable matrix, there exists a positive-definite matrix $\mathbf{P}$ that satisfies Equation (10–115). (See Problem **A–10–15**.)

Hence our procedure is to determine the elements of $\mathbf{P}$ from Equation (10–115) and see if it is positive definite. (Note that more than one matrix $\mathbf{P}$ may satisfy this equation. If the system is stable, there always exists one positive-definite matrix $\mathbf{P}$ to satisfy this equation. This means that, if we solve this equation and find one positive-definite matrix $\mathbf{P}$, the system is stable. Other $\mathbf{P}$ matrices that satisfy this equation are not positive definite and must be discarded.)

The performance index J can be evaluated as

$$J = \int_0^\infty \mathbf{x}^*(\mathbf{Q} + \mathbf{K}^*\mathbf{R}\mathbf{K})\mathbf{x}\, dt = -\mathbf{x}^*\mathbf{P}\mathbf{x}\Big|_0^\infty = -\mathbf{x}^*(\infty)\mathbf{P}\mathbf{x}(\infty) + \mathbf{x}^*(0)\mathbf{P}\mathbf{x}(0)$$

Since all eigenvalues of $\mathbf{A} - \mathbf{B}\mathbf{K}$ are assumed to have negative real parts, we have $\mathbf{x}(\infty) \to \mathbf{0}$. Therefore, we obtain

$$J = \mathbf{x}^*(0)\mathbf{P}\mathbf{x}(0) \qquad (10\text{–}116)$$

Thus, the performance index J can be obtained in terms of the initial condition $\mathbf{x}(0)$ and $\mathbf{P}$.

To obtain the solution to the quadratic optimal control problem, we proceed as follows: Since $\mathbf{R}$ has been assumed to be a positive-definite Hermitian or real symmetric matrix, we can write

$$\mathbf{R} = \mathbf{T}^*\mathbf{T}$$

where $\mathbf{T}$ is a nonsingular matrix. Then Equation (10–115) can be written as

$$(\mathbf{A}^* - \mathbf{K}^*\mathbf{B}^*)\mathbf{P} + \mathbf{P}(\mathbf{A} - \mathbf{BK}) + \mathbf{Q} + \mathbf{K}^*\mathbf{T}^*\mathbf{TK} = \mathbf{0}$$

which can be rewritten as

$$\mathbf{A}^*\mathbf{P} + \mathbf{PA} + \left[\mathbf{TK} - (\mathbf{T}^*)^{-1}\mathbf{B}^*\mathbf{P}\right]^*\left[\mathbf{TK} - (\mathbf{T}^*)^{-1}\mathbf{B}^*\mathbf{P}\right] - \mathbf{PBR}^{-1}\mathbf{B}^*\mathbf{P} + \mathbf{Q} = \mathbf{0}$$

The minimization of J with respect to $\mathbf{K}$ requires the minimization of

$$\mathbf{x}^*\left[\mathbf{TK} - (\mathbf{T}^*)^{-1}\mathbf{B}^*\mathbf{P}\right]^*\left[\mathbf{TK} - (\mathbf{T}^*)^{-1}\mathbf{B}^*\mathbf{P}\right]\mathbf{x}$$

with respect to $\mathbf{K}$. (See Problem **A–10–16**.) Since this last expression is nonnegative, the minimum occurs when it is zero, or when

$$\mathbf{TK} = (\mathbf{T}^*)^{-1}\mathbf{B}^*\mathbf{P}$$

Hence,

$$\mathbf{K} = \mathbf{T}^{-1}(\mathbf{T}^*)^{-1}\mathbf{B}^*\mathbf{P} = \mathbf{R}^{-1}\mathbf{B}^*\mathbf{P} \tag{10–117}$$

Equation (10–117) gives the optimal matrix $\mathbf{K}$. Thus, the optimal control law to the quadratic optimal control problem when the performance index is given by Equation (10–114) is linear and is given by

$$\mathbf{u}(t) = -\mathbf{Kx}(t) = -\mathbf{R}^{-1}\mathbf{B}^*\mathbf{Px}(t)$$

The matrix $\mathbf{P}$ in Equation (10–117) must satisfy Equation (10–115) or the following reduced equation:

$$\mathbf{A}^*\mathbf{P} + \mathbf{PA} - \mathbf{PBR}^{-1}\mathbf{B}^*\mathbf{P} + \mathbf{Q} = \mathbf{0} \tag{10–118}$$

Equation (10–118) is called the reduced-matrix Riccati equation. The design steps may be stated as follows:

1. Solve Equation (10–118), the reduced-matrix Riccati equation, for the matrix $\mathbf{P}$. [If a positive-definite matrix $\mathbf{P}$ exists (certain systems may not have a positive-definite matrix $\mathbf{P}$), the system is stable, or matrix $\mathbf{A} - \mathbf{BK}$ is stable.]
2. Substitute this matrix $\mathbf{P}$ into Equation (10–117). The resulting matrix $\mathbf{K}$ is the optimal matrix.

A design example based on this approach is given in Example 10–9. Note that if the matrix $\mathbf{A} - \mathbf{BK}$ is stable, the present method always gives the correct result.

Finally, note that if the performance index is given in terms of the output vector rather than the state vector, that is,

$$J = \int_0^\infty (\mathbf{y}^*\mathbf{Qy} + \mathbf{u}^*\mathbf{Ru})\,dt$$

then the index can be modified by using the output equation

$$\mathbf{y} = \mathbf{Cx}$$

to

$$J = \int_0^\infty (\mathbf{x}^*\mathbf{C}^*\mathbf{QCx} + \mathbf{u}^*\mathbf{Ru})\,dt \tag{10–119}$$

and the design steps presented in this section can be applied to obtain the optimal matrix $\mathbf{K}$.

EXAMPLE 10–9 Consider the system shown in Figure 10–36. Assuming the control signal to be

$$u(t) = -\mathbf{K}\mathbf{x}(t)$$

determine the optimal feedback gain matrix $\mathbf{K}$ such that the following performance index is minimized:

$$J = \int_0^\infty \left(\mathbf{x}^T \mathbf{Q} \mathbf{x} + u^2\right) dt$$

where

$$\mathbf{Q} = \begin{bmatrix} 1 & 0 \\ 0 & \mu \end{bmatrix} \qquad (\mu \geq 0)$$

From Figure 10–36, we find that the state equation for the plant is

$$\dot{\mathbf{x}} = \mathbf{A}\mathbf{x} + \mathbf{B}u$$

where

$$\mathbf{A} = \begin{bmatrix} 0 & 1 \\ 0 & 0 \end{bmatrix}, \qquad \mathbf{B} = \begin{bmatrix} 0 \\ 1 \end{bmatrix}$$

We shall demonstrate the use of the reduced-matrix Riccati equation in the design of the optimal control system. Let us solve Equation (10–118), rewritten as

$$\mathbf{A}^*\mathbf{P} + \mathbf{P}\mathbf{A} - \mathbf{P}\mathbf{B}\mathbf{R}^{-1}\mathbf{B}^*\mathbf{P} + \mathbf{Q} = 0$$

Noting that matrix $\mathbf{A}$ is real and matrix $\mathbf{Q}$ is real symmetric, we see that matrix $\mathbf{P}$ is a real symmetric matrix. Hence, this last equation can be written as

$$\begin{bmatrix} 0 & 0 \\ 1 & 0 \end{bmatrix}\begin{bmatrix} p_{11} & p_{12} \\ p_{12} & p_{22} \end{bmatrix} + \begin{bmatrix} p_{11} & p_{12} \\ p_{12} & p_{22} \end{bmatrix}\begin{bmatrix} 0 & 1 \\ 0 & 0 \end{bmatrix}$$

$$- \begin{bmatrix} p_{11} & p_{12} \\ p_{12} & p_{22} \end{bmatrix}\begin{bmatrix} 0 \\ 1 \end{bmatrix}[1][0 \quad 1]\begin{bmatrix} p_{11} & p_{12} \\ p_{12} & p_{22} \end{bmatrix} + \begin{bmatrix} 1 & 0 \\ 0 & \mu \end{bmatrix} = \begin{bmatrix} 0 & 0 \\ 0 & 0 \end{bmatrix}$$

This equation can be simplified to

$$\begin{bmatrix} 0 & 0 \\ p_{11} & p_{12} \end{bmatrix} + \begin{bmatrix} 0 & p_{11} \\ 0 & p_{12} \end{bmatrix} - \begin{bmatrix} p_{12}^2 & p_{12}p_{22} \\ p_{12}p_{22} & p_{22}^2 \end{bmatrix} + \begin{bmatrix} 1 & 0 \\ 0 & \mu \end{bmatrix} = \begin{bmatrix} 0 & 0 \\ 0 & 0 \end{bmatrix}$$

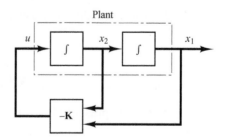

Plant

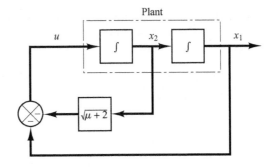

Figure 10–37
Optimal control of
the plant shown in
Figure 10–36.

from which we obtain the following three equations:

$$1 - p_{12}^2 = 0$$

$$p_{11} - p_{12}p_{22} = 0$$

$$\mu + 2p_{12} - p_{22}^2 = 0$$

Solving these three simultaneous equations for p_{11}, p_{12}, and p_{22}, requiring **P** to be positive definite, we obtain

$$\mathbf{P} = \begin{bmatrix} p_{11} & p_{12} \\ p_{12} & p_{22} \end{bmatrix} = \begin{bmatrix} \sqrt{\mu + 2} & 1 \\ 1 & \sqrt{\mu + 2} \end{bmatrix}$$

Referring to Equation (10–117), the optimal feedback gain matrix **K** is obtained as

$$\mathbf{K} = \mathbf{R}^{-1}\mathbf{B}^*\mathbf{P}$$

$$= [1][0 \quad 1]\begin{bmatrix} p_{11} & p_{12} \\ p_{12} & p_{22} \end{bmatrix}$$

$$= [p_{12} \quad p_{22}]$$

$$= [1 \quad \sqrt{\mu + 2}]$$

Thus, the optimal control signal is

$$u = -\mathbf{K}\mathbf{x} = -x_1 - \sqrt{\mu + 2}\, x_2 \tag{10–120}$$

Note that the control law given by Equation (10–120) yields an optimal result for any initial state under the given performance index. Figure 10–37 is the block diagram for this system.

Since the characteristic equation is

$$|s\mathbf{I} - \mathbf{A} + \mathbf{B}\mathbf{K}| = s^2 + \sqrt{\mu + 2}\, s + 1 = 0$$

if $\mu = 1$, the two closed-loop poles are located at

$$s = -0.866 + j\,0.5, \qquad s = -0.866 - j\,0.5$$

These correspond to the desired closed-loop poles when $\mu = 1$.

Solving Quadratic Optimal Regulator Problems with MATLAB. In MATLAB, the command

lqr(A,B,Q,R)

solves the continuous-time, linear, quadratic regulator problem and the associated Riccati equation. This command calculates the optimal feedback gain matrix $\mathbf{K}$ such that the feedback control law

$$u = -\mathbf{Kx}$$

minimizes the performance index

$$J = \int_0^\infty (\mathbf{x}^*\mathbf{Qx} + \mathbf{u}^*\mathbf{Ru})\,dt$$

subject to the constraint equation

$$\dot{\mathbf{x}} = \mathbf{Ax} + \mathbf{Bu}$$

Another command

$$[K,P,E] = lqr(A,B,Q,R)$$

returns the gain matrix $\mathbf{K}$, eigenvalue vector $\mathbf{E}$, and matrix $\mathbf{P}$, the unique positive-definite solution to the associated matrix Riccati equation:

$$\mathbf{PA} + \mathbf{A}^*\mathbf{P} - \mathbf{PBR}^{-1}\mathbf{B}^*\mathbf{P} + \mathbf{Q} = \mathbf{0}$$

If matrix $\mathbf{A} - \mathbf{BK}$ is a stable matrix, such a positive-definite solution $\mathbf{P}$ always exists. The eigenvalue vector $\mathbf{E}$ gives the closed-loop poles of $\mathbf{A} - \mathbf{BK}$.

It is important to note that for certain systems matrix $\mathbf{A} - \mathbf{BK}$ cannot be made a stable matrix, whatever $\mathbf{K}$ is chosen. In such a case, there does not exist a positive-definite matrix $\mathbf{P}$ for the matrix Riccati equation. For such a case, the commands

$$K = lqr(A,B,Q,R)$$

$$[K,P,E] = lqr(A,B,Q,R)$$

do not give the solution. See MATLAB Program 10–18.

EXAMPLE 10–10 Consider the system defined by

$$\begin{bmatrix} \dot{x}_1 \\ \dot{x}_2 \end{bmatrix} = \begin{bmatrix} -1 & 1 \\ 0 & 2 \end{bmatrix}\begin{bmatrix} x_1 \\ x_2 \end{bmatrix} + \begin{bmatrix} 1 \\ 0 \end{bmatrix}u$$

Show that the system cannot be stabilized by the state-feedback control scheme

$$u = -\mathbf{Kx}$$

whatever matrix $\mathbf{K}$ is chosen. (Notice that this system is not state controllable.)
Define

$$\mathbf{K} = \begin{bmatrix} k_1 & k_2 \end{bmatrix}$$

Then

$$\mathbf{A} - \mathbf{BK} = \begin{bmatrix} -1 & 1 \\ 0 & 2 \end{bmatrix} - \begin{bmatrix} 1 \\ 0 \end{bmatrix}\begin{bmatrix} k_1 & k_2 \end{bmatrix}$$

$$= \begin{bmatrix} -1 - k_1 & 1 - k_2 \\ 0 & 2 \end{bmatrix}$$

Hence, the characteristic equation becomes

$$|s\mathbf{I} - \mathbf{A} + \mathbf{B}\mathbf{K}| = \begin{vmatrix} s + 1 + k_1 & -1 + k_2 \\ 0 & s - 2 \end{vmatrix}$$

$$= (s + 1 + k_1)(s - 2) = 0$$

The closed-loop poles are located at

$$s = -1 - k_1, \qquad s = 2$$

Since the pole at $s = 2$ is in the right-half s plane, the system is unstable whatever $\mathbf{K}$ matrix is chosen. Hence, quadratic optimal control techniques cannot be applied to this system.

Let us assume that matrices $\mathbf{Q}$ and $\mathbf{R}$ in the quadratic performance index are given by

$$\mathbf{Q} = \begin{bmatrix} 1 & 0 \\ 0 & 1 \end{bmatrix}, \qquad R = [1]$$

and that we write MATLAB Program 10–18. The resulting MATLAB solution is

$$K = [\text{NaN} \quad \text{NaN}]$$

(NaN means 'not a number.') Whenever the solution to a quadratic optimal control problem does not exist, MATLAB tells us that matrix $\mathbf{K}$ consists of NaN.

MATLAB Program 10–18

```
% ---------- Design of quadratic optimal regulator system ----------

A = [-1 1;0  2];
B = [1;0];
Q = [1  0;0  1];
R = [1];

K = lqr(A,B,Q,R)

Warning: Matrix is singular to working precision.

K =

   NaN  NaN

% ***** If we enter the command [K,P,E] = lqr(A,B,Q,R), then *****

[K,P,E] = lqr(A,B,Q,R)

Warning: Matrix is singular to working precision.

K =

   NaN  NaN

P =

   -Inf  -Inf
   -Inf  -Inf

E =

   -2.0000
   -1.4142
```

EXAMPLE 10–11 Consider the system described by

$$\dot{x} = Ax + Bu$$

where

$$A = \begin{bmatrix} 0 & 1 \\ 0 & -1 \end{bmatrix}, \qquad B = \begin{bmatrix} 0 \\ 1 \end{bmatrix}$$

The performance index J is given by

$$J = \int_0^\infty (x'Qx + u'Ru)\,dt$$

where

$$Q = \begin{bmatrix} 1 & 0 \\ 0 & 1 \end{bmatrix}, \qquad R = [1]$$

Assume that the following control u is used.

$$u = -Kx$$

Determine the optimal feedback gain matrix K.

The optimal feedback gain matrix K can be obtained by solving the following Riccati equation for a positive-definite matrix P:

$$A*P + PA - PBR^{-1}B*P + Q = 0$$

The result is

$$P = \begin{bmatrix} 2 & 1 \\ 1 & 1 \end{bmatrix}$$

Substituting this P matrix into the following equation gives the optimal K matrix:

$$K = R^{-1}B*P$$

$$= [1][0 \quad 1]\begin{bmatrix} 2 & 1 \\ 1 & 1 \end{bmatrix} = [1 \quad 1]$$

Thus, the optimal control signal is given by

$$u = -Kx = -x_1 - x_2$$

MATLAB 10–19 also yields the solution to this problem.

MATLAB Program 10–19

```
% ---------- Design of quadratic optimal regulator system ----------
A = [0  1;0 -1];
B = [0;1];
Q = [1  0; 0  1];
R = [1];

K = lqr(A,B,Q,R)

K =

    1.0000   1.0000
```

EXAMPLE 10–12   Consider the system given by

$$\dot{\mathbf{x}} = \mathbf{A}\mathbf{x} + \mathbf{B}u$$

where

$$\mathbf{A} = \begin{bmatrix} 0 & 1 & 0 \\ 0 & 0 & 1 \\ -35 & -27 & -9 \end{bmatrix}, \quad \mathbf{B} = \begin{bmatrix} 0 \\ 0 \\ 1 \end{bmatrix}$$

The performance index J is given by

$$J = \int_0^\infty (\mathbf{x}'\mathbf{Q}\mathbf{x} + u'Ru)\, dt$$

where

$$\mathbf{Q} = \begin{bmatrix} 1 & 0 & 0 \\ 0 & 1 & 0 \\ 0 & 0 & 1 \end{bmatrix}, \quad R = [1]$$

Obtain the positive-definite solution matrix **P** of the Riccati equation, the optimal feedback gain matrix **K**, and the eigenvalues of matrix **A** − **BK**.

MATLAB Program 10–20 will solve this problem.

MATLAB Program 10–20

```
% ---------- Design of quadratic optimal regulator system ----------
A = [0 1 0;0 0 1;-35 -27 -9];
B = [0;0;1];
Q = [1 0 0;0 1 0;0 0 1];
R = [1];
[K,P,E] = lqr(A,B,Q,R)

K =

    0.0143    0.1107    0.0676

P =

    4.2625    2.4957    0.0143
    2.4957    2.8150    0.1107
    0.0143    0.1107    0.0676

E =

   -5.0958
   -1.9859 + 1.7110i
   -1.9859 - 1.7110i
```

Next, let us obtain the response **x** of the regulator system to the initial condition **x**(0), where

$$\mathbf{x}(0) = \begin{bmatrix} 1 \\ 0 \\ 0 \end{bmatrix}$$

With state feedback $u = -\mathbf{Kx}$, the state equation for the system becomes

$$\dot{\mathbf{x}} = \mathbf{Ax} + \mathbf{B}u = (\mathbf{A} - \mathbf{BK})\mathbf{x}$$

Then the system, or sys, can be given by

$$sys = ss(A\text{-}B*K, \ eye(3), \ eye(3), \ eye(3))$$

MATLAB Program 10–21 produces the response to the given initial condition. The response curves are shown in Figure 10–38.

MATLAB Program 10–21

```
% Response to initial condition.

A = [0 1 0;0 0 1;-35 -27 -9];
B = [0;0;1];
K = [0.0143 0.1107 0.0676];
sys = ss(A-B*K, eye(3),eye(3),eye(3));
t = 0:0.01:8;
x = initial(sys,[1;0;0],t);
x1 = [1 0 0]*x';
x2 = [0 1 0]*x';
X3 = [0 0 1]*x';

subplot(2,2,1); plot(t,x1); grid
xlabel('t (sec)'); ylabel('x1')

subplot(2,2,2); plot(t,x2); grid
xlabel('t (sec)'); ylabel('x2)

subplot(2,2,3); plot(t,x3); grid
xlabel('t (sec)'); ylabel('x3')
```

EXAMPLE 10–13 Consider the system shown in Figure 10–39. The plant is defined by the following state-space equations:

$$\dot{\mathbf{x}} = \mathbf{Ax} + \mathbf{B}u$$
$$y = \mathbf{Cx} + Du$$

where

$$\mathbf{A} = \begin{bmatrix} 0 & 1 & 0 \\ 0 & 0 & 1 \\ 0 & -2 & -3 \end{bmatrix}, \quad \mathbf{B} = \begin{bmatrix} 0 \\ 0 \\ 1 \end{bmatrix}, \quad \mathbf{C} = \begin{bmatrix} 1 & 0 & 0 \end{bmatrix}, \quad D = \begin{bmatrix} 0 \end{bmatrix}$$

The control signal u is given by

$$u = k_1(r - x_1) - (k_2 x_2 + k_3 x_3) = k_1 r - (k_1 x_1 + k_2 x_2 + k_3 x_3)$$

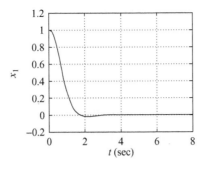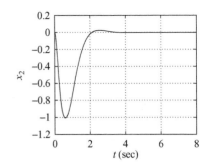

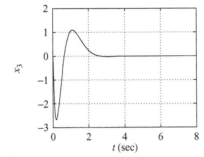

Figure 10–38
Response curves to
initial condition.

In determining an optimal control law, we assume that the input is zero, or $r = 0$.
Let us determine the state-feedback gain matrix **K**, where

$$\mathbf{K} = \begin{bmatrix} k_1 & k_2 & k_3 \end{bmatrix}$$

such that the following performance index is minimized:

$$J = \int_0^\infty (\mathbf{x}'\mathbf{Q}\mathbf{x} + u'Ru)\,dt$$

where

$$\mathbf{Q} = \begin{bmatrix} q_{11} & 0 & 0 \\ 0 & q_{22} & 0 \\ 0 & 0 & q_{33} \end{bmatrix}, \quad R = 1, \quad \mathbf{x} = \begin{bmatrix} x_1 \\ x_2 \\ x_3 \end{bmatrix} = \begin{bmatrix} y \\ \dot{y} \\ \ddot{v} \end{bmatrix}$$

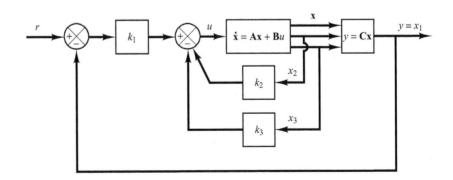

Figure 10–39
Control system.

To get a fast response, q_{11} must be sufficiently large compared with q_{22}, q_{33}, and R. In this problem, we choose

$$q_{11} = 100, \qquad q_{22} = q_{33} = 1, \qquad R = 0.01$$

To solve this problem with MATLAB, we use the command

$$K = lqr(A,B,Q,R)$$

MATLAB Program 10–22 yields the solution to this problem.

MATLAB Program 10–22

```
% ---------- Design of quadratic optimal control system ----------
A = [0 1 0;0 0 1;0 -2 -3];
B = [0;0;1];
Q = [100 0 0;0 1 0;0 0 1];
R = [0.01];

K = lqr(A,B,Q,R)

K =

   100.0000   53.1200   11.6711
```

Next we shall investigate the step-response characteristics of the designed system using the matrix **K** thus determined. The state equation for the designed system is

$$\dot{\mathbf{x}} = \mathbf{A}\mathbf{x} + \mathbf{B}u$$

$$= \mathbf{A}\mathbf{x} + \mathbf{B}(-\mathbf{K}\mathbf{x} + k_1 r)$$

$$= (\mathbf{A} - \mathbf{B}\mathbf{K})\mathbf{x} + \mathbf{B}k_1 r$$

and the output equation is

$$y = \mathbf{C}\mathbf{x} = \begin{bmatrix} 1 & 0 & 0 \end{bmatrix} \begin{bmatrix} x_1 \\ x_2 \\ x_3 \end{bmatrix}$$

To obtain the unit-step response, use the following command:

$$[y,x,t] = step(AA,BB,CC,DD)$$

where

$$AA = \mathbf{A} - \mathbf{B}\mathbf{K}, \qquad BB = \mathbf{B}k_1, \qquad CC = \mathbf{C}, \qquad DD = D$$

MATLAB Program 10–23 produces the unit-step response of the designed system. Figure 10–40 shows the response curves x_1, x_2, and x_3 versus t on one diagram.

MATLAB Program 10–23

```
% ---------- Unit-step response of designed system ----------

A = [0  1  0;0  0  1;0  -2  -3];
B = [0;0;1]
C = [1  0  0];
D = [0];
K = [100.0000  53.1200  11.6711];
k1 = K(1); k2 = K(2); k3 = K(3);

% ***** Define the state matrix, control matrix, output matrix,
% and direct transmission matrix of the designed systems as AA,
% BB, CC, and DD *****

AA = A - B*K;
BB = B*k1;
CC = C;
DD = D;
t = 0:0.01:8;
[y,x,t] = step (AA,BB,CC,DD,1,t);

plot(t,x)
grid
title('Response Curves x1, x2, x3, versus t')
xlabel('t Sec')
ylabel('x1,x2,x3')
text(2.6,1.35,'x1')
text(1.2,1.5,'x2')
text(0.6,3.5,'x3')
```

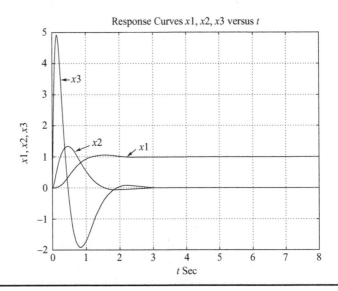

Figure 10–40
Response curves x_1
versus t, x_2 versus t,
and x_3 versus t.

Concluding Comments on Optimal Regulator Systems

1. Given any initial state $\mathbf{x}(t_0)$, the optimal regulator problem is to find an allowable control vector $\mathbf{u}(t)$ that transfers the state to the desired region of the state space and for which the performance index is minimized. For the existence of an optimal control vector $\mathbf{u}(t)$, the system must be completely state controllable.

2. The system that minimizes (or maximizes, as the case may be) the selected performance index is, by definition, optimal. Although the controller may have nothing to do with "optimality" in many practical applications, the important point is that the design based on the quadratic performance index yields a stable control system.

3. The characteristic of an optimal control law based on a quadratic performance index is that it is a linear function of the state variables, which implies that we need to feed back all state variables. This requires that all such variables be available for feedback. If not all state variables are available for feedback, then we need to employ a state observer to estimate unmeasurable state variables and use the estimated values to generate optimal control signals.

   Note that the closed-loop poles of the system designed by the use of the quadratic optimal regulator approach can be found from

   $$|s\mathbf{I} - \mathbf{A} + \mathbf{BK}| = 0$$

   Since these closed-loop poles correspond to the desired closed-loop poles in the pole-placement approach, the transfer functions of the observer controllers can be obtained from either Equation (10–74) if the observer is of full-order type or Equation (10–108) if the observer is of minimum-order type.

4. When the optimal control system is designed in the time domain, it is desirable to investigate the frequency-response characteristics to compensate for noise effects. The system frequency-response characteristics must be such that the system attenuates highly in the frequency range where noise and resonance of components are expected. (To compensate for noise effects, we must in some cases either modify the optimal configuration and accept suboptimal performance or modify the performance index.)

5. If the upper limit of integration in the performance index J given by Equation (10–114) is finite, then it can be shown that the optimal control vector is still a linear function of the state variables, but with time-varying coefficients. (Therefore, the determination of the optimal control vector involves that of optimal time-varying matrices.)

10–9 ROBUST CONTROL SYSTEMS

Suppose that given a control object (i.e., a system with a flexible arm) we wish to design a control system. The first step in the design of a control system is to obtain a mathematical model of the control object based on the physical law. Quite often the model may be nonlinear and possibly with distributed parameters. Such a model may be difficult to analyze. It is desirable to approximate it by a linear constant-coefficient system that will approximate the actual object fairly well. Note that even though the model to be used for design purposes may be a simplified one, it is necessary that such a model must include any intrinsic character of the actual object. Assuming that we can get a model that approximates the actual system quite well, we must get a simplified model for the purpose of designing the control system that will require a compensator of lowest order possible. Thus, a model of a control object (whatever it may be) will probably include an error in the modeling process. Note that in the frequency-response approach to control systems design, we use phase and gain margins to take care of the modeling errors. However, in the state-space approach, which is based on the differential equations of the plant dynamics, no such "margins" are involved in the design process.

Since the actual plant differs from the model used in the design, a question arises whether the controller designed using a model will work satisfactorily with the actual plant. To ensure that it will do so, robust control theory has been developed since around 1980.

Robust control theory uses the assumption that the models we use in designing control systems have modeling errors. We shall present an introduction to this theory in this section. Basically, the theory assumes that there is an uncertainty or error between the actual plant and its mathematical model and includes such uncertainty or error in the design process of the control system.

Systems designed based on the robust control theory will possess the following properties:

(1) *Robust stability.* The control system designed is stable in the presence of perturbation.

(2) *Robust performance.* The control system exhibits predetermined response characteristics in the presence of perturbation.

This theory requires considerations based on frequency-response analysis and time-domain analysis. Because of the mathematical complications associated with robust control theory, detailed discussion of robust control theory is beyond the scope of the senior engineering student. In this section, only introductory discussion of robust control theory is presented.

Uncertain Elements in Plant Dynamics. The term *uncertainty* refers to the differences or errors between the model of the plant and the actual plant.

Uncertain elements that may appear in practical systems may be classified as *structured* uncertainty and *unstructured* uncertainty. An example of structured uncertainty is any parametric variation in the plant dynamics, such as variations in poles and zeros of the plant transfer function. Examples of unstructured uncertainty include frequency-dependent uncertainty, such as high-frequency modes that we normally neglect in modeling plant dynamics. For example, in the modeling of a flexible-arm system, the model may include a finite number of modes of oscillation. The modes of oscillation that are not included in the modeling behave as uncertainty of the system. Another example of uncertainty occurs in the linearization of a nonlinear plant. If the actual plant is nonlinear and its model is linear, then the difference acts as unstructured uncertainty.

In this section we consider the case where the uncertainty is unstructured. In addition we assume that the plant involves only one uncertainty. (Some plants may involve multiple uncertain elements.)

In the robust control theory, we define unstructured uncertainty as $\Delta(s)$. Since the exact description of $\Delta(s)$ is unknown, we use an estimate of $\Delta(s)$ (as to the magnitude and phase characteristics) and use this estimate in the design of the controller that stabilizes the control system. Stability of a system with unstructured uncertainty can then be examined by use of the small gain theorem to be given following the definition of the H_∞ norm.

H_∞ Norm.   The H_∞ norm of a stable single-input–single-output system is the largest possible amplification factor of the steady-state response to sinusoidal excitation.

For a scalar $\Phi(s)$, $\|\Phi\|_\infty$ gives the maximum value of $|\Phi(j\omega)|$. It is called the H_∞ norm. See Figure 10–41.

In robust control theory we measure the magnitude of the transfer function by the H_∞ norm. Assume that the transfer function $\Phi(s)$ is proper and stable. [Note that a transfer function $\Phi(s)$ is called proper if $\Phi(\infty)$ is limited and definite. If $\Phi(\infty) = 0$, it is called strictly proper.] The H_∞ norm of $\Phi(s)$ is defined by

$$\|\Phi\|_\infty = \bar{\sigma}\left[\Phi(j\omega)\right]$$

$\bar{\sigma}\left[\Phi(j\omega)\right]$ means the maximum singular value of $[\Phi(j\omega)]$. ($\bar{\sigma}$ means σ_{max}.) Note that the singular value of a transfer function Φ is defined by

$$\sigma_i(\Phi) = \sqrt{\lambda_i(\Phi^*\Phi)}$$

where $\lambda_i(\Phi^*\Phi)$ is the ith largest eigenvalue of $\Phi^*\Phi$ and it is always a non-negative real value. By making $\|\Phi\|_\infty$ smaller, we make the effect of input w on the output z smaller. It is frequently the case that instead of using the maximum singular value $\|\Phi\|_\infty$, we use the inequality

$$\|\Phi\|_\infty < \gamma$$

and limit the magnitude of $\Phi(s)$ by γ. To make the magnitude of $\|\Phi\|_\infty$ small, we choose a small γ and require that $\|\Phi\|_\infty < \gamma$.

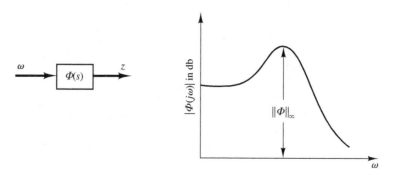

Figure 10–41
Bode diagram and
the H_∞ norm $\|\Phi\|_\infty$.

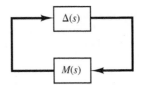

Figure 10–42
Closed-loop system.

Small-Gain Theorem.   Consider the closed-loop system shown in Figure 10–42. In the figure $\Delta(s)$ and $M(s)$ are stable and proper transfer functions.

The small-gain theorem states that if

$$\|\Delta(s)M(s)\|_\infty < 1$$

then this closed-loop system is stable. That is, if the H_∞ norm of $\Delta(s)M(s)$ is smaller than 1, this closed-loop system is stable. This theorem is an extension of the Nyquist stability criterion.

It is important to note that the small-gain theorem gives a sufficient condition for stability. That is, a system may be stable even if it does not satisfy this theorem. However, if a system satisfies the small-gain theorem, it is always stable.

System with Unstructured Uncertainty.   In some cases an unstructured uncertainty error may be considered multiplicative such that

$$\tilde{G} = G(1 + \Delta_m)$$

where $\tilde{G}$ is the true plant dynamics and G is the model plant dynamics. In other cases an unstructured uncertainty error may be considered additive such that

$$\tilde{G} = G + \Delta_a$$

In either case we assume that the norm of Δ_m or Δ_a is bounded such that

$$\|\Delta_m\| < \gamma_m, \qquad \|\Delta_a\| < \gamma_a$$

where γ_m and γ_a are positive constants.

EXAMPLE 10–14   Consider a control system with unstructured multiplicative uncertainty. We shall consider robust stability and robust performance of the system. (A system with unstructured additive uncertainty will be discussed in Problem **A–10–18**.)

Robust Stability.   Let us define

$\tilde{G}$ = true plant dynamics

G = model of plant dynamics

Δ_m = unstructured multiplicative uncertainty

We assume that Δ_m is stable and its upper bound is known. We also assume that $\tilde{G}$ and G are related by

$$\tilde{G} = G(I + \Delta_m)$$

Consider the system shown in Figure 10–43(a). Let us examine the transfer function between point A and point B. Notice that Figure 10–43(a) can be redrawn as shown in Figure 10-43(b). The transfer function between point A and point B can be given by

$$\frac{KG}{1 + KG} = (1 + KG)^{-1} KG$$

Define

$$(1 + KG)^{-1} KG = T \qquad (10\text{--}121)$$

Using Equation (10–121) we can redraw Figure 10–43(b) as Figure 10–43(c). Applying the small-gain theorem to the system consisting of Δ_m and T as shown in Figure 10–43(c), we obtain the condition for stability to be

$$\|\Delta_m T\|_\infty < 1 \qquad (10\text{--}122)$$

In general, it is impossible to precisely model Δ_m. Therefore, let us use a scalar transfer function $W_m(j\omega)$ such that

$$\overline{\sigma}\{\Delta_m(j\omega)\} < |W_m(j\omega)|$$

where $\overline{\sigma}\{\Delta_m(j\omega)\}$ is the largest singular value of $\Delta_m(j\omega)$.

Consider, instead of Inequality (10–122), the following inequality:

$$\|W_m T\|_\infty < 1 \qquad (10\text{--}123)$$

If Inequality (10–123) holds true, Inequality (10–122) will always be satisfied. By making the H_∞ norm of $W_m T$ to be less than 1, we obtain the controller K that will make the system stable.

Suppose that we cut the line at point A in Figure 10–43(a). Then we obtain Figure 10–43(d). Replacing Δ_m by $W_m I$, we obtain Figure 10–43(e). Redrawing Figure 10–43(e), we obtain Figure 10–43(f). Figure 10–43(f) is called a *generalized plant diagram*.

Referring to Equation (10–121), T is given by

$$T = \frac{KG}{1 + KG} \qquad (10\text{--}124)$$

Then Inequality (10–123) can be rewritten as

$$\left\|\frac{W_m K(s)G(s)}{1 + K(s)G(s)}\right\|_\infty < 1 \qquad (10\text{--}125)$$

Clearly, for a stable plant model $G(s)$, $K(s) = 0$ will satisfy Inequality (10–125). However, $K(s) = 0$ is not the desirable transfer function for the controller. To find an acceptable transfer function for $K(s)$, we may add another condition—for example, that the resulting system will have robust performance such that the system output follows the input with minimum error, or another reasonable condition. In what follows we shall obtain the condition for robust performance.

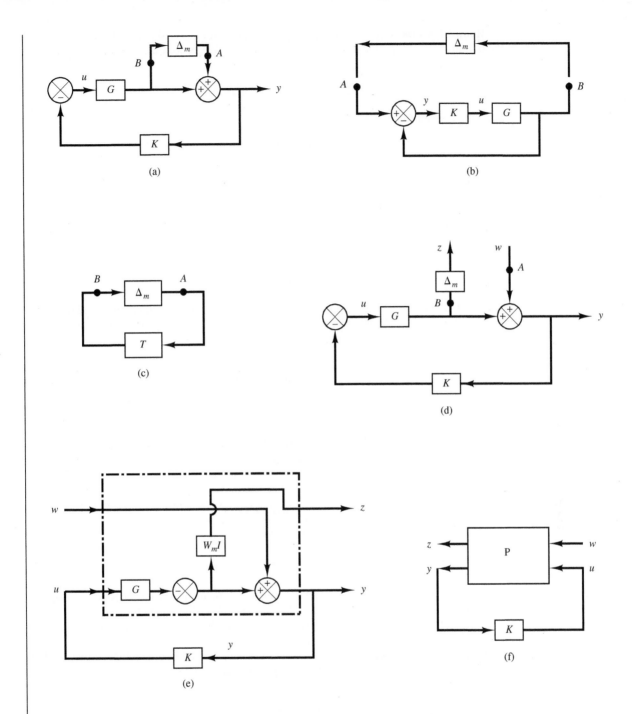

Figure 10–43
(a) Block diagram of a system with unstructured multiplicative uncertainty;
(b)–(d) successive modifications of the block diagram of (a);
(e) block diagram showing a generalized plant with unstructured multiplicative uncertainty;
(f) generalized plant diagram.

Robust Performance. Consider the system shown in Figure 10–44. Suppose that we want the output $y(t)$ to follow the input $r(t)$ as closely as possible, or we wish to have

$$\lim_{t \to \infty} [r(t) - y(t)] = \lim_{t \to \infty} e(t) \to 0$$

Since the transfer function $Y(s)/R(s)$ is

$$\frac{Y(s)}{R(s)} = \frac{KG}{1 + KG}$$

we have

$$\frac{E(s)}{R(s)} = \frac{R(s) - Y(s)}{R(s)} = 1 - \frac{Y(s)}{R(s)} = \frac{1}{1 + KG}$$

Define

$$\frac{1}{1 + KG} = S$$

where S is commonly called the sensitivity function and T defined by Equation (10–124) is called the complementary sensitivity function. In this robust performance problem we want to make the H_∞ norm of S smaller than the desired transfer function W_s^{-1} or $\|S\|_\infty < W_s^{-1}$ which can be written as

$$\|W_s S\|_\infty < 1 \tag{10–126}$$

Combining Inequalities (10–123) and (10–126), we get

$$\left\| \begin{matrix} W_m T \\ W_s S \end{matrix} \right\|_\infty < 1$$

where $T + S = 1$, or

$$\left\| \begin{matrix} W_m(s) \dfrac{K(s)G(s)}{1 + K(s)G(s)} \\ W_s(s) \dfrac{1}{1 + K(s)G(s)} \end{matrix} \right\|_\infty < 1 \tag{10–127}$$

Our problem then becomes to find $K(s)$ that will satisfy Inequality (10–127). Note that depending on the chosen $W_m(s)$ and $W_s(s)$ there may be many $K(s)$ that satisfy Inequality (10–127), or may be no $K(s)$ that satisfies Inequality (10–127). Such a robust control problem using Inequality (10–127) is called a mixed-sensitivity problem.

Figure 10–45(a) is a generalized plant diagram, where two conditions (robust stability and robust performance) are specified. A simplified version of this diagram is shown in Figure 10–45(b).

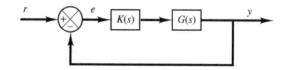

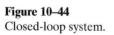

Figure 10–44
Closed-loop system.

Chapter 10 / Control Systems Design in State Space

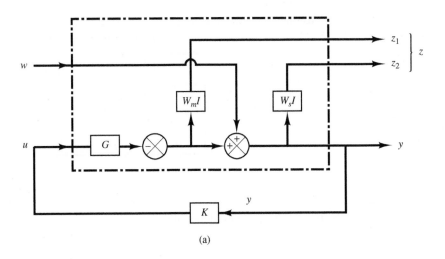

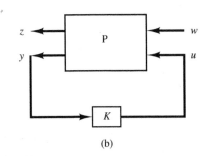

Figure 10–45
(a) Generalized
plant diagram;
(b) simplfied version
of the generalized
plant diagram
shown in (a).

Finding Transfer Function $z(s)/w(s)$ from a Generalized Plant Diagram. Consider the generalized plant diagram shown in Figure 10–46.

In this diagram $w(s)$ is the exogenous disturbance and $u(s)$ is the manipulated variable. $z(s)$ is the controlled variable and $y(s)$ is the observed variable.

Consider this control system consisting of the generalized plant $P(s)$ and the controller $K(s)$. The equation that relates the outputs $z(s)$ and $y(s)$ and the inputs $w(s)$ and $u(s)$ of the generalized plant $P(s)$ is

$$\begin{bmatrix} z(s) \\ y(s) \end{bmatrix} = \begin{bmatrix} P_{11} & P_{12} \\ P_{21} & P_{22} \end{bmatrix} \begin{bmatrix} w(s) \\ u(s) \end{bmatrix}$$

The equation that relates $u(s)$ and $y(s)$ is given by

$$u(s) = K(s)y(s)$$

Define the transfer function that relates the controlled variable z(s) to the exogenous disturbance $w(s)$ as $\Phi(s)$. Then

$$z(s) = \Phi(s)w(s)$$

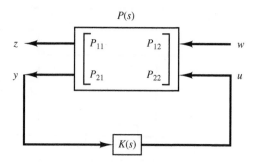

Figure 10–46
A generalized plant
diagram.

Note that $\Phi(s)$ can be determined as follows: Since

$$z(s) = P_{11}w(s) + P_{12}u(s)$$

$$y(s) = P_{21}w(s) + P_{22}u(s)$$

$$u(s) = K(s)y(s)$$

we obtain

$$y(s) = P_{21}w(s) + P_{22}K(s)y(s)$$

Hence

$$[I - P_{22}K(s)]y(s) = P_{21}w(s)$$

or

$$y(s) = [I - P_{22}K(s)]^{-1}P_{21}w(s)$$

Therefore,

$$z(s) = P_{11}w(s) + P_{12}K(s)[I - P_{22}K(s)]^{-1}P_{21}w(s)$$

$$= \{P_{11} + P_{12}K(s)[I - P_{22}K(s)]^{-1}P_{21}\}w(s)$$

Hence,

$$\Phi(s) = P_{11} + P_{12}K(s)[I - P_{22}K(s)]^{-1}P_{21} \tag{10–128}$$

EXAMPLE 10–15 Let us determine the P matrix in the generalized plant diagram of the control system considered in Example 10–14. We derived Inequality (10–125) for the control system to be robust stable. Rewriting Inequality (10–125), we have

$$\left\| \frac{W_m KG}{1 + KG} \right\|_\infty < 1 \tag{10–129}$$

If we define

$$\Phi_1 = \frac{W_m KG}{1 + KG} \tag{10–130}$$

then Inequality (10–129) can be written as

$$\|\Phi_1\|_\infty < 1$$

Referring to Equation (10–128), rewritten as

$$\Phi = P_{11} + P_{12}K(I - P_{22}K)^{-1}P_{21}$$

notice that if we choose the generalized plant P matrix as

$$P = \begin{bmatrix} 0 & W_m G \\ I & -G \end{bmatrix} \tag{10–131}$$

then we obtain

$$\Phi = P_{11} + P_{12}K(I - P_{22}K)^{-1}P_{21}$$

$$= W_m KG(I + KG)^{-1}$$

which is exactly the same as Φ_1 in Equation (10–130).

We derived in Example 10–14 that if we wished to have the output y follow the input r as close as possible, we needed to make the H_∞ norm of $\Phi_2(s)$, where

$$\Phi_2 = \frac{W_s}{I + KG} \tag{10–132}$$

less than 1. [See Inequality (10–126).]

Note that the controlled variable z is related to the exogenous disturbance w by

$$z = \Phi(s)w$$

and referring to Equation (10–128)

$$\Phi(s) = P_{11} + P_{12}K(I - P_{22}K)^{-1}P_{21}$$

Notice that if we choose the P matrix as

$$P = \begin{bmatrix} W_s & -W_s G \\ I & -G \end{bmatrix} \tag{10–133}$$

then we obtain

$$\Phi = P_{11} + P_{12}K(I - P_{22}K)^{-1}P_{21}$$

$$= W_s - W_s KG(I + KG)^{-1}$$

$$= W_s \left[1 - \frac{KG}{1 + KG} \right]$$

$$= W_s \left[\frac{1}{1 + KG} \right]$$

which is the same as Φ_2 in Equation (10–132).

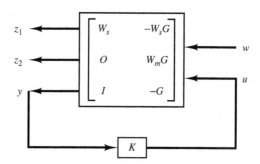

Figure 10–47
Generalized plant of the system discussed in Example 10–15.

If both the robust stability and robust performance conditions are required, the control system must satisfy the condition given by Inequality (10–127), rewritten as

$$\left\| \begin{matrix} W_m \dfrac{KG}{1 + KG} \\[2mm] W_s \dfrac{1}{1 + KG} \end{matrix} \right\|_\infty < 1 \qquad (10\text{–}134)$$

For the P matrix, we combine Equations (10–133) and (10–131) and get

$$P = \begin{bmatrix} W_s & -W_s G \\ 0 & W_m G \\ I & -G \end{bmatrix} \qquad (10\text{–}135)$$

If we construct $P(s)$ as given by Equation (10–135), then the problem of designing a control system to satisfy both robust stability and robust performance conditions can be formulated by using the generalized plant represented by Equation (10–135). As mentioned earlier, such a problem is called a mixed-sensitivity problem. By using the generalized plant given by Equation (10–135) we are able to determine the controller $K(s)$ that satisfies Inequality (10–134). The generalized plant diagram for the system considered in Example 10–15 becomes as shown in Figure 10–47.

H Infinity Control Problem. To design a controller K of a control system to satisfy various stability and performance specifications, we utilize the concept of the generalized plant.

As mentioned earlier a generalized plant is a linear model consisting of a model of the plant and weighting functions corresponding to the specifications for the required performance. Referring to the generalized plant shown in Figure 10–48, the H infinity control problem is a problem to design a controller K that will make the H_∞ norm of the transfer function from the exogenous disturbance w to the controlled variable z less than a specified value.

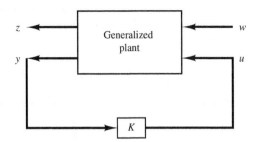

Figure 10–48
A generalized plant diagram.

The reason to use generalized plants, rather than individual block diagrams of control systems, is that a number of control systems with uncertain elements have been designed using generalized plants and, consequently, established design approaches using such plants are available.

Note that any weighting function, such as $W(s)$, is an important parameter to influence the resulting controller $K(s)$. In fact, the goodness of the resulting designed system depends on the choice of the weighting function or functions used in the design process.

Note that the controller that is the solution to the H infinity control problem is commonly called the H infinity controller.

Solving Robust Control Problems. There are three established approaches to solve robust control problems. They are

1. Solve robust control problems by deriving the Riccati equations and solving them.
2. Solve robust control problems by using the linear matrix inequality approach.
3. Solve robust control problems that involve structural uncertainties by using the μ analysis and μ synthesis approach.

Solving robust control problems by use of any of the above methods requires a broad mathematical background.

In this section we have presented only an introduction to the robust control theory. Solving any robust control problem requires mathematical background beyond the scope of the senior engineering student. Therefore, an interested reader may take a graduate-level control course at an established college or university and study this subject in detail.

EXAMPLE PROBLEMS AND SOLUTIONS

A–10–1. Consider the system defined by

$$\dot{\mathbf{x}} = \mathbf{A}\mathbf{x} + \mathbf{B}u$$

Suppose that this system is not completely state controllable. Then the rank of the controllability matrix is less than n, or

$$\text{rank}\begin{bmatrix} \mathbf{B} & \vdots & \mathbf{AB} & \vdots & \cdots & \vdots & \mathbf{A}^{n-1}\mathbf{B} \end{bmatrix} = q < n \qquad (10\text{–}136)$$

Example Problems and Solutions **817**

This means that there are q linearly independent column vectors in the controllability matrix. Let us define such q linearly independent column vectors as $\mathbf{f}_1, \mathbf{f}_2, \ldots, \mathbf{f}_q$. Also, let us choose $n - q$ additional n-vectors $\mathbf{v}_{q+1}, \mathbf{v}_{q+2}, \ldots, \mathbf{v}_n$ such that

$$\mathbf{P} = \left[\mathbf{f}_1 \;\vdots\; \mathbf{f}_2 \;\vdots\; \cdots \;\vdots\; \mathbf{f}_q \;\vdots\; \mathbf{v}_{q+1} \;\vdots\; \mathbf{v}_{q+2} \;\vdots\; \cdots \;\vdots\; \mathbf{v}_n \right]$$

is of rank n. By using matrix $\mathbf{P}$ as the transformation matrix, define

$$\mathbf{P}^{-1}\mathbf{A}\mathbf{P} = \hat{\mathbf{A}}, \qquad \mathbf{P}^{-1}\mathbf{B} = \hat{\mathbf{B}}$$

Show that $\hat{\mathbf{A}}$ can be given by

$$\hat{\mathbf{A}} = \left[\begin{array}{c|c} \mathbf{A}_{11} & \mathbf{A}_{12} \\ \hline \mathbf{0} & \mathbf{A}_{22} \end{array} \right]$$

where $\mathbf{A}_{11}$ is a $q \times q$ matrix, $\mathbf{A}_{12}$ is a $q \times (n - q)$ matrix, $\mathbf{A}_{22}$ is an $(n - q) \times (n - q)$ matrix, and $\mathbf{0}$ is an $(n - q) \times q$ matrix. Show also that matrix $\hat{\mathbf{B}}$ can be given by

$$\hat{\mathbf{B}} = \left[\begin{array}{c} \mathbf{B}_{11} \\ \hline \mathbf{0} \end{array} \right]$$

where $\mathbf{B}_{11}$ is a $q \times 1$ matrix and $\mathbf{0}$ is an $(n - q) \times 1$ matrix.

Solution. Notice that

$$\mathbf{A}\mathbf{P} = \mathbf{P}\hat{\mathbf{A}}$$

or

$$\left[\mathbf{A}\mathbf{f}_1 \;\vdots\; \mathbf{A}\mathbf{f}_2 \;\vdots\; \cdots \;\vdots\; \mathbf{A}\mathbf{f}_q \;\vdots\; \mathbf{A}\mathbf{v}_{q+1} \;\vdots\; \cdots \;\vdots\; \mathbf{A}\mathbf{v}_n \right]$$

$$= \left[\mathbf{f}_1 \;\vdots\; \mathbf{f}_2 \;\vdots\; \cdots \;\vdots\; \mathbf{f}_q \;\vdots\; \mathbf{v}_{q+1} \;\vdots\; \cdots \;\vdots\; \mathbf{v}_n \right]\hat{\mathbf{A}} \qquad (10\text{--}137)$$

Also,

$$\mathbf{B} = \mathbf{P}\hat{\mathbf{B}} \qquad (10\text{--}138)$$

Since we have q linearly independent column vectors $\mathbf{f}_1, \mathbf{f}_2, \ldots, \mathbf{f}_q$, we can use the Cayley–Hamilton theorem to express vectors $\mathbf{A}\mathbf{f}_1, \mathbf{A}\mathbf{f}_2, \ldots, \mathbf{A}\mathbf{f}_q$ in terms of these q vectors. That is,

$$\mathbf{A}\mathbf{f}_1 = a_{11}\mathbf{f}_1 + a_{21}\mathbf{f}_2 + \cdots + a_{q1}\mathbf{f}_q$$

$$\mathbf{A}\mathbf{f}_2 = a_{12}\mathbf{f}_1 + a_{22}\mathbf{f}_2 + \cdots + a_{q2}\mathbf{f}_q$$

$$\cdot$$

$$\cdot$$

$$\cdot$$

$$\mathbf{A}\mathbf{f}_q = a_{1q}\mathbf{f}_1 + a_{2q}\mathbf{f}_2 + \cdots + a_{qq}\mathbf{f}_q$$

Hence, Equation (10–137) may be written as follows:

$$\left[\mathbf{Af}_1 \mid \mathbf{Af}_2 \mid \cdots \mid \mathbf{Af}_q \mid \mathbf{Av}_{q+1} \mid \cdots \mid \mathbf{Av}_n\right]$$

$$= \left[\mathbf{f}_1 \mid \mathbf{f}_2 \mid \cdots \mid \mathbf{f}_q \mid \mathbf{v}_{q+1} \mid \cdots \mid \mathbf{v}_n\right] \begin{bmatrix} a_{11} & \cdots & a_{1q} & a_{1q+1} & \cdots & a_{1n} \\ a_{21} & \cdots & a_{2q} & a_{2q+1} & \cdots & a_{2n} \\ \cdot & & \cdot & \cdot & & \cdot \\ \cdot & & \cdot & \cdot & & \cdot \\ \cdot & & \cdot & \cdot & & \cdot \\ a_{q1} & \cdots & a_{qq} & a_{qq+1} & \cdots & a_{qn} \\ \hline 0 & \cdots & 0 & a_{q+1q+1} & \cdots & a_{q+1n} \\ \cdot & & \cdot & \cdot & & \cdot \\ \cdot & & \cdot & \cdot & & \cdot \\ \cdot & & \cdot & \cdot & & \cdot \\ 0 & \cdots & 0 & a_{nq+1} & \cdots & a_{nn} \end{bmatrix}$$

Define

$$\begin{bmatrix} a_{11} & \cdots & a_{1q} \\ a_{21} & \cdots & a_{2q} \\ \cdot & & \cdot \\ \cdot & & \cdot \\ \cdot & & \cdot \\ a_{q1} & \cdots & a_{qq} \end{bmatrix} = \mathbf{A}_{11}$$

$$\begin{bmatrix} a_{1q+1} & \cdots & a_{1n} \\ a_{2q+1} & \cdots & a_{2n} \\ \cdot & & \cdot \\ \cdot & & \cdot \\ \cdot & & \cdot \\ a_{qq+1} & \cdots & a_{qn} \end{bmatrix} = \mathbf{A}_{12}$$

$$\begin{bmatrix} 0 & \cdots & 0 \\ \cdot & & \cdot \\ \cdot & & \cdot \\ \cdot & & \cdot \\ 0 & \cdots & 0 \end{bmatrix} = \mathbf{A}_{21} = (n - q) \times q \text{ zero matrix}$$

$$\begin{bmatrix} a_{q+1q+1} & \cdots & a_{q+1n} \\ \cdot & & \cdot \\ \cdot & & \cdot \\ \cdot & & \cdot \\ a_{nq+1} & \cdots & a_{nn} \end{bmatrix} = \mathbf{A}_{22}$$

Then Equation (10–137) can be written as

$$\left[\mathbf{Af}_1 \mid \mathbf{Af}_2 \mid \cdots \mid \mathbf{Af}_q \mid \mathbf{Av}_{q+1} \mid \cdots \mid \mathbf{Av}_n\right]$$

$$= \left[\mathbf{f}_1 \mid \mathbf{f}_2 \mid \cdots \mid \mathbf{f}_q \mid \mathbf{v}_{q+1} \mid \cdots \mid \mathbf{v}_n\right] \begin{bmatrix} \mathbf{A}_{11} & \mathbf{A}_{12} \\ \hline \mathbf{0} & \mathbf{A}_{22} \end{bmatrix}$$

Thus,

$$\mathbf{AP} = \mathbf{P}\left[\begin{array}{c|c} \mathbf{A}_{11} & \mathbf{A}_{12} \\ \hline \mathbf{0} & \mathbf{A}_{22} \end{array}\right]$$

Hence,

$$\mathbf{P}^{-1}\mathbf{AP} = \hat{\mathbf{A}} = \left[\begin{array}{c|c} \mathbf{A}_{11} & \mathbf{A}_{12} \\ \hline \mathbf{0} & \mathbf{A}_{22} \end{array}\right]$$

Next, referring to Equation (10–138), we have

$$\mathbf{B} = \left[\mathbf{f}_1 \,\vdots\, \mathbf{f}_2 \,\vdots\, \cdots \,\vdots\, \mathbf{f}_q \,\vdots\, \mathbf{v}_{q+1} \,\vdots\, \cdots \,\vdots\, \mathbf{v}_n\right]\hat{\mathbf{B}} \qquad (10\text{–}139)$$

Referring to Equation (10–136), notice that vector $\mathbf{B}$ can be written in terms of q linearly independent column vectors $\mathbf{f}_1, \mathbf{f}_2, \ldots, \mathbf{f}_q$. Thus, we have

$$\mathbf{B} = b_{11}\mathbf{f}_1 + b_{21}\mathbf{f}_2 + \cdots + b_{q1}\mathbf{f}_q$$

Consequently, Equation (10–139) may be written as follows:

$$b_{11}\mathbf{f}_1 + b_{21}\mathbf{f}_2 + \cdots + b_{q1}\mathbf{f}_q = \left[\mathbf{f}_1 \,\vdots\, \mathbf{f}_2 \,\vdots\, \cdots \,\vdots\, \mathbf{f}_q \,\vdots\, \mathbf{v}_{q+1} \,\vdots\, \cdots \,\vdots\, \mathbf{v}_n\right]\begin{bmatrix} b_{11} \\ b_{21} \\ \cdot \\ \cdot \\ \cdot \\ b_{q1} \\ 0 \\ \cdot \\ \cdot \\ \cdot \\ 0 \end{bmatrix}$$

Thus,

$$\hat{\mathbf{B}} = \left[\begin{array}{c} \mathbf{B}_{11} \\ \hline \mathbf{0} \end{array}\right]$$

where

$$\mathbf{B}_{11} = \begin{bmatrix} b_{11} \\ b_{21} \\ \cdot \\ \cdot \\ \cdot \\ b_{q1} \end{bmatrix}$$

A–10–2. Consider a completely state controllable system

$$\dot{\mathbf{x}} = \mathbf{Ax} + \mathbf{B}u$$

Define the controllability matrix as $\mathbf{M}$:

$$\mathbf{M} = \left[\mathbf{B} \,\vdots\, \mathbf{AB} \,\vdots\, \cdots \,\vdots\, \mathbf{A}^{n-1}\mathbf{B}\right]$$

Show that

$$\mathbf{M}^{-1}\mathbf{A}\mathbf{M} = \begin{bmatrix} 0 & 0 & \cdots & 0 & -a_n \\ 1 & 0 & \cdots & 0 & -a_{n-1} \\ 0 & 1 & \cdots & 0 & -a_{n-2} \\ \cdot & \cdot & & \cdot & \cdot \\ \cdot & \cdot & & \cdot & \cdot \\ \cdot & \cdot & & \cdot & \cdot \\ 0 & 0 & \cdots & 1 & -a_1 \end{bmatrix}$$

where $a_1, a_2, \ldots, a_n$ are the coefficients of the characteristic polynomial

$$|s\mathbf{I} - \mathbf{A}| = s^n + a_1 s^{n-1} + \cdots + a_{n-1}s + a_n$$

Solution. Let us consider the case where $n = 3$. We shall show that

$$\mathbf{A}\mathbf{M} = \mathbf{M} \begin{bmatrix} 0 & 0 & -a_3 \\ 1 & 0 & -a_2 \\ 0 & 1 & -a_1 \end{bmatrix} \tag{10-140}$$

The left-hand side of Equation (10–140) is

$$\mathbf{A}\mathbf{M} = \mathbf{A}\begin{bmatrix} \mathbf{B} & \vdots & \mathbf{A}\mathbf{B} & \vdots & \mathbf{A}^2\mathbf{B} \end{bmatrix} = \begin{bmatrix} \mathbf{A}\mathbf{B} & \vdots & \mathbf{A}^2\mathbf{B} & \vdots & \mathbf{A}^3\mathbf{B} \end{bmatrix}$$

The right-hand side of Equation (10–140) is

$$\begin{bmatrix} \mathbf{B} & \vdots & \mathbf{A}\mathbf{B} & \vdots & \mathbf{A}^2\mathbf{B} \end{bmatrix} \begin{bmatrix} 0 & 0 & -a_3 \\ 1 & 0 & -a_2 \\ 0 & 1 & -a_1 \end{bmatrix} = \begin{bmatrix} \mathbf{A}\mathbf{B} & \vdots & \mathbf{A}^2\mathbf{B} & \vdots & -a_3\mathbf{B} - a_2\mathbf{A}\mathbf{B} - a_1\mathbf{A}^2\mathbf{B} \end{bmatrix} \tag{10-141}$$

The Cayley–Hamilton theorem states that matrix $\mathbf{A}$ satisfies its own characteristic equation or, in the case of $n = 3$,

$$\mathbf{A}^3 + a_1\mathbf{A}^2 + a_2\mathbf{A} + a_3\mathbf{I} = \mathbf{0} \tag{10-142}$$

Using Equation (10–142), the third column of the right-hand side of Equation (10–141) becomes

$$-a_3\mathbf{B} - a_2\mathbf{A}\mathbf{B} - a_1\mathbf{A}^2\mathbf{B} = (-a_3\mathbf{I} - a_2\mathbf{A} - a_1\mathbf{A}^2)\mathbf{B} = \mathbf{A}^3\mathbf{B}$$

Thus, Equation (10–141) becomes

$$\begin{bmatrix} \mathbf{B} & \vdots & \mathbf{A}\mathbf{B} & \vdots & \mathbf{A}^2\mathbf{B} \end{bmatrix} \begin{bmatrix} 0 & 0 & -a_3 \\ 1 & 0 & -a_2 \\ 0 & 1 & -a_1 \end{bmatrix} = \begin{bmatrix} \mathbf{A}\mathbf{B} & \vdots & \mathbf{A}^2\mathbf{B} & \vdots & \mathbf{A}^3\mathbf{B} \end{bmatrix}$$

Hence, the left-hand side and the right-hand side of Equation (10–140) are the same. We have thus shown that Equation (10–140) is true. Consequently,

$$\mathbf{M}^{-1}\mathbf{A}\mathbf{M} = \begin{bmatrix} 0 & 0 & -a_3 \\ 1 & 0 & -a_2 \\ 0 & 1 & -a_1 \end{bmatrix}$$

The preceding derivation can be easily extended to the general case of any positive integer n.

A–10–3. Consider a completely state controllable system

$$\dot{\mathbf{x}} = \mathbf{A}\mathbf{x} + \mathbf{B}u$$

Define

$$\mathbf{M} = \begin{bmatrix} \mathbf{B} & \vdots & \mathbf{A}\mathbf{B} & \vdots & \cdots & \vdots & \mathbf{A}^{n-1}\mathbf{B} \end{bmatrix}$$

and

$$\mathbf{W} = \begin{bmatrix} a_{n-1} & a_{n-2} & \cdots & a_1 & 1 \\ a_{n-2} & a_{n-3} & \cdots & 1 & 0 \\ \cdot & \cdot & & \cdot & \cdot \\ \cdot & \cdot & & \cdot & \cdot \\ \cdot & \cdot & & \cdot & \cdot \\ a_1 & 1 & \cdots & 0 & 0 \\ 1 & 0 & \cdots & 0 & 0 \end{bmatrix}$$

where the a_i's are coefficients of the characteristic polynomial

$$|s\mathbf{I} - \mathbf{A}| = s^n + a_1 s^{n-1} + \cdots + a_{n-1}s + a_n$$

Define also

$$\mathbf{T} = \mathbf{MW}$$

Show that

$$\mathbf{T}^{-1}\mathbf{AT} = \begin{bmatrix} 0 & 1 & 0 & \cdots & 0 \\ 0 & 0 & 1 & \cdots & 0 \\ \cdot & \cdot & \cdot & & \cdot \\ \cdot & \cdot & \cdot & & \cdot \\ \cdot & \cdot & \cdot & & \cdot \\ 0 & 0 & 0 & \cdots & 1 \\ -a_n & -a_{n-1} & -a_{n-2} & \cdots & -a_1 \end{bmatrix}, \qquad \mathbf{T}^{-1}\mathbf{B} = \begin{bmatrix} 0 \\ 0 \\ \cdot \\ \cdot \\ \cdot \\ 0 \\ 1 \end{bmatrix}$$

Solution. Let us consider the case where $n = 3$. We shall show that

$$\mathbf{T}^{-1}\mathbf{AT} = (\mathbf{MW})^{-1}\mathbf{A}(\mathbf{MW}) = \mathbf{W}^{-1}(\mathbf{M}^{-1}\mathbf{AM})\mathbf{W} = \begin{bmatrix} 0 & 1 & 0 \\ 0 & 0 & 1 \\ -a_3 & -a_2 & -a_1 \end{bmatrix} \qquad (10\text{--}143)$$

Referring to Problem **A–10–2**, we have

$$\mathbf{M}^{-1}\mathbf{AM} = \begin{bmatrix} 0 & 0 & -a_3 \\ 1 & 0 & -a_2 \\ 0 & 1 & -a_1 \end{bmatrix}$$

Hence, Equation (10–143) can be rewritten as

$$\mathbf{W}^{-1}\begin{bmatrix} 0 & 0 & -a_3 \\ 1 & 0 & -a_2 \\ 0 & 1 & -a_1 \end{bmatrix}\mathbf{W} = \begin{bmatrix} 0 & 1 & 0 \\ 0 & 0 & 1 \\ -a_3 & -a_2 & -a_1 \end{bmatrix}$$

Therefore, we need to show that

$$\begin{bmatrix} 0 & 0 & -a_3 \\ 1 & 0 & -a_2 \\ 0 & 1 & -a_1 \end{bmatrix}\mathbf{W} = \mathbf{W}\begin{bmatrix} 0 & 1 & 0 \\ 0 & 0 & 1 \\ -a_3 & -a_2 & -a_1 \end{bmatrix} \qquad (10\text{--}144)$$

The left-hand side of Equation (10–144) is

$$\begin{bmatrix} 0 & 0 & -a_3 \\ 1 & 0 & -a_2 \\ 0 & 1 & -a_1 \end{bmatrix}\begin{bmatrix} a_2 & a_1 & 1 \\ a_1 & 1 & 0 \\ 1 & 0 & 0 \end{bmatrix} = \begin{bmatrix} -a_3 & 0 & 0 \\ 0 & a_1 & 1 \\ 0 & 1 & 0 \end{bmatrix}$$

The right-hand side of Equation (10–144) is

$$\begin{bmatrix} a_2 & a_1 & 1 \\ a_1 & 1 & 0 \\ 1 & 0 & 0 \end{bmatrix} \begin{bmatrix} 0 & 1 & 0 \\ 0 & 0 & 1 \\ -a_3 & -a_2 & -a_1 \end{bmatrix} = \begin{bmatrix} -a_3 & 0 & 0 \\ 0 & a_1 & 1 \\ 0 & 1 & 0 \end{bmatrix}$$

Clearly, Equation (10–144) holds true. Thus, we have shown that

$$\mathbf{T}^{-1}\mathbf{A}\mathbf{T} = \begin{bmatrix} 0 & 1 & 0 \\ 0 & 0 & 1 \\ -a_3 & -a_2 & -a_1 \end{bmatrix}$$

Next, we shall show that

$$\mathbf{T}^{-1}\mathbf{B} = \begin{bmatrix} 0 \\ 0 \\ 1 \end{bmatrix} \tag{10–145}$$

Note that Equation (10–145) can be written as

$$\mathbf{B} = \mathbf{T}\begin{bmatrix} 0 \\ 0 \\ 1 \end{bmatrix} = \mathbf{MW}\begin{bmatrix} 0 \\ 0 \\ 1 \end{bmatrix}$$

Noting that

$$\mathbf{T}\begin{bmatrix} 0 \\ 0 \\ 1 \end{bmatrix} = \begin{bmatrix} \mathbf{B} & \vdots & \mathbf{AB} & \vdots & \mathbf{A}^2\mathbf{B} \end{bmatrix} \begin{bmatrix} a_2 & a_1 & 1 \\ a_1 & 1 & 0 \\ 1 & 0 & 0 \end{bmatrix} \begin{bmatrix} 0 \\ 0 \\ 1 \end{bmatrix} = \begin{bmatrix} \mathbf{B} & \vdots & \mathbf{AB} & \vdots & \mathbf{A}^2\mathbf{B} \end{bmatrix} \begin{bmatrix} 1 \\ 0 \\ 0 \end{bmatrix} = \mathbf{B}$$

we have

$$\mathbf{T}^{-1}\mathbf{B} = \begin{bmatrix} 0 \\ 0 \\ 1 \end{bmatrix}$$

The derivation shown here can be easily extended to the general case of any positive integer n.

A–10–4. Consider the state equation

$$\dot{\mathbf{x}} = \mathbf{A}\mathbf{x} + \mathbf{B}u$$

where

$$\mathbf{A} = \begin{bmatrix} 1 & 1 \\ -4 & -3 \end{bmatrix}, \qquad \mathbf{B} = \begin{bmatrix} 0 \\ 2 \end{bmatrix}$$

The rank of the controllability matrix $\mathbf{M}$,

$$\mathbf{M} = \begin{bmatrix} \mathbf{B} & \vdots & \mathbf{AB} \end{bmatrix} = \begin{bmatrix} 0 & 2 \\ 2 & -6 \end{bmatrix}$$

is 2. Thus, the system is completely state controllable. Transform the given state equation into the controllable canonical form.

Solution. Since

$$|s\mathbf{I} - \mathbf{A}| = \begin{vmatrix} s - 1 & -1 \\ 4 & s + 3 \end{vmatrix} = (s - 1)(s + 3) + 4$$

$$= s^2 + 2s + 1 = s^2 + a_1 s + a_2$$

we have

$$a_1 = 2, \qquad a_2 = 1$$

Define

$$\mathbf{T} = \mathbf{MW}$$

where

$$\mathbf{M} = \begin{bmatrix} 0 & 2 \\ 2 & -6 \end{bmatrix}, \qquad \mathbf{W} = \begin{bmatrix} 2 & 1 \\ 1 & 0 \end{bmatrix}$$

Then

$$\mathbf{T} = \begin{bmatrix} 0 & 2 \\ 2 & -6 \end{bmatrix}\begin{bmatrix} 2 & 1 \\ 1 & 0 \end{bmatrix} = \begin{bmatrix} 2 & 0 \\ -2 & 2 \end{bmatrix}$$

and

$$\mathbf{T}^{-1} = \begin{bmatrix} 0.5 & 0 \\ 0.5 & 0.5 \end{bmatrix}$$

Define

$$\mathbf{x} = \mathbf{T}\hat{\mathbf{x}}$$

Then the state equation becomes

$$\dot{\hat{\mathbf{x}}} = \mathbf{T}^{-1}\mathbf{A}\mathbf{T}\hat{\mathbf{x}} + \mathbf{T}^{-1}\mathbf{B}u$$

Since

$$\mathbf{T}^{-1}\mathbf{A}\mathbf{T} = \begin{bmatrix} 0.5 & 0 \\ 0.5 & 0.5 \end{bmatrix}\begin{bmatrix} 1 & 1 \\ -4 & -3 \end{bmatrix}\begin{bmatrix} 2 & 0 \\ -2 & 2 \end{bmatrix} = \begin{bmatrix} 0 & 1 \\ -1 & -2 \end{bmatrix}$$

and

$$\mathbf{T}^{-1}\mathbf{B} = \begin{bmatrix} 0.5 & 0 \\ 0.5 & 0.5 \end{bmatrix}\begin{bmatrix} 0 \\ 2 \end{bmatrix} = \begin{bmatrix} 0 \\ 1 \end{bmatrix}$$

we have

$$\begin{bmatrix} \dot{\hat{x}}_1 \\ \dot{\hat{x}}_2 \end{bmatrix} = \begin{bmatrix} 0 & 1 \\ -1 & -2 \end{bmatrix}\begin{bmatrix} \hat{x}_1 \\ \hat{x}_2 \end{bmatrix} + \begin{bmatrix} 0 \\ 1 \end{bmatrix}u$$

which is in the controllable canonical form.

A–10–5. Consider a system defined by

$$\dot{\mathbf{x}} = \mathbf{A}\mathbf{x} + \mathbf{B}u$$

$$y = \mathbf{C}\mathbf{x}$$

where

$$\mathbf{A} = \begin{bmatrix} 0 & 1 \\ -2 & -3 \end{bmatrix}, \qquad \mathbf{B} = \begin{bmatrix} 0 \\ 2 \end{bmatrix}, \qquad \mathbf{C} = \begin{bmatrix} 1 & 0 \end{bmatrix}$$

The characteristic equation of the system is

$$|s\mathbf{I} - \mathbf{A}| = \begin{vmatrix} s & -1 \\ 2 & s+3 \end{vmatrix} = s^2 + 3s + 2 = (s+1)(s+2) = 0$$

The eigenvalues of matrix $\mathbf{A}$ are -1 and -2.

It is desired to have eigenvalues at -3 and -5 by using a state-feedback control $u = -\mathbf{Kx}$. Determine the necessary feedback gain matrix $\mathbf{K}$ and the control signal u.

Solution. The given system is completely state controllable, since the rank of

$$\mathbf{M} = \begin{bmatrix} \mathbf{B} & \vdots & \mathbf{AB} \end{bmatrix} = \begin{bmatrix} 0 & 2 \\ 2 & -6 \end{bmatrix}$$

is 2. Hence, arbitrary pole placement is possible.

Since the characteristic equation of the original system is

$$s^2 + 3s + 2 = s^2 + a_1 s + a_2 = 0$$

we have

$$a_1 = 3, \qquad a_2 = 2$$

The desired characteristic equation is

$$(s+3)(s+5) = s^2 + 8s + 15 = s^2 + \alpha_1 s + \alpha_2 = 0$$

Hence,

$$\alpha_1 = 8, \qquad \alpha_2 = 15$$

It is important to point out that the original state equation is not in the controllable canonical form, because matrix $\mathbf{B}$ is not

$$\begin{bmatrix} 0 \\ 1 \end{bmatrix}$$

Hence, the transformation matrix $\mathbf{T}$ must be determined.

$$\mathbf{T} = \mathbf{MW} = \begin{bmatrix} \mathbf{B} & \vdots & \mathbf{AB} \end{bmatrix} \begin{bmatrix} a_1 & 1 \\ 1 & 0 \end{bmatrix} = \begin{bmatrix} 0 & 2 \\ 2 & -6 \end{bmatrix} \begin{bmatrix} 3 & 1 \\ 1 & 0 \end{bmatrix} = \begin{bmatrix} 2 & 0 \\ 0 & 2 \end{bmatrix}$$

Hence,

$$\mathbf{T}^{-1} = \begin{bmatrix} 0.5 & 0 \\ 0 & 0.5 \end{bmatrix}$$

Referring to Equation (10–13), the necessary feedback gain matrix is given by

$$\mathbf{K} = \begin{bmatrix} \alpha_2 - a_2 & \vdots & \alpha_1 - a_1 \end{bmatrix} \mathbf{T}^{-1}$$

$$= \begin{bmatrix} 15 - 2 & \vdots & 8 - 3 \end{bmatrix} \begin{bmatrix} 0.5 & 0 \\ 0 & 0.5 \end{bmatrix} = \begin{bmatrix} 6.5 & 2.5 \end{bmatrix}$$

Thus, the control signal u becomes

$$u = -\mathbf{Kx} = -\begin{bmatrix} 6.5 & 2.5 \end{bmatrix} \begin{bmatrix} x_1 \\ x_2 \end{bmatrix}$$

A–10–6. A regulator system has a plant

$$\frac{Y(s)}{U(s)} = \frac{10}{(s+1)(s+2)(s+3)}$$

Define state variables as

$$x_1 = y$$
$$x_2 = \dot{x}_1$$
$$x_3 = \dot{x}_2$$

By use of the state-feedback control $u = -\mathbf{Kx}$, it is desired to place the closed-loop poles at

$$s = -2 + j2\sqrt{3}, \qquad s = -2 - j2\sqrt{3}, \qquad s = -10$$

Obtain the necessary state-feedback gain matrix $\mathbf{K}$ with MATLAB.

Solution. The state-space equations for the system become

$$\begin{bmatrix} \dot{x}_1 \\ \dot{x}_2 \\ \dot{x}_3 \end{bmatrix} = \begin{bmatrix} 0 & 1 & 0 \\ 0 & 0 & 1 \\ -6 & -11 & -6 \end{bmatrix} \begin{bmatrix} x_1 \\ x_2 \\ x_3 \end{bmatrix} + \begin{bmatrix} 0 \\ 0 \\ 10 \end{bmatrix} u$$

$$y = \begin{bmatrix} 1 & 0 & 0 \end{bmatrix} \begin{bmatrix} x_1 \\ x_2 \\ x_3 \end{bmatrix} + 0u$$

Hence,

$$\mathbf{A} = \begin{bmatrix} 0 & 1 & 0 \\ 0 & 0 & 1 \\ -6 & -11 & -6 \end{bmatrix}, \qquad \mathbf{B} = \begin{bmatrix} 0 \\ 0 \\ 10 \end{bmatrix}$$

$$\mathbf{C} = \begin{bmatrix} 1 & 0 & 0 \end{bmatrix}, \qquad D = \begin{bmatrix} 0 \end{bmatrix}$$

(Note that, for the pole placement, matrices $\mathbf{C}$ and D do not affect the state-feedback gain matrix $\mathbf{K}$.)

Two MATLAB programs for obtaining state-feedback gain matrix $\mathbf{K}$ are given in MATLAB Programs 10–24 and 10–25.

MATLAB Program 10–24

```
A = [0 1 0;0 0 1;-6 -11 -6];
B = [0;0;10];
J = [-2+j*2*sqrt(3) -2-j*2*sqrt(3) -10];
K = acker(A,B,J)

K =

    15.4000    4.5000    0.8000
```

```
MATLAB Program 10–25

A = [0 1 0;0 0 1; -6 -11 -6];
B = [0;0;10];
J = [-2+j*2*sqrt(3) -2-J*2*Sqrt(3) -10];
K = place(A,B,J)
place: ndigits= 15

K =

   15.4000   4.5000   0.8000
```

A–10–7. Consider a completely observable system

$$\dot{\mathbf{x}} = \mathbf{Ax}$$

$$y = \mathbf{Cx}$$

Define the observability matrix as **N**:

$$\mathbf{N} = \left[\mathbf{C^*} \; \vdots \; \mathbf{A^*C^*} \; \vdots \; \cdots \; \vdots \; (\mathbf{A^*})^{n-1}\mathbf{C^*}\right]$$

Show that

$$\mathbf{N^*A(N^*)}^{-1} = \begin{bmatrix} 0 & 1 & 0 & \cdots & 0 \\ 0 & 0 & 1 & \cdots & 0 \\ \cdot & \cdot & \cdot & & \cdot \\ \cdot & \cdot & \cdot & & \cdot \\ \cdot & \cdot & \cdot & & \cdot \\ 0 & 0 & 0 & \cdots & 1 \\ -a_n & a_{n-1} & -a_{n-2} & \cdots & -a_1 \end{bmatrix} \qquad (10\text{–}146)$$

where $a_1, a_2, \dots, a_n$ are the coefficients of the characteristic polynomial

$$|s\mathbf{I} - \mathbf{A}| = s^n + a_1 s^{n-1} + \cdots + a_{n-1}s + a_n$$

Solution. Let us consider the case where $n = 3$. Then Equation (10–146) can be written as

$$\mathbf{N^*A(N^*)}^{-1} = \begin{bmatrix} 0 & 1 & 0 \\ 0 & 0 & 1 \\ -a_3 & -a_2 & -a_1 \end{bmatrix} \qquad (10\text{–}147)$$

Equation (10–147) may be rewritten as

$$\mathbf{N^*A} = \begin{bmatrix} 0 & 1 & 0 \\ 0 & 0 & 1 \\ -a_3 & -a_2 & -a_1 \end{bmatrix} \mathbf{N^*} \qquad (10\text{–}148)$$

We shall show that Equation (10–148) holds true. The left-hand side of Equation (10–148) is

$$\mathbf{N^*A} = \begin{bmatrix} \mathbf{C} \\ \mathbf{CA} \\ \mathbf{CA}^2 \end{bmatrix} \mathbf{A} = \begin{bmatrix} \mathbf{CA} \\ \mathbf{CA}^2 \\ \mathbf{CA}^3 \end{bmatrix} \qquad (10\text{–}149)$$

The right-hand side of Equation (10–148) is

$$\begin{bmatrix} 0 & 1 & 0 \\ 0 & 0 & 1 \\ -a_3 & -a_2 & -a_1 \end{bmatrix} \mathbf{N^*} = \begin{bmatrix} 0 & 1 & 0 \\ 0 & 0 & 1 \\ -a_3 & -a_2 & -a_1 \end{bmatrix} \begin{bmatrix} \mathbf{C} \\ \mathbf{CA} \\ \mathbf{CA^2} \end{bmatrix}$$

$$= \begin{bmatrix} \mathbf{CA} \\ \mathbf{CA^2} \\ -a_3 \mathbf{C} - a_2 \mathbf{CA} - a_1 \mathbf{CA^2} \end{bmatrix} \qquad (10\text{–}150)$$

The Cayley–Hamilton theorem states that matrix $\mathbf{A}$ satisfies its own characteristic equation, or

$$\mathbf{A}^3 + a_1 \mathbf{A}^2 + a_2 \mathbf{A} + a_3 \mathbf{I} = \mathbf{0}$$

Hence,

$$-a_1 \mathbf{CA^2} - a_2 \mathbf{CA} - a_3 \mathbf{C} = \mathbf{CA^3}$$

Thus, the right-hand side of Equation (10–150) becomes the same as the right-hand side of Equation (10–149). Consequently,

$$\mathbf{N^*A} = \begin{bmatrix} 0 & 1 & 0 \\ 0 & 0 & 1 \\ -a_3 & -a_2 & -a_1 \end{bmatrix} \mathbf{N^*}$$

which is Equation (10–148). This last equation can be modified to

$$\mathbf{N^*A(N^*)}^{-1} = \begin{bmatrix} 0 & 1 & 0 \\ 0 & 0 & 1 \\ -a_3 & -a_2 & -a_1 \end{bmatrix}$$

The derivation presented here can be extended to the general case of any positive integer n.

A–10–8. Consider a completely observable system defined by

$$\dot{\mathbf{x}} = \mathbf{Ax} + \mathbf{B}u \qquad (10\text{–}151)$$

$$y = \mathbf{Cx} + Du \qquad (10\text{–}152)$$

Define

$$\mathbf{N} = \begin{bmatrix} \mathbf{C^*} & \vdots & \mathbf{A^*C^*} & \vdots & \cdots & \vdots & (\mathbf{A^*})^{n-1}\mathbf{C^*} \end{bmatrix}$$

and

$$\mathbf{W} = \begin{bmatrix} a_{n-1} & a_{n-2} & \cdots & a_1 & 1 \\ a_{n-2} & a_{n-3} & \cdots & 1 & 0 \\ \cdot & \cdot & & \cdot & \cdot \\ \cdot & \cdot & & \cdot & \cdot \\ \cdot & \cdot & & \cdot & \cdot \\ a_1 & 1 & \cdots & 0 & 0 \\ 1 & 0 & \cdots & 0 & 0 \end{bmatrix}$$

where the a's are coefficients of the characteristic polynomial

$$|s\mathbf{I} - \mathbf{A}| = s^n + a_1 s^{n-1} + \cdots + a_{n-1}s + a_n$$

Define also

$$\mathbf{Q} = (\mathbf{WN^*})^{-1}$$

Show that

$$\mathbf{Q}^{-1}\mathbf{A}\mathbf{Q} = \begin{bmatrix} 0 & 0 & \cdots & 0 & -a_n \\ 1 & 0 & \cdots & 0 & -a_{n-1} \\ 0 & 1 & \cdots & 0 & -a_{n-2} \\ \cdot & \cdot & & \cdot & \cdot \\ \cdot & \cdot & & \cdot & \cdot \\ \cdot & \cdot & & \cdot & \cdot \\ 0 & 0 & \cdots & 1 & -a_1 \end{bmatrix}$$

$$\mathbf{C}\mathbf{Q} = \begin{bmatrix} 0 & 0 & \cdots & 0 & 1 \end{bmatrix}$$

$$\mathbf{Q}^{-1}\mathbf{B} = \begin{bmatrix} b_n - a_n b_0 \\ b_{n-1} - a_{n-1} b_0 \\ \cdot \\ \cdot \\ \cdot \\ b_1 - a_1 b_0 \end{bmatrix}$$

where the b_k's $(k = 0, 1, 2, \ldots, n)$ are those coefficients appearing in the numerator of the transfer function when $\mathbf{C}(s\mathbf{I} - \mathbf{A})^{-1}\mathbf{B} + D$ is written as follows:

$$\mathbf{C}(s\mathbf{I} - \mathbf{A})^{-1}\mathbf{B} + D = \frac{b_0 s^n + b_1 s^{n-1} + \cdots + b_{n-1} s + b_n}{s^n + a_1 s^{n-1} + \cdots + a_{n-1} s + a_n}$$

where $D = b_0$.

Solution. Let us consider the case where $n = 3$. We shall show that

$$\mathbf{Q}^{-1}\mathbf{A}\mathbf{Q} = (\mathbf{W}\mathbf{N}^*)\mathbf{A}(\mathbf{W}\mathbf{N}^*)^{-1} = \begin{bmatrix} 0 & 0 & -a_3 \\ 1 & 0 & -a_2 \\ 0 & 1 & -a_1 \end{bmatrix} \tag{10-153}$$

Note that, by referring to Problem A–10–7, we have

$$(\mathbf{W}\mathbf{N}^*)\mathbf{A}(\mathbf{W}\mathbf{N}^*)^{-1} = \mathbf{W}\big[\mathbf{N}^*\mathbf{A}(\mathbf{N}^*)^{-1}\big]\mathbf{W}^{-1} = \mathbf{W}\begin{bmatrix} 0 & 1 & 0 \\ 0 & 0 & 1 \\ -a_3 & -a_2 & -a_1 \end{bmatrix}\mathbf{W}^{-1}$$

Hence, we need to show that

$$\mathbf{W}\begin{bmatrix} 0 & 1 & 0 \\ 0 & 0 & 1 \\ -a_3 & -a_2 & -a_1 \end{bmatrix}\mathbf{W}^{-1} = \begin{bmatrix} 0 & 0 & -a_3 \\ 1 & 0 & -a_2 \\ 0 & 1 & -a_1 \end{bmatrix}$$

or

$$\mathbf{W}\begin{bmatrix} 0 & 1 & 0 \\ 0 & 0 & 1 \\ -a_3 & -a_2 & -a_1 \end{bmatrix} = \begin{bmatrix} 0 & 0 & -a_3 \\ 1 & 0 & -a_2 \\ 0 & 1 & -a_1 \end{bmatrix}\mathbf{W} \tag{10-154}$$

The left-hand side of Equation (10–154) is

$$\mathbf{W}\begin{bmatrix} 0 & 1 & 0 \\ 0 & 0 & 1 \\ -a_3 & -a_2 & -a_1 \end{bmatrix} = \begin{bmatrix} a_2 & a_1 & 1 \\ a_1 & 1 & 0 \\ 1 & 0 & 0 \end{bmatrix}\begin{bmatrix} 0 & 1 & 0 \\ 0 & 0 & 1 \\ -a_3 & -a_2 & -a_1 \end{bmatrix}$$

$$= \begin{bmatrix} -a_3 & 0 & 0 \\ 0 & a_1 & 1 \\ 0 & 1 & 0 \end{bmatrix}$$

The right-hand side of Equation (10–154) is

$$\begin{bmatrix} 0 & 0 & -a_3 \\ 1 & 0 & -a_2 \\ 0 & 1 & -a_1 \end{bmatrix}\mathbf{W} = \begin{bmatrix} 0 & 0 & -a_3 \\ 1 & 0 & -a_2 \\ 0 & 1 & -a_1 \end{bmatrix}\begin{bmatrix} a_2 & a_1 & 1 \\ a_1 & 1 & 0 \\ 1 & 0 & 0 \end{bmatrix}$$

$$= \begin{bmatrix} -a_3 & 0 & 0 \\ 0 & a_1 & 1 \\ 0 & 1 & 0 \end{bmatrix}$$

Thus, we see that Equation (10–154) holds true. Hence, we have proved Equation (10–153). Next we shall show that

$$\mathbf{CQ} = \begin{bmatrix} 0 & 0 & 1 \end{bmatrix}$$

or

$$\mathbf{C}(\mathbf{WN^*})^{-1} = \begin{bmatrix} 0 & 0 & 1 \end{bmatrix}$$

Notice that

$$\begin{bmatrix} 0 & 0 & 1 \end{bmatrix}(\mathbf{WN^*}) = \begin{bmatrix} 0 & 0 & 1 \end{bmatrix}\begin{bmatrix} a_2 & a_1 & 1 \\ a_1 & 1 & 0 \\ 1 & 0 & 0 \end{bmatrix}\begin{bmatrix} \mathbf{C} \\ \mathbf{CA} \\ \mathbf{CA}^2 \end{bmatrix}$$

$$= \begin{bmatrix} 1 & 0 & 0 \end{bmatrix}\begin{bmatrix} \mathbf{C} \\ \mathbf{CA} \\ \mathbf{CA}^2 \end{bmatrix} = \mathbf{C}$$

Hence, we have shown that

$$\begin{bmatrix} 0 & 0 & 1 \end{bmatrix} = \mathbf{C}(\mathbf{WN^*})^{-1} = \mathbf{CQ}$$

Next define

$$\mathbf{x} = \mathbf{Q\hat{x}}$$

Then Equation (10–151) becomes

$$\dot{\mathbf{x}} = \mathbf{Q}^{-1}\mathbf{AQ\hat{x}} + \mathbf{Q}^{-1}\mathbf{B}u \qquad (10\text{–}155)$$

and Equation (10–152) becomes

$$y = \mathbf{CQ\hat{x}} + Du \qquad (10\text{–}156)$$

Referring to Equation (10–153), Equation (10–155) becomes

$$\begin{bmatrix} \dot{\hat{x}}_1 \\ \dot{\hat{x}}_2 \\ \dot{\hat{x}}_3 \end{bmatrix} = \begin{bmatrix} 0 & 0 & -a_3 \\ 1 & 0 & -a_2 \\ 0 & 1 & -a_1 \end{bmatrix}\begin{bmatrix} \hat{x}_1 \\ \hat{x}_2 \\ \hat{x}_3 \end{bmatrix} + \begin{bmatrix} \gamma_3 \\ \gamma_2 \\ \gamma_1 \end{bmatrix}u$$

where

$$\begin{bmatrix} \gamma_3 \\ \gamma_2 \\ \gamma_1 \end{bmatrix} = \mathbf{Q}^{-1}\mathbf{B}$$

The transfer function $G(s)$ for the system defined by Equations (10–155) and (10–156) is

$$G(s) = \mathbf{CQ}(s\mathbf{I} - \mathbf{Q}^{-1}\mathbf{AQ})^{-1}\mathbf{Q}^{-1}\mathbf{B} + D$$

Noting that

$$\mathbf{CQ} = \begin{bmatrix} 0 & 0 & 1 \end{bmatrix}$$

we have

$$G(s) = \begin{bmatrix} 0 & 0 & 1 \end{bmatrix} \begin{bmatrix} s & 0 & a_3 \\ -1 & s & a_2 \\ 0 & -1 & s + a_1 \end{bmatrix}^{-1} \begin{bmatrix} \gamma_3 \\ \gamma_2 \\ \gamma_1 \end{bmatrix} + D$$

Note that $D = b_0$. Since

$$\begin{bmatrix} s & 0 & a_3 \\ -1 & s & a_2 \\ 0 & -1 & s + a_1 \end{bmatrix}^{-1} = \frac{1}{s^3 + a_1 s^2 + a_2 s + a_3} \begin{bmatrix} s^2 + a_1 s + a_2 & -a_3 & -a_3 s \\ s + a_1 & s^2 + a_1 s & -a_2 s - a_3 \\ 1 & s & s^2 \end{bmatrix}$$

we have

$$G(s) = \frac{1}{s^3 + a_1 s^2 + a_2 s + a_3} \begin{bmatrix} 1 & s & s^2 \end{bmatrix} \begin{bmatrix} \gamma_3 \\ \gamma_2 \\ \gamma_1 \end{bmatrix} + D$$

$$= \frac{\gamma_1 s^2 + \gamma_2 s + \gamma_3}{s^3 + a_1 s^2 + a_2 s + a_3} + b_0$$

$$= \frac{b_0 s^3 + (\gamma_1 + a_1 b_0)s^2 + (\gamma_2 + a_2 b_0)s + \gamma_3 + a_3 b_0}{s^3 + a_1 s^2 + a_2 s + a_3}$$

$$= \frac{b_0 s^3 + b_1 s^2 + b_2 s + b_3}{s^3 + a_1 s^2 + a_2 s + a_3}$$

Hence,

$$\gamma_1 = b_1 - a_1 b_0, \qquad \gamma_2 = b_2 - a_2 b_0, \qquad \gamma_3 = b_3 - a_3 b_0$$

Thus, we have shown that

$$\mathbf{Q}^{-1}\mathbf{B} = \begin{bmatrix} \gamma_3 \\ \gamma_2 \\ \gamma_1 \end{bmatrix} = \begin{bmatrix} b_3 - a_3 b_0 \\ b_2 - a_2 b_0 \\ b_1 - a_1 b_0 \end{bmatrix}$$

Note that what we have derived here can be easily extended to the case when n is any positive integer.

A–10–9. Consider a system defined by

$$\dot{\mathbf{x}} = \mathbf{A}\mathbf{x} + \mathbf{B}u$$

$$y = \mathbf{C}\mathbf{x}$$

where

$$A = \begin{bmatrix} 1 & 1 \\ -4 & -3 \end{bmatrix}, \qquad B = \begin{bmatrix} 0 \\ 2 \end{bmatrix}, \qquad C = \begin{bmatrix} 1 & 1 \end{bmatrix}$$

The rank of the observability matrix **N**,

$$N = \begin{bmatrix} C^* & \vdots & A^*C^* \end{bmatrix} = \begin{bmatrix} 1 & -3 \\ 1 & -2 \end{bmatrix}$$

is 2. Hence, the system is completely observable. Transform the system equations into the observable canonical form.

Solution. Since

$$|sI - A| = s^2 + 2s + 1 = s^2 + a_1 s + a_2$$

we have

$$a_1 = 2, \qquad a_2 = 1$$

Define

$$Q = (WN^*)^{-1}$$

where

$$N = \begin{bmatrix} 1 & -3 \\ 1 & -2 \end{bmatrix}, \qquad W = \begin{bmatrix} a_1 & 1 \\ 1 & 0 \end{bmatrix} = \begin{bmatrix} 2 & 1 \\ 1 & 0 \end{bmatrix}$$

Then

$$Q = \left\{ \begin{bmatrix} 2 & 1 \\ 1 & 0 \end{bmatrix} \begin{bmatrix} 1 & 1 \\ -3 & -2 \end{bmatrix} \right\}^{-1} = \begin{bmatrix} -1 & 0 \\ 1 & 1 \end{bmatrix}^{-1} = \begin{bmatrix} -1 & 0 \\ 1 & 1 \end{bmatrix}$$

and

$$Q^{-1} = \begin{bmatrix} -1 & 0 \\ 1 & 1 \end{bmatrix}$$

Define

$$x = Q\hat{x}$$

Then the state equation becomes

$$\dot{\hat{x}} = Q^{-1}AQ\hat{x} + Q^{-1}Bu$$

or

$$\begin{bmatrix} \dot{\hat{x}}_1 \\ \dot{\hat{x}}_2 \end{bmatrix} = \begin{bmatrix} -1 & 0 \\ 1 & 1 \end{bmatrix} \begin{bmatrix} 1 & 1 \\ -4 & -3 \end{bmatrix} \begin{bmatrix} -1 & 0 \\ 1 & 1 \end{bmatrix} \begin{bmatrix} \hat{x}_1 \\ \hat{x}_2 \end{bmatrix} + \begin{bmatrix} -1 & 0 \\ 1 & 1 \end{bmatrix} \begin{bmatrix} 0 \\ 2 \end{bmatrix} u$$

$$= \begin{bmatrix} 0 & -1 \\ 1 & -2 \end{bmatrix} \begin{bmatrix} \hat{x}_1 \\ \hat{x}_2 \end{bmatrix} + \begin{bmatrix} 0 \\ 2 \end{bmatrix} u \qquad\qquad (10\text{--}157)$$

The output equation becomes

$$y = CQ\hat{x}$$

or

$$y = \begin{bmatrix} 1 & 1 \end{bmatrix} \begin{bmatrix} -1 & 0 \\ 1 & 1 \end{bmatrix} \begin{bmatrix} \hat{x}_1 \\ \hat{x}_2 \end{bmatrix} = \begin{bmatrix} 0 & 1 \end{bmatrix} \begin{bmatrix} \hat{x}_1 \\ \hat{x}_2 \end{bmatrix} \tag{10–158}$$

Equations (10–157) and (10–158) are in the observable canonical form.

A–10–10. For the system defined by

$$\dot{\mathbf{x}} = \mathbf{Ax} + \mathbf{B}u$$

$$y = \mathbf{Cx}$$

consider the problem of designing a state observer such that the desired eigenvalues for the observer gain matrix are $\mu_1, \mu_2, \ldots, \mu_n$.

Show that the observer gain matrix given by Equation (10–61), rewritten as

$$\mathbf{K}_e = (\mathbf{WN}^*)^{-1} \begin{bmatrix} \alpha_n - a_n \\ \alpha_{n-1} - a_{n-1} \\ \cdot \\ \cdot \\ \cdot \\ \alpha_1 - a_1 \end{bmatrix} \tag{10–159}$$

can be obtained from Equation (10–13) by considering the dual problem. That is, the matrix $\mathbf{K}_e$ can be determined by considering the pole-placement problem for the dual system, obtaining the state-feedback gain matrix $\mathbf{K}$, and taking its conjugate transpose, or $\mathbf{K}_e = \mathbf{K}^*$.

Solution. The dual of the given system is

$$\dot{\mathbf{z}} = \mathbf{A}^*\mathbf{z} + \mathbf{C}^*v \tag{10–160}$$

$$n = \mathbf{B}^*\mathbf{z}$$

Using the state-feedback control

$$v = -\mathbf{Kz}$$

Equation (10–160) becomes

$$\dot{\mathbf{z}} = (\mathbf{A}^* - \mathbf{C}^*\mathbf{K})\mathbf{z}$$

Equation (10–13), which is rewritten here, is

$$\mathbf{K} = \begin{bmatrix} \alpha_n - a_n & \vdots & \alpha_{n-1} - a_{n-1} & \vdots & \cdots & \vdots & \alpha_2 - a_2 & \vdots & \alpha_1 - a_1 \end{bmatrix} \mathbf{T}^{-1} \tag{10–161}$$

where

$$\mathbf{T} = \mathbf{MW} = \begin{bmatrix} \mathbf{C}^* & \vdots & \mathbf{A}^*\mathbf{C}^* & \vdots & \cdots & \vdots & (\mathbf{A}^*)^{n-1}\mathbf{C}^* \end{bmatrix} \mathbf{W}$$

For the original system, the observability matrix is

$$\begin{bmatrix} \mathbf{C}^* & \vdots & \mathbf{A}^*\mathbf{C}^* & \vdots & \cdots & \vdots & (\mathbf{A}^*)^{n-1}\mathbf{C}^* \end{bmatrix} = \mathbf{N}$$

Hence, matrix $\mathbf{T}$ can also be written as

$$\mathbf{T} = \mathbf{NW}$$

Since $\mathbf{W} = \mathbf{W}^*$, we have

$$\mathbf{T}^* = \mathbf{W}^*\mathbf{N}^* = \mathbf{WN}^*$$

and

$$(\mathbf{T}^*)^{-1} = (\mathbf{WN}^*)^{-1}$$

Taking the conjugate transpose of both sides of Equation (10–161), we have

$$\mathbf{K}^* = \left(\mathbf{T}^{-1}\right)^* \begin{bmatrix} \alpha_n - a_n \\ \alpha_{n-1} - a_{n-1} \\ \cdot \\ \cdot \\ \cdot \\ \alpha_1 - a_1 \end{bmatrix} = \left(\mathbf{T}^*\right)^{-1} \begin{bmatrix} \alpha_n - a_n \\ \alpha_{n-1} - a_{n-1} \\ \cdot \\ \cdot \\ \cdot \\ \alpha_1 - a_1 \end{bmatrix} = \left(\mathbf{WN}^*\right)^{-1} \begin{bmatrix} \alpha_n - a_n \\ \alpha_{n-1} - a_{n-1} \\ \cdot \\ \cdot \\ \cdot \\ \alpha_1 - a_1 \end{bmatrix}$$

Since $\mathbf{K}_e = \mathbf{K}^*$, this last equation is the same as Equation (10–159). Thus, we obtained Equation (10–159) by considering the dual problem.

A–10–11. Consider an observed-state feedback control system with a minimum-order observer described by the following equations:

$$\dot{\mathbf{x}} = \mathbf{A}\mathbf{x} + \mathbf{B}u \qquad (10\text{–}162)$$

$$y = \mathbf{C}\mathbf{x}$$

$$u = -\mathbf{K}\tilde{\mathbf{x}} \qquad (10\text{–}163)$$

where

$$\mathbf{x} = \begin{bmatrix} x_a \\ \hline \mathbf{x}_b \end{bmatrix}, \qquad \tilde{\mathbf{x}} = \begin{bmatrix} x_a \\ \hline \tilde{\mathbf{x}}_b \end{bmatrix}$$

(x_a is the state variable that can be directly measured, and $\tilde{\mathbf{x}}_b$ corresponds to the observed state variables.)

Show that the closed-loop poles of the system comprise the closed-loop poles due to pole placement [the eigenvalues of matrix $(\mathbf{A} - \mathbf{BK})$] and the closed-loop poles due to the minimum-order observer [the eigenvalues of matrix $(\mathbf{A}_{bb} - \mathbf{K}_e\mathbf{A}_{ab})$]

Solution. The error equation for the minimum-order observer may be derived as given by Equation (10–94), rewritten thus:

$$\dot{\mathbf{e}} = \left(\mathbf{A}_{bb} - \mathbf{K}_e\mathbf{A}_{ab}\right)\mathbf{e} \qquad (10\text{–}164)$$

where

$$\mathbf{e} = \mathbf{x}_b - \tilde{\mathbf{x}}_b$$

From Equations (10–162) and (10–163), we obtain

$$\dot{\mathbf{x}} = \mathbf{A}\mathbf{x} - \mathbf{BK}\tilde{\mathbf{x}} = \mathbf{A}\mathbf{x} - \mathbf{BK}\begin{bmatrix} x_a \\ \hline \tilde{\mathbf{x}}_b \end{bmatrix} = \mathbf{A}\mathbf{x} - \mathbf{BK}\begin{bmatrix} x_a \\ \hline \mathbf{x}_b - \mathbf{e} \end{bmatrix}$$

$$= \mathbf{A}\mathbf{x} - \mathbf{BK}\left\{\mathbf{x} - \begin{bmatrix} 0 \\ \hline \mathbf{e} \end{bmatrix}\right\} = (\mathbf{A} - \mathbf{BK})\mathbf{x} + \mathbf{BK}\begin{bmatrix} 0 \\ \hline \mathbf{e} \end{bmatrix} \qquad (10\text{–}165)$$

Combining Equations (10–164) and (10–165) and writing

$$\mathbf{K} = \begin{bmatrix} K_a & \vdots & \mathbf{K}_b \end{bmatrix}$$

we obtain

$$\begin{bmatrix} \dot{\mathbf{x}} \\ \dot{\mathbf{e}} \end{bmatrix} = \begin{bmatrix} \mathbf{A} - \mathbf{BK} & \mathbf{BK}_b \\ 0 & \mathbf{A}_{bb} - \mathbf{K}_e\mathbf{A}_{ab} \end{bmatrix}\begin{bmatrix} \mathbf{x} \\ \mathbf{e} \end{bmatrix} \qquad (10\text{–}166)$$

Equation (10–166) describes the dynamics of the observed-state feedback control system with a minimum-order observer. The characteristic equation for this system is

$$\begin{vmatrix} s\mathbf{I} - \mathbf{A} + \mathbf{BK} & -\mathbf{BK}_b \\ 0 & s\mathbf{I} - \mathbf{A}_{bb} + \mathbf{K}_e\mathbf{A}_{ab} \end{vmatrix} = 0$$

or

$$\left|s\mathbf{I} - \mathbf{A} + \mathbf{BK}\right|\left|s\mathbf{I} - \mathbf{A}_{bb} + \mathbf{K}_e\mathbf{A}_{ab}\right| = 0$$

The closed-loop poles of the observed-state feedback control system with a minimum-order observer consist of the closed-loop poles due to pole placement and the closed-loop poles due to the minimum-order observer. (Therefore, the pole-placement design and the design of the minimum-order observer are independent of each other.)

A–10–12. Consider a completely state controllable system defined by

$$\dot{\mathbf{x}} = \mathbf{Ax} + \mathbf{B}u \qquad (10\text{–}167)$$

$$y = \mathbf{Cx}$$

where $\mathbf{x}$ = state vector (n-vector)

u = control signal (scalar)

y = output signal (scalar)

$\mathbf{A} = n \times n$ constant matrix

$\mathbf{B} = n \times 1$ constant matrix

$\mathbf{C} = 1 \times n$ constant matrix

Suppose that the rank of the following $(n + 1) \times (n + 1)$ matrix

$$\begin{bmatrix} \mathbf{A} & \mathbf{B} \\ -\mathbf{C} & 0 \end{bmatrix}$$

is $n + 1$. Show that the system defined by

$$\dot{\mathbf{e}} = \hat{\mathbf{A}}\mathbf{e} + \hat{\mathbf{B}}u_e \qquad (10\text{–}168)$$

where

$$\hat{\mathbf{A}} = \begin{bmatrix} \mathbf{A} & \mathbf{0} \\ -\mathbf{C} & 0 \end{bmatrix}, \qquad \hat{\mathbf{B}} = \begin{bmatrix} \mathbf{B} \\ 0 \end{bmatrix}, \qquad u_e = u(t) - u(\infty)$$

is completely state controllable.

Solution. Define

$$\mathbf{M} = \begin{bmatrix} \mathbf{B} & \vdots & \mathbf{AB} & \vdots & \cdots & \vdots & \mathbf{A}^{n-1}\mathbf{B} \end{bmatrix}$$

Because the system given by Equation (10–167) is completely state controllable, the rank of matrix $\mathbf{M}$ is n. Then the rank of

$$\begin{bmatrix} \mathbf{M} & \mathbf{0} \\ \mathbf{0} & 1 \end{bmatrix}$$

is $n + 1$. Consider the following equation:

$$\begin{bmatrix} \mathbf{A} & \mathbf{B} \\ -\mathbf{C} & 0 \end{bmatrix}\begin{bmatrix} \mathbf{M} & \mathbf{0} \\ \mathbf{0} & 1 \end{bmatrix} = \begin{bmatrix} \mathbf{AM} & \mathbf{B} \\ -\mathbf{CM} & 0 \end{bmatrix} \qquad (10\text{–}169)$$

Since matrix

$$\begin{bmatrix} \mathbf{A} & \mathbf{B} \\ -\mathbf{C} & 0 \end{bmatrix}$$

is of rank $n + 1$, the left-hand side of Equation (10–169) is of rank $n + 1$. Therefore, the right-hand side of Equation (10–169) is also of rank $n + 1$. Since

$$\begin{bmatrix} \mathbf{AM} & \mathbf{B} \\ -\mathbf{CM} & 0 \end{bmatrix} = \begin{bmatrix} \mathbf{A}\begin{bmatrix} \mathbf{B} & \vdots & \mathbf{AB} & \vdots & \cdots & \vdots & \mathbf{A}^{n-1}\mathbf{B} \end{bmatrix} & \mathbf{B} \\ -\mathbf{C}\begin{bmatrix} \mathbf{B} & \vdots & \mathbf{AB} & \vdots & \cdots & \vdots & \mathbf{A}^{n-1}\mathbf{B} \end{bmatrix} & 0 \end{bmatrix}$$

$$= \begin{bmatrix} \mathbf{AB} & \vdots & \mathbf{A}^2\mathbf{B} & \vdots & \cdots & \vdots & \mathbf{A}^n\mathbf{B} & \vdots & \mathbf{B} \\ -\mathbf{CB} & \vdots & -\mathbf{CAB} & \vdots & \cdots & \vdots & -\mathbf{CA}^{n-1}\mathbf{B} & \vdots & 0 \end{bmatrix}$$

$$= \begin{bmatrix} \hat{\mathbf{A}}\hat{\mathbf{B}} & \vdots & \hat{\mathbf{A}}^2\hat{\mathbf{B}} & \vdots & \cdots & \vdots & \hat{\mathbf{A}}^n\hat{\mathbf{B}} & \vdots & \hat{\mathbf{B}} \end{bmatrix}$$

we find that the rank of

$$[\hat{\mathbf{B}} \ \vdots \ \hat{\mathbf{A}}\hat{\mathbf{B}} \ \vdots \ \hat{\mathbf{A}}^2\hat{\mathbf{B}} \ \vdots \ \cdots \ \vdots \ \hat{\mathbf{A}}^n\hat{\mathbf{B}}]$$

is $n + 1$. Thus, the system defined by Equation (10–168) is completely state controllable.

A–10–13. Consider the system shown in Figure 10–49. Using the pole-placement-with-observer approach, design a regulator system such that the system will maintain the zero position $(y_1 = 0$ and $y_2 = 0)$ in the presence of disturbances. Choose the desired closed-loop poles for the pole-placement part to be

$$s = -2 + j2\sqrt{3}, \qquad s = -2 - j2\sqrt{3}, \qquad s = -10, \qquad s = -10$$

and the desired poles for the minimum-order observer to be

$$s = -15, \qquad s = -16$$

First, determine the state feedback gain matrix $\mathbf{K}$ and observer gain matrix $\mathbf{K}_e$. Then, obtain the response of the system to an arbitrary initial condition—for example,

$$y_1(0) = 0.1, \qquad y_2(0) = 0, \qquad \dot{y}_1(0) = 0, \qquad \dot{y}_2(0) = 0$$
$$e_1(0) = 0.1, \qquad e_2(0) = 0.05$$

where e_1 and e_2 are defined by

$$e_1 = y_1 - \tilde{y}_1$$
$$e_2 = y_2 - \tilde{y}_2$$

Assume that $m_1 = 1$ kg, $m_2 = 2$ kg, $k = 36$ N/m, and $b = 0.6$ N-s/m.

Solution. The equations for the system are

$$m_1\ddot{y}_1 = k(y_2 - y_1) + b(\dot{y}_2 - \dot{y}_1) + u$$
$$m_2\ddot{y}_2 = k(y_1 - y_2) + b(\dot{y}_1 - \dot{y}_2)$$

By substituting the given numerical values for $m_1, m_2, k,$ and b and simplifying, we obtain

$$\ddot{y}_1 = -36y_1 + 36y_2 - 0.6\dot{y}_1 + 0.6\dot{y}_2 + u$$
$$\ddot{y}_2 = 18y_1 - 18y_2 + 0.3\dot{y}_1 - 0.3\dot{y}_2$$

Let us choose the state variables as follows:

$$x_1 = y_1$$
$$x_2 = y_2$$
$$x_3 = \dot{y}_1$$
$$x_4 = \dot{y}_2$$

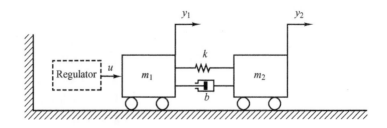

Figure 10–49
Mechanical system.

Then, the state-space equations become

$$
\begin{bmatrix} \dot{x}_1 \\ \dot{x}_2 \\ \dot{x}_3 \\ \dot{x}_4 \end{bmatrix} = \begin{bmatrix} 0 & 0 & 1 & 0 \\ 0 & 0 & 0 & 1 \\ -36 & 36 & -0.6 & 0.6 \\ 18 & -18 & 0.3 & -0.3 \end{bmatrix} \begin{bmatrix} x_1 \\ x_2 \\ x_3 \\ x_4 \end{bmatrix} + \begin{bmatrix} 0 \\ 0 \\ 1 \\ 0 \end{bmatrix} u
$$

$$
\begin{bmatrix} y_1 \\ y_2 \end{bmatrix} = \begin{bmatrix} 1 & 0 & 0 & 0 \\ 0 & 1 & 0 & 0 \end{bmatrix} \begin{bmatrix} x_1 \\ x_2 \\ x_3 \\ x_4 \end{bmatrix}
$$

Define

$$
A = \left[\begin{array}{cc|cc} 0 & 0 & 1 & 0 \\ 0 & 0 & 0 & 1 \\ \hline -36 & 36 & -0.6 & 0.6 \\ 18 & -18 & 0.3 & -0.3 \end{array} \right] = \left[\begin{array}{c|c} \mathbf{A}_{aa} & \mathbf{A}_{ab} \\ \hline \mathbf{A}_{ba} & \mathbf{A}_{bb} \end{array} \right], \qquad B = \left[\begin{array}{c} 0 \\ 0 \\ \hline 1 \\ 0 \end{array} \right] = \left[\begin{array}{c} \mathbf{B}_a \\ \hline \mathbf{B}_b \end{array} \right]
$$

The state feedback gain matrix $\mathbf{K}$ and observer gain matrix $\mathbf{K}_e$ can be obtained easily by use of MATLAB as follows:

$$
\mathbf{K} = \begin{bmatrix} 130.4444 & -41.5556 & 23.1000 & 15.4185 \end{bmatrix}
$$

$$
\mathbf{K}_e = \begin{bmatrix} 14.4 & 0.6 \\ 0.3 & 15.7 \end{bmatrix}
$$

(See MATLAB Program 10–26.)

MATLAB Program 10–26

```
A = [0 0 1 0;0 0 0 1;-36 36 -0.6 0.6;18 -18 0.3 -0.3];
B = [0;0;1;0];
J = [-2+j*2*sqrt(3) -2-j*2*sqrt(3) -10 -10];
K = acker(A,B,J)

K =

   130.4444   -41.5556   23.1000   15.4185

Aab = [1 0;0 1];
Abb = [-0.6 0.6;0.3 -0.3];
L = [-15 -16];
Ke = place(Abb',Aab',L)'
place: ndigits= 15

Ke =

   14.4000    0.6000
    0.3000   15.7000
```

Response to Initial Condition: Next, we obtain the response of the designed system to the given initial condition. Since

$$
\dot{\mathbf{x}} = \mathbf{Ax} + \mathbf{B}u
$$

$$
u = -\mathbf{K}\tilde{\mathbf{x}}
$$

$$
\tilde{\mathbf{x}} = \begin{bmatrix} \mathbf{x}_a \\ \tilde{\mathbf{x}}_b \end{bmatrix} = \begin{bmatrix} \mathbf{y} \\ \tilde{\mathbf{x}}_b \end{bmatrix}
$$

we have

$$\dot{\mathbf{x}} = \mathbf{Ax} - \mathbf{BK}\widetilde{\mathbf{x}} = (\mathbf{A} - \mathbf{BK})\mathbf{x} + \mathbf{BK}(\mathbf{x} - \widetilde{\mathbf{x}}) \qquad (10\text{--}170)$$

Note that

$$\mathbf{x} - \widetilde{\mathbf{x}} = \begin{bmatrix} \mathbf{x}_a \\ \hline \mathbf{x}_b \end{bmatrix} - \begin{bmatrix} \mathbf{x}_a \\ \hline \widetilde{\mathbf{x}}_b \end{bmatrix} = \begin{bmatrix} \mathbf{0} \\ \hline \mathbf{x}_b - \widetilde{\mathbf{x}}_b \end{bmatrix} = \begin{bmatrix} \mathbf{0} \\ \hline \mathbf{e} \end{bmatrix} = \begin{bmatrix} \mathbf{0} \\ \hline \mathbf{I} \end{bmatrix}\mathbf{e} = \mathbf{Fe}$$

where

$$\mathbf{F} = \begin{bmatrix} \mathbf{0} \\ \hline \mathbf{I} \end{bmatrix}$$

Then, Equation (10–170) can be written as

$$\dot{\mathbf{x}} = (\mathbf{A} - \mathbf{BK})\mathbf{x} + \mathbf{BKFe} \qquad (10\text{--}171)$$

Since, from Equation (10–94), we have

$$\dot{\mathbf{e}} = \left(\mathbf{A}_{bb} - \mathbf{K}_e\mathbf{A}_{ab}\right)\mathbf{e} \qquad (10\text{--}172)$$

by combining Equations (10–171) and (10–172) into one equation, we have

$$\begin{bmatrix} \dot{\mathbf{x}} \\ \hline \dot{\mathbf{e}} \end{bmatrix} = \begin{bmatrix} \mathbf{A} - \mathbf{BK} & \mathbf{BKF} \\ \hline \mathbf{0} & \mathbf{A}_{bb} - \mathbf{K}_e\mathbf{A}_{ab} \end{bmatrix}\begin{bmatrix} \mathbf{x} \\ \hline \mathbf{e} \end{bmatrix}$$

The state matrix here is a 6 × 6 matrix. The response of the system to the given initial condition can be obtained easily with MATLAB. (See MATLAB Program 10–27.) The resulting response curves are shown in Figure 10–50. The response curves seem to be acceptable.

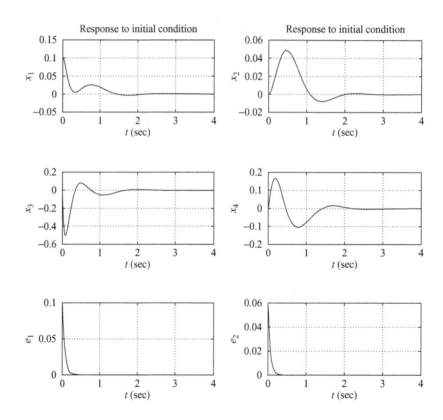

Figure 10–50
Response curves to initial condition.

```
MATLAB Program 10–27

% Response to initial condition

A = [0  0  1  0;0  0  0  1;-36  36  -0.6  0.6;18  -18  0.3  -0.3];
B = [0;0;1;0];
K = [130.4444  -41.5556  23.1000  15.4185];
Ke = [14.4  0.6;0.3  15.7];
F = [0  0;0  0;1  0;0  1];
Aab = [1  0;0  1];
Abb = [-0.6  0.6;0.3  -0.3];
AA = [A-B*K  B*K*F; zeros(2,4)  Abb-Ke*Aab];
sys = ss(AA,eye(6),eye(6),eye(6));
t = 0:0.01:4;
y = initial(sys,[0.1;0;0;0;0.1;0.05],t);
x1 = [1  0  0  0  0  0]*y';
x2 = [0  1  0  0  0  0]*y';
x3 = [0  0  1  0  0  0]*y';
x4 = [0  0  0  1  0  0]*y';
e1 = [0  0  0  0  1  0]*y';
e2 = [0  0  0  0  0  1]*y';

subplot(3,2,1); plot(t,x1); grid; title('Response to initial condition'),
xlabel('t (sec)'); ylabel('x1')
subplot(3,2,2); plot(t,x2); grid; title('Response to initial condition'),
xlabel('t (sec)'); ylabel('x2')
subplot(3,2,3); plot(t,x3); grid; xlabel('t (sec)'); ylabel('x3')
subplot(3,2,4); plot(t,x4); grid; xlabel('t (sec)'); ylabel('x4')
subplot(3,2,5); plot(t,e1); grid; xlabel('t (sec)');ylabel('e1')
subplot(3,2,6); plot(t,e2); grid; xlabel('t (sec)'); ylabel('e2')
```

A–10–14. Consider the system shown in Figure 10–51. Design both the full-order and minimum-order observers for the plant. Assume that the desired closed-loop poles for the pole-placement part are located at

$$s = -2 + j2\sqrt{3}, \qquad s = -2 - j2\sqrt{3}$$

Assume also that the desired observer poles are located at

(a) $s = -8$, $s = -8$ for the full-order observer

(b) $s = -8$ for the minimum-order observer

Compare the responses to the initial conditions specified below:

(a) for the full-order observer:

$$x_1(0) = 1, \qquad x_2(0) = 0, \qquad e_1(0) = 1, \qquad e_2(0) = 0$$

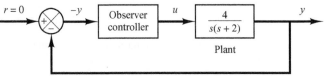

Figure 10–51
Regulator system.

(b) for the minimum-order observer:

$$x_1(0) = 1, \qquad x_2(0) = 0, \qquad e_1(0) = 1$$

Also, compare the bandwidths of both systems.

Solution. We first determine the state-space representation of the system. By defining state variables x_1 and x_2 as

$$x_1 = y$$
$$x_2 = \dot{y}$$

we obtain

$$\begin{bmatrix} \dot{x}_1 \\ \dot{x}_2 \end{bmatrix} = \begin{bmatrix} 0 & 1 \\ 0 & -2 \end{bmatrix}\begin{bmatrix} x_1 \\ x_2 \end{bmatrix} + \begin{bmatrix} 0 \\ 4 \end{bmatrix} u$$

$$y = \begin{bmatrix} 1 & 0 \end{bmatrix}\begin{bmatrix} x_1 \\ x_2 \end{bmatrix}$$

For the pole-placement part, we determine the state feedback gain matrix **K**. Using MATLAB, we find **K** to be

$$\mathbf{K} = \begin{bmatrix} 4 & 0.5 \end{bmatrix}$$

(See MATLAB Program 10–28.)

Next, we determine the observer gain matrix $\mathbf{K}_e$ for the full-order observer. Using MATLAB, we find $\mathbf{K}_e$ to be

$$\mathbf{K}_e = \begin{bmatrix} 14 \\ 36 \end{bmatrix}$$

(See MATLAB Program 10–28.)

MATLAB Program 10–28

```
% Obtaining matrices K and Ke.

A = [0 1;0 -2];
B = [0;4];
C = [1 0];
J = [-2+j*2*sqrt(3) -2-j*2*sqrt(3)];
L = [-8 -8];
K = acker(A,B,J)

K =

   4.0000  0.5000

Ke = acker(A',C',L)'

Ke =

   14
   36
```

Now we find the response of this system to the given initial condition. Referring to Equation (10–70), we have

$$\begin{bmatrix} \dot{\mathbf{x}} \\ \dot{\mathbf{e}} \end{bmatrix} = \begin{bmatrix} \mathbf{A} - \mathbf{BK} & \mathbf{BK} \\ \mathbf{0} & \mathbf{A} - \mathbf{K}_e\mathbf{C} \end{bmatrix}\begin{bmatrix} \mathbf{x} \\ \mathbf{e} \end{bmatrix}$$

This equation defines the dynamics of the designed system using the full-order observer. MATLAB Program 10–29 produces the response to the given initial condition. The resulting response curves are shown in Figure 10–52.

MATLAB Program 10–29

```
% Response to initial condition ---- full-order observer

A = [0  1;0  -2];
B = [0;4];
C = [1  0];
K = [4  0.5];
Ke = [14;36];
AA = [A-B*K  B*K; zeros(2,2)  A-Ke*C];
sys = ss(AA, eye(4), eye(4), eye(4));
t = 0:0.01:8;
x = initial(sys, [1;0;1;0],t);
x1 = [1  0  0  0]*x';
x2 = [0  1  0  0]*x';
e1 = [0  0  1  0]*x';
e2 = [0  0  0  1]*x';

subplot(2,2,1); plot(t,x1); grid
xlabel('t (sec)'); ylabel('x1')

subplot(2,2,2); plot(t,x2); grid
xlabel('t (sec)'); ylabel('x2')

subplot(2,2,3); plot(t,e1); grid
xlabel('t (sec)'); ylabel('e1')

subplot(2,2,4); plot(t,e2); grid
xlabel('t (sec)'); ylabel('e2')
```

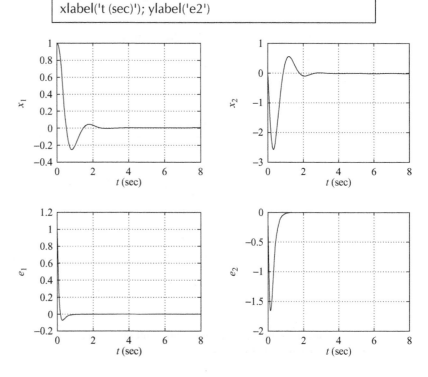

Figure 10–52
Response curves to
initial condition.

To obtain the transfer function of the observer controller, we use MATLAB. MATLAB Program 10–30 produces this transfer function. The result is

$$\frac{num}{den} = \frac{74s + 256}{s^2 + 18s + 108} = \frac{74(s + 3.4595)}{(s + 9 + j5.1962)(s + 9 - j5.1962)}$$

MATLAB Program 10–30

```
% Determination of transfer function of observer controller ---- full-order observer

A = [0  1;0  -2];
B = [0;4];
C = [1  0];
K = [4  0.5];
Ke = [14;36];
[num,den] = ss2tf(A-Ke*C-B*K, Ke,K,0)

num =

    0   74.0000   256.0000

den =

    1   18   108
```

Next, we obtain the observer gain matrix K_e for the minimum-order observer. MATLAB Program 10–31 produces K_e. The result is

$$K_e = 6$$

MATLAB Program 10–31

```
% Obtaining Ke ---- minimum-order observer

Aab = [1];
Abb = [-2];
LL = [-8];
Ke = acker(Abb',Aab',LL)'

Ke =

    6
```

The response of the system with minimum-order observer to the initial condition can be obtained as follows: By substituting $u = -\mathbf{K}\tilde{\mathbf{x}}$ into the plant equation given by Equation (10–79) we find

$$\dot{\mathbf{x}} = \mathbf{A}\mathbf{x} - \mathbf{B}\mathbf{K}\tilde{\mathbf{x}} = \mathbf{A}\mathbf{x} - \mathbf{B}\mathbf{K}\mathbf{x} + \mathbf{B}\mathbf{K}(\mathbf{x} - \tilde{\mathbf{x}})$$

$$= (\mathbf{A} - \mathbf{B}\mathbf{K})\mathbf{x} + \mathbf{B}\begin{bmatrix} K_a & K_b \end{bmatrix}\begin{bmatrix} 0 \\ e \end{bmatrix}$$

or

$$\dot{\mathbf{x}} = (\mathbf{A} - \mathbf{B}\mathbf{K})\mathbf{x} + \mathbf{B}K_b e$$

The error equation is

$$\dot{e} = \left(A_{bb} - K_e A_{ab} \right)e$$

Hence the system dynamics are defined by

$$\begin{bmatrix} \dot{\mathbf{x}} \\ \dot{e} \end{bmatrix} = \begin{bmatrix} \mathbf{A} - \mathbf{B}\mathbf{K} & \mathbf{B}K_b \\ 0 & A_{bb} - K_e A_{ab} \end{bmatrix}\begin{bmatrix} \mathbf{x} \\ e \end{bmatrix}$$

Based on this last equation, MATLAB Program 10–32 produces the response to the given initial condition. The resulting response curves are shown in Figure 10–53.

MATLAB Program 10–32

```
% Response to intial condition ---- minimum-order observer

A = [0  1;0 -2];
B = [0;4];
K = [4  0.5];
Kb = 0.5;
Ke = 6;
Aab = 1; Abb = -2;
AA = [A-B*K B*Kb; zeros(1,2) Abb-Ke*Aab];
sys = ss(AA,eye(3),eye(3),eye(3));
t = 0:0.01:8;
x = initial(sys,[1;0;1],t);
x1 = [1 0 0]*x';
x2 = [0 1 0]*x';
e = [0 0 1]*x';

subplot(2,2,1); plot(t,x1); grid
xlabel('t (sec)'); ylabel('x1')

subplot(2,2,2); plot(t,x2); grid
xlabel('t (sec)'); ylabel('x2')

subplot(2,2,3); plot(t,e); grid
xlabel('t (sec)'); ylabel('e')
```

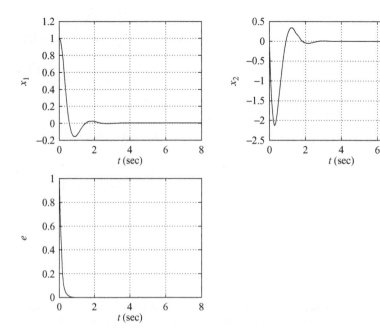

Figure 10–53
Response curves to initial condition.

The transfer function of the observer controller, when the system uses the minimum-order observer, can be obtained by use of MATLAB Program 10–33. The result is

$$\frac{num}{den} = \frac{7s + 32}{s + 10} = \frac{7(s + 4.5714)}{s + 10}$$

MATLAB Program 10–33

```
% Determination of transfer function of observer controller ---- minimum-order observer
A = [0  1;0 -2];
B = [0;4];
Aaa = 0; Aab = 1; Aba = 0; Abb = -2;
Ba = 0; Bb = 4;
Ka = 4; Kb = 0.5;
Ke = 6;
Ahat = Abb - Ke*Aab;
Bhat = Ahat*Ke + Aba - Ke*Aaa;
Fhat = Bb - Ke*Ba;
Atilde = Ahat - Fhat*Kb;
Btilde = Bhat - Fhat*(Ka + Kb*Ke);
Ctilde = -Kb;
Dtilde = -(Ka + Kb*Ke);
[num,den] = ss2tf(Atilde, Btilde, -Ctilde, -Dtilde)

num =

   7  32

den =

   1  10
```

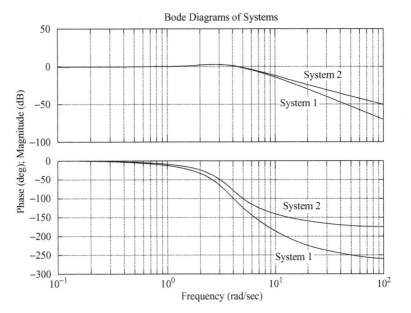

Figure 10–54
Bode diagrams of System 1
(system with full-order
observer) and System 2
(system with minimum-
order observer).
System 1 =
$(296s + 1024)/$
$(s^4 + 20s^3 + 144s^2$
$+ 512s + 1024);$
System 2 = $(28s + 128)/$
$(s^3 + 12s^2 + 48s + 128).$

The observer controller is clearly a lead compensator.

The Bode diagrams of System 1 (closed-loop system with full-order observer) and of System 2 (closed-loop system with minimum-order observer) are shown in Figure 10–54. Clearly, the bandwidth of System 2 is wider than that of System 1. System 1 has a better high-frequency noise-rejection characteristic than System 2.

A–10–15. Consider the system

$$\dot{\mathbf{x}} = \mathbf{A}\mathbf{x}$$

where $\mathbf{x}$ is a state vector (n-vector) and $\mathbf{A}$ is an $n \times n$ constant matrix. We assume that $\mathbf{A}$ is non-singular. Prove that if the equilibrium state $\mathbf{x} = \mathbf{0}$ of the system is asymptotically stable (that is, if $\mathbf{A}$ is a stable matrix), then there exists a positive-definite Hermitian matrix $\mathbf{P}$ such that

$$\mathbf{A}^*\mathbf{P} + \mathbf{P}\mathbf{A} = -\mathbf{Q}$$

where $\mathbf{Q}$ is a positive-definite Hermitian matrix.

Solution. The matrix differential equation.

$$\dot{\mathbf{X}} = \mathbf{A}^*\mathbf{X} + \mathbf{X}\mathbf{A}, \qquad \mathbf{X}(0) = \mathbf{Q}$$

has the solution

$$\mathbf{X} = e^{\mathbf{A}^*t}\mathbf{Q}e^{\mathbf{A}t}$$

Integrating both sides of this matrix differential equation from $t = 0$ to $t = \infty$, we obtain

$$\mathbf{X}(\infty) - \mathbf{X}(0) = \mathbf{A}^*\left(\int_0^\infty \mathbf{X}\, dt\right) + \left(\int_0^\infty \mathbf{X}\, dt\right)\mathbf{A}$$

Noting that $\mathbf{A}$ is a stable matrix and, therefore, $\mathbf{X}(\infty) = \mathbf{0}$, we obtain

$$-\mathbf{X}(0) = -\mathbf{Q} = \mathbf{A}^*\left(\int_0^\infty \mathbf{X}\, dt \right) + \left(\int_0^\infty \mathbf{X}\, dt \right)\mathbf{A}$$

Let us put

$$\mathbf{P} = \int_0^\infty \mathbf{X}\, dt = \int_0^\infty e^{\mathbf{A}^*t}\mathbf{Q}e^{\mathbf{A}t}\, dt$$

Note that the elements of $e^{\mathbf{A}t}$ are finite sums of terms like $e^{\lambda_i t}$, $te^{\lambda_i t} \ldots, t^{m_i-1}e^{\lambda_i t}$, where the λ_i are the eigenvalues of $\mathbf{A}$ and m_i is the multiplicity of λ_i. Since the λ_i possess negative real parts,

$$\int_0^\infty e^{\mathbf{A}^*t}\mathbf{Q}e^{\mathbf{A}t}\, dt$$

exists. Note that

$$\mathbf{P}^* = \int_0^\infty e^{\mathbf{A}^*t}\mathbf{Q}e^{\mathbf{A}t}\, dt = \mathbf{P}$$

Thus $\mathbf{P}$ is Hermitian (or symmetric if $\mathbf{P}$ is a real matrix). We have thus shown that for a stable $\mathbf{A}$ and for a positive-definite Hermitian matrix $\mathbf{Q}$, there exists a Hermitian matrix $\mathbf{P}$ such that $\mathbf{A}^*\mathbf{P} + \mathbf{P}\mathbf{A} = -\mathbf{Q}$. We now need to prove that $\mathbf{P}$ is positive definite. Consider the following Hermitian form:

$$\mathbf{x}^*\mathbf{P}\mathbf{x} = \mathbf{x}^* \int_0^\infty e^{\mathbf{A}^*t}\mathbf{Q}e^{\mathbf{A}t}\, dt\, \mathbf{x}$$

$$= \int_0^\infty (e^{\mathbf{A}t}\mathbf{x})^*\mathbf{Q}(e^{\mathbf{A}t}\mathbf{x})\, dt > 0, \qquad \text{for } \mathbf{x} \neq \mathbf{0}$$

$$= 0, \qquad \text{for } \mathbf{x} = \mathbf{0}$$

Hence, $\mathbf{P}$ is positive definite. This completes the proof.

A–10–16. Consider the control system described by

$$\dot{\mathbf{x}} = \mathbf{A}\mathbf{x} + \mathbf{B}u \tag{10–173}$$

where

$$\mathbf{A} = \begin{bmatrix} 0 & 1 \\ 0 & 0 \end{bmatrix}, \qquad \mathbf{B} = \begin{bmatrix} 0 \\ 1 \end{bmatrix}$$

Assuming the linear control law

$$u = -\mathbf{K}\mathbf{x} = -k_1 x_1 - k_2 x_2 \tag{10–174}$$

determine the constants k_1 and k_2 so that the following performance index is minimized:

$$J = \int_0^\infty \mathbf{x}^T\mathbf{x}\, dt$$

Consider only the case where the initial condition is

$$\mathbf{x}(0) = \begin{bmatrix} c \\ 0 \end{bmatrix}$$

Choose the undamped natural frequency to be 2 rad/sec.

Solution. Substituting Equation (10–174) into Equation (10–173), we obtain

$$\dot{\mathbf{x}} = \mathbf{Ax} - \mathbf{BKx}$$

or

$$\begin{bmatrix} \dot{x}_1 \\ \dot{x}_2 \end{bmatrix} = \begin{bmatrix} 0 & 1 \\ 0 & 0 \end{bmatrix}\begin{bmatrix} x_1 \\ x_2 \end{bmatrix} + \begin{bmatrix} 0 \\ 1 \end{bmatrix}[-k_1 x_1 - k_2 x_2]$$

$$= \begin{bmatrix} 0 & 1 \\ -k_1 & -k_2 \end{bmatrix}\begin{bmatrix} x_1 \\ x_2 \end{bmatrix} \tag{10–175}$$

Thus,

$$\mathbf{A} - \mathbf{BK} = \begin{bmatrix} 0 & 1 \\ -k_1 & -k_2 \end{bmatrix}$$

Elimination of x_2 from Equation (10–175) yields

$$\ddot{x}_1 + k_2 \dot{x}_1 + k_1 x_1 = 0$$

Since the undamped natural frequency is specified as 2 rad/sec, we obtain

$$k_1 = 4$$

Therefore,

$$\mathbf{A} - \mathbf{BK} = \begin{bmatrix} 0 & 1 \\ -4 & -k_2 \end{bmatrix}$$

$\mathbf{A} - \mathbf{BK}$ is a stable matrix if $k_2 > 0$. Our problem now is to determine the value of k_2 so that the performance index

$$J = \int_0^\infty \mathbf{x}^T \mathbf{x}\, dt = \mathbf{x}^T(0)\mathbf{P}(0)\mathbf{x}(0)$$

is minimized, where the matrix $\mathbf{P}$ is determined from Equation (10–115), rewritten

$$(\mathbf{A} - \mathbf{BK})^*\mathbf{P} + \mathbf{P}(\mathbf{A} - \mathbf{BK}) = -(\mathbf{Q} + \mathbf{K}^*\mathbf{RK})$$

Since in this system $\mathbf{Q} = \mathbf{I}$ and $\mathbf{R} = \mathbf{0}$, this last equation can be simplified to

$$(\mathbf{A} - \mathbf{BK})^*\mathbf{P} + \mathbf{P}(\mathbf{A} - \mathbf{BK}) = -\mathbf{I} \tag{10–176}$$

Since the system involves only real vectors and real matrices, $\mathbf{P}$ becomes a real symmetric matrix. Then Equation (10–176) can be written as

$$\begin{bmatrix} 0 & -4 \\ 1 & -k_2 \end{bmatrix}\begin{bmatrix} p_{11} & p_{12} \\ p_{12} & p_{22} \end{bmatrix} + \begin{bmatrix} p_{11} & p_{12} \\ p_{12} & p_{22} \end{bmatrix}\begin{bmatrix} 0 & 1 \\ -4 & -k_2 \end{bmatrix} = \begin{bmatrix} -1 & 0 \\ 0 & -1 \end{bmatrix}$$

Solving for the matrix $\mathbf{P}$, we obtain

$$\mathbf{P} = \begin{bmatrix} p_{11} & p_{12} \\ p_{12} & p_{22} \end{bmatrix} = \begin{bmatrix} \dfrac{5}{2k_2} + \dfrac{k_2}{8} & \dfrac{1}{8} \\ \dfrac{1}{8} & \dfrac{5}{8k_2} \end{bmatrix}$$

The performance index is then

$$J = \mathbf{x}^T(0)\mathbf{P}\mathbf{x}(0)$$

$$= \begin{bmatrix} c & 0 \end{bmatrix}\begin{bmatrix} p_{11} & p_{12} \\ p_{12} & p_{22} \end{bmatrix}\begin{bmatrix} c \\ 0 \end{bmatrix} = p_{11}c^2$$

$$= \left(\frac{5}{2k_2} + \frac{k_2}{8}\right)c^2 \tag{10–177}$$

To minimize J, we differentiate J with respect to k_2 and set $\partial J/\partial k_2$ equal to zero as follows:

$$\frac{\partial J}{\partial k_2} = \left(\frac{-5}{2k_2^2} + \frac{1}{8}\right)c^2 = 0$$

Hence,

$$k_2 = \sqrt{20}$$

With this value of k_2, we have $\partial^2 J/\partial k_2^2 > 0$. Thus, the minimum value of J is obtained by substituting $k_2 = \sqrt{20}$ into Equation (10–177), or

$$J_{\min} = \frac{\sqrt{5}}{2}c^2$$

The designed system has the control law

$$u = -4x_1 - \sqrt{20}x_2$$

The designed system is optimal in that it results in a minimum value for the performance index J under the assumed initial condition.

A–10–17. Consider the same inverted-pendulum system as discussed in Example 10–5. The system is shown in Figure 10–8, where $M = 2$ kg, $m = 0.1$ kg, and $l = 0.5$ m. The block diagram for the system is shown in Figure 10–9. The system equations are given by

$$\dot{\mathbf{x}} = \mathbf{A}\mathbf{x} + \mathbf{B}u$$

$$y = \mathbf{C}\mathbf{x}$$

$$u = -\mathbf{K}\mathbf{x} + k_I\xi$$

$$\dot{\xi} = r - y = r - \mathbf{C}\mathbf{x}$$

where

$$A = \begin{bmatrix} 0 & 1 & 0 & 0 \\ 20.601 & 0 & 0 & 0 \\ 0 & 0 & 0 & 1 \\ -0.4905 & 0 & 0 & 0 \end{bmatrix}, \quad B = \begin{bmatrix} 0 \\ -1 \\ 0 \\ 0.5 \end{bmatrix}, \quad C = \begin{bmatrix} 0 & 0 & 1 & 0 \end{bmatrix}$$

Referring to Equation (10–51), the error equation for the system is given by

$$\dot{\mathbf{e}} = \hat{\mathbf{A}}\mathbf{e} + \hat{\mathbf{B}}u_e$$

where

$$\hat{\mathbf{A}} = \begin{bmatrix} \mathbf{A} & \mathbf{0} \\ -\mathbf{C} & 0 \end{bmatrix} = \begin{bmatrix} 0 & 1 & 0 & 0 & 0 \\ 20.601 & 0 & 0 & 0 & 0 \\ 0 & 0 & 0 & 1 & 0 \\ -0.4905 & 0 & 0 & 0 & 0 \\ 0 & 0 & -1 & 0 & 0 \end{bmatrix}, \quad \hat{\mathbf{B}} = \begin{bmatrix} \mathbf{B} \\ 0 \end{bmatrix} = \begin{bmatrix} 0 \\ -1 \\ 0 \\ 0.5 \\ 0 \end{bmatrix}$$

and the control signal is given by Equation (10–41):

$$u_e = -\hat{\mathbf{K}}\mathbf{e}$$

where

$$\hat{\mathbf{K}} = \begin{bmatrix} \mathbf{K} & \vdots & -k_I \end{bmatrix} = \begin{bmatrix} k_1 & k_2 & k_3 & k_4 & \vdots & -k_I \end{bmatrix}$$

$$\mathbf{e} = \begin{bmatrix} \mathbf{x}_e \\ \xi_e \end{bmatrix} = \begin{bmatrix} \mathbf{x}(t) - \mathbf{x}(\infty) \\ \xi(t) - \xi(\infty) \end{bmatrix}$$

$$\mathbf{x} = \begin{bmatrix} x_1 \\ x_2 \\ x_3 \\ x_4 \end{bmatrix} = \begin{bmatrix} \theta \\ \dot{\theta} \\ x \\ \dot{x} \end{bmatrix}$$

Using MATLAB, determine the state feedback gain matrix $\hat{\mathbf{K}}$ such that the following performance index J is minimized:

$$J = \int_0^\infty (\mathbf{e}^* \mathbf{Q}\mathbf{e} + u^* Ru) \, dt$$

where

$$\mathbf{Q} = \begin{bmatrix} 100 & 0 & 0 & 0 & 0 \\ 0 & 1 & 0 & 0 & 0 \\ 0 & 0 & 1 & 0 & 0 \\ 0 & 0 & 0 & 1 & 0 \\ 0 & 0 & 0 & 0 & 1 \end{bmatrix}, \qquad R = 0.01$$

Obtain the unit-step response of the system designed.

Solution. A MATLAB program to determine $\hat{\mathbf{K}}$ is given in MATLAB Program 10–34. The result is

$$k_1 = -188.0799, \qquad k_2 = -37.0738, \qquad k_3 = -26.6767, \qquad k_4 = -30.5824, \qquad k_I = -10.0000$$

MATLAB Program 10–34

```
% Design of quadratic optimal control system

A = [0 1 0 0;20.601 0 0 0;0 0 0 1;-0.4905 0 0 0];
B = [0;-1;0;0.5];
C = [0 0 1 0];
D = [0];
Ahat = [A zeros(4,1);-C 0];
Bhat = [B;0];
Q = [100 0 0 0 0;0 1 0 0 0;0 0 1 0 0;0 0 0 1 0;0 0 0 0 1];
R = [0.01];
Khat = lqr(Ahat,Bhat,Q,R)

Khat =

   -188.0799  -37.0738  -26.6767  -30.5824  10.0000
```

Unit-Step Response. Once we have determined the feedback gain matrix $\mathbf{K}$ and the integral gain constant k_I, we can determine the unit-step response of the designed system. The system equation is

$$\begin{bmatrix} \dot{\mathbf{x}} \\ \dot{\xi} \end{bmatrix} = \begin{bmatrix} \mathbf{A} & \mathbf{0} \\ -\mathbf{C} & 0 \end{bmatrix} \begin{bmatrix} \mathbf{x} \\ \xi \end{bmatrix} + \begin{bmatrix} \mathbf{B} \\ 0 \end{bmatrix} u + \begin{bmatrix} 0 \\ 1 \end{bmatrix} r \tag{10–178}$$

[Refer to Equation (10–35).] Since

$$u = -\mathbf{K}\mathbf{x} + k_I\xi$$

Equation (10–178) can be written as follows:

$$\begin{bmatrix} \dot{\mathbf{x}} \\ \dot{\xi} \end{bmatrix} = \begin{bmatrix} \mathbf{A} - \mathbf{B}\mathbf{K} & \mathbf{B}k_I \\ -\mathbf{C} & 0 \end{bmatrix} \begin{bmatrix} \mathbf{x} \\ \xi \end{bmatrix} + \begin{bmatrix} 0 \\ 1 \end{bmatrix} r \tag{10–179}$$

The output equation is

$$y = \begin{bmatrix} \mathbf{C} & 0 \end{bmatrix} \begin{bmatrix} \mathbf{x} \\ \xi \end{bmatrix} + \begin{bmatrix} 0 \end{bmatrix} r$$

MATLAB Program 10–35 gives the unit-step response of the system given by Equation (10–179). The resulting response curves are presented in Figure 10–55. It shows response curves $\theta\left[= x_1(t)\right]$ versus t, $\dot{\theta}\left[= x_2(t)\right]$ versus t, $y\left[= x_3(t)\right]$ versus t, $\dot{y}\left[= x_4(t)\right]$ versus t, and $\xi\left[= x_5(t)\right]$ versus t, where the input $r(t)$ to the cart is a unit-step function $\left[r(t) = 1 \text{ m}\right]$. All initial conditions are set equal to zero. Figure 10–56 is an enlarged version of the cart position $y\left[= x_3(t)\right]$ versus t. The cart moves backward a very small amount for the first 0.6 sec or so. (Notice that the cart velocity is negative for the first 0.4 sec.) This is due to the fact that the inverted-pendulum-on-the-cart system is a nonminimum-phase system.

MATLAB Program 10–35

```
% Unit-step response

A = [0 1 0 0;20.601 0 0 0;0 0 0 1;-0.4905 0 0 0];
B = [0;-1;0;0.5];
C = [0 0 1 0];
D = [0];
K = [-188.0799 -37.0738 -26.6767 -30.5824];
kl = -10.0000;
AA = [A-B*K  B*kl; -C  0];
BB = [0;0;0;0;1];
CC= [C  0];
DD = D;
t = 0:0.01:10;
[y,x,t] = step(AA,BB,CC,DD,1,t);
x1 = [1 0 0 0 0]*x';
x2 = [0 1 0 0 0]*x';
x3 = [0 0 1 0 0]*x';
x4 = [0 0 0 1 0]*x';
x5 = [0 0 0 0 1]*x';

subplot(3,2,1); plot(t,x1); grid;
xlabel('t (sec)'); ylabel('x1')

subplot(3,2,2); plot(t,x2); grid;
xlabel('t (sec)'); ylabel('x2')

subplot(3,2,3); plot(t,x3); grid;
xlabel('t (sec)'); ylabel('x3')

subplot(3,2,4); plot(t,x4); grid;
xlabel('t (sec)'); ylabel('x4')

subplot(3,2,5); plot(t,x5); grid;
xlabel('t (sec)'); ylabel('x5')
```

Comparing the step-response characteristics of this system with those of Example 10–5, we notice that the response of the present system is less oscillatory and exhibits less maximum overshoot in the position response $\left(x_3 \text{ versus } t\right)$. The system designed by use of the quadratic optimal regulator approach generally gives such characteristics—less oscillatory and well damped.

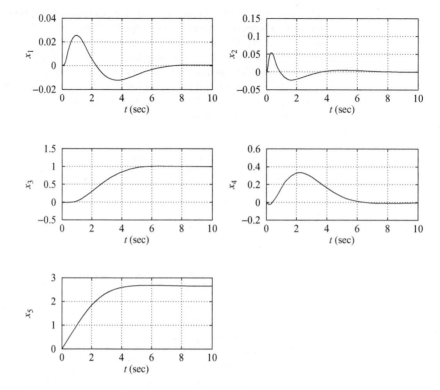

Figure 10–55
Response curves to a
unit-step input.

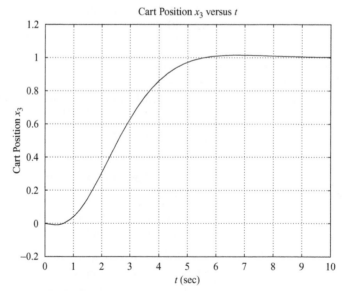

Figure 10–56
Cart position versus t
curve.

A–10–18. Consider the stability of a system with unstructured additive uncertainty as shown in Figure
10–57(a). Define

$\tilde{G}$ = true plant dynamics
G = model of plant dynamics
Δ_a = unstructured additive uncertainty

Chapter 10 / **Control Systems Design in State Space**

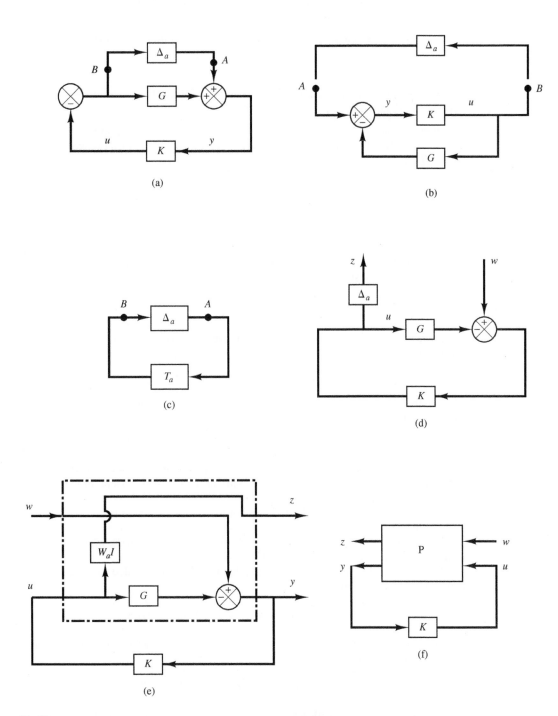

Figure 10–57
(a) Block diagram of a system with unstructured additive uncertainty;
(b)–(d) successive modifications of the block diagram of (a);
(e) block diagram showing a generalized plant with unstructured additive uncertainty;
(f) generalized plant diagram.

Example Problems and Solutions

853

Assume that Δ_a is stable and its upper bound is known. Assume also that $\tilde{G}$ and G are related by

$$\tilde{G} = G + \Delta_a$$

Obtain the condition that the controller K must satisfy for robust stability. Also, obtain a generalized plant diagram for this system.

Solution. Let us obtain the transfer function between point A and point B in Figure 10–57(a). Redrawing Figure 10–57(a), we obtain Figure 10–57(b). Then the transfer function between points A and B can be obtained as

$$\frac{K}{1 + GK} = K(1 + GK)^{-1}$$

Define

$$K(1 + GK)^{-1} = T_a$$

Then Figure 10–57(b) can be redrawn as Figure 10–57(c). By using the small-gain theorem, the condition for the robust stability of the closed-loop system can be obtained as

$$\|\Delta_a T_a\|_\infty < 1 \tag{10–180}$$

Since it is impossible to model Δ_a precisely, we need to find a scalar transfer function $W_a(j\omega)$ such that

$$\bar{\sigma}\{\Delta_a(j\omega)\} < |W_a(j\omega)| \quad \text{for all } \omega$$

and use this $W_a(j\omega)$ instead of Δ_a. Then, the condition for the robust stability of the closed-loop system can be given by

$$\|W_a T_a\|_\infty < 1 \tag{10–181}$$

If Inequality (10–181) holds true, then it is evident that Inequality (10–180) also holds true. So this is the condition to guarantee the robust stability of the designed system. In Figure 10–57(e), Δ_a in Figure 10–57(d) was replaced by $W_a I$.

To summarize, if we make the H_∞ norm of the transfer function from w to z to be less than 1, the controller K that satisfies Inequality (10–181) can be determined.

Figure 10–57(e) can be redrawn as that shown in Figure 10–57(f), which is the generalized plant diagram for the system considered.

Note that for this problem the Φ matrix that relates the controlled variable z and the exogenous disturbance w is given by

$$z = \Phi(s)w = (W_a T_a)w = [W_a K(I + GK)^{-1}]w$$

Noting that $u(s) = K(s)y(s)$ and referring to Equation (10–128), $\Phi(s)$ is given by the elements of the P matrix as follows:

$$\Phi(s) = P_{11} + P_{12}K(I - P_{22}K)^{-1}P_{21}$$

To make this $\Phi(s)$ equal to $W_a K(I + GK)^{-1}$, we may choose $P_{11} = 0$, $P_{12} = W_a$, $P_{21} = I$, and $P_{22} = -G$. Then, the P matrix for this problem can be obtained as

$$P = \begin{bmatrix} 0 & W_a \\ I & -G \end{bmatrix}$$

B–10–1. Consider the system defined by

$$\dot{x} = Ax + Bu$$

$$y = Cx$$

where

$$A = \begin{bmatrix} -1 & 0 & 1 \\ 1 & -2 & 0 \\ 0 & 0 & -3 \end{bmatrix}, \quad B = \begin{bmatrix} 0 \\ 0 \\ 1 \end{bmatrix}, \quad C = \begin{bmatrix} 1 & 1 & 0 \end{bmatrix}$$

Transform the system equations into (a) controllable canonical form and (b) observable canonical form.

B–10–2. Consider the system defined by

$$\dot{x} = Ax + Bu$$

$$y = Cx$$

where

$$A = \begin{bmatrix} -1 & 0 & 1 \\ 1 & -2 & 0 \\ 0 & 0 & -3 \end{bmatrix}, \quad B = \begin{bmatrix} 0 \\ 1 \\ 1 \end{bmatrix}, \quad C = \begin{bmatrix} 1 & 1 & 1 \end{bmatrix}$$

Transform the system equations into the observable canonical form.

B–10–3. Consider the system defined by

$$\dot{x} = Ax + Bu$$

where

$$A = \begin{bmatrix} 0 & 1 & 0 \\ 0 & 0 & 1 \\ -1 & -5 & -6 \end{bmatrix}, \quad B = \begin{bmatrix} 0 \\ 1 \\ 1 \end{bmatrix}$$

By using the state-feedback control $u = -Kx$, it is desired to have the closed-loop poles at $s = -2 \pm j4$, $s = -10$. Determine the state-feedback gain matrix K.

B–10–4. Solve Problem B–10–3 with MATLAB.

B–10–5. Consider the system defined by

$$\begin{bmatrix} \dot{x}_1 \\ \dot{x}_2 \end{bmatrix} = \begin{bmatrix} 0 & 1 \\ 0 & 2 \end{bmatrix} \begin{bmatrix} x_1 \\ x_2 \end{bmatrix} + \begin{bmatrix} 1 \\ 0 \end{bmatrix} u$$

Show that this system cannot be stabilized by the state-feedback control $u = -Kx$, whatever matrix K is chosen.

B–10–6. A regulator system has a plant

$$\frac{Y(s)}{U(s)} = \frac{10}{(s + 1)(s + 2)(s + 3)}$$

Define state variables as

$$x_1 = y$$
$$x_2 = \dot{x}_1$$
$$x_3 = \dot{x}_2$$

By use of the state-feedback control $u = -Kx$, it is desired to place the closed-loop poles at

$$s = -2 + j2\sqrt{3}, \qquad s = -2 - j2\sqrt{3}, \qquad s = -10$$

Determine the necessary state-feedback gain matrix K.

B–10–7. Solve Problem B–10–6 with MATLAB.

B–10–8. Consider the type 1 servo system shown in Figure 10–58. Matrices A, B, and C in Figure 10–58 are given by

$$A = \begin{bmatrix} 0 & 1 & 0 \\ 0 & 0 & 1 \\ 0 & -5 & -6 \end{bmatrix}, \quad B = \begin{bmatrix} 0 \\ 0 \\ 1 \end{bmatrix}, \quad C = \begin{bmatrix} 1 & 0 & 0 \end{bmatrix}$$

Determine the feedback gain constants k_1, k_2, and k_3 such that the closed-loop poles are located at

$$s = -2 + j4, \qquad s = -2 - j4, \qquad s = -10$$

Obtain the unit-step response and plot the output $y(t)$-versus-t curve.

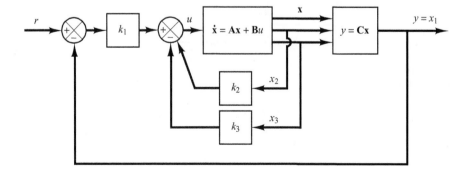

Figure 10–58
Type 1 servo system.

B–10–9. Consider the inverted-pendulum system shown in Figure 10–59. Assume that

$$M = 2 \text{ kg}, \quad m = 0.5 \text{ kg}, \quad l = 1 \text{ m}$$

Define state variables as

$$x_1 = \theta, \quad x_2 = \dot{\theta}, \quad x_3 = x, \quad x_4 = \dot{x}$$

and output variables as

$$y_1 = \theta = x_1, \quad y_2 = x = x_3$$

Derive the state-space equations for this system.

It is desired to have closed-loop poles at

$$s = -4 + j4, \quad s = -4 - j4, \quad s = -20, \quad s = -20$$

Determine the state-feedback gain matrix **K**.

Using the state-feedback gain matrix **K** thus determined, examine the performance of the system by computer simulation. Write a MATLAB program to obtain the response of the system to an arbitrary initial condition. Obtain the response curves $x_1(t)$ versus t, $x_2(t)$ versus t, $x_3(t)$ versus t, and $x_4(t)$ versus t for the following set of initial condition:

$$x_1(0) = 0, \quad x_2(0) = 0, \quad x_3(0) = 0, \quad x_4(0) = 1 \text{ m/s}$$

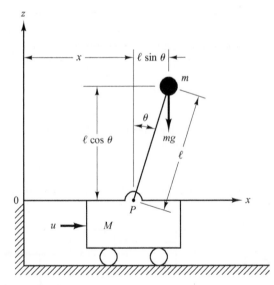

Figure 10–59
Inverted-pendulum system.

B–10–10. Consider the system defined by

$$\dot{x} = Ax$$

$$y = Cx$$

where

$$A = \begin{bmatrix} -1 & 1 \\ 1 & -2 \end{bmatrix}, \quad C = [1 \quad 0]$$

Design a full-order state observer. The desired observer poles are $s = -5$ and $s = -5$.

B–10–11. Consider the system defined by

$$\dot{x} = Ax + Bu$$

$$y = Cx$$

where

$$A = \begin{bmatrix} 0 & 1 & 0 \\ 0 & 0 & 1 \\ -5 & -6 & 0 \end{bmatrix}, \quad B = \begin{bmatrix} 0 \\ 0 \\ 1 \end{bmatrix}, \quad C = [1 \quad 0 \quad 0]$$

Design a full-order state observer, assuming that the desired poles for the observer are located at

$$s = -10, \quad s = -10, \quad s = -15$$

B–10–12. Consider the system defined by

$$\begin{bmatrix} \dot{x}_1 \\ \dot{x}_2 \\ \dot{x}_3 \end{bmatrix} = \begin{bmatrix} 0 & 1 & 0 \\ 0 & 0 & 1 \\ 1.244 & 0.3956 & -3.145 \end{bmatrix} \begin{bmatrix} x_1 \\ x_2 \\ x_3 \end{bmatrix}$$

$$+ \begin{bmatrix} 0 \\ 0 \\ 1.244 \end{bmatrix} u$$

$$y = [1 \quad 0 \quad 0] \begin{bmatrix} x_1 \\ x_2 \\ x_3 \end{bmatrix}$$

Given the set of desired poles for the observer to be

$$s = -5 + j5\sqrt{3}, \quad s = -5 - j5\sqrt{3}, \quad s = -10$$

design a full-order observer.

B–10–13. Consider the double integrator system defined by

$$\ddot{y} = u$$

If we choose the state variables as

$$x_1 = y$$

$$x_2 = \dot{y}$$

then the state-space representation for the system becomes as follows:

$$\begin{bmatrix} \dot{x}_1 \\ \dot{x}_2 \end{bmatrix} = \begin{bmatrix} 0 & 1 \\ 0 & 0 \end{bmatrix} \begin{bmatrix} x_1 \\ x_2 \end{bmatrix} + \begin{bmatrix} 0 \\ 1 \end{bmatrix} u$$

$$y = [1 \quad 0] \begin{bmatrix} x_1 \\ x_2 \end{bmatrix}$$

It is desired to design a regulator for this system. Using the pole-placement-with-observer approach, design an observer controller.

Choose the desired closed-loop poles for the pole-placement part to be

$$s = -0.7071 + j0.7071, \qquad s = -0.7071 - j0.7071$$

and assuming that we use a minimum-order observer, choose the desired observer pole at

$$s = -5$$

B–10–14. Consider the system

$$\dot{x} = Ax + Bu$$

$$y = Cx$$

where

$$A = \begin{bmatrix} 0 & 1 & 0 \\ 0 & 0 & 1 \\ -6 & -11 & -6 \end{bmatrix}, \quad B = \begin{bmatrix} 0 \\ 0 \\ 1 \end{bmatrix}, \quad C = \begin{bmatrix} 1 & 0 & 0 \end{bmatrix}$$

Design a regulator system by the pole-placement-with-observer approach. Assume that the desired closed-loop poles for pole placement are located at

$$s = -1 + j, \qquad s = -1 - j, \qquad s = -5$$

The desired observer poles are located at

$$s = -6, \qquad s = -6, \qquad s = -6$$

Also, obtain the transfer function of the observer controller.

B–10–15. Using the pole-placement-with-observer approach, design observer controllers (one with a full-order observer and the other with a minimum-order observer) for the system shown in Figure 10–60. The desired closed-loop poles for the pole-placement part are

$$s = -1 + j2, \qquad s = -1 - j2, \qquad s = -5$$

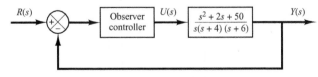

Figure 10–60
Control system with observer controller in the feedforward path.

The desired observer poles are

$$s = -10, \quad s = -10, \quad s = -10 \quad \text{for the full-order observer}$$

$$s = -10, \quad s = -10 \quad \text{for the minimum-order observer.}$$

Compare the unit-step responses of the designed systems. Compare also the bandwidths of both systems.

B–10–16. Using the pole-placement-with-observer approach, design the control systems shown in Figures 10–61(a) and (b). Assume that the desired closed-loop poles for the pole placement are located at

$$s = -2 + j2, \qquad s = -2 - j2$$

and the desired observer poles are located at

$$s = -8, \qquad s = -8$$

Obtain the transfer function of the observer controller. Compare the unit-step responses of both systems. [In System (b), determine the constant N so that the steady-state output $y(\infty)$ is unity when the input is a unit-step input.]

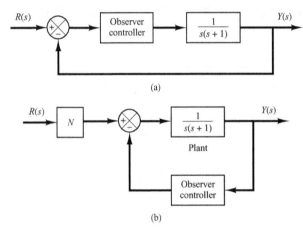

Figure 10–61
Control systems with observer controller: (a) observer controller in the feedforward path; (b) observer controller in the feedback path.

B–10–17. Consider the system defined by

$$\dot{x} = Ax$$

where

$$A = \begin{bmatrix} 0 & 1 & 0 \\ 0 & 0 & 1 \\ -1 & -2 & -a \end{bmatrix}$$

$$a = \text{adjustable parameter} > 0$$

Determine the value of the parameter a so as to minimize the following performance index:

$$J = \int_0^\infty \mathbf{x}^T \mathbf{x}\, dt$$

Assume that the initial state $\mathbf{x}(0)$ is given by

$$\mathbf{x}(0) = \begin{bmatrix} c_1 \\ 0 \\ 0 \end{bmatrix}$$

B–10–18. Consider the system shown in Figure 10–62. Determine the value of the gain K so that the damping ratio ζ of the closed-loop system is equal to 0.5. Then determine also the undamped natural frequency ω_n of the closed-loop system. Assuming that $e(0) = 1$ and $\dot{e}(0) = 0$, evaluate

$$\int_0^\infty e^2(t)\, dt$$

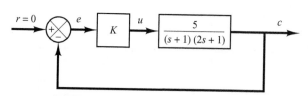

Figure 10–62
Control system.

B–10–19. Determine the optimal control signal u for the system defined by

$$\dot{\mathbf{x}} = \mathbf{A}\mathbf{x} + \mathbf{B}u$$

where

$$\mathbf{A} = \begin{bmatrix} 0 & 1 \\ 0 & -1 \end{bmatrix}, \qquad \mathbf{B} = \begin{bmatrix} 0 \\ 1 \end{bmatrix}$$

such that the following performance index is minimized:

$$J = \int_0^\infty (\mathbf{x}^T \mathbf{x} + u^2)\, dt$$

B–10–20. Consider the system

$$\begin{bmatrix} \dot{x}_1 \\ \dot{x}_2 \end{bmatrix} = \begin{bmatrix} 0 & 1 \\ 0 & 0 \end{bmatrix}\begin{bmatrix} x_1 \\ x_2 \end{bmatrix} + \begin{bmatrix} 0 \\ 1 \end{bmatrix}u$$

It is desired to find the optimal control signal u such that the performance index

$$J = \int_0^\infty (\mathbf{x}^T \mathbf{Q}\mathbf{x} + u^2)\, dt, \qquad \mathbf{Q} = \begin{bmatrix} 1 & 0 \\ 0 & \mu \end{bmatrix}$$

is minimized. Determine the optimal signal $u(t)$.

B–10–21. Consider the inverted-pendulum system shown in Figure 10–59. It is desired to design a regulator system that will maintain the inverted pendulum in a vertical position in the presence of disturbances in terms of angle θ and/or angular velocity $\dot{\theta}$. The regulator system is required to return the cart to its reference position at the end of each control process. (There is no reference input to the cart.)

The state-space equation for the system is given by

$$\dot{\mathbf{x}} = \mathbf{A}\mathbf{x} + \mathbf{B}u$$

where

$$\mathbf{A} = \begin{bmatrix} 0 & 1 & 0 & 0 \\ 20.601 & 0 & 0 & 0 \\ 0 & 0 & 0 & 1 \\ -0.4905 & 0 & 0 & 0 \end{bmatrix}$$

$$\mathbf{B} = \begin{bmatrix} 0 \\ -1 \\ 0 \\ 0.5 \end{bmatrix}, \qquad \mathbf{x} = \begin{bmatrix} \theta \\ \dot{\theta} \\ x \\ \dot{x} \end{bmatrix}$$

We shall use the state-feedback control scheme

$$u = -\mathbf{K}\mathbf{x}$$

Using MATLAB, determine the state-feedback gain matrix $\mathbf{K} = \begin{bmatrix} k_1 & k_2 & k_3 & k_4 \end{bmatrix}$ such that the following performance index J is minimized:

$$J = \int_0^\infty (\mathbf{x}^* \mathbf{Q}\mathbf{x} + u^* R u)\, dt$$

where

$$\mathbf{Q} = \begin{bmatrix} 100 & 0 & 0 & 0 \\ 0 & 1 & 0 & 0 \\ 0 & 0 & 1 & 0 \\ 0 & 0 & 0 & 1 \end{bmatrix}, \qquad R = 1$$

Then obtain the system response to the following initial condition:

$$\begin{bmatrix} x_1(0) \\ x_2(0) \\ x_3(0) \\ x_4(0) \end{bmatrix} = \begin{bmatrix} 0.1 \\ 0 \\ 0 \\ 0 \end{bmatrix}$$

Plot response curves θ versus t, $\dot{\theta}$ versus t, x versus t, and $\dot{x}$ versus t.

Appendix

Laplace Transform Tables

Appendix A first presents the complex variable and complex function. Then it presents tables of Laplace transform pairs and properties of Laplace transforms. Finally, it presents frequently used Laplace transform theorems and Laplace transforms of pulse function and impulse function.

Complex Variable.   A complex number has a real part and an imaginary part, both of which are constant. If the real part and/or imaginary part are variables, a complex quantity is called a *complex variable*. In the Laplace transformation we use the notation s as a complex variable; that is,

$$s = \sigma + j\omega$$

where σ is the real part and ω is the imaginary part.

Complex Function.   A complex function $G(s)$, a function of s, has a real part and an imaginary part or

$$G(s) = G_x + jG_y$$

where G_x and G_y are real quantities. The magnitude of $G(s)$ is $\sqrt{G_x^2 + G_y^2}$, and the angle θ of $G(s)$ is $\tan^{-1}(G_y/G_x)$. The angle is measured counterclockwise from the positive real axis. The complex conjugate of $G(s)$ is $\bar{G}(s) = G_x - jG_y$.

Complex functions commonly encountered in linear control systems analysis are single-valued functions of s and are uniquely determined for a given value of s.

A complex function $G(s)$ is said to be *analytic* in a region if $G(s)$ and all its derivatives exist in that region. The derivative of an analytic function $G(s)$ is given by

$$\frac{d}{ds} G(s) = \lim_{\Delta s \to 0} \frac{G(s + \Delta s) - G(s)}{\Delta s} = \lim_{\Delta s \to 0} \frac{\Delta G}{\Delta s}$$

Since $\Delta s = \Delta\sigma + j\Delta\omega$, Δs can approach zero along an infinite number of different paths. It can be shown, but is stated without a proof here, that if the derivatives taken along two particular paths, that is, $\Delta s = \Delta\sigma$ and $\Delta s = j\Delta\omega$, are equal, then the derivative is unique for any other path $\Delta s = \Delta\sigma + j\Delta\omega$ and so the derivative exists.

For a particular path $\Delta s = \Delta\sigma$ (which means that the path is parallel to the real axis),

$$\frac{d}{ds} G(s) = \lim_{\Delta\sigma \to 0} \left(\frac{\Delta G_x}{\Delta\sigma} + j \frac{\Delta G_y}{\Delta\sigma} \right) = \frac{\partial G_x}{\partial\sigma} + j \frac{\partial G_y}{\partial\sigma}$$

For another particular path $\Delta s = j\Delta\omega$ (which means that the path is parallel to the imaginary axis),

$$\frac{d}{ds} G(s) = \lim_{j\Delta\omega \to 0} \left(\frac{\Delta G_x}{j\Delta\omega} + j \frac{\Delta G_y}{j\Delta\omega} \right) = -j \frac{\partial G_x}{\partial\omega} + \frac{\Delta G_y}{\partial\omega}$$

If these two values of the derivative are equal,

$$\frac{\partial G_x}{\partial\sigma} + j \frac{\partial G_y}{\partial\sigma} = \frac{\partial G_y}{\partial\omega} - j \frac{\partial G_x}{\partial\omega}$$

or if the following two conditions

$$\frac{\partial G_x}{\partial\sigma} = \frac{\partial G_y}{\partial\omega} \quad \text{and} \quad \frac{\partial G_y}{\partial\sigma} = -\frac{\partial G_x}{\partial\omega}$$

are satisfied, then the derivative $dG(s)/ds$ is uniquely determined. These two conditions are known as the Cauchy–Riemann conditions. If these conditions are satisfied, the function $G(s)$ is analytic.

As an example, consider the following $G(s)$:

$$G(s) = \frac{1}{s + 1}$$

Then

$$G(\sigma + j\omega) = \frac{1}{\sigma + j\omega + 1} = G_x + jG_y$$

where

$$G_x = \frac{\sigma + 1}{(\sigma + 1)^2 + \omega^2} \quad \text{and} \quad G_y = \frac{-\omega}{(\sigma + 1)^2 + \omega^2}$$

It can be seen that, except at $s = -1$ (that is, $\sigma = -1$, $\omega = 0$), $G(s)$ satisfies the Cauchy–Riemann conditions:

$$\frac{\partial G_x}{\partial \sigma} = \frac{\partial G_y}{\partial \omega} = \frac{\omega^2 - (\sigma + 1)^2}{\left[(\sigma + 1)^2 + \omega^2\right]^2}$$

$$\frac{\partial G_y}{\partial \sigma} = -\frac{\partial G_x}{\partial \omega} = \frac{2\omega(\sigma + 1)}{\left[(\sigma + 1)^2 + \omega^2\right]^2}$$

Hence $G(s) = 1/(s + 1)$ is analytic in the entire s plane except at $s = -1$. The derivative $dG(s)/ds$, except at $s = 1$, is found to be

$$\frac{d}{ds}G(s) = \frac{\partial G_x}{\partial \sigma} + j\frac{\partial G_y}{\partial \sigma} = \frac{\partial G_y}{\partial \omega} - j\frac{\partial G_x}{\partial \omega}$$

$$= -\frac{1}{(\sigma + j\omega + 1)^2} = -\frac{1}{(s + 1)^2}$$

Note that the derivative of an analytic function can be obtained simply by differentiating $G(s)$ with respect to s. In this example,

$$\frac{d}{ds}\left(\frac{1}{s + 1}\right) = -\frac{1}{(s + 1)^2}$$

Points in the s plane at which the function $G(s)$ is analytic are called *ordinary* points, while points in the s plane at which the function $G(s)$ is not analytic are called *singular* points. Singular points at which the function $G(s)$ or its derivatives approach infinity are called *poles*. Singular points at which the function $G(s)$ equals zero are called *zeros*.

If $G(s)$ approaches infinity as s approaches $-p$ and if the function

$$G(s)(s + p)^n, \quad \text{for } n = 1, 2, 3, \dots$$

has a finite, nonzero value at $s = -p$, then $s = -p$ is called a pole of order n. If $n = 1$, the pole is called a simple pole. If $n = 2, 3, \dots$, the pole is called a second-order pole, a third-order pole, and so on.

To illustrate, consider the complex function

$$G(s) = \frac{K(s + 2)(s + 10)}{s(s + 1)(s + 5)(s + 15)^2}$$

$G(s)$ has zeros at $s = -2$, $s = -10$, simple poles at $s = 0$, $s = -1$, $s = -5$, and a double pole (multiple pole of order 2) at $s = -15$. Note that $G(s)$ becomes zero at $s = \infty$. Since for large values of s

$$G(s) \doteq \frac{K}{s^3}$$

$G(s)$ possesses a triple zero (multiple zero of order 3) at $s = \infty$. If points at infinity are included, $G(s)$ has the same number of poles as zeros. To summarize, $G(s)$ has five zeros $(s = -2,\ s = -10,\ s = \infty,\ s = \infty,\ s = \infty)$ and five poles $(s = 0,\ s = -1,\ s = -5,\ s = -15,\ s = -15)$.

Laplace Transformation.   Let us define

$f(t)$ = a function of time t such that $f(t) = 0$ for $t < 0$

s = a complex variable

$\mathcal{L}$ = an operational symbol indicating that the quantity that it prefixes is to be transformed by the Laplace integral $\int_0^\infty e^{-st}\,dt$

$F(s)$ = Laplace transform of $f(t)$

Then the Laplace transform of $f(t)$ is given by

$$\mathcal{L}[f(t)] = F(s) = \int_0^\infty e^{-st}\,dt\,[f(t)] = \int_0^\infty f(t)e^{-st}\,dt$$

The reverse process of finding the time function $f(t)$ from the Laplace transform $F(s)$ is called the *inverse Laplace transformation*. The notation for the inverse Laplace transformation is $\mathcal{L}^{-1}$, and the inverse Laplace transform can be found from $F(s)$ by the following inversion integral:

$$\mathcal{L}^{-1}[F(s)] = f(t) = \frac{1}{2\pi j} \int_{c-j\infty}^{c+j\infty} F(s)e^{st}\,ds, \qquad \text{for } t > 0$$

where c, the abscissa of convergence, is a real constant and is chosen larger than the real parts of all singular points of $F(s)$. Thus, the path of integration is parallel to the $j\omega$ axis and is displaced by the amount c from it. This path of integration is to the right of all singular points.

Evaluating the inversion integral appears complicated. In practice, we seldom use this integral for finding $f(t)$. We frequently use the partial-fraction expansion method given in Appendix B.

In what follows we give Table A–1, which presents Laplace transform pairs of commonly encountered functions, and Table A–2, which presents properties of Laplace transforms.

Table A–1   Laplace Transform Pairs

	$f(t)$	$F(s)$
1	Unit impulse $\delta(t)$	1
2	Unit step $1(t)$	$\dfrac{1}{s}$
3	t	$\dfrac{1}{s^2}$
4	$\dfrac{t^{n-1}}{(n-1)!}$ $(n = 1, 2, 3, \ldots)$	$\dfrac{1}{s^n}$
5	t^n $(n = 1, 2, 3, \ldots)$	$\dfrac{n!}{s^{n+1}}$
6	e^{-at}	$\dfrac{1}{s+a}$
7	te^{-at}	$\dfrac{1}{(s+a)^2}$
8	$\dfrac{1}{(n-1)!}t^{n-1}e^{-at}$ $(n = 1, 2, 3, \ldots)$	$\dfrac{1}{(s+a)^n}$
9	$t^n e^{-at}$ $(n = 1, 2, 3, \ldots)$	$\dfrac{n!}{(s+a)^{n+1}}$
10	$\sin \omega t$	$\dfrac{\omega}{s^2 + \omega^2}$
11	$\cos \omega t$	$\dfrac{s}{s^2 + \omega^2}$
12	$\sinh \omega t$	$\dfrac{\omega}{s^2 - \omega^2}$
13	$\cosh \omega t$	$\dfrac{s}{s^2 - \omega^2}$
14	$\dfrac{1}{a}\left(1 - e^{-at}\right)$	$\dfrac{1}{s(s+a)}$
15	$\dfrac{1}{b-a}\left(e^{-at} - e^{-bt}\right)$	$\dfrac{1}{(s+a)(s+b)}$
16	$\dfrac{1}{b-a}\left(be^{-bt} - ae^{-at}\right)$	$\dfrac{s}{(s+a)(s+b)}$
17	$\dfrac{1}{ab}\left[1 + \dfrac{1}{a-b}\left(be^{-at} - ae^{-bt}\right)\right]$	$\dfrac{1}{s(s+a)(s+b)}$

(continues on next page)

Table A–1 *(continued)*

18	$\dfrac{1}{a^2}\left(1 - e^{-at} - ate^{-at}\right)$	$\dfrac{1}{s(s + a)^2}$
19	$\dfrac{1}{a^2}\left(at - 1 + e^{-at}\right)$	$\dfrac{1}{s^2(s + a)}$
20	$e^{-at}\sin\omega t$	$\dfrac{\omega}{(s + a)^2 + \omega^2}$
21	$e^{-at}\cos\omega t$	$\dfrac{s + a}{(s + a)^2 + \omega^2}$
22	$\dfrac{\omega_n}{\sqrt{1 - \zeta^2}}e^{-\zeta\omega_n t}\sin\omega_n\sqrt{1 - \zeta^2}\,t \quad (0 < \zeta < 1)$	$\dfrac{\omega_n^2}{s^2 + 2\zeta\omega_n s + \omega_n^2}$
23	$-\dfrac{1}{\sqrt{1 - \zeta^2}}e^{-\zeta\omega_n t}\sin\left(\omega_n\sqrt{1 - \zeta^2}\,t - \phi\right)$ $\phi = \tan^{-1}\dfrac{\sqrt{1 - \zeta^2}}{\zeta}$ $(0 < \zeta < 1, \quad 0 < \phi < \pi/2)$	$\dfrac{s}{s^2 + 2\zeta\omega_n s + \omega_n^2}$
24	$1 - \dfrac{1}{\sqrt{1 - \zeta^2}}e^{-\zeta\omega_n t}\sin\left(\omega_n\sqrt{1 - \zeta^2}\,t + \phi\right)$ $\phi = \tan^{-1}\dfrac{\sqrt{1 - \zeta^2}}{\zeta}$ $(0 < \zeta < 1, \quad 0 < \phi < \pi/2)$	$\dfrac{\omega_n^2}{s\left(s^2 + 2\zeta\omega_n s + \omega_n^2\right)}$
25	$1 - \cos\omega t$	$\dfrac{\omega^2}{s\left(s^2 + \omega^2\right)}$
26	$\omega t - \sin\omega t$	$\dfrac{\omega^3}{s^2\left(s^2 + \omega^2\right)}$
27	$\sin\omega t - \omega t\cos\omega t$	$\dfrac{2\omega^3}{\left(s^2 + \omega^2\right)^2}$
28	$\dfrac{1}{2\omega}t\sin\omega t$	$\dfrac{s}{\left(s^2 + \omega^2\right)^2}$
29	$t\cos\omega t$	$\dfrac{s^2 - \omega^2}{\left(s^2 + \omega^2\right)^2}$
30	$\dfrac{1}{\omega_2^2 - \omega_1^2}\left(\cos\omega_1 t - \cos\omega_2 t\right) \quad \left(\omega_1^2 \neq \omega_2^2\right)$	$\dfrac{s}{\left(s^2 + \omega_1^2\right)\left(s^2 + \omega_2^2\right)}$
31	$\dfrac{1}{2\omega}\left(\sin\omega t + \omega t\cos\omega t\right)$	$\dfrac{s^2}{\left(s^2 + \omega^2\right)^2}$

Table A-2 Properties of Laplace Transforms

1	$$\mathscr{L}\big[Af(t)\big] = AF(s)$$
2	$$\mathscr{L}\big[f_1(t) \pm f_2(t)\big] = F_1(s) \pm F_2(s)$$
3	$$\mathscr{L}_{\pm}\left[\frac{d}{dt}f(t)\right] = sF(s) - f(0\pm)$$
4	$$\mathscr{L}_{\pm}\left[\frac{d^2}{dt^2}f(t)\right] = s^2F(s) - sf(0\pm) - \dot{f}(0\pm)$$
5	$$\mathscr{L}_{\pm}\left[\frac{d^n}{dt^n}f(t)\right] = s^nF(s) - \sum_{k=1}^{n} s^{n-k}\overset{(k-1)}{f}(0\pm)$$ $$\text{where } \overset{(k-1)}{f}(t) = \frac{d^{k-1}}{dt^{k-1}}f(t)$$
6	$$\mathscr{L}_{\pm}\left[\int f(t)\,dt\right] = \frac{F(s)}{s} + \frac{1}{s}\left[\int f(t)\,dt\right]_{t=0\pm}$$
7	$$\mathscr{L}_{\pm}\left[\int \cdots \int f(t)(dt)^n\right] = \frac{F(s)}{s^n} + \sum_{k=1}^{n}\frac{1}{s^{n-k+1}}\left[\int \cdots \int f(t)(dt)^k\right]_{t=0\pm}$$
8	$$\mathscr{L}\left[\int_0^t f(t)\,dt\right] = \frac{F(s)}{s}$$
9	$$\int_0^{\infty} f(t)\,dt = \lim_{s\to 0} F(s) \qquad \text{if } \int_0^{\infty} f(t)\,dt \text{ exists}$$
10	$$\mathscr{L}\big[e^{-\alpha t}f(t)\big] = F(s + a)$$
11	$$\mathscr{L}\big[f(t-\alpha)1(t-\alpha)\big] = e^{-\alpha s}F(s) \qquad \alpha \geq 0$$
12	$$\mathscr{L}\big[tf(t)\big] = -\frac{dF(s)}{ds}$$
13	$$\mathscr{L}\big[t^2f(t)\big] = \frac{d^2}{ds^2}F(s)$$
14	$$\mathscr{L}\big[t^nf(t)\big] = (-1)^n\frac{d^n}{ds^n}F(s) \qquad (n = 1, 2, 3, \dots)$$
15	$$\mathscr{L}\left[\frac{1}{t}f(t)\right] = \int_s^{\infty} F(s)\,ds \qquad \text{if } \lim_{t\to 0}\frac{1}{t}f(t) \text{ exists}$$
16	$$\mathscr{L}\left[f\left(\frac{1}{a}\right)\right] = aF(as)$$
17	$$\mathscr{L}\left[\int_0^t f_1(t-\tau)f_2(\tau)\,d\tau\right] = F_1(s)F_2(s)$$
18	$$\mathscr{L}\big[f(t)g(t)\big] = \frac{1}{2\pi j}\int_{c-j\infty}^{c+j\infty} F(p)G(s-p)\,dp$$

Finally, we present two frequently used theorems, together with Laplace transforms of the pulse function and impulse function.

Initial value theorem	$f(0+) = \lim_{t \to 0+} f(t) = \lim_{s \to \infty} sF(s)$
Final value theorem	$f(\infty) = \lim_{t \to \infty} f(t) = \lim_{s \to 0} sF(s)$
Pulse function $$f(t) = \frac{A}{t_0}1(t) - \frac{A}{t_0}1(t - t_0)$$	$$\mathcal{L}[f(t)] = \frac{A}{t_0 s} - \frac{A}{t_0 s}e^{-st_0}$$
Impulse function $$g(t) = \lim_{t_0 \to 0} \frac{A}{t_0}, \quad \text{for } 0 < t < t_0$$ $$= 0, \quad \text{for } t < 0, t_0 < t$$	$$\mathcal{L}[g(t)] = \lim_{t_0 \to 0} \left[\frac{A}{t_0 s}(1 - e^{-st_0}) \right]$$ $$= \lim_{t_0 \to 0} \frac{\dfrac{d}{dt_0}[A(1 - e^{-st_0})]}{\dfrac{d}{dt_0}(t_0 s)}$$ $$= \frac{As}{s} = A$$

Appendix

Partial-Fraction Expansion

Before we present MATLAB approach to the partial-fraction expansions of transfer functions, we discuss the manual approach to the partial-fraction expansions of transfer functions.

Partial-Fraction Expansion when $F(s)$ Involves Distinct Poles Only.   Consider $F(s)$ written in the factored form

$$F(s) = \frac{B(s)}{A(s)} = \frac{K(s + z_1)(s + z_2) \cdots (s + z_m)}{(s + p_1)(s + p_2) \cdots (s + p_n)}, \qquad \text{for } m < n$$

where $p_1, p_2, \ldots, p_n$ and $z_1, z_2, \ldots, z_m$ are either real or complex quantities, but for each complex p_i or z_j there will occur the complex conjugate of p_i or z_j, respectively. If $F(s)$ involves distinct poles only, then it can be expanded into a sum of simple partial fractions as follows:

$$F(s) = \frac{B(s)}{A(s)} = \frac{a_1}{s + p_1} + \frac{a_2}{s + p_2} + \cdots + \frac{a_n}{s + p_n} \qquad (\text{B--1})$$

where a_k ($k = 1, 2, \ldots, n$) are constants. The coefficient a_k is called the *residue* at the pole at $s = -p_k$. The value of a_k can be found by multiplying both sides of Equation (B–1) by $(s + p_k)$ and letting $s = -p_k$, which gives

$$\left[(s + p_k) \frac{B(s)}{A(s)} \right]_{s=-p_k} = \left[\frac{a_1}{s + p_1}(s + p_k) + \frac{a_2}{s + p_2}(s + p_k) \right.$$

$$\left. + \cdots + \frac{a_k}{s + p_k}(s + p_k) + \cdots + \frac{a_n}{s + p_n}(s + p_k) \right]_{s=-p_k}$$

$$= a_k$$

We see that all the expanded terms drop out with the exception of a_k. Thus the residue a_k is found from

$$a_k = \left[(s + p_k) \frac{B(s)}{A(s)} \right]_{s=-p_k}$$

Note that, since $f(t)$ is a real function of time, if p_1 and p_2 are complex conjugates, then the residues a_1 and a_2 are also complex conjugates. Only one of the conjugates, a_1 or a_2, needs to be evaluated, because the other is known automatically.

Since

$$\mathcal{L}^{-1} \left[\frac{a_k}{s + p_k} \right] = a_k e^{-p_k t}$$

$f(t)$ is obtained as

$$f(t) = \mathcal{L}^{-1}[F(s)] = a_1 e^{-p_1 t} + a_2 e^{-p_2 t} + \cdots + a_n e^{-p_n t}, \qquad \text{for } t \geq 0$$

EXAMPLE B–1   Find the inverse Laplace transform of

$$F(s) = \frac{s + 3}{(s + 1)(s + 2)}$$

The partial-fraction expansion of $F(s)$ is

$$F(s) = \frac{s + 3}{(s + 1)(s + 2)} = \frac{a_1}{s + 1} + \frac{a_2}{s + 2}$$

where a_1 and a_2 are found as

$$a_1 = \left[(s + 1) \frac{s + 3}{(s + 1)(s + 2)} \right]_{s=-1} = \left[\frac{s + 3}{s + 2} \right]_{s=-1} = 2$$

$$a_2 = \left[(s + 2) \frac{s + 3}{(s + 1)(s + 2)} \right]_{s=-2} = \left[\frac{s + 3}{s + 1} \right]_{s=-2} = -1$$

Thus

$$f(t) = \mathcal{L}^{-1}[F(s)]$$

$$= \mathcal{L}^{-1} \left[\frac{2}{s + 1} \right] + \mathcal{L}^{-1} \left[\frac{-1}{s + 2} \right]$$

$$= 2e^{-t} - e^{-2t}, \qquad \text{for } t \geq 0$$

EXAMPLE B–2   Obtain the inverse Laplace transform of

$$G(s) = \frac{s^3 + 5s^2 + 9s + 7}{(s + 1)(s + 2)}$$

Here, since the degree of the numerator polynomial is higher than that of the denominator polynomial, we must divide the numerator by the denominator.

$$G(s) = s + 2 + \frac{s + 3}{(s + 1)(s + 2)}$$

Note that the Laplace transform of the unit-impulse function $\delta(t)$ is 1 and that the Laplace transform of $d\delta(t)/dt$ is s. The third term on the right-hand side of this last equation is $F(s)$ in Example B–1. So the inverse Laplace transform of $G(s)$ is given as

$$g(t) = \frac{d}{dt}\delta(t) + 2\delta(t) + 2e^{-t} - e^{-2t}, \qquad \text{for } t \geq 0-$$

EXAMPLE B–3 Find the inverse Laplace transform of

$$F(s) = \frac{2s + 12}{s^2 + 2s + 5}$$

Notice that the denominator polynomial can be factored as

$$s^2 + 2s + 5 = (s + 1 + j2)(s + 1 - j2)$$

If the function $F(s)$ involves a pair of complex-conjugate poles, it is convenient not to expand $F(s)$ into the usual partial fractions but to expand it into the sum of a damped sine and a damped cosine function.

Noting that $s^2 + 2s + 5 = (s + 1)^2 + 2^2$ and referring to the Laplace transforms of $e^{-\alpha t}\sin\omega t$ and $e^{-\alpha t}\cos\omega t$, rewritten thus,

$$\mathcal{L}\left[e^{-\alpha t}\sin\omega t\right] = \frac{\omega}{(s + \alpha)^2 + \omega^2}$$

$$\mathcal{L}\left[e^{-\alpha t}\cos\omega t\right] = \frac{s + \alpha}{(s + \alpha)^2 + \omega^2}$$

the given $F(s)$ can be written as a sum of a damped sine and a damped cosine function:

$$F(s) = \frac{2s + 12}{s^2 + 2s + 5} = \frac{10 + 2(s + 1)}{(s + 1)^2 + 2^2}$$

$$= 5\frac{2}{(s + 1)^2 + 2^2} + 2\frac{s + 1}{(s + 1)^2 + 2^2}$$

It follows that

$$f(t) = \mathcal{L}^{-1}\left[F(s)\right]$$

$$= 5\mathcal{L}^{-1}\left[\frac{2}{(s + 1)^2 + 2^2}\right] + 2\mathcal{L}^{-1}\left[\frac{s + 1}{(s + 1)^2 + 2^2}\right]$$

$$= 5e^{-t}\sin 2t + 2e^{-t}\cos 2t, \qquad \text{for } t \geq 0$$

Partial-Fraction Expansion when $F(s)$ Involves Multiple Poles.   Instead of discussing the general case, we shall use an example to show how to obtain the partial-fraction expansion of $F(s)$.

Consider the following $F(s)$:

$$F(s) = \frac{s^2 + 2s + 3}{(s + 1)^3}$$

The partial-fraction expansion of this $F(s)$ involves three terms,

$$F(s) = \frac{B(s)}{A(s)} = \frac{b_1}{s + 1} + \frac{b_2}{(s + 1)^2} + \frac{b_3}{(s + 1)^3}$$

where b_3, b_2, and b_1 are determined as follows. By multiplying both sides of this last equation by $(s + 1)^3$, we have

$$(s + 1)^3 \frac{B(s)}{A(s)} = b_1(s + 1)^2 + b_2(s + 1) + b_3 \qquad \text{(B–2)}$$

Then letting $s = -1$, Equation (B–2) gives

$$\left[(s + 1)^3 \frac{B(s)}{A(s)} \right]_{s=-1} = b_3$$

Also, differentiation of both sides of Equation (B–2) with respect to s yields

$$\frac{d}{ds} \left[(s + 1)^3 \frac{B(s)}{A(s)} \right] = b_2 + 2b_1(s + 1) \qquad \text{(B–3)}$$

If we let $s = -1$ in Equation (B–3), then

$$\frac{d}{ds} \left[(s + 1)^3 \frac{B(s)}{A(s)} \right]_{s=-1} = b_2$$

By differentiating both sides of Equation (B–3) with respect to s, the result is

$$\frac{d^2}{ds^2} \left[(s + 1)^3 \frac{B(s)}{A(s)} \right] = 2b_1$$

From the preceding analysis it can be seen that the values of b_3, b_2, and b_1 are found systematically as follows:

$$b_3 = \left[(s + 1)^3 \frac{B(s)}{A(s)} \right]_{s=-1}$$
$$= (s^2 + 2s + 3)_{s=-1}$$
$$= 2$$

$$b_2 = \left\{ \frac{d}{ds} \left[(s + 1)^3 \frac{B(s)}{A(s)} \right] \right\}_{s=-1}$$
$$= \left[\frac{d}{ds} (s^2 + 2s + 3) \right]_{s=-1}$$
$$= (2s + 2)_{s=-1}$$
$$= 0$$

$$b_1 = \frac{1}{2!} \left\{ \frac{d^2}{ds^2} \left[(s + 1)^3 \frac{B(s)}{A(s)} \right] \right\}_{s=-1}$$
$$= \frac{1}{2!} \left[\frac{d^2}{ds^2} (s^2 + 2s + 3) \right]_{s=-1}$$
$$= \frac{1}{2} (2) = 1$$

We thus obtain

$$f(t) = \mathcal{L}^{-1}[F(s)]$$

$$= \mathcal{L}^{-1}\left[\frac{1}{s+1}\right] + \mathcal{L}^{-1}\left[\frac{0}{(s+1)^2}\right] + \mathcal{L}^{-1}\left[\frac{2}{(s+1)^3}\right]$$

$$= e^{-t} + 0 + t^2 e^{-t}$$

$$= (1 + t^2)e^{-t}, \qquad \text{for } t \geq 0$$

Comments. For complicated functions with denominators involving higher-order polynomials, partial-fraction expansion may be quite time consuming. In such a case, use of MATLAB is recommended.

Partial-Fraction Expansion with MATLAB. MATLAB has a command to obtain the partial-fraction expansion of $B(s)/A(s)$. Consider the following function $B(s)/A(s)$:

$$\frac{B(s)}{A(s)} = \frac{\text{num}}{\text{den}} = \frac{b_0 s^n + b_1 s^{n-1} + \cdots + b_n}{s^n + a_1 s^{n-1} + \cdots + a_n}$$

where some of a_i and b_j may be zero. In MATLAB row vectors num and den specify the coefficients of the numerator and denominator of the transfer function. That is,

$$\text{num} = [b_0 \ b_1 \ \dots \ b_n]$$
$$\text{den} = [1 \ a_1 \ \dots \ a_n]$$

The command

$$[r,p,k] = \text{residue(num,den)}$$

finds the residues (r), poles (p), and direct terms (k) of a partial-fraction expansion of the ratio of two polynomials $B(s)$ and $A(s)$.

The partial-fraction expansion of $B(s)/A(s)$ is given by

$$\frac{B(s)}{A(s)} = \frac{r(1)}{s - p(1)} + \frac{r(2)}{s - p(2)} + \cdots + \frac{r(n)}{s - p(n)} + k(s) \qquad \text{(B–4)}$$

Comparing Equations (B–1) and (B–4), we note that $p(1) = -p_1$, $p(2) = -p_2, \ldots,$ $p(n) = -p_n; r(1) = a_1, r(2) = a_2, \ldots, r(n) = a_n$. [$k(s)$ is a direct term.]

EXAMPLE B–4 Consider the following transfer function,

$$\frac{B(s)}{A(s)} = \frac{2s^3 + 5s^2 + 3s + 6}{s^3 + 6s^2 + 11s + 6}$$

For this function,

$$\text{num} = [2 \quad 5 \quad 3 \quad 6]$$
$$\text{den} = [1 \quad 6 \quad 11 \quad 6]$$

The command

$$[r,p,k] = \text{residue(num,den)}$$

gives the following result:

```
[r,p,k] = residue(num,den)

r =

    -6.0000
    -4.0000
     3.0000

p =

    -3.0000
    -2.0000
    -1.0000

k =

     2
```

(Note that the residues are returned in column vector r, the pole locations in column vector p, and the direct term in row vector k.) This is the MATLAB representation of the following partial-fraction expansion of $B(s)/A(s)$:

$$\frac{B(s)}{A(s)} = \frac{2s^3 + 5s^2 + 3s + 6}{s^3 + 6s^2 + 11s + 6}$$

$$= \frac{-6}{s + 3} + \frac{-4}{s + 2} + \frac{3}{s + 1} + 2$$

Note that if $p(j) = p(j + 1) = \cdots = p(j + m - 1)$ [that is, $p_j = p_{j+1} = \cdots = p_{j+m-1}$], the pole $p(j)$ is a pole of multiplicity m. In such a case, the expansion includes terms of the form

$$\frac{r(j)}{s - p(j)} + \frac{r(j + 1)}{[s - p(j)]^2} + \cdots + \frac{r(j + m - 1)}{[s - p(j)]^m}$$

For details, see Example B–5.

EXAMPLE B–5 Expand the following $B(s)/A(s)$ into partial fractions with MATLAB.

$$\frac{B(s)}{A(s)} = \frac{s^2 + 2s + 3}{(s + 1)^3} = \frac{s^2 + 2s + 3}{s^3 + 3s^2 + 3s + 1}$$

For this function, we have

$$\text{num} = [1 \ \ 2 \ \ 3]$$
$$\text{den} = [1 \ \ 3 \ \ 3 \ \ 1]$$

The command

$$[r,p,k] = \text{residue(num,den)}$$

gives the result shown next:

```
num = [1  2  3];
den = [1  3  3  1];
[r,p,k] = residue(num,den)

r =

    1.0000
    0.0000
    2.0000

p =

   -1.0000
   -1.0000
   -1.0000

k =

   []
```

It is the MATLAB representation of the following partial-fraction expansion of $B(s)/A(s)$:

$$\frac{B(s)}{A(s)} = \frac{1}{s + 1} + \frac{0}{(s + 1)^2} + \frac{2}{(s + 1)^3}$$

Note that the direct term k is zero.

Appendix

Vector-Matrix Algebra

In this appendix we first review the determinant of a matrix, then we define the adjoint matrix, the inverse of a matrix, and the derivative and integral of a matrix.

Determinant of a Matrix. For each square matrix, there exists a determinant. The determinant of a square matrix $\mathbf{A}$ is usually written as $|\mathbf{A}|$ or det $\mathbf{A}$. The determinant has the following properties:

1. If any two consecutive rows or columns are interchanged, the determinant changes its sign.

2. If any row or any column consists only of zeros, then the value of the dererminant is zero.

3. If the elements of any row (or any column) are exactly k times those of another row (or another column), then the value of the determinant is zero.

4. If, to any row (or any column), any constant times another row (or column) is added, the value of the determinant remains unchanged.

5. If a determinant is multiplied by a constant, then only one row (or one column) is multiplied by that constant. Note, however, that the determinant of k times an $n \times n$ matrix $\mathbf{A}$ is k^n times the determinant of $\mathbf{A}$, or

$$|k\mathbf{A}| = k^n|\mathbf{A}|$$

This is because

$$kA = \begin{bmatrix} ka_{11} & ka_{12} & \cdots & ka_{1m} \\ ka_{21} & ka_{22} & \cdots & ka_{2m} \\ \vdots & \vdots & & \vdots \\ ka_{n1} & ka_{n2} & \cdots & ka_{nm} \end{bmatrix}$$

6. The determinant of the product of two square matrices A and B is the product of determinants, or

$$|AB| = |A|\,|B|$$

If $B = n \times m$ matrix and $C = m \times n$ matrix, then

$$\det(I_n + BC) = \det(I_m + CB)$$

If $A \neq 0$ and $D = m \times m$ matrix, then

$$\det\begin{bmatrix} A & B \\ C & D \end{bmatrix} = \det A \cdot \det S$$

where $S = D - CA^{-1}B$.
If $D \neq 0$, then

$$\det\begin{bmatrix} A & B \\ C & D \end{bmatrix} = \det D \cdot \det T$$

where $T = A - BD^{-1}C$.
If $B = 0$ or $C = 0$, then

$$\det\begin{bmatrix} A & 0 \\ C & D \end{bmatrix} = \det A \cdot \det D$$

$$\det\begin{bmatrix} A & B \\ 0 & D \end{bmatrix} = \det A \cdot \det D$$

Rank of Matrix.   A matrix A is said to have rank m if there exists an $m \times m$ sub-matrix M of A such that the determinant of M is nonzero and the determinant of every $r \times r$ submatrix (where $r \geq m + 1$) of A is zero.
As an example, consider the following matrix:

$$A = \begin{bmatrix} 1 & 2 & 3 & 4 \\ 0 & 1 & -1 & 0 \\ 1 & 0 & 1 & 2 \\ 1 & 1 & 0 & 2 \end{bmatrix}$$

Note that $|\mathbf{A}| = 0$. One of a number of largest submatrices whose determinant is not equal to zero is

$$\begin{bmatrix} 1 & 2 & 3 \\ 0 & 1 & -1 \\ 1 & 0 & 1 \end{bmatrix}$$

Hence, the rank of the matrix $\mathbf{A}$ is 3.

Minor M_{ij}.   If the ith row and jth column are deleted from an $n \times n$ matrix $\mathbf{A}$, the resulting matrix is an $(n-1) \times (n-1)$ matrix. The determinant of this $(n-1) \times (n-1)$ matrix is called the minor M_{ij} of the matrix $\mathbf{A}$.

Cofactor A_{ij}.   The cofactor A_{ij} of the element a_{ij} of the $n \times n$ matrix $\mathbf{A}$ is defined by the equation

$$A_{ij} = (-1)^{i+j} M_{ij}$$

That is, the cofactor A_{ij} of the element a_{ij} is $(-1)^{i+j}$ times the determinant of the matrix formed by deleting the ith row and the jth column from $\mathbf{A}$. Note that the cofactor A_{ij} of the element a_{ij} is the coefficient of the term a_{ij} in the expansion of the determinant $|\mathbf{A}|$, since it can be shown that

$$a_{i1}A_{i1} + a_{i2}A_{i2} + \cdots + a_{in}A_{in} = |\mathbf{A}|$$

If $a_{i1}, a_{i2}, \ldots, a_{in}$ are replaced by $a_{j1}, a_{j2}, \ldots, a_{jn}$, then

$$a_{j1}A_{i1} + a_{j2}A_{i2} + \cdots + a_{jn}A_{in} = 0 \qquad i \ne j$$

because the determinant of $\mathbf{A}$ in this case possesses two identical rows. Hence, we obtain

$$\sum_{k=1}^{n} a_{jk}A_{ik} = \delta_{ji}|\mathbf{A}|$$

Similarly,

$$\sum_{k=1}^{n} a_{ki}A_{kj} = \delta_{ij}|\mathbf{A}|$$

Adjoint Matrix.   The matrix $\mathbf{B}$ whose element in the ith row and jth column equals A_{ji} is called the adjoint of $\mathbf{A}$ and is denoted by adj $\mathbf{A}$, or

$$\mathbf{B} = (b_{ij}) = (A_{ji}) = \text{adj } \mathbf{A}$$

That is, the adjoint of $\mathbf{A}$ is the transpose of the matrix whose elements are the cofactors of $\mathbf{A}$, or

$$\text{adj } \mathbf{A} = \begin{bmatrix} A_{11} & A_{21} & \cdots & A_{n1} \\ A_{12} & A_{22} & \cdots & A_{n2} \\ \vdots & \vdots & & \vdots \\ A_{1n} & A_{2n} & \cdots & A_{nn} \end{bmatrix}$$

Note that the element of the jth row and ith column of the product $\mathbf{A}(\text{adj } \mathbf{A})$ is

$$\sum_{k=1}^{n} a_{jk} b_{ki} = \sum_{k=1}^{n} a_{jk} A_{ik} = \delta_{ji} |\mathbf{A}|$$

Hence, $\mathbf{A}(\text{adj } \mathbf{A})$ is a diagonal matrix with diagonal elements equal to $|\mathbf{A}|$, or

$$\mathbf{A}(\text{adj } \mathbf{A}) = |\mathbf{A}| \, \mathbf{I}$$

Similarly, the element in the jth row and ith column of the product $(\text{adj } \mathbf{A})\mathbf{A}$ is

$$\sum_{k=1}^{n} b_{jk} a_{ki} = \sum_{k=1}^{n} A_{kj} a_{ki} = \delta_{ij} |\mathbf{A}|$$

Hence, we have the relationship

$$\mathbf{A}(\text{adj } \mathbf{A}) = (\text{adj } \mathbf{A})\mathbf{A} = |\mathbf{A}| \, \mathbf{I} \qquad\qquad \text{(C–1)}$$

Thus

$$\mathbf{A}^{-1} = \frac{\text{adj } \mathbf{A}}{|\mathbf{A}|} = \begin{bmatrix} \dfrac{A_{11}}{|\mathbf{A}|} & \dfrac{A_{21}}{|\mathbf{A}|} & \cdots & \dfrac{A_{n1}}{|\mathbf{A}|} \\[2mm] \dfrac{A_{12}}{|\mathbf{A}|} & \dfrac{A_{22}}{|\mathbf{A}|} & \cdots & \dfrac{A_{n2}}{|\mathbf{A}|} \\[1mm] \vdots & \vdots & & \vdots \\[1mm] \dfrac{A_{1n}}{|\mathbf{A}|} & \dfrac{A_{2n}}{|\mathbf{A}|} & \cdots & \dfrac{A_{nn}}{|\mathbf{A}|} \end{bmatrix}$$

where A_{ij} is the cofactor of a_{ij} of the matrix $\mathbf{A}$. Thus, the terms in the ith column of $\mathbf{A}^{-1}$ are $1/|\mathbf{A}|$ times the cofactors of the ith row of the original matrix $\mathbf{A}$. For example, if

$$\mathbf{A} = \begin{bmatrix} 1 & 2 & 0 \\ 3 & -1 & -2 \\ 1 & 0 & -3 \end{bmatrix}$$

then the adjoint of $\mathbf{A}$ and the determinant $|\mathbf{A}|$ are respectively found to be

$$\text{adj } \mathbf{A} = \begin{bmatrix} \begin{vmatrix} -1 & -2 \\ 0 & -3 \end{vmatrix} & -\begin{vmatrix} 2 & 0 \\ 0 & -3 \end{vmatrix} & \begin{vmatrix} 2 & 0 \\ -1 & -2 \end{vmatrix} \\[4mm] -\begin{vmatrix} 3 & -2 \\ 1 & -3 \end{vmatrix} & \begin{vmatrix} 1 & 0 \\ 1 & -3 \end{vmatrix} & -\begin{vmatrix} 1 & 0 \\ 3 & -2 \end{vmatrix} \\[4mm] \begin{vmatrix} 3 & -1 \\ 1 & 0 \end{vmatrix} & -\begin{vmatrix} 1 & 2 \\ 1 & 0 \end{vmatrix} & \begin{vmatrix} 1 & 2 \\ 3 & -1 \end{vmatrix} \end{bmatrix}$$

$$= \begin{bmatrix} 3 & 6 & -4 \\ 7 & -3 & 2 \\ 1 & 2 & -7 \end{bmatrix}$$

and

$$|\mathbf{A}| = 17$$

Hence, the inverse of $\mathbf{A}$ is

$$\mathbf{A}^{-1} = \frac{\text{adj } \mathbf{A}}{|\mathbf{A}|} = \begin{bmatrix} \frac{3}{17} & \frac{6}{17} & -\frac{4}{17} \\ \frac{7}{17} & -\frac{3}{17} & \frac{2}{17} \\ \frac{1}{17} & \frac{2}{17} & -\frac{7}{17} \end{bmatrix}$$

In what follows, we give formulas for finding inverse matrices for the 2×2 matrix and the 3×3 matrix. For the 2×2 matrix

$$\mathbf{A} = \begin{bmatrix} a & b \\ c & d \end{bmatrix} \qquad \text{where } ad - bc \neq 0$$

the inverse matrix is given by

$$\mathbf{A}^{-1} = \frac{1}{ad - bc} \begin{bmatrix} d & -b \\ -c & a \end{bmatrix}$$

For the 3×3 matrix

$$\mathbf{A} = \begin{bmatrix} a & b & c \\ d & e & f \\ g & h & i \end{bmatrix} \qquad \text{where } |\mathbf{A}| \neq 0$$

the inverse matrix is given by

$$\mathbf{A}^{-1} = \frac{1}{|\mathbf{A}|} \begin{bmatrix} \begin{vmatrix} e & f \\ h & i \end{vmatrix} & -\begin{vmatrix} b & c \\ h & i \end{vmatrix} & \begin{vmatrix} b & c \\ e & f \end{vmatrix} \\ -\begin{vmatrix} d & f \\ g & i \end{vmatrix} & \begin{vmatrix} a & c \\ g & i \end{vmatrix} & -\begin{vmatrix} a & c \\ d & f \end{vmatrix} \\ \begin{vmatrix} d & e \\ g & h \end{vmatrix} & -\begin{vmatrix} a & b \\ g & h \end{vmatrix} & \begin{vmatrix} a & b \\ d & e \end{vmatrix} \end{bmatrix}$$

Note that

$$(\mathbf{A}^{-1})^{-1} = \mathbf{A}$$

$$(\mathbf{A}^{-1})' = (\mathbf{A}')^{-1}$$

$$(\mathbf{A}^{-1})^* = (\mathbf{A}^*)^{-1}$$

There are several more useful formulas available. Assume that $\mathbf{A} = n \times n$ matrix, $\mathbf{B} = n \times m$ matrix, $\mathbf{C} = m \times n$ matrix, and $\mathbf{D} = m \times m$ matrix. Then

$$[\mathbf{A} + \mathbf{BC}]^{-1} = \mathbf{A}^{-1} - \mathbf{A}^{-1}\mathbf{B}[\mathbf{I}_m + \mathbf{CA}^{-1}\mathbf{B}]^{-1}\mathbf{CA}^{-1}$$

If $|\mathbf{A}| \neq 0$ and $|\mathbf{D}| \neq 0$, then

$$\begin{bmatrix} \mathbf{A} & \mathbf{B} \\ \mathbf{0} & \mathbf{D} \end{bmatrix}^{-1} = \begin{bmatrix} \mathbf{A}^{-1} & -\mathbf{A}^{-1}\mathbf{B}\mathbf{D}^{-1} \\ \mathbf{0} & \mathbf{D}^{-1} \end{bmatrix}$$

$$\begin{bmatrix} \mathbf{A} & \mathbf{0} \\ \mathbf{C} & \mathbf{D} \end{bmatrix}^{-1} = \begin{bmatrix} \mathbf{A}^{-1} & \mathbf{0} \\ -\mathbf{D}^{-1}\mathbf{C}\mathbf{A}^{-1} & \mathbf{D}^{-1} \end{bmatrix}$$

If $|\mathbf{A}| \neq 0$, $\mathbf{S} = \mathbf{D} - \mathbf{C}\mathbf{A}^{-1}\mathbf{B}$, $|\mathbf{S}| \neq 0$, then

$$\begin{bmatrix} \mathbf{A} & \mathbf{B} \\ \mathbf{C} & \mathbf{D} \end{bmatrix}^{-1} = \begin{bmatrix} \mathbf{A}^{-1} + \mathbf{A}^{-1}\mathbf{B}\mathbf{S}^{-1}\mathbf{C}\mathbf{A}^{-1} & -\mathbf{A}^{-1}\mathbf{B}\mathbf{S}^{-1} \\ -\mathbf{S}^{-1}\mathbf{C}\mathbf{A}^{-1} & \mathbf{S}^{-1} \end{bmatrix}$$

If $|\mathbf{D}| \neq 0$, $\mathbf{T} = \mathbf{A} - \mathbf{B}\mathbf{D}^{-1}\mathbf{C}$, and $|\mathbf{T}| \neq 0$, then

$$\begin{bmatrix} \mathbf{A} & \mathbf{B} \\ \mathbf{C} & \mathbf{D} \end{bmatrix}^{-1} = \begin{bmatrix} \mathbf{T}^{-1} & -\mathbf{T}^{-1}\mathbf{B}\mathbf{D}^{-1} \\ -\mathbf{D}^{-1}\mathbf{C}\mathbf{T}^{-1} & \mathbf{D}^{-1} + \mathbf{D}^{-1}\mathbf{C}\mathbf{T}^{-1}\mathbf{B}\mathbf{D}^{-1} \end{bmatrix}$$

Finally, we present the MATLAB approach to obtain the inverse of a square matrix. If all elements of the matrix are given as numerical values, this approach is best.

MATLAB Approach to Obtain the Inverse of a Square Matrix. The inverse of a square matrix $\mathbf{A}$ can be obtained with the command

$$inv(A)$$

For example, if matrix $\mathbf{A}$ is given by

$$\mathbf{A} = \begin{bmatrix} 1 & 1 & 2 \\ 3 & 4 & 0 \\ 1 & 2 & 5 \end{bmatrix}$$

then the inverse of matrix $\mathbf{A}$ is obtained as follows:

```
A = [1  1  2;3  4  0;1  2  5];
inv(A)

ans =

        2.2222      -0.1111      -0.8889
       -1.6667       0.3333       0.6667
        0.2222      -0.1111       0.1111
```

That is

$$\mathbf{A}^{-1} = \begin{bmatrix} 2.2222 & -0.1111 & -0.8889 \\ -1.6667 & 0.3333 & 0.6667 \\ 0.2222 & -0.1111 & 0.1111 \end{bmatrix}$$

MATLAB Is Case Sensitive. It is important to note that MATLAB is case sensitive. That is, MATLAB distinguishes between upper- and lowercase letters. Thus, x and X are not the same variable. All function names must be in lowercase, such as inv(A), eig(A), and poly(A).

Differentiation and Integration of Matrices. The derivative of an $n \times m$ matrix $\mathbf{A}(t)$ is defined to be the $n \times m$ matrix, each element of which is the derivative of the corresponding element of the original matrix, provided that all the elements $a_{ij}(t)$ have derivatives with respect to t. That is,

$$\frac{d}{dt}\mathbf{A}(t) = \left(\frac{d}{dt}a_{ij}(t)\right) = \begin{bmatrix} \dfrac{d}{dt}a_{11}(t) & \dfrac{d}{dt}a_{12}(t) & \cdots & \dfrac{d}{dt}a_{1m}(t) \\ \dfrac{d}{dt}a_{21}(t) & \dfrac{d}{dt}a_{22}(t) & \cdots & \dfrac{d}{dt}a_{2m}(t) \\ \vdots & \vdots & & \vdots \\ \dfrac{d}{dt}a_{n1}(t) & \dfrac{d}{dt}a_{n2}(t) & \cdots & \dfrac{d}{dt}a_{nm}(t) \end{bmatrix}$$

Similarly, the integral of an $n \times m$ matrix $\mathbf{A}(t)$ is defined to be

$$\int \mathbf{A}(t)\,dt = \left(\int a_{ij}(t)\,dt\right) = \begin{bmatrix} \displaystyle\int a_{11}(t)\,dt & \displaystyle\int a_{12}(t)\,dt & \cdots & \displaystyle\int a_{1m}(t)\,dt \\ \displaystyle\int a_{21}(t)\,dt & \displaystyle\int a_{22}(t)\,dt & \cdots & \displaystyle\int a_{2m}(t)\,dt \\ \vdots & \vdots & & \vdots \\ \displaystyle\int a_{n1}(t)\,dt & \displaystyle\int a_{2n}(t)\,dt & \cdots & \displaystyle\int a_{nm}(t)\,dt \end{bmatrix}$$

Differentiation of the Product of Two Matrices. If the matrices $\mathbf{A}(t)$ and $\mathbf{B}(t)$ can be differentiated with respect to t, then

$$\frac{d}{dt}[\mathbf{A}(t)\mathbf{B}(t)] = \frac{d\mathbf{A}(t)}{dt}\mathbf{B}(t) + \mathbf{A}(t)\frac{d\mathbf{B}(t)}{dt}$$

Here again the multiplication of $\mathbf{A}(t)$ and $d\mathbf{B}(t)/dt$ [or $d\mathbf{A}(t)/dt$ and $\mathbf{B}(t)$] is, in general, not commutative.

Differentiation of $\mathbf{A}^{-1}(t)$. If a matrix $\mathbf{A}(t)$ and its inverse $\mathbf{A}^{-1}(t)$ are differentiable with respect to t, then the derivative of $\mathbf{A}^{-1}(t)$ is given by

$$\frac{d\mathbf{A}^{-1}(t)}{dt} = -\mathbf{A}^{-1}(t)\frac{d\mathbf{A}(t)}{dt}\mathbf{A}^{-1}(t)$$

The derivative may be obtained by differentiating $\mathbf{A}(t)\mathbf{A}^{-1}(t)$ with respect to t. Since

$$\frac{d}{dt}[\mathbf{A}(t)\mathbf{A}^{-1}(t)] = \frac{d\mathbf{A}(t)}{dt}\mathbf{A}^{-1}(t) + \mathbf{A}(t)\frac{d\mathbf{A}^{-1}(t)}{dt}$$

and

$$\frac{d}{dt}[\mathbf{A}(t)\mathbf{A}^{-1}(t)] = \frac{d}{dt}\mathbf{I} = \mathbf{0}$$

we obtain

$$\mathbf{A}(t)\frac{d\mathbf{A}^{-1}(t)}{dt} = -\frac{d\mathbf{A}(t)}{dt}\mathbf{A}^{-1}(t)$$

or

$$\frac{d\mathbf{A}^{-1}(t)}{dt} = -\mathbf{A}^{-1}(t)\frac{d\mathbf{A}(t)}{dt}\mathbf{A}^{-1}(t)$$

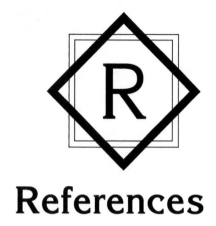

References

A–1   Anderson, B. D. O., and J. B. Moore, *Linear Optimal Control*. Upper Saddle River, NJ: Prentice Hall, 1971.

A–2   Athans, M., and P. L. Falb, *Optimal Control: An Introduction to the Theory and Its Applications*. New York: McGraw-Hill Book Company, 1965.

B–1   Barnet, S., "Matrices, Polynomials, and Linear Time-Invariant Systems," *IEEE Trans. Automatic Control*, **AC-18** (1973), pp. 1–10.

B–2   Bayliss, L. E., *Living Control Systems.* London: English Universities Press Limited, 1966.

B–3   Bellman, R., *Introduction to Matrix Analysis*. New York: McGraw-Hill Book Company, 1960.

B–4   Bode, H. W., *Network Analysis and Feedback Design*. New York: Van Nostrand Reinhold, 1945.

B–5   Brogan, W. L., *Modern Control Theory*. Upper Saddle River, NJ: Prentice Hall, 1985.

B–6   Butman, S., and R. Sivan (Sussman), "On Cancellations, Controllability and Observability," *IEEE Trans. Automatic Control*, **AC-9** (1964), pp. 317–8.

C–1   Campbell, D. P, *Process Dynamics*. New York: John Wiley & Sons, Inc., 1958.

C–2   Cannon, R., *Dynamics of Physical Systems*. New York: McGraw-Hill Book Company, 1967.

C–3   Chang, P. M., and S. Jayasuriya, "An Evaluation of Several Controller Synthesis Methodologies Using a Rotating Flexible Beam as a Test Bed," *ASME J. Dynamic Systems, Measurement, and Control*, **117** (1995), pp. 360–73.

C–4   Cheng, D. K., *Analysis of Linear Systems*. Reading, MA: Addison-Wesley Publishing Company, Inc., 1959.

C–5   Churchill, R. V., *Operational Mathematics*, 3rd ed. New York: McGraw-Hill Book Company, 1972.

C–6   Coddington, E. A., and N. Levinson, *Theory of Ordinary Differential Equations*. New York: McGraw-Hill Book Company, 1955.

C–7   Craig, J. J., *Introduction to Robotics, Mechanics and Control*. Reading, MA: AddisonWesley Publishing Company, Inc., 1986.

C–8   Cunningham, W J., *Introduction to Nonlinear Analysis*. New York: McGraw-Hill Book Company, 1958.

D–1   Dorf, R. C., and R. H. Bishop, *Modern Control Systems*, 9th ed. Upper Saddle River, NJ: Prentice Hall, 2001.

E–1   Enns, M., J. R. Greenwood III, J. E. Matheson, and F. T. Thompson, "Practical Aspects of State-Space Methods Part I: System Formulation and Reduction," *IEEE Trans. Military Electronics*, **MIL-8** (1964), pp. 81–93.

E–2   Evans, W. R., "Graphical Analysis of Control Systems," *AIEE Trans. Part II*, **67** (1948), pp. 547-51.

E–3   Evans, W. R., "Control System Synthesis by Root Locus Method," *AIEE Trans Part II*, **69** (1950), pp. 66–9.

E–4   Evans, W. R., "The Use of Zeros and Poles for Frequency Response or Transient Response," *ASME Trans*. **76** (1954), pp. 1135–44.

E–5   Evans, W. R., *Control System Dynamics*. New York: McGraw-Hill Book Company, 1954.

F–1   Franklin, G. F, J. D. Powell, and A. Emami-Naeini, *Feedback Control of Dynamic Systems*, 3rd ed. Reading, MA: Addison-Wesley Publishing Company, Inc., 1994.

F–2   Friedland, B., *Control System Design*. New York: McGraw-Hill Book Company, 1986.

F–3   Fu, K. S., R. C. Gonzalez, and C. S. G. Lee, Robotics: Control, Sensing, Vision, and Intelligence. New York: McGraw-Hill Book Company, 1987.

G–1   Gantmacher, F. R., *Theory of Matrices*, Vols. I and II. NewYork: Chelsea Publishing Company, Inc., 1959.

G–2   Gardner, M. F, and J. L. Barnes, *Transients in Linear Systems*. New York: John Wiley & Sons, Inc., 1942.

G–3   Gibson, J. E., *Nonlinear Automatic Control*. New York: McGraw-Hill Book Company, 1963.

G–4   Gilbert, E. G., "Controllability and Observability in Multivariable Control Systems," *J. SIAM Control*, ser. A, **1** (1963) , pp. 128–51.

G–5   Graham, D., and R. C. Lathrop, "The Synthesis of Optimum Response: Criteria and Standard Forms," *AIEE Trans. Part II*, **72** (1953), pp. 273–88.

H–1   Hahn, W., *Theory and Application of Liapunov's Direct Method*. Upper Saddle River, NJ: Prentice Hall, 1963.

H–2   Halmos, P. R., *Finite Dimensional Vector Spaces*. New York: Van Nostrand Reinhold, 1958.

H–3   Higdon, D. T., and R. H. Cannon, Jr., "On the Control of Unstable Multiple-Output Mechanical Systems," *ASME Paper no. 63-WA-148*, 1963.

I–1   Irwin, J. D., *Basic Engineering Circuit Analysis*. New York: Macmillan, Inc., 1984.

J–1   Jayasuriya, S., "Frequency Domain Design for Robust Performance Under Parametric, Unstructured, or Mixed Uncertainties," *ASME J. Dynamic Systems, Measurement, and Control*, **115** (1993), pp. 439–51.

K–1   Kailath, T., *Linear Systems*. Upper Saddle River, NJ: Prentice Hall, 1980.

K–2   Kalman, R. E., "Contributions to the Theory of Optimal Control," *Bol. Soc Mat. Mex.*, **5** (1960), pp. 102–19.

K–3   Kalman, R. E., "On the General Theory of Control Systems," *Proc. First Intern. Cong. IFAC, Moscow*, 1960, *Automatic and Remote Control*. London: Butterworths & Company Limited, 1961, pp. 481–92.

K–4   Kalman, R. E., "Canonical Structure of Linear Dynamical Systems," *Proc. Natl. Acad. Sci., USA*, **48** (1962), pp. 596–600.

K–5   Kalman, R. E., "When Is a Linear Control System Optimal?" *ASME J. Basic Engineering*, ser. D, **86** (1964), pp. 51–60.

K–6   Kalman, R. E., and J. E. Bertram, "Control System Analysis and Design via the Second Method of Lyapunov: I Continuous-Time Systems," *ASME J. Basic Engineering*, ser. D, **82** (1960), pp. 371–93.

K–7   Kalman, R. E., Y. C. Ho, and K. S. Narendra, "Controllability of Linear Dynamic Systems," in *Contributions to Differential Equations*, Vol. 1. New York: Wiley-Interscience Publishers, Inc., 1962.

K–8   Kautsky, J., and N. Nichols, "Robust Pole Assignment in Linear State Feedback," *Intern. J. Control*, **41** (1985), pp 1129–55.

K–9   Kreindler, E., and P. E. Sarachick, "On the Concepts of Controllability and Observability of Linear Systems," *IEEE Trans. Automatic Control*, **AC-9** (1964), pp. 129–36.

K–10 Kuo, B. C., *Automatic Control Systems*, 6th ed. Upper Saddle River, NJ: Prentice Hall, 1991.

L–1   LaSalle, J. P, and S. Lefschetz, *Stability by Liapunov's Direct Method with Applications*. New York: Academic Press, Inc., 1961.

L–2   Levin, W. S., *The Control Handbook*. Boca Raton, FL: CRC Press, 1996.

L–3   Levin, W. S. *Control System Fundamentals*. Boca Raton, FL: CRC Press, 2000.

L–4   Luenberger, D. G., "Observing the State of a Linear System," *IEEE Trans. Military Electr.*, **MIL-8** (1964), pp. 74–80.

L–5   Luenberger, D. G., "An Introduction to Observers," *IEEE Trans. Automatic Control*, **AC-16** (1971), pp. 596–602.

L–6   Lur'e, A. I., and E. N. Rozenvasser, "On Methods of Constructing Liapunov Functions in the Theory of Nonlinear Control Systems," *Proc. First Intern. Cong. IFAC*, Moscow, 1960, *Automatic and Remote Control*. London: Butterworths & Company Limited, 1961, pp. 928–33.

M–1   MathWorks, Inc., *The Student Edition of MATLAB*, version 5. Upper Saddle River, NJ: Prentice Hall, 1997.

M–2   Melbourne, W. G., "Three Dimensional Optimum Thrust Trajectories for Power-Limited Propulsion Systems," *ARS J.*, **31** (1961), pp. 1723–8.

M–3   Melbourne, W. G., and C. G. Sauer, Jr., "Optimum Interplanetary Rendezvous with Power-Limited Vehicles," *AIAA J.*, **1** (1963), pp. 54–60.

M–4   Minorsky, N., *Nonlinear Oscillations*. New York: Van Nostrand Reinhold, 1962.

M–5   Monopoli, R. V., "Controller Design for Nonlinear and Time-Varying Plants," *NASA* **CR152**, Jan., 1965.

N–1   Noble, B., and J. Daniel, *Applied Linear Algebra*, 2nd ed. Upper Saddle River, NJ: Prentice Hall, 1977.

N–2   Nyquist, H., "Regeneration Theory," *Bell System Tech. J.*, **11** (1932), pp. 126–47.

O–1   Ogata, K., *State Space Analysis of Control Systems*. Upper Saddle River, NJ: Prentice Hall, 1967.

O–2   Ogata, K., *Solving Control Engineering Problems with MATLAB*. Upper Saddle River, NJ: Prentice Hall, 1994.

O–3   Ogata, K., *Designing Linear Control Systems with MATLAB*. Upper Saddle River, NJ: Prentice Hall, 1994.

O–4   Ogata, K., *Discrete-Time Control Systems*, 2nd ed. Upper Saddle River, NJ: Prentice Hall, 1995.

O–5   Ogata, K., *System Dynamics*, 4th ed. Upper Saddle River, NJ: Prentice Hall, 2004.

O–6   Ogata, K., *MATLAB for Control Engineers.* Upper Saddle River, NJ: Pearson Prentice Hall, 2008.

P–1   Phillips, C. L., and R. D. Harbor, *Feedback Control Systems.* Upper Saddle River, NJ: Prentice Hall, 1988.

P–2   Pontryagin, L. S., V. G. Boltyanskii, R. V. Gamkrelidze, and E. F. Mishchenko, *The Mathematical Theory of Optimal Processes.* New York: John Wiley & Sons, Inc., 1962.

R–1   Rekasius, Z. V., "A General Performance Index for Analytical Design of Control Systems," *IRE Trans. Automatic Control,* **AC-6** (1961), pp. 217–22.

R–2   Rowell, G., and D. Wormley, *System Dynamics.* Upper Saddle River, NJ: Prentice Hall, 1997.

S–1   Schultz, W. C., and V. C. Rideout, "Control System Performance Measures: Past, Present, and Future," *IRE Trans. Automatic Control,* **AC-6** (1961), pp. 22–35.

S–2   Smith, R. J., *Electronics: Circuits and Devices,* 2d ed. New York: John Wiley & Sons, Inc., 1980.

S–3   Staats, P. F. "A Survey of Adaptive Control Topics," *Plan B paper,* Dept. of Mech. Eng., University of Minnesota, March 1966.

S–4   Strang, G., *Linear Algebra and Its Applications.* New York: Academic Press, Inc., 1976.

T–1   Truxal, J. G., *Automatic Feedback Systems Synthesis.* New York: McGraw-Hill Book Company, 1955.

U–1   Umez-Eronini, E., *System Dynamics and Control.* Pacific Grove, CA: Brooks/Cole Publishing Company, 1999.

V–1   Valkenburg, M. E., *Network Analysis.* Upper Saddle River, NJ: Prentice Hall, 1974.

V–2   Van Landingham, H. F., and W. A. Blackwell, "Controller Design for Nonlinear and Time-Varying Plants," *Educational Monograph,* College of Engineering, Oklahoma State University, 1967.

W–1   Webster, J. G., *Wiley Encyclopedia of Electrical and Electronics Engineering,* Vol. 4. New York: John Wiley & Sons, Inc., 1999.

W–2   Wilcox, R. B., "Analysis and Synthesis of Dynamic Performance of Industrial Organizations—The Application of Feedback Control Techniques to Organizational Systems," *IRE Trans. Automatic Control,* **AC-7** (1962), pp. 55–67.

W–3   Willems, J. C., and S. K. Mitter, "Controllability, Observability, Pole Allocation, and State Reconstruction," *IEEE Trans. Automatic Control,* **AC-16** (1971), pp. 582–95.

W–4   Wojcik, C. K., "Analytical Representation of the Root Locus," *ASME J. Basic Engineering,* ser. D, **86** (1964), pp. 37–43.

W–5   Wonham, W. M., "On Pole Assignment in Multi-Input Controllable Linear Systems," *IEEE Trans. Automatic Control,* **AC-12** (1967), pp. 660–65.

Z–1   Zhou, K., J. C. Doyle, and K. Glover, *Robust and Optimal Control.* Upper Saddle River, NJ: Prentice Hall, 1996.

Z–2   Zhou, K., and J. C. Doyle, *Essentials of Robust Control,* Upper Saddle River, NJ: Prentice Hall, 1998.

Z–3   Ziegler, J. G., and N. B. Nichols, "Optimum Settings for Automatic Controllers," *ASME Trans.* **64** (1942), pp. 759–68.

Z–4   Ziegler, J. G., and N. B. Nichols, "Process Lags in Automatic Control Circuits," *ASME Trans.* **65** (1943), pp. 433–44.

Index

A

Absolute stability, 160
Ackermann's formula:
    for observer gain matrix, 756–57
    for pole placement, 730–31
Actuating error, 8
Actuator, 21–22
Adjoint matrix, 876
Air heating system, 150
Aircraft elevator control system, 156
Analytic function, 860
Angle:
    of arrival, 286
    of departure, 280, 286
Angle condition, 271
Asymptotes:
    Bode diagram, 406–07
    root loci, 274–75, 284–85
Attenuation, 165
Attitude-rate control system, 386
Automatic controller, 21
Automobile suspension system, 86
Auxiliary polynomial, 216

B

Back emf, 95
    constant, 95

Bandwidth, 474, 539
Basic control actions:
    integral, 24
    on-off, 22
    proportional, 24
    proportional-plus-derivative, 25
    proportional-plus-integral, 24
    proportional-plus-integral-plus-
        derivative, 25
    two-position, 22–23
Bleed-type relay, 111
Block, 17
Block diagram, 17–18
    reduction, 27–28, 48
Bode diagram, 403
    error in asymptotic expression of, 403
    of first-order factors, 406–07, 409
    general procedure for plotting, 413
    plotting with MATLAB, 422–25
    of quadratic factors, 410–12
    of system defined in state space,
        426–27
Branch point, 18
Break frequency, 406
Breakaway point, 275–76, 285–86, 351
Break-in point, 276, 281, 285–86, 351
Bridged-T networks, 90, 520
Business system, 5

C

Canonical forms:
    controllable, 649
    diagonal, 650
    Jordan, 651, 653
    observable, 650
Capacitance:
    of pressure system, 107–09
    of thermal system, 137
    of water tank, 103
Cancellation of poles and zeros, 288
Cascaded system, 20
Cascaded transfer function, 20
Cauchy–Riemann conditions, 860–61
Cauchy's theorem, 526
Cayley–Hamilton theorem, 668, 701
Characteristic equation, 652
Characteristic polynomial, 34
Characteristic roots, 652
Circular root locus, 282
Classical control theory, 2
Classification of control systems, 225
Closed-loop control system, 8
Closed-loop system, 20
Closed-loop frequency response, 477
Closed-loop frequency response curves:
    desirable shapes of, 492
    undesirable shapes of, 492
Closed-loop transfer function, 19–20
Cofactor, 876
Command compensation, 630
Compensation:
    feedback, 308
    parallel, 308
    series, 308
Compensator:
    lag, 323, 503–04
    lag–lead, 332–34, 511–13
    lead, 312–13, 495–96
Complete observability, 683–84
    conditions for, 684–85
    in the s plane, 684
Complete output controllablility, 714
Complete state controllability, 676–81
    in the s plane, 680–81
Complex-conjugate poles:
    cancellation of undesirable, 520
Complex function, 859
Complex impedence, 75
Complex variable, 859
Computational optimization approach to
    design PID controller, 583–89
Conditional stability, 299–300, 510–11
Conditionally stable system, 299–300,
    458, 510–11
Conduction heat transfer, 137

Conformal mapping, 447, 462–64
Conical water tank system, 152
Constant-gain loci, 302–03
Constant-magnitude loci (M circles),
    478–79
Constant phase-angle loci (N circles),
    480–81
Constant ω_n loci, 296
Constant ζ lines, 298
Constant ζ loci, 296
Control actions, 21
Control signal, 3
Controllability, 675–81
    matrix, 677
    output, 681
Controllable canonical form, 649, 688
Controlled variable, 3
Controller, 22
Convection heat transfer, 137
Conventional control theory, 29
Convolution, integral, 16
Corner frequency, 406
Critically damped system, 167
Cutoff frequency, 474
Cutoff rate, 475

D

Damped natural frequency, 167
Damper, 64, 132
Damping ratio, 165
    lines of constant, 296
Dashpot, 64, 132–33
Dead space, 43
Decade, 405
Decibel, 403
Delay time, 169–70
Derivative control action, 118–20, 222
Derivative gain, 84
Derivative time, 25, 61
Detectability, 688
Determinant, 874
Diagonal canonical form, 694
Diagonalization of $n \times n$ matrix, 652
Differential amplifier, 78
Differential gap, 23, 24
Differentiating system, 231
Differentiation:
    of inverse matrix, 881
    of matrix, 880
    of product of two matrices, 880
Differentiator:
    approximate, 617
Direct transmission matrix, 31
Disturbance, 3, 26
Dominant closed-loop poles, 182
Duality, 754

E

$e^{\mathbf{A}t}$:
   computation of, 670–71
Eigenvalue, 652
   invariance of, 655
Electromagnetic valve, 23
Electronic controller, 77, 83
Engineering organizational system, 5–6
Equivalent moment of inertia, 234
Equivalent spring constant, 64
Equivalent viscous-friction coefficient,
   65, 234
Evans, W. R., 2, 11, 269
Exponential response curve, 162

F

Feedback compensation, 308–09, 342, 519
Feedback control, 3
Feedback control system, 7
Feedback system, 20
Feedforward transfer function, 19
Final value theorem, 866
First-order lag circuit, 80
First-order system, 161–64
   unit-impulse response of, 163
   unit-ramp response of, 162–63
   unit-step response of, 161–62
Flapper, 110
   valve, 156
Fluid systems:
   mathematical modeling of, 100
Free-body diagram, 69–70
Frequency response, 398
   correlation between step response
      and, 471–74
   lag compensation based on, 502–11
   lag–lead compensation based on,
      511–17
   lead compensation based on, 493–502
Full-order state observer, 752–53
Functional block, 17

G

Gain crossover frequency, 467–69
Gain margin, 464–67
Gas constant, 108
   for air, 142
   universal, 108
Gear train, 232
   system, 232–34
Generalized plant, 813, 815–17
   diagram, 810–16, 853–54

H

H infinity control problem, 816
H infinity norm, 6, 808

Hazen, 2, 11
High-pass filter, 495
Higher-order systems, 179
   transient response of, 180–81
Hurwitz determinants, 252–58
Hurwitz stability criterion, 252–53, 255–58
   equivalence of Routh's stability
      criterion and, 255–57
Hydraulic controller:
   integral, 130
   jet-pipe, 147
   proportional, 131
   proportional-plus-derivative, 134–35
   proportional-plus-integral, 133–34
   proportional-plus-integral-plus-
      derivative, 135–36
Hydraulic servo system, 124–25
Hydraulic servomotor, 128, 130, 156
Hydraulic system, 106, 123–39, 149
   advantages and disadvantages of, 124
   compared with pneumatic system, 106

I

Ideal gas law, 108
Impedance:
   approach to obtain transfer function,
      75–76
Impulse function, 866
Impulse response, 163, 178–79, 195–97
   function, 16–17
Industrial controllers, 22
Initial condition:
   response to, 203–11
Initial value theorem, 866
Input filter, 261, 630
Input matrix, 31
Integral control, 220
Integral control action, 24–25, 218
Integral controller, 22
Integral gain, 61
Integral time, 25, 61
Integration of matrix, 880
Inverse Laplace transform:
   partial-fraction expansion method for
      obtaining, 867–73
Inverse Laplace transformation, 862
Inverse of a matrix:
   MATLAB approach to obtain, 879
Inverse polar plot, 461–62, 537–38
Inverted-pendulum system, 68–72, 98
Inverted-pendulum control system,
   746–51
Inverting amplifier, 78
I-PD control, 591–92
I-PD-controlled system, 592, 628–29, 643
   with feedforward control, 642

J

Jet-pipe controller, 146–47
Jordan blocks, 679
Jordan canonical form, 651, 695, 706–07

K

Kalman, R. E., 12, 675
Kirchhoff's current law, 72
Kirchhoff's loop law, 72
Kirchhoff's node law, 72
Kirchhoff's voltage law, 72

L

Lag compensation, 321
Lag compensator, 311, 321, 502
  Bode diagram of, 503
  design by frequency-response method, 502–11
  design by root-locus method, 321, 323
  polar plot of, 503
Lag network, 82, 542
Lag–lead compensation, 330, 335, 338, 377, 511–18
Lag–lead compensator:
  Bode diagram of, 558
  design by frequency-response method, 513–17
  design by root-locus method, 331–32, 380–82
  electronic, 330–32
  polar plot of, 512
Lag–lead network:
  electronic, 330–32
  mechanical, 366
Lagrange polynomial, 708
Lagrange's interpolation formula, 708
Laminar-flow resistance, 102
Laplace transform, 862
  properties of, 865
  table of, 863–64
Lead compensator, 311, 493
  Bode diagram of, 494
  design by frequency-response method, 493–502
  design by root-locus method, 311–18
  polar plot of, 494
Lead, lag, and lag–lead compensators:
  comparison of, 517–18
Lead network, 542
  electronic, 82
  mechanical, 365
Lead time, 5
Linear approximation:
  of nonlinear mathematical models, 43
Linear system, 14
  constant coefficient, 14

Linear time-invariant system, 14, 164
Linear time-varying system, 14
Linearization:
  of nonlinear systems, 43
Liquid-level control system, 157
Liquid-level systems, 101, 103–04, 140–41
Log-magnitude curves of quadratic transfer function, 411
Logarithmic decrement, 237
Logarithmic plot, 403
Log-magnitude versus phase plot, 403, 443–44
LRC circuit, 72–73

M

M circles, 478–79
  a family of constant, 479
Magnitude condition, 271
Manipulated variable, 3
Mapping theorem, 448–49
Mathematical model, 13

MATLAB commands:

  MATLAB:
  obtaining maximum overshoot with, 194
  obtaining peak time with, 194
  obtaining response to initial condition with, 266
  partial-fraction expansion with, 871–73
  plotting Bode diagram with, 422–23
  plotting root loci with, 290–91
  writing text in diagrams with, 188–89
  [A,B,C,D] = tf2ss(num,den), 40, 656, 698
  bode(A,B,C,D), 422, 426
  bode(A,B,C,D,iu), 426–27
  bode(A,B,C,D,iu,w), 422
  bode(A,B,C,D,w), 422
  bode(num,den), 422
  bode(num,den,w), 422, 425, 551
  bode(sys), 422
  bode(sys,w), 552
  c = step(num,den,t), 190
  for loop, 243, 249, 584
  [Gm,pm,wcp,wcg,] = margin(sys), 468–69
  gtext ('text'), 189
  impulse(A,B,C,D), 195
  impulse(num, den), 195
  initial(A,B,C,D,[initial condition],t), 209
  inv(A), 879
  K = acker(A,B,J), 736
  K = lqr(A,B,Q,R), 798
  K = place(A,B,J), 736

MATLAB commands (*Cont.*)

K_e = acker(A',C',L)', 773
K_e = acker(Abb,Aab,L)', 773
K_e = place(A',C',L)', 773
K_e = place(Abb',Aab',L)', 773
[K,P,E] = lqr(A,B,Q,R), 798
[K,r] = rlocfind(num,den), 303
logspace(d1,d2), 422
logspace(d1,d2,n), 422–23, 534
lqr(A,B,Q,R), 797
lsim(A,B,C,D,u,t), 201
lsim(num,den,r,t), 201
magdB = 20*log10(mag), 422
[mag,phase,w] = bode(A,B,C,D), 422
[mag,phase,w] = bode(A,B,C,D,iu,w), 422
[mag,phase,w] = bode(A,B,C,D,w), 422
[mag,phase,w] = bode(num,den), 422
[mag,phase,w] = bode(num,den,w), 422, 476
[mag,phase,w] = bode(sys), 422
[mag,phase,w] = bode(sys,w), 476
mesh, 192
mesh(y), 192, 249
mesh(y'), 192, 249
[Mp,k] = max(mag), 476
NaN, 799
[num,den] = feedback(num1,den1, num2,den2), 20–21
[num,den] = parallel(num1,den1, num2,den2), 20–21
[num,den] = series(num1,den1, num2,den2), 20–21
[num,den] = ss2tf(A,B,C,D), 41, 657
[num,den] = ss2tf(A,B,C,D,iu), 41–42, 58, 657
[NUM,den] = ss2tf(A,B,C,D,iu), 59, 659
nyquist(A,B,C,D), 436, 441–42
nyquist(A,B,C,D,iu), 441
nyquist(A,B,C,D,iu,w), 436, 441
nyquist(A,B,C,D,w), 436
nyquist(num,den), 436
nyquist(num, den,w), 436
nyquist(sys), 436
polar(theta,r), 545
printsys(num,den), 20–21, 189
printsys(num,den,'s'), 189
r = abs(z), 544
[r,p,k] = residue(num,den), 239, 871–72
[re,im,w] = nyquist(A,B,C,D), 436
[re,im,w] = nyquist(A,B,C,D,iu,w), 436
[re,im,w] = nyquist(A,B,C,D,w), 436
[re,im,w] = nyquist(num,den), 436
[re,im,w] = nyquist(num,den,w), 436
[re,im,w] = nyquist(sys), 436
residue, 867
resonant_frequency = w(k), 476
resonant_peak = 20*log10(Mp), 476
rlocfind, 303
rlocus(A,B,C,D), 295
rlocus(A,B,C,D,K), 290, 295
rlocus(num,den), 290–91
rlocus(num,den,K), 290
sgrid, 297
sortsolution, 584
step(A,B,C,D), 184, 186
step(A,B,C,D,iu), 184
step(num,den), 184
step(num,den,t), 184
step(sys), 184
sys = ss(A,B,C,D), 184
sys = tf(num,den), 184
text, 188
theta = angle(z), 544
w = logspace(d2,d3,100), 425
y = lsim(A,B,C,D,u,t), 201
y = lsim(num,den,r,t), 201
[y, x, t] = impulse(A,B,C,D), 195
[y, x, t] = impulse(A,B,C,D,iu), 195
[y, x, t] = impulse(A,B,C,D,iu,t), 195
[y, x, t] = impulse(num,den), 195
[y, x, t] = impulse(num,den,t), 195
[y, x, t] = step(A,B,C,D,iu), 184
[y, x, t] = step(A,B,C,D,iu,t), 184
[y, x, t] = step(num,den,t), 184, 190
z = re+j*im, 544

End of MATLAB commands

Matrix exponential, 661, 669–674
    closed solution for, 663
Matrix Riccati equation, 798, 800
Maximum overshoot:
    in unit-impulse response, 179
    in unit-step response, 170, 172
    versus ζ curve, 174
Maximum percent overshoot, 170
Maximum phase lead angle, 494, 498
Measuring element, 21
Mechanical lag–lead system, 366
Mechanical lead system, 365
Mechanical vibratory system, 236
Mercury thermometer system, 151
Minimal polynomial, 669, 704–06
Minimum-order observer, 767–77
    based controller, 777
Minimum-order state observer, 752
Minimum-phase system, 415–16
Minimum-phase transfer function, 415
Minor, 876
Modern control theory, 2, 29
    versus conventional control theory, 29

Motor torque constant, 95
Motorcycle suspension system, 87
Multiple-loop system, 458–59

N

N circles, 480–81
  a family of constant, 481
Newton's second law, 66
Nichols, 2, 11, 398
Nichols chart, 482–85
Nichols plots, 403
Nonbleed-type relay, 111
Nonhomogeneous state equation:
  solution of, 666–67
Noninverting amplifier, 79
Nonlinear mathematical models:
  linear approximation of, 43–45
Nonlinear system, 43
Nonminimum-phase systems, 300–01,
  415, 417
Nonminimum-phase transfer function,
  415, 488
Nonuniqueness:
  of a set of state variables, 655
Nozzle-flapper amplifier, 110
Number-decibel conversion line, 404
Nyquist, H., 2, 11, 398
Nyquist path, 545
Nyquist plot, 403, 439–40, 443
  of positive-feedback system, 535–37
  of system defined in state space, 440–43
Nyquist stability analysis, 454–62
Nyquist stability criterion, 445–54
  applied to inverse polar plots, 461–62

O

Observability, 675, 682–88
  complete, 683–85
  matrix, 653
Observable canonical form, 650, 692
Observation, 752
Observed-state feedback control system,
  761
Observer, 753
  design of control system with, 786–93
  full-order, 753
  mathematical model of, 752
  minimum-order, 767–73
Observer-based controller:
  transfer function of, 761
Observer controller:
  in the feedback path of control system,
  787, 790–93
  in the feedforward path of control
  system, 787–90
Observer-controller matrix, 762

Observer-controller transfer function,
  761–62
Observer error equation, 753
Observer gain matrix, 755
  MATLAB determination of, 773
Octave, 405
Offset, 258
On-off control action, 22–23
On-off controller, 22
One-degree-of-freedom control system,
  593
op amps, 78
Open-loop control system, 8
  advantages of, 9
  disadvantages of, 9
Open-loop frequency response curves:
  reshaping of, 493
Open-loop transfer function, 19
Operational amplifier, 78
Operational amplifier circuits, 93–94
  for lead or lag compensator:
  table of, 85
Optimal regulator problem, 806
Ordinary point, 861
Orthogonality:
  of root loci and constant gain loci,
  301–02
Output controllability, 681
Output equation, 31
Output matrix, 31
Overdamped system, 168–69
Overlapped spool valve, 146
Overlapped valve, 130

P

Parallel compensation, 308–09, 342–43
Partial-fraction expansion, 867–73
  with MATLAB, 871–73
PD control, 373
PD controller, 614–15
Peak time, 170, 172, 193
Performance index, 793
Performance specifications, 9
Phase crossover frequency, 467–69
Phase margin, 464–67
  versus ζ curve, 472
PI controller, 2, 614–15
PI-D control, 590–92
PID control system, 572–77, 583, 587,
  617–21, 628–29, 642–43
  basic, 590
  with input filter, 629
  two-degrees-of-freedom, 592–95
PID controller, 567, 577, 614–16, 620, 632
  modified, 616
  using operational amplifiers, 83–84

Pilot valve, 124, 130
PI-PD control, 592
PID-PD control, 592
Plant, 3
Pneumatic actuating valve, 117–18
Pneumatic controllers, 144–45, 154–55
Pneumatic nozzle-flapper amplifier, 110
Pneumatic on-off controller, 115
Pneumatic pressure system, 142
Pneumatic proportional controller, 112–16
   force-balance type, 115–16
   force-distance type, 112–15
Pneumatic proportional-plus-derivative
   controller, 119–20
Pneumatic proportional-plus-integral
   control action, 120–22
Pneumatic proportional-plus-integral-
   plus-derivative control action,
   122–23
Pneumatic relay, 111
   bleed type, 111
   nonbleed type, 111
   reverse acting, 112
Pneumatic systems, 106–23, 153
   compared with hydraulic system, 106
Pneumatic two-position controller, 115
Polar grids, 297
Polar plot, 403, 427–28, 430, 432
Pole: 861
   of order n, 861
   simple, 861
Pole assignment technique, 723
Pole-placement:
   necessary and sufficient conditions for
     arbitrary, 725
Pole placement problem, 723–35
   solving with MATLAB, 735–36
Positive-feedback system:
   Nyquist plot for, 536–37
   root loci for, 303–07
Positional servo system, 95–97
Pressure system, 107, 109
Principle of duality, 687
Principle of superposition, 43
Process, 3
Proportional control, 219
Proportional control action, 24
Proportional controller, 22
Proportional gain, 25, 61
Proportional-plus-derivative control:
   of second-order system, 224
   of system with inertia load, 223
Proportional-plus-derivative control
   action, 25
Proportional-plus-derivative controller,
   22, 542

Proportional-plus-integral control action,
   24
Proportional-plus-integral controller, 22,
   121, 542
Proportional-plus-integral-plus-
   derivative control action, 25
Proportional-plus-integral-plus-
   derivative controller, 22
Pulse function, 866

Q
Quadratic factor, 410
   log-magnitude curves of, 411
   phase-angle curves of, 411
Quadratic optimal control problem:
   MATLAB solution of, 804
Quadratic optimal regulator system,
   793–95
   MATLAB design of, 797

R
Ramp response, 197
Rank of matrix, 875
Reduced-matrix Riccati equation, 795–97
Reduced-order observer, 752
Reduced-order state observer, 752
Reference input, 21
Regulator system with observer
   controller, 778–86, 789
Relative stability, 160, 217, 462
Residue, 867
Residue theorem, 527
Resistance:
   gas-flow, 107
   laminar-flow, 101–02
   of pressure system, 107, 109
   of thermal system, 137
   turbulent-flow, 102
Resonant frequency, 430, 470
Resonant peak, 413, 430, 470
   versus ζ curve, 413
Resonant peak magnitude, 413, 470
Response:
   to arbitrary input, 201
   to initial condition, 203–11
   to torque disturbance, 221
Reverse-acting relay, 112
Riccati equation, 795
Rise time, 169–171
   obtaining with MATLAB, 193–94
Robust control:
   system, 16, 806–17
   theory, 2, 7
Robust performance, 7, 807, 812
Robust pole placement, 735
Robust stability, 7, 807, 809

Root loci:
    general rules for constructing, 283–87
    for positive-feedback system, 303–07
Root locus, 271
    method, 269–70
Routh's stability criterion, 212–18

S

Schwarz matrix, 268
Second-order system, 164
    impulse response of, 178–79
    standard form of, 166
    step response of, 165–75
    transient-response specification of, 171
    unit-step response curves of, 169
Sensor, 21
Series compensation, 308–09, 342
Servo system, 95, 164–65
    design of, 739–51
    with tachometer feedback, 268
    with velocity feedback, 175–77
Servomechanism, 2
Set point, 21
Set-point kick, 590
Settling time, 170, 172–73
    obtaining with MATLAB, 194
    versus ζ curve, 174
Sign inverter, 79
Simple pole, 861
Singular points, 861
Sinusoidal signal generator, 486
Sinusoidal transfer function, 401
Small gain theorem, 809
Space vehicle control system, 367, 538–39
Speed control system, 4, 148
Spool valve:
    linealized mathematical model of, 127
Spring-loaded pendulum system, 98
Spring-mass-dashpot system, 66
Square-law nonlinearity, 43
S-shaped curve, 569
Stability analysis, 454–62
    in the complex plane, 182
Stabilizability, 688
Stack controller, 115
Standard second-order system, 189
State, 29
State controllability:
    complete, 676, 678, 680
State equation, 31
    solution of homogeneous, 660
    solution of nonhomogeneous, 666–67
    Laplace transform solution of, 663
State-feedback gain matrix, 724
    MATLAB approach to determine,
        735–36

State matrix, 31
State observation:
    necessary and sufficient conditions for,
        754–55
State observer, 751–77
    design with MATLAB, 773
    type 1 servo system with, 746
State observer gain matrix: 755
    Ackermann's formula to obtain, 756–57
    direct substitution approach to obtain,
        756
    transformation approach to obtain, 755
State space, 30
State-space equation, 30
    correlation between transfer function
        and, 649, 656
    solution of, 660
State-space representation:
    in canonical forms, 649
    of nth order system, 36–39
State-transition matrix, 664
    properties of, 665
State variable, 29
State vector, 30
Static acceleration error constant,
        228, 421
    determination of, 421–22
Static position error constant,
        226, 419
Static velocity error constant,
        227, 420
Steady-state error, 160, 226
    for unit parabolic input, 229
    for unit ramp input, 228
    in terms of gain K, 230
Steady-state response, 160
Step response, 699–700
    of second-order system, 165–69
Summing point, 18
Suspension system:
    automobile, 86–87
    motorcycle, 87
Sylvester's interpolation formula, 673,
        709–713
System, 3
Sytem types, 419
    type 0, 225, 230, 419, 433, 487–88
    type 1, 225, 230, 420, 433, 487–88
    type 2, 225, 230, 421, 433, 487–88
System response to initial condition:
    MATLAB approach to obtain, 203–11

T

Tachometer, 176
    feedback, 343
Taylor series expansion, 43–45

Temperature control systems, 4–5
Test signals, 159
Text:
    writing on the graphic screen, 188
Thermal capacitance, 137
Thermal resistance, 137
Thermal systems, 100, 136–39
Thermometer system, 151–52
Three-degrees-of-freedom system, 645
Three-dimensional plot, 192
    of unit-step response curves with
        MATLAB, 191–93
Traffic control system, 8
Transfer function, 15
    of cascaded elements, 73–74
    of cascaded systems, 20
    closed-loop, 20
    of closed-loop system, 20
    experimental determination of, 489–90
    expression in terms of $\mathbf{A}$, $\mathbf{B}$, $\mathbf{C}$, and D, 34
    of feedback system, 19
    feedforward, 19
    of minimum-order observer-based
        controller, 777
    of nonloading cascaded elements,
        77
    observer-controller, 762, 780–82
    open-loop, 19
    of parallel systems, 20
    sinusoidal, 401
Transfer matrix, 35
Transformation:
    from state space to transfer function,
        41–42, 657
    from transfer function to state space,
        40–41, 656
Transient response, 160
    analysis with MATLAB, 183–211
    of higher-order system, 180
    specifications, 169, 171
Transport lag, 417
    phase angle characteristics of, 417
Turbulent-flow resistance, 102
Two-degrees-of-freedom control system,
    593–95, 599–614, 636–41, 646–47
Two-position control action, 22–23
Two-position controller, 22
Type 0 system, 225, 230, 488
    log-magnitude curve for, 419, 488
    polar plot of, 433
Type 1 servo system:
    design of, 743–51
    pole-placement design of, 739–46
Type 1 system, 420
    log-magnitude curve for, 420, 488
    polar plot of, 433

Type 2 system, 421
    log-magnitude curve for, 421, 488
    polar plot of, 433

U

Uncontrollable system, 681
Undamped natural frequency, 165
Underdamped system, 166–67
Underlapped spool valve, 146
Unit acceleration input, 247
Unit-impulse response:
    of first-order system, 163
    of second-order system, 178
Unit-impulse response curves:
    a family of, 178
    obtained by use of MATLAB, 196–97
Unit-ramp response:
    of first-order system, 162–63
    of second-order system, 197–200
    of system defined in state space,
        199–200
Unit-step response:
    of first-order system, 161
    of second-order system, 163, 167, 169
Universal gas constant, 108
Unstructured uncertainty:
    additive, 852–53
    multiplicative, 809
    system with, 809

V

Valve:
    overlapped, 130
    underlapped, 130
    zero-lapped, 130
Valve coefficient, 127
Vectors:
    linear dependence of, 674
    linear independence of, 674
Velocity error, 227
Velocity feedback, 176, 343, 519

W

Watt's speed governor, 4
Weighting function, 17

Z

Zero, 861
    of order m, 862
Zero-lapped valve, 130
Zero placement, 595, 597, 612
    approach to improve response charac-
        teristics, 595–97
Ziegler–Nichols tuning rules, 11, 568–77
    first method, 569–70
    second method, 570–71